The Earth: Its Origin, Structure and Evolution

The Earth: Its Origin, Structure and Evolution

edited by

M. W. McELHINNY

Research School of Earth Sciences,
Australian National University,
Canberra, Australia

1979

ACADEMIC PRESS

London New York San Francisco

A Subsidiary of Harcourt Brace Jovanovich, Publishers

ACADEMIC PRESS INC. (LONDON) LTD.
24/28 Oval Road,
London NW1

United States Edition published by
ACADEMIC PRESS INC.
111 Fifth Avenue,
New York, New York 10003

British Library Cataloguing in Publication Data
The earth.
1. Earth sciences
I. McElhinny, Michael William
551.1 QE26.2 79–66701

ISBN 0–12–482750–0

Printed in England by J. W. Arrowsmith Ltd., Bristol.

Preface and Dedication

This volume was written to honour two outstanding scientists who have been responsible for the development of Earth Sciences at the Australian National University. John Conrad Jaeger became the foundation Professor of Geophysics in January 1952, and established the Department of Geophysics. Under his guidance, the Department expanded greatly and achieved a wide degree of international recognition in a remarkably short time, firstly in several fields of geophysics, and subsequently, in areas of geochemistry. Following Jaeger's retirement in December, 1972, the decision was reached to separate the Department from its parental Research School of Physical Sciences and establish a new, autonomous, and more broadly based Research School of Earth Sciences. Anton Hales was the first Director of the new School. Under his leadership, existing research activities were strengthened and new research groups established in geophysical fluid dynamics, economic geology and global geodynamics.

Jaeger and Hales were scientists of great distinction and breadth prior to joining the A.N.U. Despite their administrative responsibilities, their continuing ability to produce good science was an inspiration to their younger colleagues. These men possessed other qualities to a rare degree—deep concern and interest in the problems, both scientific and personal, of all their staff, creative leadership, unselfishness and also a degree of courage that enabled them to take difficult decisions, both within and outside the School, where necessary. It was this combination of human qualities and professional distinction that won the affection, loyalty and respect of all their colleagues, subordinates and students.

The book has been written exclusively by present and past members of the Faculty of the Research School of Earth Sciences and associates. The topics cover aspects of the major research interests of the Research School in a way that enables them to be placed together in a volume about the Earth. Indeed in many respects the contents of this volume reflects the view of the Earth as seen from the A.N.U. and represents an overview of the developments planned and pursued by Jaeger and Hales. Starting from the beginning it is appropriate to discuss the origin of the Earth and its composition as seen from a geochemical standpoint. A chapter on the rotation of the Earth is followed by aspects relating to the core through to the crust in sequence. The various chapters demonstrate the successful merging of geophysics and geochemistry that has been one of the hallmarks of the Research School.

As Faculty members under Jaeger and Hales the authors are proud to present this volume in their honour. Writing about the science that has been carried out under their leadership is undoubtedly the sincerest compliment that can be paid to them. As editor I could not have asked for a more pleasant or easier task.

M. W. McElhinny *Canberra, June 1978.*

John Conrad Jaeger

Anton L. Hales

List of Contributors

R. S. Anderssen, *Computing Research Group, The Australian National University, P.O. Box 4, Canberra 2600, Australia*

B. W. Chappell, *Department of Geology, School of General Studies, The Australian National University, P.O. Box 4, Canberra 2600, Australia*

J. R. Cleary, *Research School of Earth Sciences, Institute of Advanced Studies, The Australian National University, P.O. Box 4, Canberra 2600, Australia*

W. Compston, *Research School of Earth Sciences, Institute of Advanced Studies, The Australian National University, P.O. Box 4, Canberra 2600, Australia*

T. J. Fitch, *Lincoln Laboratory, Massachussetts Institute of Technology, 42 Carleton Street, Cambridge, Mass. 02142, USA*

D. H. Green, *Department of Geology, University of Tasmania, Hobart, Tasmania, Australia*

L. B. Gustafson, *Research School of Earth Sciences, Institute of Advanced Studies, The Australian National University, P.O. Box 4, Canberra 2600, Australia*

W. O. Hibberson, *Research School of Earth Sciences, Institute of Advanced Studies, The Australian National University, P.O. Box 4, Canberra 2600, Australia*

E. Irving, *Geomagnetic Laboratory, Earth Physics Branch, Department of Energy, Mines and Resources, Ottawa, Canada*

L. A. Jaques, *Bureau of Mineral Resources, Geology and Geophysics, Canberra, Australia*

A. H. Lachenbruch, *Geothermal Studies, U.S. Geological Survey, 345 Middlefield Road, Menlo Park, California 94025, USA*

K. Lambeck, *Research School of Earth Sciences, Institute of Advanced Studies, The Australian National University, P.O. Box 4, Canberra 2600, Australia*

R. C. Liebermann, *Department of Earth and Space Science, State University of New York at Stony Brook, Stony Brook, N.Y. 11790, USA*

F. E. M. Lilley, *Research School of Earth Sciences, Institute of Advanced Studies, The Australian National University, P.O. Box 4, Canberra 2600, Australia*

Lin-Gun Liu, *Research School of Earth Sciences, Institute of Advanced Studies, The Australian National University, P.O. Box 4, Canberra 2600, Australia*

I. McDougall, *Research School of Earth Sciences, Institute of Advanced Studies, The Australian National University, P.O. Box 4, Canberra 2600, Australia*

M. W. McElhinny, *Research School of Earth Sciences, Institute of Advanced Studies, The Australian National University, P.O. Box 4, Canberra 2600, Australia*

M. S. Paterson, *Research School of Earth Sciences, Institute of Advanced Studies, The Australian National University, P.O. Box 4, Canberra 2600, Australia*

A. E. Ringwood, *Research School of Earth Sciences, Institute of Advanced Studies, The Australian National University, P.O. Box 4, Canberra 2600, Australia*

J. H. Sass, *Geothermal Studies, U.S. Geological Survey,* 345 *Middlefield Road, Menlo Park, California 94025, USA*

D. J. Stevenson, *Department of Earth and Space Sciences, University of California, Los Angeles, California 90024, USA*

S. R. Taylor, *Research School of Earth Sciences, Institute of Advanced Studies, The Australian National University, P.O. Box 4, Canberra 2600, Australia*

J. S. Turner, *Research School of Earth Sciences, Institute of Advanced Studies, The Australian National University, P.O. Box 4, Canberra 2600, Australia*

Contents

1. Composition and Origin of the Earth

A. E. RINGWOOD

Abstract

The composition and constitution of the mantle are reviewed in detail. The pyrolite chemical composition of the upper mantle is found to be applicable throughout the entire mantle when the effects of phase transformations on density and seismic velocity distributions are considered. The abundance patterns of major and minor elements in pyrolite provide key boundary conditions relating to the cosmochemical environment in which the Earth accreted. Compared to the primordial abundances, pyrolite is depleted in a wide range of volatile elements to degrees which are correlated with the volatilities of these elements in the solar nebula. The abundances of many siderophile elements (e.g. Ni, Co, Re, Au) in pyrolite are 10–1000 times higher than would be expected if pyrolite had ever equilibrated with a metallic iron phase which had subsequently sunk into the core. Likewise, pyrolite is much more oxidized than would be expected if it had accreted in equilibrium with metallic iron. The abundances of many crystallographically compatible elements (e.g. Mg, Fe, Ni, Co, Mn, Zn) appear to display remarkably low dispersions in mantle rocks. These elements display extremely diverse degrees of volatile, involatile and siderophile behaviour in the cosmochemical environment of the solar nebula. The uniformity of these distributions within the mantle implies that accretion was accompanied by, or preceded by, a very efficient mixing process, and also that the mantle was not extensively melted and differentiated during accretion.

Although the core is mainly composed of iron, its anomalously low density requires the presence of 5–15 per cent of a light element in solution. If the light element is sulphur, it would imply that the Earth accreted about half of the primordial abundance of this extremely volatile element, whilst at the same time becoming much more strongly depleted in a wide range of elements which are much less volatile than sulphur. This appears highly improbable and we conclude that sulphur is not the principal light element in the core. Calculations based upon experimental data show that large quantities of FeO would dissolve in molten iron under the *P*, *T* conditions prevailing in the core. It seems most likely that the anomalously low density of the core is caused by the presence of about 10 per cent of oxygen.

The bulk composition of the Earth is estimated from inferred compositions of mantle plus core. It is found that the relative abundances of major elements, Mg, Fe, Si, Ca and Al, in the Earth are similar to the primordial relative abundances of these elements in Type 1 carbonaceous chondrites. There is no evidence which requires that significant degrees of fractionation of metallic iron from silicates, or of high-temperature condensates (CaO, Al_2O_3) from intermediate-temperature condensates (MgO, Fe) occurred during the formation of the Earth. It appears that formation of the Earth was a more simple cosmochemical process than is proposed by many current theories.

The origin of the Earth is discussed in the light of the boundary conditions obtained from its present composition. A wide range of geochemical evidence is found to contradict hypotheses involving heterogeneous accretion. The composition of the Earth can be explained satisfactorily if it accreted homogeneously from a mixture of 15 per cent of low-temperature solar nebula condensate similar in composition to Type 1 carbonaceous chondrites, and 85 per cent of a high temperature, devolatilized, metal-rich condensate. The entry of ferrous oxide into solution in the core combined with disproportionation of Fe^{2+} into $Fe^{3+} + Fe^0$ deep in the Earth's interior causes a large increase in oxygen fugacity of the mantle–core system. This may be responsible for the high abundances of siderophile elements and the high oxidation state of the mantle. The bulk composition of the Earth including its depletions in volatile elements can be readily interpreted in terms of accretion from planetesimals which became chemically isolated from the solar nebula over a wide range of temperatures in an environment somewhat resembling that of a modern infra-red cocoon protostar.

1. Introduction

The last thirty years has witnessed an enormous advance in our knowledge of the composition and constitution of the Earth's interior. It is now possible to provide reasonable estimates of the chemical and mineralogical composition of the mantle and to place constraints upon the extent of possible chemical zoning. Corresponding advances have been made in our understanding of the structure and composition of the Earth's core. As a result of these developments, many new and important boundary conditions for the origin of the Earth have emerged. I do not believe that these boundary conditions, mainly of a geochemical nature, have been adequately recognized in many recent discussions of the origin of terrestrial planets in general and of the Earth in particular.

This paper will present an outline of current knowledge relating to the internal composition and constitution of the Earth, with particular emphasis upon key properties which are relevant to its origin. We will then discuss the bearing of these properties upon theories of origin of the Earth.

Because of the complexity of the subject, it is not possible, in a review of this length, to provide complete documentation and references for many of the conclusions stated. For further detailed discussion of matters related to the composition and structure of the mantle, the reader is referred to my book (Ringwood, 1975). The composition of the core is considered in a recent detailed paper (Ringwood, 1977). Finally, the entire subject matter covered in this paper is treated more exhaustively in another paper now in press (Ringwood, 1978).

2. Constitution of the Mantle

The *crust* represents the product of extremely complex and efficient differentiation processes acting over geologic time, which have concentrated a large portion of the Earth's inventory of certain key incompatible elements (e.g. U, Ba, La) into this thin surficial region. Since we are primarily interested in the geochemical properties of the mantle in its primitive state, before being affected by the differentiation processes which formed the crust, detailed discussion of the properties of the crust is not necessary.

We will be concerned chiefly with the mantle, representing about 68 per cent of the mass of the Earth, and the core, about 32 per cent. The principal physical information on the structure of the mantle and core is given by the variation of seismic P- and S-wave velocities with depth (Fig. 1). These provide the basis for subdividing the Earth's interior into a series of layers (Fig. 2).

2.1. The Upper Mantle

The upper mantle extends from the Moho to a depth of 300–400 km (Figs 1, 2). It is characterized generally by rather low P- and S-wave velocity–depth gradients, except perhaps in the vicinity of the low-velocity zone, and by the existence of pronounced lateral variations in velocity distributions. In most regions of the Earth, the S-wave velocity and possible the P-wave velocity pass through a minimum between depths of about 70–150 km. This minimum may be caused by the occurrence of a small degree (~ 1 per cent) of partial melting along grain boundaries in this region. It is of considerable importance as a zone of low "viscosity" and tectonic decoupling of the overlying rigid lithosphere from underlying more mobile asthenosphere regions of the mantle.

The P-wave velocity of most regions of the uppermost mantle beneath both continents and oceans is in the range 8.2 ± 0.2 km s^{-1}. This property, together with certain broad petrological and chemical limitations, effectively restricts the mineralogical composition of this region to some combination of olivine, pyroxene(s) and garnet. The principal rock types carrying these minerals are peridotite (olivine–pyroxene) and eclogite (pyroxene–garnet). Complete mineralogical transitions between these two rock types are rare.

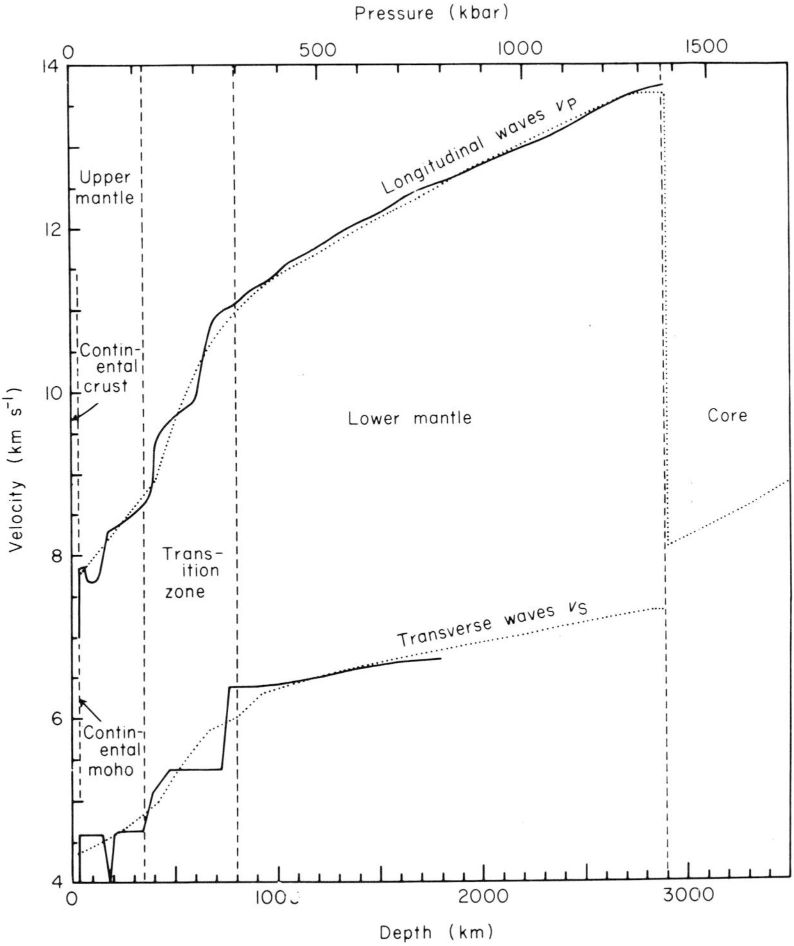

Fig. 1. Variation of seismic P- and S-wave velocities throughout mantle and outermost core. P-waves: solid line, Johnson (1967, 1969). S-waves: solid line, Nuttli (1969); dotted lines, Jeffreys (1939).

A wide range of evidence, reviewed by Ringwood (1975), shows that the uppermost mantle or lithosphere is dominantly composed of peridotite, with eclogite widely distributed as local segregations but relatively small in total amount. The vast majority of naturally occurring peridotites are strongly depleted in low-melting point components, so that they are unable to produce basaltic magmas when partially melted. On the other hand, we know that magmas of the basaltic suite have been erupted in copious volumes throughout geological time at localities scattered all over the Earth's surface, both continental and oceanic. It appears, therefore, that

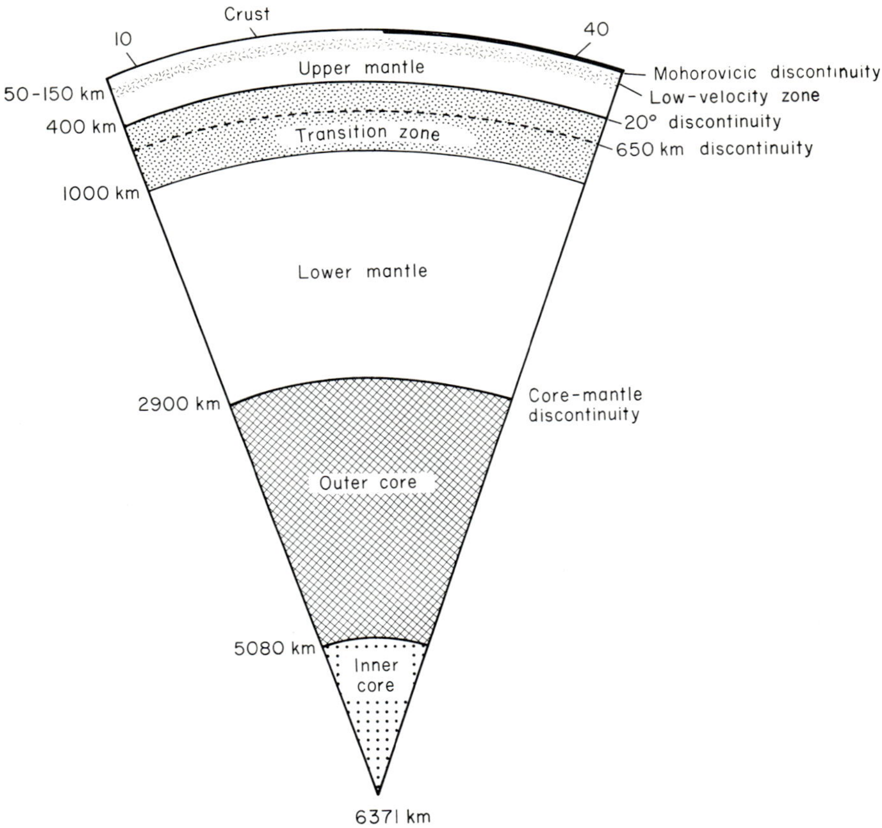

Fig. 2. Principal internal subdivisions of the Earth.

beneath the refractory peridotite layer there must exist a more primitive source region from which basaltic magmas have not been extracted. This primitive source material has been denoted by the term "pyrolite", implying a non-specific pyroxene–olivine rock, capable of yielding basaltic magmas on partial melting. Peridotite is interpreted to represent the refractory residue remaining behind after the basaltic magma has been removed. We thus arrive at a chemically zoned model for the upper mantle, as indicated in Fig. 3. Material of pyrolite composition is believed to be the dominant constituent of the upper mantle (Ringwood, 1975).

In a general sense, the composition of pyrolite must lie between those of basalt and peridotite. This composition may be obtained from the study of chemical equilibria between the various classes of basaltic magmas and their respective peridotitic residues, using the techniques of experimental petrology at high pressures and temperatures. This approach is combined

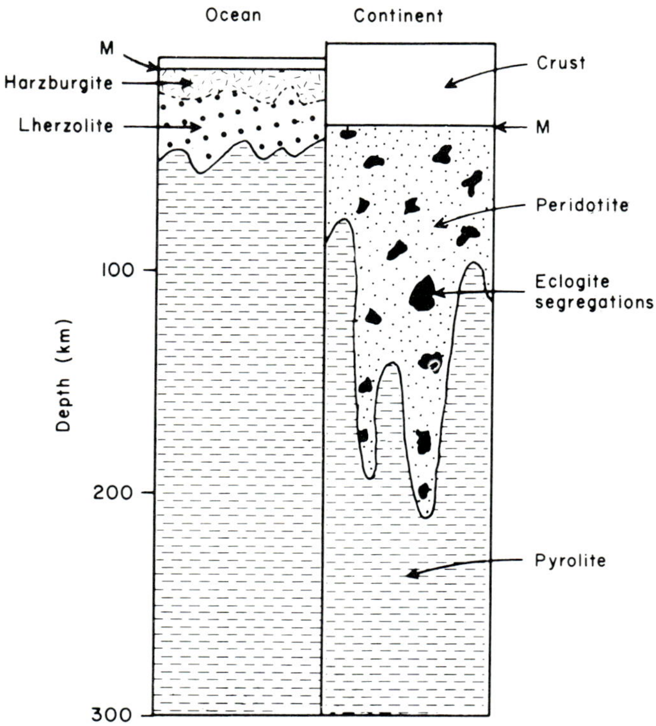

Fig. 3. Chemically zoned model for the upper mantle. Harzburgite (olivine + orthopyroxene + chromite) and lherzolite (olivine + orthopyroxene + clinopyroxene + spinel) are varieties of peridotite.

with the use of trace element distributions between basaltic and peridotitic components in the light of experimentally determined crystal–liquid partition coefficients. By the use of these and other methods, reviewed in detail by Ringwood (1975, 1978), it has been possible to provide an estimate of the pyrolite composition, as given in Table 1. It is emphasized that this is not meant to be taken as a *unique* composition. There is probably a significant composition range for pyrolite in the upper mantle. Present indications are that this range is probably quite small for major elements and compatible† trace elements, but may be very large for incompatible‡ trace elements.

† Compatible elements are defined as those possessing ionic radii and charges which permit them to substitute readily in the principal minerals of the mantle (e.g. Mg, Fe, Ni, Co, Mn, Zn).

‡ Incompatible elements are defined as those possessing ionic radii and/or ionic charges which do not permit their ions to substitute readily into the principal phases of the mantle. As a result, they are strongly partitioned into magmas during partial melting processes. Examples are Cs, Rb, K, Ba, Pb, La, Ce, U, Th, Ti, Nb, B.

Table 1. Primordial model mantle composition derived from Type 1 carbonaceous chondrites† compared with pyrolite composition.

	Primordial model mantle (weight per cent)	Pyrolite (weight per cent)
SiO_2	48.2	45.1
TiO_2	0.15	0.2
Al_2O_3	3.5	3.3
Cr_2O_3	0.7	0.4
MgO	34.0	38.1
FeO	8.1	8.0
MnO	0.5	0.15
CaO	3.3	3.1
Na_2O	1.6	0.4
K_2O	0.15	0.03
$CaO/(MgO+SiO_2)$	0.040	0.037
$Al_2O_3/(MgO+SiO_2)$	0.043	0.040

† Based on average of chemical analyses of Orgueil and Ivuna (Wiik, 1956) on a volatile and sulphur-free basis. Siderophile components (NiO and most of the FeO) have been removed and assigned to the model "core" following the procedure of Ringwood (1966), leaving sufficient FeO in mantle to produce an MgO/(MgO+FeO) ratio of 0.88.

2.2. The Transition Zone

It is seen from Figs 1 and 2 that, on the average, seismic velocities and density increase very rapidly with depth between 350 and 900 km. Major seismic and density discontinuities occur around 400 and 650 km and there is also evidence of smaller discontinuities around 350 km, 500 km and deeper than 650 km. Bullen (1936) first demonstrated that density increased with depth more rapidly throughout the transition zone than could be explained by self-compression of homogeneous material within the Earth's gravitational field. He concluded that chemical and/or phase changes must occur in this region. Birch (1952) inferred from his study of the elastic properties of the mantle that phase changes were primarily responsible for the inhomogeneity between depths of 350 and 900 km, although chemical changes were also possible.

Birch's hypothesis was verified in principle by direct experimentation at very high pressures mainly by Ringwood and co-workers in Canberra during the period 1958–72 and by Akimoto and co-workers in Tokyo from 1965 onwards. Comprehensive reviews of this work are given in Ringwood (1975), Akimoto (1972) and in the chapter by Liu in this book (Chapter 6).

It is of interest to consider the series of phase transformations which are believed to occur with increasing depth in a mantle of pyrolite composition, based upon the experimental data referred to above. These are summarized in Fig. 4. At a depth of about 350 km†, pyroxene forms a complex solid solution with pre-existing garnet. The solid solution involves the formation of a novel class of garnet-type end-members in which one quarter of the silicon atoms are octahedrally coordinated. The transformation involves an increase in the effective density of the pyroxene component of pyrolite of about 10 per cent.

At a slightly greater depth, around 400 km, a major transformation occurs in olivine, the most abundant mineral in the upper mantle, and is probably responsible for the major seismic and density discontinuity occurring around this depth. Olivine transforms to the β-Mg_2SiO_4 structure, which is about 8 per cent denser than the olivine and is closely related to the spinel structure.

At greater depths, probably around 500–550 km, the calcium silicate ($CaSiO_3$) component of the garnet transforms to the extremely dense perovskite structure while β-Mg_2SiO_4 transforms to the γ-Mg_2SiO_4 structure, accompanied by a density increase of about 2 per cent.

Liu (1975) has shown that at a pressure of about 250 kbar Mg_2SiO_4 spinel disproportionates to an assemblage of $MgSiO_3$ possessing an orthorhombic perovskite structure plus MgO (rocksalt). This transformation is probably responsible for the major seismic discontinuity around 650 km. Another transformation in this region is of $MgSiO_3$-Al_2O_3 garnet first to the ilmenite structure, followed by transformation to the perovskite structure. Also in this region, sodium is expected to be present in a high-pressure modification of $NaAlSiO_4$, possessing the extremely dense calcium ferrite structure.

Ringwood (1975) has shown that the above series of phase transformations, discovered in the laboratory and occurring in material possessing the bulk chemical composition of pyrolite, provides a satisfactory explanation of the positions and magnitudes of the seismic discontinuities in the transition zone and also provides an adequate explanation of the density change through this region.

2.3. The Lower Mantle

Birch (1952) concluded that the variation of seismic velocities and density between depths of 900 and 2700 km could be explained by self-compression of homogeneous material within the Earth's gravitational field, and that

† Based on the results of Ringwood (1972). However Akaogi and Akimoto (1977) have since concluded that the pyroxene–garnet transition may occur at a greater depth than the olivine to β-Mg_2SiO_4 transition. Further work is necessary to clarify the situation.

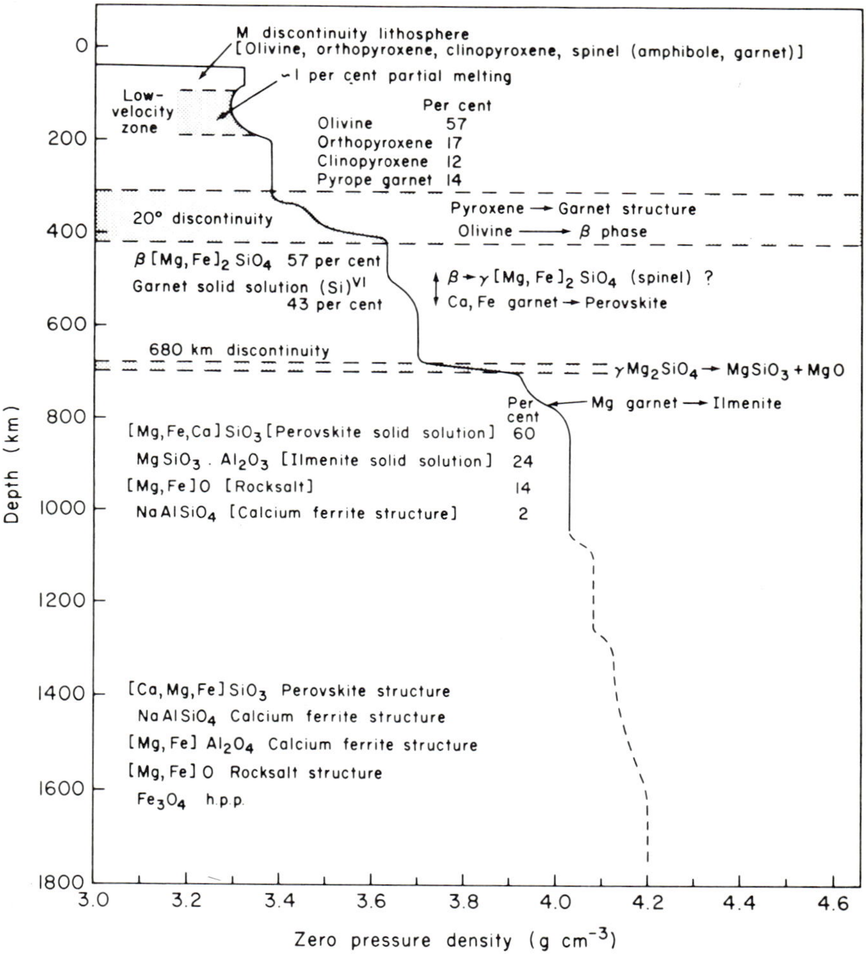

Fig. 4. Mineral assemblages and corresponding zero pressure densities for a model mantle of pyrolite composition based upon the results of static high pressure–high temperature experimentation.

there was no evidence for the occurrence of *major* phase transformations or of substantial changes of chemical composition in this region.

Many previous discussions of the constitution of the lower mantle have been based upon a comparison of the observed velocities and densities in this region with the corresponding properties of mixtures of oxides: MgO + FeO + SiO_2 (stishovite) as inferred from elasticity and shock-wave data (e.g. Anderson, 1967, 1968). This was justified, in part, by a widespread belief that the lower mantle indeed consisted of a physical mixture of oxide phases formed by the disproportionation of transition-zone minerals under pressure (e.g. Birch, 1952), and also by the convenience of using simple oxide phases as a comparative yardstick.

Studies of this kind suggested that the density of the lower mantle was a few per cent greater than would be expected for a mixture of oxides isochemical with pyrolite (e.g. Davies and Dziewonski, 1975). Although this discrepancy might appear rather minor, it has been the subject of a great deal of discussion and is important to the dynamics of the mantle and also to cosmochemical considerations.

In principle, the density differential might be caused by an increased FeO/MgO ratio in the lower mantle, as strongly argued by Anderson (1967, 1970), or by additional phase transformations to minerals which were intrinsically denser (i.e. closer-packed) than the mixed-oxide assemblage, the bulk composition of the lower mantle being similar to that of the upper mantle (Ringwood, 1970, 1975). In the former case, a stable stratification of the mantle would be implied, preventing mass transport, e.g. via convection currents, between upper and lower mantle. Moreover, Anderson's models required the overall iron abundance in the Earth (relative to Mg and Si) to be substantially higher than in Type 1 carbonaceous chondrites and in the Sun, implying that large-scale fractionation of iron from silicates had occurred in the solar nebula prior to accretion of planets, as suggested by Urey (1952). On the other hand, if the second alternative were correct, mass transfer processes between the upper and lower mantles would be possible, whilst the total abundance of iron in the Earth (relative to lithophile elements) could be similar to that in Type 1 carbonaceous chondrites and in the Sun.

Much of the earlier ambiguity arising from attempts to interpret the properties of the lower mantle in terms of mixed oxides arose from uncertainties in the elastic properties of stishovite and wüstite. These have since been accurately determined by several workers. Using these improved values, Davies and Dziewonski (1975) and Liebermann *et al.* (1976) showed that the properties of the lower mantle cannot be explained by oxide mixtures possessing stoichiometries ranging from those of olivine to pyroxene composition. In particular, the properties of the $(Mg_{0.64}Fe_{0.36})O$ + SiO_2 mixture proposed to represent the composition of the lower mantle by

Anderson (1970) are in serious disagreement with the corresponding properties of the lower mantle derived by Davies and Dziewonski (1975). The unsatisfactory nature of Anderson's composition is also demonstrated directly by shock-wave data on the densities of a pyroxenite composition, $(Mg_{0.86}Fe_{0.14})SiO_3$, by McQueen *et al.* (1967). This rock attains densities of 5.15–5.3 g cm^{-3} at pressures of 950–1100 kbar, which are in close agreement with the densities actually obtained for the mantle in this pressure range. The much more iron-rich pyroxenite composition $(Mg_{0.64}Fe_{0.36})SiO_3$ advocated by Anderson would be about 0.3 g cm^{-3} denser at the same pressures and well above the mantle density range.

Ringwood (1962, 1970) had suggested that the lower mantle was probably composed, not of mixed oxides, but of normal binary compounds like $MgSiO_3$ and Mg_2SiO_4, which occurred in crystal structures intrinsically denser than the isochemical mixed oxides. In particular, he suggested that $MgSiO_3$ might adopt the perovskite structure at very high pressure, thereby achieving a density a few per cent higher than that of isochemical periclase + stishovite.

Discovery of the predicted orthorhombic perovskite form of $MgSiO_3$ by Liu (1974), using diamond anvil techniques at a pressure of about 250 kbar, was a milestone. This was followed by his demonstration (Liu, 1975) that at around 250 kbar Mg_2SiO_4 spinel disproportionated directly into $MgSiO_3$ (perovskite) plus MgO (rocksalt) without an intervening episode of disproportionation into $MgO + SiO_2$ (stishovite), which had previously been believed to occur. The density of $MgSiO_3$ perovskite was 3.5 per cent higher than the isochemical mixture of $MgO + SiO_2$ (stishovite). Moreover, several other binary phases intrinsically denser than isochemical mixed oxides were synthesized (Ringwood, 1975). Shock-wave data (McQueen and Marsh, 1966; Jackson *et al.*, 1977) also indicate that pure Mg_2SiO_4 adopts a structure intrinsically denser than $MgSiO_3$ perovskite in the deep mantle.

In the light of this data, it now seems certain that the lower mantle consists of an assemblage of mineral phases which are a few per cent denser than isochemical mixed oxides. By a depth of 1000 km, experimental data indicate that pyrolite would crystallize in the following mineral assemblage: $MgSiO_3$ perovskite, $CaSiO_3$ perovskite, (Mg, Fe)O rocksalt, and $NaAlSiO_4$ (calcium ferrite). Further transitions to slightly denser states appear quite likely to occur between 1000–2700 km and may be responsible for the very small seismic discontinuities observed in the lower mantle by Johnson (1967) and others.

The conclusion that the mineral assemblage present in the lower mantle is more closely packed than an isochemical mixture of component oxides effectively resolves the density discrepancy referred to earlier. There seems little doubt that material of pyrolite composition occurring in a mineral assemblage a few per cent denser than isochemical mixed oxides is capable of explaining the density and seismic velocity distributions observed throughout most of the lower mantle (Ringwood, 1970, 1975; Wang and Simmons, 1972; Mao, 1974; Davies and Dziewonski, 1975).

3. Geochemistry of the Mantle

3.1. Bulk Composition

The probable model composition of the upper mantle was derived by direct studies of upper mantle materials. In the previous section, we concluded that experimentally discovered phase transformations occurring in material of pyrolite composition were capable of providing a quantitative explanation of the distribution of seismic velocities and densities throughout the entire mantle. In particular, there is no evidence which requires the existence of major chemical zoning (e.g. an increase of FeO/MgO ratio with depth) throughout the mantle, except in the uppermost and lowermost 200 km. It seems quite likely, therefore, that the pyrolite bulk composition of the upper mantle applies throughout nearly all of the mantle. At the very least, this would be a reasonable assumption in attempting to estimate the bulk composition of the mantle.

This conclusion receives support from our interpretation of the dynamical behaviour of the mantle as manifested in plate tectonic processes. It is clear that there has been a considerable degree of convective redistribution of material throughout the upper mantle and transition zones. Thus, the pyrolite now present in the low-velocity zone which is believed to represent the immediate source material of modern basalts may ultimately have been derived from considerable depths in the mantle. There are now some important indications (Jordan, 1975, 1977; Fitch, 1977) that subducted lithosphere plates sink below the 650 km discontinuity into the deep mantle. If it can finally be demonstrated that convective interchange of material occurs throughout the *entire* mantle, then our previous inferences concerning the relevance of the pyrolite composition to the bulk composition of the entire mantle will have been further strengthened.

3.2. Chemical Homogeneity vs. Heterogeneity in the Mantle

Strictly speaking, the above conclusion is valid only over scales exceeding those of seismic wavelengths. At smaller scales, various kinds of local differentiation processes have affected large volumes of the mantle. The most important of these arise from differentiation associated with the generation and subduction of lithosphere plates. These processes represent a two-stage, irreversible differentiation of the mantle (Ringwood, 1969, 1975). Pyrolite partially melts beneath mid-oceanic ridges to form a basaltic crust overlying a residual peridotitic lithosphere. The differentiated lithosphere is then subducted into the mantle beneath trenches where a second stage of partial melting of the oceanic crust occurs, leading to the removal of a small amount of siliceous liquid, rich in incompatible elements. Via a complex series of petrochemical processes, this liquid participates in the formation of andesitic magmas which are ultimately incorporated into continents. The refractory differentiated lithosphere sinks deeply into the mantle. Its composition is such that it can never again serve as a source of basaltic magmas. Rather, it displaces primitive pyrolite upwards into the lower-velocity zone to serve as a continuing source of basaltic magmas.

Since the bulk composition of the differentiated lithosphere averaged over dimensions of 50 km or so is still very similar (except for incompatible elements) to that of pyrolite, our earlier conclusion concerning the applicability of the pyrolite composition to the bulk mantle is not affected. Moreover, at a very early stage of Earth history, the amount of differentiation of this kind which had occurred must have been quite small.

The mantle has also been subjected to very complex and pervasive chemical differentiation processes involving highly incompatible elements (Gast, 1968). Typical examples are the processes which have caused depletions and enrichments of barium and the light rare earths and fractionation of Rb from Sr, and Pb from U, in the source regions of basalts. Nevertheless, these affect elements which, in the main, are present only in small concentrations, and appear to be decoupled from the major element chemistry. The results of Sun and Nesbitt (1977) imply that the degree of such minor element heterogeneity was much smaller early in the Archaean. Apparently, the fractionation processes have been evolutionary in nature.

It is hazardous to make general statements about the degree of chemical heterogeneity or homogeneity of the mantle without defining both the scales and components involved. Although the differentiation processes referred to above are of considerable geochemical interest and importance, their scales are unlikely to invalidate our estimates of the bulk major element mantle composition, based upon the inference that this approximates the mean composition of the upper mantle pyrolite.

Compatible Elements

In contrast to the wide dispersions of incompatible elements in the upper mantle, the dispersions of abundances of many compatible elements, on the basis of currently available data, appear to be remarkably low (Ringwood, 1978). For example, in fresh ultramafic rocks of mantle origin, the vast majority of Ni, Co and Mn abundances are within ±50 per cent of the mean values. Although the data are more limited, a similar pattern seems to apply to Ga and Ge in ocean floor basalts and presumably also to their source regions. The dispersion of zinc abundances in ultramafic rocks and basalts also appears to be low. Moreover, the vast majority of mantle-derived alpine ultramafics and lherzolite xenoliths in alkali basalts possess $FeO/(FeO+MgO)$ ratios within the range 0.07–0.11 (molecular). These displays of uniformity on a world-wide basis suggest that a corresponding degree of uniformity extends throughout the low-velocity zone. Because of the large degree of convective circulation believed to have occurred within the mantle, uniformity of this type may well extend to great depths. It will be seen later that these abundance patterns possess considerable cosmochemical significance.

3.3. Limitation on Large-scale Melting and Differentiation

When large bodies of magma cool within the Earth's crust, they display strong fractionation caused by crystallization differentiation. In general the efficiency of this process is expected to increase with the dimensions of the body of magma. If the mantle had been totally or extensively melted during formation of the Earth, the inevitable operation of crystallization differentiation should have resulted in gross petrological and chemical diversity on a very large scale. This is not observed. The evidence previously discussed, implying a large degree of chemical homogeneity throughout the mantle, more especially at an early stage of its evolution before the smaller-scale differentiation processes connected with plate tectonic and dynamical processes in the mantle had become prevalent, provides strong evidence that the mantle did not experience a large degree of melting during the formation of the Earth.

This conclusion is further supported by studies of the abundances of many relatively incompatible elements in pyrolite (Sun and Nesbitt, 1977). Several of these elements, including Ti, Y, Zr, Nb, Sc and the heavy rare earth elements, are present in pyrolite in primordial ratios, when compared to major elements such as Mg, Si, Ca and Al. If the mantle had ever been totally or largely melted, intense fractionations of these elements would have occurred.

3.4. Cosmochemical Aspects of the Pyrolite Composition

It is now generally agreed that, except for elements more volatile than xenon, the elemental abundances in Type I carbonaceous (C1) chondrites closely approach the primordial abundances of the solar nebula, in which the Earth was formed. The close agreement between C1 and solar abundances is shown in Fig. 5.

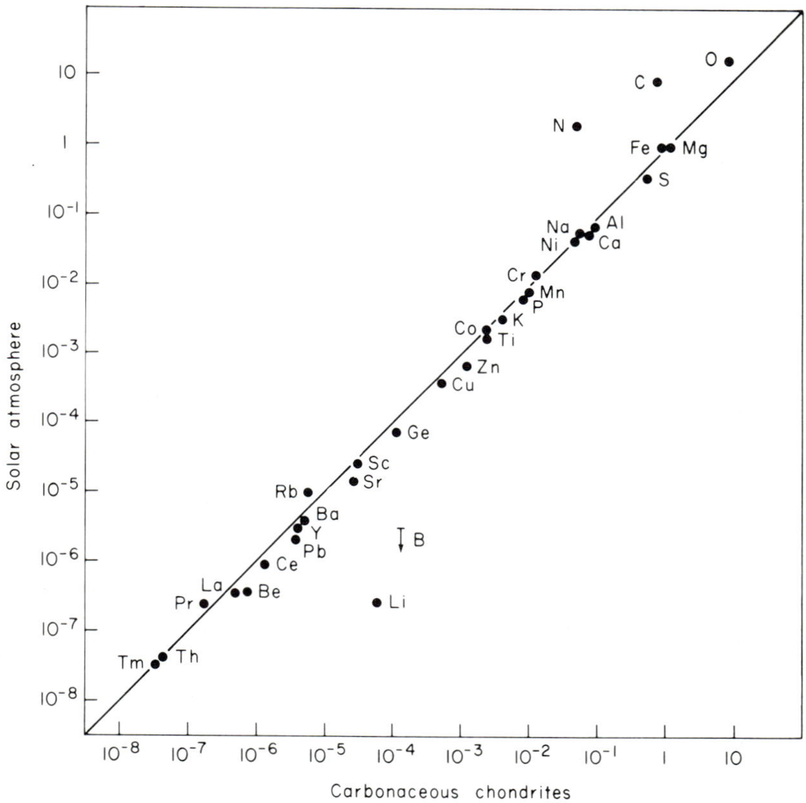

Fig. 5. Comparison of elemental abundances in Type 1 carbonaceous chondrites with those in the solar photosphere. All abundances normalized on the basis of Si $= 10^6$. (From Ross and Aller, 1976.)

A model mantle composition may be derived from the primordial C1 composition by removing siderophile elements (assigning them to the notional "core") but leaving sufficient FeO in the mantle composition to yield an MgO/(MgO + FeO) ratio of 0.89 which corresponds to the pyrolite ratio. This "primordial mantle" composition is compared with pyrolite in Table 1.

The overall agreement in abundances of most major elements is very close. In particular, the abundances of the key components of the "high-temperature condensate"† group (CaO and Al_2O_3) relative to the major components (MgO and SiO_2) is very close, implying that the high-temperature condensates, as a group, were not strongly fractionated from magnesium silicates during the formation of the Earth. The principal differences between pyrolite and the primordial mantle composition are that silica is modestly depleted in pyrolite (by 17 per cent) relative to MgO, whilst Cr_2O_3 (43 per cent), MnO (70 per cent), Na_2O (75 per cent) and K_2O (80 per cent) are depleted by increasingly large factors. It is seen in Table 5 that this is also the order of increasing volatility of these elements. We will return to this topic in the next section.

Further studies of the abundances in pyrolite of other elements belonging to the high-temperature condensate group have shown that they are present at levels (relative to the major components MgO and SiO_2) which are similar to their corresponding relative abundances as in Type 1 carbonaceous chondrites (Ringwood, 1975; Sun and Nesbitt, 1977). Elements in this category include Ti, Y, heavy rare earth elements, Zr, Nb, Sc, U and Th. It seems that fractionation of the entire group of lithophile high-temperature condensates (relative to MgO and SiO_2 during the formation of the Earth has been decidedly minor.

3.5. Volatile Elements

Ringwood and Kesson (1977) have compared the abundances of many elements in pyrolite with their abundances in Type 1 carbonaceous chondrites, normalized to magnesium in both cases (Fig. 6). As noted earlier, representative lithophile elements which are less volatile than silicon are present in primordial proportions. However, a wide range of other elements which are more volatile than magnesium are seen to be depleted by variable factors. The least volatile element to be significantly depleted in the Earth (by 17 per cent) is silicon. If we consider only the non-siderophile elements, the degrees of depletion generally increase with increasing volatility as Si, Cr, Mn, Na, K, As, Ga, Rb and F become progressively more depleted, until a plateau is reached consisting of Cs, In, Zn and Cl, possessing depletion factors in the range 0.05–0.08. There can thus be little doubt that the terrestrial depletions are primarily a consequence of selective condensation/volatilization processes. In some cases, e.g. Cu, Ag, Bi, Pb, Tl, Cd, Ge, Au, S and Se, the observed depletions in the mantle have been enhanced

† The term refers to a group of involatile elements and components which condense before Mg_2SiO_4 and metallic iron during the cooling from high temperatures of a gas of solar composition (Table 5). The high-temperature condensates include the silicates and oxides of Ca, Al, Ti, Sc, Zr, U, Th, Ba, Sr, rare earth elements, platinum metals and elemental W, Mo, Ta and Nb.

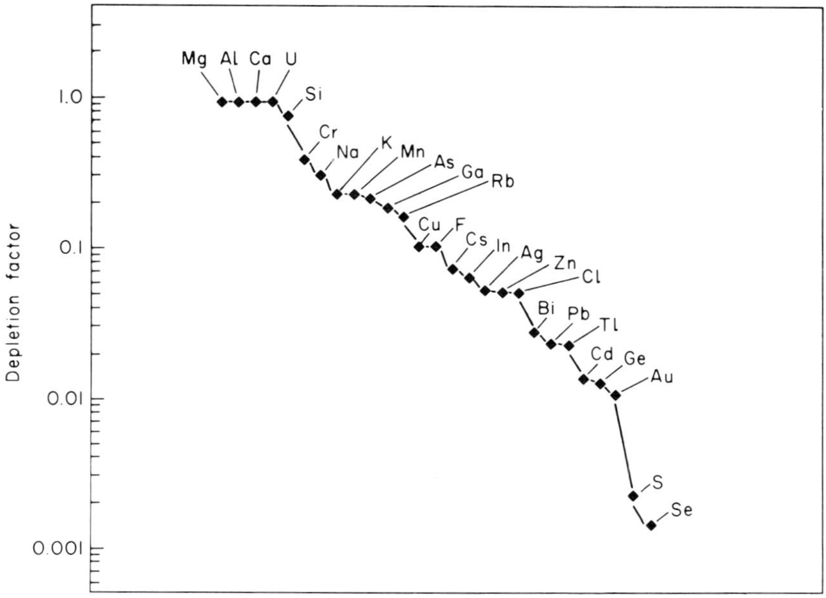

Fig. 6. Abundances of elements in the Earth's mantle compared to primordial abundances as given by C1 chondrites. (From Ringwood and Kesson, 1977.)

by their siderophile tendencies, which have caused variable degrees of entry into the Earth's core. However, the observed depletions of these elements are much too high to have been caused solely by siderophile behaviour. Elements which are much less volatile and much more strongly siderophile (e.g. Ni, Co) have not been depleted in the mantle to the same extent as Bi, Pb, Cd, Tl and Ge.

Although the depletion of silicon in pyrolite is relatively small, (~17 per cent) when compared to the primordial abundances as given by C1 chondrites, it is firmly established. Silicon is substantially more volatile than magnesium in a system of solar composition (Grossman and Larimer, 1974). In view of the strong depletions of Cr, Mn and Na in the mantle, which may reasonably be attributed to volatility, it seems likely that the small depletion of silicon is also caused by this factor.

3.6. Siderophile Elements and Redox Conditions in the Mantle

Siderophile elements are those which become preferentially partitioned in the metal phase when metallic iron (or nickel–iron) are equilibrated. Such elements are spoken of as being more "noble" than iron. Some, including Ni, Co, Ir, Pt, Re, are less volatile than silicon under appropriate cosmochemical conditions either as metals or oxides, so that they are likely to be present in the Earth in their primordial abundances. Others, such as Cu, Ag, Au, Ge and As are more volatile than silicon, so that they are likely to be depleted in the Earth compared to the primordial abundances.

Table 2. Abundances of some siderophile elements in pyrolite compared to C1 abundances together with relevant silicate/metal distribution coefficients, *K*. (From Ringwood and Kesson, 1977.)

Element	Pyrolite abundance† / C1 abundance	$K \frac{\text{Silicate}}{\text{Fe metal}}$
Involatile		
Co	0.08	0.005
Ni	0.07	0.0006
Re	0.001	5×10^{-4}–10^{-6}
Os	0.004	Probably similar to Au and Re
Ir	0.002	Probably similar to Au and Re
Pt	0.01	Probably similar to Au and Re
Volatile		
S	0.0023	
Cu	0.11	0.02–0.003
Ge	0.013	0.001
As	0.23	<0.01
Se	0.0015	
Pd	0.007	
Ag	0.055	
Au	0.01	$<3\times10^{-5}$

† Normalized to 10^6 Mg atoms.

In Table 2 we have compared the abundances of some siderophile elements in pyrolite with their primordial abundances. These elements are all depleted in the upper mantle and it may confidently be assumed that they have become dominantly segregated into the core. The partition coefficients for many of these elements between iron and appropriate silicate phases have been measured or estimated (mostly from meteoritic data) and are also given. The remarkable feature is that the elements for which partition coefficients are available are more abundant in the upper mantle by factors varying from 10 to 1000, than would be expected if these elements had been partitioned into the core under (low-pressure) equilibrium conditions. From what is known of the general thermodynamic properties of Ir, Os, Pt, Pd and Ag compounds, the same conclusion probably applies to them. The discrepancies in the cases of Cu, Au, Pd, Ag, Ge and As are probably greater than indicated in Table 2, since these elements are moderately volatile and may well have been intrinsically depleted in the bulk Earth by factors of 3–10 (Fig. 6), so that the ratios in column 1 should be multiplied by factors in this range. Likewise, sulphur and selenium are volatile, so that appropriate correction factors should be introduced.

It is clear that these relationships must have a vital bearing upon the conditions under which the Earth's core was formed, a topic discussed in the next section.

The oxidation state of iron in the mantle raises problems which are related to those discussed in connection with the "overabundances" of siderophile elements in the mantle. In order to explain the Fe^{3+}/Fe^{2+} ratios of fresh oceanic basaltic glasses and of unaltered primary mantle minerals (from xenoliths and peridotites), a Fe^{3+}/Fe^{2+} of 0.05–0.1 is required for primary pyrolite. On the other hand, the Fe^{3+}/Fe^{2+} ratios of basalts, pyroxenes and spinels which have equilibrated with metallic iron at high temperatures are much lower, probably by an order of magnitude (Ringwood, 1971).

The difference in oxidation states is reflected in the nature of the volatiles developed by degassing of the mantle. It is well known that the volatiles degassed from the Earth are dominantly composed of H_2O and CO_2, rather than H_2 and CO. Thus, the gas phase in equilibrium with an average Hawaiian tholeiite at 1200 °C would have $H_2O/H_2 \sim 120$ and $CO_2/CO \sim 35$ (Holland, 1963). These ratios agree closely with Rubey's (1951) estimates for the composition of volatiles degassed from the Earth. The observed composition of the gas phase is thus compatible with the oxidation state of the pyrolite mantle, which indeed constitutes the ultimate redox state buffer. On the other hand, if the mantle had equilibrated with metallic iron, the gas phase would be dominated by CO and H_2, rather than CO_2 and H_2O (at 1200 °C, $CO_2/CO \sim 0.05$, $H_2O/H_2 \sim 0.1$).

4. Composition of the core

4.1. Light Elements in the Core

A wide range of evidence now implies rather conclusively that the Earth's core is dominantly composed of metallic iron (e.g. Birch, 1963; Jacobs, 1975; Ringwood, 1978). However, when the density of the outer core is compared with that of iron under equivalent *P*, *T* conditions, it is found to be 8–10 per cent less dense than pure iron (Fig. 7). This discrepancy implies that

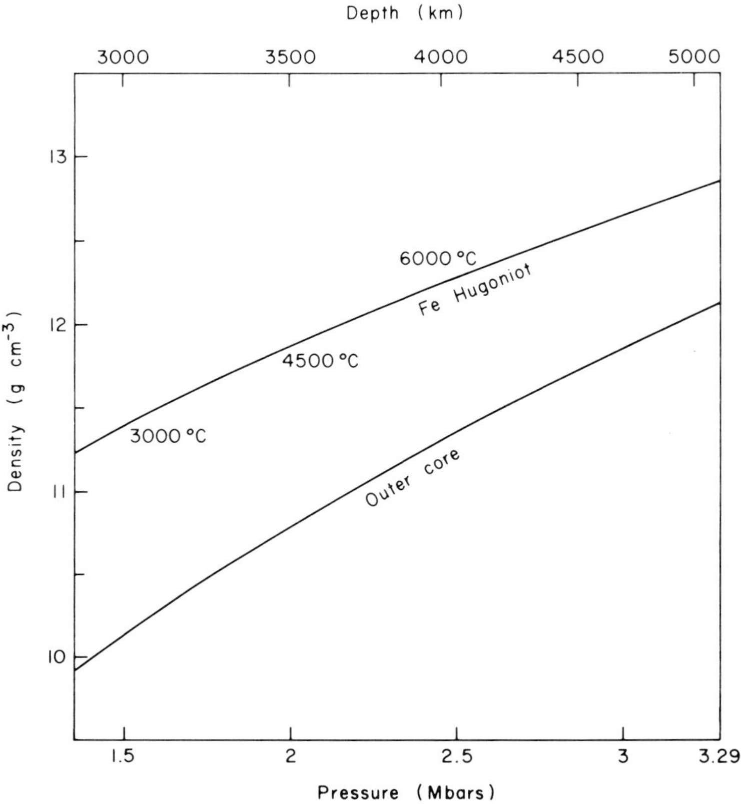

Fig. 7. Variation of density with pressure in the outer core (Dziewonski *et al.*, 1975) and along the Hugoniot for pure iron (McQueen and Marsh, 1966). Temperatures on the Hugoniot are also indicated.

the outer core contains about 5–15 per cent of an element or elements of low atomic weight. The identification, both of these light element(s) and the manner in which they become incorporated in the core, has important implications for the origin of the Earth.

In recent years, considerable support has developed in favour of the view that sulphur is most likely to be the principal light element in the outer core (e.g. Murthy and Hall, 1970). Ahrens (1976) has examined the composition of the outer core in the light of shock-wave data on the densities of FeS, Fe and Ni under very high pressures. Assuming that the additional light element in the core is sulphur, he finds that about 14 weight per cent of this element would be necessary to account for the lower density of the outer core as compared to that of a pure iron–nickel alloy of appropriate composition. This implies that the Earth must have accreted about half of the primordial abundance of sulphur, assuming the latter to be given by Type 1 (C1) carbonaceous chondrites.

In the parental solar nebula, sulphur would be condensed as FeS at temperatures below 650 K (Table 5). Nearly all metals are fully condensed at temperatures higher than this (exceptions are Pb, Bi, In and Tl). Sulphur is thus an extremely volatile element in a nebular of solar composition.

Studies of the composition of the Earth's mantle, as discussed in Section 3.5, demonstrate that it is depleted by factors ranging from 1.3 to over 100, in a wide range of elements, as compared to the primordial abundances as given by C1 chondrites. The general sequence of increasing depletion factors (Fig. 6) corresponds quite well to the sequence of condensation of elements and phases from a cooling gas of solar composition, strongly implying that the terrestrial depletions are primarily a consequence of selective condensation/volatilization processes.

A formidable problem confronting the hypothesis that sulphur is the principal light element in the Earth's core lies in explaining why such a highly volatile element is not as depleted in the bulk Earth as several *non-siderophile* elements which are much less volatile than sulphur in the solar nebula. For example, Cr, Mn, Na, K, Rb, F, Cs, Zn and Cl are all less volatile than S and yet appear to be depleted in the Earth by much larger factors. Depletion factors of these elements compared to the primordial C1 and solar abundances range from 0.3 to 0.03 as compared to the implied terrestrial depletion factor for sulphur of about 0.5.

The problem of retaining sulphur whilst losing less volatile elements is highlighted by the compositions of meteorites. In the chemical processing by which the ordinary chondrites were formed from primordial material, about 80 per cent of the sulphur was lost, but Na, K, Rb, Mn and Cr remained essentially intact. The highly oxidized, metal-free assemblage of the Karoonda C3 equilibrated chondrite indicates that it condensed at lower temperatures than ordinary chondrites, yet its sulphur content is smaller. Moreover, iron meteorites contain, on the average, only about 1 per cent of sulphur, yet many of them evidently formed in an environment (prior to metal segregation) where Na, K and Rb were undepleted, as indicated by the compositions of silicate inclusions.

We must, therefore, acknowledge the very severe difficulties encountered by models which propose that the Earth, accreting in a nebula of solar composition, condensed about half of the primordial complement of sulphur, but nevertheless lost much larger proportions of several elements which are much less volatile than sulphur. It does not seem possible to devise possible conditions of condensation and accretion within a nebula of solar composition by which sulphur is retained and these other elements lost.

There is, accordingly, a strong incentive to explore the possibility that the principal additional compound in the core is an element other than sulphur.

4.2. Iron Oxide in the Core

Dubrovskiy and Pan'kov (1972) suggested that the density of the outer core could be explained if it consisted of a mixture of about 60 per cent liquid metallic FeO, plus 40 per cent of nickel–iron. A related suggestion was put forward by Bullen (1977*a*,*b*), who proposed that the outer core consists of iron suboxide Fe_2O. These suggestions derived from efforts to interpret the density distributions within the Earth and terrestrial planets, and chemical evidence bearing upon these hypotheses was largely ignored.

Because of the technological importance of the Fe–O system, a considerable amount of chemical information relevant to this hypothesis exists, particularly in the metallurgical literature. Ringwood (1977) investigated the feasibility of the hypothesis in the light of this data. A key issue relates to the solubility of FeO in molten iron. This is very small, in the vicinity of 0.6 per cent near the liquidus at atmospheric pressure. However, the solubility of molten FeO in molten Fe increases rapidly with increasing temperatures in the range 1500–2000 °C. Distin *et al.* (1971) showed that solubility of oxygen in molten iron in this temperature interval is accurately described by the following relationship

$$\log_{10}(\text{weight per cent oxygen}) = -\frac{6380}{T} + 2.765,$$

where T is the temperature in kelvins. The high quality of their experimental results is shown by Fig. 8. They imply that the solubility of oxygen in molten iron reaches 3.1 atom per cent (equivalent to 6.2 mol per cent of $F_{1/2}O_{1/2}$) at 2000 °C. The above relationship obtained by Distin *et al.* (1971) has been used to calculate the solubility of FeO in iron at higher temperatures outside the experimental range. Data from the same workers were also used to calculate the solubilities of Fe in FeO. Results are shown in Fig. 9. It is seen that approximately 40 mol per cent of $Fe_{1/2}O_{1/2}$ would dissolve in liquid iron at 3000 °C, whilst at 3500 °C the proportion of dissolved wüstite is expected to be in the vicinity of 70 mol per cent. The temperature at the base of the mantle probably exceeds 3000 °C (Jacobs, 1975; Higgins and Kennedy, 1971), so that extensive solubility of $Fe_{1/2}O_{1/2}$ in iron should be possible in the core.

The effect of high pressures upon this solubility was also considered by Ringwood (1977). The partial molar volume of FeO when in solution in molten Fe was found to be about 18 per cent smaller than the molar volume of pure, near-stoichiometric FeO. Therefore, high pressures will tend to *increase* the solubility of FeO in molten iron. Semiquantitative calculations of this effect, including the effect of the $P\Delta v$ term on the free energy of the solution equilibria, show that high pressure will depress and reduce the extent of the liquid immiscibility field between the (Fe–FeO) metal solution and the oxide solution. Such calculations indicate that above about 500 kbar, complete miscibility would occur between metallic iron and iron oxide above the solidus. The resultant phase diagram would be qualitatively similar to that displayed by the analogous system Fe–FeS (Hansen and Anderko, 1958).

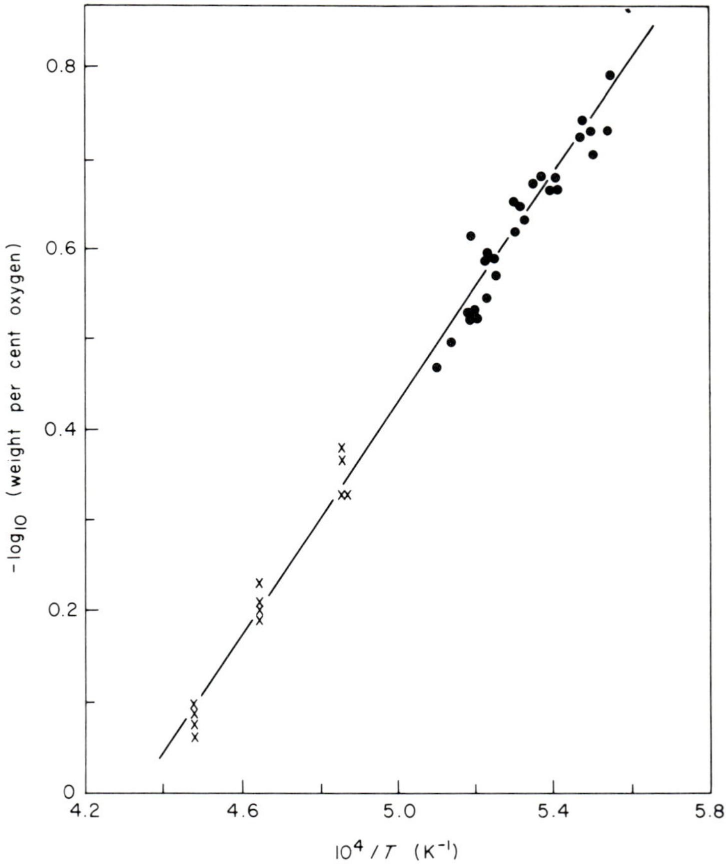

Fig. 8. Solubility (in $\log_{10}$ weight per cent) of oxygen in liquid iron vs. reciprocal of absolute temperature in range 1800–2230 K. Dots refer to data of Distin *et al.* (1971) and crosses to Taylor and Chipman (1943). (From Distin *et al.*, 1971.)

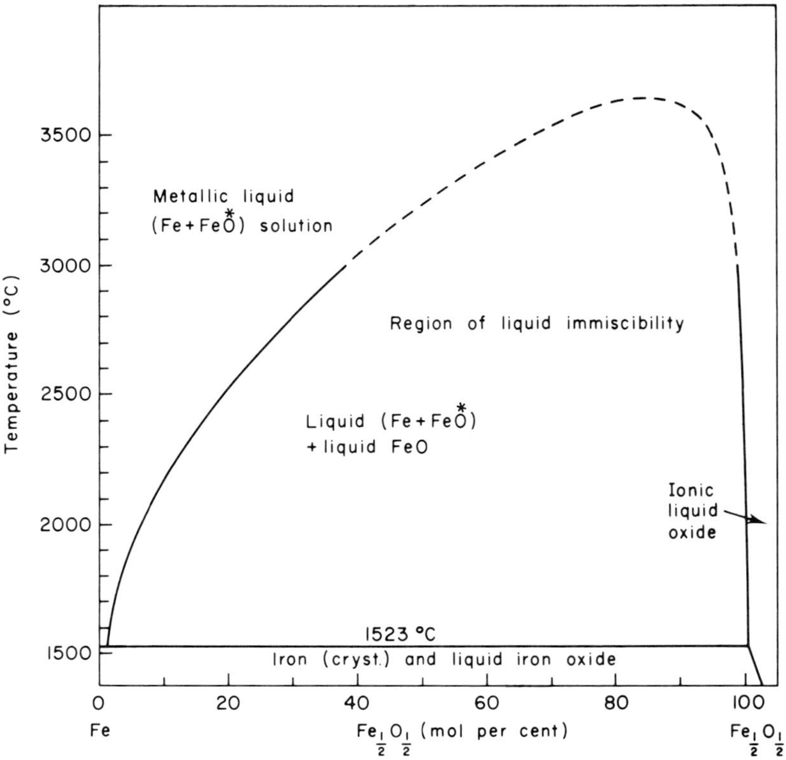

Fig. 9. Calculated phase diagram for the system Fe–$Fe_{1/2}O_{1/2}$ at 1 atm based upon experimental measurement of solubility relationships between 1500 and 2000 °C (from Fig. 8). Asterisk over FeO refers to a metallic solution.

Iron oxide is an important component of the Earth's mantle; however, its thermodynamic activity may be reduced by compound formation and solid solution in other minerals. Ringwood (1977) made a semiquantitative estimate of the significance of this effect by calculating the solubilities of FeO in molten iron in equilibrium with a possible oxide component of the lower mantle, $(Mg_{0.88}Fe_{0.12})O$, at a temperature of 2500 °C as a function of pressure. Results are given in Table 3. We see that, at this temperature, the metal phase in equilibrium with the possible lower mantle mineral assemblage would contain large amounts of FeO at pressures as low as 500 kbar and temperatures as low as 2500 °C, and that solubility increases very rapidly with pressure in the interval 500–700 kbar.

Since FeO is already *known* to be an important component of the Earth's mantle, there can be little doubt, in the light of this evidence, that oxygen (in the form of the component FeO) is also an important constituent of the core.

Table 3. Calculated solubilities of FeO in molten iron in equilibrium with $(Mg_{0.88}Fe_{0.12})O$ at 2500 °C as a function of pressure *P*. (After Ringwood, 1977.)†

P(kbars)	Solubility (mol per cent $Fe_{1/2}O_{1/2}$)
0	2
100	4
200	7
300	11
400	18
500	31
600	52
700	87

† The pressures calculated in Table 3 of the Ringwood (1977) paper are too low by a factor of two. The error arose because the molar volume of ferrous oxide used in the calculations was that of FeO rather than $Fe_{1/2}O_{1/2}$. The pressure calculated for the non-metal to metal transition in FeO at 2500 °C in that paper should accordingly be increased from 158 to 316 kbar.

The amount of FeO in the core necessary to account for its observed density is difficult to estimate. A calculation on the basis of the density of the covalent-metallic form of FeO implies that the core could contain as much as 65 weight per cent FeO (Ringwood, 1977). Alternative estimates by Dubrovskiy and Pan'kov (1972) yield FeO contents in the vicinity of 57–65 per cent. Another estimate is obtained from a comparison of the observed core density at 1.4 Mbar (Fig. 7) with the densities of pure Fe and of a mixture of $(Fe_3O_4 + Fe)$ as obtained from shock-wave data at 1.4 Mbar. This method yields a value of 28 weight per cent FeO in the core.† In view of the uncertainties, we take a value of 44 ± 16 weight per cent FeO, which is equivalent to 10 ± 4 weight per cent of oxygen.

4.3. Conditions for Core Segregation

Anticipating the argument of Section 8, we assume that the Earth accreted as an intimate mixture of metallic iron and silicate particles. Formation of the core thus required melting and segregation of iron into bodies which

† Incorrectly given as 44 weight per cent FeO on page 123 of Ringwood (1977).

were large enough to sink through the mantle into the Earth's deep interior. Urey (1952) first pointed out that this major redistribution of matter to a state of lower potential energy within the Earth's gravitational field would yield a large amount of thermal energy, largely evolved via viscous dissipation. He calculated that this would amount to about 700 cal g^{-1}, enough to heat the entire Earth by over 2000 °C. Urey suggested that segregation of much of the core occurred gradually, throughout geological time, thus driving mantle convection and providing a tectonic engine, continuing to the present day. However, Ringwood (1960) noted that the core formation process as postulated by Urey was likely to be intrinsically unstable, and once started, would proceed to completion in a very short time. The evolution of heat resulting from metal segregation would cause the effective viscosity of the mantle to fall exponentially with temperature, thereby increasing the rate of core formation, leading to a runaway process. A similar conclusion was reached by Birch (1965).

Ringwood (1960) also pointed out that segregation of the core would cause a significant fractionation of lead from uranium, since the molten iron would be expected to carry down substantial quantities of lead, but not uranium. Since the Earth's "age" of 4.55 billion years, as measured by the lead isotope method, refers to the time when, to a first approximation (and neglecting second-order differentiation effects), the present overall Pb/U ratio of the upper mantle/crust system was established, the "age of the Earth" as determined in this manner actually dates the time of formation of the core.

This argument was subsequently quantified by Oversby and Ringwood (1971), who measured the partition coefficient for lead between a series of relevant iron alloys and a basaltic melt, and obtained a relationship between the partition coefficient and the time which would have elapsed between accretion of the Earth and segregation of the core. They concluded that the time interval between accretion and core formation was probably smaller than 10^8 years and almost certainly was smaller than 5×10^8 years. It was pointed out by Ringwood (1960) and Oversby and Ringwood (1971) that this early period of core formation represents a basic boundary condition for all hypotheses relating to the origin and early evolution of the Earth.

The time interval between accretion and core formation was reconsidered by Vollmer (1977) on the basis of small subsequent revisions to the uranium decay time constants and isotopic composition of primordial lead. Vollmer concluded that core formation was not completed within 10^8 years after accretion, but more probably occurred over a period of about 5×10^8 years after accretion. However, Vollmer assumed that deviations in isotopic compositions of oceanic basaltic leads from the lead growth curve of the terrestrial crust were caused not by minor fractionations of lead from uranium in the source region of basalts (Gast, 1968) but by the core formation process itself. A wide range of geochemical evidence, including neodymium–samarium isotope systematics, cast doubt upon the first part of this assumption.

The problem was also examined by Gancarz and Wasserburg (1977) in relation to the lead isotopic composition of the 3.6 billion-year-old Amitsoq Greiss of West Greenland. They found that providing the Earth's age was close to 4.55 billion years, and that most of the lead entered the metal phase during core formation, as indicated by the experimental partition coefficient data cited earlier, the most probable time of core formation was around 4.4 billion years ago, i.e. about 1.5×10^8 years after accretion. For a wide range of initial assumptions, core formation occurred within 4×10^8 years after accretion.

5. Boundary Conditions for the Origin of the Earth

In the previous sections, we have reviewed much evidence relating to the composition and internal constitution of the Earth. We will now attempt to utilize this evidence in formulating a set of boundary conditions for theories of origin.

5.1. Abundance of Iron in the Earth

The major element compositions of the mantle and core may be combined in their known proportions to obtain the total abundances of the principal metallic elements Mg, Si, Fe, Al and Ca in the Earth. These are compared with the primordial abundances of these elements in Table 4. Apart from minor depletion of Si in the Earth (Section 3.5), the relative abundances of these major elements are identical within their uncertainty limits. In particular, there is no evidence for significant fractionation of iron from silicates, a basic postulate in the cosmogonic hypotheses of Urey (1952), Anders (1968, 1971) and many others.

5.2. Abundance of "High-temperature Condensates" in the Earth

From Table 4, we see that the terrestrial abundances of the principal members of this group, Ca and Al, are similar (relative to Fe, Mg and Si) to their primordial abundances. As discussed in Section 3.4, this characteristic applies also to many less abundant elements of this group, e.g. Ti, Zr, Nb, Sc, Y, rare earth elements and probably Ba, Sr, U and Th.

Table 4. Comparison of major element bulk composition of Earth (as obtained in this paper) with corresponding compositions for Type 1 carbonaceous chondrites.†

Atom fractions (Mg = 1.0)	C1 chondrites	Earth‡
Si	0.95	0.80
Mg	1.00	1.00
Fe§	0.90	0.92
Al	0.08	0.07
Ca	0.07	0.06

† Average of Orgueil and Ivuna (Wiik, 1956).
‡ Core assumed to contain 10 weight per cent of oxygen.
§ Ni has been included with Fe.

It is clear that the bulk composition of the Earth is related very simply to the primordial composition given by C1 chondrites and the Sun. There is no need to employ the complex seven-component fractionation models employed by Anders (1968, 1971) and co-workers in discussing the composition of the chondritic meteorites in order to explain the composition of the Earth. Our planet yields no evidence that these processes played a significant role in its genesis. This suggests that the formation of the Earth may have been much more simple, in a cosmochemical sense, than envisaged in the theories of Anders and co-workers.

5.3. Depletion of Volatile Elements in the Earth

The Earth has been depleted in a wide range of elements which are more volatile than silicon under appropriate cosmochemical conditions. For non-siderophile elements, there is a strong correlation between the degree of depletion and the degree of volatility. The overall composition of the Earth can be matched by a mixture of about 10 per cent of a low-temperature nebula condensate (similar to C1 chondrites) which has retained the primordial abundance of volatile metals, with 90 per cent of a high-temperature condensate which has been subjected to varying degrees of loss of volatile metals, in amounts which correlated with their volatilities.

Boundary conditions given in Sections 5.1–5.3 above imply that the present composition of the Earth can be derived very simply from the primordial composition (as given, for example, by C1 chondrites), simply by reduction of most of the oxidized iron and nickel in C1 chondrites to the metallic phase, accompanied by varying degrees of loss of volatile components. This relationship has often been *assumed* in the past, as in the "chondritic Earth model". It is only comparatively recently, however, that this assumption has become independently justified by our increased understanding of the structure and composition of the Earth, as reviewed in previous sections.

5.4. Chemical Homogeneity of the Primordial Mantle

Differentiation processes connected with the dynamical behaviour of the mantle as manifested ultimately in the formation and subduction of lithospheric plates have caused a large degree of local heterogeneity within the mantle throughout geological time. However, source regions of basaltic magmas which have not been subjected to these processes appear to be remarkably homogeneous in major element composition and, particularly, in their distributions of compatible elements. This degree of homogeneity was more prevalent early in Earth's history before the operation of these local differentiation processes.

It seems clear that when the Earth was formed, a high degree of chemical homogeneity prevailed throughout the mantle in the cases of major components (MgO, SiO_2, FeO, Al_2O_3, CaO) and, in particular, for crystallographically compatible groups of elements such as (Mg, Fe, Ni, Co, Mn, Zn), (Ge, Si) and (Al, Ga). The compatible elements seem to have been remarkable well mixed. They include elements exhibiting highly variable degrees of volatility (e.g. Mg, Fe, Mn, Zn; Al, Ga; Si, Ge), differing siderophile nature (Mg, Ni, Co, Mn; Si, Ge; Al, Ga) and also components which are indicators of redox conditions (FeO/MgO).

The high degree of homogeneity in the mantle of groups of elements which possess extremely different cosmochemical behaviour in the solar nebula provides strong constraints on the nature of the accretion process.

5.5. Thermal State of Mantle after Formation of the Earth

The degree of homogeneity of major element and compatible minor element distributions throughout the mantle, as inferred to be present early in the Earth's history, demonstrates that the mantle was not completely or extensively melted during formation of the Earth.

5.6. Siderophile Elements in the Mantle

Many siderophile elements (e.g. Ni, Co, Cu, Au, Re) are present in the upper mantle at levels from 10 to 100 times higher than would be expected if this region had equilibrated with metallic iron, which had subsequently settled into the core. Moreover, the upper mantle is too oxidized, as manifested by Fe^{3+}/Fe^{2+} ratios and by the CO_2/CO and H_2O/H_2 ratios of gases evolved, to have been in equilibrium with metallic iron.

5.7. Composition of the Earth's Core

Although the core is dominantly composed of iron, its density requires that it contain 5–15 per cent of an element of relatively low atomic weight. The principal light element component is probably oxygen, although a significant amount of sulphur may also be present.

5.8. Time of Formation of the Core

The core formed during or very soon after the accretion of the Earth had been essentially completed. This requires a source of heat sufficient to cause temperatures in large regions of the Earth's interior to exceed the melting point of the metal phase during or very soon after accretion. The most likely source of this energy is partial retention of gravitational energy of accretion.

6. Origin I: Homogeneous Accretion

6.1. The Parental Nebula

The solar system is believed to have formed during the gravitational collapse of portion of an interstellar cloud of dust and gas. Theoretical studies of the collapse phase of star formation (Larson, 1972; Bodenheimer, 1972) show that it is highly non-homologous, leading to the generation of a dense hydrostatic central core surrounded by an extended nebula of infalling material. This phase of development may have been observed in some infra-red stars consisting of a circumstellar shell of silicate particles radiating in the infra-red at temperatures of 300–1000 K and surrounding a young, luminous star (Bodenheimer, 1972; Werner *et al.*, 1977).

We will assume that the Earth formed by accretion within a corresponding rotating circumstellar shell of dust and gas left behind during contraction of the protosun. An impression of the rotating solar nebula at this infra-red "cocoon star" stage is given in Fig. 10. The protosun is seen to be surrounded by a series of shells of dust and gas with temperatures decreasing radially outwards.

The processes by which dust particles began to adhere to one another and to form a hierarchy of planetesimals of varying sizes are poorly understood. It is clear, however, that for accretion to commence, the relative velocities of dust grains had to be minimized so that they approached each other very slowly, thereby allowing surface chemical forces to cause them to adhere during gentle impacts. The presence of gas at this stage played a crucial role in damping random chaotic motions by viscous dissipation, thereby causing dust grains to rotate in nearly coplanar circular orbits with small relative velocities. Dust grains began to adhere, building up a first generation of small aggregations or planetesimals, perhaps with dimensions up to a few centimetres, which would become chemically isolated from the gas phase, where they formed. These small planetesimals would fall quite rapidly through the gas, reaching the equatorial plane in a period which could be as short as 1–1000 years (Safronov, 1972). Thus, a thin, relatively dense disc of dust and small planetesimals would collect in the equatorial plane (Fig. 10).

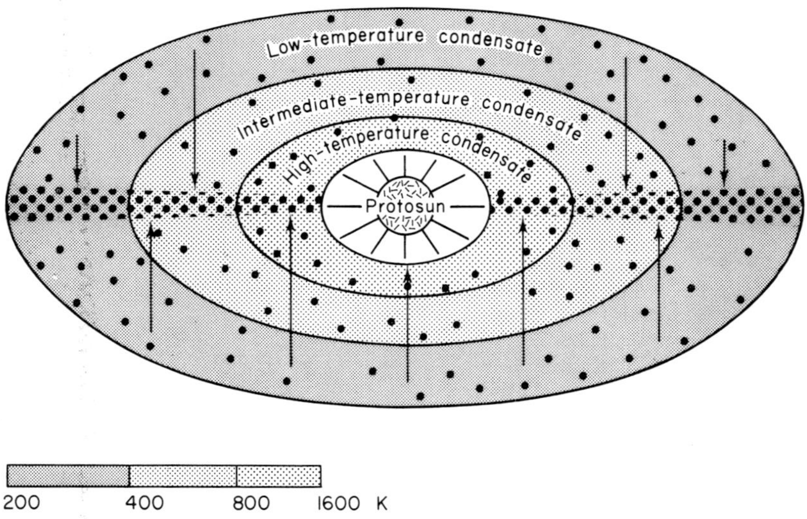

Fig. 10. Sketch showing notional temperature distribution in "cocoon nebula" surrounding sun, and prior to collapse into discoidal configuration. Small first-generation planetesimals form within shell and spiral towards ecliptic plane. Actual paths taken by sinking planetesimals are more complex than indicated and are controlled by a combination of gravity acting perpendicular to ecliptic plane and gas–solid dissipation.

It is seen from Fig. 10 that the solid material which ultimately settled into the ecliptic plane would have reached equilibrium with surrounding gas over a wide range of temperatures. Since the composition of the solid condensate is sensitively dependent upon temperature (Table 5), the condensate would display a high degree of chemical heterogeneity, varying from highly oxidized, volatile-rich planetesimals resembling C1 chondrites which formed in cool regions of the nebula, far away from the ecliptic plane, to highly reduced, devolatilized, silicate-iron planetesimals, formed at higher temperatures, close to the ecliptic plane. Notice also, that the overall ratio of low-temperature to high-temperature condensates increases with increasing distance from the Sun in the ecliptic plane (Fig. 10). Thus, there will be a tendency for the proportion of low-temperature condensates ultimately incorporated in planets to increase with distance from the Sun.

Consider, now, the formation of a planet, e.g. Earth, from planetesimals which have collected in the ecliptic plane. As emphasized by Safronov (1972), the temperature in the ecliptic plane will be low because of the high opacity of the gas–solid system which shields the planetesimals from solar radiation. In the region where the Earth accreted, temperatures would have been in the range 50–300 K, according to the particular model chosen (Safronov, 1972; Kusaka *et al.*, 1970).

Goldreich and Ward (1973) demonstrated that the disc of small first-generation planetesimals would be characterized by an acute gravitational instability which would cause it to collapse rapidly ($\sim 10^4$ years) to form a series of self-gravitating "cloudlets" which would then contract to form a second generation of planetesimals with dimensions of a few kilometres.

The accumulation of planets from kilometre-sized planetesimals which have formed in the ecliptic plane according to the theory of Schmidt (1958) and Safronov (1972) is depicted in Fig. 11. Planetesimals were initially moving in circular, coplanar orbits so that approach velocities were low and a significant proportion of collisions led to effective fusion of the colliding bodies. A hierarchy of planetesimals with different sizes developed. Beyond a certain critical radius, on the order of a few hundred kilometres, the bodies would develop gravity fields which were sufficiently large to perturb the orbits of approaching planetesimals, increasing their eccentricities and inclinations (Fig. 11). This represents a key stage in planetary accumulation because, at this critical radius, the planetesimal begins to sweep up surrounding dust and small planetesimals not with its *geometric cross-section*, which is proportional to R^2, but with its *gravitational cross-section*, which is proportional to R^4 (Hartmann, 1969).

Thus, in a given zone of the solar system, if one planetesimal exceeded the critical radius, it would grow very rapidly, sweeping up mass at the expense of surrounding planetesimals. In such a zone, this particular planetesimal would grow much more rapidly than all others and would become a *planetary nucleus.* This nucleus would rapidly develop a large gravitational field and would dominate the subsequent accretional processes in its "feeding-zone." Other planetesimals may have exceeded the critical radius for gravitational growth, but in general would have been much smaller than the planetary nucleus. Thus, Safronov estimates that, for most planets, the mass

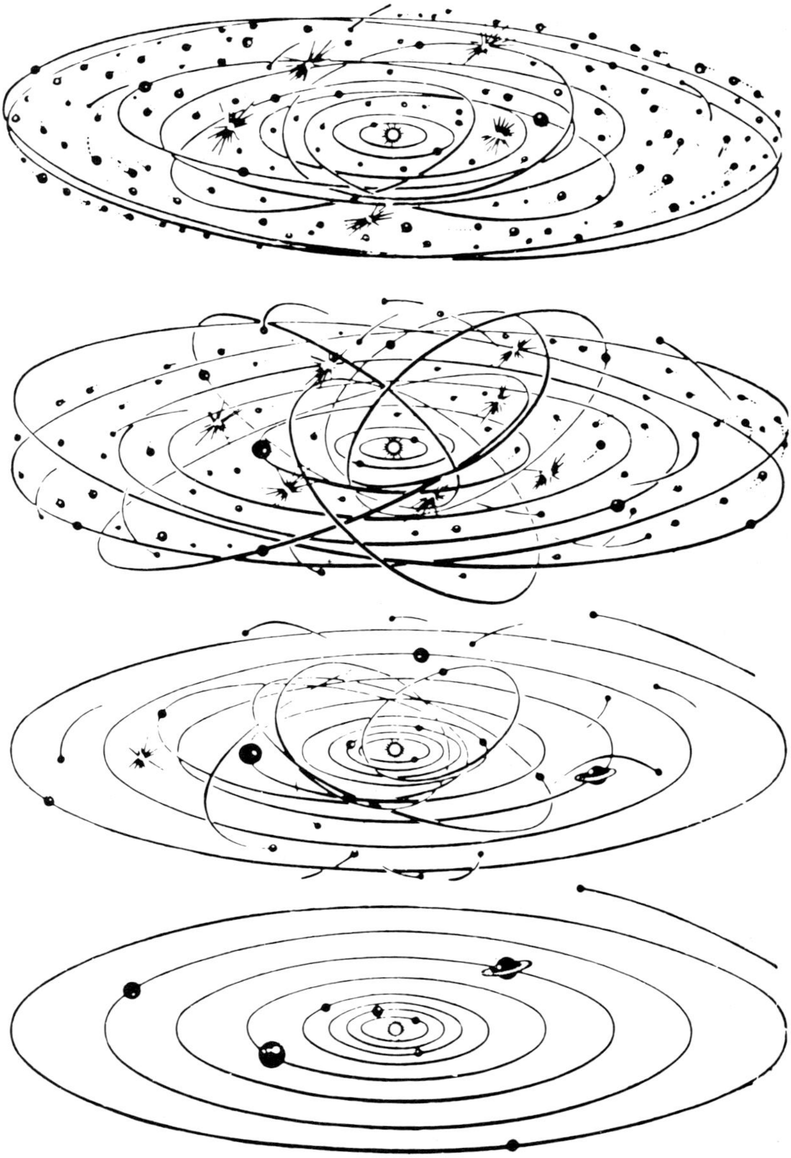

Fig. 11. Stages in the accretion of a planet from a disc of planetesimals in the ecliptic plane as envisaged in the models of Schmidt and Safronov. (From Levin, 1972.)

of the largest impacting planetesimals was from 10^{-2} to 10^{-3} of the mass of the planet.† We see from Fig. 11 that, although accretion commenced in a highly flattened disc of planetesimals rotating in coplanar circular orbits, the perturbations by growing planetary nuclei caused large increases of orbital eccentricities and inclinations of planetesimals, so that the effective thickness of the disc increased considerably during the accretion process. The present nearly circular, coplanar orbits of the planets result from an effective averaging of the chaotic motions of very large numbers of individual planetesimals. Safronov explains the present inclinations of the axes of rotation of planets as being caused by a random component of angular momentum contributed by impact of the "next largest planetesimal" upon the planetary nucleus.

As noted above, gravitational perturbations would often lead to highly elliptical orbits of planetesimals, so that there would be a considerable degree of radial mixing of planetesimals initially formed in different regions of the nebula. The effective "feeding zones" of planetesimals would thus overlap considerably. We have already seen that at the stage when the equatorial disc first formed (Fig. 10), the proportion of low-temperature C1 condensate among the planetesimals increased with distance from the sun. The subsequent radial mixing of planetesimals originally formed at varying distances from the sun by gravitational perturbations during accretion would also make a major contribution to the chemical heterogeneity of the condensates swept up by a given planet.

6.2. Early Homogeneous Accretion Hypotheses

The most widely supported class of hypotheses of the Earth's origin during the period 1950–70 maintained that the Earth accreted from an intimate mixture of silicate particles and metal particles, generally resembling chondritic meteorites (e.g. Urey, 1952, 1958, 1962; Kuiper, 1952; Vinogradov, 1961; Elsasser, 1963; Birch, 1965). The chondritic material was assumed to have been formed in the solar nebula by a complex series of chemical and physical processes which had occurred prior to the accretion of planets.

According to this class of hypotheses, accretion of the Earth occurred over a sufficiently long period (10^7–10^8 years) so that its gravitational potential energy was efficiently radiated away and it formed in an initially "cool" and unmelted condition, with an average temperature less than 1000 °C. Subsequently, heating by long-lived radioactive elements occurred, leading to melting of the metal phase and its segregation into the core, much later

† Uranus, with its rotational obliquity of 98° may have been hit by a "planetesimal" amounting to 7 per cent of its mass (Safronov, 1972).

(e.g. 10^9 years) than the primary accretion of the Earth. Thus, according to these models, not only did the Earth accrete initially from relatively homogeneous, well-mixed material, but also, for a considerable period thereafter, the Earth itself was approximately homogeneous, consisting of an intimate mixture of metal and silicate phases.

Homogeneous accretion hypotheses can explain, in principle, the major element composition of the Earth and its depletions in volatile elements, which are assumed to have resulted from the high-temperature processing which occurred in the solar nebula, prior to accretion. These hypotheses also account for the inferred approximate chemical uniformity of the mantle subsequent to core formation and the small dispersions in abundances of many compatible siderophile, lithophile and volatile minor elements in the mantle.

These early homogeneous accretion hypotheses did not, however, explain our inference that the core formed during, or very soon after the accretion of the Earth. They also encountered some serious chemical difficulties:

(i) If the Earth had once consisted of an intimate mixture of iron and silicate phases which had been slowly heated until the metal melted and segregated, it is inevitable that local chemical equilibrium would have been established between metal and silicate phases. Yet the abundances of Ni, Co, Cu, Au and many other siderophile elements in pyrolite are 10–1000 times higher than can be explained by equilibrium partition between metal and silicate phases (Section 3.6).

(ii) The Fe^{3+}/Fe^{2+} ratio of the mantle is at least ten times higher than would be expected if the mantle had equilibrated with metallic iron which subsequently separated into the core. In consequence, the volatiles degassed from the mantle are dominantly composed of CO_2 and H_2O, whereas if the mantle had once equilibrated with metallic iron, the gases evolved should be composed dominantly of CO and H_2 (Section 3.6).

(iii) The homogeneous accretion hypotheses discussed above do not account for the presence of large amounts (10–15 per cent) of a light element in the Earth's core. Hypotheses which have addressed this problem generally maintain that sulphur is the principal light element in the Earth's core (e.g. Murthy and Hall, 1970). According to these models, the depletions of volatile elements now observed in the Earth occurred in the solar nebula, prior to accretion. However, it is not possible to devise conditions whereby half of the primordial complement of highly volatile sulphur remained condensed, whereas far larger proportions of many elements which are much less volatile than sulphur (e.g. Cr, Mn, Ka, K, F, Zn) were lost (Section 3.5).

Analogous problems are encountered if other light elements (e.g. H, C, N) are postulated to be present in the core in large quantities.

These and other difficulties encountered by homogeneous accretion models caused several authors to abandon them and to explore other possibilities. Not the least of these difficulties was their complexity and contrived nature, particularly regarding the multistage chemical and physical fractionation processes which were assumed to have occurred in the solar nebula prior to accretion (e.g. Urey, 1952, 1956, 1957, 1962, 1963; Anders, 1968, 1971, 1977; Ganapathy and Anders, 1974). Our conclusion (Section 5) that the proportions of "high-temperature condensates", iron and magnesium silicates, in the Earth are similar to those in the Sun and in Type 1 carbonaceous chondrites casts considerable doubt on the need for invoking the operation of these complex processes. Moreover, they are not needed to explain existing data bearing upon the compositions of Mars and Venus (Ringwood and Clark, 1971; Ringwood and Anderson, 1977).

7. Origin II: Heterogeneous Accretion Hypotheses

7.1. Single-stage Model

In an attempt to avoid some of the difficulties encountered by early versions of homogeneous accretion hypotheses, Ringwood (1960, 1966, 1975) proposed that the Earth formed by accretion from planetesimals having compositions similar to Type 1 carbonaceous chondrites. Formation of metallic iron and loss of volatiles were believed to have occurred *during accretion* by reaction of carbon in the accreting material with oxidized iron. The temperature and thus the chemical composition of the accreting Earth was controlled by the gravitational potential energy liberated during accretion, which increases as R^2. In order to obtain sufficiently high temperatures to evaporate and lose volatile components, it was necessary to assume an accretion time smaller than 10^6 years.

According to this model (Fig. 12), the Earth formed "inside out", with a cool, oxidized, volatile-rich nucleus containing siderophile elements, and becoming progressively more metal-rich and devolatilized towards the outside. Because of these characteristics, it can be classed in the heterogeneous accretion category of models. Eventually, gravitational instability occurred, leading to rapid core segregation. Oxidized primordial material from the central nucleus was rapidly mixed in with the iron-free silicates of the rest of the Earth to form the mantle under conditions where equilibrium with metallic iron was not achieved, thereby explaining the oxidation state of the mantle and its high abundance of siderophile elements. The reduction

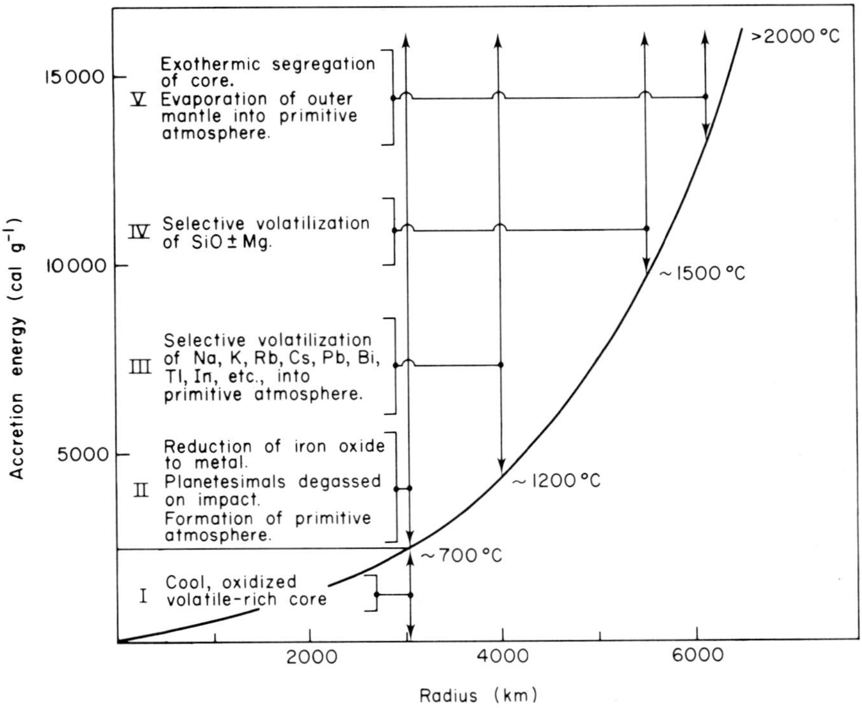

Fig. 12. Relationship between energy of accretion and radius of the growing Earth according to Ringwood's (1970) single-stage model. The principal stages of accretion are also shown in relation to the energy of accretion and approximate surface temperature.

products and volatiles escaped from the Earth continuously during accretion via turbulent mixing with hydrogen from the surrounding nebula. The principal criticisms levelled against this model were:

(i) difficulty of disposing of gaseous reduction products (e.g. Urey, 1960);
(ii) problems of forming a core out of equilibrium with mantle (e.g. Grossman, 1972);
(iii) short time scale required for accretion (Levin, 1972).

In my view, these difficulties can be overcome, as in the revised model by Ringwood (1975). However, the resultant model is rather contrived.

A more serious problem concerns the inferred original homogeneity of the mantle and the low dispersions in abundances of many siderophile, volatile and lithophile elements. It seems rather difficult to explain these small dispersions in terms of physical mixing of oxidized, volatile-rich material from the nucleus into the surrounding devolatilized, reduced mantle during core formation. It is scarcely credible that such a mixing process occurring in the solid state could be as efficient as is required. A further difficulty is that the siderophile and volatile elements from the nucleus should have been introduced into the mantle in their primordial ratios. However, the observed relative abundances (e.g. Ni/Ir, Ni/Ge) of these elements in the mantle diverge widely from primordial ratios.

7.2. Heterogeneous Accretion/Nebula Condensation Hypothesis

According to another version of heterogeneous accretion, the composition of the material from which the Earth accreted changed with time so that the layered structure of the Earth resulted, to a considerable degree from the primary accretion process (Eucken, 1944; Turekian and Clark, 1969; Clark *et al.*, 1972). The latter workers proposed specifically that most of the Earth accreted simultaneously with cooling of the solar nebula on a time scale of 10^3–10^4 years, according to Cameron's (1969) model, and that the core segregated during accretion (Fig. 13). The upper mantle is believed to have

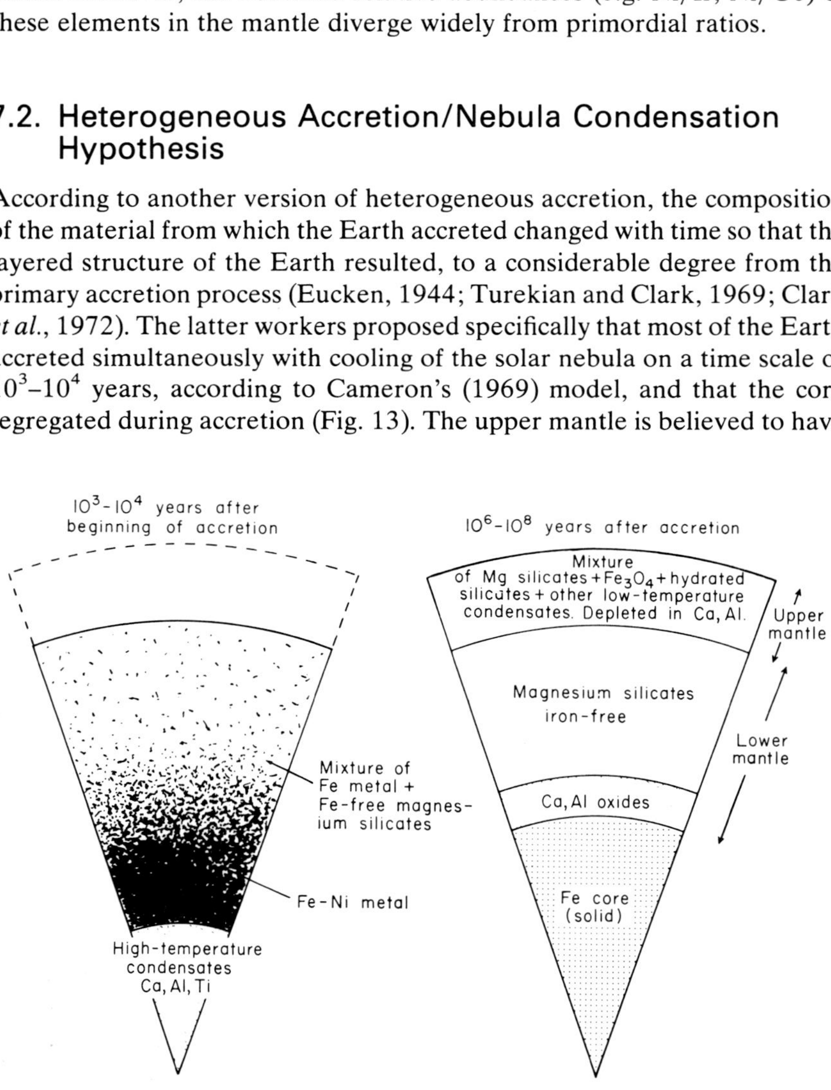

Fig. 13. Zonal structure of Earth suggested by heterogeneous accretion hypothesis of Turekian and Clark (1969) and Clark *et al.* (1977).

accreted later, as a mixture of high-temperature and low-temperature condensates over a time scale of 10^6–10^8 years. A key feature of this model is that the upper mantle has never been in contact with the core, thereby explaining the presence of siderophile elements and volatiles in the upper mantle and also its oxidation state.

This model has achieved a degree of favourable attention in the literature. Nevertheless it is beset by numerous difficulties, most of them fatal.

Primary Zoning

The model leads to an Earth zoned according to volatility (Fig. 13 and Table 5). Moreover, internal temperatures after accretion are well below the solidus at all depths. Since most of the metallic iron is concentrated towards the deep interior, the gravitational energy liberated by final segregation of the core is quite small and insufficient to cause strong heating or to drive intense convective mixing processes within the mantle. It is a source of considerable mystery as to how this configuration of the Earth, after completion of accretion, could have evolved to its present structure. The two configurations could hardly be more different.

The strongly zoned initial structure implied in this heterogeneous accretion model cannot be reconciled with our conclusion concerning the homogeneity of the mantle with respect to the distribution of siderophile, lithophile and volatile elements possessing widely different condensation temperatures in the solar nebula (Section 3.3). Once a large-scale stratification is established, it is extremely difficult to remove. Even if initially unstable, e.g. a dense layer overlying a less dense layer, the most likely result would be a simple convective overturn, leading to reversal of the layers and a stable stratification. The model does not provide an adequate means of homogenizing the mantle to the required degree.

The model also implies that the lower mantle is iron-free, but this is clearly contradicted by its elastic properties and density (Ringwood, 1975). One might expect the parental upper mantle (pyrolite) to be depleted in Ca, Al and other involatile elements relative to primordial abundances, but this is not the case.

Further difficulties encountered by this heterogeneous accretion model are summarized below. These, and others, are discussed in greater detail by Ringwood (1975, 1978).

Upper Mantle Chemistry

If siderophile and volatile elements were introduced into the upper mantle via a C1-type condensate and had never equilibriated with metallic iron, many of these volatile and siderophile elements (e.g. Ni, Ir, Ge) should be present in their primordial proportions. However, the Ni/Ir and Ni/Ge ratios of the upper mantle are 10–50 times higher than the primordial ratios.

Table 5. Condensation temperatures and sequence of phases and elements separating from gas of solar composition at 10^{-4} atmospheres total pressure. Temperatures correspond to 50 per cent condensation of a given element. (Based on Grossman and Larimer, 1974, Fig. 2; Grossman, 1972; Anders, 1968, Fig. 5; Wasson and Wai, 1976; and Wai and Wasson, 1977.)†

Element or compound	Temperature (K)
Ca, Al, Ti oxides and silicates; platinum metals, W, Mo, Ta, Zr, rare earth elements	>1400
Mg_2SiO_4	~1360
Fe–Ni metal	~1360
Remaining SiO_2 (as $MgSiO_3$)	1200–1350
Cr_2O_3	—
P	1290
Au	1230
Li	1225
Mn_2SiO_4	1190
As	1135
Cu	1118
Ga	1075
$(K,Na)AlSi_3O_8$	~1000
Ag	952
Sb	910
F	855
Ge	812
Sn	720
Zn	660–760
Se	684
Te	680
Cd	680
S (as FeS)	648
FeO (10 per cent solid solution as $(Mg_{0.9}Fe_{0.1})_2SiO_4$)	~600
Pb	520
Bi	470
In	460
Tl	440
Fe_3O_4	400
NiO	—
H_2O (as hydrated silicates)	~300

† Significant differences appear in calculations by the authors cited. We have selected what appear to be the most reliable estimates based upon recent thermodynamic data.

Composition of Core
The model does not provide a means of introducing light elements into the core which is insulated from the FeO and FeS of the upper mantle by an iron- and sulphur-free lower mantle.

Accretion Time Scale
The model assumes that most of the Earth accreted on a time scale of 10^3–10^4 years, in accordance with Cameron's (1969) model for the formation of the solar system. Such a short time scale raises severe dynamical problems which are unsolved (Levin, 1972). Moreover, Cameron (1975) has since abandoned his model. To the extent to which the heterogeneous accretion hypothesis is based upon Cameron's model, it would appear to be without foundation.

8. Homogeneous Accretion Revisited

8.1. Some General Considerations

The difficulties encountered by the heterogeneous accretion models are sufficiently serious as to justify reconsideration of the homogeneous accretion hypothesis, particularly in the light of more recent developments. For example, the studies of Safronov (1972) and Wetherill (1976) have shown that, providing a substantial proportion of accreting bodies were large, ranging perhaps up to 1000 km in radius, sufficient accretional thermal energy could be trapped within the Earth to permit early formation of the core, even for an accretion time scale as long as 10^8 years. This removes one of the difficulties faced by early versions of homogeneous accretion.

In our discussion of the infra-red cocoon protostar as a model for the early development of the solar system (Section 6.1), it was noted that solids may have become chemically isolated (as planetesimals) from the gas phase over a very wide range of temperatures (Fig. 10). The planetesimals which spiral down to the ecliptic plane would thus show a great diversity of compositions, controlled by the condensation sequence of Table 5. This diversity is increased by the gravitational perturbations of planetesimal orbits during accretion, which leads to a considerable degree of radial mixing. There would be a complete continuum from low-temperature condensates, similar to C1 chondrites, through planetesimals displaying increasing degrees of reduction of metallic iron and loss of volatiles, to relatively high-temperature condensates, totally reduced and strongly devolatilized, even to the extent of losing about 17 per cent of silicon, as SiO.

The subsequent homogeneous accretion of the Earth from this mixture of planetesimals readily explains some of our first-order conclusions concerning its composition, namely, the approximate chondritic relative abundances of major elements, Mg, Fe, Si, Ca, Al, and of elements less volatile than Si. It also explains the depletions of volatile elements in the Earth and our conclusion that the degree of depletion increases approximately with volatility. Finally, it explains the inferred high degree of homogeneity of the primordial mantle with respect to the distribution of crystallographically compatible lithophile, siderophile and volatile elements.

Although it is important to recognize that the Earth accreted from a mixture of planetesimals possessing widely different thermal histories and, hence, compositions, some aspects of our discussion of the subsequent accretion of the Earth can be simplified if we consider the Earth to have accreted as a mixture of two principal components: A, an oxidized, volatile-rich primordial component similar to C1 chondrites, and B, a highly reduced, metal-rich devolatilized material formed at high temperatures. Compositions of components A and B are given in Table 6.

Table 6. Compositions of Orgueil C1 chondrite and of reduced, devolatilized component condensing at elevated temperatures (~1000 °C) in solar nebula.

	Orgueil† A	High-temperature‡ component B
Metallic nickel–iron (~5 per cent Ni)	—	34.1
SiO_2	21.7	32.8
TiO_2	0.1	0.2
Al_2O_3	1.6	2.8
Cr_2O_3	0.35	0.2
MnO	0.2	0.1
FeO	22.9	—
NiO	1.2	—
MgO	15.2	27.7
CaO	1.2	2.3
Na_2O	0.7	—
K_2O	0.07	—
P_2O_5	0.3	—
Water	19.2	—
Organic compounds	9.7	—
Sulphur	5.7	—

† Wiik (1956).

‡ Based on bulk composition of Earth (Ringwood, 1975) minus volatile alkalis and with all oxidized iron reduced to the metallic state.

In order to account for the abundances of non-siderophile volatile elements (e.g. Zn, Cl, Cs, In) in the mantle, Ringwood (1977) estimated that about 10 per cent of component A would need to be mixed with 90 per cent of component B. On the other hand, in order to account for the large amount of oxygen believed to be present in the core, about 15 per cent of component A would be necessary. The discrepancy (if real) is not serious, and plausible explanations can be advanced (Ringwood, 1977, 1978).

8.2 Accretion of the Earth

We now consider accretion of the Earth from a mixture of initially small planetesimals comprising about 15 per cent of component A and 85 per cent of component B. The accretion time scale is not a key factor in the following discussion, which could accommodate a "long" time scale of 10^7–10^8 years. The sizes of the initial planetesimal population in the ecliptic plane of the nebula ranged from a few kilometres downwards. Accretion is believed to have proceeded as discussed in Section 6.1, whereby a population of intermediate-sized bodies with radii extending perhaps up to 1000 km was produced. The Earth thus formed from a hierarchy of intermediate bodies, planetesimals and dust, as discussed by Safronov (1972) and Wetherill (1976).

During the early stages of accretion, the accretion energy was low (Fig. 11). This would permit the trapping of a significant proportion of highly volatile components introduced by component A at this stage. After the mass of the growing nucleus had increased to about one tenth of the present Earth mass, planetesimals and the intermediate-sized bodies would have arrived at the Earth's surface with sufficiently high velocities to cause strong transient heating and evaporation of volatiles. Moreover, the impact velocities would have caused total disruption and fragmentation of incoming projectiles, accompanied by intense and repeated comminution and mixing. This would have caused intimate mixing of components A and B, probably at the centimetre (or smaller) scale. This state of intimate mixing of A and B thus prevailed throughout most of the accretion of the Earth.

The temperature distribution during accretion depended upon several variables, especially the time scale of accretion and the opacity of the atmosphere. For a wide range of plausible initial conditions, the *average* near-surface temperatures throughout most of the accretion of Earth were likely to have exceeded the boiling point of water. Thus, water evaporated from component A during impact of planetesimals with the Earth would have accumulated in the primitive atmosphere, and, together with hydrogen (see below), would be its principal component. Smaller amounts of other gases, including N_2, NH_3, inert gases and CO (or CO_2, dependent upon temperature), would also accumulate in the primitive atmosphere. Moreover, transient high temperatures exceeding 2000 °C would be reached during the entry of planetesimals and dust with high velocities into the primitive atmosphere. The interaction of these smaller bodies and particles with the primitive atmosphere would resemble that of meteors in the present atmosphere. Most would be totally evaporated and recondensed before they reached the surface, but would have produced extremely high transient temperatures. Intermediate-sized bodies such as the larger meteorites would have been extensively oblated at high temperatures, losing most of their mass before reaching the surface. The largest bodies would penetrate the atmosphere without large proportionate loss of mass, becoming intensely shock heated and degassed upon impact with the solid earth. A large proportion of these bodies would be vaporized and recondensed. Thus, only a small proportion of the incoming component A would retain its volatiles during atmospheric entry and impact with the growing Earth.

Our chosen mixture of 15 per cent component A and 85 per cent component B would contain the equivalent of 3 per cent H_2O and 29 per cent metallic nickel–iron (5 per cent Ni). The preponderance of metal phase is to be noted. Thus, the principal chemical interaction of infalling planetesimals with the primitive atmosphere will be between metallic iron and water vapour. The equilibrium constants for the relevant reaction are given in Table 7.

Table 7. Equilibrium constants for the simplified reaction $MgSiO_3 + Fe + H_2O = \frac{1}{2}Fe_2SiO_4 + Mg_2SiO_4 + H_2$ corresponding to olivine and pyroxene compositions with $FeO/(FeO+MgO) = 0.22$. (Based on Mueller, 1964.)

Temperature (°C)	"K" $= P_{H_2}/P_{H_2O}$
400	132
600	35
800	17
1000	11
1200	7.6
1400	10

It is immediately seen that most of the water vapour in the primitive atmosphere will be reduced by infalling excess metallic iron to form FeO (which enters ferromagnesian silicates in the accreting Earth) and hydrogen.

In the presence of a young and active sun, the exosphere temperature of the Earth is likely to have been significantly higher than at present. Even now, hydrogen is known to escape by thermal evaporation from the Earth's exosphere. It seems likely, then, that the hydrogen produced by the above reaction would continue to escape from the atmosphere whilst the Earth accreted. This effect would be greatly enhanced by very high temperatures transiently reached when dust and planetesimals entered the primitive atmosphere at very high velocities.

A simple mass-balance shows that most of the FeO finally incorporated in the Earth would have been produced by reaction of metallic iron in component B with water vapour from component A. The proportion of FeO introduced directly from component A was much smaller (Ringwood, 1977).

Although most of the volatile components from component A were transiently evaporated into the atmosphere during entry into the atmosphere, only the most highly volatile components—e.g. H_2, H_2O, CH_4, CO_2, CO, N_2, NH_3, some H_2S and inert gases—remained in the atmosphere. Components of lesser volatility, e.g. most of the depleted metals (alkali metals, Pb, Tl, Bi, etc.) from component A simply recondensed in the primitive atmosphere and were incorporated into the solid Earth. It seems likely that most of the sulphur in component A was likewise ultimately condensed as FeS, given the redox state of the primitive atmosphere—see further discussion by Ringwood (1977).

A relatively small proportion of component A would have reached the Earth's surface in a cool state, retaining its volatiles, and would thereby become incorporated in the growing Earth. However, most of the water in this component would eventually be reduced by excess metallic iron as temperatures and pressure rose:

$$Fe + H_2O = FeO + H_{2\,(Fe)}; \qquad \Delta v = -\text{ve}.$$

The hydrogen would form interstitial solutions with metallic iron and would ultimately be carried into the core.

8.3. Formation of the Core

The temperature distributions within the Earth after accretion are given in Fig. 14 for two very different accretional models, one for a time scale of 10^6 years and the other for a time scale of 10^8 years. Although they are based upon radically different assumptions, the two estimates in Fig. 14 are not grossly dissimilar, the maximum temperature difference between the curves being in the vicinity of 1000 °C.

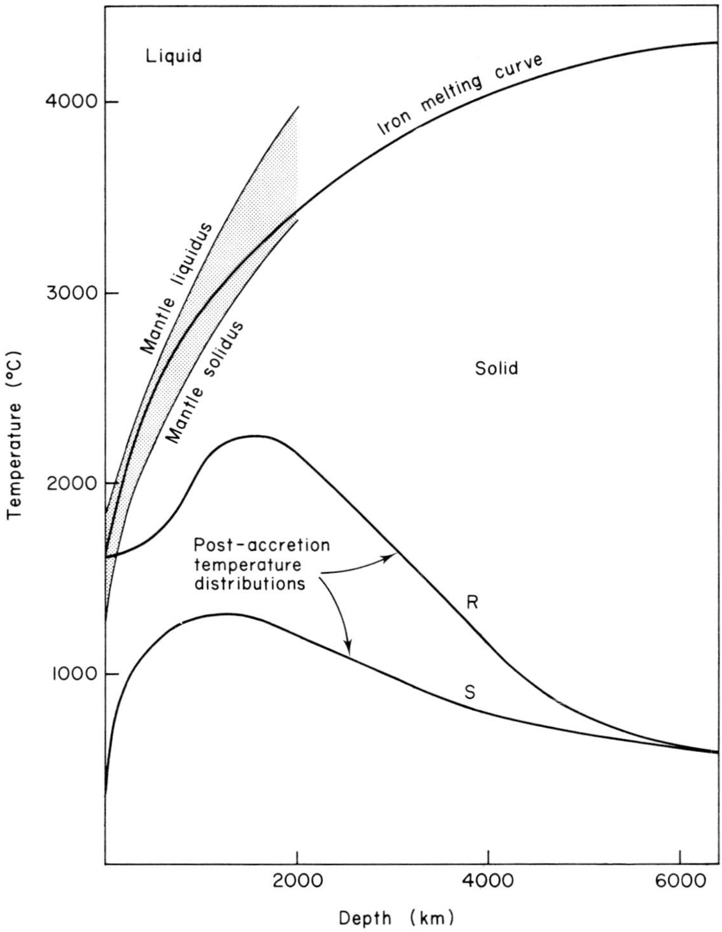

Fig. 14. Estimates of melting temperatures of iron and melting interval of mantle (Higgins and Kennedy, 1971; Kennedy and Higgins, 1973) together with estimates of post-accretional temperature distributions within the Earth, prior to core formation. Curve S = Safronov (1972) and is based upon the assumptions that the Earth accreted on a time scale of 10^8 years and that a substantial proportion of the accreting bodies were large objects exceeding 100 km in size. Curve R = Ringwood (1975) and requires a short accretion time scale so that surface temperatures reached during most of the accretion period were sufficiently high to evaporate the volatile elements inferred to be depleted in the mantle.

The estimated melting point of iron with depth is shown in Fig. 14 in relation to estimates of the mantle liquidus and solidus to a depth of 2000 km. It appears likely that the melting point of pure iron with depth lies between the solidus and liquidus of the silicate mantle at least to a depth of

1000 km, and probably to 2000 km. This corresponds to a region containing about half of the mass of the Earth. We note, moreover, that the melting curve for iron and the mantle liquidus are much higher than either of the two post-accretional temperature profiles given in Fig. 14. This is also true of the mantle solidus, except within the uppermost 100 km.

We have concluded earlier that segregation of the core occurred close to 4.5 billion years ago and that most of the Earth's mantle remained unmelted during the process. In order to begin the segregation of metal from silicate, it is necessary to exceed the melting point of the metal phase so as to permit it to segregate into large bodies. However, the similarities of iron and mantle melting intervals down to great depths would seem to make it difficult to achieve this without melting the silicate phase. Moreover, the fact that the melting curves of iron and mantle materials are much higher than the expected post-accretional temperature distribution increases the difficulty of early core formation.

The most plausible solution to these problems is the proposal by Murthy and Hall (1970) that the melting point of iron (but not of silicates) was lowered strongly by another component. The presence of large amounts of a light element in the core is well established (Fig. 7). According to the arguments developed in this paper, the additional component is believed to consist mainly of FeO, though significant quantities of sulphur could also be present. This is expected to have a drastic effect in reducing the melting point of iron. Thus, 1 per cent of FeO in solution is known to decrease the melting point of iron by 11 °C, whereas 1 per cent FeS reduces the melting point by only 6 °C (Hansen and Anderko, 1958). It has been suggested (Section 4.2) that, under high pressures, complete miscibility may occur between Fe and FeO (as in the Fe–FeS system), and we would therefore expect FeO to greatly reduce the melting point of iron at pressures sufficiently high for miscibility to occur. These are probably attained at depths of about 1300 km.

The commencement of core segregation at great depths would have caused acute convective instability in the overlying regions. This was accentuated because of the decrease in FeO/(FeO+MgO) ratio of the silicate–oxide phases with depth as FeO entered the metal phase. After segregation of metal from a localized region, the residual silicate–oxide phase rose convectively or diapirically upward, to be replaced by a new batch of dense FeO-rich silicate and metal mixture from the upper mantle. This new batch quickly re-equilibrated. FeO migrated from silicate–oxide phase into the molten metal phase, which segregated and sank into the core. The FeO content of the residual silicate–oxide phase was thus established by high-pressure equilibrium with (Fe–FeO) metal phase deep within the Earth. Ultimately, the entire volume of the Earth was convectively recycled in this manner and all of the metal entered the core. One of the principal results of this process was to produce a fairly uniform mean FeO/(FeO+MgO) ratio in the silicate–oxide phase, this ratio reflecting equilibria established ultimately in the deep mantle. A more detailed discussion of the processes responsible for the small dispersion of FeO/(FeO+MgO) ratios in the mantle is given by Ringwood (1978).

Mao (1974) has suggested that, in the deep mantle, ferrous iron is likely to disproportionate into ferric iron plus metallic iron, e.g.

$$4FeO = Fe_3O_4\ (h.p.p.) + \epsilon Fe; \qquad \Delta v = -5.8\ cm^3.$$

This reaction appears quite plausible. Because of entropy considerations, some undisproportionated FeO would have remained in solid solution in the rocksalt and perovskite phases of the lower mantle. We suggest that this region attained a final composition corresponding to a FeO/(FeO+MgO) ratio close to 0.12. However, whilst this assemblage would be stable under subsolidus conditions, the effect of Fe_3O_4 (h.p.p.) upon the melting point of iron would be identical to that of FeO. In both cases, we are considering, ultimately, the solubility of *oxygen*, in molten iron. Indeed, this is the form in which metallurgists usually express their phase diagrams. Thus, any Fe_3O_4 (h.p.p.) produced by this subsolidus reaction would greatly lower the melting point of pure iron in the deep mantle and would dissolve to form a liquid metallic (Fe–O) alloy as the internal temperature rose above the solidus of this system.

It has sometimes been assumed that once the temperatures in the Earth's interior have risen sufficiently to permit metal segregation, the additional 600–700 cal g^{-1} liberated during rapid formation of the core will cause extensive or complete melting of the mantle (e.g. Ringwood, 1960). This, however, is not correct. The process of core formation necessarily results in strong convection throughout the Earth, leading rapidly to the establishment of an adiabatic gradient. Upper mantle temperatures (required for metal segregation) are probably in the vicinity of 1500 °C, and the adiabatic temperature distribution would prevail at greater depths. Because of the drastic effect of pressure upon melting point of silicates so that the adiabatic gradient is much smaller than the melting point gradient, the mantle below depths of 400 km or more is solid (Fig. 15).

As core-forming energy is pumped into the Earth, the mean temperature of the convecting mantle depends upon the rate at which energy can be dissipated at the Earth's surface. If the core formation process takes more than a million years, most of this energy will be radiated away by a thin layer of magma near the surface, and the net temperature distribution within the solid Earth at greater depths will remain unchanged. However, even if core formation were a catastrophic process (e.g. Ringwood, 1960, 1975) occur-

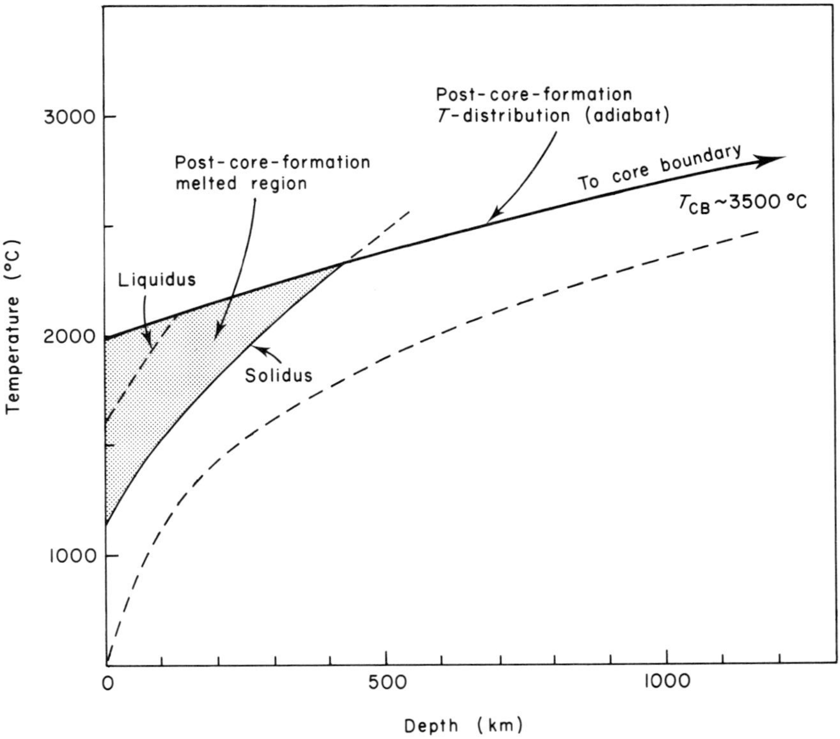

Fig. 15. Maximum temperature distribution in mantle following catastrophic core formation, in relation to pyrolite solidus and liquidus. The mantle temperature distribution follows an adiabat from 2000 °C, based upon data of Birch (1952) and Verhoogen (1956). Note extensive molten and partly molten zone extending to a depth of 400 km. Also shown is an estimate of present mantle temperature distribution (broken line). (From Ringwood, 1975, with permission).

ring, say, in a period on the order of 10^3–10^4 years, the temperature at the Earth's surface would be unlikely to exceed about 2000 °C. Above this temperature, silicates will be evaporated from the surface, a process which provides an effective thermostat. Even in this case, the molten layer would not extend below about 400 km (Fig. 15).

8.4. Siderophile Elements and Oxidation State within the Mantle

The relatively high abundances of siderophile elements and oxidized species (Fe^{3+}, CO_2, H_2O) in the upper mantle show clearly that this region could not have equilibrated with metallic iron (Section 3.6). This was a principal reason for rejecting early homogeneous accretion models. Ringwood (1966, 1971) and Brett (1971) considered the possibility that high pressures deep in the mantle might have caused increases in the distribution coefficients of some siderophile elements between metal and silicates. Although this is possible, even likely in specific cases, Ringwood rejected this as a general explanation since it would have been highly coincidental for the Δv terms in the appropriate exchange equilibria to have had the same sign, thereby permitting *all* of these partition coefficients to be changed in the same sense.

It now seems more likely that the high oxidation state and siderophile element abundances in the mantle arise from changes in the oxygen fugacity of the mantle–core system, caused partly by the extensive solubility of oxygen in the core (Ringwood, 1977). High pressures are relevant to the extent that they are responsible for solubility of FeO in the metal phase, not because of their contribution to the $P\Delta v$ terms in the free energies of the relevant exchange equilibria. Solubility of FeO decreases the activity of metallic iron in the solution, causing the oxygen fugacity of coexisting mantle silicates to increase proportionally to $1/a_{\mathrm{Fe}}^2$ (Ringwood, 1977). It seems quite possible that the oxygen fugacity in regions of the Earth's mantle which had equilibrated with Fe–FeO solutions prior to and during core formation could have been 1–1.5 log units higher than in the case where solution of FeO did not occur.

This increase in oxidation state may have been enhanced by another effect. We have earlier considered Mao's suggestion that, in the deep mantle, FeO may partially disproportionate to form a high-pressure phase of Fe_3O_4 plus metallic iron. This assemblage would be stable only under subsolidus conditions. Once the melting temperature of the iron–oxygen solution in the deep mantle had been exceeded substantially, the Fe_3O_4 (h.p.p.) phase would have been totally dissolved in excess liquid metallic iron and would ultimately have been carried into the core. However, under subsolidus conditions *prior* to core segregation, the lower mantle phases, e.g. $(Mg_{0.88}Fe_{0.12})\ SiO_3$ perovskite, would have equilibrated with Fe_3O_4 (h.p.p.) and would, therefore, contain a small proportion of Fe_2O_3 which would possess limited solid solubility in the perovskite phase. After the metallic (Fe–O) solution had segregated into the core, the Fe^{3+}-bearing perovskite would be returned to the upper mantle via convection, thereby accounting for the presence of a small amount of Fe^{3+} in this region and a greatly increased oxygen fugacity.

It seems possible that under the much higher oxygen fugacities in the mantle, caused by a combination of the effects discussed above, oxidized species such as OH^{-1} and CO_3^{-2} may have been stable throughout most of the Earth during the formation of the core. Likewise, the higher f_{O_2} conditions would be expected to cause considerable increases in the equilibrium amounts of siderophile elements which entered the silicate phases of the mantle. The high abundances of these elements might then be attributed to equilibrium partition between (Fe–O) metallic liquid and silicates.

The apparent uniformity in distribution of several compatible siderophile elements in the upper mantle may thus represent a combination of equilibrium partitions between metal and silicate phases in the deep mantle combined with convective mixing during core formation and also subsequent to core formation. The equilibrium model might also provide an explanation for other puzzling characteristics of the mantle, not readily explained by the heterogeneous accretion model, e.g. the observations that the relative abundances of many siderophile elements in the upper mantle, e.g. Ni/Ge, Ni/Ir, deviate strongly from the primordial relative abundances, as discussed earlier.

8.5. Escape of Primitive Atmosphere

Most of the low-temperature C1 component A accreting upon the Earth was completely degassed upon impact. Assuming that the mixture from which the Earth accreted contained 15 per cent of component A, the Earth should contain about 400 p.p.m. of nitrogen, most of which would have entered the primitive atmosphere. However, the actual amount of nitrogen in the present atmosphere corresponds to less than 1 p.p.m. in the bulk earth. It is clear, therefore, that a sizeable primitive atmosphere must have escaped from the Earth during or soon after accretion, if a substantial amount of component A had been accreted. Consideration of the mass balances of other volatiles in the accreting material and in the present Earth would lead to the same conclusion.

It can readily be shown that atmosphere loss on the scale required cannot occur via the Jeans–Spitzer mechanism of evaporative escape of single atoms from the exosphere. The key physical process required is that of continuous atmospheric blow-off during accretion. Öpik (1963) pointed out that blowing-off of a planetary atmosphere is governed by an escape parameter $\bar{B}$ which represents the ratio of gravitational potential energy to thermal energy of the molecules

$$\bar{B} = GM\bar{m}/RkT,$$

where G is the gravitational constant, M is the mass of the planet inside the spherical surface with radius R, $\bar{m}$ is the *mean* mass of the gas molecules, k is Boltzman's constant and T is absolute temperature. For a diatomic gas, the thermal energy equals 2.5 kT per molecule; if this exceeds the gravitational energy, the top of the atmosphere blows off into interplanetary space. Thus, for escape of this kind, which is indiscriminate as to molecular species and depends only on the mean properties of the gas,

$$GM\bar{m}/RkT = \bar{B} < 2.5.$$

In the case of the primitive Earth, rotating with angular velocity ω, the effective gravitational energy is reduced and the condition for blow-off becomes

$$\bar{B} = \frac{\bar{m}}{kT}\left(\frac{GM}{R} - \omega^2 R^2\right) < 2.5.$$

Clearly, blow-off is favoured by increasing R, T and ω, and decreasing $\bar{m}$.

Ringwood (1975) estimated that blow-off of a primitive terrestrial atmosphere would occur during accretion providing the Earth was rotating with a period of 4–5 h (as implied if the Moon were derived from the Earth), the atmospheric temperature was 1800–2000 K (comparable to the present exosphere temperature when the Sun is in an active phase) and the atmosphere possessed a mean molecular weight $\bar{m}$ of 3.5. According to his model, this low value of $\bar{m}$ was produced by turbulent mixing of the primitive terrestrial atmosphere with hydrogen derived from the solar nebula. This model requires a time scale for accretion shorter than one million years.

In the present model, the required supply of hydrogen is produced by the reaction of metallic iron in component B with water from component A:

$$Fe + H_2O = FeO + H_2.$$

This reaction occurs mostly in the primitive terrestrial atmosphere as metallic iron from component B falls with high velocity into an atmosphere of water vapour plus other volatiles evaporated from component A. The relevant equilibrium constants for the reaction (Table 7) show that, in the lower regions of the atmosphere, the H_2/H_2O ratio was probably in the vicinity of 100. Because of the great preponderance of hydrogen in the system, the mean molecular weight $\bar{m}$ is only slightly greater than 2. Under these circumstances, providing high temperatures (1500–2000 K) can be maintained in the upper atmosphere, the primitive atmosphere will blow-off continually. Thus, hydrogen produced by this reaction continually ‘washes’ other volatile components, including N_2, CO, CO_2, NH_3, inert gases and volatile metals, back into the nebula during accretion. For this reason, the total amount of component A which was accreted by the Earth may have been significantly higher than estimates based upon the content of volatile elements Zn, In, Cs, Cl now present in the mantle (Section 3.5). A substantial amount of these elements may have been lost from the Earth by the above mechanism. This may account for the discrepancy between the amount of component A implied by the present inventory of volatile elements in the mantle (10 per cent) and the amount required to explain the total FeO present in the mantle and core (15 per cent).

The required high temperature in the upper atmosphere could readily have been produced by the early Sun providing that it was a somewhat more active state than today. This appears quite likely. Transient high temperatures would also be produced by the falling of small planetesimals into the atmosphere. It is possible that atmosphere escape may have occurred via a large number of small 'puffs' as limited volumes were heated to extremely high temperatures by planetesimal infalls. The phenomena would be similar to those produced by the infall of meteors and meteorites, but on a vastly greater scale.

We conclude from this discussion that the loss of a sizeable primitive atmosphere from the Earth does not appear to present any severe problems, whether accretion occurs on a time scale of 10^6 or 10^8 years.

9. Conclusion

The present extension of the homogeneous accretion hypothesis appears capable of explaining several difficulties encountered by earlier versions of this model. These include the high abundances of siderophile elements in the mantle, the redox state of the mantle, the nature of the light elements in the core and the escape of a primitive atmosphere during accretion. It also has the major advantages of explaining the homogeneity of distribution of lithophile, siderophile and volatile compatible elements in the mantle and the bulk chemical composition of the Earth.

The viability of the present model depends to a considerable extent upon our conclusions that the compatible elements are indeed relatively uniformly distributed throughout the mantle and that oxygen is the principal light element in the core. The evidence in favour of these conclusions is strong but not yet finally decisive. Fortunately, this key evidence is in principle procurable. If the above conclusions are finally confirmed, the homogeneous accretion model will become most attractive, being simple in concept, economical of assumptions and consistent with a wide range of geochemical and geophysical data.

Acknowledgements

This paper was written whilst the author was the recipient of the Andrew D. White Professorship in the Department of Geological Sciences at Cornell University. The support and hospitality of Cornell University are gratefully acknowledged. Helpful comments and constructive criticisms from Dr R. Arculus and Dr S. Kesson are also appreciated.

References

Ahrens, T. J. (1976). Shock wave data for pyrrhotite and constraints on the composition of the outer core (abstract). In *U.S.–Japan Seminar on High-Pressure Research Applications in Geophysics*, p. 31.

Akaogi, M. and Akimoto, S. (1977). Pyroxene–garnet solid solution equilibria in the systems $Mg_4Si_4O_{12}$–$Mg_3Al_2Si_3O_{12}$ and $Fe_4Si_4O_{12}$–$Fe_3Al_2Si_3O_{12}$ at high pressures and temperatures. *Inst. Solid State Physics, Univ. Tokyo tech. Rep.* A800, 1–52.

Akimoto, S. I. (1972). The system MgO–FeO–SiO_2 at high pressures and temperatures—phase equilibria and elastic properties. *Tectonophysics* **13,** 161–187.

Anders, E. (1968). Chemical processes in the early solar system as inferred from meteorites. *Acc. chem. Res.* **1**, 289–298.

Anders, E. (1971). Meteorites and the early solar system. *A. Rev. Astron. Astrophys.* **9**, 1–34.

Anders, E. (1977). Chemical compositions of the Moon, Earth and eucrite parent body. *Phil. Trans. R. Soc. Lond. A* **285**, 23–40.

Anderson, D. L. (1967). Phase changes in the upper mantle. *Science, N.Y.* **157,** 1165–1173.

Anderson, D. L. (1968). Chemical heterogeneity of the mantle. *Earth planet. Sci. Lett.* **5**, 89–94.

Anderson, D. L. (1970). Petrology of the mantle. *Mineral Soc. Am. spec. Paper* **3,** 85–93.

Anderson, D. L. (1977). Composition of the mantle and core. *A. Rev. Earth planet. Sci.* **5**, 179–202.

Birch, F. (1952). Elasticity and constitution of the Earth's interior. *J. geophys. Res.* **57**, 227–286.

Birch, F. (1963). Some geophysical applications of high pressure research. In *Solids under Pressure* (W. Paul and D. Warschauer, eds.), McGraw-Hill, New York, pp. 137–162.

Birch, F. (1965). Speculations on the earth's thermal history. *Bull. geol. Soc. Am.* **76**, 133–154.

Bodenheimer, P. (1972). Stellar evolution toward the main sequence. *Rep. Progress Phys.* **35**, 1–54.

Brett, R. (1971). The earth's core: speculations on its chemical equilibrium with the mantle. *Geochim. cosmochim. acta* **35**, 203–221.

Bullen, K. E. (1936). The variation of density and the ellipticities of strata of equal density within the earth. *Mon. Not. R. astronom. Soc., Geophys. Suppl.* **3**, 395–401.

Bullen, K. E. (1973*a*). Cores of the terrestrial planets. *Nature, Lond.* **243,** 68–70.

Bullen, K. E. (1973*b*). On planetary cores. *The Moon* **7**, 384–395.

Cameron, A. G. W. (1969). Physical conditions in the primitive solar nebula. In *Meteorite Research* (P. Millman, ed.), Reidel, Dordrecht, Holland, pp. 7–15.

Cameron, A. G. W. (1975). The original and evolution of the solar system. *Scient. Am.* **233**, 32–41.

Clark. S. P., Turekian, K. and Grossman, L. (1972). Model for early history of the earth. In *The Nature of the Solid Earth* (E. C. Robertson, ed.), McGraw-Hill, New York, pp. 3–18.

Davies, G. F. and Dziewonski, A. M. (1975). Homogeneity and constitution of the Earth's lower mantle and core. *Phys. Earth planet. Interiors* **10**, 330–343.

Distin, P. A., Whiteway, S. and Mason, C. (1971). Solubility of oxygen in liquid iron from 1785° to 1960°. A new technique for the study of slag-metal equilibria. *Can. Metal. Qtly* **10**, 13–18.

Dubrovskiy, V. A. and Pan'kov, V. (1972). On the composition of the earth's core. *Izv. Phys. solid Earth*, No. 7, 452–455.

Dizewonski, A., Hales, A. and Lapwood, E. (1975). Parametrically simple earth models consistent with geophysical data. *Phys. Earth planet. Interiors* **10**, 12–48.

Elsasser, W. M. (1963). Early history of the Earth. In *Earth Science and Meteorites* (J. Geiss and E. Goldberg, eds.), North Holland, Amsterdam, pp. 1–30.

Eucken, A. (1944). Physikalisch-chemische Betrachtungen über der früeste Entwicklungsgesichte der Erde. *Nach. Akad. Wiss. Göttingen, Math.-Phys. K1.* **1**, 1-25.

Fitch, T. J. (1977). *In situ* P-wave velocities in deep earthquake zones of the SW Pacific: Evidence for a phase boundary between the upper and lower mantle. In *Island Arcs, Deep Sea Trenches and Back-Arc Basins* (M. Talwani and W. C. Pitman III, eds.), Maurice Ewing Series **1**, American Geophysical Union, Washington D.C., pp. 123–136.

Ganapathy, R. and Anders, E. (1974). Bulk compositions of the Moon and Earth, estimated from meteorites. *Proc. 5th Lunar Sci. Conf.* 1181–1206.

Gancarz, A. J. and Wasserburg, G. J. (1977). Initial Pb of the Amitsoq gneiss, West Greenland, and implications for age of the Earth. *Geochim. cosmochim. acta.* **41**, 1283–1301.

Gast, P. W. (1968). Trace element fractionation and the origin of tholeiitic and alkaline magma types. *Geochim. cosmochim. acta* **32**, 1057–1086.

Goldreich, P. and Ward, W. R. (1973). The formation of planetesimals. *Astrophys. J.* **183**, 1051–1061.

Grossman, L. (1972). Condensation in the primitive solar nebula. *Geochim. cosmochim. acta* **36**, 597–619.

Grossman, L. and Larimer, J. (1974). Early chemical history of the solar system. *Rev. geophys. Space Phys.* **17**, 71–101.

Hansen, M. and Anderko, K. (1958). *Constitution of Binary Alloys*, 2nd ed. McGraw-Hill, New York.

Hartmann, W. K. (1969). Angular momentum of Icarus. *Icarus* **10**, 445–446.

Higgins, G. H. and Kennedy, G. C. (1971). The adiabatic gradient and the melting point gradient in the core of the earth. *J. geophys. Res.* **76**, 1870–1878.

Hölland, H. D. (1963). On the chemical evolution of the terrestrial and Cytherean atmospheres. In *The Origin and Evolution of Atmospheres and Oceans* (P. Brancazio and A. G. W. Cameron, eds.), Wiley, New York, pp. 86–101.

Jackson, I. A., Ahrens, T. J. and Richeson, H. (1977). Dynamic compression of forsterite to 150 GPa. *Trans. Am. geophys. Union* **6**, 518.

Jacobs, J. A. (1975). *The Earth's Core*, Academic Press, London.

Jeffreys, H. (1939). The times of P, S and SKS and the velocities of P and S. *Mon. Not. R. astronom. Soc., Geophys. Suppl.* **4**, 498–533.

Johnson, L. R. (1967). Array measurements of P velocities in the upper mantle. *J. geophys. Res.* **72**, 6309–6325.

Johnson, L. R. (1969). Array measurements of P velocities in the lower mantle. *Bull. seismol. Soc. Am.* **59**, 973–1008.

Jordan, T. (1975). Lateral heterogeneity and mantle dynamics. *Nature, Lond.* **257**, 745–750.

Jordan, T. (1977). Lithospheric slab penetration into the lower mantle beneath the Sea of Okhotsk. *J. Geophys.* **43**, 473–496.

Kennedy, G. C. and Higgins, G. H. (1973). Temperature gradients at the core–mantle interface. *The Moon* **7**, 14–21.

Kuiper, G. P. (ed.) (1952). *The Atmospheres of the Earth and Planets*, 2nd ed., Chicago University Press, Chicago, pp. 306–405.

Kusaka, T., Nakano, T. and Hayashi, C. (1970). Growth of solid particles in the primordial solar nebula. *Prog. theor. Phys.* **44**, 1580–1595.

Larson, R. B. (1972). The evolution of spherical protostars with masses 0.25 $M_\odot$ to 10 $M_\odot$. *Mon. Not. R. astronom. Soc.* **157**, 121–145.

Levin, B. J. (1972). Origin of the earth. *Tectonophysics* **13**, 7–29.

Liebermann, R. C., Ringwood, A. E. and Major, A. (1976). Elasticity of polycrystalline stishovite. *Earth planet. Sci. Lett.* **32**, 127–140.

Liu, L. G. (1974). Silicate perovskite from phase transformation of pyrope-garnet at high pressures and temperatures. *Geophys. Res. Lett.* **1**, 277–280.

Liu, L. G. (1975). Post-oxide phases of olivine and pyroxene and mineralogy of the mantle. *Nature, Lond.* **258**, 510–512.

McQueen, R. G. and Marsh, S. P. (1966). In *Handbook of Physical Constants* (S. P. Clark, Jr., ed.), *Geol. Soc. Am. Mem.* 97.

McQueen, R. G., Marsh, S. P. and Fritz, J. N. (1967). Hugoniot equation of state of twelve rocks. *J. geophys. Res.* **72**, 4999–5036.

Mao, N. (1974). Velocity–density systematics and iron content of the mantle. *EOS* **55**, 416.

Mueller, R. F. (1964). A comparison of oxidative equilibria in meteorites and terrestrial rocks. *Geochim. cosmochim. acta* **27**, 273–278.

Murthy, R. V. and Hall, H. (1970). The chemical composition of the earth's core: possibility of sulphur in the core. *Phys. Earth planet. Interiors* **2**, 276–282.

Nuttli, O. W. (1969). Travel times and amplitudes of S waves from nuclear explosions in Nevada. *Bull. seismol. Soc. Am.* **59**, 385–398.

Öpik, E. J. (1963). Selective escape of gases. *Geophys. J.* **7**, 490–509.

Oversby, V. M. and Ringwood, A. E. (1971). Time of formation of the earth's core. *Nature, Lond.* **234**, 463–465.

Ringwood, A. E. (1960). Some aspects of the thermal evolution of the earth. *Geochim. cosmochim. acta* **20**, 241–259.

Ringwood, A. E. (1962). Mineralogical constitution of the deep mantle. *J. geophys. Res.* **67**, 4005–4010.

Ringwood, A. E. (1966). The chemical composition and origin of the earth. In *Advances in Earth Sciences* (P. M. Hurley, ed.), MIT Press, Cambridge, Mass., pp. 287–356.

Ringwood, A. E. (1969). Composition and evolution of the upper mantle. In *The Earth's Crust and Upper Mantle*, Geophysics Monograph **13**, American Geophysical Union, Washington D.C., pp. 1–17.

Ringwood, A. E. (1970). Phase transformations and the constitution of the mantle. *Phys. Earth planet. Interiors* **3**, 109–155.

Ringwood, A. E. (1971). Core–mantle equilibrium: comments on a paper by R. Brett. *Geochim. cosmochim. acta* **35**, 223–230.

Ringwood, A. E. (1975). *Composition and Petrology of the Earth's Mantle*, McGraw-Hill, New York.

Ringwood, A. E. (1977). Composition of the core and implications for origin of the earth. *Geochem. J.* **11**, 111–135.

Ringwood, A. E. (1978). Composition and origin of the Earth. The Vernadsky Lecture, U.S.S.R. Academy of Sciences, in press.
Ringwood, A. E. and Anderson, D. L. (1977). Earth and Venus, a comparative study. *Icarus* **30**, 243–253.
Ringwood, A. E. and Clark, S. P. (1971). Internal constitution of Mars. *Nature, Lond.* **234**, 89–92.
Ringwood, A. E. and Kesson, S. E. (1977). Basaltic magmatism and the composition of the moon. Part II. Siderophile and volatile elements in moon, earth and chondrites: implications for lunar origin. *The Moon*, **16**, 425–464.
Ross, J. E. and Aller, L. H. (1976). The chemical composition of the sun. *Science, N.Y.* **191**, 1223–1229.
Rubey, W. W. (1951). Geologic history of sea water. *Bull. geol. Soc. Am.* **62**, 1111–1147.
Safronov, V. S. (1972). *Evolution of the Protoplanetary Cloud and Formation of the Earth and Planets.* (English transl.), IPST, Jerusalem, pp. 1–206.
Schmidt, O. Y. (1958). *A Theory of the Origin of the Earth: Four Lectures*, Foreign Languages Publishing House, Moscow; Lawrence and Wishart, London, pp. 58–59.
Sun, S. S. and Nesbitt, R. W. (1977). Chemical heterogeneity of the Archaean mantle, composition of the earth and mantle evolution. *Earth planet. Sci. Lett.* **35**, 429–448.
Taylor, C. R. and Chipman, J. (1943). Equilibria of liquid iron and simple basic and acid slags in a rotating iron furnace. *Am. Inst. metal. Engrs Trans.* **154**, 228–245.
Turekian, K. and Clark. S. P. (1969). Inhomogeneous accumulation of the earth from the primitive solar nebula. *Earth planet. Sci. Lett.* **6**, 346–348.
Urey, H. C. (1952). *The Planets*, Yate University Press, New Haven, Conn.
Urey, H. C. (1956). Diamonds, meteorites and the origin of the solar system. *Astrophys. J.* **124**, 623–637.
Urey, H. C. (1957). Boundary conditions for the origin of the solar system. In *Physics and Chemistry of the Earth* (L. Ahrens, F. Press, K. Rankama and S. Runcorn, eds.), Pergamon Press, London, pp. 46–76.
Urey, H. C. (1958). The early history of the solar system as indicated by the meteorites. *Proc. chem. Soc.*, **1958**, 67–78.
Urey, H. C. (1960). On the chemical evolution and densities of the planets. *Geochim. cosmochim. acta* **18**, 151–153.
Urey, H. C. (1962). Evidence regarding the origin of the earth. *Geochim. cosmochim. acta* **26**, 1–13.
Urey, H. C. (1963). The original and evolution of the solar system. In *Space Science* (D. LeGalley, ed.), Wiley, New York, pp. 123–168.
Verhoogen, J. (1956). Temperatures within the earth. In *Physics and Chemistry of the Earth* (L. Ahrens, F. Press, K. Rankama and S. Runcorn, eds), Pergamon Press, London, pp. 17–43.
Vinogradov, A. P. (1961). The origin of the material of the earth's crust. Communication 1. *Geochemistry USSR* (English transl.), 1–32.
Vollmer, R. (1977). Terrestrial lead isotopic evolution and formation time of Earth's core. *Nature, Lond.* **270**, 144–147.
Wai, C. M. and Wasson, J. T. (1977). Nebular condensation of moderately volatile elements and their abundances in ordinary chondrites. *Earth planet. Sci. Lett.* **36**, 1–13.

Wang, H. and Simmons, G. (1972). FeO and SiO_2 in the lower mantle. *Earth planet. Sci. Lett.* **14**, 83–86.
Wasson, J. T. and Wai, C. M. (1976). Explanation for the very low Ga and Ge concentrations in some iron meteorite groups. *Nature, Lond.* **261**, 114–116.
Werner, M. W., Becklin, E. E., and Neugebaur, G. (1977). Infrared studies of star formation. *Science, N.Y.* **197**, 723–732.
Wetherill, G. W. (1976). The role of large bodies in the formation of the Earth and Moon. *Proc. 7th Lunar Sci Conf.* **3**, 3245–3257
Wiik, H. B. (1956). The chemical composition of some stony meteorites. *Geochim. cosmochim. acta* **2**, 91–117.

2. The History of the Earth's Rotation

KURT LAMBECK

Abstract

Observational evidence for the secular acceleration of the Earth's spin and lunar motion throughout the geological past is reviewed. The Earth's average acceleration for the past 500×10^6 years appears to be of the order 5×10^{-22} s^{-2}, only marginally lower than the present rate. There is no strong evidence for any non-tidal secular acceleration and this presents a limit on $\dot{G}/G$ of about $2\text{–}3 \times 10^{-11}$. Consequences and implications of the tidal evolution of the lunar motion are reviewd. The observational data do not place stringent constraints on theories of lunar evolution. Polar wander throughout the past does not appear to have been significant, despite the prediction by Goldreich and Toomre's model that polar wander should exceed the average rate of plate motions. This does not imply a high viscosity for the lower mantle but is a consequence of an absence of any isostatic compensation of the plates in their model.

1. Introduction

The discussion of the Earth's rotation is conveniently separated into three aspects: polar motion, changes in the length of day and precession and nutation. Polar motion is the movement of the axis of rotation with respect to the Earth's crust. Variations in the speed of rotation about this instantaneous axis cause changes in the length of day. Precession is the motion of the rotation axis with respect to inertial space; an oscillation of the rotation axis about the pole of the ecliptic with a period of some 26 000 years. Superimposed on this are the much shorter-period nutations. We are concerned here with the first two components of the motion, and in particular in the variations over very long time periods. The polar motion over geological time is usually referred to as polar wander while the main long-term behaviour of the length of day is characterized by a secular deceleration. Studies of these phenomena have a twofold interest. First, one would like to know what causes them. What geophysical phenomena cause the pole to wander and the Earth to slow down; what geophysically useful properties can be deduced from a knowledge of these changes. Second, both polar wander and the secular deceleration have consequences on, amongst other subjects, paleomagnetism, paleoclimatology, paleobiology and on the dynamical evolution of the Moon's orbit throughout geological time. Only some of these aspects can be touched upon in this review.

2. Length of Day

2.1. The Tidal Dissipation Problem

Telescope observations and ancient records of solar eclipses indicate that the Earth has been slowing down at a rate of about 5×10^{-22} rad s^{-1}, or that the length of the day has been increasing at a rate of about 1.5 ms per century (Muller, 1976). These observations also indicate that the Moon's mean motion is changing secularly, that the Moon is moving away from the Earth at a rate of about 4–5 cm y^{-1}. Paleontological evidence suggests that these accelerations have persisted for at least the last 500 million years. The reason for the secular evolution of the Earth's spin and the Moon's acceleration in its orbit lies in the tidal torques exerted by the Moon on the Earth and vice versa. The subject has been recently reviewed by Lambeck (1977). See also the volume edited by Brosche and Sündermann (1978). When the motion is extrapolated into the geological past, the consequence of the tidal accelerations is to bring the Moon closer to the Earth along a well-defined path and any theory of the lunar formation and subsequent evolution must comply with these orbital constraints. The Earth's spin was also faster in the past than it is now, due mainly to the tidal torques, but part may be due to a changing mass distribution inside the Earth. Hence observations of the rotation throughout the geological past may also provide some constraints on the evolution of the Earth.

The Moon's gravitational attraction raises a tide on the non-rigid Earth that can be approximated to a high degree of accuracy by an ellipsoidal bulge. Due to imperfections in elasticity, the Earth's response is not immediate and in the time, Δt, it takes for it to respond, the planet will have rotated through an angle $\Omega\, \Delta t$, where Ω is the speed of rotation. Likewise, the Moon will be displaced in longitude by an angle $n_{☾}\, \Delta t$ where $n_{☾}$ is the mean motion of the Moon. As $\Omega \gg n_{☾}$, the tidal bulge will not be aligned with the Earth–Moon axis and the Moon's attraction on this bulge results in the tidal torque,

$$L_1 = \int_V \rho \mathbf{x} \operatorname{grad} U \, dv, \tag{1}$$

where U is the potential of the lunar attraction and the integral is over the volume v of the Earth whose density at $\mathbf{x}$ is $\rho(\mathbf{x})$. Integrated over one period of the Moon's motion, this integral does not vanish if there is a lag in the response. The equal but opposite torque, created by the potential, ΔU, of this bulge on the Moon is

$$L_2 = \int_V \rho \mathbf{x} \operatorname{grad} \Delta U \, \mathrm{d}v, \tag{2}$$

where the integral is over the Moon's volume. The torque (1) leads to a modification of the Earth's spin, since

$$L_1 = \dot{H} = \frac{\mathrm{d}}{\mathrm{d}t}(C\Omega) = C\dot{\Omega}, \tag{3}$$

where H is the angular momentum of the Earth's spin and C is its polar moment of inertia. The torque (2) adds to the Moon's orbital momentum an amount equal to that lost by the Earth, or

$$L_2 = \dot{H}_☾ = \frac{\mathrm{d}}{\mathrm{d}t}\left[\frac{Mm_☾}{M+m_☾} a_☾^2 (1-e_☾^2)^{1/2} n_☾\right], \tag{4}$$

where M is the mass of the Earth, $m_☾$ the mass of the Moon, $a_☾$ the semimajor axis of the lunar orbit and $e_☾$ its eccentricity. The mean motion $n_☾$ relates to $a_☾$ by Kepler's law:

$$G(M+m_☾) = n_☾^2 a_☾^3, \tag{5}$$

where G is the gravitational constant. The component of $H_☾$ parallel to the Earth's spin is $H_☾ \cos i$. Conservation of angular momentum requires that $\dot{H} + \dot{H}_☾ \cos i = 0$. (We ignore the small contribution from the Moon's spin and assume that G, C, M, $m_☾$ remain constant). Then

$$\dot{\Omega}_{\mathrm{T}} = -\frac{1}{C}\frac{Mm_☾}{M+m_☾}\frac{\mathrm{d}}{\mathrm{d}t}\left[a_☾^2 n_☾ (1-e_☾^2)^{1/2} \cos i_☾\right], \tag{6}$$

establishes a relation between the change in the Earth's spin and changes in the lunar orbit due to the tidal torques. The contributions to $\dot{\Omega}_T$ due to $\dot{e}_☾$ and $\mathrm{d}i_☾/\mathrm{d}t$ are small and the dominant part comes from $\dot{n}_☾$ or $\dot{a}_☾$. Once $\dot{n}_☾$ is observed, these smaller contributions can be readily estimated (Lambeck, 1977) and the total lunar tidal acceleration $(\dot{\Omega}_{\mathrm{T}})_☾$ can be deduced from (6). The solar tides must also be considered for, while $\dot{n}_\odot$ will be very small, the torque by the Sun on the solar tide is significant. As $\dot{n}_\odot$ is too small to be observable, even when integrated over millions of years, $(\dot{\Omega}_{\mathrm{T}})_\odot$ can be estimated only by considering a specific model for the solar tide. At present, $(\dot{\Omega}_{\mathrm{T}})_\odot < (\dot{\Omega}_{\mathrm{T}})_☾$. Any difference between $\dot{\Omega}_{\mathrm{T}} = (\dot{\Omega}_{\mathrm{T}})_☾ + (\dot{\Omega}_{\mathrm{T}})_\odot$ and the observed acceleration determined the non-tidal acceleration $\dot{\Omega}_{\mathrm{NT}}$.

The rotational energy associated with the Earth's spin is

$$E_1 = \tfrac{1}{2} C\Omega^2$$

and the energy associated with the orbital motion is

$$E_2 = -GMm_{☾}/2a_{☾}.$$

This includes the potential and kinetic energies. The total rate of energy dissipations due to the tidal interaction is

$$\frac{\mathrm{d}E}{\mathrm{d}t} = \frac{\mathrm{d}E_1}{\mathrm{d}t} + \frac{\mathrm{d}E_2}{\mathrm{d}t} = C\Omega\dot{\Omega}_{\mathrm{T}} - \frac{1}{3} m_{☾} n_{☾} a_{☾}^2 \dot{n}_{☾}$$

$$\simeq (5.86 \dot{\Omega}_{\mathrm{T}} - 9.58\, \dot{n}_{☾})\, 10^{40} \quad \mathrm{erg\, s^{-1}}. \tag{7}$$

The rate of energy dissipation can therefore be deduced from the tidal accelerations without requiring a knowledge of the energy sink. The tidal force grad ΔU entering into the torque (2) accelerates the Moon. Of interest here are the changes in $a_{☾}$, $e_{☾}$, $i_{☾}$ due to this force. For the former we can write, approximately,

$$\dot{a} = Aa^{-11/2}, \tag{8}$$

where

$$A \simeq \frac{3Gm_{☾}R^5 k_2}{[G(M+m_{☾})]^{1/2}} \frac{1}{Q} \simeq 7.8 \times 10^{52} \frac{k_2}{Q} \quad \mathrm{cm^{13/2}\, s^{-1}}.$$

k_2 is the Love number of the Earth and Q^{-1} is the specific dissipation. Both k_2 and Q are tidal effective parameters, including the combined solid Earth and ocean response. Integrating (8) from the present time T_0 to some time T into the past gives

$$(a_{☾}|_T)^{13/2} = (a_{☾}|_{T_0})^{13/2} + \frac{13}{2} A(T - T_0). \tag{9}$$

The effective lag angle of the Earth is some 5–6° (Lambeck, 1977) or $Q \simeq 10$. If this has remained constant, the Moon would have been within 10 Earth radii about 1.3×10^9 years ago. From (6), the length of day then would have been about 6.7 h. If the average lag throughout geological time has been one half of the above value, close approach would have occurred some 2.6×10^9 years ago.

If the Moon is initially in an eccentric orbit, the transfer of angular momentum and energy is greater at perigee than at apogee. Thus, the satellite is accelerated more in the former position than in the latter and the eccentricity increases with time; the lunar orbit spirals outwards from the Earth in an increasingly eccentric orbit until other forces dominate. The third element of interest is the inclination of the lunar orbit. In an inclined orbit the displacement of the tidal bulge due to the delay Δt is a function of the position of the Moon according to $\Omega \Delta t \cos \phi$, where ϕ is the latitude. This displacement occurs in a plane approximately parallel to the equator. At the maximum latitude attained by the Moon, $\phi = i_☾$, the displacement is less than at the equator and the torque is a minimum. The torque is a maximum when the Moon is over the equator. Integrating over one orbital period gives a non-zero average torque in a plane normal to the intersection of the orbit and equator, with a component directed towards the point of maximum orbital elevation. This torque tends to bring together the angular momentum vector and the rotation axis and the inclination of the orbit evolves slowly with time. At distances beyond about 20 Earth radii the orbit will move towards the ecliptic. Thus the inclination on the ecliptic in the past was greater than the present 5°. For bodies much closer to the Earth, $a_☾ < 15$–20 Earth radii, the orbital plane moves towards the equator. (See Fig. 1 and Goldreich, 1966, for a more complete discussion.)

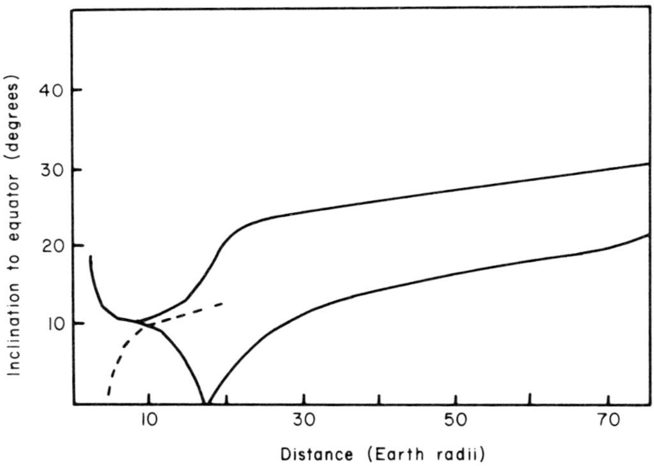

Fig. 1. Inclination of the Moon's orbit to the equator as calculated by Goldreich. The two branches of the curve delimit the oscillation in inclination due to precession. Note that for $a_☾ < 10R_E$ the inclination is significantly non-zero. The dashed curve represents Rubincam's result of the change in inclination if the Moon is perturbed out of the equator by 3° when at 3.8 R_E. An effective viscosity of 10^{18} poise is adopted (see Section 2.4).

Recent studies of the integration of the equations of motion for the Moon back into time have been made by Goldreich (1966) and by Gerstenkorn (1967, 1969). These and other investigations all agree in that (i) there has been a minimum Earth–Moon distance at some time in the past, (ii) the inclination of the lunar orbit was substantially greater in the past than it is now, and (iii) the eccentricity of the lunar orbit increases as the distance increases. In these integrations the tides raised by the Earth on the Moon and by the Sun on the Earth must also be considered as must the mutual attractions of the Earth, Sun and Moon. Important in these integrations are the assumptions made about the delay in the tidal response, for dissipation occurs mainly in the oceans (Lambeck, 1975, 1977) and there is no assurance that this has remained constant throughout geological time. Thus the time scale of the past motion cannot be uniquely established once the lag in response is permitted to vary with time.

In the calculations by Gerstenkorn (1969) the minimum Earth–Moon distance was close to, or within the Roche limit of about 2.8 Earth radii. At this point the self attraction of the satellite is exceeded by the tidal attraction of the planet; the satellite becomes gravitationally unstable and breaks up. Even if disintegration does not occur, the tides will be so large that extensive melting of the crust and of the Moon will occur. The geological consequences of this close approach—or Gerstenkorn event—are severe indeed and, if it occurred in the last 4×10^9 years, it should be firmly imprinted on the geological record of both the Moon and Earth.

2.2. Observational Evidence

Table 1 summarizes the accelerations as determined from telescope observations and eclipse data based on the work of Stephenson (1972), Muller (1975, 1976) and Muller and Stephenson (1975). Numerous estimates for $\dot{n}_{☾}$ and $\dot{\Omega}$ have been made in recent years with considerable fluctuations in results (see Lambeck, 1977, for a review) but there is now a general consensus that these recent values are reliable. The eclipse records establish

Table 1. Accelerations of the Earth and Moon from the eclipse records, representing average conditions over the last 3000 years, and from the coral and bivalve growth rhythms, representing average conditions over the last 300×10^6 years.

Data	$\dot{\Omega}$ $(10^{-22}\ s^{-2})$	$\dot{\Omega}_{NT}$ $(10^{-22}\ s^{-2})$	$\dot{n}_{☾}$ $(10^{-23}\ s^{-2})$
Eclipse records and telescope observations (Muller, 1976; Lambeck, 1977)	-5.5 ± 0.5	1.6 ± 0.6	-1.35 ± 0.10
Coral and bivalve data (Lambeck, 1978)	-5.3 ± 0.6	1.0 ± 0.6	-1.2 ± 0.2
Coral and bivalve data with $\dot{\Omega}_{NT} = 0$ (Lambeck, 1978)	-5.2 ± 0.5	0	-1.0 ± 0.2

average accelerations for only the last 3000 years. Of potentially much greater importance for the orbital evolution studies are the fossil records of growth cycles in certain invertebrates, mainly corals and bivalves. These growth rhythms have been interpreted in terms of astronomical cycles (see the papers in Rosenberg and Runcorn, 1975) and appear to provide information on: (i) N_1 = the number of solar days per year, (ii) N_2 = the number of solar days per synodic month and (iii) N_3 = the number of synodic months per year. Observations of changes of these quantities with time enable the accelerations $\dot{\Omega}$ and $\dot{n}_☾$ to be estimated. Any results must be considered, however, with caution. The evidence for the interpretations is largely circumstantial rather than based on an understanding of the environmental factors controlling the formation of the growth lines, and there is little assurance that the environmental factors have remained constant through time. It is not even certain whether the solar day and synodic month are the appropriate units visible in the growth patterns.

Scrutton (1978) has reviewed the present state of the art of the periodic growth features in fossil organisms. Scrutton (1978) and Lambeck (1978*a*) have independently reviewed the published data and came to similar results as to which values are more acceptable than others. Lambeck (1978*a*, *b*) estimated the accelerations from these data (see Table 1) using the following relations:

$$N_1 = [\Omega(t) - n_\odot]/n_\odot,$$

$$N_2 = [\Omega(t) - n_\odot]/[n_☾(t) - n_\odot],$$

$$N_3 = [n_☾(t) - n_\odot]/n_\odot,$$

where $n_\odot$ is the mean motion of the Sun and is assumed to be constant with time. Values for N_1, N_2 and N_3 at different epochs permit $\Omega(t)$ and $n_☾(t)$, or $\dot{\Omega}$ and $\dot{n}_☾$, to be determined. The coral and bivalve data give quite comparable results, perhaps permitting one to have some confidence in the results unless all the published results are influenced by *a priori* assumptions on what the accelerations should be.

The acceleration $\dot{\Omega}(t)$ is the total acceleration of the Earth and may consist of two parts: (i) a tidal part $\dot{\Omega}_T(t)$, which follows from (6) once $\dot{n}_☾$ is determined, and (ii) a non-tidal part $\dot{\Omega}_{NT}(t)$ due to other forces acting on the Earth or to changes in the inertia tensor. Allowing for the solar tides as well, the present relation between $\dot{n}_☾$ and $\dot{\Omega}_T$ is (Lambeck 1977)

$$\dot{\Omega}_T \simeq 51.4\, \dot{n}_☾ \tag{10}$$

and is unlikely to have changed very significantly over the last 5×10^8 years, so that

$$\dot{\Omega}_{NT} = \dot{\Omega} - 51.4\, \dot{n}_☾.$$

Fossil remains of corals and bivalves do not occur much beyond the Ordovician and only some 10 per cent of the Earth's age is represented. Stromatolites are the only hope for extending the fossil record further back but, after the initial optimism by Pannella (1972), the more careful appraisals of the data by Pannella (1976) and Scrutton (1978) are cautious.

Both the coral and bivalve data indicate that the average accelerations for the last 400×10^6 years have been somewhat less than the present values. These data are also compatible with there not having been any significant non-tidal secular acceleration of the Earth. Perhaps the best solution is to combine the coral and bivalve data with the assumption ((6) or (10)) that angular momentum is conserved. This gives the accelerations in the last line of Table 1. The associated precision estimates are only as good as the validity of the underlying assumptions made about the growth rhythms.

2.3. Consequences and Implications

Comparing the present rate of dissipation with that computed from the paleontological data indicates that present values exceed the average values since the Ordovician by a factor of about 1.4. Such a change can be readily attributed to changes in the ocean configuration and to changes in ocean volume. Brosche and Sündermann (1977) have attempted to model the ocean tide in the Permian and find a reduction in dissipation, compared with the present rate, by a factor of two or three. A simple argument concerning the periods of the free oscillations of the oceans also suggests a reduction in the rate of dissipation. The free oscillation period is very approximately $2L/\sqrt{gh}$, where L is the typical horizontal length scale of the ocean basin, g is gravity and h is the depth. In the Pangea configuration, L would have been greater than now and hence the periods longer and further away from the period of the forcing function. There is evidence of widespread shallow seas having occurred at some epochs in the geological past and it has been suggested that the tidal dissipation may have been much greater than it is now. This need not be so, for now the free oscillation periods will increase and the tidal amplitudes decrease. Variation in sea level due to changes in the extent of glaciations may also lead to changes in the rate of dissipation. While little is still known about the manner in which the energy is dissipated (see Wunsch, 1975; Schott, 1977) it is inopportune to speculate further.

Assuming that the average values of the accelerations found since the Ordovician are representative of all times for which there were oceans, leads, with (9), to a close approach of the Moon some 2 billion years ago, compared with 1.3 billion years ago based on the present value. Another interpretation may be that dissipation has increased slowly with time due to a gradual evolution of the ocean basins so that at the Ordovician, dissipation was only about half of the average value. With this approximation, close approach occurred some 3.8 billion years ago. This is somewhat more satisfactory and illustrates that it will not be very difficult to accommodate the situation in which close approach never occurred at all.

In the absence of reliable fossil records, any further extrapolation backwards into time can only be tested against less direct observations. The stromatolite records are indicative of tidal activity at least as long ago as 3.0–3.3 billion years and do not require the excessively high tides that would characterize a close approach (Pannella, 1976). Sedimentological evidence is also suggestive of lunar tides as far back as 3.3 billion years ago (Brunn and Hobday, 1976; Klein, 1972).

The catastrophic consequences of the close approach, as already indicated, should be present in the geologies of the Earth and Moon. The early part of the geological record is, however, uncertain and subject to much controversy and perhaps the only two points upon which there is general agreement are (i) close approach has not occurred during the last 2×10^9 years and (ii) speculation on its possible occurrence prior to about 4×10^9 years is futile as the geological record is entirely absent for the first $0.5–0.8 \times 10^9$ years of the Earth's lifetime. Extensive studies of Precambrian rock ages indicate groupings of radiometric ages at several epochs, around approximately 3.8–3.5, 2.8–2.5 and 1.9–1.6 billion years ago. Other events occurred later in the Precambrian. These ages are indicative of major stages in the crustal evolution, during which various mantle isotopes were homogenized. Several attempts have been made to relate one or another of these epochs to the Gerstenkorn event. Cloud (1968), for example, suggested a date around 3.8 billion years ago, while Turcotte *et al.* (1977) suggested 2.85 ± 0.25 billion years ago. There is no compelling case for associating either one of these dates, or any other epoch of major upheavals in the crustal evolution, with close approach of the Moon. Turcotte *et al.*'s date is based on one stromatolite point obtained by Pannella from the Bulawayan formation. They see no reason to have less confidence in this result than in the other stromatolite results, even though Pannella (1976) considers that only his value for the later Gunflint stromatolite may be significant. Even this value is not in agreement with Mohr's (1975) results for the same formation (see Lambeck, 1978*a*). Turcotte *et al.* stress that the date of 2.85 billion years ago coincides with (i) the end of the mobile period of Archean tectonics and the beginning of crustal thickening and stable continent formation, and (ii) the end of the mare volcanism on the Moon. They propose that significant tidal interactions between the Earth and Moon occurred over an extended period between about 3.8–2.8 billion years ago, and that the tidal energy dissipated in both the Earth and in the Moon was responsible for the Archean and mare volcanism. The obvious problem with this argument is how to store the Moon close to the Earth for such a long period; all the extrapolation calculations show that once the Moon is within about 10 Earth radii the orbital evolution is extremely rapid. Recourse to vague resonances has to be made (see also Alfvén and Arrhenius, 1969), but Kaula and Harris (1973, 1975) did not find any stable configurations that could maintain the Moon close to the Earth.

The relatively quiescent nature of the evolution of the lunar surface during the last three aeons is good evidence that close approach must have occurred earlier than 3.0 billion years ago, if it occurred at all. The flooding of the basins between 3.9 and 3.1 billion years ago could have been initiated by internal heating of the Moon during a rapid close approach stage, but this explanation is not without its own difficulties. In particular, the large-scale relief of the lunar highlands has been preserved and the crystallization ages of many rocks are much older than this: melting could not have been Moon-wide.

One observation that may place an important constraint on the evolution of the orbit is the non-hydrostatic shape of the Moon. If the Moon were in an equilibrium configuration, its shape could be determined from the self-attraction and its spin and orbital motions. If this shape differs from its observed position then it can be argued that the observed shape represents the equilibrium condition at some time in the past when the Moon was closer to the Earth. Jeffreys (1915, 1937) attempted such calculations. Assuming that the Moon is essentially a homogeneous body and that, in its initial equilibrium state, it was already rotating with an angular velocity equal to the mean motion about the Earth, the theoretical coefficients in the lunar gravity field describing the dynamical ellipticities of the visible lunar face and of the lunar equator, are

$$C_{20}^{\mathrm{E}} \equiv -\frac{I_{33}-\frac{1}{2}(I_{1}+I_{22})}{m_{☾}R_{☾}^{2}} = \frac{-15}{12}\frac{M}{m_{☾}}\left(\frac{R_{☾}}{a_{☾}}\right)^{3}$$

and

$$C_{22}^{\mathrm{E}} \equiv -\frac{1}{4}\frac{I_{11}-I_{22}}{m_{☾}R_{☾}^{2}} = +\frac{9}{24}\frac{M}{m_{☾}}\left(\frac{R_{☾}}{a_{☾}}\right)^{3}.$$

Equating these theoretical values with the observed values (Ferrari, 1977), $C_{20}^{\bigcirc} = -2.046\times10^{-4}$ and $C_{22}^{\bigcirc} = 0.217\times10^{-4}$, and solving for $a_{☾}$ gives $a_{☾} \simeq 25\,R_{\mathrm{E}}$. This suggests that the Moon attained its present shape when it was about 20–30 R_{E} from the Earth. However, the ratio C_{22}/C_{20} is independent of $a_{☾}$ and, if this interpretation is valid, then

$$\frac{C_{22}^{\mathrm{E}}}{C_{20}^{\mathrm{E}}} = \frac{C_{22}^{\bigcirc}}{C_{20}^{\bigcirc}} = \frac{3}{10}.$$

But the observed ratio gives 0.10. This discrepancy may be a consequence of the fact that the Moon is not homogeneous, but Jeffrey's (1915) calculations using a Wiechert model do not give very different results. One other reason may be that the Moon's spin was not yet locked to the Earth at the time it attained its equilibrium state; but in this case $I_{11}-I_{22}=0$ and the expected $C^{\circ}_{22}/C^{\circ}_{20}$ ratio would be zero. An easy way out of this would be to assume, as has sometimes been done for the Earth, that the lunar C°_{20} and C°_{22} terms have different origins; that the former is due to the earlier rotation but that C_{22} is a consequence of other density anomalies within the satellite. Further study of this problem may be worthwhile. To provide a useful result, the time at which the equilibrium shape was attained is required. The most plausible date would be soon after the termination of the period of intense surface bombardment, or about 3.9 billion years ago. Another limit is that it occurred prior to the final stages of the flooding of the basins, or before 3.0 billion years ago, since the large gravity anomalies over many of the mare are indicative that the lunar mantle can support non-hydrostatic stresses over long periods. Lambeck and Pullan (1979) discuss this problem further.

In conclusion, there is no compelling geological evidence to suggest that the Gerstenkorn event occurred during the last 4.0 billion years. Most likely the Moon was already well outside the Roche limit at this time.

2.4. Constraints on Theories of Lunar Evolution

In the absence of any clear landmarks to fix the lunar orbit in the geologically distant past, a variety of conjectures have been proposed to explain the origin and evolution of the Moon. Traditionally these hypothesis fall into four main categories: (i) capture of the Moon by the Earth, (ii) fission of the Moon from the Earth, (iii) simultaneous but separate formation of the Earth and Moon in an orbit about each other, and (iv) coagulation or accretion of numerous small moons in orbit around the Earth. Various combinations of these hypotheses have also been proposed. Kaula (1971), Kaula and Harris (1975), and Smith (1976) review the currently more acceptable postulates. Once the two bodies are formed in more or less their present forms, the tidal evolution of the Moon's orbit must be such that at some point it falls into the evolutionary scheme sketched out above. The number of proposed or implied connection points is great and the available observational data do not provide stringent controls, particularly while the time scale remains uncertain. Formation times for the Moon earlier or later than that for the Earth, a formation close to the Earth, in the asteroid belt, near Venus, inside the orbit of Mercury or in the Earth–Sun Lagrangian point, have all been proposed. In the absence of more stringent dynamical controls, the choice depends mainly on the various authors' interpretation of the chemical and petrological evidence. One dynamical constraint is stringent, and that is the observation that the Moon's orbit could not have lain in the equatorial plane when close to the Earth unless some mechanisms wrenched it out of this plane. If the Moon had formed on the equatorial plane close to the Earth, its subsequent motion—due to the attractions of the Earth's bulge and of the Sun on the Moon—would have been such to keep it in the equator since the Earth's bulge dominates the attraction. Later, when the Sun's attraction dominates over the Earth's flattening, the Moon would have been kept in the ecliptic. Instead, the orbit plane is now inclined at some 5° to the ecliptic. Kaula (1971) has suggested that the removal of a primordial atmosphere would have caused the plane to be tilted to the equator. Some have suggested that an infall of planetesimals during the final stages of accretion tipped the Earth's rotation axis. Yet other authors have argued that any non-uniformity of tidal dissipation may significantly modify the orbital evolution and that the scenarios such as Fig. 1 will be invalid in such a case (e.g. Munk, 1968; Alfvén and Arrhenius, 1969). However, any enhanced localized dissipation can only modify the time scale; it cannot modify the relations $\mathrm{d}i/\mathrm{d}a$ and $\mathrm{d}e/\mathrm{d}a$ (Lambeck, 1977). One possible way of modifying the inclination is by postulating past situations in which the relative importance of the tide components is different from what it is at present. For example, the main contributors to $\mathrm{d}i/\mathrm{d}t$ come from the M_2, K_1 and O_1 frequencies, with the first two bringing the Moon into the equator and the last moving the Moon out of the equator with time. Situations may have existed in the past where the free oscillation periods of some of the oceans resonated with the O_1 tide but not with the others. Such a situation is unlikely today since K_1 and O_1 have quite similar frequencies, but this has not always been so in the past, when $n_☾$ was of the same magnitude as Ω. A solution like this is, however, unsatisfactory, as it is unlikely to have been more than a transient phenomena without any long-term consequence on the lunar orbit. Possibly a more satisfactory explanation is the one by Rubincam (1975) who, following O'Keefe (1972) and G. H. Darwin, considers the evolution of the lunar orbit under the assumption that dissipation occurs in the mantle by linear viscous damping. In such a model, the tangent of the lag angle is proportional to the frequency of the tide component and to the viscosity. For viscosities greater than about 10^{16} poise, the lag angles go to 90° when the tidal frequencies are significantly non-zero and the amplitude of the tidal bulge becomes vanishingly small. When $n_☾ = 2\Omega$ the M_2 and K_1 frequencies are non-zero, but the O_1 frequency is close to zero, and this tide may dominate the evolution. Persuing Darwin's hypothesis, Rubincam finds that if there is an initial perturbation of some 3° in the inclination of the lunar orbit on the equator at the time that $\Omega \simeq 2n_☾$—at 3.8 Earth radii from the Earth—the present 5°

inclination to the ecliptic can be explained, provided that the Earth's viscosity at that time exceeded 10^{18} poise. In this model, the evolution of the orbit is quite different from that proposed by Gerstenkorn and Goldreich (Fig. 1) and stresses the rheology dependence of the evolution once the Moon approaches the Earth.

2.5. Non-tidal Acceleration

If we accept the paleoacceleration solution for the combined coral and bivalve data (Table 1), we obtain

$$\dot{\Omega}_{\mathrm{NT}} = (1.0 \pm 0.5)\, 10^{-22} \quad \mathrm{s}^{-2} \tag{11}$$

which is hardly significant. Thus it is not very useful to speculate on processes that may lead to an acceleration of the Earth, except in that this result does exclude some of the more dramatic ones. Dicke (1966) proposed that a secular change in the gravitational constant will lead to a secular acceleration $\dot{\Omega}_{\mathrm{NT}}$. Allowing for gravitational compression of the Earth, a change in G results in an acceleration of (Dicke, 1969)

$$\dot{\Omega}/\Omega = -1.8\, \dot{G}/G$$

and, with the above result for $\dot{\Omega}_{\mathrm{NT}}$,

$$\dot{G}/G \simeq -(2.4 + 1.2)\, 10^{-11} \quad \mathrm{y}^{-1}.$$

From an analysis of the ancient eclipse data, and comparison of atomic and ephemeris time scales, Muller (1976) obtains $\dot{G}/G \simeq -(5 \pm 3)10^{-11}\,\mathrm{y}^{-1}$. Dicke (1969) has also proposed that $\dot{\Omega}_{\mathrm{NT}}$ may be a consequence of the isostatic post-glacial rebound. A third proposal, by Yukutake (1972), is that it is caused by core–mantle electromagnetic coupling. These last two mechanisms are both irregular when viewed over geologic time and will not contribute to the average $\dot{\Omega}_{\mathrm{NT}}$ for the last 400 million years. Other proposals involve a growth of the core, variations with time in the depth of the principal mantle transition zones, or an expanding Earth. Without examining the validity of these mechanisms, the following rough estimates provide estimates of the maximum permissible changes.

Core Growth

Consider a homogeneous mantle of radius R_{m} and density ρ_{m}, surrounding a core of radius R_{c} and density ρ_{c}. We assume that the core grows due to a precipitation of heavier elements in the mantle. Thus both R_{c} and ρ_{m} will vary with time. When mass is conserved, the rate of change of the inertia tensor is

$$\frac{\mathrm{d}C}{\mathrm{d}t} = \frac{8\pi}{3}(\rho_{\mathrm{c}} - \rho_{\mathrm{m}}) R_c^4 \left\{ 1 - \frac{3}{5}\left(\frac{R_{\mathrm{m}}^5 - R_{\mathrm{c}}^5}{R_{\mathrm{m}}^3 - R_{\mathrm{c}}^3} \right) \frac{1}{R_{\mathrm{c}}^2} \right\} \frac{\mathrm{d}R_{\mathrm{c}}}{\mathrm{d}t}$$

and

$$\dot{\Omega} = 2.8 \times 10^{-14} \frac{\mathrm{d}R_c}{\mathrm{d}t}.$$

For $\dot{\Omega}_{\mathrm{NT}} \simeq 10^{-22}$ rad s^{-2},

$$\frac{\mathrm{d}R_c}{\mathrm{d}t} \simeq 0.1 \text{ cm y}^{-1}.$$

Migration of Transition Zones

A cooling Earth could result in an upward movement of the transition zone if the phase transitions are exothermic. Consider only the 400 km discontinuity, then

$$\frac{\mathrm{d}C}{\mathrm{d}t} \simeq -2.5 \times 10^{35} \frac{\mathrm{d}R_{400}}{\mathrm{d}t},$$

where R_{400} is the radius of the transition zone. Alternatively,

$$\dot{\Omega}_{\mathrm{NT}} \simeq 2.3 \times 10^{-14} \frac{\mathrm{d}R_{400}}{\mathrm{d}t}$$

and, with (11),

$$\frac{\mathrm{d}R_{400}}{\mathrm{d}t} \sim 0.1 \text{ cm y}^{-1}.$$

Expanding Earth

If we can consider that the core is not subject to change, then the change in the inertia tensor due to a uniform expansion of the mantle is

$$\frac{\mathrm{d}C}{\mathrm{d}t} = \frac{8\pi}{3} \rho_{\mathrm{m}} R_{\mathrm{m}}^{4} \left\{ 1 - \frac{3}{5} \left(\frac{R_{\mathrm{m}}^{5} - R_{\mathrm{c}}^{5}}{R_{\mathrm{m}}^{3} - R_{\mathrm{c}}^{3}} \right) \frac{1}{R_{\mathrm{m}}^{2}} \right\} \frac{\mathrm{d}R_{\mathrm{m}}}{\mathrm{d}t}$$

$$\simeq 4.3 \times 10^{35} \frac{\mathrm{d}R_{\mathrm{m}}}{\mathrm{d}t}$$

and

$$\dot{\Omega}_{\mathrm{NT}} = -3.9 \times 10^{-14} \frac{\mathrm{d}R_{\mathrm{m}}}{\mathrm{d}t}.$$

In the simple model, an expanding Earth leads to a further deceleration. For the observed value of $\dot{\Omega}_{\mathrm{NT}} \simeq 10^{-22}$ s^{-2},

$$\frac{\mathrm{d}R_{\mathrm{m}}}{\mathrm{d}t} \lesssim -0.1 \text{ cm y}^{-1}.$$

The above rates are no more than very rough order-of-magnitude estimates. First, the observational evidence for $\dot{\Omega}_{NT}$ is quite uncertain, it being the small difference between two quantities that are only precise to about 10 per cent. Second, any more realistic calculation will be complicated by the fact that several mechanisms will contribute to the overall change in inertia tensor, but in view of the uncertainty in $\dot{\Omega}_{NT}$ more detailed calculations are probably premature. These results do, however, indicate that expansion of the Earth at a rate of a few centimetres per year can be ruled out. This has also been concluded from a recent examination of the paleomagnetic data (McElhinny *et al.*, 1978).

3. Polar Wander

The concept of polar wander, the large-scale wandering of the Earth's axis of rotation throughout geological time, goes back more than 100 years and has its origin in observations of fossil plant and animal distributions and in the scars, tillites and moraines of past glaciations. For example, the discovery that a subtropical climate existed in Spitsbergen at a time when central and southern Europe was subject to a tropical climate and when extensive glaciations occurred in southern Africa, led to the conclusion that in the western hemisphere the carboniferous equator must have lain far to the north of the present equator. A further degree of freedom to the interpretation of the paleontological and paleoclimatic data was introduced by Wegener and by F. B. Taylor with their concept of continental drift, in which the continents are postulated to have moved relative to each other over large distances throughout geologic time. Two important paleomagnetic observations—that large changes have occurred in the mean direction of the geomagnetic field and that this field has periodically reversed itself—have revived interest in the polar wander hypothesis. The separation of drift and wander has, however, remained problematical, and this has plagued much of the subsequent discussions. Presumably if drift occurs, so will the pole wander; the rotation axis moving along with the principal axis of maximum inertia as a redistribution of mass occurs. Drift and wander will have a common explanation yet the available data does not permit a clear separation of the two.

Kelvin, in 1863, apparently could see little objection to the polar wander hypothesis; the formation of the world's major mountains being sufficient to shift the Earth's axis by some tens of degrees. G. H. Darwin also concluded that, if the Earth's rheology resembles that of a plastic body, polar wander may well occur on an extensive scale, but that it would not exceed a few degrees if the Earth's behaviour is essentially that of a rigid body. Gold (1955) came to a similar conclusion; that large-scale polar wander is inevitable if the long-term rheology is anelastic. Munk and MacDonald (1960) and Goldreich and Toomre (1969) further discuss the polar wander model.

The problem is to determine the motion of the rotation axis of a body that evolves slowly with time. Goldreich and Toomre refer to such a body as a "quasi-rigid" body and they illustrate it with an extension of Gold's beetle model in which a colony of N beetles roam a rigid rotating sphere. The solution of the equations of motion is such that the rotation axis, once it lies close to the principal axis of either maximum or minimum inertia, continues to follow this axis. As the beetles crawl slowly along random paths, the positions of the inertia axis of the combined sphere and beetle population evolve with time and the rotation axis follows it. This is for the motion viewed from the sphere itself. In an absolute frame, it is the angular momentum axis that remains fixed in space, and for the present purpose this can be identified with the rotation axis. An important aspect of Goldreich and Toomre's model is that they find that the speed of the motion of the rotation axis exceeds the average speed of the beetles by a factor of $N^{1/2}$. Thus, for large N, even modest displacements on the sphere lead to large-scale polar wandering.

Not surprisingly, the actual Earth differs from this beetle colony in some important ways. Most important is that the Earth possesses a rotational bulge. If the Earth were in a truly hydrostatic equilibrium state, this will not significantly modify the above model since the bulge will reorient itself relative to the rotational axis, should the latter shift in position. Should the bulge not adjust, any polar wander will be small indeed. A question raised by Munk and others is, to what extent does the bulge follow this hydrostatic equilibrium behaviour? The importance of this is illustrated by Gold's calculation in which he estimates that if South America is suddenly raised by 30 m the change in the inertia tensor is such as to change the pole position by 0.01°. But if the rotational bulge follows the rotation axis, the total displacement is some 10°.

The Earth's rotational bulge is described by a second-degree zonal coefficient in the geopotential. The observed value of this term is $C_{20} = -4.841 \times 10^{-4}$ compared with -4.798×10^{-4} if the planet were in complete hydrostatic equilibrium. The difference between these two estimates has sometimes been thought to be sufficiently important, compared with other non-hydrostatic terms in the geopotential, to warrant a special interpretation. MacDonald (1963, 1965) and McKenzie (1966) concluded that this difference was a consequence of a delay in the Earth's response to a change in the speed of rotation. As the Earth slows down under the influence of tidal forces, it takes time for the bulge to adjust to the new angular velocity and the new equilibrium state. The observed bulge is then interpreted as a fossil bulge corresponding to the equilibrium state at some time earlier. The relaxation of the oblateness involves a change in the overall mass distribution of the mantle and it should be controlled by the flow properties of the mantle as a whole. McKenzie concludes that this fossil bulge hypothesis leads to a lower mantle viscosity of the order 10^{26} poise, compared with values of 10^{21}–10^{22} poise for the upper mantle deduced from post-glacial uplift studies. Such high values in the lower mantle effectively confine convection to the upper mantle and probably excludes large-scale polar wander. The difference between the observed and equilibrium values of C_{20} is 4.3×10^{-6} compared with $C_{22}\simeq2.4\times10^{-6}$ and $S_{22}\simeq-1.4\times10^{-6}$. In other words,

$$\Delta C_{20}/(C_{22}^2+S_{22}^2)^{1/2}\simeq1.6$$

and there is approximately 2.5 times more potential energy contained in the non-hydrostatic bulge than there is in the coefficients C_{22}, S_{22}. Goldreich and Toomre, however, have shown that neither this interpretation nor the resulting conclusion are justified, and that the discrepancies in the bulge are not anomalous but are a consequence of the choice of coordinate system adopted (see also O'Connell, 1971). Thus, if we consider a new set of axes x', y', z' such that z' is aligned with the equatorial axis of smallest moment of inertia and the x' axis is aligned with the present pole, the anomalous potential coefficient $\Delta C'_{20}$ becomes

$$(\Delta C'_{20})^2=\frac{3}{4}\left[C_{22}^2+S_{22}^2\right]+\frac{1}{4}\Delta C_{20}^2-\frac{\sqrt{3}}{2}\Delta C_{20}\left[C_{22}^2+S_{22}^2\right]^{1/2}. \quad (12)$$

The degrees variances

$$\Sigma_m\,(C_{lm}^2+S_{lm}^2)$$

are invariant to a rotation of the axes, so that once $\Delta C'_{20}$ is determined from (12), $(C'^2_{22}+S'^2_{22})$ is also known and

$$\Delta C'_{20}/(C'^2_{22}+S'^2_{22})^{1/2}\simeq1.6,$$

a ratio equal to that obtained before, but now the $\Delta C'_{20}$ is associated with the ellipticity of the equator while the C'_{22}, S'_{22} are associated with the ellipiticity of the bulge: the non-hydrostatic bulge is no more anomalous than the equatorial bulge and if a special explanation is to given to the one, it must also be given to the other. But, as Goldreich and Toomre remark, no one has yet proposed that the large negative geoid anomaly on the equator south of India has anything to do with some past rotation about the corresponding equatorial axis.

The main point of Goldreich and Toomre's study is to emphasize that, for a quasi-rigid body, large-scale displacements of the pole can be readily explained by small relative horizontal displacements of material in the mantle. Second, they stress that the non-hydrostatic part of the rotational bulge is not anomalously large and that it is not necessary to suppose that the Earth's response to long-wavelength deformations is much different from its response to shorter-wavelength deformations. The Earth can be considered as a quasi-rigid body whose rotational bulge flows in conformity with the changing position of the rotation axis. If mantle convection causes the inertia axis to shift, so will the rotation axis wander; both drift and wander can be expected to occur but the rate of wander may be much larger than the rate of drift.

The history of the Earth's polar wander lies in part in the record of fossil magnetization of rocks (see, for example, McElhinny, 1973) and paleomagnetic observations permit the position of the magnetic pole to be estimated, if the field can be assumed to correspond to that of an axial geocentric dipole. In general, virtual magnetic poles for a tectonically stable region, with ages ranging over some millions of years, cluster clearly about a mean pole suggesting that the time averaged field approximates a dipole field. Archaeomagnetic observations give virtual pole positions that cluster about the geographic pole indicating that the time average field is representative of an axial dipole field. Further back into time the general agreement with magnetic pole positions and paleoclimatic determinations of the equator position also suggests that the field, on the average, is axial (Stehli, 1968; Briden, 1968). Curie's principle of symmetry argues for this parallelism (Stacey, 1977).

Assuming that, on the average, the past field has remained that of an axial dipole, the apparent pole path for a given continent indicates the relative motion of that continent with respect to the rotation axis. Tectonic motions of the continent or plate cannot be separated from any drift of this axis. Marine magnetic anomalies, which permit the relative motion of the plates to be determined throughout the Cainozoic and Mesozoic, also do not help in separating the two without further assumptions, such as adopting a reference frame fixed by hotspots (for example, Minister *et al.* 1974) or by minimizing the motions of the plate margins (Kaula, 1975). To fix any one plate requires three parameters and, if there are N moving plates and a wandering pole, there will be a total of $(3N-1)+2$ unknowns (if one longitude is arbitrarily fixed) for any one geological epoch. As independent observations we have the relative motions between the plates and the position of the apparent pole relative to one plate, or a total of $2+3(N-1)$. Thus the number of unknowns always exceeds the number of observations. A unique solution is impossible. Polar wander becomes an unnecessary concept in paleomagnetism and in plate tectonics (Munk and MacDonald, 1960; McKenzie, 1972; McElhinny, 1973). McKenzie suggests, however, that the concept of polar wander may nevertheless be a useful one if the motion of the pole relative to any one plate is very much faster than the motion between the plates themselves. The linear velocity of the pole relative to a plate A can be expressed by

$$ {}_{A}v_z = R({}_{A}\Omega_z \times \hat{z}) $$

where ${}_{A}\Omega_z$ describes the motion of the plate A relative to the geographic axis $\hat{z}$ by an angular velocity vector in the geographic equator (see McKenzie, 1972). The mean velocity of the N plates is

$$ v_m = \frac{1}{N} \sum_{n=1}^{N} {}_{n}v_z. $$

If

$$ \sum_{n=1}^{N} {}_{n}v_z \gg \sum_{n=1}^{N} ({}_{n}v_z - v_m), \tag{13} $$

it may be useful to consider the v_m as polar wander and the differences ${}_{n}v_z - v_m$ as drift. This procedure has been followed by McElhinny, after weighing the velocities by the plate areas so as to reduce the influence of the small fast-moving oceanic plates. He finds no evidence that large-scale wander has occurred during the last 40×10^6 years. Jurdy and Van der Voo (1975) have extended the analysis back to the Early Cretaceous and conclude that there is no compelling evidence for large-scale polar wander during the last 110×10^6 years. Further back into time, the evidence for both polar wander and plate motions becomes much less certain, in particular since marine magnetic anomalies are not found further back in time than the Mesozoic. With perhaps the exception of the Ordovician (McElhinny and Briden, 1971), the ensemble of the paleozoic data suggests that drift was predominant and that wander alone cannot explain the apparent polar wander.

The paleomagnetic evidence appears to be at variance with Goldreich and Toomre's model, which predicts large-scale wandering for even a modest drift. One reason could be that the lower mantle, after all, is too viscous to permit the rotational bulge to adjust itself to changes in the position of the rotation axis. But there is little evidence for this. Cathles's (1975) work on the post-glacial rebound problem suggests that there is no need for a rapid increase in viscosity with depth and that the sea level data may in fact exclude such a model. This conclusion is confirmed by Peltier and Andrews (1976). O'Connell (1977) reviews the flow properties of mantle materials and also believes that there is no compelling argument for adopting a high viscosity for the lower mantle. An alternative and more acceptable explanation is that, whereas the beetles on Goldreich and Toomre's planet are uncompensated, the mass redistributions associated with the drift and concomitant mantle flow will result in a much reduced modification of the inertia tensor. In their model, gravity anomalies will be strongly correlated with the positions of the beetles, but such a relation between gravity and topography is not seen on the Earth. Also, convection within the mantle may only have a minor impact on the inertia tensor since lateral seismic anomalies for the mantle appear to be a magnitude larger than predicted from the gravity field and equations of state; one way of reconciling the disparity is to assume some form of compensation whereby positive density anomalies are overlain by negative anomalies and vice versa (Lambeck, 1976; Kaula, 1977). Changes in the inertia tensor due to continental motions and mantle convection will therefore be much smaller than implied in the Goldreich and Toomre model. Polar wander will, in consequence, also be much less and the approximate separation of drift and wander indicated by (13) is not useful. Such a separation becomes feasible only by adopting some "absolute" reference frame such as the hotspot hypothesis, by introducing paleoclimate indicators or by modelling the changing mass distributions associated with plate motions. All three approaches have been attempted but the results are still inadequate.

References

Alfvén, H. and Arrhenius, G. (1969). Two alternatives for the history of the Moon. *Science, N.Y.* **165**, 11–17.

Briden, J. C. (1968). Paleoclimatic evidence of geocentric axial dipole field. In *History of the Earth's Crust* (R.. A. Phinney, ed.), Princeton University Press, Princeton, N.J., pp. 178–194.

Brosche, P. and Sünderman, J. (1977). Effect of oceanic tides on the rotation of the Earth. In *Scientific Applications of Lunar Laser Ranging* (J. D. Mulholland, ed.), D. Reidel Publishing Co., New York, pp. 131–141.

Brosche, P. and Sünderman, J. (eds) (1978). *Tidal Friction and Earth's Rotation*, Springer-Verlag, Berlin, Heidelberg, New York (in press).

Brunn, V. and Hobday, D. K. (1976). Early Precambrian tidal sedimentation in the Pongola supergroup of South Africa. *J. Sediment. Petrol.* **46**, 670–679.

Cathles, L. M. (1975). *The Viscosity of the Earth's Mantle*, Princeton University Press, Princeton, N.J.

Cloud, P. E. (1968). Atmospheric and hydrospheric evolution on the primitive Earth. *Science*, N.Y. **160**, 729–736.

Dicke, R. H. (1966). The secular acceleration of the Earth's rotation and cosmology. In *The Earth–Moon System* (B. G. Marsden and A. G. W. Cameron, eds), Plenum Press, New York, pp. 98–163.

Dicke, R. H. (1969). Average acceleration of the Earth's rotation and the viscosity of the deep mantle. *J. geophys. Res.* **74**, 5895–5902.

Ferrari, A. J. (1977). Lunar gravity: a harmonic analysis. *J. geophys. Res.* **82**, 3065–3084.

Gerstenkorn, H. (1967). On the controversy over the effect of tidal friction upon the history of the Earth–Moon system. *Icarus* **7**, 160–167.
Gerstenkorn, H. (1969). The earliest past of the Earth–Moon system. *Icarus* **11**, 189–207.
Gold, T. (1955). Instability of the Earth's axis of rotation. *Nature, Lond.* **175**, 526–529.
Goldreich, P. (1966). History of the lunar orbit. *Rev. Geophys.* **4**, 411–439.
Goldreich, P. and Toomre, A. (1969). Some remarks on polar wandering. *J. geophys. Res.* **74**, 2555–2567.
Jeffreys, H. (1915). Certain hypotheses as to the internal structures of the Earth and Moon. *R. astronom. Soc. Mem.* **60**, 197–217.
Jeffreys, H. (1937). On the figures of the Earth and Moon. *Mon. Not. R. astronom. Soc., Geophys. Suppl.* **4**, 1–13.
Jurdy, D. M. and Van der Voo, R. (1975). True polar wander since the Early Cretaceous. *Science, N.Y.* **187**, 1193–1196.
Kaula, W. M. (1971). Dynamical aspects of lunar origin. *Rev. Geophys. Space Phys.* **9**, 217–238.
Kaula, W. M. (1975). Absolute plate motions by boundary velocity minimizations. *J. geophys. Res.* **80**, 244–248.
Kaula, W. M. (1977). Geophysical inferences from statistical analysis of the gravity field. In *The Changing World of Geodetic Science*, Dept. Geod. Sci. Rept 250, Ohio State University, Columbus, pp. 119–141.
Kaula, W. M. and Harris, A. W. (1973). Dynamically plausible hypotheses of lunar origin. *Nature, N.Y.* **245**, 367–369.
Kaula, W. M. and Harris, A. W. (1975). Dynamics of lunar origin and orbital evolution. *Rev. Geophys. Space Phys.* **13**, 363–371.
Klein, G. (1972). Determination of paleotidal range in clastic sedimentary rocks. *Int. Geol. Congr.* **24**, sec. 6, 397–405.
Lambeck, K. (1975). Effects of tidal dissipation in the oceans on the Moon's orbit and the Earth's rotation. *J. geophys. Res.* **80**, 2917–2925.
Lambeck, K. (1976). Lateral density anomalies in the mantle. *J. geophys. Res.* **81**, 6333–6340.
Lambeck, K. (1977). Tidal dissipation in the oceans astronomical, geophysical and oceanographic consequences. *Phil. Trans. R. Soc. Lond. A* **287**, 545–594.
Lambeck, K. (1978*a*). *The Earth's Rotation; Geophysical Causes and Consequences*, Cambridge University Press, Cambridge (in press).
Lambeck, K. (1978*b*). Paleorotations of the Earth. In *Tidal Friction and the Rotation of the Earth* (P. Brosche and J. Sünderman, eds), Springer-Verlag, Berlin, Heidelberg, New York (in press).
Lambeck, K. and Pullan, S. (1979). Comments on the lunar gravity and topography. Submitted to *Geophys. J. R. astronom. Soc.*
MacDonald, G. J. F. (1963). The deep structure of the oceans and continents. *Rev. Geophys.* **1**, 587–655.
MacDonald, G. J. F. (1965). The figure and long-term mechanical properties of the Earth. In *Advances in Earth Sciences* (P. M. Hurley, ed), MIT Press, Cambridge, Mass., pp. 199–245.
McElhinny, M. W. (1973). *Palaeomagnetism and Plate Tectonics*, Cambridge University Press, Cambridge.
McElhinny, M. W. and Briden, J. C. (1971). Continental drift during the Palaeozoic. *Earth planet. Sci. Lett.* **10**, 407–416.

McElhinny, M. W., Taylor, S. R. and Stevenson, D. J. (1977). Limits to the expansion of the Earth, Moon, Mars and Mercury and to changes in the gravitational constant. *Nature, Lond.* **279**, 316–321.

McKenzie, D. P. (1966). The viscosity of the lower mantle. *Geophys. J. R. astronom. Soc.* **14**, 297–305.

McKenzie, D. P. (1972). Plate tectonics. In *The Nature of the Solid Earth* (E. Robertson, ed.), McGraw-Hill, New York, pp. 323–360.

Minister, J. B., Jordan, T. H., Molnar, P. and Haines, E. (1974). Numerical modelling of instantaneous plate tectonics. *Geophys. J. R. astronom. Soc.* **36**, 541–576.

Mohr, R. E. (1975). Measured periodicities of the Biwabik Stromatolites and their geophysical significance. In *Growth Rhythms and the History of the Earth's Rotation* (G. D. Rosenberg and S. K. Runcorn, eds), John Wiley, New York, pp. 43–55.

Muller, P. M. (1975). An analysis of the ancient astronomical observations with the implications for geophysics and cosmology. Thesis, University of Newcastle.

Muller, P. M. (1976). Determination of the cosmological rate of change of G and the tidal accelerations of Earth and Moon. Report SP43–36, Jet Propulsion Lab., Pasadena.

Muller, P. M. and Stephenson, F. R. (1975). The acceleration of the Earth and Moon from early astronomical observations. In *Growth Rhythms and History of the Earth's Rotation* (G. D. Rosenberg and S. K. Runcorn, eds), John Wiley, New York, pp. 459–534.

Munk, W. H. (1968). Once again—tidal friction. *Q. J. R. astronom. Soc.* **9**, 352–375.

Munk, W. H. and MacDonald, G. J. F. (1960). *The Rotation of the Earth*, Cambridge University Press, Cambridge, p. 323.

O'Connell, R. J. (1971). Pleistocene glaciation and the viscosity of the lower mantle. *Geophys. J. R. astronom. Soc.* **23**, 299–327.

O'Connell, R. J. (1977). On the scale of mantle convection. *Tectonophysics* **38**, 119–136.

O'Keefe, J. A. (1972). Inclination of the Moon's orbit: The early history. *Irish astronom. J.* **10**, 241–250.

Pannella, G. (1972). Paleontological evidence on the Earth's rotational history since Early Precambrian. *Astrophys. Space Sci.* **16**, 212–237.

Pannella, G. (1976). Geophysical inferences from stromatolite lamination. In *Stromatolites* (M. R. Walter, ed.), Elsevier, New York, pp. 673–685.

Peltier, W. R. and Andrews, J. T. (1976). Glacial-isostatic adjustment. I. The forward problem. *Geophys. J. R. astronom. Soc.* **46**, 605–646.

Rosenberg, G. D. and Runcorn, S. K. (1975). *Growth Rhythms and the History of the Earth's Rotation*, John Wiley, Chichester.

Rubincam, D. P. (1975). Tidal friction and the early history of the Moon's orbit. *J. geophys. Res.* **80**, 1537–1548.

Schott, F. (1977). On the energetics of baroclinic tides in the North Atlantic. *Ann. Geophys.* **33**, 41–62.

Scrutton, C. T. (1978). Periodic growth features in fossil organisms and the length of day and month. In *Tidal Friction and the Rotation of the Earth* (P. Brosche and J. Sünderman, eds), Springer Verlag, Berlin, Heidelberg, New York (in press).

Smith, J. V. (1976). Development of the Earth–Moon system with implications for the geology of the early Earth. In *The Early History of the Earth* (B. F. Windley, ed.), John Wiley, Chichester, pp. 3–19.

Stacey, F. D. (1977). *Physics of the Earth*, 2nd ed., John Wiley, New York.

Stehli, F. G. (1968). A paleoclimatic test of the hypothesis of an axial depolar magnetic field. In *The History of the Earth's Crust* (R. A. Phinney, ed.), Princeton University Press, Princeton, N.J., pp. 195–207.

Stephenson, F. R. (1972). Some geophysical, astrophysical and chronological deductions from early astronomical records. Thesis, University of Newcastle.

Turcotte, D. L., Cisne, J. L. and Nordmann, J. C. (1977). On the evolution of the lunar orbit. *Icarus* **30**, 251–266.

Wunsch, C. (1975). Internal tides in the ocean. *Rev. Geophys. Space Phys.* **13**, 167–182.

Yukutake, T. (1972). The effect of change in the geomagnetic dipole moment on the rate of the Earth's rotation. *J. Geomag. Geoelect.* **24**, 19–47.

3. Geomagnetism and the Earth's Core

F. E. M. LILLEY

Abstract

That the Earth has a magnetic field is one of the basic physical facts concerning the planet, and the history of geomagnetic observation forms a substantial part of the early history of science. Palaeomagnetism has contributed much information to the knowledge of the geomagnetic field over past geological time, especially in demonstrating geomagnetic reversals.

The most satisfactory process to invoke in explanation of the geomagnetic field is dynamo action in the liquid core of the Earth. The crucial flow pattern required and many other details of the process remain unknown, though various feasible models have been constructed for them. It appears possible to resolve the physical and chemical circumstances deduced for the core with a convection dynamo, driven at least partially by a power source of order 5×10^{12} W arising from cooling.

Current research aimed at increasing the precision of knowledge about the core is centred on increasing the resolution of seismic interpretation techniques, and on exploring the possibilities of detecting the "gravitational undertones" and inertial modes of core oscillation.

1. Introduction

1.1. History

The study of the magnetic field of the Earth has ancient origins, lost at some time in antiquity when man first discovered that certain rocks tended to align themselves geographically if they were free to rotate. Such were the beginnings of the magnetic compass, which as a basic instrument of navigation for exploration and trade has played a fundamental part in the progress of civilization over the last millenium and possibly for much longer.

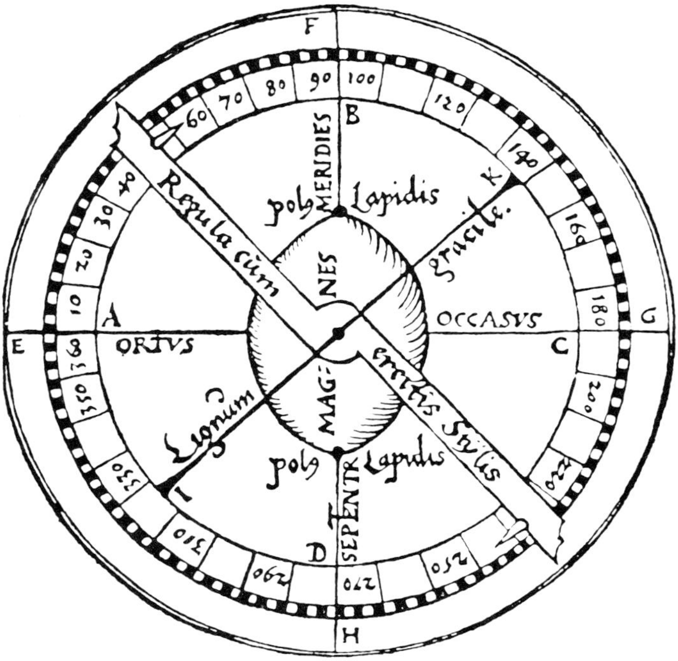

Fig. 1. Petrus Peregrinus' diagram for a compass which uses a naturally magnetized stone (*magnes*), supported to float in water. (From Harradon, 1943.)

Figure 1 shows the design of the first known documented compass to originate in Europe. The figure is taken from the article on thirteenth century Petrus Peregrinus by Harradon (1943). Smith (1970) points out that Peregrinus' work was significant not only for magnetism, but indeed for the development of the whole philosophy of experimental science.

The practical application of the compass in navigation caused the magnetic field of the Earth to be a subject of continued research. It is now measured all over the globe by ships, aircraft and satellites, and its fluctuations with time are monitored simultaneously by observatories throughout the world.

1.2. Deceptive Simplicity of the Dipole Field

The common picture of the magnetic field of the Earth as being dipolar, with poles effectively coincident with the geographic poles of the Earth, has a simplicity which is perhaps deceptive in view of the complexities of the subject which become evident upon closer examination: for analysis of the geomagnetic field in all its detail involves the application of physics ranging from the properties of a highly rarified plasma leaving the Sun to the properties of ultra-compressed material in the core of the Earth. This article will be restricted to the basic problem of solid Earth geomagnetism: the physical processes in the Earth's core giving rise to what is known as the "main magnetic field" of the Earth.

1.3. Gauss' Theorem

The nineteenth century mathematician C. F. Gauss used the mathematical method of spherical harmonics to analyse the existing measurements of the global magnetic field. Several centuries earlier, Gilbert had concluded that the Earth's magnetic field arose from within the planet (as part of his theory that the Earth itself was like a great magnet), at a time when it had been common to attribute the directional properties of compasses to causes external to the Earth: one popular theory (since most observations were made in the northern hemisphere) was that the pole star exercised some attractive effect. Gauss showed in 1839 that the main magnetic field did indeed originate inside the planet, and this demonstration was perhaps the founding step of modern geomagnetism.

The essence of Gauss' method lies in the theoretical result that the magnetic potential of the observations may be expressed in terms with radial dependence r^n, where r is the distance from the centre of the Earth and n is some integer. Terms with negative n then necessarily indicate an internal origin, for which the potential must go to zero at large radial distances. Similarly, terms with positive n necessarily indicate an external origin, for which the radial derivative of the potential must be finite at small radial distances. In his analysis, Gauss showed that the terms corresponding to internal origin were overwhelmingly dominant.

1.4. The Magnetization of Crustal Rocks

Rocks can support magnetic fields through the induced and permanent magnetization of certain of their constituent minerals, and the phenomenon of rock magnetization is the basis of all magnetic techniques of geologic mapping over both land and sea. Crustal magnetization is responsible for the magnetic stripes on the ocean floors, which demonstrate sea-floor spreading. The principles of palaeomagnetism, central to the articles in this volume by M. W. McElhinny and E. Irving (Chapters 4 and 17), depend upon magnetization in crustal rocks.

However, rock magnetization is limited to crustal depths, as the phenomenon disappears when the temperature of a mineral reaches its "Curie temperature". This temperature is 580 °C for magnetite, so with typical thermal gradients at the Earth's surface being of order 25 °C km^{-1}, it can be seen that the base of the magnetized layer must be at depths of order 20 km. This is a very thin layer on the scale of the radius of the Earth, and magnetization of such a layer is not strong enough to account for the Earth's main magnetic field. Thus the dynamo theory has arisen, which places the origin of the main magnetic field in the earth's core, and attributes it to the flow of macroscopic electric currents, supported by a process of electromagnetic induction.

1.5. Electromagnetic Induction in the Earth

Electromagnetic induction takes place in two different regions of the Earth, in two different processes. Near the Earth's surface, primary fields originating external to the Earth induce secondary fields to arise within the Earth. These fluctuating fields combined are the "natural magnetic variations", which, measured at the Earth's surface, appear as small perturbations of the main magnetic field. The physics of such surface induction is relatively well understood, though the mathematics of applying it to general geological situations is involved. The study of natural magnetic variations is currently yielding much information on the electrical conductivity structure of the crust and upper mantle of the Earth.

Much deeper in the Earth, motional electromagnetic induction in the core is thought to generate the main magnetic field of the Earth in a dynamo process. Here the physics and mathematics are complex and the "geology" of necessity simple, in attempting to explain an induction process with the fascinating property that the primary and secondary fields are the same. Of all the contributions to the geomagnetic field, this core originating dynamo process is by far the most important.

1.6. Plan of this Article

The second part of this article introduces the physical elements of dynamo theory. A third part discusses miscellaneous aspects of the core and the geomagnetic field, chosen largely on the basis of their fascination for the present author. It should be emphasized in advance that much of the relevant knowledge concerning the core is very indefinite, and with its many degrees of freedom core science has been a fertile ground for creative ideas. As far as he can in these matters the author has adopted the philosophy of seeking to represent a consensus where he judges that one exists, or, if not, the majority view. It should of course be borne in mind that the majority view need not be the correct one.

2. Main Field Production by Dynamo Action in the Core

Thirty years ago discussion centred on whether a dynamo process in the core of the Earth was physically possible (cf. Elsasser, 1946, and a series of succeeding papers; Bullard and Gellman, 1954). By ten years ago it had essentially been established that this was so, but there was some fascination in investigating whether reasonably simple flow patterns could act as dynamos from a mathematical point of view (e.g. Lilley, 1970; Roberts, 1970; Gubbins, 1972). This demonstration has now been achieved for a number of cases, and the term "geodynamo" has come into the literature (Busse, 1975; Jacobs, 1975). Attention now concentrates particularly on the physics of the process in the Earth: what is the energy source, what is the flow pattern, how do reversals occur? A topical discussion also centres on the possibility that the core might be stably stratified, at least in part.

In leading up to one possible description for the dynamo, pertinent facts about the core shall be grouped according to whether they are (i) well established, (ii) under debate or (iii) hardly known at all. More informally, it will be seen that this grouping corresponds to the evidence for the facts being (i) beyond reasonable doubt, (ii) debatable, and (iii) so scant that it is hardly possible even to argue about them.

2.1. Well-established Facts about the Core

Liquid Nature

That the Earth has a liquid core is one of the great historic discoveries of seismology, and followed from the observation by Oldham in 1906 that the inner part of the Earth does not transmit seismic shear waves, as liquids do not.

Size

From seismology also comes evidence that the outer boundary of the core is clearly defined. Jeffreys (1939) determined the core radius to be 3473 ± 4.2 km, and Dziewonski and Haddon (1974) recently obtained 3485 ± 3 km. That such values are quoted to four significant figures is evidence of the precise nature of this branch of seismology, and, in the context of the uncertainties of dynamo theory, the core radius is known exactly. The possibility remains, however, of relatively very small undulations existing on the core–mantle interface; the subject of these is discussed in the contributions to this volume by M. W. McElhinny, J. R. Cleary and R. S. Anderssen (Chapters 4 and 5).

Solid Inner Core

Though the outer core does not transmit shear waves, certain seismic observations are best accounted for by an inner core which does transmit shear waves and so is "solid". This inner core is of radius 1220–1230 km (Bullen, 1975).

Composition Mainly Iron

The main constituent of the core is held to be iron, which has approximately the correct density and elastic properties to match seismic information and information on the Earth's total mass and moment of inertia, and which needs to form most of the core if the chemical abundances of the whole Earth are to be consistent with those of chondritic meteorites. There is considerable debate over the minor constituents of the core, which will be discussed below in Section 2.2, "Impurities in the iron" and "The potassium question".

A Dynamo Exists in the Earth's Core

The author has little hesitation in grouping the existence of a dynamo in the core as a well-established fact. While historically there have been other theories for the origin of the main geomagnetic field, the wealth of recently accumulated evidence especially on the time-dependent behaviour of the Earth's magnetic field leaves a core dynamo as really the only feasible source for it.

Rotation Influences Dipole Field Generation

The present circumstance of near coincidence of the geomagnetic and geographic poles is taken as evidence that the daily rotation of the core (as part of the Earth) in some way strongly influences the geodynamo, through the effect of Coriolis forces on the fluid flow in the core. While this is not understood in detail (as in fact the core flow itself is not known, see "Fluid flow pattern in the core" in Section 2.3), there is little doubt that some process of this nature occurs. That it has always been so over geologic time is a basic principle of palaeomagnetism, in which determinations of time-averaged magnetic pole position are taken to be determinations of geographic pole position, following early work on this fundamental problem by Irving (1956) and Blackett (1961) (see also Briden 1968).

This work showed, for the cases examined, that the geographic latitudes in which the rocks had formed (as deduced from palaeoclimatic data) agreed with the geomagnetic latitudes in which they had formed (as deduced from palaeomagnetic data).

A clear example of such a coincidence is given in Fig. 2 for the palaeomagnetic poles of reconstructed Gondwanaland (from McElhinny, 1977). The Ordovician–Silurian (O–S) palaeomagnetic pole falls in the present Sahara Desert of North Africa, where there is also geological evidence of Silurian glaciation, implying that at that time the present North Africa was near a geographic pole. Similarly the Permo-Carboniferous (PC)

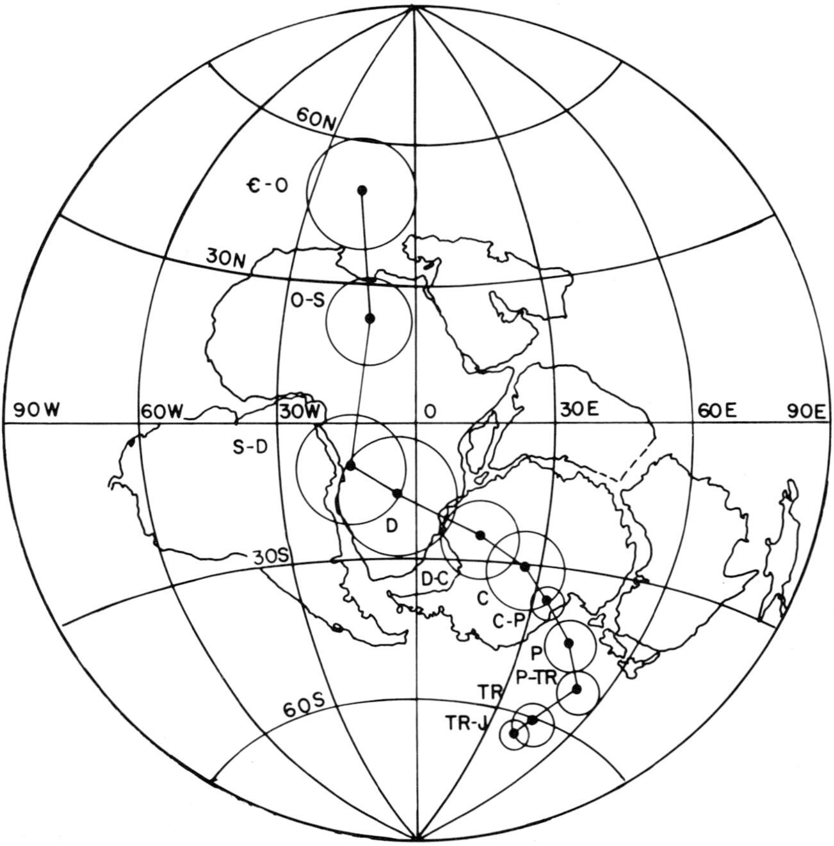

Fig. 2. Palaeomagnetic pole positions for Gondwanaland. (From McElhinny, 1977.)

palaeomagnetic pole falls near Tasmania, where independent geological evidence is also found of Permian glaciation.

Most palaeomagnetic pole determinations are a statistical average for many different rock samples, spread out in time so that their mean result will smooth out small excursions of the magnetic pole from the geographic pole such as occurs for the Earth at present.

2.2. Facts in Debate about the Core

The previous section listed facts which have been convincingly demonstrated. This section now approaches upon the more delicate ground of facts at present under debate, in some instances under heated debate. There is no significance to the order in which they are discussed.

Impurities in the Iron

The inner core may indeed be nearly pure iron, but the outer core must hold a component which is less dense and which increases the compressional seismic velocity. Upon the basis of geochemical abundance arguments (cosmic, solar and meteoritic) the most likely companion of iron in the core is nickel. However, adding nickel to iron in a model core does not improve the match of expected model properties with seismic Earth observations, and the further addition of a lighter element (or elements) must be made to the model. The main present contenders are oxygen, potassium, silicon and sulphur (Brett, 1976), with hydrogen also being reconsidered (Stevenson, 1977).

These impurities are important because their presence could affect the electrical conductivity of the iron, and also its melting characteristics. The question of potassium is particularly important for dynamo theory, because potassium in the core would form a radioactive heat source, crucial in discussing the energetics of dynamos (see "Energy sources for the geodynamo" in Section 2.4). Potassium is therefore now discussed separately.

The Potassium Question

Recently the question of potassium in the core has been revived. The argument has been basically that if the Earth is presently to have the appropriate chondritic abundance of potassium, then this potassium must be in the lower mantle or core as it is not found in the upper mantle or crust. Further, it could be incorporated in the core if it has a tendency to combine with sulphur which might be there (see Goettel, 1976).

The point has been discussed widely. Ringwood (1977) has recently reasserted rejection of this hypothesis on the grounds that there is no reason to expect the Earth to not be depleted in potassium: as are the Moon and the class of meteorites known as eucrites. Indeed, in this recent paper, Ringwood re-examines whether oxygen, not sulphur, might be the main "light" element in the Earth's core.

Electrical Conductivity

Early estimates of the electrical conductivity of a molten iron core by Elsasser (1946) and Bullard (1949) gave a mean value of 3×10^5 S m^{-1}, and this value was used by Bullard and Gellman (1954) in discussing core energetics, with the reservation that it was uncertain by a factor of three.

A number of investigations have been carried out on the problem since, from different points of view. Gardiner and Stacey (1971) and Stacey (1972) also arrive at the value of 3×10^5 S m^{-1}, in a range $(2 \rightarrow 10) \times 10^5$ S m^{-1}. There is uncertainty due to the unknown effect of the impurities in the core (Johnston and Strens, 1973), especially as the composition of these impurities is itself unknown.

Temperatures within the Core

The basic structure from seismology of a liquid outer core and solid inner core was explained by Jacobs (1953) as being due to a relationship between melting point and actual temperature, as shown in Fig. 3. This has remained the most straightforward explanation. Thus determining the melting point of iron (with its possible impurities) at the known pressure of the inner core–outer core boundary gives a temperature estimate for that part of the interior of the Earth.

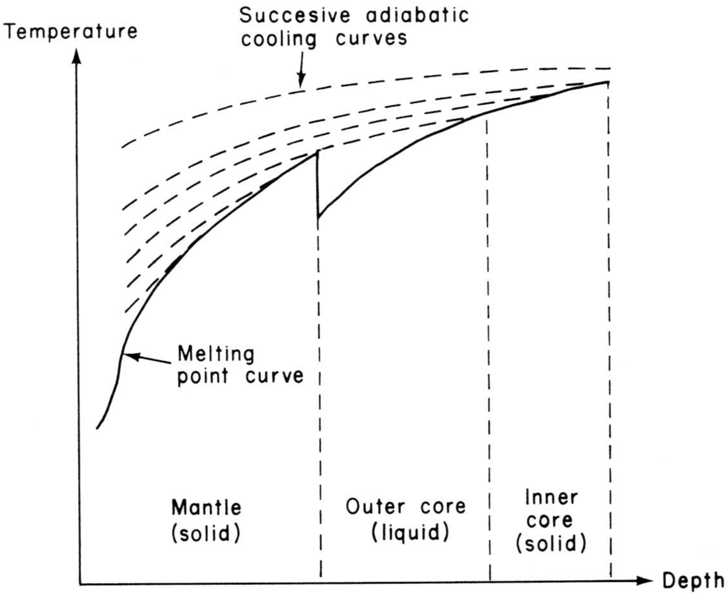

Fig. 3. Formation of a liquid outer core between a solid mantle and a solid inner core. The outer core becomes "trapped" as the inner core solidifies from its centre outwards, and the mantle subsequently solidifies from its base outwards. (From Jacobs, 1953.)

One of the first such estimates was 3900 K by Simon (1953) and one of the most recent is 4160 K (Stacey and Irvine, 1977), and indeed a figure of about 4000 K has not recently been much questioned.

Determination of the temperature at the mantle–core boundary is a more subtle matter. Traditionally the whole outer core is taken to be completely in convection, so that the actual temperature gradient should be the appropriate adiabatic temperature gradient for the material of the molten outer core. Then, given a temperature at the inner core–outer core boundary, it is possible to move radially outwards along the adiabatic gradient until the core–mantle boundary is reached, and thus to obtain a temperature there. An early such estimate was given by Jacobs (1954) as 3600 K, and Stacey and Irving (1977) recently obtained 2940 K, with several hundred degrees uncertainty.

A major difficulty lies in estimating the adiabatic gradient for the material of the outer core. Higgins and Kennedy (1971) in a controversial determination found that the adiabatic gradient was less than the melting point gradient, which if so would mean that the core was stable against convection (except for the innermost outer core; Kennedy and Higgins, 1973). Stronger evidence for this state of affairs appears to be needed, however, before the other estimates of adiabatic gradient (which allow convection) have to be abandoned, and with them a convection process driving the geodynamo. Liu (1975) has pointed out that considering the full phase diagram for iron at core pressures appears to avoid the difficulty.

Increased awareness of the possibility of stability in the core has, nevertheless, led to ideas which, depending on the distribution of heat sources, allow either the inner or the outer part of the fluid core to be stable against convection. In such parts heat flow must then take place by conduction, and the thermal conductivity of the core material becomes a key parameter. This is usually estimated using the Wiedemann–Franz relationship between electrical and thermal conductivities. Stacey (1972) thus obtains a core thermal conductivity estimate of 28 W m^{-1} K^{-1}.

With such a value for thermal conductivity, the heat flow out from the core by conduction only along the adiabatic gradient is of order 3×10^{12} W (Stacey, 1977). This must be a minimum estimate of heat flow from a convecting core since the convection would transport heat in excess of that transported by conduction alone.

2.3. Hardly-known Facts about the Core

The following questions arise in considering the core. Their nature is such, however, that it is not possible to predict at present how they might ever be clarified or resolved.

Change with Time of Core Conditions

Theories on core science depend crucially on the extent to which the core is developing with time. Information on whether the present inner core is cooling or growing, or both, would be invaluable as means of testing a whole range of possibilities. The search for such information rightly falls in the domain of palaeogeophysics, but there seems little hope of a direct answer regarding core evolution. There is a most significant body of data in palaeomagnetism (see "Palaeomagnetic evidence for the geodynamo",

Section 2.4), and this can be expected to increasingly constrain theories for the geodynamo, but the data apply to the end results of a very complicated process, and it is not clear that they will be able to clarify the origins of the process.

The Toroidal Magnetic Field within the Earth

One property of a toroidal magnetic field in an electrically conducting sphere is that the field lines close on themselves entirely within the sphere, and never emerge outside. Toroidal fields play a most important role in most dynamo theories, but virtually by definition any toroidal magnetic field in the Earth's core can not be detected by surface observations. A number of investigations regarding the possible detection of toroidal fields in the Earth by their physical effects have been carried out, such as the following:

(i) the possibility that the observed secular variation of the geomagnetic field represents waves in a core toroidal field (Hide and Stewartson, 1972);
(ii) the possibility that seismic wave propagation through the core would suffer magnetoelastic effects (see Section 3.1);
(iii) the possibility that the electric currents supporting a core field would leak to the Earth's surface (Runcorn, 1954).

The results of such investigations have on the whole, however, so far proved inconclusive, and some independent estimate of the Earth's toroidal core magnetic field would be a great advance for dynamo theory.

Stability of the Fluid Core against Convection

The density distribution of a homogeneous material under an adiabatic temperature gradient follows the Williamson–Adams (or Adams–Williamson) equation:

$$\frac{d\rho}{dz} = \frac{g\rho}{\phi}$$

(Bullen, 1975), where ρ denotes density, g local gravitational acceleration, z depth into the material, and $\phi = k/\rho$, where k is the adiabatic incompressibility. Such a material, if a fluid, is neutrally stable as a particle displaced within it will stay put, neither seeking to travel further nor to return to its original position.

Generally, seismic observations have been consistent with material in the outer core of the Earth obeying this equation (Dziewonski *et al.*, 1975), indicating that the outer core is neutrally stable or else convecting: sufficiently rapid convection will maintain a temperature gradient which is nearly adiabatic. However, the increasing resolution of seismic data has given rise to the possibility of checking the consistency of the core material with the Williamson–Adams condition more closely (Jacobs and Masters, 1976). If, following Bullen, one writes

$$\frac{\mathrm{d}\rho}{\mathrm{d}z} = \eta \frac{g\rho}{\phi}$$

for the core, then the departure of η from unity is a measure of the departure of core conditions from those of chemical homogeneity and temperature adiabaticity. In particular $\eta > 1$ will indicate stability against convection, and $\eta < 1$ instability.

The parameter η has the practical advantage that it can be estimated inside much of the Earth by applying values obtained from the interpretation of seismic data to the above equation or to

$$\eta = \frac{\mathrm{d}k}{\mathrm{d}p} - g^{-1} \frac{\mathrm{d}\phi}{\mathrm{d}z},$$

where p denotes pressure. There is also a close connection between η and the squared "Brunt–Väisälä" frequency $N^2(z)$, given by

$$N^2(z) = (\eta - 1)\frac{g^2}{\phi},$$

where $N(z)$ is the angular frequency at which a slightly displaced fluid particle will oscillate adiabatically about its rest position, due to buoyancy forces. Thus departures of N^2 from zero correspond to departures of η from unity and indicate departures from Williamson–Adams conditions in the core: $N^2 > 0$ ($\eta > 1$) indicates stability against convection, and $N^2 < 0$ ($\eta < 1$) instability.

An estimate of the distribution of N^2 for a recent model of the Earth's outer core is shown in Fig. 4 from Smith (1976). The validity of such results depends upon the resolution of the seismically-determined parameters and the density distribution used. The indications of core stability or instability from such diagrams must be notional until the resolution question is clarified, but are given in the figure to illustrate the principle involved. There may in fact be physical grounds for disallowing negative values of N^2, so that the problem would become the converse one of using marginal core instability as an extra constraint on the interpretation of seismic data.

Fluid Flow Pattern in the Core

The fluid flow pattern in the core is the key basis for the geodynamo, and the author is not a little melancholy about grouping it here as a "hardly-known fact". But while postulated flows range from oscillations of a stably stratified fluid through smooth global circulation to small-scale turbulence, the question seems completely open. Analysis of the secular variation of the

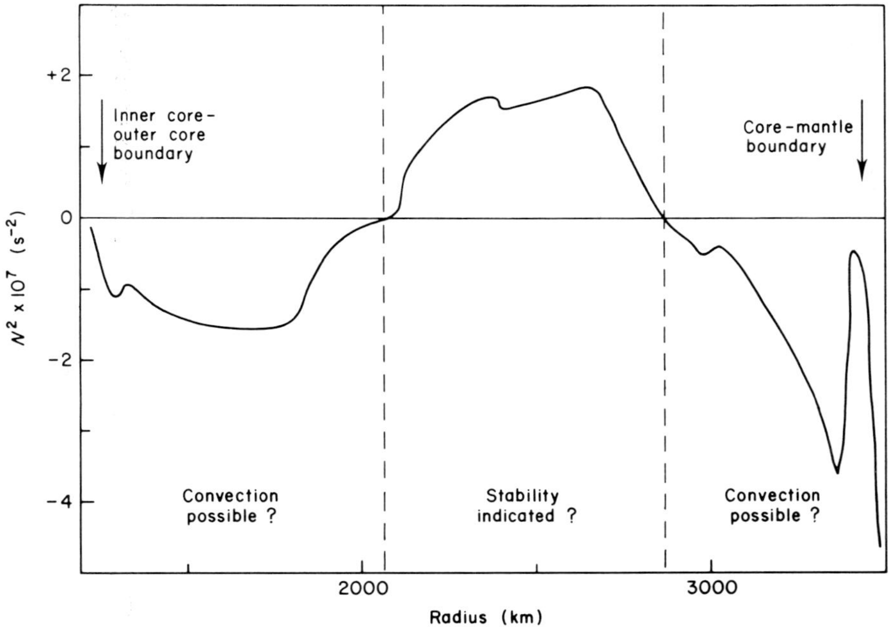

Fig. 4. The quantity N^2 estimated using the outer-core parameters of a particular Earth model (DG579). (After Smith, 1976.) The erratic behaviour of the curve reflects the dependence of N^2 on the first radial derivative of density: see text for discussion.

geomagnetic field observed on the Earth's surface over recent historic time has given rise to some estimates of fluid flow at the core–mantle boundary, and these are smooth, but the method would not be expected to resolve turbulence in the core fluid.

2.4. The Dynamo Equation

The description given here of dynamo theory will be physical rather than mathematical. Dynamo theory seeks to find a fluid motion which will perpetuate, by motional electromagnetic induction, an existing magnetic field. For reasons of simplicity, and because there is little justification in doing otherwise, dynamo regions are taken to be of homogeneous and isotropic electrical conductivity: profoundly different from the common dynamos of electrical engineering.

The basic equation for dynamo action is then quite straightforward to derive from electromagnetic induction theory. It is

$$\frac{\partial \mathbf{B}}{\partial t} = \nabla \times (\mathbf{v} \times \mathbf{B}) + \frac{1}{\sigma\mu}\nabla^2 \mathbf{B}, \qquad (1)$$

where **B** denotes magnetic induction, **v** velocity, and t, σ and μ denote time, electrical conductivity and magnetic permeability, respectively. In models for the Earth's core, **v** and σ are considered to be non-zero only within a sphere, though the **B** field exists over all space (vanishing at great distances). A number of important aspects to be expected in dynamo behaviour may be seen simply by examining this equation.

(i) *Variation with time.* Since time enters the equation, there should be no surprise at the observations, over the last several hundred years, of secular change in the magnetic field of the Earth; nor at the much greater time fluctuations demonstrated by palaeomagnetism. Indeed, a dynamo which produces a steady magnetic field requires a rather special balance between the two terms of the right-hand side of equation (1), so that the left-hand side can be zero. Because equation (1) is a vector equation, this balance must occur in each of its three constituent scalar equations, at all points within the sphere.

(ii) *No preferred sign for* **B**. Because the dynamo equation is homogeneous in **B**, if it is satisfied by a particular **B** field it will also be satisfied by the negative of that **B** field. Thus a dynamo which supports a magnetic field in one direction can equally easily, by the same process, support a field which is everywhere reversed. This feature of dynamo action agrees in a satisfying manner with the palaeomagnetic evidence for no preferred polarity (either normal or reversed) of the geomagnetic field over geologic time.

(iii) *Balance of creation against diffusion.* The term $\nabla\times(\mathbf{v}\times\mathbf{B})$ in the dynamo equation is the source term, and represents the physical process by which magnetic induction is "created" through the flow of fluid across lines of force. By contrast, the term $(1/\sigma\mu)\nabla^2\mathbf{B}$ represents the tendency for field to decay away through ohmic energy loss by the electric currents supporting the field. The balance of these two terms, at a particular point, determines how the magnetic field changes with time at that point.

(iv) *Magnetohydrodynamic effect.* A magnetohydrodynamic effect enters the dynamo process as **v** may itself be affected by **B**: the general equation of motion for the fluid will contain an electromagnetic term for the Lorentz force on the fluid, of form $\mathbf{J}\times\mathbf{B}$, where **J** is electric current density. Since this term may alternatively be written $(1/\mu)(\nabla\times\mathbf{B})\times\mathbf{B}$, it will have the same value regardless of the sign of **B**, so that the indifference of the dynamo equation to magnetic field polarity carries through also into the equation of motion.

However including this non-linear effect of **B** in the process, though necessary physically, makes mathematical analysis of the dynamo problem very much more complicated, especially as the general equation of motion for the fluid (the "Navier–Stokes" equation), will also contain a non-linear term in the velocity **v**.

The simplest level of attack on the dynamo problem is to see whether a steady or growing **B** field can be found satisfying equation (1) for a given (steady) **v** field. This is called the kinematic dynamo problem, and may be attempted for unbounded or bounded regions. Attempts for unbounded regions have generally met with more success, although of course attempts for bounded regions come closer to geophysical application. Physical effects of boundaries in the dynamo problem have been discussed recently by Bullard and Gubbins (1977), who point out that strong electric current sheets tend to form near insulating boundaries. Because equation (1) is linear in **B**, it does not give any information on the magnitude of the **B** field to be expected in a kinematic dynamo.

The next level of complication is to solve the equation of motion simultaneously for **v**, putting in some appropriate body force. This is very difficult, but a number of attempts at the problem have been made on the basis of a time-stepping numerical method. Starting from some set of initial conditions, instantaneous values are calculated for the rates of change of **v** and **B** with time at all points. Increments to the **v** and **B** fields are made according to these rates of change operating for a short time interval, and the whole process is then repeated using the new **v** and **B** fields. In this way, time fluctuations of **v** and **B** have been clearly demonstrated numerically.

Dynamo theory is a subject of advanced mathematics and the interested reader is referred to reviews by Roberts (1967, 1971), Gubbins (1974), Levy (1976) and Krause (1977) for more details. There is, however, one important development appropriate to describe here: it has the name of "mean field electrodynamics" and is useful for treating problems involving turbulence. The concept is that the flow shall have two length scales: a small length scale for the turbulence and a large length scale for the regional flow in which the turbulence occurs. The viscosity must be sufficiently low that turbulence is not inhibited. The dynamo process can then be examined on these two different scales with valuable simplifying approximations, and the interactions between the scales analysed. The interesting physical effect to emerge is that small-scale motions can sustain a large-scale magnetic field.

Palaeomagnetic Evidence for the Geodynamo

Palaeomagnetic studies have yielded invaluable evidence regarding the history of the earth's magnetic field. In particular:

(i) The geomagnetic field has existed from the earliest geological times. Even the oldest pre-Cambrian rocks examined give evidence that they acquired their magnetization in field strengths of a similar order of magnitude to the present geomagnetic field strength (McElhinny and Evans, 1968).

(ii) The Earth's magnetic field has fluctuated with time, reversing irregularly. Its general behaviour is known for the past 500 million years, with better accuracy for more recent periods. In this volume, the paper by I. McDougall (Chapter 16) describes the history of reversals since the Tertiary Period, and the contribution from M. W. McElhinny (Chapter 4) deals with reversal rates. Palaeomagnetic resolution has been sufficient to trace some reversals through in detail (Dagley and Lawley, 1974). It has been pointed out by Jones (1977) that the long-term behaviour of the geomagnetic field may be influenced by convection in the lower mantle.

(iii) The evidence for long-term average coincidence of the magnetic poles and the rotation poles of the Earth was discussed in "Rotation influences dipole field generation" in Section 2.1. A new way of demonstrating that the field at Earth's surface has been predominantly dipolar was given recently by Evans (1976). Evans' demonstration is repeated in Fig. 5.

Energy Needs for the Geodynamo

The dynamo theory requires that the magnetic field of the Earth be supported by a flow of electric currents in the Earth's core (in association with a fluid flow of the material there). These electric currents, even for a steady magnetic field, will be continuously dissipating energy into heat by ohmic loss, due to the finite electrical conductivity of the core material. It is relevant to estimate what this ohmic power loss might be. The electrical conductivity of the core enters the calculations, and its uncertainty is carried through. To support the present field of the Earth as measured on the surface, Lowes (1970) estimated the minimum ohmic dissipation in the core to be of order 10^8 W. Gubbins (1976) estimated $7.6 \times 10^{13}/\sigma$W, which, taking σ as 5×10^5 S m^{-1} (see "Electrical conductivity" in Section 2.2), also gives order 10^8 W. If there is a strong hidden toroidal field in the core, then the ohmic power dissipation must be much higher, of order $10^{10} \rightarrow 10^{11}$ W for a toroidal field of say 10 mT.

These are estimates of the power dissipated by the ohmic currents of the geodynamo. However, the power the dynamo will need to be supplied with will be greater, depending on the efficiency of the dynamo process. Various possible energy supplies for the dynamo process and their efficiencies will be discussed in the next section, but for one of the most obvious, heat, the efficiency is necessarily rather low. A heat engine operating between two temperatures T_1 and T_2 ($T_1 > T_2$) has a maximum possible efficiency of

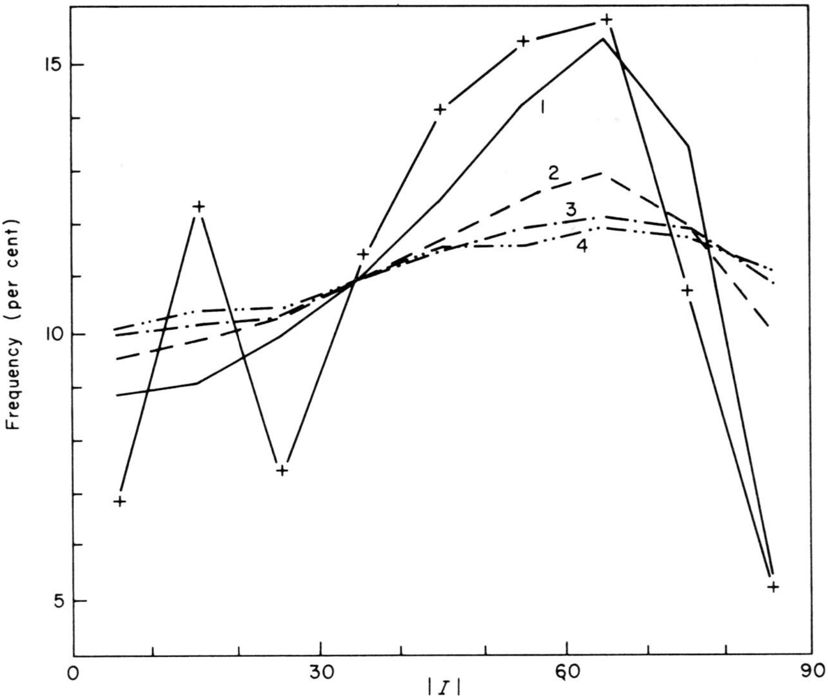

Fig. 5. Distribution of measured inclinations (I) of world-wide palaeomagnetic data. (From Evans, 1976.) Crosses show actual data. Curve 1 shows theoretical distribution for a dipole geomagnetic field; curve 2 for a quadrupole field, and curves 3 and 4 for corresponding higher orders. The data demonstrate that the geomagnetic field has been predominantly dipolar.

order $(T_1 - T_2)/T_1$ (Bullard and Gellman, 1954; Metchnik *et al.*, 1974), and estimates of the efficiency of a convection dynamo in the core are of order 10 per cent maximum. Thus, to sustain a dynamo with ohmic dissipation in the range $10^8 \rightarrow 10^{11}$ W, the total energy flow through the process must be of order $10^9 \rightarrow 10^{12}$ W. The total heat flow out of the surface of the Earth, about 2×10^{13} W, itself places a restriction on the heat flux out of the core and thus leads to a restriction in the maximum size of a toroidal field in the core (see Gubbins, 1976).

An interesting aspect of a thermally-driven core dynamo model is that the energy dissipated through ohmic loss by electric current flow is not lost to the system but reappears as a heat source in the fluid. Thus, viewed from outside, it is not possible to tell what fraction of the total power flowing out of the system has been "side-tracked" through the dynamo process.

Energy Sources for the Geodynamo

(i) *Cooling of core.* A general cooling of the core produces (a) a heat source uniformly distributed through the outer core, plus (b) heat emerging at the inner core–outer core boundary from the cooling inner core, plus (c) latent heat released at the inner core–outer core boundary by the inner core freezing there.

Verhoogen (1961) estimated the amount of heat produced by a cooling core and there seems little reason to modify his estimates. He found the simple cooling source and the freezing source to be of the same order of magnitude; and for them together to produce an energy flow of 10^{12} W out of the core, the Earth's interior should be cooling at the rate of some tens of degrees over 10^9 years.

For an Earth with a history of cooling this is a modest change; so that cooling easily provides power flow for the geodynamo.

(ii) *Radioactivity in the core.* As mentioned in Section 2.2), the case for potassium in the core may need some special geochemical pleading: otherwise radioactive heating of the geodynamo would be a physically reasonable possibility. As a thermal process it too would suffer the general inefficiency of such processes discussed above in "Energy needs for the geodynamo".

(iii) *Precession of the Earth.* This was proposed particularly by Malkus (1968) as a power source for the geodynamo, but its adequacy as a power supply if flow in the core is laminar has recently been questioned by Rochester *et al.* (1975) and Loper (1975).

(iv) *Sedimentation.* There are several different ways in which the core fluid could be kept in motion by heavier material sinking and lighter material rising. Pondering the formation of the core itself, Urey (1952) suggested core motions could be set up by iron continuously falling into the core from the mantle. Later, Braginskii (1964) found that due to the inefficiency of thermal convection his dynamo models required too great a flow of heat out of the core, and he developed a theory in which silicon is released at the inner core–outer core boundary as the inner core (of effectively pure iron) is continuously formed by freezing the impure outer core material. The silicon, being lighter than the outer core fluid, floats upwards and may thus drive core motions.

The energy released by such a chemical differentiation process depends crucially upon the existence of a density increase at the inner core–outer core boundary. Modern seismological techniques are approaching the point of establishing whether or not a significant density increase occurs, and, on the basis of evidence that it indeed may, Gubbins (1976) has also favoured gravitational settling as a major contribution to the dynamo process. Gubbins (1977) points out that such a process does not have a fundamental inefficiency like thermal convection, but can in fact be very efficient.

(v) *Seismic energy*. The possibility that sufficient of the energy given off by earthquakes might be converted in the core to drive the dynamo was suggested by Mullan (1973), on the basis of an earthquake energy supply rate of 3×10^{11}–3×10^{12} W. Comparison of this power figure with the requirements of the geodynamo discussed above indicates that an unreasonably efficient seismic to dynamo energy transfer process would be required, though it is possible that the seismic energy figure may be an underestimate with regard to long period disturbances. Crossley and Smylie (1975) found that core oscillations suffered little damping and so might couple with a dynamo process, though Gubbins (1975) calculated that they would have insufficient amplitude. The case for seismic energy driving the dynamo seems at best to be marginal on present evidence.

An Outline of the Geodynamo

Repeating the caveat that core theories are generally indefinite, the author will here synthesize the foregoing ideas into a description of one way in which a dynamo process could be taking place in the core of the Earth. For the newcomer to dynamo theory the description may illustrate concepts and provide a model to think in terms of. The description which follows is not profoundly different from that of Bullard and Gellman (1954), whose dynamo model has been a working basis of earth scientists for more than twenty years, including the crucial period when development and acceptance of the sea-floor spreading hypothesis depended upon understanding the behaviour of the geomagnetic field.

First, the Earth is taken to be cooling from a hot origin (see A. E. Ringwood's contribution to this volume, Chapter 1), so that the power flow out of the cooling core is comfortably adequate to drive the dynamo process and has been so since early pre-Cambrian times. This power flow is of order 5×10^{12} W: a modest fraction of the surface terrestrial heat flow, but sufficient to include the heat conducted along an adiabatic gradient in the outer core; a lesser flow would mean that the core was stable against convection (at least in part). This power flow comes both from general cooling and also from latent heat and possibly buoyancy forces at the inner core boundary, and so is enough to support an ohmic dissipation of perhaps 10^{11} W, thus allowing moderate toroidal fields within the Earth. The cooling rate is of order some tens of degrees per 10^9 years, and the whole mantle also cooling at about this rate provides for another 5×10^{12} W of the terrestrial heat flow. In the mantle, it is necessary for the heat given off by cooling to be transported by convection. The other 10^{13} W of surface heat flow comes from radioactive elements, mostly concentrated in the crust.

The actual flow patterns in the core, aligned by Earth's rotation and thus in some unspecified way producing an axial field, may be like those determined experimentally by Hide (1953) (see also Runcorn 1954) and Busse (1975), from where Fig. 6 is taken. They will be modified by the radial

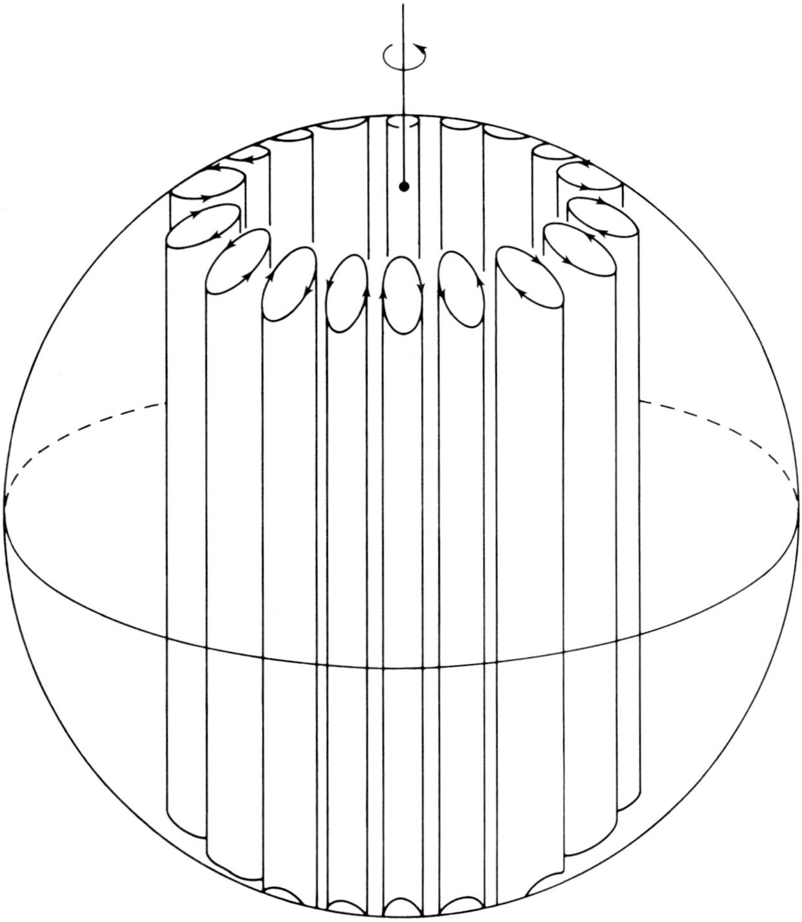

Fig. 6. A sketch for the onset of convection in a rotating sphere. (From Busse, 1975.)

buoyancy occurring in the central gravitational force field of the Earth's core and also (possibly extensively) by magnetohydrodynamic forces. Time fluctuations of the magnetic field will occur even when the flow pattern is relatively steady, due to the very stringent requirement for stability that the right-hand side of equation (1) must balance in all three components at all points in the core. The magnetic time fluctuations will be exacerbated when the fluid flow changes with time due to magnetohydrodynamic forces, and even due to the simple developments which occur in convective flows when the area above an uprising column is heated in excess of its surroundings, and an area below a sinking column is cooled in excess of its surroundings. The delicate state of core thermal conditions may well cause by themselves a quite natural migration of convection patterns with time.

The determination and observation of new core-dependent parameters may enable such a model to be modified and refined. But the exact nature of the geodynamo may remain one of the Earth's best kept secrets.

3. Miscellaneous Processes

The core may demonstrate a whole range of processes which could not be modelled under laboratory conditions and which, if not unique to the Earth, could nevertheless only at present be reasonably expected to be studied in any detail on this planet: foremost amongst these is, of course, the dynamo process itself, just described. The final part of this article now briefly discusses three others.

3.1. Interaction of Magnetic Fields and Seismic Waves

As noted in Section 2.3, any evidence on the toroidal magnetic field strength within the core would be most valuable. A number of authors have therefore examined the possibility (first suggested it seems by Cagniard, 1952), that the magnetic field in the core would interact with seismic waves passing through it, to an extent detectable in surface seismic observations. Such phenomena in which magnetic and elastic forces interact have been called "magnetoelastic" (a term also used to describe the somewhat different phenomena of the elastic properties of actual magnetic material).

Elastic waves can be strongly damped by non-uniform magnetic fields in laboratory experiments (Lilley and Carmichael, 1968, 1970), but it appears that in the core of the Earth the effect would be negligible (Lilley and Smylie, 1968). Crossley and Smylie (1975) have shown, in fact, that the electromagnetic damping of core oscillations is likely to be so slight that the converse possibility arises of core oscillations contributing significantly to a dynamo process (see "Energy sources for the geodynamo" in Section 2.4).

3.2. Oscillations of a Core Stable against Convection

Slichter (1961) pointed out a possible oscillation of the whole Earth in which the inner core moves about its equilibrium central position. The restoring force is not predominantly elastic but gravitational, so that the inner core floats (or more precisely, sinks), backwards and forwards. Fundamental to this process is the central gravitational force field experienced by a particle in the Earth's core.

The inner core represents a basic stratification within the core. It has to be slightly denser than the outer core (more precisely to depart from the Williamson–Adams equation with positive η, see "Stability of the fluid core against convection" in Section 2.3), or else it would be neutrally buoyant and not confined to the Earth's centre. If the fluid in the outer core is also stable against convection a whole range of extra oscillations are allowed, similar to the basic Slichter mode in that the dominant restoring forces are due to buoyancy. These modes have become known as the "gravitational undertones" of Earth oscillation; although arising in the core they will in fact affect the whole Earth because the mantle–core boundary is not rigid. Streamlines for the basic Slichter mode and one undertone are shown in Fig. 7.

Such modes have been the subject of much study (e.g. Crossley, 1975; Smith, 1976). In a recent paper, Johnson and Smylie (1977) discuss the theory for computing all different oscillation modes of a model Earth. Also important are the inertial modes, arising in a fluid core simply as a result of its rotation. These inertial modes do not depend on stable stratification for their existence, but are influenced by self-gravitation and compression.

The general effect of rotation is to couple different modes together, in particular into two distinct chains. The coupling is weak at periods of 1 h or less (for the normal elastic modes) but strong at periods of several hours or more (for the gravitational undertones and inertial oscillations).

A very weak gravity signal of appropriate frequency and amplitude should occur at the Earth's surface if these modes can exist and if they are excited by a strong-enough earthquake (Jackson and Slichter, 1974). There is currently much interest in developing ultra-sensitive gravimetric techniques because of this possibility of collecting new data pertaining to the Earth's core.

3.3. Biological Effects of Geomagnetic Reversals

The solar wind, a stream of particles continually flowing off the Sun's surface, interacts with the geomagnetic field to produce two main effects: (i) the solar wind is deflected clear of the Earth's surface, and (ii) the geomagnetic field is contained within a region about the Earth known as the magnetosphere. This situation is depicted in Fig. 8.

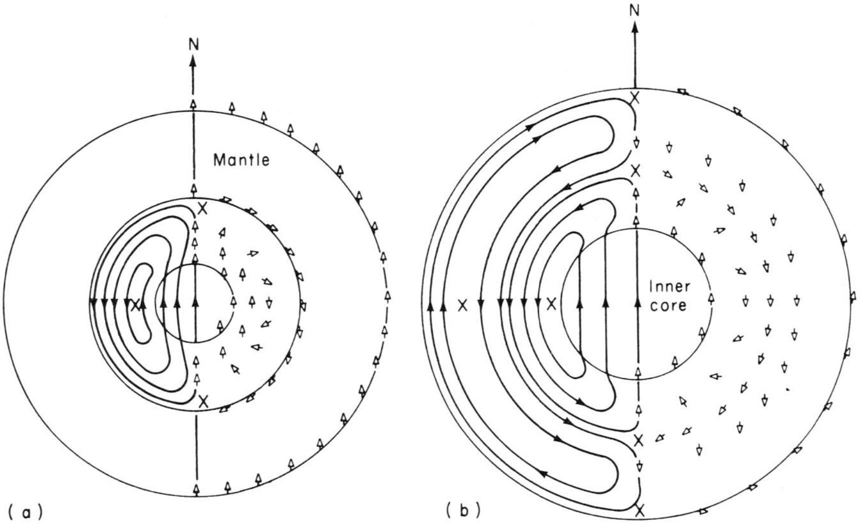

Fig. 7. (a) The Slichter mode of inner core oscillation (often denoted ${}_{-1}S_1^\circ$); (b) the second undertone (${}_{-2}S_1^\circ$). (From Crossley 1975). Note different scales of (a) whole Earth and (b) the core only. The right half of each figures shows instantaneous motion and the left half the corresponding streamlines.

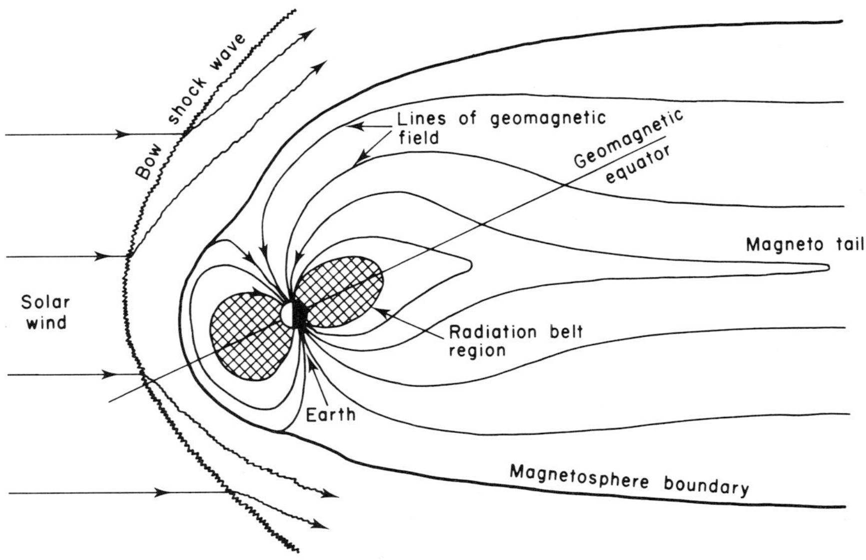

Fig. 8. The magnetosphere formed by flow of the solar wind around the Earth's magnetic field. (After Ness, 1970.)

A reversal of the main geomagnetic field will cause gross changes in the magnetosphere. If the poles of a weakened geomagnetic field migrate along lines of geographic longitude (as palaeomagnetic studies have shown that they may, see "Palaeomagnetic evidence for the geodynamo" in Section 2.4), then as the magnetic poles approach the geographic equator substantial changes in the magnetosphere will occur during each day. If the field reverses by weakening to near zero in one axial direction and then growing in the other there will be a time of near-zero field when the magnetosphere will be greatly reduced, if not collapsed completely (cf. McCormac and Evans, 1969; Siscoe *et al.*, 1976).

Uffen (1963) first pointed out that by thus removing or at any rate much reducing the magnetosphere, a geomagnetic reversal could have biological effects. He identified two mechanisms during a reversal: (i) charged particles would be dumped on Earth's surface from the Van Allen radiation belts in the magnetosphere where they are normally trapped, and (ii) with the magnetospheric "solar-wind shield" removed, the charged particles of the solar wind would be able to reach Earth's surface directly. Both sources of charged particles could induce genetic mutations in living organisms at the Earth's surface, and thus exert a pressure on evolution.

Uffen's model was qualitative, and more quantitative estimates (Sagan, 1965; Waddington, 1967; Black, 1967) did not support his particular mechanisms as being likely. However, detailed evidence of coincidences of faunal extinctions and geomagnetic reversals began to accumulate, especially from analyses of deep-ocean sediment cores (Harrison and Funnell, 1964; Opdyke *et al.*, 1966; Watkins and Goodell, 1967; Steuerwald *et al.*, 1968; Hays, 1971). In seeking an intermediate process in a possible chain of causes it was noticed that climate patterns also correlated in some instances (Harrison, 1968; Wollin *et al.*, 1971; Harrison and Prospero, 1974; Fairbridge, 1977).

Most recently, Reid *et al.* (1976) have re-examined the question of ultra-violet light received at the Earth's surface during a geomagnetic reversal. Knowledge of the Earth's stratospheric ozone layer which normally absorbs much of the ultra-violet radiation from the Sun has increased recently, due to its relevance in various topical environmental matters. Reid *et al.* now estimate that during a geomagnetic reversal the number of solar protons reaching the stratosphere may be greatly increased and consequently the ozone concentration there greatly reduced. Such a mechanism is only one step from Uffen's original hypothesis: the solar protons may not cause mutations directly but they affect the atmosphere and so allow in an increased flux of potentially harmful ultra-violet radiation.

In as much as it involves the Earth's atmosphere, the solar proton process may also be expected to affect climate. Indeed, on the shorter time scale of recorded history there is much current interest in the possible influences which the geomagnetic field and its solar relationships may have on terrestrial weather (Wilcox *et al.*, 1974; King, 1974; Wollin *et al.*, 1973).

Acknowledgements

For advice on topical matters concerning the geomagnetic field and the core I am grateful to many people, especially J. R. Cleary, D. Gubbins, J. A. Jacobs, M. W. McElhinny, A. E. Ringwood, M. G. Rochester, D. E. Smylie, F. D. Stacey and D. J. Stevenson: several of these I thank also for comments on a first version of this manuscript.

The diagrams in this review have come from a variety of sources, and I thank the authors of the original articles concerned for permission to reproduce their diagrams here.

Finally, in this article, I should like to acknowledge the pleasure it has been to work for the last ten years in the stimulating atmosphere for earth science engendered at the Australian National University by J. C. Jaeger and A. L. Hales.

References

Black, D. I. (1967). Cosmic ray effects and faunal extinctions at geomagnetic field reversals. *Earth planet. Sci. Lett.* **3**, 225–236.

Blackett, P. M. S. (1961). Comparison of ancient climates with the ancient latitudes deduced from rock magnetic measurements. *Proc. R. Soc. A* **263**, 1–30.

Braginskii, S. I. (1964). Magnetohydrodynamics of the earth's core. *Geomag. Aeron., Engl. Transl.* **4**, 698–712.

Brett, R. (1976). The current status of speculations on the composition of the core of the Earth. *Rev. Geophys. Space Phys.* **14**, 375–383.

Briden, J. C. (1968). Paleoclimatic evidence of a geocentric axial dipole field. In *The History of the Earth's Crust* (R. A. Phinney, ed.), Princeton University Press, Princeton, N.J., pp. 178–194.

Bullard, E. C. (1949). The magnetic field within the earth. *Proc. R. Soc. A* **197**, 433–453.

Bullard, E. C. and Gellman, H. (1954). Homogeneous dynamos and terrestrial magnetism. *Phil. Trans. R. Soc. A* **247**, 213–278.

Bullard, E. C. and Gubbins, D. (1977). Generation of magnetic fields by fluid motions of global scale. *Geophys. astrophys. Fluid Dynamics* **8**, 43–56.

Bullen, K. E. (1975). *The Earth's Density*, Chapman and Hall, London.

Busse, F. H. (1975). A model of the geodynamo, *Geophys. J. R. astronom. Soc.* **42**, 437–459.

Cagniard, M. L. (1952), Sur la nature des ondes séismiques capables de traverser le noyau terrestre. *C.r. hebd Séanc. Acad. Sci., Paris* **234**, 1705–1706.

Crossley, D. J. (1975). Core undertones with rotation. *Geophys. J. R. astronom. Soc.* **42**, 477–488.

Crossley, D. J. and Smylie, D. E. (1975). Electromagnetic and viscous damping of core oscillations. *Geophys. J. R. astronom. Soc.* **42**, 1011–1034.

Dagley, P. and Lawley, E. (1974). Palaeomagnetic evidence for the transitional behaviour of the geomagnetic field. *Geophys. J. R. astronom. Soc.* **36**, 577–598.

Dziewonski, A. M., Hales, A. L. and Lapwood, E. R. (1975). Parametrically simple earth models consistent with geophysical data. *Phys. Earth planet. Interiors* **10**, 12–48.

Dziewonski, A. M. and Haddon, R. A. W. (1974). The radius of the core-mantle boundary inferred from travel time and free oscillation data: a critical review. *Phys. Earth planet. Interiors* **9**, 28–35.

Elsasser, W. M. (1946). Induction effects in terrestrial magnetism. *Phys. Rev.* **69**, 106–116.

Evans, M. E. (1976). Test of the dipolar nature of the geomagnetic field throughout Phanerozoic time. *Nature, Lond.* **262**, 676–677.

Fairbridge, R. W. (1977). Global climate change during the 13,500 b.p. Gothenburg geomagnetic excursion. *Nature, Lond.* **265**, 430–431.

Gardiner, R. B. and Stacey, F. D. (1971). Electricity resistivity of the core. *Phys. Earth planet. Interiors* **4**, 406–410.

Goettel, K. A. (1976). Models for the origin and composition of the earth, and the hypothesis of potassium in the earth's core. *Geophys. Surveys* **2**, 369–397.

Gubbins, D. (1972). Kinematic dynamos and geomagnetism. *Nature phys. Sci.* **238**, 119–122.

Gubbins, D. (1974). Theories of the geomagnetic and solar dynamos. *Rev. Geophys. Space Phys.* **12**, 137–154.

Gubbins, D. (1975). Can the earth's magnetic field be sustained by core oscillations? *Geophys. Res. Lett.* **2**, 409–412.

Gubbins, D. (1976). Observational constraints on the generation process of the earth's magnetic field. *Geophys. J. R. astronom. Soc.* **47**, 19–39.

Gubbins, D. (1977). Energetics of the earth's core, *J. Geophys.* **43**, 453–464.

Harradon, H. D. (1943). Some early contributions to the history of geomagnetism—part 1, *Terrest. Mag. atmosp. Elect.* **48**, 3–17.

Harrison, C. G. A. (1968). Evolutionary processes and reversals of the earth's magnetic field, *Nature, Lond.* **217**, 46–47.

Harrison, C. G. A. and Funnell, B. M. (1964). Relationship of palaeomagnetic reversals and micropalaeontology in two late Caenozoic cores from the Pacific Ocean. *Nature, Lond.* **204**, 566.

Harrison, C. G. A. and Prospero, J. M. (1974). Reversals of the earth's magnetic field and climate changes. *Nature, Lond.* **250**, 563–565.

Hays, J. D. (1971). Faunal extinctions and reversals of the earth's magnetic field, *Bull. geol. Soc. Am.*, **82**, 2433–2447.

Hide, R. (1953). Some experiments of thermal convection in a rotating liquid. *Q. Jl R. meteor. Soc.* **79**, 161.

Hide, R. and Stewartson, K. (1972). Hydromagnetic oscillations of the earth's core. *Rev. Geophys. Space Phys.* **10**, 579–598.

Higgins, G. and Kennedy, G. C. (1971). The adiabatic gradient and the melting point gradient in the core of the earth. *J. geophys. Res.* **76**, 1870–1878.

Irving, E. (1956). Palaeomagnetic and palaeoclimatological aspects of polar wandering. *Geofis. pura appl.* **33**, 23–41.

Jackson, B. V. and Slichter, L. B. (1974). The residual daily earth tides at the South Pole. *J. geophys. Res.* **79**, 1711–1715.

Jacobs, J. A. (1953). The earth's inner core, *Nature, Lond.* **172**, 297–298.
Jacobs, J. A. (1954). Temperature distribution within the earth's core, *Nature, Lond.* **173**, 258.
Jacobs, J. A. (1975). *The Earth's Core*, Academic Press, London.
Jacobs, J. A. and Masters, G. (1976). The stability of the earth's core and the geodynamo, *Nature, Lond.* **261**, 483–484.
Jain, A. and Evans, R. (1972). Calculation of the electrical resistivity of liquid iron in the earth's core. *Nature, Lond.* **235**, 165–167.
Jeffreys, H. (1939). The times of P, S and SKS, and the velocities of P and S, *Mon. Not. R. astronom. Soc., Geophys. Suppl.* **4**, 498–533.
Johnson, I. M. and Smylie, D. E. (1977). A variational approach to whole-earth dynamics, *Geophys. J. R. astronom. Soc.* **50**, 35–54.
Johnston, M. J. S. and Strens, R. G. J. (1973). Electrical conductivity of molten Fe–Ni–S–C core mix, *Phys. Earth planet. Interiors* **7**, 217–218.
Jones, G. M. (1977). Thermal interaction of the core and the mantle and long-term behaviour of the geomagnetic field, *J. geophys. Res.* **82**, 1703–1709.
Kennedy, G. C. and Higgins, G. H. (1973). The core paradox, *J. geophys. Res.* **78**, 900–904.
King, J. W. (1974). Weather and the earth's magnetic field. *Nature, Lond.* **247**, 131–134.
Krause, F. (1977). Mean-field electrodynamics and dynamo theory of the earth's magnetic field, *J. Geophys.* **43**, 421–440.
Levy, E. H. (1976). Generation of planetary magnetic fields. *A. Rev. Earth planet. Sci.* **4**, 159–185.
Lilley, F. E. M. (1970). On kinematic dynamos. *Proc. R. Soc. A* **316**, 153–167.
Lilley, F. E. M. and Carmichael, C. M. (1968). Electromagnetic damping of elastic waves: experimental results. *Can. J. Earth Sci.* **5**, 825–829.
Lilley, F. E. M. and Carmichael, C. M. (1970). Electromagnetic damping of elastic waves: a simple theory. *Can. J. Earth Sci.* **7**, 1304–1307.
Lilley, F. E. M. and Smylie, D. E. (1968). Elastic wave motion and a nonuniform magnetic field in electrical conductors. *J. geophys. Res.* **73**, 6527–6533.
Liu, L. (1975). On the (γ, ϵ, l) triple point of iron and the earth's core. *Geophys. J. R. astronom. Soc.* **43**, 697–706.
Loper, D. E. (1975). Torque balance and energy budget for precessionally driven dynamos. *Phys. Earth planet. Interiors* **11**, 43–60.
Lowes, F. J. (1970). Possible evidence on core evolution from geomagnetic dynamo theories. *Phys. Earth planet. Interiors* **2**, 382–385.
Malkus, W. V. R. (1968). Precession of the earth as the cause of geomagnetism. *Science, N.Y.* **160**, 259–264.
McCormac, B. M. and Evans, J. E. (1969). Consequences of very small planetary magnetic moments. *Nature, Lond.* **223**, 1255.
McElhinny, M. W. (1977). A closer look at Laurasia, Gondwanaland and Pangea. Plenary lecture, 4th International Palynological Congress, Lucknow, India.
McElhinny, M. W. and Evans, M. E. (1968). An investigation of the strength of the geomagnetic field in the early pre-Cambrian. *Phys. Earth planet. Interiors* **1**, 485–497.
Metchnik, V. I., Gladwin, M. T. and Stacey, F. D. (1974). Core convection as a power source for the geomagnetic dynamo—a thermodynamic argument. *J. Geomag. Geoelect.* **26**, 405–415.

Mullan, D. J. (1973). Earthquake waves and the geomagnetic dynamo. *Science, N.Y.* **181**, 553–554.
Ness, N. F. (1970). Magnetometers for space research. *Space Sci. Revs* **11**, 459–554.
Opdyke, N. D., Glass, B., Hays, J. D. and Foster, J. (1966). Palaeomagnetic study of Antarctic deep-sea cores. *Science, N.Y.* **154**, 349–357.
Reid, G. C., Isaksen, I. S. A., Holzer, T. E. and Crutzen, P. J. (1976). Influence of ancient solar-proton events on the evolution of life, *Nature, Lond.* **259**, 177–179.
Ringwood, A. E. (1977). Composition of the core and implications for origin of the earth, *Geochem. J.* **11**, 111–135.
Roberts, G. O. (1970). Spatially periodic dynamos, *Phil. Trans. R. Soc. A* **266**, 535–558.
Roberts, P. H. (1967). *An Introduction to Magnetohydrodynamics*, Longmans, London.
Roberts, P. H. (1971). Dynamo theory. In *Mathematical Problems in the Geophysical Sciences* (W. H. Reid, ed.), American Mathematics Society, Providence, R.I., pp. 129–206.
Rochester, M. G., Jacobs, J. A., Smylie, D. E. and Chong, K. F. (1975). Can precession power the geomagnetic dynamo? *Geophys. J. R. astronom. Soc.* **43**, 661–678.
Runcorn, S. K. (1954). The earth's core. *Trans. Am. geophys. Union* **35**, 49–63.
Sagan, C. (1965). Is the early evolution of life related to the development of the earth's core? *Nature, Lond.* **206**, 448.
Simon, F. E. (1953). The melting of iron at high pressures. *Nature, Lond.* **172**, 746–747.
Siscoe, G. L., Chen, C. and Harel, M. (1976). On the altitude of the magnetopause during geomagnetic reversals. *J. atmos. terrest. Phys.* **38**, 1327–1331.
Slichter, L. B. (1961). The fundamental free mode of the earth's inner core. *Proc. natn. Acad. Sci. U.S.A.* **47**, 186–190.
Smith, M. L. (1976). Translational inner core oscillations of a rotating, slightly elliptical earth. *J. geophys. Res.* **81**, 3055–3065.
Smith, P. J. (1970). Petrus Peregrinus' Epistola: the beginning of experimental studies of magnetism in Europe. *Earth Sci. Rev. Atlas. Suppl.* **6**, A11–A17.
Stacey, F. D. (1972). Physical properties of the earth's core, *Geophys. Surveys* **1**, 99–119.
Stacey, F. D. (1977). Applications of thermodynamics to fundamental earth physics. *Geophys. Surveys*, **3**, 175–204.
Stacey, F. D. and Irvine, R. D. (1977). Theory of melting: thermodynamic basis of Lindemann's law. *Australian J. Phys.* **30**, 631–640.
Steuerwald, B. A., Clark, D. L. and Andrew, J. A. (1968). Magnetic stratigraphy and faunal patterns in Arctic Ocean sediments. *Earth planet. Sci. Lett.* **5**, 79–85.
Stevenson, D. J. (1977). Hydrogen in the earth's core. *Nature, Lond.* **268**, 130.
Uffen, R. J. (1963). Influence of the earth's core on the origin and evolution of life. *Nature, Lond.* **198**, 143–144.
Urey, H. C. (1952). *The Planets: Their Origin and Development*, Yale University Press, New Haven, Conn.
Verhoogen, J. (1961). Heat balance of the earth's core. *Geophys. J. R. astronom. Soc.* **4**, 276–281.
Waddington, C. J. (1967). Paleomagnetic field reversals and cosmic radiation. *Science, N.Y.* **158**, 913–915.

Watkins, N. D. and Goodell, H. G. (1967). Geomagnetic polarity change and faunal extinction in the Southern ocean. *Science, N.Y.* **156**, 1083–1087.
Wilcox, J. M., Scherrer, P. H., Svalgaard, L., Roberts, W. O., Olsen, R. H. and Jenne, R. L. (1974). Influence of solar magnetic sector structure on terrestrial atmospheric vorticity. *J. atmos. Sci.* **31**, 581–588.
Wollin, G., Ericson, D. B., Ryan, W. B. F. and Foster, J. H. (1971). Magnetism of the earth and climatic changes. *Earth planet. Sci. Lett.* **12**, 175–183.
Wollin, G., Kukla, G. J., Ericson, D. B., Ryan, W. B. F. and Wollin, J. (1973). Magnetic intensity and climatic changes 1925–1970. *Nature, Lond.* **242**, 34–37.

4. Palaeogeomagnetism and the Core–Mantle Interface

M. W. McELHINNY

Abstract

Studies of the geomagnetic secular variation and theoretical considerations suggest that the time constants of the geomagnetic field generation extend up to 10^4 years. This view is supported by the similar time taken for the field to change polarity. The long-term palaeomagnetic record of the reversal history and palaeosecular variation show that there are time constants associated with the geomagnetic field that extend to 10^8 years or longer. Correlation and spectral analysis of the reversal record confirms that polarity intervals are independent and may be treated as a random process using stochastic models, thus eliminating any requirement for long time constants in the core dynamo processes. The core therefore contains no memory of past configurations beyond the average life time of the magnetohydrodynamic process. The long-term changes must then reflect changes in conditions under which the dynamo operates and changes at the core–mantle boundary appear most likely. Correlations between the large-scale non-axisymmetric features of the gravitational and geomagnetic potentials suggest the existence of topographic bumps on the core–mantle interface. Analysis of global seismic P-wave velocity anomalies in the lower mantle shows a negative correlation between long-wavelength gravity anomalies observed at the surface and those computed from the velocity anomalies under the assumption of proportionality between velocity and density perturbations. The gross features of these velocity anomalies in the lower mantle correlate with similar gross features of the anomalous time-averaged palaeomagnetic field over the past 5 million years. Furthermore there appears to be a correlation between major plate tectonic changes at the surface and changes in reversal frequency, suggesting a relation between the tectonic evolution of the upper mantle and processes in the core. The interrelation of all these phenomena strongly supports the existence of mantle-wide convection and the presence of large-scale topographic bumps at the core–mantle interface.

1. Introduction

Until the development of studies in archaeomagnetism and palaeomagnetism, the investigation of the geomagnetic field and its time variation has been based on the analysis of observatory measurements made over the past few centuries. The short-term variations arise from currents flowing in the ionosphere giving rise to magnetic storms and the more or less regular daily variation. Variations at this frequency originating in the core are attenuated by electromagnetic induction in the mantle. Cox and Doell (1964) estimated the cut-off point to be periods between 1 and 10 years. From a global analysis of magnetic records over a 50-year period (Currie (1966, 1967, 1968) found no significant differences in geomagnetic spectra from different regions of the Earth and estimated that for periods up to 3.7 years the spectrum originates entirely from outside the Earth.

The power spectrum of the geomagnetic field of internal origin is illustrated in Fig. 1. The power level increases sharply with increasing period up to 30 years following the analysis of Currie (1968). The steep rise in spectral power with increasing period indicates that most of the power is present in periods very much longer than the time of historical observations. The most significant of these longer period variations is the westward drift of the geomagnetic field first noticed by E. Halley in 1692 but not analysed on a global scale until the work of Bullard *et al.* (1950). For the non-dipole field the drift rate was found to be $0.18 \pm 0.015° \text{ y}^{-1}$ and for the secular variation $0.32 \pm 0.067° \text{ y}^{-1}$. These drift rates are independent of latitude and correspond to periods of 2000 and 1125 years, respectively. Archaeomagnetic measurements (Burlatskaya *et al.*, 1965, 1970; Burlatskaya, 1972; Yukutake, 1967) suggest similar drift rates for the main features of the declination and inclination indicating that the westward drift has persisted over at least the past 1000 years. Since the westward drift produces a substantial part of the present secular variation, it is clear that the power continues to increase to periods of at least 10^3 years (Fig. 1).

Changes in the Earth's magnetic moment deduced from archaeomagnetic data (Cox, 1968; Burlatskaya, 1970; Smith, 1970) indicate a period of about 10^4 years and amplitudes of 50 per cent. Bullard *et al.* (1950) and Vestine *et al.* (1963) have noted that there is no evidence supporting a change in position of the geomagnetic pole since 1829. McDonald and Gunst (1968) showed that the average rate of westward drift of the geomagnetic poles during the past 130 years is $0.042° \text{ y}^{-1}$ in azimuth angle and no progressive motion in polar angle. This would appear to indicate a period of about 10^4 years for changes in orientation of the main dipole (dipole wobble). A similar conclusion was reached by Nagata (1965). The free-decay period of the Earth's magnetic field was initially estimated to be about 50 000 years

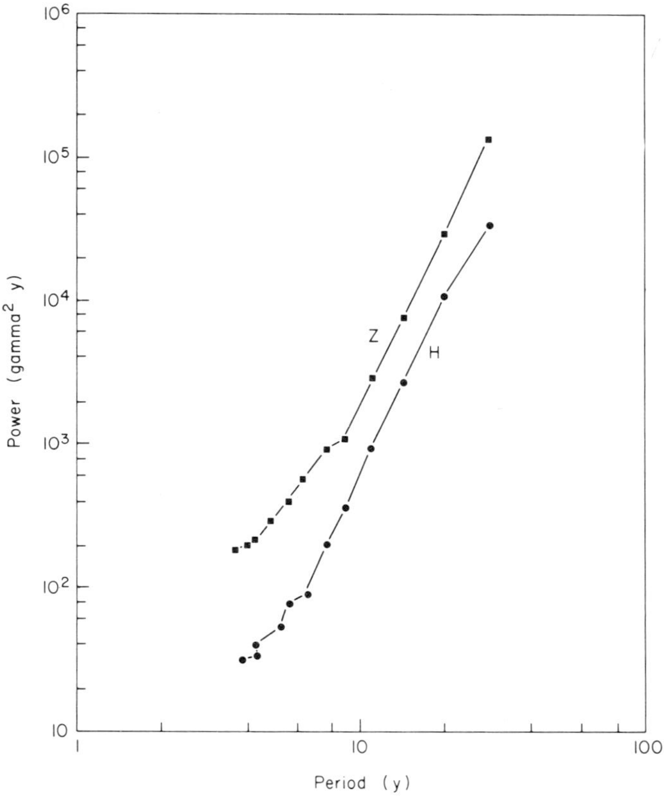

Fig. 1. Power spectrum of the geomagnetic field of internal origin from data covering approximately 50 years for the first half of this century after Currie (1968). Composite H and Z spectra derived from eight magnetic observatories in the northern hemisphere are shown.

for an arrested core fluid (Elsasser, 1946). With allowance for eddy diffusion this period should be reduced by an order of magnitude, so that periods of 10^3–10^4 years are to be associated with the field generation mechanism (see also Braginskiy, 1972).

Evidence that the Earth's magnetic field has reversed its polarity many times in the past has been one of the major discoveries of palaeomagnetic investigations. For the past 5 million years the pattern of reversals is now well established (Cox, 1969; McDougall, this volume, Chapter 16). The time constants associated with the reversal process are, however, very much longer than those predicted by dynamo theories of the field generation mechanism. The time required for the field to change polarity is in the range 10^3–10^4 years (Ninkovich *et al.*, 1966; Harrison and Somayajulu, 1966; Cox and Dalrymple, 1967; McElhinny, 1971*a*). This is consistent with dynamo theory that the field generation mechanism has time constants in this range. However the time between successive polarity changes may be as short as 3×10^4 years and longer than 10^7 years (Cox, 1975). Furthermore, changes in the average frequency of reversals occur at intervals of 5×10^7 years or longer (McElhinny, 1971*b*; Ulrych, 1972; Irving and Pullaiah, 1976). Unless the time constants predicted by dynamo theory are incorrect, these phenomena are not related to the field generation process but must have some other cause.

Cox (1975) has drawn an analogy between meteorological phenomena and the spectrum of geomagnetic changes. The short period fluctuations correspond to daily pressure and temperature changes due to the Earth's rotation either directly or indirectly. In geomagnetism the analogues are micropulsations, daily and annual variations and effects of the sun-spot cycle. Medium period variations due to individual storms or seasonal changes correspond to processes occurring in the air masses themselves, the relevant time constants being those of the underlying meteorological processes. Here the geomagnetic analogue is the secular variation reflecting the time constants of dynamo processes in the core.

The very long period part of the spectrum in meteorology corresponds to changes in climate. The underlying physical process has some other cause, such as changes in the rotation axis of the Earth, the formation of mountain ranges or the effects of plate tectonics involving the opening and closing of oceans. These latter effects are essentially changes in the boundary conditions at the fluid–surface interface. By analogy the very long period geomagnetic changes are probably due to boundary conditions changing at the core–mantle interface.

2. Time Constants of Dynamo Processes

Analysis of marine magnetic anomalies in terms of sea-floor spreading very soon extended the reversal time scale back to 76 million years ago (Heirtzler *et al.*, 1968). Subsequently, the details of the reversal time scale from 76 to 153 million years was clarified by Larson and Pitman (1972). The changes in reversal frequency for the past 153 million years deduced from these time scales are illustrated in Fig. 2 as seen through a sliding window 10 million years long (Cox, 1975). The variations in average reversal frequency occur in discontinuous steps. From 153 to 107 million years the average frequency is about 1 per million years. Between 107 and 85 million years this drops to zero and then abruptly increases to 1 per million years between 85 and 45 million years. At 45 million years there is an abrupt increase to 3 per million years.

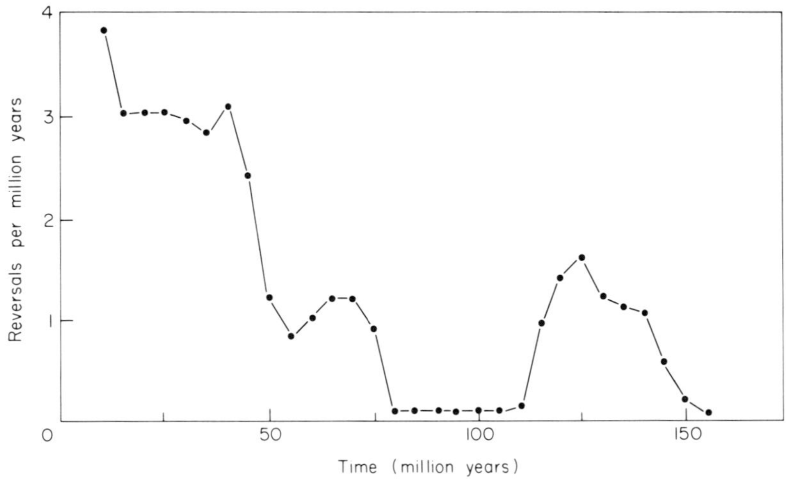

Fig. 2. Variations in the average frequency of reversals as seen through a sliding window 10 million years long after Cox (1975). From 76 million years ago to the present the rates are from Heirtzler *et al.* (1968), and prior to 76 million years ago they are from Larson and Pitman (1972).

It should be stressed that these abrupt changes refer to the *average* frequency of reversals. The reversals themselves still take place at widely varying intervals from 3×10^4 years to 10^7 years or more (Cox, 1975). The time constants of fluid motions in the core as deduced from secular variation and from the time taken to change polarity range up to a maximum of 10^4 years. If this time constant is characteristic of the longest time constants of magnetohydrodynamic processes in the core, the implication is that the magnetic memory of the core is short relative to the average length of polarity intervals. On the other hand, if the dynamo process is coherent over times much longer than the average length of polarity intervals, the lengths of these intervals would not be independent. To determine whether successive polarity intervals are independent thus provides an indirect means of measuring the time constants of dynamo processes.

The independence of sequences of geomagnetic polarity intervals has been tested statistically by Phillips *et al.* (1975) using correlation analysis to test the hypothesis that reversals are generated by a renewal process. Let t be the time elapsed since a previous polarity change and $p(t)\mathrm{d}t$ the probability that a reversal will occur between t and $t + \mathrm{d}t$. For a renewal process, $p(t)$ depends *only* on the time since the previous reversal and *not* on the length of prior polarity intervals (i.e. the lengths of polarity intervals will be independent). For a Poisson process, $p(t)$ is a constant, but for a general renewal process $p(t)$ varies.

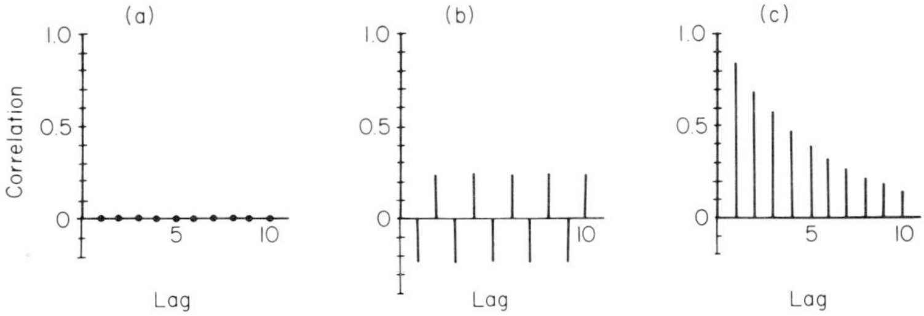

Fig. 3. Theoretical correlation coefficients as a function of lag for various types of processes after Phillips *et al.* (1975). (a) Renewal process, intervals are independent; (b) alternating renewal process, normal and reversed polarity intervals have different mean lengths; (c) first-order Markov process, the length of each polarity interval depends on the length of the previous interval.

Figure 3 shows the predictions of a correlation analysis for three possibilities, a renewal process, an alternating process in which normal and reversed polarity intervals have different mean lengths and a first-order Markov process in which the lengths of polarity intervals depend on the length of one or more previous polarity intervals. Using four sections of the geomagnetic polarity time scale for the past 153 million years, Phillips *et al.* (1975) derived the correlation coefficients shown in Fig. 4. The correlation coefficients all drop below the level of significance for the three data sets showing stationary behaviour (the average interval length is constant throughout). The 166–153 million year section shows non-stationary behaviour. In a later paper, Phillips and Cox (1976) applied spectral analysis to determine whether periodic components were present in these sections of the time scale. Maximum entropy spectral analysis showed no evidence for periodicities (Fig. 4). Both analyses lead to the conclusion that polarity intervals are indeed independent and support the hypothesis that observed reversals occur at random times. The analysis of Phillips and Cox (1976) is compatible with a gamma renewal process or some similar random process.

For gamma distribution polarity intervals, the probability density function $f(x)$ that an interval will have a length between x and $x+\mathrm{d}x$ is given by

$$f(x)=\frac{(k/\mu)^k x^{k-1}\,\mathrm{e}^{-kx/\mu}}{\Gamma(k)},$$

where μ is the mean interval length and k is a parameter known as the gamma index. For $k=1$, the distribution of polarity intervals is exponential and the process is Poisson. When $k>1$, the density function peaks away from the origin so that the probability of finding short polarity intervals is small. When $k \gg 1$, $f(x)$ becomes normal about the mean and when $k \to \infty$ all

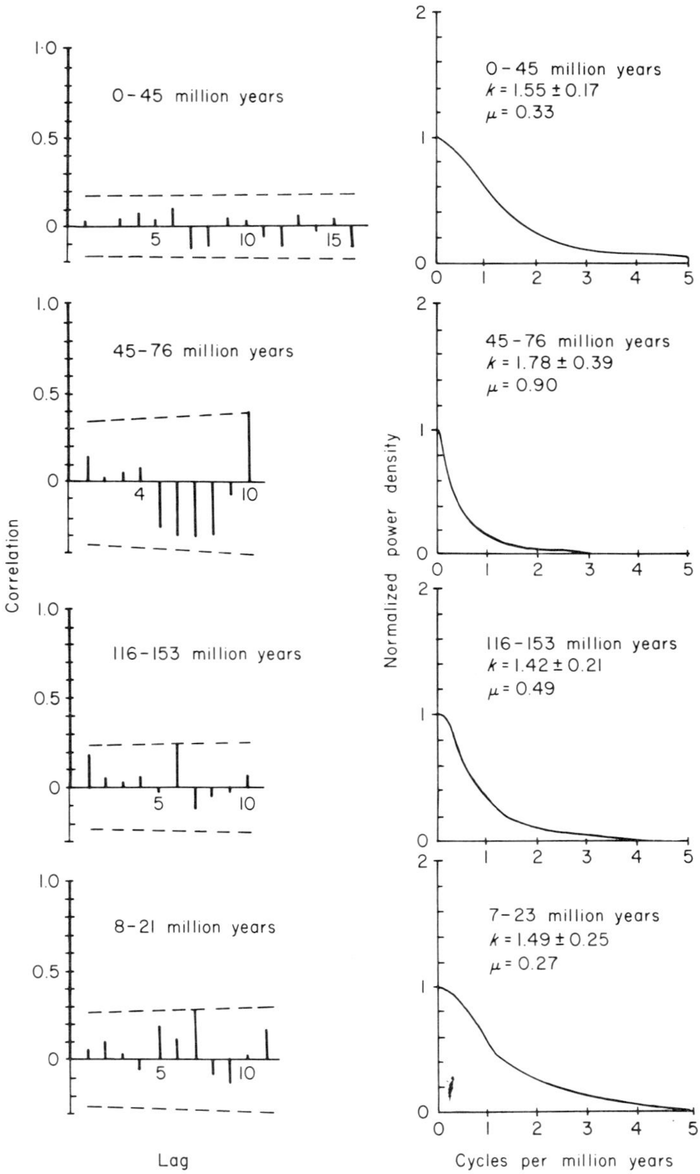

Fig. 4. Analysis of polarity intervals for four sections of the reversal time scale. Left-hand diagrams show correlation coefficients (see Fig. 3) with the dashed lines defining 95 per cent confidence regions about the expected values for a renewal process (Phillips *et al.*, 1975). Right-hand diagrams show maximum entropy spectral estimates (Phillips and Cox, 1976). Estimates of the gamma index k and its standard error together with the mean interval length μ are indicated.

intervals have equal lengths. The gamma index k thus indicates how close the observed distribution is to being exponential and how dispersed they are about their mean.

Phillips (1977) has examined the four reversal sections to test that they are generated by a gamma renewal process. The gross statistical properties of the observations support the model but the statistics of normal polarity intervals are found to differ significantly from those of reversed polarity intervals. The various parameters are listed in Table 1. The gamma indices for *all* polarity intervals are not significantly different for the four sequences. So whereas the mean interval length is quite different for all four sequences, the gamma index k is constant. This seems to suggest that the reversal process operating through the Cenozoic did so even through sudden changes in average reversal frequency. However the most surprising result of Table 1

Table 1. Statistical parameters for polarity intervals (after Phillips, 1977).

Age (million years)	Number			Mean length (million years)			Gamma index		
	N	N_n	N_r	μ	μ_n	μ_r	k	k_n	k_r
0–44.77	136	36	68	0.33	0.33	0.33	1.55 ± 0.17	2.28 ± 0.37	1.19 ± 0.18
44.77–76.33	35	18	17	0.90	0.78	1.03	1.78 ± 0.39	2.79 ± 0.88	1.35 ± 0.42
116.4–153.0	75	37	38	0.49	0.58	0.40	1.42 ± 0.21	1.90 ± 0.41	1.22 ± 0.25
7.39–22.69	57	29	28	0.27	0.27	0.26	1.49 ± 0.25	1.77 ± 0.43	1.28 ± 0.30

is the clear differences in the values of k between the normal and reversed intervals. The gamma indices estimated for normal intervals are consistently larger than those estimated for the reversed intervals. This implies that they have different distributions and that reversals cannot be described by a simple renewal process. A more complicated alternating renewal process is required. Phillips (1977) suggests that the persistent differences in k for normal and reversed states is evidence that the dynamo is more stable after a transition to normal polarity than it is after a transition to reversed polarity. The time constants of this stability are similar to the time constants of fluid motions in the core.

Cyclonic convection dynamo models of the geomagnetic field (Parker, 1955, 1969; Levy, 1972*a,b,c*) assume a dynamo that is spherical and axisymmetric, irrespective of polarity. The asymmetry between the normal and reversed states described may therefore be related to departures from sphericity. At the present time the meaning of this asymmetry is not clear. However, the wide range of time intervals between successive polarity changes and the clear evidence that successive polarity intervals are independent of one another shows that they may be treated as a random process using stochastic models. This eliminates any requirement for long time constants in the dynamo processes in the core. Polarity changes can be presumed to be due to instabilities in the magnetohydrodynamic process and that an instability may take place at any time *t* following a previous instability. Possible forms of instability that have been suggested include too much symmetry in the fluid motions in the core (Nagata, 1969) or some critical configuration of cyclonic convection cells (Parker, 1969; Levy, 1972*a*,*b*,*c*). These models presume the core contains no memory of past configurations beyond the average lifetime of the magnetohydrodynamic process. This presumption is now clearly supported by the reversal record.

3. Very Long Geomagnetic Time Constants

Sudden changes in the average frequency of reversals in the past became apparent in the analysis of Heirtzler *et al.* (1968) of marine magnetic anomalies for the early Cenozoic. The most noticeable of these changes occurred about 45 million years ago (Fig. 2). McElhinny (1971*b*) first attempted to analyse long-term trends in reversal history by examining the distribution of normal and reversed measurements in land-based palaeomagnetic results. Assuming that typical palaeomagnetic studies cover roughly equal time intervals then the reversal rate will have been high if studies persistently observed both polarities and will have been low if only one polarity was persistently observed. A plot of the number of mixed polarity studies as a fraction of the whole is therefore a rough approximation to changes in reversal frequency. Jacobs (1975) interpreted this as roughly equivalent to a logarithmic plot of reversal rate vs. time. This interpretation is illustrated in Fig. 5. It shows that the reversal rate was low especially during the late Palaeozoic and during the Cretaceous. These correspond to the well-known late Palaeozoic reversed interval that lasted about 50 million years (Irving and Parry, 1963) and the Cretaceous normal interval that lasted about 25 million years (Helsley and Steiner, 1969). Figure 5 indicates that changes in reversal rate occur at intervals of 5×10^7 years or longer.

Similar long period fluctuations are observed in the polarity bias of the field in the past. McElhinny (1971*b*) first noticed a cyclic behaviour with period of about 350 million years. The field was heavily biased to normal during the Mesozoic and to reversed during the late Palaeozoic. A more recent analysis by Irving and Pullaiah (1976) confirms the same general pattern (Fig. 6). Maximum entropy spectral analysis of this variation (Fig. 7) shows a predominant peak around 300 million years ago with minor peaks

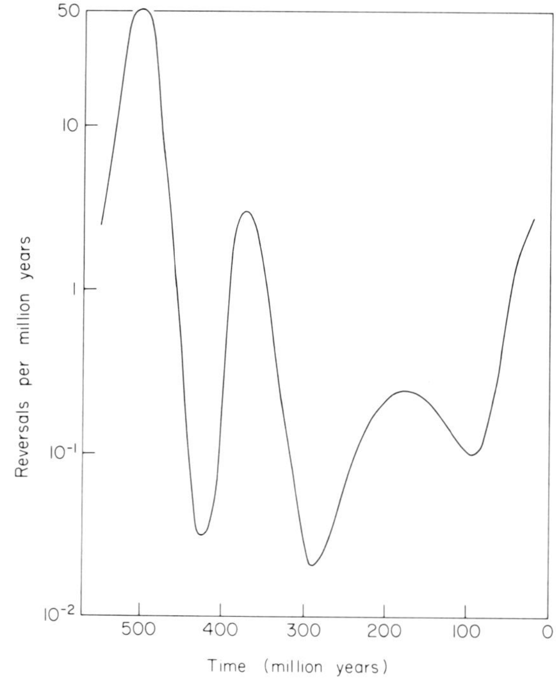

Fig. 5. Reversal rate of the geomagnetic field during the Phanerozoic as interpreted by Jacobs (1975) from the analysis of McElhinny (1971*b*).

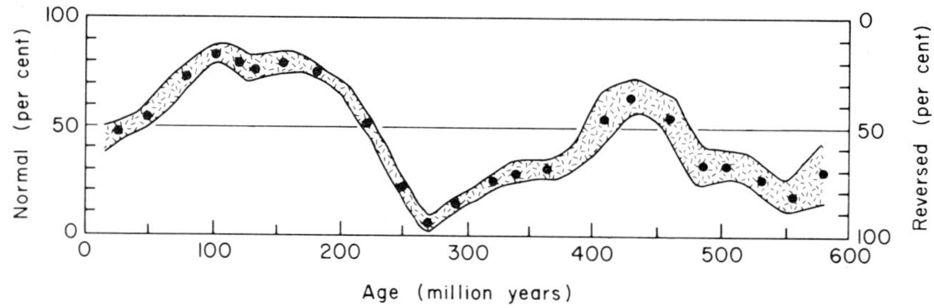

Fig. 6. Polarity bias of the geomagnetic field during the Phanerozoic after Irving and Pullaiah (1976). Overlapping 50 million years averages of polarity ratios as observed in palaeomagnetic results are shown together with the limits of the standard errors.

around 110 and 60 million years ago. The previous suggestion of a 700 million year periodicity (Ulrych, 1972) is not seen in these data.

Other long-term trends in the geomagnetic field have been noted by Brock (1971). He noticed that the magnitude of palaeosecular variation (as

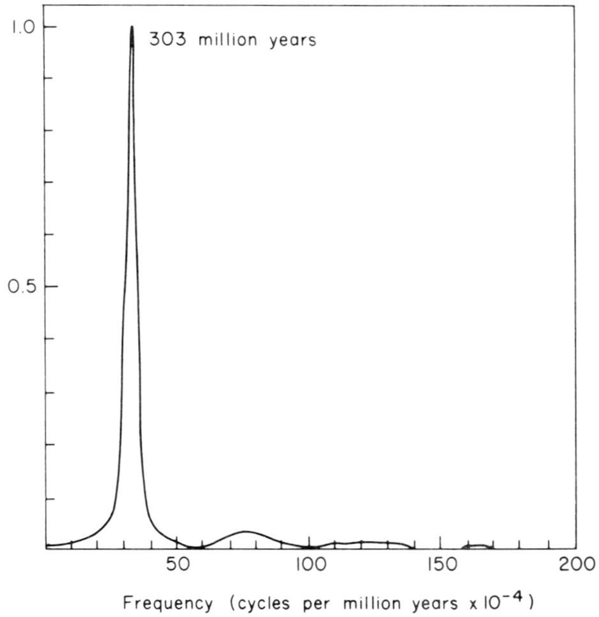

Fig. 7. Maximum entropy power spectrum of the polarity ratios of Fig. 6.

measured by the scatter in palaeomagnetic directions for any epoch) was much less in pre-Cenozoic times than during the Cenozoic. He suggested that this was also related to changes in reversal frequency, the lower palaeosecular variation in the late Palaeozoic and Mesozoic reflecting the much lower reversal rate. An analysis of angular standard deviations (**S**) observed in palaeomagnetic results has been made at 25 million year intervals covering the past 350 million years by Irving and Pullaiah (1976). Irving and Ward (1964) have observed that the angular standard deviation of palaeomagnetic results and of field directions of the present geomagnetic field are a function of latitude represented by the relation

$$S = S_e(1 + 3\sin^2\lambda)^{-1/2}.$$

In order to compare results from different latitudes the values of S were equator normalized to produce values of S_e using the palaeolatitude values for each formation studied. The variation deduced by Irving and Pullaiah (1976) is shown in Fig. 8. It confirms the suggestion of Brock (1971) that secular variation is apparently related to changes in reversal frequency.

The relationship between palaeosecular variation and reversal frequency follows the model of Cox (1968) for geomagnetic reversals in which the

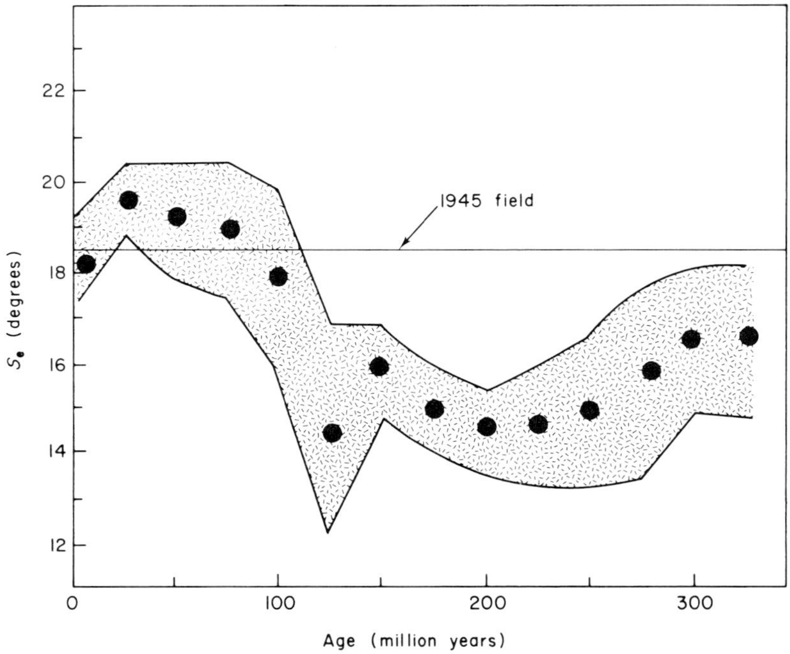

Fig. 8. Palaeosecular variation as a function of time. S_e represents the equator normalized value of the angular standard deviation observed in palaeomagnetic data. Fifty million year overlapping averages with standard errors are shown.

reversal frequency varies according to the ratio of non-dipole field to the main dipole field. Clearly, palaeosecular variation as measured by the scatter of palaeomagnetic results is also a measure of this same ratio. Because the time scales of all these phenomena are very much longer than those of the dynamo processes in the core, these changes must reflect a change in conditions under which the dynamo operates. Of the possible ways in which such a change might be accomplished, a change in the conditions at the core–mantle boundary appears most likely.

The source of the non-dipole field is generally ascribed to eddies in the outer core close to the core–mantle boundary. Hide (1967) has suggested that the form of these eddies is likely to be sensitive to changes in temperature and/or topography at the core–mantle interface. Such changes may result from processes that occur at the base of the mantle. Since the fundamental period associated with these processes is 300 million years and this is also the time scale of plate tectonics, it seems probable that there is a relation between the tectonic evolution of the upper mantle and processes at the base of the lower mantle.

The very long period of 300 million years reflects both the gross changes in reversal frequency, polarity bias and the time scale of plate tectonics. However the shorter periods observed in the polarity bias spectrum and the abrupt changes in average reversal frequency occurring at intervals of about 50 million years also seem to be related to times of tectonic change at the surface of the Earth. Vogt (1975) has noted that major plate tectonic changes at about 42–45, 70–80 and 110–120 million years ago coincide also with changes in reversal frequency (see Fig. 2). All of these correlations strongly suggest that the tectonics of the upper mantle and the behaviour of the core reflected in long term geomagnetic changes are coupled in some way. The most obvious way in which this could occur is through mantle-wide convection.

If the geomagnetic dynamo is maintained by thermal convection in the outer core, convection in the mantle may be required to transport the heat that crosses the core–mantle boundary through the mantle. Numerical investigations of thermal convection at high Rayleigh numbers by Jones (1977) admit the possibility that this convection will be intermittent, consisting of rising plumes of material which are occasionally emitted from a thermal boundary layer at the base of the mantle. During the convection part of the intermittent cycle, temperatures at the core–mantle boundary decrease, leading to an increase in the Rayleigh number of the core. When the lower mantle again becomes quiescent, temperatures at the core–mantle boundary rise and the Rayleigh number of the core decreases. These fluctuations in the Rayleigh number of the core are likely to disturb the flow field in the core and thus affect the frequency of reversals.

4. Correlation of the Gravity and Magnetic Field

Although the long-term geomagnetic field variations consistently suggest that they are related to changes in temperature and/or topography at the core–mantle interface, analysis of seismic travel times shows that short-wavelength topographic features there cannot exceed a few kilometres in height (Cox and Cain, 1972). Long-wavelength features of this height cannot be resolved. However, there is no necessity for these topographic features or bumps to be any more than kilometre-sized features. Hide (1969) has suggested that bumps of kilometre size could provide topographic coupling between the core and the mantle. This would account for the decade fluctuations in the length of the day since neither viscous nor electromagnetic coupling seem adequate.

Hide and Horai (1968) have shown that because of the high-density contrast at the core–mantle boundary (5.58 g cm^{-3} at the base of the mantle to 9.90 g cm^{-3} in the outer core, following Jordan and Anderson (1974)), bumps with horizontal dimensions of 1000 km and height about 1 km would make a significant contribution to gravity field variations at the Earth's surface. However, such a contribution would only be to those global-scale features associated with spherical harmonics of degree not greater than four. Also Hide (1966, 1967) has argued that, owing to Coriolis forces associated with the Earth's rotation, topographic bumps might interact strongly with magnetohydrodynamic motions throughout the core and thus measurably distort the geomagnetic field. Therefore any correlation between gravity and magnetic anomalies should reflect on the nature of the core–mantle boundary. With this in mind, Hide and Malin (1970) studied the degree of correlation of those large-scale non-axisymmetric features of the gravitational potential, which could be described by a spherical harmonic series truncated at terms of degree four, with corresponding features of the geomagnetic potential. Qualitative analysis of the manner in which bumps at the core–mantle interface might distort the geomagnetic field suggests the distortions could be displaced relative to the bumps by distances comparable with their horizontal scale—thousands of kilometres.

Hide and Malin (1970) established that the correlation coefficient for epoch 1965 attains a maximum value of 0.84 when the magnetic field is displaced 160° eastwards in longitude λ, compared with only 0.1 when $\lambda = 0$. Since the gravitational field is unlikely to change appreciably on the time scale of the geomagnetic secular variation, the variation of λ with time can be calculated back to 1835—the date of the first reliable spherical harmonic analysis of the geomagnetic field by Gauss. This variation is shown in Fig. 9, indicating a linear increase of λ with time according to

$$\lambda = 126.2° + 0.273\ (t - 1835),$$

where t is the epoch (year AD). The time dependence of λ is associated with the westward drift of the geomagnetic field.

5. Anomalies in the Time-averaged Palaeomagnetic Field

Palaeomagnetists have long taken as the basic assumption of their discipline that the palaeomagnetic field averaged over sufficient time is that of an axial geocentric dipole.

Wilson (1970, 1971, 1972) and Wilson and McElhinny (1974) have shown that, although the assumption is true to first order, there are second-order departures which can be modelled as though the geocentric dipole were displaced slightly northwards axially from the geocentre. This suggests that there is some asymmetry in the time-averaged palaeomagnetic field.

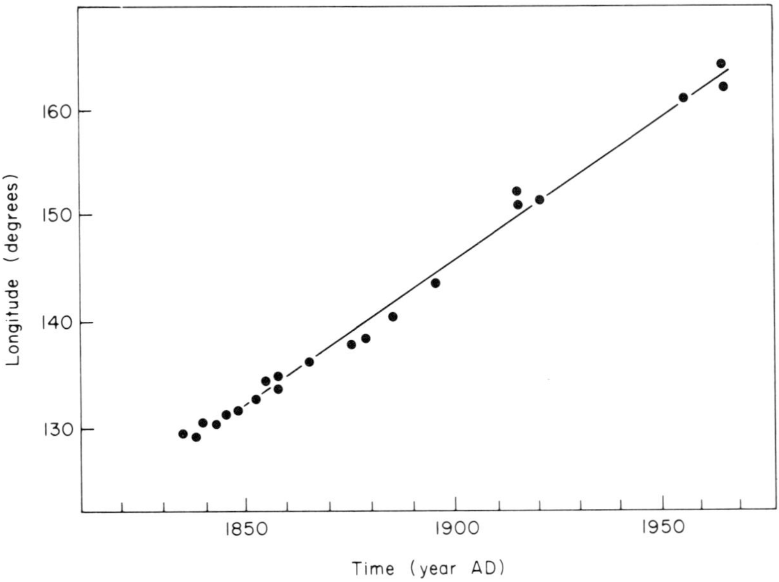

Fig. 9. The variation with time of the eastward displacement in longitude λ between the Earth's magnetic and gravitational fields (Hide and Malin, 1970).

Although our present understanding of the origin of the geomagnetic field is poor, dynamo theories suggest the following simple analysis.

The magnetic induction equation is fundamental to all dynamo theories:

$$\partial \mathbf{H}/\partial t = k\nabla^2\mathbf{H} + \nabla \times (\boldsymbol{\nu} \times \mathbf{H}),$$

where k is the magnetic diffusivity and $\boldsymbol{\nu}$ is the fluid velocity. Replacement of $\mathbf{H}$ with $-\mathbf{H}$ does not change this equation and it is therefore symmetric with respect to the magnetic field. In considering hydromagnetic dynamos it is necessary to consider the Navier–Stokes equation. The magnetic field enters this last equation only through the body force in terms of the Lorentz force, a force which is antisymmetric in $\mathbf{H}$. However, recent progress in hydromagnetic dynamos (Busse, 1975, Gubbins, 1974) suggests it is reasonable to assume the Lorentz force as having only a minor effect on the velocity field. If this is so, then to a first-order approximation, the dynamo problem is symmetric with respect to the magnetic field.

Merrill and McElhinny (1977) have made a spherical harmonic analysis of palaeomagnetic data covering the past 5 million years to see if there are any major asymmetries in the time-averaged field. A 5 million year time-span was used because a previous analysis of palaeosecular variation by McElhinny and Merrill (1975) suggested there were long enough periods present

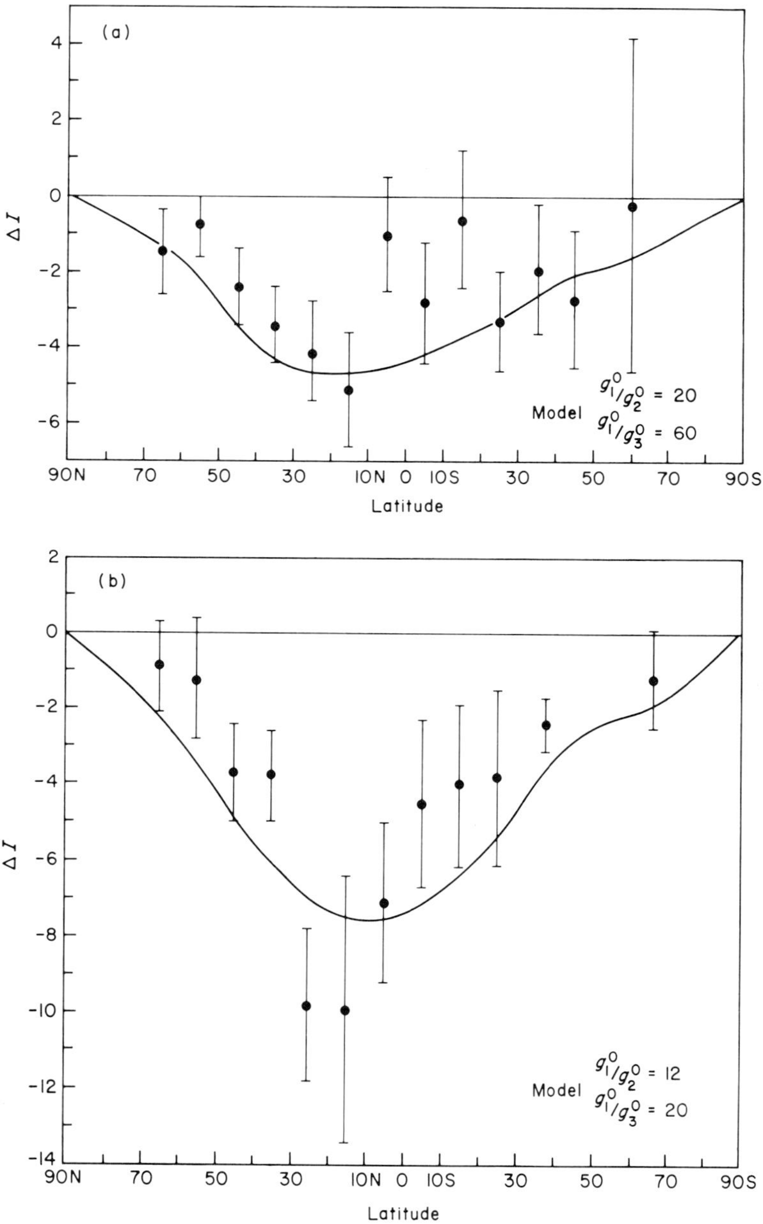

Fig. 10. ΔI anomaly averaged by 10° latitude strips using world-wide palaeomagnetic data for the past 5 million years together with their standard errors. The solid line is a model variation using the g_2^0 and g_3^0 terms indicated separately for the normal (a) and reversed (b) data. (After Merrill and McElhinny, 1977.)

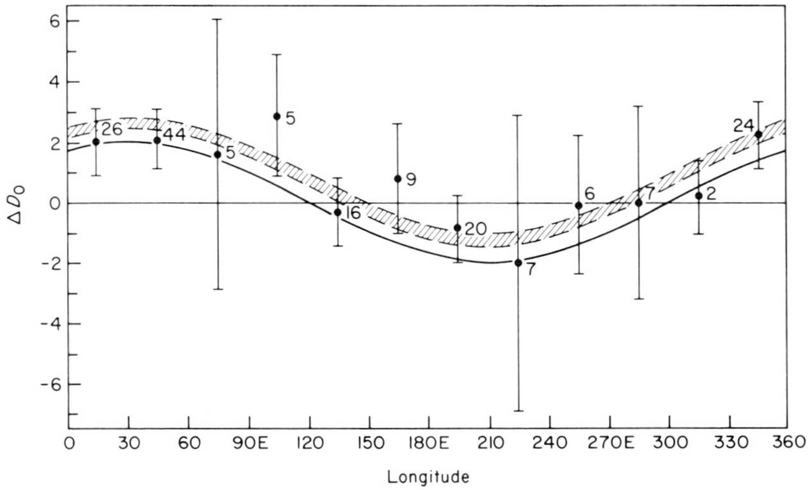

Fig. 11. ΔD anomaly averaged by 30° longitude sectors for normal polarity palaeomagnetic data for the past 5 million years. Simple sinusoidal curves that best fit the data lie within the batched region. The best symmetric curve is also shown for a ratio of equatorial dipole to geocentric axial dipole of 0.035. (After Merrill and McElhinny, 1977.)

in the geomagnetic secular variation to require several million years to obtain a truely time-averaged field. The procedure adopted was to analyse the inclination anomaly ΔI (departure of I from the expectations of an axial geocentric dipole) around latitude strips to determine the zonal harmonics. Analysis of the declination anomaly ΔD (departure of D from zero) around longitude sectors determines the first non-zonal harmonics. The results are illustrated in Figs 10 and 11. The ΔI analysis shows that there are clear differences between the normal and reversed fields and best fitting models using g_2^0 and g_3^0 terms are shown. Apart from the normal-reversed differences there is a clear north–south asymmetry as represented by the g_3^0 term.

The ΔD analysis suffers from lack of sufficient data. Whereas results from deep-sea cores can be used in a ΔI analysis, declinations are not available so that data are restricted to land based measurements only. No meaningful pattern arises from the few reversed data but the normal data give the results of Fig. 11. Simple sinusoidal curves that best fit these results lie in the hatched area of Fig. 11. If a symmetrical curve is drawn as should be expected (the mean ΔD should be zero unless there is a toroidal field at the Earth's surface) the lower curve in Fig. 11 results. From this curve the non-zonal terms g_1^1 and h_1^1 can be estimated. The major feature of Fig. 11 is the tendency for a negative ΔD in the Pacific hemisphere and a positive ΔD

in the opposite continental hemisphere. The Gaussian coefficients are listed in Table 2.

Table 2. Gaussian coefficients (normalized to the geocentric axial dipole term g_1^0) for the time-averaged palaeomagnetic field for the past 5 million years (after Merrill and McElhinny, 1977).

Time	Present field	Normal	Reversed
g_1^1/g_1^0	+0.067	−0.017	−?
h_1^1/g_1^0	−0.190	+0.030	+?
g_2^0/g_1^0	+0.063	+0.050	+0.083
g_3^0/g_1^0	−0.043	+0.017	+0.034

Note that the g_1^0 term is negative for the present (normal field) and is positive for the reversed field. The table therefore indicates that the sign of all the coefficients changes with g_1^0.

The suggestion that the geographical variations in the non-dipole field result from large-scale temperature or topographic variations at the core–mantle interface can be applied to the observations above. In particular, the development of bumps at the core–mantle boundary could affect the magnetic field through alteration of the convection pattern. The occurrence of these bumps may be related to other observed hemispherical differences. A large proportion of subduction taking place over the past few million years has occurred preferentially in the northern hemisphere. Also the northern hemisphere contains the majority of continents. Similar differences exist between the oceanic (Pacific) and continental hemispheres. Besides the obvious difference in continental distribution, almost all subduction zones are confined to the boundaries of the Pacific Ocean, and the majority of these dip under nearby continents.

The significant differences between the normal and reversed polarity results must provide some insight into the mechanism for generating the geomagnetic field and the origin of geomagnetic reversals. These differences are not manifested in differences between the time spent in a given polarity epoch in the Cenozoic or in significant differences in palaeointensity values between normal and reversed polarity times. Although the existing explanations for the origin of reversals of the geomagnetic field are inadequate, the work of Parker (1969) and Levy (1972*a,b,c*) appears consistent with the differences observed. The mean angular velocity of the core is believed to be the same as that of the mantle. The mean angular velocities at the core–mantle and the inner core boundaries are probably also the same as those for the Earth as a whole because of frictional coupling (material and/or magnetic friction) at these boundaries. However, one expects that there are regions in the core where there is non-uniform rotation that produces toroidal fields. Of course, a decrease in the mean rotation rate at some region in the core must be balanced with an increase in some other region. Therefore, even in the very simple situation where the initial magnetic field has uniform direction throughout the core, toroidal fields of opposite sign will be produced in different regions of the core. Although in the real Earth the situation is far more complicated, it seems reasonable to assume that there are regions where toroidal fields are predominately of one sign and other regions where they are predominately of the opposite sign. Although the precise geometry of such regions may be controversial, Parker (1969) and Levy (1972*a,b,c*) have shown that a reversal could occur when there is a shift in the majority of cyclones from regions which have toroidal fields predominately of one sign to regions which have toroidal fields predominately of the opposite sign. (For the particular geometries invoked by Parker and Levy, a reversal occurs when the majority of cyclones shift from low latitudes to high latitudes.) Thus there is the possibility that normal and reversed magnetic polarities occur according to which region of the core convection is dominant. A geographical variation of the temperature at the core–mantle interface or of the positions of topographic bumps could affect such regions differently and produce all the observed palaeomagnetic results.

6. Discussion

Dziewonski *et al.* (1977) have examined P-wave velocity anomalies through the Earth's mantle from an analysis of nearly 700 000 P-wave travel time residuals. Relatively large wavelength features were resolved below 1500 km depth. In the region between 2200 and 2900 km (the core–mantle boundary), highly significant large-scale anomalies exist (Fig. 12). Statistically a positive P-wave velocity anomaly exists in the northern hemisphere and a negative P-wave wave anomaly in the southern hemisphere. There is also a positive P-wave anomaly associated with much of the Pacific hemisphere.

Dziewonski *et al.* present statistical evidence for a negative correlation between the long wavelength gravity anomalies observed at the surface and those computed from velocity anomalies at depths greater than 1100 km under the assumption of proportionality between velocity and density perturbations. The proportionality coefficient $\Delta v/\Delta\rho$ has been determined as −4.45 or −6.02 using two different methods. The fact that this coefficient is negative means that regions of high velocity are associated with low

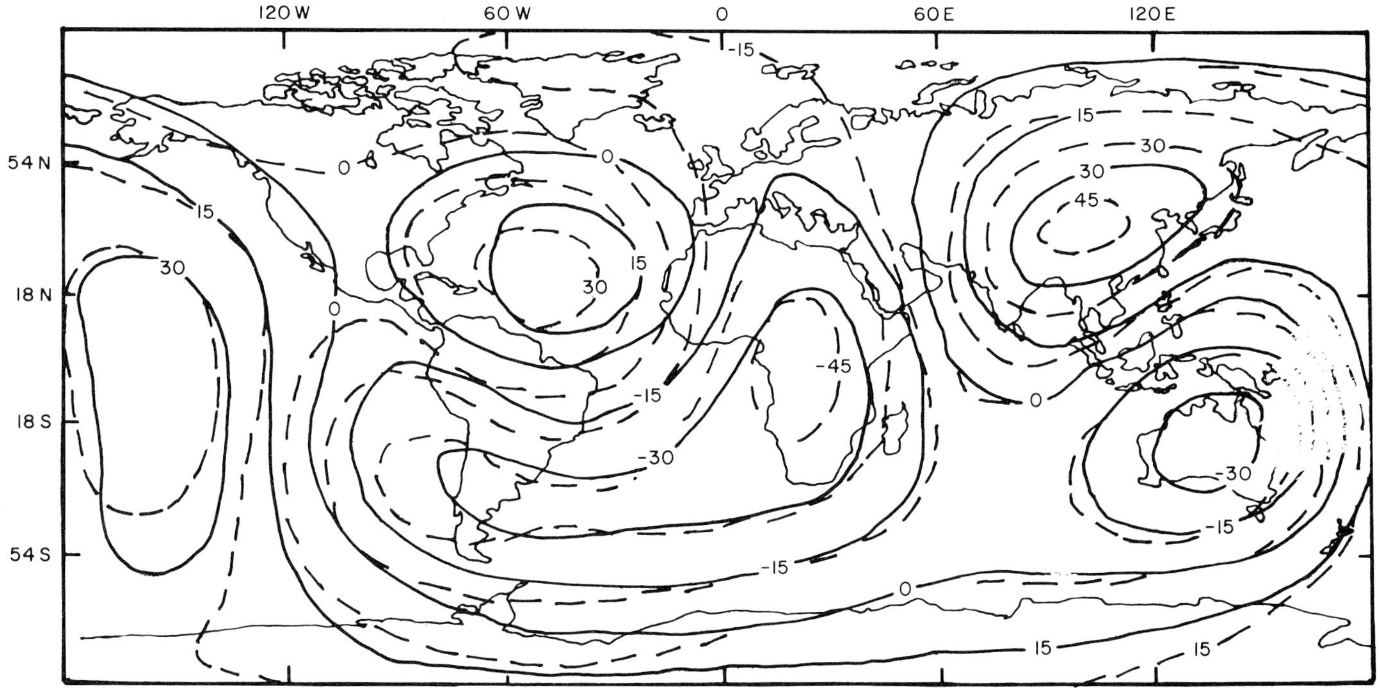

Fig. 12. Seismic P-wave velocity anomalies (dashed lines) in the region 2200 km to the core–mantle boundary after Dziewonski *et al.* (1977). Solid contours corresponds to the solution perturbed to satisfy the gravity data assuming velocity and density perturbations are proportional.

density and vice versa. This essentially rules out interpretation simply in terms of local temperature differences. Possibly origins of this correlation include (1) sinking of eclogite-rich material into the lower mantle from regions of lithospheric subduction following Ringwood (1975), (2) chemical plumes of light high-velocity material originating near the core–mantle boundary following Anderson (1975), (3) static chemical heterogeneities in a non-convecting mantle, or (4) temperature differences and perturbations of the core–mantle boundary associated with mantle-wide convection.

Only the last explanation effectively ties the seismic observations with the magnetic observations. It must be remembered that the density anomalies are not directly observed but are inferred from the gravity anomalies. The correlation may not be with density anomalies at all but with some other phenomenon that affects the geopotential. This could arise if both anomalies were related to mantle-wide convection. Temperature differences associated with convection would give rise to a positive correlation between velocity and density anomalies, the opposite to that observed. However, the gravity field will not reflect the local density anomalies alone. Over a rising current the surface of the Earth is elevated owing to the dynamical effects of the flow. This effect alone would cause a positive gravity anomaly over a rising region in spite of the higher temperature and lower density there (Runcorn, 1967; McKenzie *et al.*, 1974).

Convection in the lower mantle may cause a similar distortion of the core–mantle boundary, which would have a similar effect on the net gravity anomalies. The velocity anomalies may be due to temperature differences associated with mantle-wide convection, and the gravity anomalies result from perturbations of the core–mantle boundary associated with the flow. Thus large-scale inhomogeneities in the lower mantle resulting from dynamic processes that have occurred in the lower mantle are manifest in both the seismic and palaeomagnetic data. The common theme that ties all these observations together is the supposition that there exist topographic bumps on the core–mantle boundary.

References

Anderson, D. L. (1975). Chemical plumes in the mantle. *Bull. geol. Soc. Am.* **86**, 1593–1600.

Braginskiy, S. I. (1972). Analytical description of the geomagnetic field of past epochs and determination of the spectrum of magnetic waves in the core of the earth. *Geomag. Aeron.* **12**, 947–957.

Brock, A. (1971). An experimental study of palaeosecular variation. *Geophys. J. R. Astronom. Soc.* **24**, 303–317.

Bullard, E. C., Freedman, C., Gellman, H. and Nixon, J. (1950). The westward drift of the earth's magnetic field. *Phil. Trans. R. Soc. A* **243**, 67–92.

Burlatskaya, S. P. (1970). Changes in geomagnetic field intensity in the last 8500 years, according to global archaeomagnetic data. *Geomag. Aeron.* **10**, 544–548.
Burlatskaya, S. P. (1972). Secular geomagnetic field variations, according to archaeomagnetic and palaeomagnetic data. *Geomag. Aeron.* **12**, 582–592.
Burlatskaya, S. P., Nechayeva, T. B. and Petrova, G. N. (1965). The westward drift of the secular variation of magnetic inclination and variations of the earth's magnetic moment according to archaeomagnetic data. *Izv. Acad. Sci. USSR Phys. Solid Earth*, no. 6, 380–385.
Burlatskaya, S. P., Kruchik, Ya., Malovskiy, Z., Nechayeva, T. B. and Petrova, G. N. (1970). Comparison of secular variations in inclination in Poland and the Ukraine. *Geomag. Aeron.* **10**, 243–245.
Busse, F. H. (1975). A model of the geomagnetic dynamo. *Geophys. J. R. Astronom. Soc.* **42**, 437–459.
Cox, A. (1968). Lengths of geomagnetic polarity intervals. *J. geophys. Res.* **73**, 3247–3260.
Cox, A. (1969). Geomagnetic reversals. *Science, N.Y.* **163**, 237–245.
Cox, A. (1975). The frequency of geomagnetic reversals and the symmetry of the nondipole field. *Rev. Geophys. Space Phys.* **13**, 35–51.
Cox, A. and Cain, J. C. (1972). International conference on the core–mantle interface. *Trans. Am. geophys. Union* **53**, 591–623.
Cox, A. and Dalrymple, G. B. (1967). Statistical analysis of geomagnetic reversal data and the precision of potassium-argon dating. *J. geophys. Res.* **72**, 2603–2614.
Cox, A. and Doell, R. R. (1964). Long period variations of the geomagnetic field. *Bull. seismol. Soc. Am.* **54**, 2243–2270.
Currie, R. G. (1966). The geomagnetic spectrum—40 days to 5.5 years. *J. geophys. Res.* **71**, 4579–4598.
Currie, R. G. (1967). Magnetic shielding properties of the earth's mantle. *J. geophys. Res.* **72**, 2623–2633.
Currie, R. G. (1968). Geomagnetic spectrum of internal origin and lower mantle conductivity. *J. geophys. Res.* **73**, 2779–2786.
Dziewonski, A. M., Hager, B. H. and O'Connell, R. J. (1977). Large-scale heterogeneities in the lower mantle. *J. geophys. Res.* **82**, 239–255.
Elsasser, W. M. (1946). Induction effects on terrestrial magnetism. I. Theory. *Phys. Rev.* **69**, 106–116.
Gubbins, D. (1974). Theories of the geomagnetic and solar dynamos. *Rev. Geophys. Space Phys.* **12**, 137–154.
Harrison, C. G. A. and Somayajulu, B. L. K. (1966). Behaviour of the earth's magnetic field during a reversal. *Nature, Lond.* **212**, 1193–1195.
Heirtzler, J. R., Dickson, G. O., Herron, E. M., Pitman, W. C., III, and Le Pichon, X. (1968). Marine magnetic anomalies, geomagnetic field reversals and motions of the ocean floor and continents. *J. geophys. Res.* **73**, 2119–2136.
Helsley, C. E. and Steiner, M. D. (1969). Evidence for long intervals of normal polarity during the Cretaceous period. *Earth planet. Sci. Lett.* **5**, 325–332.
Hide, R. (1966). Free hydromagnetic oscillations of the earth's core and the theory of geomagnetic secular variation. *Phil. Trans. R. Soc. A* **259**, 615–650.
Hide, R. (1967). Motions of the earth's core and mantle, and variations of the geomagnetic field. *Science, N.Y.* **157**, 55–56.
Hide, R. (1969). Interaction between the earth's liquid core and solid mantle. *Nature, Lond.* **222**, 1055–1056.

Hide, R. and Horai, K. (1968). On the topography of the core–mantle interface. *Phys. Earth planet. Interiors* **1**, 305–308.
Hide, R. and Malin, S. R. C. (1970). Novel correlations between global features of the earth's gravitational and magnetic fields. *Nature, Lond.* **225**, 605–609.
Irving, E. and Parry, L. G. (1963). The magnetism of some Permian rocks from New South Wales. *Geophys. J. R. astronom. Soc.* **7**, 395–411.
Irving, E. and Pullaiah, G. (1976). Reversals of the geomagnetic field, magnetostratigraphy, and the relative magnitude of palaeosecular variation in the Phanerozoic. *Earth Sci. Rev.* **12**, 35–64.
Irving, E. and Ward, M. A. (1964). A statistical model of the geomagnetic field. *Pure appl. Geophys.* **57**, 47–52.
Jacobs, J. A. (1975). *The Earth's Core*, Academic Press, London, New York and San Francisco.
Jones, G. M. (1977). Thermal interaction of the core and the mantle and long-term behaviour of the geomagnetic field. *J. geophys. Res.* **82**, 1703–1709.
Jordan, T. H. and Anderson, D. L. (1974). Earth structure from free oscillations and travel times. *Geophys. J. R. astronom. Soc.* **36**, 411–450.
Larson, R. L. and Pitman, W. C., III (1972). World-wide correlation of Mesozoic magnetic anomalies and its implications. *Bull. geol. Soc. Am.* **83**, 3645–3662.
Levy, E. H. (1972*a*). Effectiveness of cyclonic convection for producing the geomagnetic field. *Astrophys. J.* **171**, 621–633.
Levy, E. H. (1972*b*). Kinematic reversal schemes for the geomagnetic dipole. *Astrophys. J.* **171**, 635–642.
Levy, E. H. (1972*c*). On the state of the geomagnetic field and its reversals. *Astrophys. J.* **175**, 573–581.
McDonald, K. L. and Gunst, R. H. (1968). Recent trends in the earth's magnetic field. *J. geophys. Res.* **73**, 2057–2067.
McElhinny, M. W. (1971*a*). Geomagnetic polarity transitions. *Comments Earth Sci. Geophys.* **1**, 150–158.
McElhinny, M. W. (1971*b*). Geomagnetic reversals during the Phanerozoic. *Science, N.Y.* **172**, 157–159.
McElhinny, M. W. and Merrill, R. T. (1975). Geomagnetic secular variation over the past 5 m.y. *Rev. Geophys. Space Phys.* **13**, 687–708.
McKenzie, D. P., Roberts, J. M. and Weiss, N. O. (1974). Convection in the earth's mantle: towards a numerical simulation. *J. Fluid Mech.* **62**, 465–538.
Merrill, R. T. and McElhinny, M. W. (1977). Anomalies in the time-averaged palaeomagnetic field and their implications for the lower mantle. *Rev. Geophys. Space Phys.* **15**, 309–323.
Nagata, T. (1965). Main characteristics of recent geomagnetic secular variation. *J. Geomag. Geoelect.* **17**, 263–276.
Nagata, T. (1969). Length of geomagnetic polarity intervals [discussion of papers by A. Cox (1968, 1969)]. *J. Geomag. Geoelect.* **21**, 701–704.
Ninkovich, D., Opdyke, N. D., Heezen, B. C. and Foster, J. H. (1966). Paleomagnetic stratigraphy, rates of deposition and tephrachronology in North Pacific deep-sea sediments. *Earth planet. Sci. Lett.* **1**, 476–492.
Parker, E. N. (1955). Hydromagnetic dynamo models. *Astrophys. J.* **122**, 293–314.
Parker, E. N. (1969). The occasional reversal of the geomagnetic field. *Astrophys. J.* **158**, 815–827.
Phillips, J. D. (1977). Time variation and asymmetry in the statistics of geomagnetic reversal sequences. *J. geophys. Res.* **82**, 835–843.

Phillips, J. D. and Cox, A. (1976). Spectral analysis of geomagnetic reversal time scales. *Geophys. J. R. astronom. Soc.* **45**, 19–33.
Phillips, J. D., Blakely, R. J. and Cox, A. (1975). Independence of geomagnetic polarity intervals. *Geophys. J.R. astronom. Soc.* **43**, 747–754.
Ringwood, A. E. (1975). *Composition and Petrology of the Earth's Mantle*, McGraw-Hill, New York.
Runcorn, S. K. (1967). Flow in the mantle inferred from the low degree harmonics of the geopotential. *Geophys. J. R. astronom. Soc.* **14**, 375–384.
Smith, P. J. (1970). The intensity of the geomagnetic field: A summary of conclusions, In *Palaeogeophysics* (S. K. Runcorn, ed.), Academic Press, New York, pp. 79–90.
Ulrych, T. (1972). Maximum entropy power spectrum of long-period geomagnetic reversals. *Nature, Lond.* **235**, 218–219.
Vestine, E. H., Sibley, W. L., Kern, J. W. and Carlstedt, J. L. (1963). Integral and spherical harmonic analyses of the geomagnetic field for 1955.0, Part 1. *J. Geomag. Geoelect.* **15**, 47–72.
Vogt, P. R. (1975). Changes in geomagnetic reversal frequency at times of tectonic change: evidence for coupling between core and upper mantle processes. *Earth planet. Sci. Letters*, **25**, 313–321.
Wilson, R. L. (1970). Permanent aspects of the earth's non-dipole magnetic field over Upper Tertiary times. *Geophys. J.R. astronom. Soc.* **19**, 413–437.
Wilson, R. L. (1971). Dipole offset—the time-average palaeomagnetic field over the past 25 million years. *Geophys. J. R. astronom. Soc.* **22**, 491–504.
Wilson, R. L. (1972). Palaeomagnetic differences between normal and reversed field sources, and the problem of far-sided and right-handed pole positions. *Geophys. J.R. astronom. Soc.* **28**, 295–304.
Wilson, R. L. and McElhinny, M. W. (1974). Investigation of the large scale palaeomagnetic field over the past 25 million years; eastward shift of the Icelandic spreading ridge. *Geophys. J. R. astronom. Soc.* **39**, 570–586.
Yukutake, T. (1967). The westward drift of the earth's magnetic field in historic times. *J. Geomag. Geoelect.* **19**, 103–116.

5. Seismology and the Internal Structure of the Earth

J. R. CLEARY and R. S. ANDERSSEN

There was a limit, however, to what could be learned about the detailed structure of the Earth's interior until seismology, by providing direct information on certain mechanical properties at specific depths below the surface, led to a further great leap forward in knowledge of solid-Earth geophysics in the twentieth century.

Bullen, *The Earth's Density*, p. 11.

Abstract

Almost from the time that seismic data became available, close to the beginning of this century, they were instrumental in effecting a spectacular increase in knowledge of the Earth's internal structure. By 1940 the velocity distribution in the Earth had been determined to first order using seismic body wave times, with consequent interpretations in terms of the elastic parameters and density. Since then, refinements to seismological techniques, including the use of arrays and the analysis of surface wave dispersion and free oscillation modes, have enabled a more precise delineation of Earth structure. Regional variations are known to exist in the upper mantle, and probably in the lower mantle as well, but the detail of such variations is not yet clear. Recent improvements in techniques are expected to assist greatly in resolving the complexities associated with this problem.

Free oscillation studies have been invaluable in providing consistency checks on models derived from travel-time data and associated assumptions about chemical homogeneity, and are currently being used to investigate the attenuation properties of the mantle. They have been less successful in resolving the non-uniqueness in the derivation of the Earth's density distribution, or in providing information about upper mantle structure. Before further major advances can be expected using free oscillation data inversion, an improvement in background theory is required.

1. Introduction

Numerous ways exist for defining what is meant by "the internal structure of the Earth"; but, in the context indicated by the title of this book, the basic structural determinant is chemical composition as a function of geographic coordinates and depth, with temperature and pressure determining the positions of phase changes. Properties such as density, elasticity and electrical conductivity are of indirect significance, since they do not provide direct insight into the "origin, evolution and composition of the Earth's interior". On the other hand, the available geophysical data (e.g. gravitational, electromagnetic and seismic) provide information about these properties rather than about chemical composition, temperature and pressure.

Thus, there is no *direct* means by which the compositional structure of the Earth's interior can be determined from available geophysical data. Such information must be obtained via a two-stage process. Initially, models of the Earth's density, elasticity and electrical conductivity must be derived from the data. Then, on the basis of theoretical and experimental results concerning the properties of terrestrial materials, models of the Earth's chemical composition can be obtained. The necessity for this indirect approach is the key to understanding the nature of geophysical and geochemical inference processes.

It is known that, at least to a first-order approximation, electrical conductivity is controlled by composition and temperature, density by composition and pressure, and the elastic parameters by composition, temperature and pressure. Clearly, then, density and the elastic parameters are key properties involved with the determination of the Earth's internal structure.

Since these are the properties which directly control the form and nature of seismic data, the importance of such data in the inference processes leading to insight into the "origin, evolution and constitution of the Earth's interior" is immediately obvious. Furthermore, they are the only available data which sample directly and in detail the elastic parameters of the Earth's interior—the Earth's figure sees them only as an integrated effect.

Even more importantly, compared with gravity and (electro)magnetic induction data, seismic data provide the most direct and explicit information about the Earth's interior. This is a consequence of the natures of gravitational attraction, electromagnetic induction and wave propagation and can be directly related mathematically to the classification of partial differential equations according to whether they are elliptic, parabolic or hyperbolic. The importance of this classification is that it precisely indicates where it is necessary to change to different mathematical and computational concepts and tools when analysing partial differential equations (see Weinberger, 1965). The mathematical differences between gravitational attraction, electromagnetic induction and wave propagation correspond directly to this classification.

The simplest physical processes usually identified with the elliptic, parabolic and hyperbolic classification for partial differential equations are the gravitational attraction of a (spherically symmetric) sphere, heat flow in a finite steel bar with its end maintained at a constant temperature, and the transverse vibrations of a clamped violin string under tension. The first represents a static situation where the data (the external gravitational field) depend only on the formal specification of the process (the density distribution of the material in the sphere); while the second and third represent dynamic situations where the data (temperature distribution as a function of time and transverse vibrations as a function of time, respectively) depend not only on the formal specification of the process (the heat conduction properties and length of the steel bar along with the constant temperature end conditions for the second, and the elastic properties, length and tension of the violin string for the third), but also on initial conditions which can be varied (the initial temperature in the steel bar, and the initial shape of the violin string). Thus, both these processes potentially allow independent experiments to be performed by which additional data about the nature of the underlying formal specification can be obtained.

This is one of the basic reasons why electromagnetic induction and seismic data potentially contain greater information than gravity data, since the initial conditions for electromagnetic induction data are the various phases of the Earth's transient (external) magnetic field, while the initial conditions for seismic data are seismic pulses generated by earthquakes or explosions. In fact, the apparent superiority of seismic data over electromagnetic induction data can in part be explained by the great variety of seismic waves which can penetrate the entire Earth, while the current systems induced by the longer period phases of the Earth's transient magnetic field only effectively penetrate to a depth of about 1000 km.

A further important point is that available gravity and electromagnetic induction data must be classified as *cumulative* or *displacement field* data in the sense that these data only sample the cumulative (displacement) effects of the wave phenomena associated with gravity and electromagnetic induction. Since the speed of propagation of such waves through matter approximates the speed of light, available instrumentation does not allow an analysis of (is unable to record) the travel times along the ray paths of such waves in the material through which they pass. On the other hand, because of the relatively low velocities of propagation of elastic waves, travel times can be obtained with reasonable accuracy and efficiency from instrumental records of the (local) displacement fields generated by seismic waves from earthquakes and explosions.

In seismic analysis and interpretation to date, it has been travel-time data, rather than the local and global behaviour of the displacement field, which have been most effective in deriving information about the behaviour of density and elastic parameters in the Earth's interior. In addition, the largest earthquakes contain sufficient energy to set the Earth into observable free vibration. Suitable analyses of the instantaneous global displacement field can be used to determine the periods of some of the modes of free oscillation of the Earth, which in turn can be defined, using the equations of motion of a SNREI (spherically symmetric, non-rotating, elastic and isotropic) Earth, in terms of density and the elastic parameters. This corresponds to the mode aspect of the ray-mode duality for SNREI objects and thereby yields an (in general) mathematically equivalent but computationally different strategy for obtaining such information.

We aim to reinforce these points by making a quasi-historic examination of the development of models of the density and the elastic parameters of the Earth's interior. The criterion by which we judge this development will be the success with which the non-uniqueness of such models has been reduced.

2. Theory without Data

In most discussions of seismology, what is not clearly explained is the source of its great momentum, i.e. the reasons why seismic data have such constraining power and why a well-established mathematical framework for the analysis of such data existed before their observation. The explanation of the former is related to the special nature of seismic data, as discussed in Section 1; while that of the latter is relevant because, without a solid mathematical foundation, it is not possible to do much with observational data other than exploratory data analysis, smoothing and some elementary statistical analyses.

The reason for the early development of a well-established mathematical framework for seismology relates to the fact that seismology is, in certain respects, a second generation science. The motivation for its initial mathematical development did not come from seismology itself, but from the development of mathematical frameworks for the study of waves in liquids and elastic media. As a study of the history of science shows, the development of mathematical foundations for most disciplines was motivated by the application of the scientific method to available data. In the case of seismology, the process was initially reversed. The theory and predictions came before the data, though the justification is the scientific method itself: extend and generalize available mathematical models to hypothetical applications. Many, such as Laplace, Poisson, Love, Maxwell and Rayleigh, associated with the development of mathematical theories for waves in fluids, observed that, with appropriate modifications, their results carried over to waves in elastic media, and hence to waves in solids.

Some of the key contributions of this nature are:

1. Though P- and S-waves were not unambiguously identified until 1900 by Oldham (1899, 1900) from the records of the Assam earthquake of 1897, their existence had been predicted theoretically by Poisson in 1829. The same is true of the waves predicted by Rayleigh in 1887 which now bear his name. On the other hand, the surface waves now called Love waves were not predicted before they were observed. Love's explanation followed in 1911, but only after he had extended the earlier work to Poisson and Rayleigh by replacing their homogeneous model by a justifiably more complex non-homogeneous one with a surface layer (cf. Jeffreys, 1962, p. 69).
2. Following a suggestion of Maxwell, Butcher introduced in 1876 quantitative stress–strain relations into the study of fracture (cf. Jeffreys, 1962, p. 7). This later became the basis for the study of earthquake mechanisms, which is an aspect of seismology not specifically pursued in this chapter.
3. Though evidence that free oscillations of the Earth could be recorded was not obtained until 1952 (by Benioff following the Kamchatka earthquake of November 4) and confirmed seven years later in 1960 (following the Chilean earthquake of May 22), Lamb predicted in 1882 that two distinct types of free oscillations of the Earth were possible.

Clearly, then, the early history of many of the major developments in seismology belongs to the theoreticians. In fact, prior to the recording of seismic data, the only data available for constraining Earth structure in terms of density and elastic wave parameters were:

(i) radius R (accurate estimates date from Fernel in 1527—cf. Bullen, 1975, p. 3);
(ii) mass M (accurate estimates date from Cavendish in 1798);
(iii) moment of inertia (a really reliable estimate dates only from artificial satellite data of 1963, though fairly accurate estimates were available during the nineteenth century);
(iv) Love's numbers h and k (accurate estimates date from Kelvin in 1863—cf. Bullen, 1975, p. 83);
(v) ellipticity—oblateness (accurate estimates date from Newton's application of Huygens' 1673 description of centrifugal force);
(vi) period of precession (though accurate astronomical estimates were not available until the beginning of the century, the fact that it was

extremely long (~26 000 years) compared with the period of rotation (one day) was known much earlier).

Even with this rather limited data set, it was possible to confirm conclusively certain of the key global features of Earth structure such as

(a) The unambiguous conclusion that the density of the Earth must increase with depth since its moment of inertia is considerably less than 0.4, the value corresponding to a spherically symmetric Earth model with uniform density. This in turn was the basis for much of the popularity of the early two-component density models of Thomson and Tait (1879), Radau (1885) and Weichart (1897), as well as their mathematical simplicity.
(b) The use of Love's numbers by Kelvin in 1863 to conclude that, on the whole, the Earth is substantially more rigid than steel.
(c) The tentative use by Hopkins (1839–42), followed by the rigorous analyses of Kelvin (1876) and Hough (1895), of the huge difference between the periods of precession and rotation of the Earth to infer that the Earth was not a thin solid shell with a liquid interior.

A detailed discussion of what was achieved by the theoreticians using this limited data set can be found in Bullen (1975, Chapters 5 and 6).

3. Body and Surface Waves

As indicated in the previous sections, the wealth of information potentially available in seismic waves was at least implicitly recognized in the first half of the nineteenth century. It was not until 1880, however, that the first practical seismographs were developed, and then only to record local events in Japan. In 1887 a large Japanese earthquake was inadvertently recorded by means of a horizontal pendulum at Potsdam. The implications of this observation were immediately understood, and following the development of the Milne seismograph in 1892 the era of "theory without data" came to a rapid conclusion.

3.1. Seismology to 1940

The identification by Oldham of P-, S- and surface waves on seismograms, as already mentioned, was followed in 1906 by his observation of a delay of about 4 min in the arrival times of P-waves beyond about 120°, confirming earlier speculation that the Earth consisted of an outer "mantle" surrounding a denser, possibly metallic, core. At about the same time, it was found that S-waves were not transmitted beneath the mantle, suggesting that the core (or part of it) is molten. In 1909 Morhorovičić found evidence of a sharp increase in velocity some 50 km below the Earth's surface. Within about 20 years, therefore, seismology had succeeded in identifying the major structural divisions of the Earth's interior: crust, mantle and core. A further subdivision was later proposed by Lehmann (1936) who interpreted the branching of PKP phases as indicating the presence of a sharp velocity increase within the core itself.

The usefulness of seismic data depends on the ability to derive P and S velocity distributions from travel time observations, and various mathematical "inversion" techniques were developed in the first decade of the twentieth century.

These all hinge, in one way or another, on transforming the basic travel-time equation (for its derivation see Jeffreys, 1962, Section 2.05)

$$\tfrac{1}{2}\Delta(p) = \int_{r=pv}^{r=R} \frac{p\,\mathrm{d}r}{(r^2/v^2 - p^2)^{1/2}}, \qquad v = v(r), \tag{1}$$

where p denotes the ray parameter, r the radius at which the ray with parameter p bottoms, R the radius of the Earth, $\Delta(p)$ the angular epicentral distance traversed by the ray with parameter p, and $v = v(r)$ the unknown velocity as a function of radius. The ray parameter p is equivalent to $\mathrm{d}T/\mathrm{d}\Delta$, the slope of the travel time curve (T, Δ) at distance Δ, where T is the travel time from source to receiver. Since, as a function of the unknown v, (1) is clearly non-linear, the aim is to transform it to some version of a linear first kind Abel-type equation of the form

$$s(p) = \int_{p}^{a} \frac{qy(q)}{(q^2 - p^2)^{1/2}}\,\mathrm{d}q \tag{2}$$

for which inversion formulas (derived by Abel, 1823, 1826) are known; in particular,

$$y(q) = -\frac{2}{\pi}\frac{\mathrm{d}}{\mathrm{d}q}\left\{ \int_{q}^{a} \frac{s(p)}{(p^2 - q^2)^{1/2}}\,\mathrm{d}p \right\}. \tag{3}$$

In fact, via the use of the Bateman–Knott transformation $u = r/v$ with u as the new independent variable, (1) becomes (for background details see Macelwane, 1951)

$$s(p) = \tfrac{1}{2}\Delta(p) = \int_{p}^{R/V} \frac{uy(u)}{(u^2 - p^2)^{1/2}}\,\mathrm{d}u, \qquad u(p) = \frac{\mathrm{d}\log r}{\mathrm{d}u} \tag{4}$$

for which the inversion formula (3) yields

$$\frac{d}{du}(\log r) = -\frac{2}{\pi}\frac{d}{du}\left(\int_u^{R/V} \frac{s(p)}{(p^2-u^2)^{1/2}}\, dp\right). \tag{5}$$

where $V = v(R)$. Since, via continuity considerations, the constant of integration must equal log R, (5) yields

$$\log\left(\frac{r}{R}\right) = -\frac{2}{\pi}\int_u^{R/V} \frac{s(p)}{(p^2-u^2)^{1/2}}\, dp, \tag{6}$$

where r is the radius at which the ray with parameter $p = u$ bottoms, which is the basis of all the various inversion formulas proposed in the literature—in one way or another they can be derived from it. The usual form is the Herglotz–Wiechert equation (Jeffreys, 1962, pp. 45–46):

$$\log\left(\frac{R}{r}\right) = \frac{1}{\pi}\int_0^{(\Delta)_u} q\, d\Delta,$$

where $(\Delta)_u$ denotes the epicentral distance associated with rays with ray parameter $p = u$, and

$$\cosh q = \left(\frac{dT}{d\Delta}\right)_u \Big/ \left(\frac{dT}{d\Delta}\right)$$

is evaluated using tables of travel time T as a function of Δ.

A review of most of the currently used methods is contained in Kennett (1976). The major difficulties associated with using any of them to invert given data arise from the data themselves, especially when they include (or should include) retrograde branches (caused by sharp velocity increases) and gaps (caused by low-velocity zones). Such difficulties include the non-uniqueness which potentially arises as a consequence of the possible existence of low-velocity zones (gaps in the data) and the uncertainties associated with the positions of the turning points of retrograde branches and the positions of the end points of gaps. These problems will be discussed further in the final section.

Following the early successes mentioned above, the quality and quantity of data improved to the point where Jeffreys (1939*a,c*) and Gutenberg and Richter (1939) independently derived P and S velocity (α and β) distributions for the whole Earth which were in substantial agreement with one another (Fig. 1). The success of these inversions may be judged from the fact that even now the Jeffreys–Bullen (JB) travel-time tables, derived from the

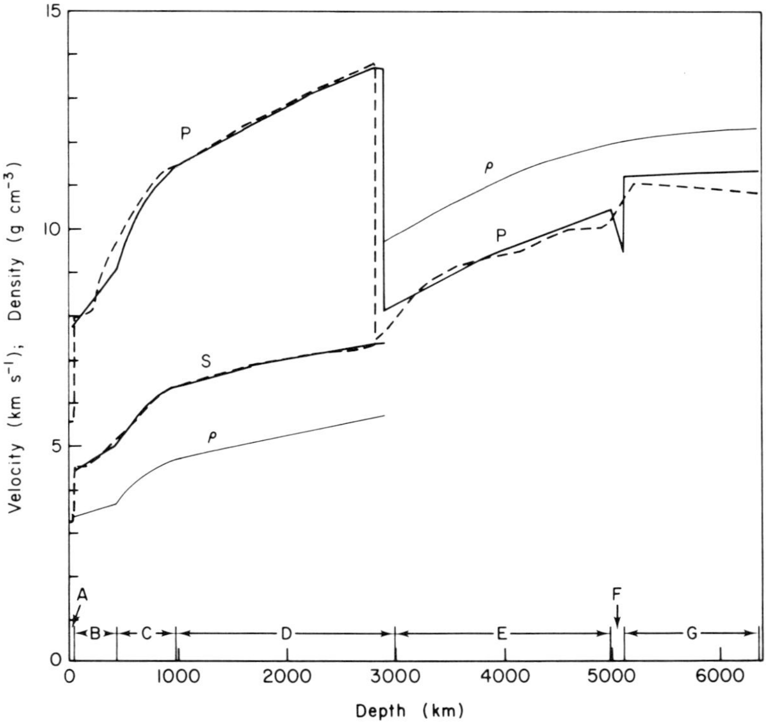

Fig. 1. P and S velocity models by Jeffreys (1939*a*,*c*) (———) and Gutenberg and Richter (1939) (– – – –), and a density model by Bullen (1940, 1942) (———). For the sake of clarity, only the P-velocity discontinuity at the core–mantle boundary has been drawn. The divisions A to G (depth scale) are due to Bullen.

It is interesting to compare these models with the recent models shown in Fig. 5. Note that although Gutenberg first postulated the presence of an upper mantle low-velocity zone in 1926, this was not quantitatively formulated until after the publication of the Gutenberg and Richter model, which shows instead a region of almost constant gradient.

Jeffreys model, are still (despite some known deficiencies) in almost universal use.

The next step (already undertaken in preliminary work by Bullen in 1936) was to use the velocity distributions to calculate corresponding distributions for the elastic parameters k (incompressibility) and μ (rigidity), and especially the density ρ, from the relationships given by elasticity theory:

$$\alpha = \sqrt{(k + \tfrac{4}{3}\mu)/\rho}, \qquad \beta = \sqrt{\mu/\rho}.$$

The solution is obviously indeterminate without additional information. Bullen approached the problem by making use of the Adams–Williamson relation for a chemically homogeneous medium:

$$-\frac{d\rho}{dr} = \frac{Gm}{r^2} \cdot \frac{\rho}{\phi}, \tag{7}$$

where $d\rho/dr$ is the density gradient, G the gravitational constant, m the mass inside a sphere of radius r, and $\phi = \alpha^2 - 4\beta^2/3$. Assuming a density of 3.32 g cm^{-3} at the top of the mantle (mainly on the basis of laboratory experiments on basic rocks at high pressures), he integrated equation (7) to obtain a density distribution for the mantle. The moment of inertia for the mantle calculated from this distribution, when compared with the known moment of inertia for the Earth, would have required a substantial decrease of density with depth in the core, which was untenable. Bullen concluded that the assumption of chemical homogeneity was probably unjustified for the region C (Fig. 1), where high velocity gradients existed. By assuming a quadratic law of density variation with depth in this region, he was able to derive a satisfactory density distribution for the mantle. The analysis was later extended to the core, the main uncertainty being the Earth's central density. A full account of these and subsequent investigations is given by Bullen (1975).

3.2. The Low-velocity Channel

With respect to the above assumptions, a first-order solution to the problem of the variation of k, μ and ρ within the Earth had been found by 1940. Certain details of fine structure remained to be resolved, especially in the region of the upper mantle. Gutenberg had for many years claimed the existence of a low-velocity zone (LVZ) in the uppermost mantle, in contradiction to the Jeffreys model. The practical difficulty is that if the velocity gradient $-dv/dr < v/r$, no rays bottom in that region. Since information about the velocity at a particular depth is provided by the "slowness" $dT/d\Delta$ of the bottoming ray ($dT/d\Delta = r/v$, where v is the velocity at the bottoming radius r of the ray emerging at distance Δ), the evidence for an LVZ rests on the *absence* of data (i.e. a shadow zone) at the corresponding epicentral distances. The presence of such a zone may be masked by other arrivals, or, alternatively, gaps may exist for other reasons.

The situation is different, however, for surface waves, where the wave velocity is controlled by the velocity structure down to a depth which increases with wavelength (and hence with period). The dispersion characteristics of surface waves are consequently a function of velocity structure, and a dispersion curve can be "inverted" to give a velocity profile. In a layered Earth two kinds of surface waves exist: Love waves, which are of SH type (transverse particle motion in the horizontal plane), and Rayleigh waves, of P-SV type (retrograde elliptical particle motion in the vertical plane). Both types are especially suited to the determination of S-velocity structure. The effective use of surface wave data for this purpose owes much to an iterative technique adapted by Haskell (1953) from a mathematical formulation developed by Thomson (1950).

Dorman *et al.* (1960) used computer techniques (novel at that time) to model observations of group velocities of Rayleigh waves with periods of 1–7 min which had undergone multiple passages round the Earth. Their results showed that models with an LVZ in the upper mantle between depths of about 100 and 200 km fitted the dispersion data better than did the Jeffreys' model at periods less than 200 s, confirming an earlier suggestion by Takeuchi *et al.* (1959). Later techniques for the processing of surface wave data (for example, the multiple filter analysis reported by Dziewonski *et al.*, 1969) permitted the accurate calculation of dispersion characteristics over specific geographical or tectonic regions, and led to "pure path" studies (e.g. Dziewonski, 1971; Knopoff, 1972) which revealed significant differences in the character of the LVZ between oceans, shields and tectonic provinces. In particular, it was shown that the evidence for the presence of an LVZ beneath shields was much less strong than for other regions (see also Alexander, 1974, Goncz and Cleary, 1976). Some of the relevant models are shown in Fig. 2.

This result was in conformity with studies of body wave travel times by Cleary and Hales (1966) and Herrin (1968), which not only showed systematic deviations from JB times but also provided estimates of station travel time anomalies which were distributed according to a regional pattern, according to which P arrivals at shield stations were up to 2 s earlier than arrivals at the same distance in continental areas of recent uplift. In a subsequent study of S times, Doyle and Hales (1967) found that S station anomalies were about four times greater than the corresponding P anomalies. Hales and Doyle (1967) interpreted this as due to partial melting in the upper mantle, thus identifying the LVZ and its regional variations as the main source of the station anomalies.

3.3. The Upper Mantle Transition Zone

Jeffreys' velocity model included a sharp change in velocity gradient at a depth of 413 km (the "20° discontinuity") whereas in Gutenberg and Richter's model the increase occurs at about 200 km (Fig. 1). This, and the failure of region C to meet the criterion of chemical homogeneity, made the "transition zone" an obvious subject for further investigation. Niazi and Anderson (1965) published one of the first array studies (from TFSO, Arizona), in which directly-measured P slownesses were plotted against

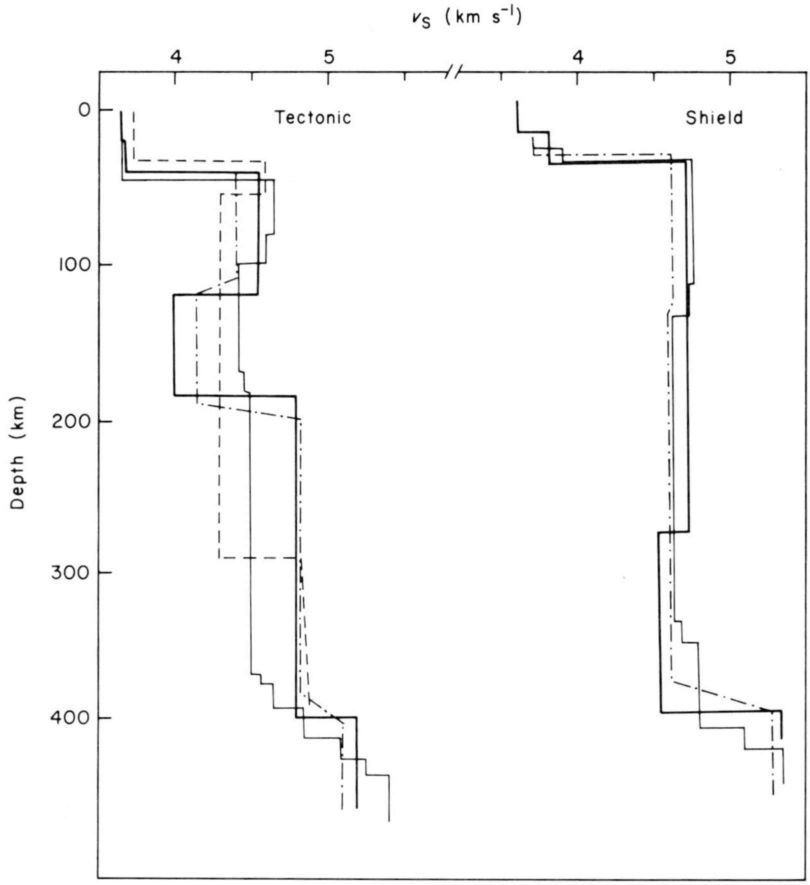

Fig. 2. S-velocity distributions in "tectonic" and shield regions, from surface wave studies by Toksöz *et al.* (1967) (———), Kanamori (1970) (- - - - -), Dziewonski (1971) (— · — · —) and Goncz and Cleary (1976) (———), showing the differences between the low velocity zones inferred for the two types of regions. The "tectonic" solutions also exemplify the non-uniqueness inherent in surface wave analysis, in that the data tend to be equally satisfied by broad, low-contrast and narrow, high-contrast low velocity zones.

distance in the range 12–30°. Their data showed relatively abrupt changes in slowness gradient at about 17° and 24°, indicating increases in velocity gradients near depths of 320 and 640 km. Many additional studies have since been made using closely-spaced stations (e.g. Green and Hales, 1968; Archambeau *et al.*, 1969; Wiggins and Helmberger, 1973) and arrays (e.g. Johnson, 1967; Simpson *et al.*, 1974), which vary in detail but which show velocity increases at average depths of about 400 and 650 km. Wiggins and

Helmberger (1973) used a technique of comparing synthetic seismograms from velocity models with the observed P train. Helmberger and Engen (1974), using a similar technique, have also found S-velocity increases at these depths. Both sets of authors include another small velocity increase at about 500 km depth.

Evidence now seems to be accumulating for the existence of a larger number of rapid velocity changes in the upper mantle, and perhaps even beyond. Hales *et al.* (1970) found velocity increases at depths of 57 km and 190 km beneath the Gulf of Mexico, while on the continents increases have been found at 80–100 km (Green and Hales, 1968; Mereu and Hunter, 1969; Hirn *et al.*, 1973, Muirhead *et al.*, 1977). From a traverse in Central Australia, Hales *et al.* (1979) suggest the presence of quasi-discontinuous changes at 75, 200, 325, 411, 512, 610, 630, 645 and 722 km. Ram Datt and Muirhead (1978), using slowness data from the Australian WRA array, find velocity increases at 85, 175, 330, 440, 520, 640, 750, 920 and 1065 km. The correspondence between these independently derived models is striking (Fig. 3). The differences between them nevertheless underline the non-uniqueness present in body wave analysis as mentioned earlier.

3.4. The Outer Core

In his determination of the P velocity distribution in the outer core, Jeffreys (1939*c*) had found it necessary to incorporate a large velocity decrease just above the inner core–outer core boundary (region F, Fig. 1) in order to bring the PKIKP branch (ray paths through the inner core) back to a distance of 110°, as the observed data seemed to indicate (Fig. 4(a)). Elsewhere the outer core velocity in his model increases with depth.

Gutenberg (1957) had observed "precursors" to the PKIKP phase back to distances of about 124°, although the PKP1 and PKP2 branches terminated at a caustic (B) at about 143°, and had attempted to explain the phenomenon in terms of wave dispersion at the inner core–outer core boundary. The problem was revived when Bullen and Burke-Gaffney (1958) observed similar precursors in nuclear explosion records, and suggested diffraction from the caustic as a possible cause.

Bolt (1962, 1964) postulated that the precursors were refraction phases from a discontinuity (transition region) in the outer core. This interpretation received wide acceptance, although variations involving multiple discontinuities were proposed by Adams and Randall (1964) and others. Difficulties arose, however, when Sacks and Saa (1969) and later Buchbinder (1971) showed that precursor amplitudes were inconsistent with the refraction interpretation. They suggested instead that the precursors were reflections from small discontinuities within the outer core (Fig. 4(b)).

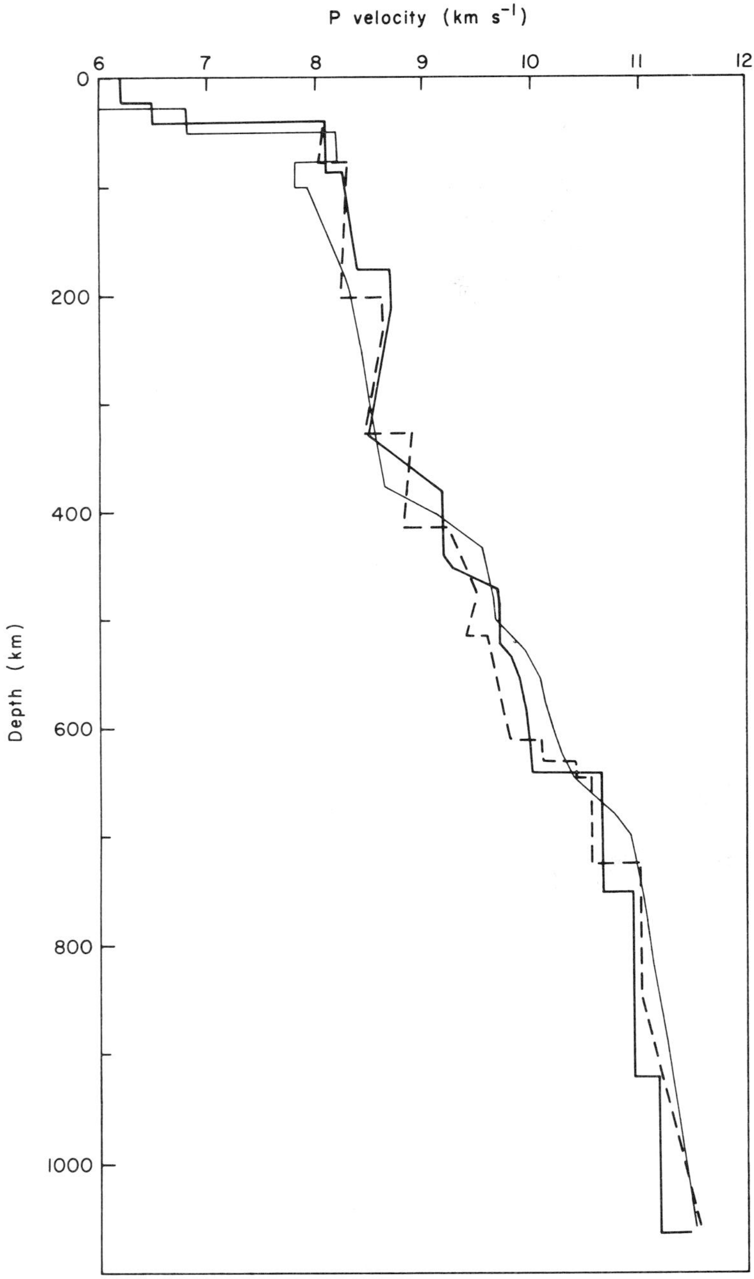

Fig. 3. Recent determinations of the P-velocity distribution beneath north-central Australia from an array study by Ram Datt and Muirhead (1978) (———) and a long-range profile by Hales *et al.* (1979) (– – – – –), compared with a distribution tabulated by Helmberger and Engen (1974) (———).

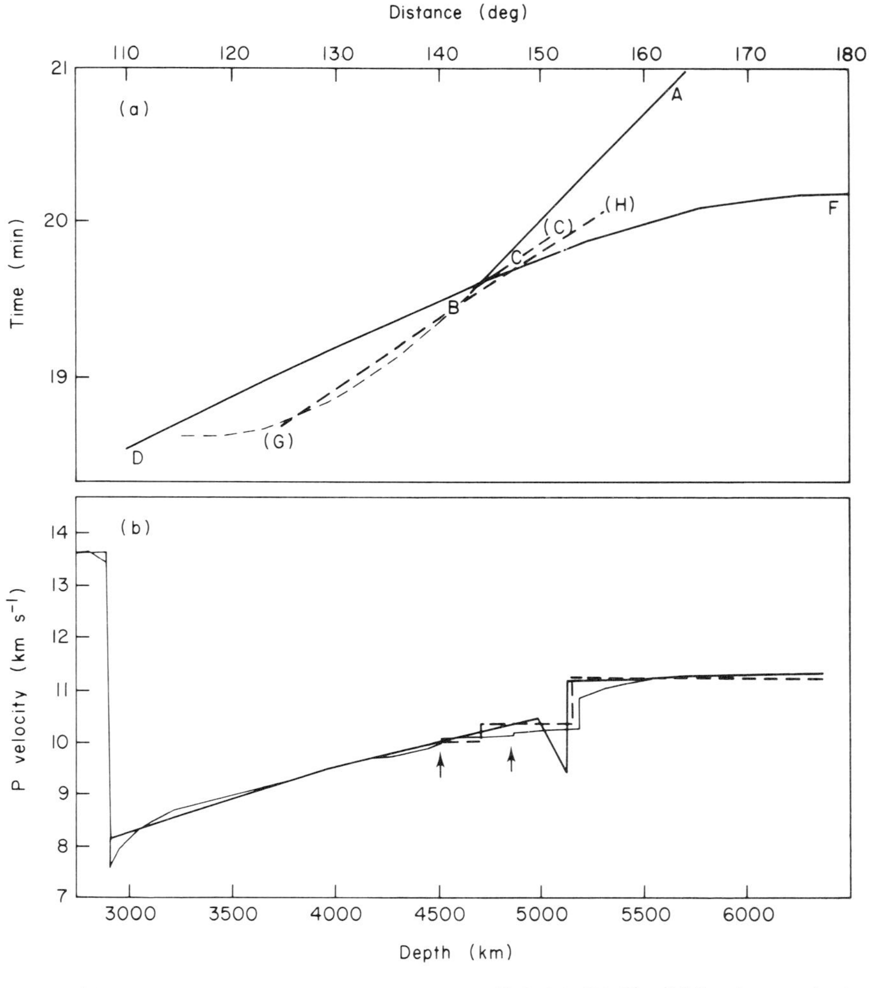

Fig. 4. (a) Two interpretations of PKP precursors: (i) Bolt's (1962) addition (— — —) of a refracted GH branch to the JB travel-time curve (———) to account for precursor arrivals at distances less than 140°. The bracketed symbols mark Bolt's additional branches. Reflection branches have been omitted; the branch sequence is ABCDF for JB, ABCGHDF for the Bolt curve. (ii) The Cleary–Haddon "minimum time curve" (— — —) for precursor arrivals, based on a hypothesis of scattering of PKP waves near the core–mantle boundary.

(b) Comparisons of the Jeffreys (1939*c*) (———) core model with models by (i) Bolt (1962) (- - - -), in which the discontinuity in the outer core is responsible for the GH branch in (a); (ii) Buchbinder (1971) (———), in which the two small discontinuities marked by arrows give reflected branches in the precursor range. The scattering hypothesis (see above) does not require the presence of outer core discontinuities.

Haddon (1972) observed that the precursor curve had a concave-upward curvature incompatible with the reflection hypothesis, and proposed that the precursors were caused by scattering near the core–mantle boundary. Cleary and Haddon (1972) showed that a theoretical minimum time curve for precursors based on this hypothesis was in excellent agreement with the observations. In an array study, Doornbos and Husebye (1972) also concluded that the precursor data were incompatible with geometrical ray theory interpretations, but favoured a diffraction interpretation. Jeffreys (1939*b*) had already shown, however, that short-period waves diffracted from the caustic would be unobservable at distances less than about 139°. Subsequent studies (King *et al.* 1973*a*,*b*; Doornbos and Vlaar, 1973; Haddon and Cleary, 1974) showed the scattering interpretation to be consistent with slowness and amplitude observations. The interpretation removes the need to postulate discontinuities or transition zones in the outer core, of the type previously envisaged. Kind and Müller (1977) in a recent study of SKS and SKKS times and amplitudes, have postulated a zone of increased velocities in the outer core at a depth of about 3750 km. Their model, however, does not produce a PKP precursor branch.

Some doubt remains about the precise origin of the scattering. The above proponents of the scattering hypothesis had preferred a model of a scattering medium above the core–mantle boundary, while differing about its probable thickness. Haddon and Cleary (1974) had noted the possibility that the scattering was caused by topography on the boundary itself, and this possibility has been strengthened by the observation of Chang and Cleary (1978) that precursors to PKKP recorded at LASA at a distance of about 60° can be interpreted as due to scattering on *underside reflection* from the boundary.

It may be added here that array observations of reflections from the inner core boundary (PKiKP) at short epicentral distances ($10° < \Delta < 50°$) enabled Engdahl *et al.* (1970) to make a fairly precise determination of about 1215 km for the radius of the inner core, close to that proposed by Bolt (1964) and in contrast to the Jeffreys (1939*c*) value of 1250 km.

3.5. Lateral Heterogeneity

Although the work of Jeffreys and Bullen was directed towards the construction of a laterally homogeneous Earth model, Jeffreys (1939*a*) considered that evidence existed for minor regional differences, especially between oceanic and continental paths. As the resolution of fine structure increased, attention was naturally directed towards the extent of lateral heterogeneity in the Earth, especially with the impetus given by the development of the theory of plate tectonics.

In the latter context mention should be made of the work of Sykes (1967) in establishing that fault plane solutions of earthquakes at oceanic ridges were in conformity with the transform fault model advanced by Wilson (1965) as a correlative of sea-floor spreading from the ridges. The associated concept of trench regions as lithospheric sinks was supported by the seismological identification of high-Q, high-velocity slabs of material dipping beneath the trench-arc systems (Oliver and Isacks, 1967; Cleary, 1967). The important paper by Isacks *et al.* (1968) placed the plate tectonics model firmly within the framework of seismology, and used seismic data to estimate the relative motions of adjoining plates, correlate rates of underthrusting with the lengths of deep seismic zones and study the stress distributions within the slabs.

On the usual plate tectonics model, the moving lithospheric plate is effectively decoupled from the underlying mantle by the asthenosphere (usually assumed to correspond to the LVZ). Although, as mentioned previously, some surface wave studies do not require the presence of an LVZ beneath the shields, they lack sufficient resolution to establish, independently of other data, whether one is actually absent in shield regions. It has been suggested from such studies, however, that lateral differences in upper mantle structure extend to depths of at least 400 km (Knopoff, 1972, Alexander, 1974), whereas the usual plate tectonics model requires a lithospheric thickness of about 200 km. Studies of ScS and multiple ScS travel time residuals (Sipkin and Jordan, 1975, 1976) seemed to confirm that lateral differences extended to depths of several hundred kilometres, and consequently to require a reinterpretation of sea-floor spreading in terms of whole-mantle convection (Jordan, 1975*a*,*b*). This has been disputed by Okal and Anderson (1975) and Hales (1976), who contend that regional differences in the ScS and multiple ScS residuals can be accommodated in the upper 250 km of the mantle. The question will be discussed further in the next section with respect to the so-called "baseline problem".

There has also been considerable discussion about the existence of lateral heterogeneity near the base of the mantle, especially in connection with the possible occurrence of mantle "plumes" which could provide a mechanism for driving the lithopheric plates. Davies and Sheppard (1972) and Kanasewich *et al.* (1973) claimed from array measurements of slowness that lateral variations did occur in the lowest 300 km of the mantle, and that at least some of them were beneath regions where plumes had been postulated (in particular, Hawaii). This conclusion was supported from travel time studies by Julian and Sengupta (1973). Wright (1974) and Berteussen (1976) cast doubt on this interpretation of the array measurements, and presented evidence that the observed slowness anomalies were more likely to be the result of structure beneath the arrays. Wright and Lyons (1975) nevertheless found from an array study that significant lateral heterogeneity existed above 800 km and below 2500 km, and proposed that convection is confined to those two regions. Sengupta and Toksöz (1976), using body wave travel time anomalies from deep focus earthquakes, also found that lateral heterogeneity was most pronounced in the upper mantle and near the core–mantle boundary. More recently, Dziewonski *et al.* (1977), in an analysis of about 700 000 P-wave travel time residuals, have contoured the lateral P-velocity variation in the lower mantle, and found an apparent correlation with large-wavelength gravity anomalies.

4. Free Oscillations

Following Benioff's 1952 evidence that free oscillations of the Earth appeared to be recordable for major earthquakes, the identification of free oscillation modes with periods of up to about 1 h from the 1960 Chilean earthquake, and the subsequent determination of eigenfrequencies of a large number of additional modes (especially by Dziewonski and Gilbert, 1972, 1973) had a great impact on recent seismology. The basic reason for the initial impact is encapsulated in the following comment of Bullen (1975, p. 310):

> Each observational period of a mode of free Earth oscillation in effect contributes an equation of condition restricting the possible distributions of (ρ, k, μ) or (ρ, α, β) in the Earth's interior.

Even though Bullen's comment is unquestionably correct, its potential implication, that the use of free oscillation data will yield a straightforward method for generating independent estimates of ρ, α and β, is not. When free oscillation data initially became available, they were thought to provide such a method. By the time of publication of *The Earth's Density* (Bullen, 1975) it had become clear that their importance in fact rested on their use as a basis for generating comparisons with and consistency checks against conclusions derived from travel-time data and associated assumptions about chemical homogeneity. The importance of free oscillation data in this sense (and not in the other) is argued strongly by Bullen in Chapter 14 of his book.

The flaw in interpreting the use of free oscillation data along the obvious lines of the above quotation is the tacit assumption that, since the oscillation equations explicitly involve ρ, α and β, the indeterminacy associated with the determination of these parameters from travel-time data (as discussed in the previous section) has (potentially) been removed. This point will be pursued in some detail in the next section.

The overall impact of free oscillation data on seismology and hence on the determinations of the internal structure of the Earth, is best described in terms of the sequence of refinements imposed on our understanding by the problems which appeared as discrepancies between the information contained in observed free oscillation periods and the interpretation of body wave data. In fact, the first discrepancies appeared soon after the identification of free oscillation modes from the Chilean earthquake.

We therefore examine the importance of free oscillation data in terms of some of the major difficulties posed by their interpretation.

4.1. S Velocity Models and the Radius of the Core

MacDonald and Ness (1961) devised an S velocity model for the mantle based on torsional oscillation data, in which the S velocity in the lower half of the mantle differed systematically by about 1 per cent from that of the Jeffreys model. As pointed out by Landisman *et al.* (1965), however, this implied that the JB S times were inaccurate by as much as 25 s at the far end of the S distance range, which was inconsistent with observational travel-time data. The latter authors devised another interpretation of the discrepancy, which required a constant density between depths of 1600 and 2800 km. Such an interpretation was also generally thought to be improbable.

Bullen and Haddon (1967*a*,*b*) found that the discrepancy could be resolved by increasing the value for the core radius by 15–20 km from the Jeffreys' value of 3473 ± 3 km. Muirhead and Cleary (1969) pointed out that a determination of the core radius by Taggart and Engdahl (1968), using PcP data, gave a value of 3477 ± 2 km. They showed that the difficulty could be overcome by reducing the shear velocity in the lowest 200 km of the mantle, in accordance with a derivation by Cleary (1969) of a shear velocity of 6.8 km s^{-1} at the base of the mantle, from observations of S waves diffracted around the Earth's core, in contrast to the Jeffreys value of 7.3 km s^{-1}.

Later workers have preferred the interpretation of a larger core radius on the grounds that (a) observed PcP-P times fit the Herrin (1968) P model with a core radius of 3485 km (Jordan and Anderson, 1974); and (b) the ScS-S times of Hales and Roberts (1970) fit their S mantle model with a similar core radius (see also the review article by Dziewonski and Haddon, 1974). More recently, however, an analysis by Mondt (1977) of SH waves at distances beyond 90°, using full wave theory, has suggested that the S velocity decreases to about 7.0 km s^{-1} at the bottom of the mantle. It seems possible, therefore, that the original discrepancy may have resulted from a combination of errors in core radius and in S velocity immediately above the core–mantle boundary. On the other hand, Mondt's result is apparently controverted by recent work of Kind and Müller (1977). The true situation may be obscured by the presence of lateral inhomogeneity in this region, as noted earlier.

4.2. A Plethora of Models

The original list of free oscillation periods obtained from recordings of the Chilean earthquake corresponded mainly to estimates for the fundamental spheroidal modes (${}_0S_l$) up to angular order number $l = 42$, and the fundamental torsional modes (${}_0T_l$) up to $l = 11$. Using improved data analysis techniques and additional data, this list was considerably extended by Pekeris (1966), Derr (1969) and Dziewonski and Landisman (1970). An important theoretical advance during this period was the calculation by Madariaga (1972) of the splitting of eigenfrequencies of Earth oscillations by lateral heterogeneities. Through the work of Mendiguren (1973), Dziewonski and Gilbert (1972, 1973), Brune and Gilbert (1974) and Gilbert and Dziewonski (1975), the number of modes, including numerous spheroidal overtones (${}_nS_l$, $n > 1$) for which accurate estimates of periods are available, has reached a number in excess of 1400.

The identification and estimation of new and more accurate periods of free oscillation has invariably been accompanied by the construction of new models to take account of the additional constraining power they contain. The resulting plethora of models has been a source of some confusion, especially since different authors have not always obtained close agreement between the models they generated to satisfy more or less the same data. The reasons for lack of consistency between models is a direct consequence of the facts that

(i) Available free oscillation data can support "globally distinct" solutions ("$\underset{=}{G}$-far" solutions in the terminology of Backus and Gilbert, 1968); i.e. solutions which differ from each other by more than fine scale variations. The potential existence of such solutions is in part a direct consequence of the incomplete nature of free oscillation data. The available data have poor resolution in the upper mantle.
(ii) Different computational strategies (in the sense that their underlying assumptions differ) have been used to generate the models.

Press (1970, 1972) used a Monte Carlo method, which depends heavily on the choice of *a priori* bounds and on the strategy by which the random realistic solutions are generated. Following the work of Backus and Gilbert (1967, 1968), many, including Dziewonski and Gilbert (1973), Jordan and Anderson (1974) and Gilbert and Dziewonski (1975), used a model refinement strategy which depends heavily on the choice of a starting solution and the algorithm used to solve the resulting non-linear equations.

Following in some respects the earlier work of Haddon and Bullen (1969), Dziewonski *et al.* (1975) used a parameterization technique which was directed towards finding the simplest model consistent with the available data. In Fig. 5 their average Earth model PEM-A is compared with model C2 of Anderson and Hart (1976) derived from a similar, but culled, data set, and generically similar starting models.

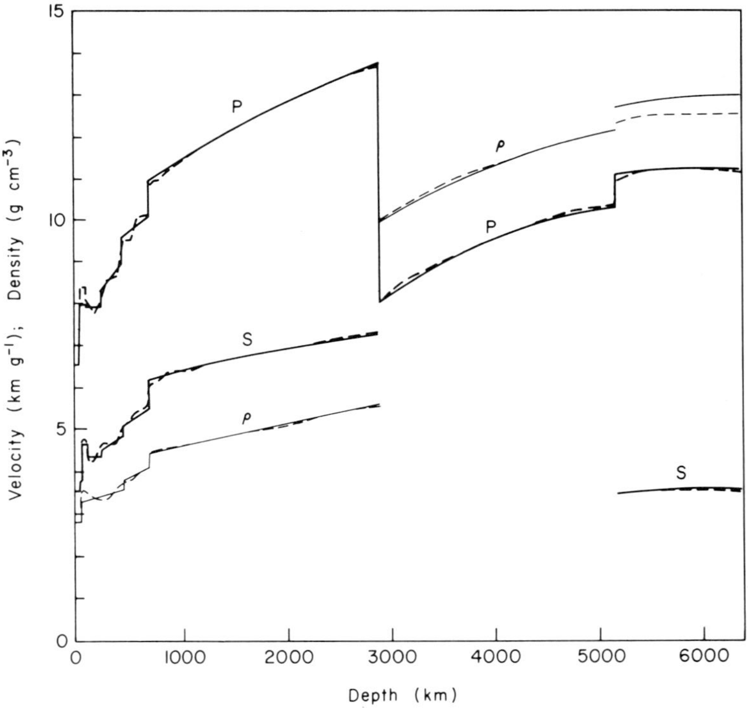

Fig. 5. Recent velocity and density models PEM-A (Dziewonski *et al.*, 1975) (———) and C2 (Anderson and Hart, 1976), (----), obtained by refinement modelling applied to free oscillation data. The starting model for C2 was B1 (Jordan and Anderson, 1974), modified in the upper mantle and core; that for PEM-A was 1066B (Gilbert and Dziewonski, 1975), which itself was derived from B1. On the scale of the figure, both sets of models below a depth of about 400 km are almost indistinguishable from their respective starting models.

One of the notable achievements at this time was the identification by Dziewonski and Gilbert (1973) of overtone periods which could be fitted adequately only by the incorporation of a solid inner core into their models, thereby providing evidence for what had previously been inferred from the seismic evidence for a sharp velocity increase within the core (Lehmann, 1936), coupled with equation of state arguments.

4.3. The Density Inversion Controversy

One of the major difficulties associated with the use of free oscillation data to generate models for ρ, α and β has been that such models, which satisfy similar data to similar confidence limits (e.g., model HB2 of Bullen and Haddon, 1970, and the Monte Carlo models of Press, 1970, 1972), can differ markedly in the region of the upper mantle. This difficulty was the background of one of the major recent controversies in geophysics; namely, the use by Press (1970, 1972) of the Monte Carlo method to generate from given free oscillation data models of ρ, α and β consistent with bounds obtained from body wave data. All the models obtained by Press supported the conclusion that "the density of the mantle between the depth of 70 and 150 km cannot be less than 3.40 g cm^{-3} and is more likely to be close to 3.50 g cm^{-3}"—a conclusion with which many disagreed on independent grounds, both geophysical (e.g., model HB2 of Bullen and Haddon, 1970, and the models of Wang, 1970, 1972) and geochemical (e.g., Ringwood, 1975).

The reasons for such lack of consistency between models has been mentioned in (i) and (ii) above. They imply that there is no guarantee that different computational strategies will identify the same subset of possible $\underset{\sim}{G}$-far solutions. In fact, for any one implementation (e.g., a starting solution for model refinement), most of the computational strategies in use identify only one solution. The exception is the Monte Carlo method. The justification for its use is that, if appropriately implemented, it can potentially identify all possible $\underset{\sim}{G}$-far solutions.

In the controversy arising out of Press's conclusion, this latter point was initially overlooked. Many argued that it was in fact the Monte Carlo method itself which was at fault. The defect in such a point of view was the motivation for Anderssen and Seneta's (1971) efforts to formalize the Monte Carlo method as a strategy for geophysical data inversion. It is clear from their work that a Monte Carlo method fails only if its implementation is invalid; that is, only if the initial *a priori* bounds were incorrectly chosen or the randomness of the generation of the realistic solutions is suspect.

The possibility that some defect in implementation was the explanation for Press's conclusion was explored by Anderssen *et al.* (1972) and Worthington *et al.* (1972). By the adoption of a procedure which favoured the generation of random globally distinct realistic solutions between given *a priori* bounds, and by working with different choices of the bounds, they were able to verify an argument previously advanced by Dziewonski (1970), that an unwarranted narrowness between 300 and 600 km in Press's shear velocity *a priori* bounds (see Fig. 6) was directly responsible for the high densities in his models at depths between 70 and 150 km.

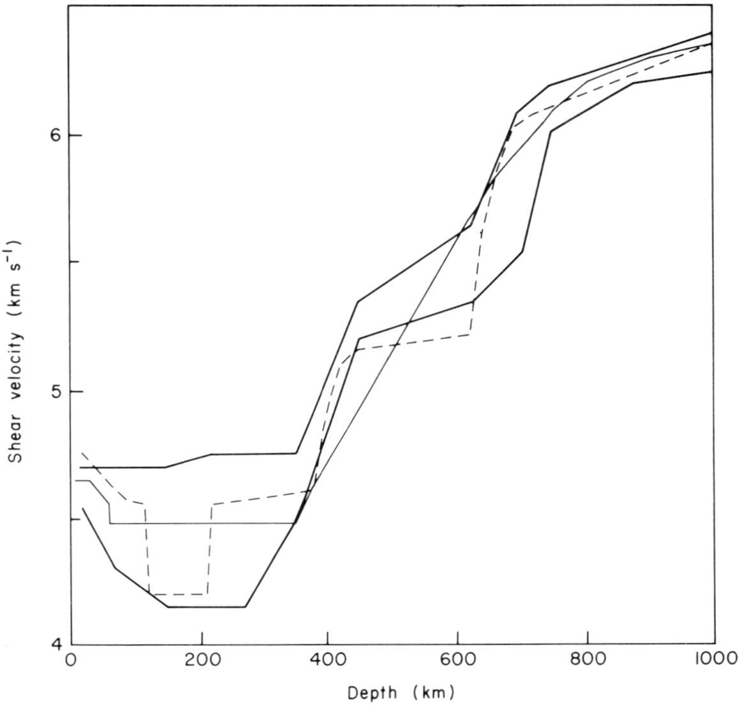

Fig. 6. S-velocity profiles from model HB2 (Bullen and Haddon, 1970) (———) and a model by Wang (1972) (– – – –), compared with initial upper and lower bounds used in Monte Carlo modelling by Press (1972). As a result of his constraints on S-velocity between depths of 300 and 600 km, all Press's models included a high-density lid in the uppermost mantle, a feature absent from HB2, Wang and other models (cf. Dziewonski, 1970; Worthington *et al.*, 1972).

In spite of this, Press's work clearly indicates the potential of the Monte Carlo method as a powerful strategy for the inversion of (geo)physical data.

4.4. Monte Carlo vs. Model Refinement

The possible existence of other globally distinct ($\underline{\underline{G}}$-far) solutions is usually overlooked when geophysical inferences are made from models by their authors. The danger is that undue stress may be placed on the apparent uniqueness of a derived model when it is used for inference purposes.

A major contributing factor has been the framework-free manner in which some authors have generated and interpreted models. The fact is ignored that, when constructing models, the key problem is how to cope with the *indeterminacy* of ρ. For example, it is assumed in refinement modelling strategies that a geophysically justifiable starting model for ρ, α and β is available. But the generation of such starting models reflects the indeterminacy of ρ. Such models are obtained by using geophysically justifiable starting models for α and β to generate a model for ρ, using additional assumptions such as an equation of state or some petrological model.

But use of such starting models potentially contradicts any claim that refinement modelling generates a model for ρ, α and β for which the dependence of ρ on α and β has been removed. This cannot be guaranteed unless the final model for ρ, α and β is $\underset{=}{G}$-far from the starting model. But this contradicts the basic justification for refinement modelling, namely: with respect to a given starting model for ρ, α and β, we only require a model which is $\underset{=}{G}$-near to the starting model.

Though the dependence of ρ on α and β may be broken for some of the solutions $\underset{=}{G}$-near to the starting solution, it will not be broken for all $\underset{=}{G}$-near solutions. This does not affect the above argument, since the implicit nature of the $\underset{=}{G}$-near solutions which a refinement modelling strategy generates cannot be controlled in advance. In fact, in refinement modelling, the corrections applied to ρ, α and β are usually constrained to satisfy the same or related equations as those satisfied by ρ, α and β—cf. the construction of the Fréchet differentiable kernels in Backus and Gilbert (1967, 1968). From a statistical point of view, we have a situation where only tentative assumptions can or must be made about the initial specifications, and thereby are involved with "inferences based on conditional specifications". A bibliography for this subject can be found in Bancroft and Han (1977).

It follows that, without the use of a suitable (statistical) framework for estimating the effect of any explicit dependence of ρ on α and β, only models constructed so that they contain no inbuilt dependence of ρ on α and β should be used for general geophysical inference purposes. Two strategies exist for generating such models:

(i) The Monte Carlo method as applied by Press (1970, 1972) and formalized by Anderssen and Seneta (1971): with respect to given *a priori* non-uniqueness bounds for ρ, α and β, and given free oscillation data and associated error bounds, generate and test random realistic models of ρ, α and β to obtain acceptable models of ρ, α and β which satisfy the data up to the level of the error bounds.

(ii) The restricted Monte Carlo method: with respect to assumed models for α and β, given *a priori* non-uniqueness bounds for ρ and given free oscillation data and associated error bounds, generate and test random realistic models of ρ to obtain acceptable models of ρ which satisfy the data up to the level of the error bounds. The aim is to see if the resulting models for ρ are consistent with independent geophysical and geochemical conclusions about ρ. If they are not, then this represents justification for concluding that the assumed models for α and β are invalid. This is more or less the strategy behind the work of Anderssen *et al.* (1972), and Worthington *et al.* (1972).

Strategies for the analysis of such models of ρ, α and β can be based on

(i) a search for inner consistency (or lack of it) among these models;
(ii) a search for inner consistency (or lack of it) between these models and independently-derived geophysical results (i.e. generally accepted geophysical/geochemical beliefs).

The goal is to find models which are consistent with independently derived geophysical results of comparable resolution and reliability, and which exhibit inner consistency which can be used to extend our understanding of the internal structure of the Earth. A possible outcome, as Press's (1970, 1972) work illustrates, is that the derived models will not be consistent with independently derived geophysical results. In this case, the problem reduces to having to decide whether it is the *implementation* of the Monte Carlo method (i.e. *a priori* non-uniqueness bounds, etc.) or the independently derived geophysical results which are in error.

4.5. The Smoothness of Models

Even though the number of free oscillation modes for which estimates of periods are available is now in excess of 1400, there are large gaps in the information about ρ, α and β which they constrain. To illustrate, we restrict attention to a consideration of the torsional modes of oscillation ${}_nT_l$ with angular order number l and overtone number n.

If the substitution $U = rV$ is applied to the equations of Alterman *et al.* (1959) for the eigenfrequencies ${}_n\sigma_l$ of the torsional oscillations of an SNREI Earth, the following Sturm–Liouville system is obtained for ${}_n\sigma_l$:

$$\frac{\mathrm{d}}{\mathrm{d}r}\left(r^4\mu\frac{\mathrm{d}V}{\mathrm{d}r}\right)+[r^4\rho\,{}_n\sigma_l^2-(l^2+l-2)\mu r^2]V=0, \quad a\leqslant r\leqslant b, \tag{8}$$

with boundary conditions

$$\mu\frac{\mathrm{d}V}{\mathrm{d}r}=0, \quad r=a, \quad r=b,$$

where a and b denote the radius of the core–mantle boundary and the surface of the Earth respectively, ρ denotes density and μ denotes rigidity.

In most published investigations of the asymptotic distribution of the eigenvalues λ_n of Sturm–Liouville systems, it is assumed that their coefficients are continuously differentiable. As a consequence, the asymptotic behaviour of λ_n (${}_n\sigma_l^2$ in (8)) always takes the form

$$\lambda_n^{1/2} = n + a_0 + a_1/n + \ldots,$$

where a_0, $a_1 \ldots$ are constants independent of n (see, for example, Fix, 1967, for a rigorous analysis). Although the (asymptotic) behaviour of λ_n for discontinuous coefficients is obviously of interest, few studies have existed until recently. For example, the problem of finding the eigenfrequencies of free oscillation modes of a heterogeneous Earth model has been considered by Saito (1971), Usami (1971), Madariaga (1972), Anderssen and Cleary (1974), Luh (1974), and Geller and Stein (1978), while general results about the asymptotic behaviour of such eigenfrequencies are contained in Lapwood (1975), McNabb *et al.* (1976), Wang *et al.* (1977) and Anderssen (1977).

In an examination of the asymptotic behaviour of the eigenvalues of a general Sturm–Liouville system associated with the second-order differential equation for the torsional free oscillations of SNREI Earth models with N discontinuities between the Earth's surface and the core–mantle boundary, McNabb *et al.* (1976) showed that the asymptotic distribution of these eigenvalues satisfied, for suitably large n (cf. Wang *et al.*, 1978),

$$\sin(c\sigma) = \sum_{j=1}^{2N-1} A_j \sin(d_j\sigma), \qquad \sigma = \lambda_n^{1/2}, \tag{9}$$

where c is the shear wave radial travel time from the Earth's surface to the core–mantle boundary, and A_j and d_j are constants which depend on the parameters defining the discontinuities. For Earth models with only one discontinuity in either the mantle or crust, $A_1 = A$ corresponds to the reflection coefficient for normally incident waves at the discontinuity and $d_1 = d$ to the difference between shear wave radial travel times from the Earth's surface and the core–mantle boundary to the discontinuity (Anderssen, 1977). Equation (9) thereby becomes

$$\sin c\sigma = A \sin d\sigma, \qquad \sigma = \lambda_n^{1/2}. \tag{10}$$

If, in addition, it is assumed that $A \ll 1$ (A is positive) and $|d| \ll 1$, then, for suitably large n,

$$\lambda_n^{1/2} \doteqdot \frac{n\pi}{c} + (-1)^n \frac{A}{c} \sin \frac{d}{c} n\pi. \tag{11}$$

As mentioned in McNabb *et al.* (1976), equations (10) and (11) are important because many of the qualitative features of the asymptotic distribution of the λ_n are immediate consequences of them. In particular:

(i) In the limit as $A \to 0$, (11) yields the well-known result that, when $\rho(x)$ and $\mu(x)$ contain no discontinuities, the asymptotic spacing between the $\lambda_n^{1/2}$ values tends to a constant value π/c.

(ii) If $\rho(x)$ and $\mu(x)$ contain a single discontinuity (at the same location), then the form of the second term on the right-hand side of (11) indicates that the asymptotic spacing between the $\lambda_n^{1/2}$ values does not tend to a constant value π/c but changes in an "irregular" and "persistent" manner—sometimes referred to as the *solotone effect.* In fact, even for a small discontinuity (i.e. small A), the persistent irregular behaviour of the asymptotic spacing is present.

(iii) If $\rho(x)$ and $\mu(x)$ contain more than one discontinuity, then the results of (11) continue to hold unless the contributions from the different discontinuities to the persistent irregular behaviour of the asymptotic spacing cancel.

From a geophysical point of view, (i), (ii) and (iii) clearly indicate that information about rapid changes in (and hence, the non-smoothness of) ρ and β in the mantle are contained in the asymptotic distribution of the eigenfrequencies ${}_n\sigma_l$. Consequently, available torsional free oscillation data (i.e. small overtone numbers) only constrain models of ρ and β for the mantle which are suitably smooth (it should be remembered that simple (limited parameterization) models do not necessarily define smooth models). Therefore, the use of non-smooth models must be independently justified.

Work by Anderssen *et al.* (1975) and Gilbert (1975) confirm that these results also hold for spheroidal free oscillation eigenfrequencies.

4.6. The Baseline Problem

One of the major difficulties associated with the reconciliation of body wave and free oscillation data has been the existence of a systematic difference of some 4–8 s between one-way vertical S-wave travel times from mantle models satisfying free oscillation data and those from body wave studies such as that of Hales and Roberts (1970) (see Jordan and Anderson, 1974, and Gilbert and Dziewonski, 1975). Since most body-wave observations are derived from continental stations, it seemed likely that the discrepancies arose because of large differences in the upper mantle S velocity structure between continents and oceans, as the observations of Sipkin and Jordan (1975, 1976) appeared to indicate. In their "continental" model PEM-C, based on free oscillation and "pure path" surface wave dispersion data, Dziewonski *et al.* (1975) were able to reduce the S travel-time discrepancy to within reasonable limits, but the differences between their continental and oceanic models extend down to 420 km.

A new approach to the problem was initiated by Randall (1976), who verified an earlier contention by Jeffreys (1965) that the effect of attenuation on dispersion was first order, and that the free oscillation periods, especially those of the torsional modes, should be corrected for this. A number of studies have followed which take the effect of attenuation into account. In particular, Hart *et al.* (1977) have found that, by applying the Q model MM8 (Anderson *et al.*, 1965) to an Earth model constructed from uncorrected data, they were able to remove effectively the baseline discrepancy.

An alternative treatment of the free oscillation data, in this context, is to use them as a constraint on Q structure within the Earth. Anderson and Hart (1978*a*) showed that the Q model MM8 used by Hart *et al.* (1977) was inadequate for certain of the longer-period mode data, and obtained a set of Q models for the mantle by trial-and-error inversion of surface wave and free oscillation Q data, with additional constraints from body wave data. Their preferred model incorporated a high-Q (~ 2000) region in the lower mantle, with low-Q zones (down to ~ 100) at top and bottom of the mantle. Sailor and Dziewonski (1977), on the other hand, were able to satisfy Q data from a "representative set" of free oscillation modes with simpler models; for example, one with $Q = 353$ below 670 km and $Q = 110$ in the upper mantle (all Q values given here specify Q in shear). Anderson and Hart included in their paper a model not very different from the above model of Sailor and Dziewonski, but considered that it did not satisfy adequately the observed low-order torsional mode data or the body wave (ScS) data. In a further study, Anderson and Hart (1978*b*) have refined the preferred model of their earlier paper. Nevertheless much of the Q data presently available are strongly scattered and subject to systematic error, and it may be some time before the question is satisfactorily resolved.

5. State of the Art

By 1939, as indicated earlier, the internal structure of the Earth had been resolved to first order. Almost 40 years later, refinements to the determination of such structure have probably reached the limits imposed by the presence of lateral and vertical inhomogeneities. The next step, that of precisely determining the structure of particular regions of the Earth, is still very much in its initial stages.

That this should be so, after many years of painstaking and expensive experimentation, may seem disappointing. What *has* been achieved is at least an approximate definition of the problem itself, in terms both of the complexity of the structure to be resolved and of the limitations inherent in the means available for its resolution. In the process (as will have been apparent from the foregoing sections), considerable advances have been made towards the ideal of using the totality of information available in the seismic record in order to provide optimum constraints on its interpretation.

5.1. Travel-time Data and Inversion

One of the major obstacles to the precise determination of Earth structure has been revealed by array studies (e.g. those of Berteussen, 1975*a*,*b*) showing considerable fluctuations in signal amplitudes (with smaller but significant fluctuations in travel times) between sensors only a few kilometres apart. Whatever the correct interpretation of these phenomena (see for example Aki *et al.*, 1977) the observations themselves emphasize the practical difficulty of interpreting record sections obtained from traverses of sparsely-spaced instruments.

The spectacular record sections obtained from long lines of closely-spaced instruments (e.g. Hirn *et al.*, 1973) or from aggregations of array data (King and Calcagnile, 1976), show that experiments of this kind are capable of clearly delineating travel-time branches in spite of the fluctuations imposed by local inhomogeneities. Once the branches have been defined with precision and (as far as possible) completeness, the data can be most effectively inverted by means of the "tau" method developed by Bessonova *et al.* (1974) from the theory of Gerver and Markushevich (1966). This method makes use of the function

$$\tau(p) = T(p) - pX(p),$$

where $p = \mathrm{d}T/\mathrm{d}X$ is the ray parameter, X is half the epicentral distance, and T is half the travel time from source to receiver. Bessonova *et al.* showed that limits of the function $\tau(p)$ could be mapped into limits for the velocity–depth function; so that with $\tau(p)$ estimated from (T, X) observations an envelope of possible velocity–depth solutions may be obtained. The efficacy of the method when good data are available has been demonstrated by Kennett (1976) and by Bates and Kanasewich (1976).

Nevertheless, as Kennett points out, "all methods of travel-time inversion suffer from the limitations imposed by the conditions for the validity of classical ray methods". The first of these limitations is the impossibility of resolving the velocity structure of a low-velocity zone within which no rays bottom. This source of uncertainty can be minimized by means of the strategy employed by Hales *et al.* (1978) of using data from sources occurring throughout the depth range of the upper mantle, in accordance with the theoretical findings of Gerver and Markushevich (1966). A second important limitation results from the failure of ray theory under certain circumstances in regions of very steep velocity gradient and at the edges of the shadow zones produced by low-velocity zones. This limitation can be overcome by comparison of the data with synthetic seismograms generated from the postulated models, enabling the amplitudes and waveforms of the signals to be used as constraints in the inversion. Various techniques have been developed in recent years for the generation of synthetic seismograms (see, for example, Fuchs and Müller, 1971; Gilbert and Helmberger, 1972). Of particular interest at present is a method incorporating the "Langer approximation" (Richards, 1976), used by Cormier and Richards (1977) in an application of full wave theory to a discontinuous velocity increase.

Supplementary methods available include the use of "pure path" surface wave dispersion data. The resolving power of such analyses would be much enhanced by the addition of higher mode data, as described by Alexander (1963), but little used since then. Recent studies of S to P and P to S conversions at upper mantle discontinuities (see for example Marshall *et al.*, 1975; Sacks and Snoke, 1975) suggest that these also may provide useful constraints on upper mantle structure.

Combinations of these methods should make it feasible to undertake a quantitative determination of the structures underlying discrete regions of the Earth, and of the differences between them (aside from logistical difficulties, which are serious enough for shields and other continental regions, but much more so if an intensive study of suboceanic structure is contemplated). The investigation of deep mantle inhomogeneity poses more subtle problems: apart from the methods already in use, as described earlier, it is to be hoped that resolution of near-surface structure will enable a more efficient use of arrays for this purpose.

5.2. Free Oscillation Data and Inversion

It is clear from the discussion in the previous sections that through

(i) the supplementation of present free oscillation data with higher overtone (rather than angular order) number eigenfrequencies in order to resolve more accurately rapid changes in global internal structure,

(ii) the use of Monte Carlo inversion techniques to sample more thoroughly the underlying non-uniqueness in ρ, α and β,

better models for and information about ρ, α and β are possible. What appears to be needed most, however, before another major advance is made with free oscillation data inversion, is an improvement in background theory. To illustrate this, we return to the point in Section 4 concerning the possible assumption that "since the free oscillation equations explicitly involve ρ, α and β, the indeterminacy associated with the determination of these parameters from travel-time data has (potentially) been removed".

Such an assumption ignores the fact that not only is the mathematical formalism for the evaluation of ρ, α and β from free oscillation data an inverse eigenvalue problem of considerable complexity, but also that theoretical work such as that of Borg (1946) (and more recently Barcilon, 1974) on such problems, shows explicitly that, at least for the vibrating stretched string with a non-symmetric continuous rigidity, one (full) spectrum of eigenfrequencies is insufficient to determine the rigidity uniquely.

Though the free oscillation data involve a different spectrum for each of the angular order numbers (i.e. an angular order spectrum), the nature of the additional constraining power they carry for a spherically symmetric Earth has not been determined theoretically, although Barcilon (1974) has made some tentative attempts to come to grips with this problem. A simple non-rigorous and intuitive argument (which does not agree with a conjecture of Backus and Gilbert, 1967, p. 265—see also Barcilon, 1974) is that, without additional assumptions, any finite set of angular order spectra does not uniquely determine ρ, α and β as functions of depth since they arise as a consequence of the mathematical framework used to examine the oscillations of a sphere rather than as an expression of geophysical reality. Independent examples of redundancy induced by spherical symmetry in geophysical contexts exist: for example, the redundancy contained in observations of the gravitational potential of a spherically symmetric sphere at more than one location on its surface (cf. Clemins̆, 1957).

Up to this point, the discussion has ignored the important practical point that available free oscillation data will always be inaccurate and incomplete. Although some have argued that the bulk of the constraining power of free oscillation data is contained in the lower eigenfrequencies (longer periods), the work of Anderssen and Cleary (1974), Lapwood (1975), McNabb *et al.* (1976), Anderssen *et al.* (1975) and Gilbert (1975) clearly indicates that this is true only if ρ, α and β, as functions of depth, are smooth slowly-varying functions with no rapid changes in properties. Even if we ignore the fact that transition zones, involving rapid changes in properties, occur in the mantle, such a restriction on the form of ρ, α and β would be completely inconsistent with the changes that ρ, α and β undergo at the core discontinuities.

Geophysically, the importance of the above mentioned work on overtone eigenfrequencies is the implication that when ρ, α and β are constrained to fit given free oscillation data, it is more the spacing between the values of the eigenfrequencies, than their actual numerical values, which should be approximated most closely. It is from this point of view that the acquisition of more higher-mode data is likely to prove useful.

Acknowledgement

We wish to thank Professor Kurt Lambeck and Dr David King for critically reading the manuscript.

See note added in proof, p. 175.

References

Abel, N. H. (1823, 1826). *Oeuvres Complètes*, 2 vols., Christiana, Oslo, 1881; Vol 1: pp. 11–27 (1823); pp. 97–101 (1826).

Adams, R. D. and Randall, M. J. (1964). The fine structure of the Earth's core. *Bull. seismol. Soc. Am.* **54**, 1299–1313.

Aki, K., Christofferson, A. and Husebye, E. S. (1977). Determination of the three-dimensional seismic structure of the lithosphere. *J. geophys. Res.* **82**, 277–296.

Alexander, S. S. (1963). Surface wave propagation in the western United States. Ph.D. Thesis, California Institute of Technology, Pasadena, California.

Alexander, S. (1974). Comparison of crust and upper mantle structure beneath shields and the influence of shields on the dynamics of plate motions. *Trans. Am. geophys. Union* **55**, 358–359.

Alterman, Z., Jarosch, H. and Pekeris, C. L. (1959). Oscillation of the Earth. *Proc. R. Soc. A.* **252**, 80–95.

Anderson, D. L. and Hart, R. S. (1976). An Earth model based on free oscillations and body waves. *J. geophys. Res.* **81**, 1461–1475.

Anderson, D. L. and Hart, R. S. (1978*a*). Attenuation models for the Earth. *Phys. Earth planet. Interiors* **16**, 289–306.

Anderson, D. L. and Hart, R. S. (1978*b*). The *Q* of the Earth. *J. geophys. Res.* **83**, 5869–5882.

Anderson, D. L., Ben-Menahem, A. and Archambeau, C. B. (1965). Attenuation of seismic energy in the upper mantle. *J. geophys. Res.* **70**, 1441–1448.

Anderssen, R. S. (1977). The effect of discontinuities in density and shear velocity on the asymptotic overtone structure of torsional eigenfrequencies of the Earth. *Geophys. J. R. astronom. Soc.* **50**, 303–310.

Anderssen, R. S. and Cleary, J. R. (1974). Asymptotic structure in torsional free oscillations of the Earth—I. Overtone structure. *Geophys. J. R. astronom. Soc.* **39**, 241–268.

Anderssen, R. S. and Seneta, E. (1971). A simple statistical estimation procedure for Monte Carlo Inversion in geophysics. *Pure appl. Geophys.* **91**, 5–13.

Anderssen, R. S., Cleary, J. R. and Dziewonski, A. M. (1975). Asymptotic structure in eigenfrequencies of spheroidal normal modes of the Earth. *Geophys. J. R. astronom. Soc.* **43**, 1001–1006.

Anderssen, R. S., Worthington, M. H. and Cleary, J. R. (1972). Density modelling by Monte Carlo Inversion—I, Methodology. *Geophys. J. R. astronom. Soc.* **29**, 433–444.

Archambeau, C. B., Flinn, E. A. and Lambert, D. G. (1969). Fine structure of the upper mantle. *J. geophys. Res.* **74**, 5825–5865.

Backus, G. E. and Gilbert, F. J. (1967). Numerical application of a formalism for geophysical inverse problems. *Geophys. J. R. astronom. Soc.* **13**, 247–276.

Backus, G. E. and Gilbert, F. J. (1968). The resolving power of gross Earth data. *Geophys. J. R. astronom. Soc.* **16**, 169–205.

Bancroft, T. A. and Han, C.-P. (1977). Inferences based on conditional specification: a note and a bibliography. *Int. statist. Rev.* **45**, 117–127.

Barcilon, V. (1974). On the uniqueness of inverse eigenvalue problems. *Geophys. J. R. astronom. Soc.* **38**, 287–298.

Bates, A. C. and Kanasewich, E. R. (1976). Inversion of seismic travel times using the Tau method. *Geophys. J. R. astronom. Soc.* **47**, 59–72.

Berteussen, K. A. (1975*a*). Crustal structure and P-wave travel time anomalies at Norsar. *J. Geophys.* **41**, 71–84.

Berteussen, K. A. (1975*b*). P-wave amplitude variability at Norsar. *J. Geophys.* **41**, 595–613.

Berteussen, K. (1976). The origin of slowness and azimuth anomalies at large arrays. *Bull. seismol. Soc. Am.* **66**, 719–741.

Bessonova, E. N., Fishman, V. M., Ryaboyi, V. Z. and Sitnikova, G. A. (1974). The tau method for the inversion of travel times—I: Deep seismic sounding data. *Geophys. J. R. astronom. Soc.* **36**, 377–398.

Bolt, B. A. (1962). Gutenberg's early PKP observations. *Nature, Lond.* **196**, 121–124.

Bolt, B. A. (1964). The velocity of seismic waves near the Earth's centre. *Bull. seismol. Soc. Am.* **54**, 191–208.

Borg, G. (1946). Eine Umbehrung der Sturm–Liouvilleschen Eigenwertaufgabe. *Acta math.* **78**, 1–96.

Brune, J. N. and Gilbert, F. (1974). Torsional overtone dispersion from correlations of S waves to SS waves. *Bull. seismol. Soc. Am.* **64**, 313–320.

Buchbinder, G. G. R. (1971). A velocity structure of the Earth's core. *Bull. seismol. Soc. Am.* **61**, 429–456.

Bullen, K. E. (1936). The variation of density and the ellipticities of strata of equal density within the Earth. *Mon. Not. R. astronom. Soc., Geophys. Suppl.* **3**, 395–401.

Bullen, K. E. (1940). The problem of the Earth's density variation. *Bull. seismol. Soc. Am.* **30**, 235–250.

Bullen, K. E. (1942). The density variation of the Earth's central core. *Bull. seismol. Soc. Am.* **32**, 19–29.

Bullen, K. E. (1975). *The Earth's Density*, Chapman and Hall, London.

Bullen, K. E. and Burke-Gaffney, T. N. (1958). Diffracted seismic waves near the PKP caustic. *Geophys. J. R. astronom. Soc.* **1**, 9–17.

Bullen, K. E. and Haddon, R. A. W. (1967*a*). Earth oscillations and the Earth's interior. *Nature, Lond.* **213**, 574–576.

Bullen, K. F. and Haddon, R. A. W. (1967*b*). Derivation of an Earth model from free oscillation data. *Proc. natn. Acad. Sci. U.S.A.* **58**, 846–852.

Bullen, K. E. and Haddon, R. A. W. (1970). Evidence from seismology and related sources on the Earth's present internal structure. *Phys. Earth planet Interiors* **2**, 342–349.

Cavendish, H. (1798). Experiments to determine the density of the Earth. *Phil. Trans. R. Soc.* **58**, 469–526.

Chang, A. and Cleary, J. R. (1978). Precursors to PKKP. *Bull. seismol. Soc. Am.* **68**, 1059–1079.

Cleary, J. R. (1967). Azimuthal variation of the Longshot source term. *Earth planet. Sci. Lett.* **3**, 29–37.

Cleary, J. R. (1969). The S velocity of the core-mantle boundary from observations of diffracted S. *Bull. seismol. Soc. Am.*, **59**, 1399–1405.
Cleary, J. R. and Haddon, R. A. W. (1972). Seismic wave scattering near the core–mantle boundary: a new interpretation of precursors to PKP. *Nature*, Lond. **240**, 549–551.
Cleary, J. R. and Hales, A. L. (1966). An analysis of the travel times of P waves to North American stations, in the distance range 32° to 100°. *Bull. seismol. Soc. Am.* **56**, 467–489.
Clemins̆, A. (1957). Direkte Verfahren zur Auswertung von Schweremessungen bei zweidimensionaler Massenverteilung. *Geofis. pura appl.* **38**, 81–122.
Cormier, V. F. and Richards, P. G. (1977). Full wave theory applied to a discontinuous velocity increase: the inner core boundary. *J. Geophys.* **43**, 3–31.
Davies, D. and Sheppard, R. M. (1972). Lateral heterogeneity in the Earth's mantle. *Nature, Lond.* **239**, 318–323.
Derr, J. S. (1969). Free oscillation observations through 1968. *Bull. seismol. Soc. Am.* **59**, 2079–2099.
Doornbos, D. J. and Husebye, E. S. (1972). Array analysis of PKP phases and their precursors. *Phys. Earth planet. Interiors* **5**, 387–399.
Doornbos, D. J. and Vlaar, N. J. (1973). Regions of seismic wave scattering in the Earth's mantle and precursors to PKP. *Nature phys. Sci.* **243**, 58–61.
Dorman, J., Ewing, M. and Oliver, J. (1960). Study of shear-velocity distribution in the upper mantle by mantle Rayleigh waves. *Bull. seismol. Soc. Am.* **50**, 87–115.
Doyle, H. A. and Hales, A. L. (1967). An analysis of the travel times of S waves to North American stations, in the distance range 28° to 82°. *Bull. seismol. Soc. Am.* **57**, 761–771.
Dziewonski, A. M. (1970). Correlation properties of free period partial derivatives and their relation to the resolution of gross Earth data. *Bull. seismol. Soc. Am.* **60**, 741–768.
Dziewonski, A. M. (1971). Upper mantle models from 'pure path' dispersion data. *J. geophys. Res.* **76**, 2587–2601.
Dziewonski, A. M. and Gilbert, F. (1972). Observations of normal modes from 84 recordings of the Alaskan earthquake of 1964 March 28. *Geophys. J. R. astronom. Soc.* **27**, 393–446.
Dziewonski, A. M. and Gilbert, F. (1973). Observations of normal modes from 84 recordings of the Alaskan earthquake of 1964 March 28, Part II. Further remarks based on new spheroidal overtone data, *Geophys. J. R. astronom. Soc.* **35**, 401–437.
Dziewonski, A. M. and Haddon, R. A. W. (1974). The radius of the core-mantle boundary inferred from travel time and free oscillation data; a critical review. *Phys. Earth planet. Interiors* **9**, 28–35.
Dziewonski, A. M. and Landisman, M. (1970). Great circle Rayleigh and Love wave dispersion from 100–900 seconds. *Geophys. J. R. astronom. Soc.* **19**, 37–91.
Dziewonski, A. M., Bloch, S. and Landisman, M. (1969). A technique for the analysis of transient seismic signals. *Bull. seismol. Soc. Am.* **59**, 427–444.
Dziewonski, A. M., Hagar, B. H. and O'Connell, R. J. (1977). Large-scale heterogeneities in the lower mantle. *J. geophys. Res.* **82**, 239–255.
Dziewonski, A. M., Hales, A. L. and Lapwood, E. R. (1975). Parametrically simple Earth models consistent with geophysical data. *Phys. Earth planet. Interiors* **10**, 12–48.
Engdahl, E. R., Flinn, E. A. and Romney, C. (1970). Seismic waves reflected from the Earth's inner core. *Nature, Lond.* **228**, 852–853.

Fix, G. (1967). Asymptotic eigenvalues of Sturm-Liouville systems. *J. math. Anal. Applics* **19**, 519–525.

Fuchs, K. and Müller, G. (1971). Computation of synthetic seismograms with the reflectivity method and comparison with observations. *Geophys. J. R. astronom. Soc.* **23**, 417–433.

Geller, R. J. and Stein, S. (1978). Normal modes of a laterally heterogeneous body: a one-dimensional example. *Bull. seismol. Soc. Am.* **68**, 103–116.

Gerver, M. and Markushevich, V. (1966). Determination of a seismic wave velocity from the travel-time curve. *Geophys. J. R. astronom. Soc.* **11**, 165–173.

Gilbert, F. (1975). Some asymptotic properties of the normal modes of the Earth. *Geophys. J. R. astronom. Soc.* **43**, 1007–1011.

Gilbert, F. and Dziewonski, A. M. (1975). An application of normal mode theory to the retrieval of structural parameters and source mechanisms from seismic spectra. *Phil. Trans. R. Soc.* **278**, 187–269.

Gilbert, F. and Helmberger, D. V. (1972). Generalized ray theory for a layered sphere. *Geophys. J. R. astronom. Soc.* **27**, 57–80.

Gilbert, F., Dziewonski, A. M. and Brune, J. N. (1973). An informative solution to a seismological inverse problem. *Proc. natn. Acad. Sci. U.S.A.* **70**, 1410–1413.

Goncz, J. H. and Cleary, J. R. (1976). Variations in the structure of the upper mantle beneath Australia, from Rayleigh wave observations. *Geophys. J. R. astronom. Soc.* **44**, 507–516.

Green, R. W. E. and Hales, A. L. (1968). The travel times of P waves to 30° in the central United States and upper mantle structure. *Bull. seismol. Soc. Am.* **58**, 267–289.

Gutenberg, B. (1957). The 'boundary' of the Earth's inner core. *Trans. Am. Geophys. Union* **38**, 750–753.

Gutenbert, B. and Richter, C. F. (1939). On seismic waves (fourth paper). *Beitr. Geophys.* **54**, 94–136.

Haddon, R. A. W. (1972). Corrugations on the mantle–core boundary or transition layers between inner and outer cores? (abstract) *Trans. Am. Geophys. Union* **53**, 600.

Haddon, R. A. and Bullen, K. E. (1969). An Earth Model incorporating free Earth oscillation data. *Phys. Earth planet. Interiors* **2**, 35–49.

Haddon, R. A. W. and Cleary, J. R. (1974). Evidence for scattering of seismic PKP waves near the mantle–core boundary. *Phys. Earth planet. Interiors*, **8**, 211–234.

Hales, A. L. (1976). ScS travel time residuals and the difference between continents and oceans. *Earth planet. Sci. Lett.* **33**, 249–254.

Hales, A. L. and Doyle, H. A. (1967). P. and S travel time anomalies and their interpretation. *Geophys. J. R. astronom. Soc.* **13**, 403–415.

Hales, A. L. and Roberts, J. L. (1970). Shear velocities in the lower mantle and the radius of the core. *Bull. seismol. Soc. Am.* **60**, 1427–1436.

Hales, A. L., Helsey, C. E. and Nation, J. B. (1970). P travel times for an oceanic path. *J. geophys. Res.* **75**, 7362–7381.

Hales, A. L., Muirhead, K. J. and Rynn, J. M. W. (1979). A compressional velocity distribution for the upper mantle. *Tectonophysics* (in press).

Hart, R. S., Anderson, D. L. and Kanamori, H. (1977). The effect of attenuation on gross Earth models. *J. geophys. Res.* **82**, 1647–1654.

Haskell, N. A. (1953). The dispersion of surface waves in multilayered media. *Bull. seismol. Soc. Am.* **44**, 17–34.

Helmberger, D. V. and Engen, G. R. (1974). Upper mantle shear structure. *J. geophys. Res.* **79**, 4017–4028.

Herrin, E. (1968). 1968 seismological tables for P phases. *Bull. seismol. Soc. Am.* **58**, 1193–1241.

Hirn, A., Steinmetz, L., Kind, R. and Fuchs, K. (1973). Long range profiles in western Europe: II. Fine structure of the lower lithosphere in France (Southern Bretagne). *Z. Geophys.* **39**, 363–384.

Hopkins, W. (1839–42). Researches in physical geology. *Phil. Trans. R. Soc.* **129**, 381; **130**, 193; **132**, 43.

Hough, S. S. (1895). The oscillations of a rotating ellipsoidal shell containing fluid. *Phil. Trans. R. Soc.* **186**, 469.

Isacks, B., Oliver, J. and Sykes, L. R. (1968). Seismology and the new global tectonics. *J. geophys. Res.* **73**, 5855–5899.

Jeffreys, H. (1939*a*). The times of P, S and SKS, and the velocities of P and S. *Mon. Not. R. astronom. Soc., Geophys. Suppl.* **4**, 498–533.

Jeffreys, H. (1939*b*). The times of the core waves. *Mon. Not. R. astronom. Soc., Geophys. Suppl.* **4**, 548–561.

Jeffreys, H. (1939*c*). The times of the core waves (second paper). *Mon. Not. R. astronom. Soc., Geophys. Suppl.* **4**, 594–615.

Jeffreys, H. (1962). *The Earth*, 4th ed., Cambridge University Press, Cambridge.

Jeffreys, H. (1965). Damping of S waves. *Nature, Lond.* **208**, 675.

Johnson, L. R. (1967). Array measurements of P velocities in the upper mantle. *J. geophys. Res.* **72**, 6309–6325.

Jordan, T. H. (1975*a*). The continental tectosphere. *Rev. Geophys. Space Phys.* **13**, 1–12.

Jordan, T. H. (1975*b*). Lateral heterogeneity and mantle dynamics. *Nature, Lond.* **255**, 745–750.

Jordan, T. H. and Anderson, D. L. (1974). Earth structure from free oscillations and travel times. *Geophys. J. R. astronom. Soc.* **36**, 411–459.

Julian, B. R. and Sengupta, M. K. (1973). Seismic travel time evidence for lateral inhomogeneity in the deep mantle. *Nature, Lond.* **242**, 243–247.

Kanamori, H. (1970). Velocity and *Q* of mantle waves. *Phys. Earth planet Interiors* **2**, 259–275.

Kanasewich, E. R., Ellis, R. M., Chapman, C. H. and Gutowski, P. R. (1973). Seismic array evidence of a core boundary source for the Hawaiian linear volcanic chain. *J. geophys. Res.* **78**, 1361–1371.

Kelvin, Lord (Thomson, W.) (1876). Effects of elastic yielding on precession and nutation, Opening Address to Sec. A. of the British Association, Glasgow meeting, 1876. In *Collected Mathematical and Physical Papers of Sir William Thomson*, Clay and Sons, Cambridge University Press Warehouse, London, 1890, Vol. III, pp. 320–336.

Kennett, B. L. N. (1976). A comparison of travel-time inversions. *Geophys. J. R. astronom. Soc.* **44**, 517–536.

Kind, R. and Müller, G. (1977). The structure of the outer core from SKS amplitudes and travel times. *Bull. seismol. Soc. Am.*, **67**, 1541–1554.

King, D. W. and Calcagnile, G. (1976). P-wave velocities in the upper mantle beneath Fennoscandia and Western Russia. *Geophys. J. R. astronom. Soc.* **46**, 407–432.

King, D. W., Haddon, R. A. W. and Cleary, J. R. (1973*a*). Evidence for seismic wave scattering in the D″ layer. *Earth planet Sci. Lett.* **20**, 353–356.

King, D. W., Haddon, R. A. W. and Cleary, J. R. (1973*b*). Array analysis of precursors to PKIKP in the distance range 128° to 142°. *Geophys. J. R. astronom. Soc.* **37**, 157–173.

Knopoff, L. (1972). Observation and inversion of surface-wave dispersion. In *The Upper Mantle* (A. R. Ritsema, ed.), *Tectonophysics* **13**, 497–519.

Lamb, H. (1882). On the vibrations of an elastic sphere. *Proc. Lond. math. Soc.* **13**, 189–212.

Landisman, M., Satô, Y. and Nafe, J. (1965). Free vibrations of the Earth and the properties of its deep interior regions. Part 1: Density. *Geophys. J. R. astronom. Soc.* **9**, 439–502.

Lapwood, E. R. (1975). The effect of discontinuities in density and rigidity on torsional eigenfrequencies of the earth. *Geophys. J. R. astronom. Soc.* **40**, 453–464.

Lehmann, I. (1936). P′. *Bur. Centr. seismol. Internat. A.* **14**, 3–31.

Luh, P. C. (1974). Normal modes of a rotating self-gravitating inhomogeneous Earth. *Geophys. J. R. astronom. Soc.* **38**, 187–224.

MacDonald, G. J. F. and Ness, N. F. (1961). A study of free oscillations of the Earth. *J. geophys. Res.* **66**, 1865–1911.

Macelwane, J. B. (1951). Evidence on the interior of the Earth derived from seismic sources. In *Internal Constitution of the Earth* (B. Butenberg, ed.), Dover, New York.

McNabb, A., Anderssen, R. S. and Lapwood, E. R. (1976). Asymptotic behaviour of the eigenvalues of a Sturm–Liouville system with discontinuous coefficients. *J. math. Anal. Applics* **54**, 741–751.

Madariaga, R. (1972). Toroidal free oscillations of the laterally heterogeneous Earth. *Geophys. J. R. astronom. Soc.* **27**, 81–100.

Marshall, P. D., Douglas, A., Barley, B. J. and Hudson, J. A. (1975). Short period teleseismic S waves. *Nature, Lond.* **253**, 181–182.

Mendiguren, J. A. (1973). Identification of free oscillation spectral peaks for 1970 July 31, Colombian deep shock using the excitation criterion. *Geophys. J. R. astronom. Soc.* **33**, 281–321.

Mereu, R. F. and Hunter, J. A. (1969). Crustal and upper mantle structure under the Canadian shield from Project Early Rise data. *Bull. seismol. Soc. Am.* **59**, 147–165.

Mondt, S. C. (1977). SH waves: theory and observations for epicentral distances greater than 90 degrees. *Phys. Earth planet. Interiors* **15**, 46–59.

Muirhead, K. J. and Cleary, J. R. (1969). Free oscillations of the Earth and the D″ layer. *Nature, Lond.* **223**, 1146.

Muirhead, K. J., Cleary, J. R. and Finlayson, D. M. (1977). A long range seismic profile in southeastern Australia. *Geophys. J. R. astronom. Soc.* **48**, 509–519.

Niazi, M. and Anderson, D. L. (1965). Upper mantle structure of western North America, from apparent velocities of P waves. J. geophys. Res. **70**, 4633–4640.

Okal, E. A. and Anderson, D. L. (1975). A study of lateral inhomogeneities in the upper mantle by multiple ScS travel time residuals. *Geophys. Res. Lett.* **2**, 313–316.

Oldham, R. D. (1899). *Mem. geol. Surv. India* **29**, 29.

Oldham, R. D. (1900). On the propagation of earthquake motion to great distances. *Phil. Trans. R. Soc.* **194**, 135–174.

Oliver, J. and Isacks, B. (1967). Deep earthquake zones, anomalous structure in the mantle, and the lithosphere. *J. geophys. Res.* **72**, 4259–4275.

Pekeris, C. L. (1966). Internal constitution of the Earth. *Geophys. J. R. astronom. Soc.* **11**, 85–132.

Press, F. (1970). Earth models consistent with geophysical data. *Phys. Earth planet. Interiors* **3**, 3–22.

Press, F. (1972). The Earth's interior as inferred from a family of models. In *The Nature of the Solid Earth* (E. R. Robertson, ed.), McGraw-Hill, New York.

Radau, R. R. (1885). Sur la loi des densités à l'intérieur de la Terre. *C.r. hebd. Séanc. Acad. Sci. Paris C*, 972.

Ram Datt and Muirhead, K. J. (1978). P-wave velocities in the mantle (preprint).

Randall, M. J. (1976). Attenuative dispersion and frequency shifts of the Earth's free oscillations. *Phys. Earth planet. Interiors* **12**, 1–4.

Richards, P. G. (1976). On the adequacy of plane-wave reflection—transmission coefficients in the analysis of seismic body waves. *Bull. seismol. Soc. Am.* **66**, 701–718.

Ringwood, A. E. (1975). *Composition and Petrology of the Earth's Mantle*, McGraw-Hill, New York.

Sacks, I. S. and Saa, G. (1969). The structure of the transition zone between the inner core and the outer core. *Carnegie Inst. Washington Year Book* **69**, 419–426.

Sacks, I. S. and Snoke, J. A. (1975). Determination of the lithosphere-asthenosphere boundary using converted waves. *Carnegie Inst. Washington Year Book* **75**, 223–229.

Sailor, R. V. and Dziewonski, A. M. (1977). *Q* models of the mantle consistent with free oscillation data (abstract). *Trans. Am. geophys. Union* **58**, 440.

Saito, M. (1971). Theory for the elastic-gravitational oscillation of a laterally heterogeneous Earth. *J. Phys. Earth* **19**, 259–270.

Sengupta, M. K. and Toksöz, M. N. (1976). Three dimensional model of seismic velocity variation in the Earth's mantle. *Geophys. Res. Lett.* **3**, 84–86.

Simpson, D. W., Mereu, R. F. and King, D. W. (1974). An array study of P-wave velocities in the upper mantle transition zone beneath northeastern Australia. *Bull. seismol. Soc. Am.* **64**, 1757–1788.

Sipkin, S. A. and Jordan, T. H. (1975). Lateral heterogeneity of the upper mantle determined from the travel times of ScS. *J. geophys. Res.* **80**, 1474–1484.

Sipkin, S. A. and Jordan, T. H. (1976). Lateral heterogeneity of the upper mantle determined from the travel times of multiple ScS. *J. geophys. Res.* **81**, 6307–6320.

Sykes, L. R. (1967). Mechanism of earthquakes and nature of faulting on the mid-oceanic ridge. *J. geophys. Res.* **72**, 2131–2153.

Taggart, J. N. and Engdahl, E. R. (1968). Estimation of PcP travel times and the depth to the core. *Bull. seismol. Soc. Am.* **58**, 1293–1304.

Takeuchi, H., Press, F. and Kobayashi, N. (1959). Rayleigh wave evidence for the low velocity zone in the mantle. *Bull. seismol. Soc. Am.* **49**, 355–364.

Thomson, W. (1950). Transmission of elastic waves through a stratified medium. *J. appl. Phys.* **21**, 89–93.

Thomson, W. (later Lord Kelvin) and Tait, P. G. (1879). *Treatise on Natural Philosophy*, Cambridge University Press, Cambridge.

Toksöz, M. N., Chinnery, M. A. and Anderson, D. L. (1967). Inhomogeneities in the Earth's mantle. *Geophys. J. R. astronom. Soc.* **13**, 31–59.

Usami, T. (1971). Effect of horizontal heterogeneity on the torsional oscillation of an elastic sphere. *J. Phys. Earth* **19**, 175–180.

Wang, C.-Y. (1970). Density and constitution of the mantle. *J. geophys. Res.* **75**, 3264–3284.

Wang, C.-Y. (1972). A simple Earth model. *J. geophys. Res.* **77**, 4318–4329.

Wang, C., Anderssen, R. S. and Cleary, J. R. (1978). Ray-mode duality for SH waves in Earth models with crust and mantle discontinuities: I. The case of one discontinuity. *Geophys. J. R. astronom. Soc.* **54**, 297–308.

Wang, C., Gettrust, J. F. and Cleary, J. R. (1977). Asymptotic overtone structure in eigenfrequencies of torsional normal modes of the Earth: a model study. *Geophys. J. R. astronom. Soc.* **50**, 289–302.

Weinberger, H. F. (1965). *A First Course in Partial Differential Equations*, Blaisdell, London.

Wiechert, E. (1897). Ueber die Massenverteilung in Innern der Erde. *Nachr. ges. Wiss. Göttingen, math.-phys. Kl.* 221–243.

Wiggins, R. A. and Helmberger, D. V. (1973). Upper mantle structure of the United States. *J. geophys. Res.* **87**, 1870–1880.

Wilson, J. T. (1965). A new class of faults and their bearing on continental drift. *Nature, Lond.* **207**, 343–347.

Worthington, M. H., Cleary, J. R. and Anderssen, R. S. (1972). Density modelling by Monte Carlo inversion—II. Comparison of recent Earth models. *Geophys. J. R. astronom. Soc.* **29**, 445–457.

Wright, C. (1974). Comments on a paper by E. R. Kanasewich *et al.*, "Seismic array evidence of a core-boundary source for the Hawaiian linear volcanic chain". *J. geophys. Res.* **80**, 1915–1919.

Wright, C. and Lyons, J. A. (1975). Seismology, $dT/d\Delta$ and deep mantle convection. *Geophys. J. R. astronom. Soc.* **40**, 115–138.

Notes added in proof

1. (p. 157). Müller (*J. geophys. Res.* **78**, 3469–3490, 1973) obtained additional evidence for the solidity of the inner core by comparing observed long-period core phase amplitudes with theoretical amplitudes derived from a suite of models.
2. (p. 162). In a series of recent papers, the influence of discontinuities on the free oscillations of the Earth has been studied by approaches different from that discussed here (Nolet and Kennett, *Geophys. J. R. astronom. Soc.* **53**, 219–226, 1978; Woodhouse, *Geophys. J. R. astronom. Soc.* **54**, 263–280, 1978; Kennett and Woodhouse, *Geophys. J. R. astronom. Soc.* **55**, 330–350, 1978).
3. (p. 167). The point raised about the validity of the Backus and Gilbert (1967) conjecture concerning angular order spectra has been clarified by Anderssen and Chandler (*Geophys. J. R. astronom. Soc.* **55**, 311–315, 1973). They show that the conjecture is false at least for certain nongeophysical models of ρ and μ.

6. Phase Transformations and the Constitution of the Deep Mantle

LIN-GUN LIU

Abstract

Recent experimental investigations on the phase transformations of mantle silicates at high pressure and high temperature conditions are reviewed. The mineralogy and structure of the Earth's mantle, especially at depths below 500 km, are proposed in the light of these studies. The seismic velocity discontinuity around 400 km is entirely attributable to the transformation of olivine into a spinel-like phase (β-phase). The region between 400 and 600 km is more or less homogeneous and is primarily composed of phases possessing the spinel and garnet structures. The most prominent seismic discontinuity centred around 650 km in the mantle is due mainly to the combination of the phase changes from spinel to perovskite plus rocksalt and from garnet to ilmenite structures. The 650 km discontinuity is thus characterized by an increase in the primary coordination of silicon from 4 to 6. A further minor discontinuity at a still greater depth may be associated with the ilmenite to perovskite transition. If a "pyrolite" composition is adopted, the Earth's deep mantle is probably composed of more than 70 percent perovskite phase of (Mg, Fe, Al) (Si, Al)O_3, about 20 per cent rocksalt phase of (Mg, Fe) O and less than 10 per cent other phases containing calcium, sodium and potassium.

1. Introduction

The mantle may be divided into three major regions on the basis of the variation of seismic velocities with depth (see Figs 1 and 2 in Chapter 1 of this volume). The upper mantle, extending from the Moho to about 350 km, is generally characterized by small velocity gradients, except in the vicinity of the low-velocity zone. The low-velocity zone is generally believed to result from a small degree of partial melting along the grain boundaries of the mineral phases in the upper mantle. It is widely accepted that the most abundant mineral phases in this region are olivine, pyroxene and garnet.

Velocity gradients in the region between about 350 and 900 km in depth (the transition zone) are far greater than those expected due to self-compression and thermal expansion of the mineral phases in the upper mantle. These anomalous velocity gradients could be explained by a gradual change in the bulk chemistry of the upper mantle and/or by a series of phase transitions in the upper mantle minerals. On the basis of his classic studies on the density of the mantle, Bullen (1936) proposed that the transition zone is chemically inhomogeneous. On the other hand, Bernal (1936) and Jeffreys (1937) suggested that the beginning of the transition zone might be caused by the phase transformation of olivine to the spinel structure, by analogy with the first discovery of olivine–spinel polymorphism in Mg_2GeO_4 by Goldschmidt (1931). Using an equation of state based upon finite strain theory, Birch (1952) proposed that the transition zone was characterized by a series of major phase transformations resulting from the instability of the upper mantle minerals at high pressure, although minor chemical inhomogeneities were also possible. He further concluded that the transformations were complete by about 1000 km, and that new high-pressure phases were present in the mantle at depths below 1000 km (the lower mantle), this region being essentially homogeneous, with no further major transformations occurring. He realized that closely packed oxide structures such as corundum, periclase, rutile and spinel have the required elastic ratios of the lower mantle (Birch, 1952).

Since the proposal that the velocity gradients in the transition zone might be caused mainly by chemical inhomogeneity cannot be directly tested by experiments in the laboratory, it seems reasonable first to study the possible phase transformations of olivine, pyroxene and garnet at high pressures and high temperatures. Only if the high-pressure phase transformations were incapable of accounting for the velocity and density distributions in both the transition zone and the lower mantle would it then be necessary to invoke chemical inhomogeneity in the Earth's mantle.

Many of Birch's conclusions were theoretical, and experimental verification was essential. It was necessary to demonstrate that olivine, pyroxene and garnet were indeed unstable at high pressure, and would transform to new polymorphs at pressure and temperature conditions equivalent to those between 400 and 1000 km, and further, that the elastic properties and densities of these new polymorphs were capable of explaining quantitatively the inferred properties of the transition zone and the lower mantle. The difficulty with experimental testing of the hypothesis of phase transformations was that, until 1963, static high-pressure–temperature apparatus was incapable of reproducing the P–T conditions in the Earth at depths greater than about 300 km. Studies of phase transformations in the preceding period therefore employed indirect experimental methods based on thermodynamics, comparative crystal chemistry and upon the study of germanate isotypes of silicates (Ringwood, 1956, 1958*a*). Because of the similarity in the chemical relationships between crystals of germanates and silicates, it appears that germanates often behave as high-pressure models for the corresponding silicates. Furthermore, because of the larger size of the germanium cation, phase transformations in germanates tend to occur at lower pressures than in the analogous silicates. The thermodynamic investigation of solid solution equilibria between silicate and germanate end-members provided sufficient information to calculate the pressure required for the silicate to transform to a high-pressure modification.

The results of indirect study of phase transformations in this period may be summarized as follows. All of the germanate olivines and pyroxenes which were studied up to pressures of 90 kbar were found to be unstable and transformed to denser phases, strongly suggesting that the corresponding silicates would transform similarly at higher pressures, see Ringwood (1970) for a detailed review. It was calculated that forsterite should transform to the spinel structure in the pressure range 130–175 kbar and temperature range 600–1500 °C, and that Mg_2SiO_4 spinel would be 11 ± 3 per cent denser than forsterite (Ringwood, 1956, 1958*a*). Another important discovery was made in this period by Stishov and Popova (1961) who showed that at about 100 kbar (~1600 °C) silica could crystallize in the rutile structure, demonstrating a change from 4 to 6 for the coordination of silicon under pressure. The rutile form of SiO_2 was subsequently discovered by Chao *et al.* (1962) as a product of shock metamorphism of sandstone caused by a meteorite impact and was named stishovite. This immediately raised the possibility of making silicates in which silicon is sixfold coordinated.

In the period between 1963 and 1972, apparatus capable of achieving pressures and temperatures equivalent to those at depths between 300 and 600 km were developed by Ringwood and Major (1966*a*, 1968). Numerous high-pressure modifications or assemblages for the olivine, pyroxene and garnet structures were discovered in this period, see Ringwood (1970) for a detailed review. Among them, the most noteworthy are the olivine to β-phase transformation in Mg_2SiO_4, reported by Ringwood and Major (1966*a*), and the pyroxene to garnet transformation in solid solutions of $MgSiO_3$–Al_2O_3 by Ringwood (1967). The 400-km seismic discontinuity was thus demonstrated to be attributable to these two phase transformations. Our endeavours to understand phase transformations which might occur at depths below 600 km in this period relied heavily upon results derived from silicate analogues under static high pressure (Ringwood, 1970) and from the behaviour of silicate minerals and rocks in shock-wave experiments (e.g. McQueen *et al.*, 1967). The former method has provided valuable information on the possible high-pressure modifications of olivine, pyroxene and garnet, and the latter has suggested a number of phase transformations in silicates and provided information on the equations of state of the high-pressure phases.

A new era of phase transformation studies commenced in 1972, owing to the development of a double-stage split-sphere apparatus, developed mainly by Kawai in Japan (Fig. 1), and also the diamond-anvil pressure cell coupled with laser heating developed mainly by Bassett in the USA (Fig. 2). Both techniques are capable of achieving pressures in excess of 300 kbar and at temperatures ranging from 1000 to 2000 °C. These techniques have extended experimentally accessible *P–T* conditions to the equivalent of the upper part of the lower mantle (about 1200 km). Thus, we have passed beyond the stage when our knowledge of phase transformations in the deep mantle relied heavily upon inference, such as studies of analogues under static high pressure and Hugoniots obtained by shock-wave experiments. Phase transformations of mantle silicates can be directly investigated in the laboratories. An important new high-pressure modification of forsterite, enstatite and pyrope (the orthorhombic perovskite structure $MgSiO_3 \cdot xAl_2O_3$, where $0 = x \leqslant \frac{1}{3}$) has been found by means of direct experiments using the diamond-anvil pressure cell (Liu, 1974, 1975*a*).

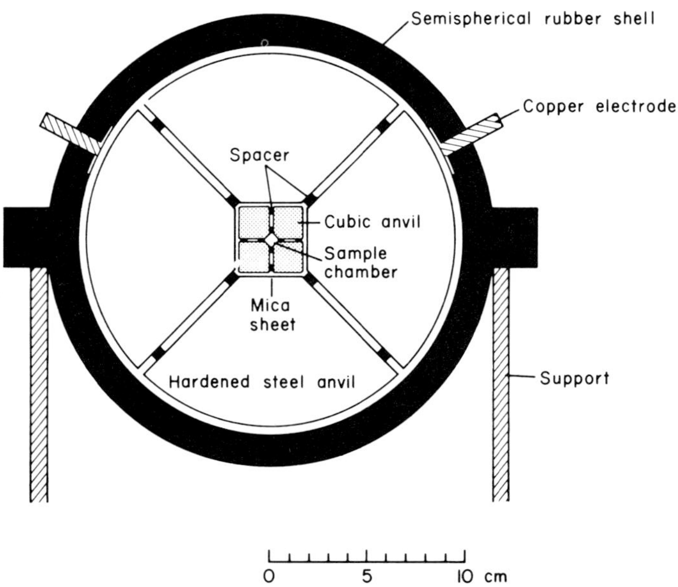

Fig. 1. Schematic drawing of a double-stage split-sphere high-pressure apparatus. (After Suito, 1972.)

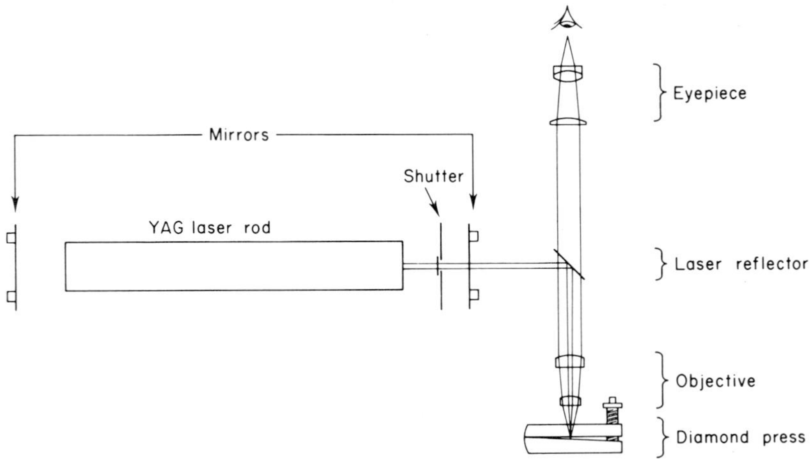

Fig. 2. Schematic diagram of a system for heating a sample under pressure in a diamond-anvil press by means of a focused laser beam. (After Liu, 1976*b*.)

2. Review of Phase Transformations in Mantle Silicates

A wide range of evidence (see Chapter 1 of this volume, for example) suggests that the MgO/(MgO + FeO) ratio for the Earth's mantle is in the vicinity of 0.9. Hence, phase transformations displayed by magnesium-rich silicates are most representative of possible major phase changes in the mantle.

2.1. The Mg_2SiO_4–Fe_2SiO_4 System

The first spinel phase in this system was synthesized by Ringwood (1958*b*) in pure Fe_2SiO_4. Sclar and Carrison (1966) reported the synthesis of a series of spinel solid solutions containing up to 35 mol per cent of Mg_2SiO_4. Ringwood and Major (1966*a*) described the synthesis of spinel solid solutions containing up to 80 mol per cent of Mg_2SiO_4. Akimoto and Fujisawa (1966) carried out a comprehensive investigation which established the nature and width of the two-phase region for compositions containing up to 30 mol per cent of Mg_2SiO_4, at pressures up to 95 kbar and at a temperature of 800 °C. Akimoto and Fujisawa (1968) then extended their earlier studies to compositions containing 40 mol per cent Mg_2SiO_4 at temperatures of 800, 1000 and 1200 °C. The nature of the olivine to spinel transformation in compositions close to pure Mg_2SiO_4 has been the subject of some debate in the past. However, forsterite was found to transform completely to a phase resembling spinel (β-Mg_2SiO_4) at pressures greater than 150 kbar and

1000 °C by Ringwood and Major (1966*a*). They also reported that the same phase was found in some runs at the $(Mg_{0.85}Fe_{0.15})_2SiO_4$ composition. An isothermal (~1000 °C) phase diagram for the system Mg_2SiO_4–Fe_2SiO_4 in the region 50–200 kbar was published and reviewed in some detail by Ringwood and Major (1970). β-Mg_2SiO_4 is about 8 per cent denser than forsterite. The phase boundary for the olivine to β-phase transition of Mg_2SiO_4 was first reported by Suito (1972), who also reported the partial transformation of β-Mg_2SiO_4 to a true spinel at pressures greater than 220 kbar and at 1000 °C. Success in synthesizing a spinel form of Mg_2SiO_4 was also reported by Ito *et al.* (1974). Spinel is about 2.5 per cent denser than β-Mg_2SiO_4. The first natural occurrence of $(Mg_{0.74}Fe_{0.26})_2SiO_4$ spinel, ringwoodite, was discovered by Binns *et al.* (1969) in a chondritic meteorite, where it had been formed under shock-wave conditions.

Studies of post-spinel phase transformations in the system Mg_2SiO_4–Fe_2SiO_4 are of considerable geophysical importance for the following reasons.

(a) This transition could be the key to explaining the 650-km discontinuity in the transition zone.
(b) The post-spinel phase could well make up the major phase in the lower mantle, which constitutes a vast portion of the Earth's interior (~50 per cent by volume).

The first direct experimental study of post-spinel phase transformations in silicates is the work of Bassett and Takahashi (1970) on Fe_2SiO_4. They found that Fe_2SiO_4 (spinel) disproportionated into constituent mixture of oxides. Subsequent work by Mao and Bell (1971) and Bassett and Ming (1972) confirmed these results. However, none of the above investigations produced X-ray patterns for well-crystallized stishovite. The same transformation in compositions extending to 80 mol per cent of Mg_2SiO_4 was studied by Ming and Bassett (1974). Finally, pure forsterite was also reported to disproportionate into a mixture of periclase plus stishovite (Kumazawa *et al.*, 1974; Ming and Bassett, 1975*a*). However, Liu (1976*a*) did not agree with the above results. He had instead found that the post-spinel phase of forsterite consisted of the assemblage of periclase plus the orthorhombic perovskite phase of $MgSiO_3$, which he had previously discovered (Liu, 1975*a*), without the intervention of the mixture of oxides. The absence of the assemblage of mixed oxide disproportionation products of forsterite was later supported by Ito (1977). The assemblage of mixed oxides was not observed in $(Mg_{0.9}Fe_{0.1})_2SiO_4$ composition either (Liu, 1976*b*). Thus, the post-spinel phase of Mg_2SiO_4 olivine is the assemblage $MgSiO_3$ (perovskite)+MgO (periclase). The latter assemblage is about 11 per cent denser than the spinel form of Mg_2SiO_4, and the transition occurs in the vicinity of 260 kbar and at about 1000 °C (Liu, 1976*a*).

The experimental results of phase transformations with fosterite are summarized as follows:

$$\underset{\text{forsterite}}{Mg_2SiO_4} \xrightarrow{\text{7.9 per cent}} \underset{\beta\text{-phase}}{Mg_2SiO_4} \xrightarrow{\text{2.4 per cent}} \underset{\text{spinel}}{Mg_2SiO_4} \xrightarrow{\text{10.7 per cent}} \underset{\text{perovskite}}{MgSiO_3} + \underset{\text{periclase}}{MgO}$$

with increasing pressure, where the percentages represent the density increase for each transition at zero pressure. The oxygen coordination of silicon does not change in the above phase transformations to spinel. It changes from 4 to 6 when spinel disproportionates into the assemblage perovskite plus periclase. The coordination number of half of Mg^{2+} also increases from 6 to 8 in the latter disproportionation reaction.

2.2. The $MgSiO_3$–$FeSiO_3$ System

Ringwood and Major (1966*b*) found that $FeSiO_3$ (pyroxene) disproportionated into a mixture of Fe_2SiO_4 (spinel) plus SiO_2 (stishovite) at about 130 kbar and at 1000 °C. The same reaction has been observed at higher pressures in compositions containing at least 75 mol per cent, and probably 85 mol per cent $MgSiO_3$ by Ringwood and Major (1966*c*), who also found that $MgSiO_3$ (enstatite) is stable to pressures of at least 200 kbar. Ito *et al.* (1972) reported that at about 1000 °C $MgSiO_3$ (enstatite) disproportionated into an assemblage of Mg_2SiO_4 (β-phase) plus SiO_2 (stishovite) at about 250 kbar,† and into Mg_2SiO_4 (spinel) plus SiO_2 (stishovite) at about 280 kbar.† A hexagonal form of $MgSiO_3$ was synthesized at about 500 kbar† by Kawai *et al.* (1974). Ming and Bassett (1975*b*) carried out experimental studies of phase transformations in this system at pressures up to 250 kbar and at 1500 ± 200 °C. They concluded that the assemblage (Mg, Fe)$_2SiO_4$ (spinel) + SiO_2 (stishovite) of this system disproportionated directly into the constituent oxide mixture with the exception of compositions containing more than 60 mol per cent of $MgSiO_3$, where the hexagonal phases of (Mg, Fe)SiO_3 intervened.

Detailed investigations of high-pressure phase transformations in pure $FeSiO_3$ and $MgSiO_3$ were made by Liu (1976*c*, *d*). The results for pure $FeSiO_3$ (Liu, 1976*c*) were generally in accordance with those reported by Ringwood and Major (1966*b*) and Ming and Bassett (1975*b*). In pure

† These pressures have been overestimated by these authors. For details refer to Liu (1976*d*).

$MgSiO_3$, however, Liu (1976*d*) produced data superior to those reported by Ito *et al.* (1972), confirming the assemblage Mg_2SiO_4 (β-phase)+SiO_2 (stishovite), and data superior to those reported by Kawai *et al.* (1974), confirming the hexagonal form of $MgSiO_3$. Liu (1976*d*, 1977*a*) showed that the hexagonal form reported by Kawai *et al.* (1974) was an ordered ilmenite structure. Moreover, Liu (1976*d*) also found that the ilmenite phase of $MgSiO_3$ transformed directly to an orthorhombic perovskite structure without the intervention of a mixed-oxide assemblage. The perovskite structure had been first synthesized by Liu (1975*a*), confirming that $MgSiO_3$ adopted the perovskite structure at very high pressures, as first suggested by Ringwood (1962). Bassett has since revised his earlier beliefs to agree with Liu that the intervening mixed oxide assemblage does not exist (personal communication, 1976). The absence of a mixed-oxide field was also later confirmed by Ito (1977). This result is consistent with the post-spinel phase transformation in Mg_2SiO_4 described in the previous section. Thus, the experimental results of phase transformations with enstatite are as follows:

$$\underset{\text{clinoenstatite}}{MgSiO_3} \xrightarrow{\text{15.2 per cent}} \tfrac{1}{2}\underset{\beta\text{-phase}}{Mg_2SiO_4} + \tfrac{1}{2}\underset{\text{stishovite}}{SiO_2} \xrightarrow{\text{1.8 per cent}} \tfrac{1}{2}\underset{\text{spinel}}{Mg_2SiO_4}$$

$$+\tfrac{1}{2}\underset{\text{stishovite}}{SiO_2} \xrightarrow{\text{2.0 per cent}} \underset{\text{ilmenite}}{MgSiO_3} \xrightarrow{\text{7.0 per cent}} \underset{\text{perovskite}}{MgSiO_3}$$

with increasing pressure, where the percentages are the increase of the zero-pressure density from one to the other. The coordination number of half of Si^{4+} increases from 4 to 6 in the disproportionation of clinoenstatite to β-phase plus stishovite, the other half increases to 6 when spinel combines with stishovite to form the ilmenite. The coordination number of silicon does not change in the ilmenite to perovskite transition, but that of Mg^{2+} increases from 6 to 8 accompanying the transition. A schematic isothermal phase diagram for the system $MgO–SiO_2$ at about 1000 °C is shown in Fig. 3.

2.3 The $Mg_3Al_2Si_3O_{12}$–$Fe_3Al_2Si_3O_{12}$ System

Two pure end-members and two naturally occurring garnets with intermediate composition of this system have been studied by Liu (1975*b*). The iron-rich portions of the system were found to transform to a griphite-type structure at a loading pressure of approximately 120 kbar and at 1500 °C (Liu, 1975*c*). This type of phase transformation is accompanied by a 1–2 per cent increase in the zero-pressure density. Griphites further disproportionated into a mixture of wüstite (FeO) plus an as yet unidentified phase (or phases) at about 155 kbar. The latter phase (or phases) may transform

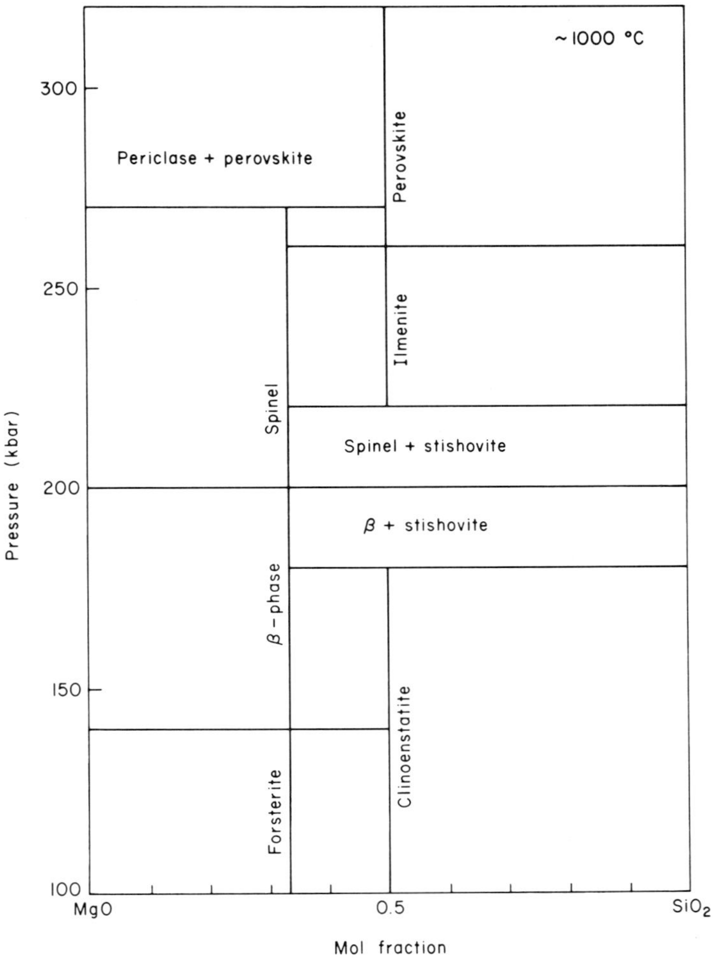

Fig. 3. Schematic isothermal phase diagram for the system $MgO–SiO_2$. (After Liu, 1976*a*.)

further to a still unknown phase (or phases) at pressures greater than 180 kbar (Liu, 1975*c*).

By analogy to the systems of olivine and pyroxene, phase transformations in the magnesium-rich portions of this system are completely different from those in the iron-rich portion. Pure $Mg_3Al_2Si_3O_{12}$ (pyrope garnet) was found to transform directly to the stoichiometric ilmenite structure at a loading pressure between 240 and 250 kbar and at about 1000 °C (Liu, 1977*b*). Ilmenite-type $Mg_3Al_2Si_3O_{12}$ further transformed to an orthorhombic perovskite modification at pressures greater than about 300 kbar

(Liu, 1977*b*,*c*). Natural and synthetic pyrope garnets were the first pure silicates found to transform to the perovskite structure (Liu, 1974). High-pressure phase transformations in this system, which are relevant to the Earth's mantle, occur as follows with increasing pressure:

$$\underset{\text{garnet}}{Mg_3Al_2Si_3O_{12}} \xrightarrow{7.9 \text{ per cent}} \underset{\text{ilmenite}}{3(Mg_{0.75}Al_{0.25})(Si_{0.75}Al_{0.25})O_3}$$

$$\xrightarrow{7.7 \text{ per cent}} \underset{\text{perovskite}}{3(Mg_{0.75}Al_{0.25})(Si_{0.75}Al_{0.25})O_3},$$

where the percentages are the same as in the previous systems. The fourfold coordinated Si^{4+} and the eightfold coordinated Mg^{2+} in garnet all occupy the sixfold coordinated sites in ilmenite. The coordination numbers of Mg^{2+} and half of Al^{3+} increase from 6 to 8 in the ilmenite–perovskite transition.

2.4. The $MgSiO_3$–$Mg_3Al_2Si_3O_{12}$ System

Pyroxene and garnet are identical in stoichiometry. Hence pyrope ($Mg_3Al_2Si_3O_{12}$) can simply be regarded as an intermediate composition between $MgSiO_3$ and Al_2O_3. The Al_2O_3 content in the system $MgSiO_3$–$Mg_3Al_2Si_3O_{12}$ is everywhere less than or equal to 25 mol per cent or 25.3 weight per cent.

This system was first studied by Boyd and England (1964) at pressures ≤50 kbar at 1100 and 1600 °C, and at somewhat higher pressures (≤60 kbar) by Boyd (1970). At 30 kbar, their results show that almost 5 weight per cent Al_2O_3 dissolves in enstatite ($MgSiO_3$) at 1100 °C, and the solubility increases to about 16 weight per cent at 1650 °C. At higher pressures and at constant temperature (about 1000 °C), these aluminous enstatites progressively disproportionated into mixtures of pyrope plus an alumina-poor aluminous enstatite, i.e.

$$3\,MgSiO_3 \cdot xAl_2O_3 \rightarrow yMg_3Al_2Si_3O_{12} + 3\,(1-y)Mg_3SiO_3 \cdot (x-y)Al_2O_3,$$

where $1 > x > y > 0$. The same reaction has also been confirmed by MacGregor and Ringwood (1964) using natural enstatite and pyrope. Boyd and England (1964) concluded that there is no detectable solid solution of excess $MgSiO_3$ in pyrope. Ringwood (1967) reported crystallization experiments on glasses of 95 per cent $MgSiO_3 \cdot 5$ per cent Al_2O_3 and 90 per cent $MgSiO_3 \cdot 10$ per cent Al_2O_3 compositions at pressures between 25 and 250 kbar at about 1000 °C. At pressures above about 100 kbar, a dramatic combination of pyrope and enstatite to form a garnet structure was indicated, with the amount of garnet increasing with pressure until, at pressures greater than 150 kbar, a single-phase homogeneous garnet of composition 90 per cent $MgSiO_3 \cdot$ 10 per cent Al_2O_3 was synthesized. Some recent data of Akaogi and Akimoto (1977) indicated that about 10 mol per cent of $MgSiO_3$ dissolves in pyrope at pressures between 40 and 70 kbar at 1000 °C and more than 60 mol per cent of $MgSiO_3$ dissolves in pyrope at pressures greater than 180 kbar and at 1000 °C. Discoveries of garnets of $(Mg_{0.75}Fe_{0.25})SiO_3$ with a few per cent of alumina and other minor components, majorites, were reported by Smith and Mason (1970) in the Coorara chondrite and by Binns *et al.* (1969) in the Tenham chondrite.

Further phase transformations in this system were studied by Liu (1977*c*), who found that the garnet field did not extend to pure $MgSiO_3$ and that the phase assemblages found for pure $MgSiO_3$, β-Mg_2SiO_4 + stishovite and Mg_2SiO_4(spinel) + stishovite, did not extend to compositions beyond 95 per cent $MgSiO_3 \cdot$ 5 per cent Al_2O_3. Moreover, complete solid solutions with the ilmenite and perovskite structures were found for the entire system (Liu, 1977*a,c*). It is generally accepted that the Earth's mantle contains about 4 weight per cent Al_2O_3. Thus, experimental results on high-pressure phase transformations of this system relevant to the Earth's mantle would be:

$$\text{pyroxene} \rightarrow (\text{pyroxene} + \text{garnet}) \rightarrow \text{garnet} \rightarrow (\text{garnet} + \text{ilmenite}) \rightarrow \text{ilmenite} \rightarrow (\text{ilmenite} + \text{perovskite}) \rightarrow \text{perovskite}$$

with increasing pressure. It is obvious that phase transformations as illustrated by this system would not be responsible for the abrupt seismic velocity discontinuity observed in the mantle if the two-phase regions of the above phase transformations are wide. Moreover, high-pressure phase transformations revealed in pure $MgSiO_3$ and $Mg_3Al_2Si_3O_{12}$ would lose their geophysical significances (see Fig. 4).

2.5. Miscellaneous Systems

The phase transformations reviewed so far are capable of accounting for phase changes involving SiO_2, MgO, FeO and Al_2O_3 in the Earth's mantle. The behaviour of other chemical components such as CaO, Na_2O and K_2O, which were not accounted for in the discussion of the above systems, is therefore discussed below.

Calcium-bearing Silicates

Calcium is expected to combine with silica to form calcium silicates. Anorthite ($CaAl_2Si_2O_8$) is known to disproportionate into a mixture of quartz, kyanite and grossularite (Boyd and England, 1961). The same assemblage, with the exception of stishovite instead of quartz, was found to occur at pressures of about 150 kbar (Ringwood, 1970). It was proposed that at still

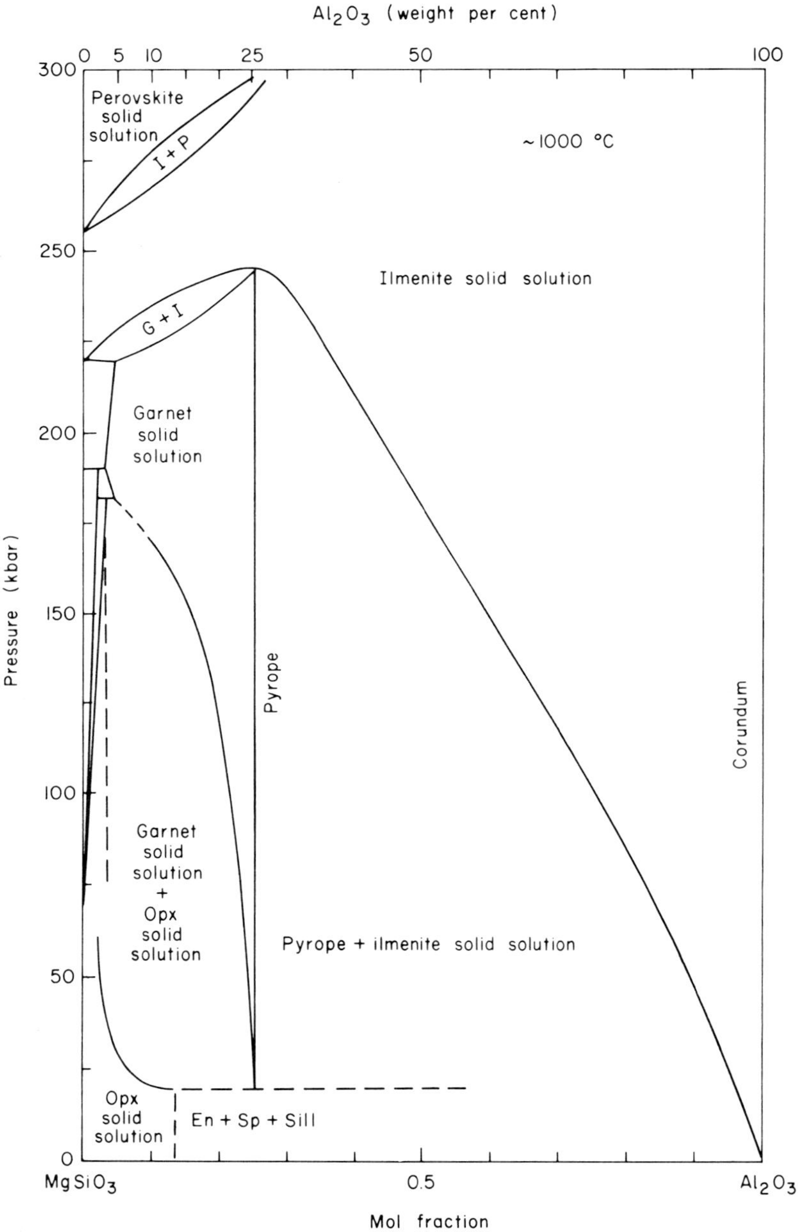

Fig. 4. Schematic isothermal phase diagram for the system $MgSiO_3$–Al_2O_3. En = $MgSiO_3$ (enstatite); Sp = $MgAl_2O_4$ (spinel); Sill = Al_2SiO_5 (sillimanite); Opx = orthopyroxene; G = garnet; I = ilmenite; P = perovskite. (After Liu, 1977*c*.)

higher pressures grossularite ($Ca_3Al_2Si_3O_{12}$) disproportionated into a mixture of perovskite plus corundum, by analogy with a phase transformation in germanium anorthite ($CaAl_2Ge_2O_8$) (Ringwood, 1970). Although transformation of anorthite to a hollandite structure was postulated by Ahrens *et al.* (1969) and Ringwood (1970), the preliminary results of static experimental study of Liu (unpublished) do not substantiate this idea.

It has been proposed that diopside ($CaMgSi_2O_6$) disproportionates into a mixture of garnet plus ilmenite by analogy with germanates (Ringwood and Seabrook, 1963); but no direct confirmation has been reported yet. Diopside is stable to pressures of at least 180 kbar (Liu, unpublished data). Aluminous diopside ($CaMgSi_2O_6 \cdot 10$ per cent Al_2O_3) was found to transform to the garnet structure at about 150 kbar (Ringwood and Major, 1966*c*). Up to 50 mol per cent of diopside was found to dissolve in garnet $(CaMg)_3Al_2Si_3O_{12}$ at about 150 kbar (Ringwood, 1970).

All of the above calcium-bearing silicates might be expected ultimately to transform to phase assemblages containing $CaSiO_3$ in the perovskite structure, as suggested by studies of the stabilities of $CaGeO_3$–$CaSiO_3$ and $CaTiO_3$–$CaSiO_3$ perovskite-type solid solutions by Ringwood and Major (1967). They concluded that they had probably synthesized a perovskite modification of pure $CaSiO_3$ at pressures exceeding 100 kbar, but that it had been impossible to retain it in the high-pressure state. It was subsequently confirmed by Liu and Ringwood (1975) that $CaSiO_3$ crystallized with the cubic perovskite structure at pressures greater than 160 kbar, using *in situ* X-ray diffraction techniques. The perovskite modification of $CaSiO_3$ is about 9 per cent denser than its constituent oxide mixture.

Sodium-bearing Silicates

Albite ($NaAlSi_3O_8$), jadeite ($NaAlSi_2O_6$) and nepheline ($NaAlSiO_4$) are with little doubt the three most important sodium-bearing rock-forming minerals. Detailed high-pressure phase transformations in these three minerals have been reviewed and studied by Liu (1978), who concluded that albite and nepheline are the stable mineral phases in the system $NaAlO_2$–SiO_2 at pressures below about 25 kbar. In the pressure region between 25 and 200 kbar, jadeite is the main stable phase. The calcium ferrite-type $NaAlSiO_4$ (Liu, 1977*d*), jadeite and $NaAlSi_3O_8$ hollandite (Liu, 1978) are all stable in the pressure region between 200 and 240 kbar. The calcium ferrite-type $NaAlSiO_4$ is the only known sodium-bearing aluminosilicate which is stable at pressures greater than 240 kbar. These results all parallel the suggestions made by Ringwood (1970) based on germanium analogues.

Potassium-bearing Silicates

Orthoclase ($KAlSi_3O_8$), leucite ($KAlSi_2O_6$) and kalsilite ($KAlSiO_4$) are potentially important potassium-bearing silicates in the mantle. Ringwood *et al.* (1967) succeeded in transforming natural orthoclase to the hollandite structure at about 120 kbar and at 1000 °C. However, orthoclase has since been found to first disproportionate into a mixture of a wadeite-type phase $K_2Si_4O_9$ + kyanite + coesite, in the pressure region between 50 and 100 kbar, before transforming to hollandite (Kinomura *et al.*, 1977). Hollandite $KAlSi_3O_8$ is stable to pressures up to at least 300 kbar at about 1000 °C (Liu, unpublished data). Leucite disproportionated into hollandite $KAlSi_3O_8$ plus a second phase believed to be a high-pressure form of $KAlO_2$ by Ringwood *et al.* (1967). Similar transformations have also been found in kalsilite (Liu, unpublished data).

3. The Constitution of the Mantle

Recent velocity–depth distributions in the mantle (see Fig. 1, Chapter 1 in this volume) show that the velocity increase in the transition zone is concentrated in two and/or three regions marked by seismic discontinuities at around 400, 650 and 800 km. The latter discontinuity is minor and not firmly established. The stepwise characteristic of the velocity distributions are best explained in terms of phase transformations displayed by mantle minerals.

3.1. The Olivine Mantle

To a first approximation, the Earth's mantle is assumed to be made up of pure olivine by many investigators (e.g. Anderson, 1968; Graham, 1970; Liu, 1973; Burdick and Anderson, 1975). This is because olivine is probably the most abundant mineral in the upper mantle. On the basis of a schematic average, Liu (1977*e*) estimated the isothermal compression curves for the various high-pressure phases or phase assemblages of forsterite described in the previous section. He further assumed that a small amount of replacement of Mg by Fe (~10 mol per cent) in these mineral phases had little effect on the elastic behaviour of the pure magnesian silicates. He then compared the calculated density–pressure curves of olivine and its various high pressure modifications with the three recent density profiles of the mantle (Jordan and Anderson, 1974; Dziewonski *et al.*, 1975; Anderson and Hart, 1976). This has been shown in Fig. 5, which indicates that the upper mantle (above *c.* 400 km) is in accordance with a composition of $(Mg_{0.97}Fe_{0.03})_2SiO_4$ at room temperature; that the general features of the transition zone (*c.* 400–650 km) can be attributed to changes of mineral phases in olivine only; and that the lower mantle (below *c.* 650 km) is consistent with a composition of $(Mg_{0.94}Fe_{0.06})_2SiO_4$ at room temperature. Corrections for the temperature rise in these regions would require a compensation increase in the

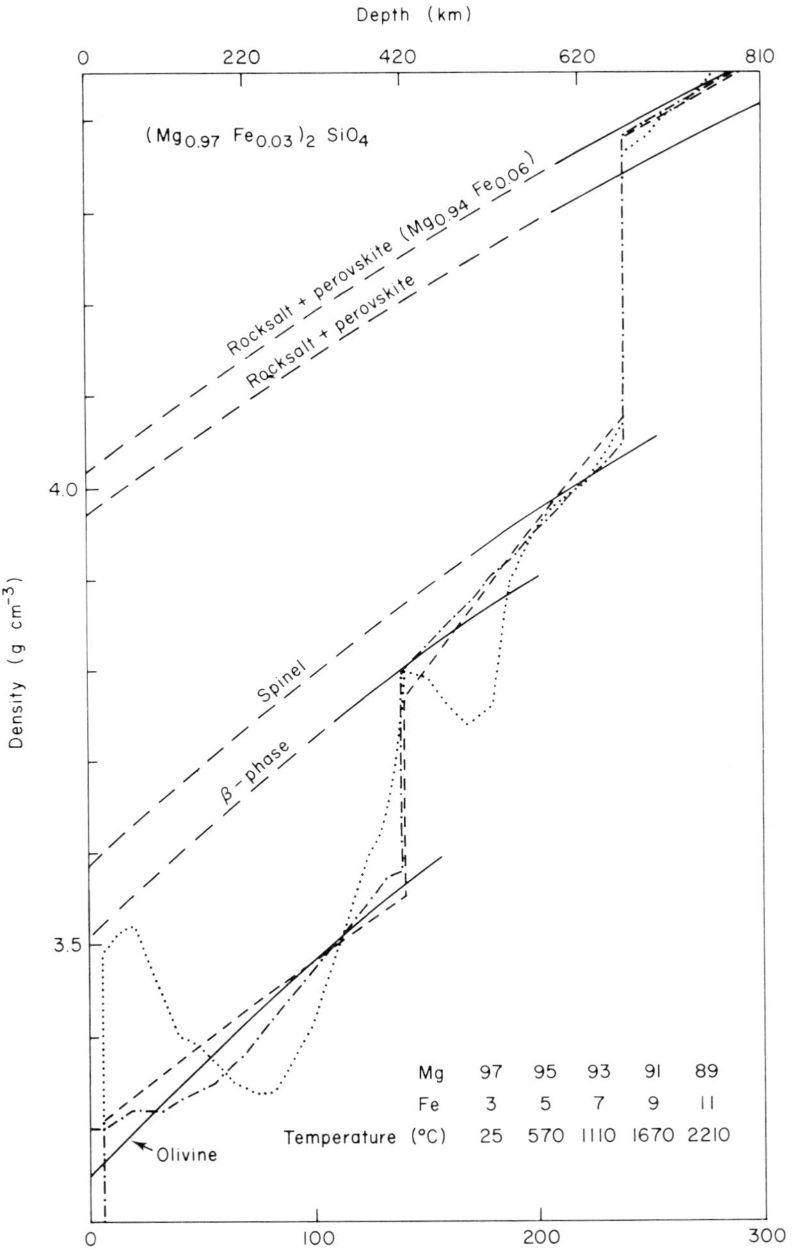

Fig. 5. Comparison of the calculated room temperature density–pressure (or depth) curves for phases of olivine, β, spinel, and the rocksalt phase plus perovskite mixture possessing an olivine stoichiometry of $(Mg_{0.97}Fe_{0.03})_2SiO_4$ with the recent density profiles of the Earth's mantle. Density–pressure curve for the mixture rocksalt plus perovskite with a stoichiometry of $(Mg_{0.94}Fe_{0.06})_2$ SiO_4 is also displayed. Variation of the FeO/(MgO + FeO) ratio of an olivine upper mantle at *c.* 400 km with temperature is also listed. (After Liu, 1977*e.*) —·—, Jordan and Anderson (1974); ––––, Dziewonski *et al.* (1975); ······, Anderson and Hart (1976).

FeO/(MgO + FeO) ratio for olivine and its various high-pressure phases. It has been shown in Fig. 5 that at *c.* 400 km an increase of temperature from 25 °C to about 2000 °C requires a change in composition of olivine from $(Mg_{0.97}Fe_{0.03})_2SiO_4$ to about $(Mg_{0.9}Fe_{0.1})_2SiO_4$. Temperatures of 2000 °C and compositions of $(Mg_{0.9}Fe_{0.1})_2SiO_4$ at *c.* 400 km are consistent with data derived from other geophysical and geochemical constraints. If composition changes due to temperature corrections at depths above *c.* 400 km are comparable to those at depths below 650 km, the model olivine mantle illustrated in Fig. 5 requires an increase of iron content by about 3 per cent, going from the upper mantle to the lower mantle. This conclusion parallels those arrived at earlier by Anderson (1968). However, a 3 per cent increase in iron content is too small to be significant when the uncertainties of estimating the various compression curves are taken into consideration. The olivine mantle is an oversimplified model and conclusions thus reached should be regarded as tentative. Pyroxene, at least, should be incorporated in the model mantle mineralogy.

3.2. The Olivine–Pyroxene Mantle

Next to olivine, pyroxene is the second major mineral constituent of the upper mantle. Using an arbitrary upper mantle with 60 mol per cent olivine plus 40 mol per cent pyroxene as a model, Liu (1977*e*) demonstrated (see Fig. 6) that the recent density profiles in the upper mantle require an FeO/(MgO + FeO) ratio of about 0.03 at room temperature. At about 400 km, temperature corrections for about 2000 °C would bring this ratio to about 0.1. The general features of the transition zone can also be represented by phase transformations in both olivine and pyroxene, as described in the previous sections. These produce mineral assemblages of β-phase plus clinopyroxene, β-phase plus stishovite, spinel phase plus stishovite, spinel plus ilmenite phases, spinel plus perovskite phases with increasing depth in the transition zone. Finally, the lower mantle is composed of the mixture perovskite plus rocksalt phases. Most importantly, this model does not require an increase in the iron content of the lower mantle, in contrast to the case of a pure olivine model mantle. The latter conclusion that the iron content of the whole mantle is probable to remain unchanged has also been reached earlier by Ringwood (1970, 1975) from a different approach. The match between the perovskite + rocksalt curve and the density profiles at depths greater than 650 km in Fig. 6 would improve if the relative amount of pyroxene were increased in the olivine–pyroxene mantle.

As was mentioned earlier, any plausible compositional model of the mantle contains about 4 weight per cent of Al_2O_3. This quantity of Al_2O_3 would drastically modify the mineral assemblages discussed above.

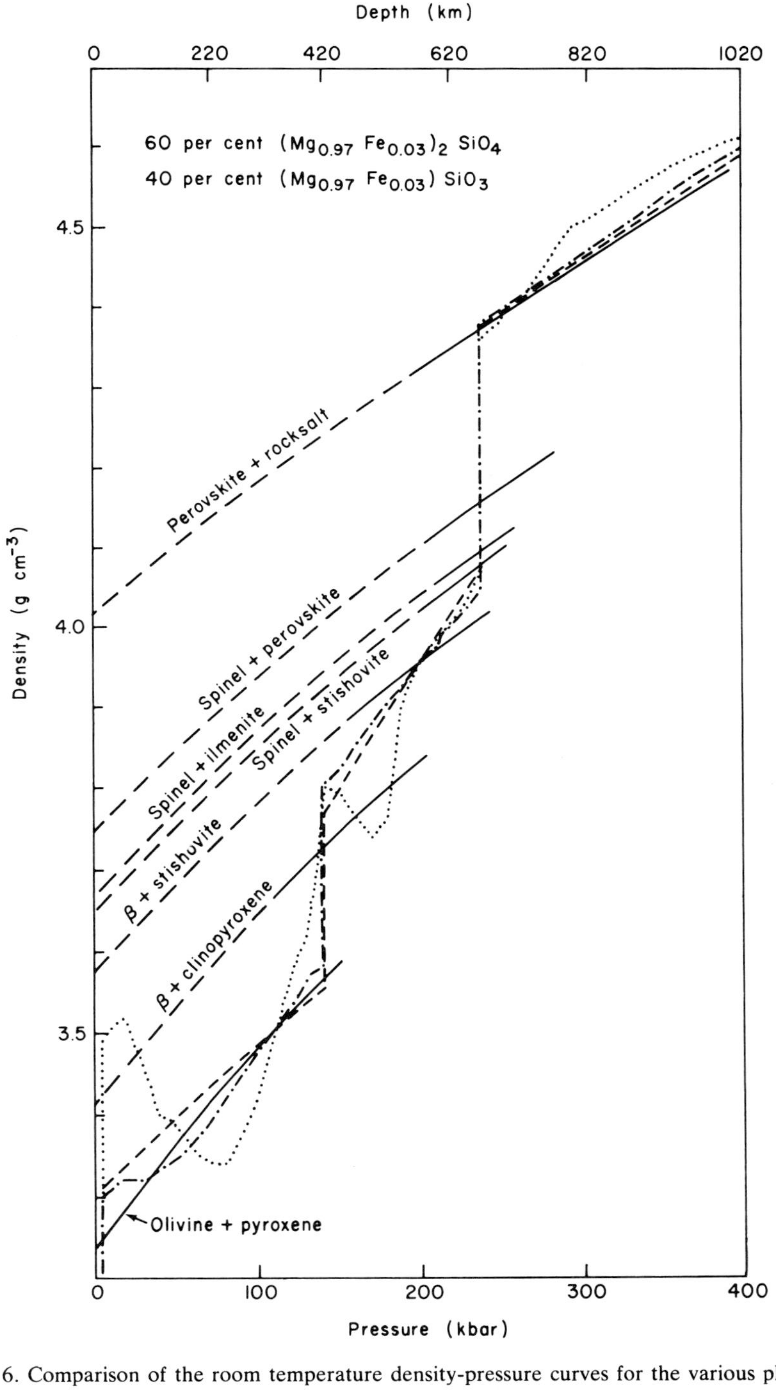

Fig. 6. Comparison of the room temperature density-pressure curves for the various phase assemblages of a mixture 60 mol per cent olivine plus 40 mol per cent pyroxene possessing the ratio FeO/(MgO + FeO) = 0.03 with the recent density profiles of the Earth's mantle above *c*. 1000 km. (After Liu, 1977*e*.) Legend in Fig. 5.

3.3. The Olivine–Pyroxene–Garnet Mantle

Liu (1977*c*) has proposed a mineralogical model of the mantle starting with a mixture of forsterite and aluminous pyroxene in 1 : 1 molecular ratio. On the basis of that approach, a similar model is now proposed for the upper mantle with the FeO/(MgO+FeO) ratio being 0.1, i.e. a mixture with olivine $(Mg_{0.9}Fe_{0.1})_2SiO_4$ and orthopyroxene 90 per cent $(Mg_{0.9}Fe_{0.1})SiO_3 \cdot$ 10 per cent Al_2O_3 in 1 : 1 molecular ratio and assuming no interaction between them. The composition of this model is equivalent (in weight per cent) to:

SiO_2	45.6
MgO	42.0
FeO	8.3
Al_2O_3	4.1
	100.0

Note that this composition is very similar to that of the "pyrolite" composition listed in Table 1 in Chapter 1 of this volume, which can be regrouped as follows:

$\sum SiO_2 + TiO_2$	45.3
$\sum MgO + CaO$	41.2
$\sum FeO + MnO$	8.2
$\sum Al_2O_3 + Cr_2O_3$	3.7
$\sum Na_2O + K_2O$	0.4
	98.8

Hence, the present model can be regarded as a simplified "pyrolite" mantle.

Figure 7 shows the effect of phase transformations on the zero-pressure density of an aluminous orthopyroxene 90 per cent $(Mg_{0.9}Fe_{0.1})SiO_3 \cdot$ 10 per cent Al_2O_3 (denoted by line 1), of an olivine $(Mg_{0.9}Fe_{0.1})_2SiO_4$ (denoted by line 2), and of their mixture in 1 : 1 ratio (denoted by the bold lines). Both the pressure and the depth scales are displayed only for reference. The effects of compression and thermal expansion on the densities are not included. This model begins with a mixture of olivine plus pyroxene at depths above 70 km, which might be called the enstatite–peridotite "crust" of the model. The "upper mantle" (70 ~ 420 km) is composed of olivine, pyroxene and garnet (garnet–peridotite). The assemblage pyroxene plus garnet is equivalent to eclogite. The relative ratio of pyroxene to garnet decreases with increasing depth in this region. The 420 km discontinuity is

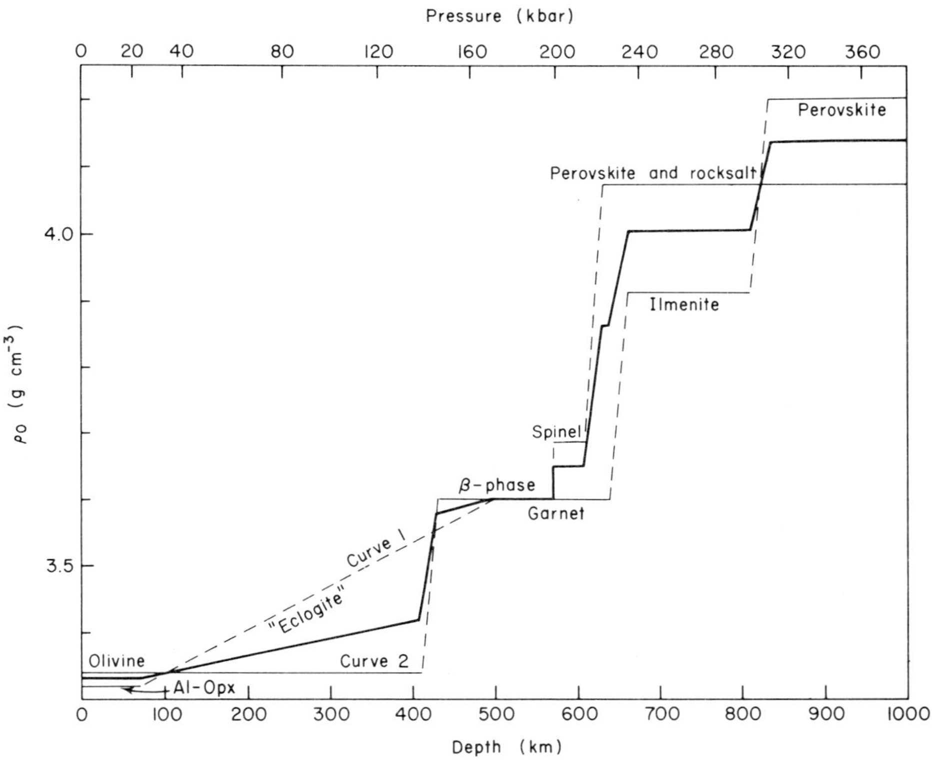

Fig. 7. Effects of phase transformations upon densities of aluminous pyroxene (90 per cent $(Mg_{0.9}Fe_{0.1})SiO_3 \cdot$ 10 per cent Al_2O_3) (curve 1); olivine ($(Mg_{0.9}Fe_{0.1})_2SiO_4$) (curve 2); and their 1 : 1 molecule mixture (the bold lines). Pressure and depth scales are shown only for reference. Effects of compression and thermal expansion on each phase are not included.

due mainly to the olivine to β-phase transition and is accompanied by a 4.7 per cent increase in the zero-pressure density. The region between 420 and 500 km is dominated by β-phase and garnet with a minor amount of pyroxene. Mixtures of β-phase plus garnet and of spinel plus garnet are present in the regions between 500 and 570 km and between 570 and 620 km, respectively. The small transition at 570 km is accompanied by a 1.4 per cent increase in the zero-pressure density which could correspond to the P-wave velocity jump of about 0.3 km s^{-1} around 550 km, recognized by Helmberger and Wiggins (1971). Hence, the "transition zone" of the model mantle is primarily composed of β-phase (or spinel) and garnet. Within a small depth region, say between 610 and 660 km, both spinel and garnet, respectively, transform into a mixture of perovskite plus rocksalt and into ilmenite. The overall increase in the zero-pressure density in this region is 9.8 per cent. Both transitions are characterized by an increase in the oxygen coordination of Si^{4+} from 4 to 6. The coordination number of half of the $(Mg_{0.9}Fe_{0.1})^{2+}$ ions increases from 6 to 8 in the spinel–postspinel transition, but decreases from 8 to 6 in the garnet–ilmenite transition. Hence, the 650 km discontinuity of the Earth's mantle may well correspond to the coordination change in Si^{4+}. This conclusion certainly seriously conflicts with Anderson's (1976) recent interpretation of the 650 km discontinuity. He proposed that this discontinuity is primarily due to phase transitions in pyroxenes and garnets, and *not* in olivines, and he based his conclusions on the *hypothetical* bulk moduli of various silicate and oxide solid solutions.

In the present model, the mineral assemblage at depths below 650 km is probably perovskite, rocksalt and ilmenite. A further discontinuity in the Earth's mantle is suggested by the present model. This is the ilmenite–perovskite transition, observed in aluminous pyroxene, which is arbitrarily shown at the depth of about 820 km in Fig. 7. The zero-pressure density increase resulting from the ilmenite–perovskite transition in the present model is 3.3 per cent. An increase of about 2 per cent in P-wave velocity at a depth of about 770 km has recently been observed by Datt and Muirhead (1976). Discontinuities near this depth (and deeper) have previously been observed (e.g. Whitcomb and Anderson, 1970). Finally, the "lower mantle" of the model is composed of perovskite and rocksalt phases.

3.4. The "Pyrolite" Mantle

Although the primary chemical composition of the above olivine–pyroxene–garnet mantle is very similar to the "pyrolite" upper mantle discussed in Chapter 1 of this volume, the "pyrolite" upper mantle is a non-specific pyroxene–olivine rock which is capable of producing basaltic magmas upon partial melting as reviewed by Ringwood (1975). In pyrolite, peridotite is interpreted as the refractory residue and basalt as a partial melting product. This implies that the composition of pyrolite lies closer to that of peridotite than to basalt.

Based on the earlier mineralogical models of the pyrolite mantle proposed by Ringwood (1970, 1975), I will now attempt to depict a deep mantle which corresponds the pyrolite composition and is in harmony with the recent observations of phase transformations described in Section 2.

The Region between Depths of 150–350 km

According to Ringwood, this region is homogeneous and the detailed mineralogy for the pyrolite composition is as follows:

Constituent	Chemical formulation	Weight per cent
Olivine (Fo_{89})	$(Mg, Fe)_2SiO_4$	57
Orthopyroxene (En_{89}?)	$(Mg, Fe)SiO_3$	17
Omphacitic pyroxene	$(Ca, Mg, Fe)_2Si_2O_6$–$NaAlSi_2O_6$	12
Pyrope garnet	$(Mg, Fe, Ca)_3(Al, Cr)_2Si_3O_{12}$	14

The density of this mineral assemblage at atmospheric pressure and room temperature is 3.38 g cm^{-3}, which is about 1.5 per cent denser than the mixture olivine plus Al-pyroxene shown in Fig. 7.

The 350–420-km Seismic Discontinuity

Two major phase transformations were suggested for this region by Ringwood:

(a) the aluminous pyroxene–garnet transition
(b) the olivine–β-phase transition.

The former involves partial sixfold coordination of silicon in the garnet structure. This transition would effectively cause only about a 2 per cent increase in the zero-pressure density for the whole mantle, but the increase is likely to spread over a relatively large depth interval owing to the effect of solid solutions. The olivine–β-phase transformation is accompanied by an overall increase in the zero-pressure density of about 4.6 per cent for the bulk pyrolite mantle.

If a discrete potassium-bearing phase should occur at depths below this, a hollandite structure of $KAlSi_3O_8$ would be expected.

The Region Between 420 and 600 km

The mineral assemblage in this region begins with 57 weight per cent β-$(Mg, Fe)_2SiO_4$ and 43 weight per cent complex garnet solid solution (Ringwood, 1975). The β-phase transforms to the spinel structure in the vicinity of 550 km. The mineral assemblages in this region are essentially the same as those shown in Fig. 7.

If excess alumina is available, the diopside portion of the omphacitic pyroxene is unstable in this region and may dissolve into the complex garnet solid solution as suggested by Ringwood (1975). Otherwise, it transforms to a phase (or phases) which has not yet been identified in this region. The cubic perovskite phase of $CaSiO_3$ is stable in this region. Therefore, the calcium component of the complex garnet solid solution, and perhaps of diopside too, might occur as the perovskite form of $CaSiO_3$.

The 650-km Discontinuity

This is the most prominent seismic discontinuity occurring in the mantle. For a pyrolite mantle model, this discontinuity should be due mainly to the combination of phase changes from spinel to the mixture of perovskite plus rocksalt in the olivine component and from garnet to ilmenite in the pyroxene–garnet system, as shown previously in Fig. 7. The overall increase in the zero-pressure density may amount to nearly 9.5 per cent within a depth interval of a few tens of kilometres.

Minor phase transformations such as the appearance of the $NaAlSi_3O_8$ hollandite and the disproportionation reaction of jadeite or omphacitic pyroxene to a phase assemblage containing the calcium ferrite structure would contribute about less than 1 per cent increase in the zero-pressure density of the whole pyrolite mantle.

Below 700 km

If phase transformations at the 650-km discontinuity are complete by about 700 km, the mineralogy below this depth would consist of $(Mg, Fe)SiO_3$ (orthorhombic perovskite), (Mg, Fe)O, ilmenite-type solid solution of $(Mg, Fe)SiO_3$–Al_2O_3, cubic perovskite $CaSiO_3$, and calcium ferrite $NaAlSiO_4$. The last two components would attribute less than 10 per cent to the total mass. The aluminous ilmenite-type solid solution transforms to the orthorhombic perovskite structure at still higher pressure. This transition may correspond to the minor seismic discontinuity in the vicinity of 800 km. The transition itself involves a 7.4 per cent increase in the zero-pressure density, hence it may cause a 2 per cent density jump for the whole pyrolite mantle at the discontinuity.

Following the present discussion, there is very little doubt that the Earth's deep mantle is composed mainly of the orthorhombic perovskite phase of $(Mg, Fe, Al) (Al, Si)O_3$, which may amount to more than 70 per cent of the total mass of the lower mantle. The rocksalt phase of (Mg, Fe)O is probably the second major mineral constituent (~20 per cent) of the lower mantle. Phase transformations in both olivine and pyroxene to a phase (or assemblage) intrinsically denser than the orthorhombic perovskite structure was also inferred by shock-wave data (McQueen and Marsh, 1966; Jeanloz and Ahrens, 1977). Hence, further transitions to phases slightly denser than the orthorhombic perovskite structure are not completely improbable in the deep mantle, but confirmation from studies by static experiments is necessary in the future.

In conclusion, several mantle models are presented in which the phase transformations, previously determined in static high-pressure experiments, can readily provide excellent qualitative explanations of the principal features of the seismic velocity distribution in the Earth's mantle.

Acknowledgement

I am grateful to S. Kesson and A. E. Ringwood for reading and commenting on the manuscript.

References

Ahrens, T. J., Anderson, D. L. and Ringwood, A. E. (1969). Equations of state and crystal structures of high-pressure phases of shocked silicates and oxides. *Rev. Geophys.* **7**, 667–701.

Akaogi, M. and Akimoto, S. (1977). Pyroxene–garnet solid solution equilibria in the systems $Mg_4Si_4O_{12}$–$Mg_3Al_2Si_3O_{12}$ and $Fe_4Si_4O_{12}$–$Fe_3Al_2Si_3O_{12}$. *Phys. Earth planet. Interiors* **15**, 90–106.

Akimoto, S. and Fujisawa, H. (1966). Olivine–spinel transition in system Mg_2SiO_4–Fe_2SiO_4 at 800 °C. *Earth planet. Sci. Lett.* **1**, 237–240.

Akimoto, S. and Fujisawa, H. (1968). Olivine–spinel solid solution equilibria in the system Mg_2SiO_4–Fe_2SiO_4. *J. geophys. Res.* **73**, 1467–1479.

Anderson, D. L. (1968). Chemical inhomogeneity of the mantle. *Earth planet. Sci. Lett.* **5**, 89–94.

Anderson, D. L. (1976). The 650 km mantle discontinuity. *Geophys. Res. Lett.* **3**, 347–349.

Anderson, D. L. and Hart, R. S. (1976). An earth model based on free oscillations and body waves. *J. geophys. Res.* **81**, 1461–1475.

Bassett, W. A. and Ming, L. (1972). Disproportionation of Fe_2SiO_4 to $2FeO + SiO_2$ at pressures up to 250 kbar and temperatures up to 3000 °C. *Phys. Earth planet. Interiors* **6**, 154–160.

Bassett, W. A. and Takahashi, T. (1970). Disproportionation of Fe_2SiO_4 to $2FeO +$ SiO_2 at high pressure and high temperature (abstract). *Trans. Am. geophys. Union* **51**, 828.

Bernal, J. D. (1936). Discussion. *Observatory* **59**, 268.

Binns, R. A., Davis, R. J. and Reed, S. B. J. (1969). Ringwoodite, natural (Mg, Fe)$_2$SiO$_4$ spinel in the Tenham meteorite. *Nature, Lond.* **221**, 943–944.

Birch, F. (1952). Elasticity and constitution of the Earth's interior. *J. geophys. Res.* **57**, 227–286.

Boyd, F. R. (1970). Garnet periodotites and the system $CaSiO_3$–$MgSiO_3$–Al_2O_3. *Min. Soc. Am. spec. Paper* **3**, 63–75.

Boyd, F. R. and England, J. L. (1961). Melting of silicates at high pressures. *Carnegie Inst. Washington Yearbook* **60**, 113–125.

Boyd, F. R. and England, J. L. (1964). The system enstatite-pyrope. *Carnegie Inst. Washington Yearbook* **63**, 157–161.

Bullen, K. E. (1936). The variation of density and the ellipticities of strata of equal density within the earth. *Mon. Not. R. astronom. Soc., Geophys. Suppl.* **3**, 395–401.

Burdick, L. J. and Anderson, D. L. (1975). Interpretation of velocity profiles of the mantle. *J. geophys. Res.* **80**, 1070–1074.

Chao, E. C. T., Fahey, J. J., Littler, J. and Milton, D. J. (1962). Stishovite, SiO_2, a very high pressure new mineral from Meteor crater, Arizona. *J. geophys. Res.* **67**, 419–421.

Datt, R. and Muirhead, K. J. (1976). Evidence for a sharp increase in P-wave velocity at about 770 km depth. *Phys. Earth planet. Interiors* **13**, 37–46.

Dziewonski, A. M., Hales, A. L. and Lapwood, E. R. (1975). Parametrically simple Earth models consistent with geophysical data. *Phys. Earth planet. Interiors* **10**, 12–48.

Goldschmidt, V. M. (1931). Zur Kristallchemie des Germaniums. *Nach. Akad. wiss. Göttingen, math. phys. Kl. 1*, No. 2, 184–190.

Graham, E. K. (1970). Elasticity and composition of the upper mantle. *Geophys. J. R. astronom. Soc.* **20**, 285–302.

Helmberger, D. and Wiggins, R. A. (1971). Upper mantle structure of mid-western states. *J. geophys. Res.* **76**, 3229–3245.

Ito, E. (1977). The absence of oxide mixture in high-pressure phases of Mg-silicates. *Geophys. Res. Lett.* **4**, 72–74.

Ito, E., Matsumoto, T., Suito, K. and Kawai, N. (1972). High-pressure break-down of enstatite. *Proc. Japan Acad.* **48**, 412–415.

Ito, E., Matsui, Y., Suito, K. and Kawai, N. (1974). Synthesis of γ-Mg_2SiO_4 *Phys. Earth planet. Interiors* **8**, 342–344.

Jeanloz, R. and Ahrens, T. J. (1977). Pyroxenes and olivines: structural implications of shock-wave data for high pressure phases. In *High-Pressure Research; Applications in Geophysics* (M. H. Manghnani and S. Akimoto, eds), Academic Press, New York, pp. 439–461.

Jeffreys, J. (1937). On the materials and density of the Earth's crust. *Mon. Not. R. astronom. Soc., Geophys. Suppl.* **4**, 50–61.

Jordan, T. H. and Anderson, D. L. (1974). Earth structure from free oscillations and travel times. *Geophys. J. R. astronom. Soc.* **36**, 411–459.

Kawai, N., Tachimori, M. and Ito, E. (1974). A high-pressure hexagonal form of $MgSiO_3$. *Proc. Japan Acad.* **50**, 378–380.

Kinomura, N., Koizumi, M. and Kume, S. (1977). Crystal structures of phases produced by disproportionation of K-feldspar under pressure. In *High-Pressure Research: Applications in Geophysics* (M. H. Manghnani and S. Akimoto, eds). Academic Press, New York, pp. 183–189.

Kumazawa, M., Sawamoto, H., Ohtani, E. and Masaki, K. (1974). Postspinel phase of forsterite and evolution of the Earth's mantle. *Nature*, Lond. **247**, 356–358.

Liu, L. (1973). Chemistry of the Earth's lower mantle. *J. geophys. Res.* **77**, 4934–4937.

Liu, L. (1974). Silicate perovskite from phase transformations of pyrope–garnet at high pressure and temperature. *Geophys. Res. Lett.* **1**, 277–280.

Liu, L. (1975*a*). Post-oxide phases of forsterite and enstatite. *Geophys. Res. Lett.* **2**, 417–419.

Liu, L. (1975*b*). High pressure reconnaissance investigation in the system $Mg_3Al_2Si_3O_{12}$–$Fe_3Al_2Si_3O_{12}$. *Earth planet. Sci. Lett.* **26**, 425–433.

Liu, L. (1975*c*). Chemistry and structure of the high-pressure phase of garnets rich in almandite. *Nature, Lond.* **255**, 213–215.

Liu, L. (1976*a*). The post-spinel phase of forsterite. *Nature, Lond.* **262**, 770–772.

Liu, L. (1976*b*). Orthorhombic perovskite phases observed in olivine, pyroxene, and garnet at high pressures and temperatures. *Phys. Earth planet. Interiors* **11**, 289–298.

Liu, L. (1976*c*). The high-pressure phases of $FeSiO_3$ with implications for Fe_2SiO_4 and FeO. *Earth planet. Sci. Lett.* **33**, 101–106.

Liu, L. (1976*d*). The high-pressure phases of $MgSiO_3$. *Earth planet. Sci. Lett.* **31**, 200–208.

Liu, L. (1977*a*). Ilmenite-type solid solutions between $MgSiO_3$ and Al_2O_3 and some structural systematics among ilmenite compounds. *Geochim. cosmochim. acta* **41**, 1355–1361.
Liu, L. (1977*b*). First occurrence of the garnet–ilmenite transition in silicates. *Science, N.Y.* 195, 990–991.
Liu, L. (1977*c*). The system enstatite-pyrope at high pressures and temperatures and the mineralogy of the Earth's mantle. *Earth planet. Sci. Lett.* **36**, 237–245.
Liu, L. (1977*d*). High-pressure $NaAlSiO_4$: the first silicate calcium ferrite isotype. *Geophys. Res. Lett.* **4**, 183–186.
Liu, L. (1977*e*). Mineralogy and chemistry of the Earth's mantle above 1000 km. *Geophys. J. R. astronom. Soc.* **48**, 53–62.
Liu, L. (1978). High-pressure phase transformations of albite, jadeite and nepheline. *Earth planet. Sci. Lett.* **37**, 438–444.
Liu, L. and Ringwood, A. E. (1975). Synthesis of a perovskite-type polymorph of $CaSiO_3$. *Earth planet. Sci. Lett.* **28**, 209–211.
MacGregor, I. D. and Ringwood, A. E. (1964). The natural system enstatite-pyrope. *Carnegie Inst. Washington Yearbook* **63**, 161–163.
McQueen, R. G. and Marsh, S. P. (1966). See Birch, F., Compressibility: elastic constants. In *Handbook of Physical Constants* (S. P. Clark, Jr., ed.), *Geol. Soc. Am. Mem.* 97, 153–159.
McQueen, R. G., Marsh, S. P. and Fritz, J. N. (1967). Hugoniot equation of state of twelve rocks. *J. geophys. Res.* **72**, 4999–5036.
Mao, H. and Bell, P. M. (1971). High-pressure decompositin of spinel (Fe_2SiO_4). *Carnegie Inst. Washington Yearbook* **70**, 176–177.
Ming, L. and Bassett, W. A. (1974). Post-spinel phases in Mg_2SiO_4–Fe_2SiO_4 system up to 80% Mg_2SiO_4 (abstract). *Trans. Am. geophys. Union* **55**, 416–417.
Ming, L. and Bassett, W. A. (1975*a*). The postspinel phases in the Mg_2SiO_4–Fe_2SiO_4 system. *Science, N.Y.* **187**, 66–68.
Ming, L. and Bassett, W. A. (1975*b*). High-pressure phase transformations in the system of $MgSiO_3$–$FeSiO_3$. *Earth planet. Sci. Lett.* **27**, 85–89.
Ringwood, A. E. (1956). The olivine–spinel transition in the Earth's mantle. *Nature, Lond.* **178**, 1303–1304.
Ringwood, A. E. (1958*a*). Constitution of the mantle I: Thermodynamics of the olivine–spinel transition. *Geochim. cosmochim. acta* **13**, 303–321.
Ringwood, A. E. (1958*b*). Constitution of the mantle II: Further data on the olivine–spinel transition. *Geochim. cosmochim. acta* **15**, 18–29.
Ringwood, A. E. (1962). Mineralogical constitution of the deep mantle. *J. geophys. Res.* **67**, 4005–4010.
Ringwood, A. E. (1967). The pyroxene–garnet transformation in the Earth's mantle. *Earth planet. Sci. Lett.* **2**, 255–263.
Ringwood, A. E. (1970). Phase transformations and the constitution of the mantle. *Phys. Earth planet. Interiors* **3**, 109–155.
Ringwood, A. E. (1975). *Composition and Petrology of the Earth's Mantle.* McGraw-Hill, New York.
Ringwood, A. E. and Major, A. (1966*a*). Synthesis of Mg_2SiO_4–Fe_2SiO_4 solid solutions. *Earth planet. Sci. Lett* **1**, 241–245.
Ringwood, A. E. and Major A. (1966*b*). High pressure transformation of $FeSiO_3$ pyroxene to spinel plus stishovite. *Earth planet. Sci. Lett.* **1**, 135–136.
Ringwood, A. E. and Major, A. (1966*c*). High pressure transformations in pyroxenes. *Earth planet. Sci. Lett.* **1**, 351–357.

Ringwood, A. E. and Major, A. (1967). Some high-pressure transformations of geophysical significance. *Earth planet. Sci. Lett.* **2**, 106–110.

Ringwood, A. E. and Major, A. (1968). Apparatus for phase transformation studies at high pressures and temperatures. *Phys. Earth planet. Interiors* **1**, 164–168.

Ringwood, A. E. and Major, A. (1970). The system Mg_2SiO_4–Fe_2SiO_4 at high pressures and temperatures. *Phys. Earth planet. Interiors* **3**, 89–108.

Ringwood, A. E. and Seabrook, M. (1963). High pressure phase transformations in germanate pyroxenes and related compounds. *J.* geophys. Res. **68**, 4601–4609.

Ringwood, A. E., Reid, A. F. and Wadsley, A. D. (1967). High pressure $KAlSi_3O_8$, an aluminosilicate with 6-fold coordination. *Acta crystallogr.* **23**, 1093–1095.

Sclar, C. B. and Carrison, L. C. (1966). High pressure synthesis of a spinel on the join Mg_2SiO_4–Fe_2SiO_4 (abstract). *Trans. Am. geophys. Union* **41**, 207.

Smith, J. V. and Mason, B. (1970). Pyroxene–garnet transformation in Coorara meteorite. *Science, N.Y.* **168**, 832–833.

Stishov, S. M. and Popova, S. V. (1961). New dense polymorphic modification of silica. *Geochemistry* **10**, 923–926.

Suito, K. (1972). Phase transformations of pure Mg_2SiO_4 into a spinel structure under high pressures and temperatures. *J. Phys. Earth* **20**, 225–243.

Whitcomb, J. H. and Anderson, D. L. (1970). Reflection of P′P′ seismic waves from discontinuities in the mantle. *J. geophys. Res.* **75**, 5713–5728.

7. Elasticity of the Mantle

R. C. LIEBERMANN

Abstract

Elasticity data for a wide variety of mantle minerals or their analog compounds as a function of pressure, temperature, composition and crystal structure are used to evaluate systematic elasticity laws and to discuss the composition and mineralogy of various regions of the Earth's mantle. Velocity–density (v–ρ) relationships across polymorphic phase transformations depend critically on crystallographic details, and are not always equivalent to v–ρ changes caused by compression or expansion of a homogeneous material, or by varying composition at constant mean atomic weight $\bar{M}$. For the Earth, these data imply that offsets on Birch v–ρ diagrams for the transition zone of the mantle cannot be ascribed to changes of $\bar{M}$ due to variations in iron content between the upper mantle and the lower mantle. For the lower mantle, only phase assemblages dominated by (Mg, Fe)SiO_3 perovskite have sufficiently high densities and seismic parameters to be compatible with the Earth models. Future work in mineral physics should emphasize the single-crystal elastic properties and their pressure dependence for high-pressure phases, the elasticity of polyphase mixtures and the effect of attenuation on comparison of laboratory and seismic elasticity data.

1. Introduction

The most direct evidence about the physical state of the Earth's interior comes from seismological investigations of the variation in elastic wave velocity and density as a function of depth. This variation may be due to changes of pressure, temperature, composition or phase and one of the principal tasks of solid-state geophysics is the interpretation of such geophysical models in terms of these different variables.

A unique interpretation of the seismic velocity and density models of the Earth's mantle in terms of composition and mineralogy remains an unattained goal. Part of this ambiguity arises from the non-uniqueness and/or lack of resolution of the geophysical models and from the complexity of petrological and mineralogical models of the mantle proposed on the basis of field and laboratory studies (see discussions in Chapters 1, 5 and 6 of this volume). However, a substantial portion of the uncertainty is still due to our insufficient knowledge and understanding of the variation of elasticity with pressure, temperature, composition and phase. This is particularly true for the effect of phase transformations and the interpretation of the transition zone (350–1000 km depth) and lower mantle (1000–2885 km depth), and is particularly unfortunate since this interpretation is critical to the question of whether the upper and lower mantle are chemically homogeneous.

In this paper, we shall summarize the state of the art in laboratory studies of the elasticity of minerals and then discuss interpretations of the velocity–depth and density–depth profiles for the upper mantle, the transition zone and the lower mantle. We shall place less emphasis in this discussion on the upper mantle since it is the region most well constrained by direct petrological evidence (e.g. Chapter 1 in this volume) and least well characterized by seismological data due to lateral heterogeneities and inadequate resolving power (e.g. Chapter 5 in this volume). This paper is complementary to Chapters 1, 5 and 6 of this volume and to two papers currently in preparation: (a) in Liebermann and Ringwood (1978), we evaluate the effects of polymorphic phase transformations, pressure and temperature on elasticity systematics; and (b) in Liebermann and Leitner (1978), we review all the extant data on the pressure and temperature derivatives of the elastic properties of minerals and discuss the variation of these properties within isostructural series of compounds.

2. Elasticity of Minerals

In the absence of reliable theoretical models or sufficient experimental data, it has been customary in geophysics to use various systematic relationships between the elastic and crystallographic properties of solids to estimate or predict velocity v and density ρ changes due to variations of pressure P, temperature T, crystal structure or chemical composition. The elasticity systematics which have enjoyed the most widespread usage are Birch's velocity–density law (Birch, 1961*a*), the bulk modulus–volume (K_s–V) law (originally discussed by O. L. Anderson and Nafe, 1965) and the seismic equation of state (D. L. Anderson, 1967*b*).

2.1. Systematic Laws

Birch (1961*a*,*b*) demonstrated a linear relationship between compressional velocity v_p and density ρ for rocks and minerals of common mean atomic weight $\bar{M}$:

$$v_p = a(\bar{M}) + b\rho. \tag{1}$$

A similar relationship has subsequently been suggested for the bulk sound velocity v_ϕ with different empirical coefficients (e.g. Wang, 1972). This parameterization of the linear v–ρ curves on the basis of intercepts which depend on $\bar{M}$ (and slopes which do not) has become known as "Birch's law" and plots of velocity versus density as "Birch diagrams".

Several investigators have shown that Birch's law may be considered as a linearization, over a limited range of density, of a power law relating velocity to density:

$$v_i = C(\bar{M})\rho^{\lambda i} \tag{2}$$

This relationship for the bulk sound velocity v_ϕ follows directly from the law of corresponding states, $K_s = AV_0^{-x}$ (O. L. Anderson and Nafe, 1965); the seismic equation of state, $\rho/\bar{M} = A(\bar{M})\phi^n$, where $\phi = K_s/\rho$ is the seismic parameter (D. L. Anderson, 1967*b*); and the power law formulations of Birch's law (Shankland, 1972; Chung, 1972). All of these semi-empirical laws are really equivalent representations of the relationship of bulk modulus to volume for materials of common $\bar{M}$ (O. L. Anderson, 1973): $n = (X-1)^{-1}$, $\lambda_\phi = \frac{1}{2}(X-1)$. Least squares fits to the existing empirical data have led to the following estimates of the parameters in these laws: $X \simeq 4$ (e.g. O. L. Anderson, 1973), $n \simeq 1/3$ (D. L. Anderson and Kanamori, 1968), $\lambda_\phi \simeq 1.25$–1.7 depending on $\bar{M}$ (Shankland and Chung, 1974).

There have been many applications of such elasticity systematics in interpretations of geophysical models for the Earth's mantle in terms of composition and mineralogy:

(a) to construct density models from the observed seismic velocities for various regions of the mantle which are assumed to be homogeneous (e.g., Birch, 1964, D. L. Anderson, 1967*b*; Wang, 1970, 1972);

(b) to calculate the magnitude of velocity increases which should accompany density increases due to predicted polymorphic phase transformations in model compositions within the transition zone of the mantle (e.g. D. L. Anderson, 1967*a*; Ringwood, 1970; Liu, chapter 6 in this volume);

(c) to evaluate the chemical homogeneity of various regions of the mantle and, in particular, to estimate any changes in composition (i.e. mean atomic weight $\bar{M}$) across the transition zone from the upper to the lower mantle (e.g. Birch, 1961*b*; McQueen *et al.*, 1964; D. L. Anderson *et al.*, 1971; Press, 1970; Soga, 1971; Chung, 1971; Mizutani *et al.*, 1972; Graham, 1973).

Birch and the originators of the other systematic relationships have speculated that, for a given chemical compound, equations (1) and (2) would describe changes of v and ρ produced by either isothermal compression, isobaric expansion or polymorphic phase transformation. In the absence of data to confirm or contradict these speculations, many subsequent authors have assumed their validity, thereby elevating them from lore to law.

2.2. Effect of Pressure and Temperature

After more than fifteen years of extensive activity in mineral physics laboratories using ultrasonic techniques, we now have a considerable body of experimental data on the elastic properties of minerals with which to test the validity of these assumptions. Measurements of the elastic properties as a function of pressure and temperature (usually not simultaneously) using various high-precision ultrasonic techniques are now available for a large number of mineral constituents of crustal and upper mantle rocks, including olivine, orthopyroxene, garnet, spinel, corundum, rutile, quartz and periclase. Investigations of the effect of chemical composition on the elasticity within a particular crystal structure have been limited by the availability of natural or synthetic single crystals of sufficient size for ultrasonic measurements, but the effect of substituting Fe for Mg or Al is at least qualitatively understood for some of the crystallographic structures represented above. Studies on polycrystalline specimens recovered from high pressures in the laboratories of S. Akimoto in Tokyo and A. E. Ringwood in Canberra since 1970 have begun to elucidate the effect of polymorphic phase transformations on the isotropic elastic properties. Finally, Weidner and Carleton's (1977) successful determination of the single-crystal elastic moduli of monoclinic coesite by Brillouin scattering techniques promises that we shall soon have these more comprehensive data on a wide variety of high-pressure phases (e.g. Leitner and Weidner, 1978).

It is now apparent that none of the systematic relationships adequately account for the observed v–ρ behavior in specific oxygen-bearing compounds produced by pressure, temperature or phase transformation. Ultrasonic data for the pressure and temperature derivatives of the velocities enable us to calculate the v–ρ trajectories due to isothermal compression, $(\partial \ln v_i/\partial \ln \rho)_T = (K_T/v_i)(\partial v_i/\partial P)_T$, or isobaric expansion, $(\partial \ln v_i/\partial \ln \rho)_P = -(\alpha_v v_i)^{-1}(\partial v_i/\partial T)_P$, where K_T = isothermal bulk modulus and α_v = volume thermal expansion. These 'thermodynamic'' values of λ_i

have been calculated by a number of previous authors (e.g. Wang, 1970; D. L. Anderson *et al.*, 1971; O. L. Anderson, 1973; Davies, 1974). In Table 1 we present values of λ_Φ for a number of important mantle minerals of $\bar{M} \simeq 21$. The wide variability of λ_Φ belies the universality of any of the elasticity laws.

Table 1. Velocity–density systematics due to changes of pressure, temperature for materials of mean atomic weight $M \simeq 21$.†

		$(\partial \ln v_\Phi/\partial \ln \rho) = \lambda_\Phi$	
Compound	Structure	Isothermal compression‡	Isobaric expansion§
Mg_2SiO_4	Olivine	2.2	1.9
$Mg_{0.80}Fe_{0.20}SiO_3$	Orthopyroxene	4.3	2.8
SiO_2	Quartz	2.7	3.2
$MgAl_2O_4$	Spinel	1.9	1.4
MgO	Rocksalt	1.4	1.4
Al_2O_3	Corundum	1.5	1.2

† Data from compilation by Liebermann and Leitner (1978).
‡ $(\partial \ln v_\Phi/\partial \ln \rho)_T = 1/2(K_0' - 1)$, where $K_0' = (\partial K_s/\partial P)_{T'}$ at $P = 1$ bar.
§ $(\partial \ln v_\Phi/\partial \ln \rho)_p = 1/2(\delta_s - 1)$, where $\delta_s = (-1/\alpha_v K_s)(\partial K_s/\partial T)_P$.

There are, however, some new patterns emerging from the enlarged data base (Liebermann and Ringwood, 1978). For each mineral, values of λ_Φ from compression or expansion are comparable which confirms the inverse dependence of elasticity on pressure and temperature. Recent advances in high-temperature, high-pressure crystallography also suggest that P and T are structurally analogous variables (e.g. Hazen, 1977). We note that for the relatively open-packed structures such as olivine, pyroxene and quartz (in all of which Si is tetrahedrally coordinated by oxygen) the elastic properties change much more rapidly with changes in density due to compression or expansion than for the more close-packed oxides such as spinel, periclase and corundum.

2.3. Effects of Polymorphic Phase Transformations

Mineralogical investigations have demonstrated that most upper mantle minerals (olivines, pyroxenes, garnets, etc.) transform to more densely packed structures when subjected to pressures and temperatures in the laboratory comparable to those existing in the Earth's mantle to depths of up to 1000 km (e.g. see summaries in Ringwood, 1975; and Liu, Chapter 6 in this volume). These studies provide crystallographic data on the density variation across such polymorphic phase transformations. Since the most directly observable parameters of the Earth's interior are the seismic wave velocities, the usefulness to geophysics of these laboratory data would be greatly enhanced if measurements of the elastic properties of the high-pressure phases were available.

The experimental difficulties involved in producing sufficient quantities of the high-pressure minerals to make ultrasonic measurements of their elastic properties are formidable. Nevertheless, such data are so vital to geophysics that considerable efforts have been made to produce such specimens in the laboratories of Akimoto in Tokyo and Ringwood in Canberra since 1970 (e.g. Mizutani *et al.*, 1972; Liebermann *et al.*, 1976). These elasticity studies cannot be performed in the diamond-anvil systems such as those employed for the mineralogical and static compression investigations described by Liu (Chapter 6 in this volume) because of the limitation of small sample volume. We must, therefore, employ large-volume high-pressure devices such as the piston-cylinder, girdle, tetrahedral-anvil and Bridgman-anvil systems.

Laboratory studies of the elastic properties of the low-pressure and high-pressure polymorphs of various compounds undergoing phase transformations may thus be viewed as providing the missing link in the chain connecting high-pressure mineralogical studies to geophysical models derived from seismological data. Ideally, we would like to perform these elasticity measurements on single-crystal specimens as a function of pressure and temperature under *in situ* conditions within the P–T stability field of the high-pressure phase. In practice, such work has been restricted to the use of ultrasonic techniques to investigate polycrystalline specimens recovered metastably at STP conditions often for analog compositions: ultrasonic techniques because until recently other techniques such as Brillouin scattering have not been adapted to polycrystalline or small single-crystal specimens (less than 0.2 mm); polycrystalline specimens because it is not possible to grow single crystals of high-pressure phases large enough for ultrasonic work (greater than 1 mm); on recovery at STP because of the technical difficulties of performing and interpreting ultrasonic measurements at high pressures inside solid media apparatuses; and for analog compositions because of the extreme conditions necessary to produce the requisite phase for mantle compositions. To date, we have been successful at the Australian National University in hot-pressing polycrystalline specimens at pressures up to 120 kbar for such elasticity studies (for experimental details, see Liebermann *et al.*, 1975, 1976).

In Liebermann and Ringwood (1978), we have combined the crystallographic and elastic data to compare density and velocity jumps across polymorphic phase transformations in the context of the systematic relationships described above. Representative samples of these calculations are given in Fig. 1, which is a Birch diagram of v_ϕ vs. ρ, and in Table 2, in which are presented average values of λ_ϕ across these phase transformations.

It is clear from Fig. 1 and Table 2 that relatively high slopes (linear or logarithmic) on v–ρ diagrams are exhibited by the olivine–spinel and olivine–β-phase transformations in germanates and silicates and for the quartz–coesite transition in SiO_2. These transitions involve no change in coordination of Ge or Si with respect to oxygen and no change in cation–anion nearest-neighbor distances and the observed v–ρ slopes are comparable to those produced by isothermal compression or isobaric expansion of the low-pressure phase (Table 1). By contrast, the coesite–rutile, pyroxene–garnet, pyroxene–ilmenite, ilmenite–perovskite transformations are characterized by low v–ρ slopes and involve major increases in cation

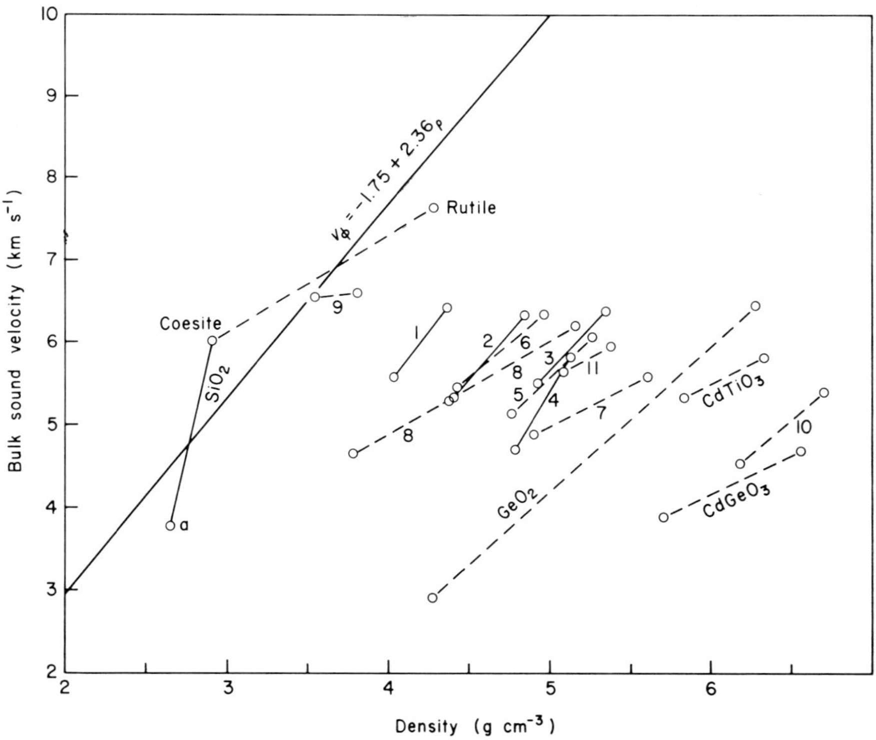

Fig. 1. Bulk sound velocity vs. density for polymorphic phase transformations in which coordination of cations and anions is unchanged (solid tie lines) or increases (dashed tie lines). Heavy solid line $v_\phi = -1.75 + 2.36\rho$ from Wang (1970, 1972) is empirical fit to data for rocks and minerals of $\bar{M} = 20$–22. Key: 1, Mg_2GeO_4; 2, Fe_2SiO_4; 3, Ni_2SiO_4; 4, Mn_2GeO_4; 5, Mg_2SnO_4; 6, $MgGeO_3$; $MnGeO_3$; 8, $CaGeO_3$; 9, Mg_2TiO_4; 10, Co_2SnO_4; 11, Co_2TiO_4.

Table 2. Velocity–density systematics across polymorphic phase transformations.†

Phase transformation	Cation–anion Coordination	Cation–anion Distance	$\lambda_\phi = (\partial \ln v_\phi / \partial \ln \rho)$
Quartz–coesite	Same	Same	5.0
Coesite–rutile	Increase	Increase	0.5
Olivine–spinel	Same	Same	1.7
Olivine–β phase	Same	Same	3.2
Pyroxene–ilmenite	Increase	Increase	1.2
Pyroxene–garnet	Increase	Increase	1.2
Garnet–Perovskite	Increase	Increase	0.9
Ilmenite–perovskite	Increase	Increase	1.1
Spinel–rocksalt + rutile	Increase	Same	1.7
Spinel–rocksalt + ilmenite	Increase	Increase	0.5

† Data from Liebermann and Ringwood (1978).

coordination and cation–anion distance (Table 3). This dependence of the v–ρ slopes on coordination has also been noted by Davies (1974) for transitions which involve isochemical mixed oxides as the transformation product.

A more quantitative generalization of distances as observations must certainly involve changes in the nearest-neighbor cation–anion distances as

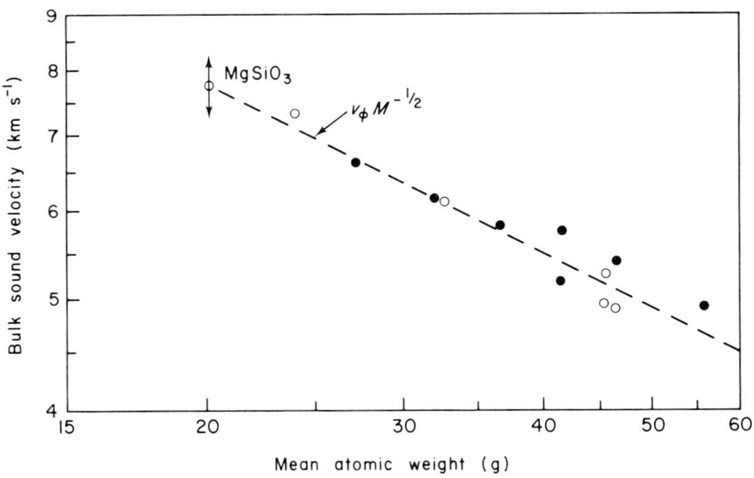

Fig. 2. Log–log plot of bulk sound velocity v_ϕ vs. mean atomic weight $\bar{M}$ for compounds with the perovskite structure (from Liebermann, Jones and Ringwood, 1977). Dashed line $v_\phi \bar{M}^{1/2}$ = constant is theoretical prediction for isostructural series (following Shankland, 1972; Chung, 1972). ○, Al; ●, Ti, Sn, Ge.

Table 3. Coordination changes across phase transformations.

$$^{IV}Si^{II}O_2 \rightarrow {}^{VI}Si^{III}O_2$$

Coesite Rutile

$$^{VI}Fe_2{}^{IV}Si^{IV}O_4 \rightarrow 2\,{}^{VI}Fe^{VI}O + {}^{VI}Si^{III}O_2$$

Normal spinel Rocksalt Rutile

$$^{VI}Mg^{IV}Mg\ {}^{VI}Sn^{IV}O_4 \rightarrow 2\,{}^{VI}Mg^{VI}O + {}^{VI}Sn^{III}O_2$$

Inverse spinel Rocksalt Rutile

$$^{VI}Mg^{IV}Mg\ {}^{VI}Ti^{IV}O_4 \rightarrow {}^{VI}Mg^{VI}O + {}^{VI}Mg^{VI}Ti^{IV}O_3$$

Inverse spinel Rocksalt Ilmenite

$$^{VI}Mn_2^{IV}Ge^{IV}O_2^{V}O^{III}O \rightarrow {}^{VII}Mn_2^{VI}Ge^{V}O_2^{IV}O_2$$

β Sr_2PbO_4

$$^{VI}Ca_3^{IV}Ge_3^{IV}O_4^{III}O_4^{II}O \rightarrow {}^{VIII}Ca_3^{VI}(CaGe)^{IV}Ge_3^{IV}O_{12}$$

Wollastonite Garnet

$$^{VIII}Ca_3^{VI}(CaGe)^{IV}Ge_3^{IV}O_{12} \rightarrow {}^{XII?}Ca^{VI}Ge^{VI}O_3$$

Garnet Perovskite

$$^{VI}Mg^{IV}Ge^{IV}O^{III}O_2 \rightarrow {}^{VI}Mg^{VI}Ge^{IV}O_3$$

Clinopyroxene Ilmenite

$$^{VI}Cd^{VI}Ti^{IV}O_3 \rightarrow {}^{VIII}Cd^{VI}Ti^{V}O_2^{IV}O$$

Ilmenite Perovskite

$$^{VIII}Na^{IV}Al^{IV}Ge_3^{III}O_8 \rightarrow {}^{VIII}Na^{VI}Al^{VI}Ge^{IV}O_8$$

Feldspar Hollandite

Reference: Shannon and Prewitt (1969).

well as the coordination changes. The critical role of the interionic distances is illustrated in the disproportionation of $A_2^{2+}B^{4+}O_4$ (spinel) compounds to 2A0–(rocksalt) + BO_2–(rutile) or AO–(rocksalt) + $AB0_3$–(ilmenite) phase assemblages (Liebermann, Jackson and Ringwood, 1977). The disproportionation of either normal (all B cations in fourfold coordination) or inverse (all B cations in sixfold coordination) spinels involves an increase in coordination from fourfold to sixfold for one third of the cations. For the inverse stannate spinels, however, the mean interionic distances are unchanged and the v–ρ slopes are high. In contrast, the mean interionic distances increase on disproportionation for the inverse titanate and normal silicate, germanate, aluminate and ferrite spinels so that their v–ρ slopes are low (D. L. Anderson, 1969; Davies, 1974). It is also important to note that the nearest-neighbor distances often dominate the repulsive interionic potential in most lattice models and, thus, the elastic moduli.

From laboratory studies such as those described above we are beginning to acquire elasticity data for the high-pressure phases of oxides and silicates. We now have a reasonable understanding of the effect of cation substitution on the elastic properties of the rocksalt, rutile, garnet, spinel and ilmenite lattices: in general, the compositional details of this substitution are of secondary importance with the velocity (modulus) depending to first-order only on the density (volume) within a given isostructural sequence. Thus, germanate, stannate and titanate compounds, which have been so important as models of the crystal chemical properties of their silicate analogs, are

Table 4. Elasticity of various model mantle mineral assemblages at zero-pressure.† Sequence of phase transformations from Liu (1977 and Chapter 6 of this volume).

Stoichiometry	Mineralogy	ρ_0 (g cm^{-3})	K_0 (Mbar)	ϕ_0 (km^2 s^{-2})
Olivine (Mg_2SiO_4)	olivine	3.21	1.3	41
	beta	3.47	2.0	58
	spinel	3.56	2.1	59
	rs + ilm	3.75	1.9	51
	rs + ru	3.85	1.9	49
	rs + per	3.93	2.2	56
Pyroxene ($MgSiO_3$)	orthopyroxene	3.21	1.2	37
	beta + ru	3.68	2.1	57
	spinel + ru	3.74	2.2	59
	ilm	3.82	2.1	55
	rs + ru	3.97	2.0	50
	per	4.08	2.5	61

† Calculations performed using the Voigt–Reuss–Hill procedure and elastic moduli measured or predicted for the various phases:

Mg_2SiO_4–olivine (Kumazawa and O. L. Anderson, 1969)
Mg_2SiO_4–β (Liebermann, 1975)
Mg_2SiO_4–spinel (Liebermann, 1975)
$MgSiO_3$–orthopyroxene (Liebermann, 1974; Weidner *et al.*, 1978)
$MgSiO_3$–ilmenite (Liebermann, 1976)
$MgSiO_3$–perovskite (Liebermann, Jones and Ringwood, 1977)
SiO_2–rutile (Liebermann *et al.*, 1976)
MgO–rocksalt (Spetzler, 1970)

Uncertainties in the moduli of the mixed phase assemblages due to the Voigt–Reuss bounds are less than 0.1 Mbar.

proving useful models for the elastic properties (e.g. Fig. 2). In previous papers, we have employed these isostructural trends to estimate the elastic properties of possible polymorphs of Mg_2SiO_4 and Mg_2SiO_3. Current estimates are given in Table 4 along with calculated properties of various model mineral assemblages. These data are critical to discussions of the elasticity and mineralogical identity of various regions of the transition zone and lower mantle, as we shall see in the next section.

3. Interpretations of Geophysical Models of Mantle Elasticity

Concurrent with recent progress in high-pressure mineralogy and mineral physics have been complementary developments in seismological studies of the velocity and density structure of the Earth's mantle. Sophisticated mathematical inversion techniques applied to gross Earth geophysical data have enabled direct determination of the density and velocity distribution with depth (e.g. Cleary and Anderssen, Chapter 5 in this volume). In Fig. 3

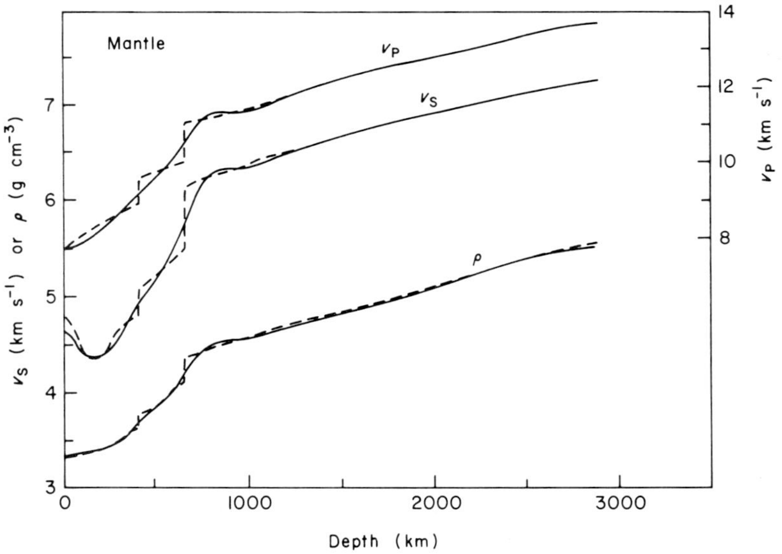

Fig. 3. Velocity and density models for the Earth's mantle from Gilbert and Dziewonski (1975). ———, Model 1066A; - - - -, model 1066B.

we give examples of two such models, both of which were obtained by fitting the eigenperiods of 1064 normal modes and the Earth's mass and moment of inertia. Other contemporary models have taken explicit account of the different properties of the continental and oceanic upper mantles (Dziewonski *et al.*, 1975) and the effect of absorption (finite Q) on the eigenperiods (Hart *et al.*, 1977). Most of the variations between models occur in the upper mantle (to depths of 350 km) in which lateral heterogeneities, high attenuation and large and variable thermal gradients are more prevalent.

3.1. Upper Mantle

The upper mantle low-velocity zone (LVZ) is a required feature of such average Earth models. D. L. Anderson and Sammis (1970) demonstrated that a wide variety of geophysical data could be explained if the LVZ were a region of incipient partial melt. Although critical thermal gradients (e.g. Liebermann and Schreiber, 1969) and grain-boundary relaxation (e.g. Goetze, 1977) are also possible explanations of the LVZ, partial melting is consistent with available petrological evidence. Green and Liebermann (1976) showed that a pyrolite petrological model and its velocity profile satisfactorily account for most of the geophysical data for various age provinces in oceanic regions of the upper mantle. If the water content of the upper mantle peridotite is <0.4 percent approximately, then its solidus has a distinctive P, T character such that the geotherm for older oceanic regions ($t>80$ million years) will enter a zone of incipient (>2 percent) melting—the LVZ—at depths of 85–95 km.

Although partial melting is an attractive explanation of the LVZ and has received wide acceptance currently, many of the arguments supporting this interpretation are based on several assumptions (e.g. the geometrical configuration of the small melt fraction) and limited experimental data (e.g. the NaCl–ice study of Spetzler and Anderson, 1968). As emphasized by Goetze (1977), the critical experiments in realistic petrological systems which would permit us to distinguish between the alternative explanations of the LVZ have not yet been performed.

The base of the LVZ occurs about 250 km and the mantle appears to be relatively homogeneous between that depth and approximately 375 km (D. L. Anderson, 1977). For such homogeneous regions, velocity and density profiles can be calculated for specified mineral assemblages via equations of state if the pressure and temperature profiles are known and the following data are available: v_p, v_s and ρ at atmospheric pressure and room temperature, the pressure and temperature derivatives of v_p and v_s and thermal expansion. These data are available (see Liebermann and Leitner, 1978, for compilation) for olivine, pyroxene and garnet, which dominate petrological models of the upper mantle, and a number of authors have demonstrated that the velocities in this region (250–375 km) may be adequately explained by realistic combination of these component minerals with FeO contents of about 10 percent (e.g. Graham, 1970; D. L. Anderson *et al.*, 1971; Ahrens, 1973; Burdick and Anderson, 1975).

3.2. The Transition Zone

Unfortunately, the normal mode data now available lack the resolving power to delineate the fine structure of the transition zone (350–1000 km) between the upper and lower mantle. Gilbert and Dziewonski (1975) have shown that both continuous (1066A) and discontinuous (1066B) functions for the velocity–depth and density–depth profiles in this region give equally good fits to the gross Earth geophysical data (Fig. 3). This is one of the most striking demonstrations of the non-linear nature of the problem of inverting such data. Solutions derived by perturbing smooth starting materials (no discontinuities) to fit all of the data contain no discontinuities (e.g. 1066A); conversely, solutions derived from discontinuous starting models retain their discontinuous character (e.g. 1066B). A comparable result was evident in models A1 and B1 of Jordan's (1972) thesis, but only B1 was published in Jordan and Anderson (1974).

However, the combined weight of other seismic data from surface wave studies, $dT/d\Delta$ studies using large seismic arrays, detailed studies of wave forms, and precursors to the phase P′P′ support the likelihood of there being two major first-order discontinuities in velocity and density near 400 km and 650 km in depth. There have also been suggestions for some time of minor discontinuities near 550 km depth (e.g. Wiggins and Helmberger, 1973; Simpson *et al.*, 1974; but see Burdick and Helmberger, 1978) and at depths greater than 700 km (e.g. Archambeau *et al.*, 1969, Ram Datt and Muirhead, 1976; Hales *et al.* 1978).

Evidence from both geophysics (e.g. D. L. Anderson, 1977) and geochemistry (e.g., Ringwood, 1975, and Chapter 1 in this volume) indicates that most, if not all, of these rapid increases in velocity and density in the transition zone are due to phase transformations. Silicon is largely tetrahedrally coordinated in mineral constituents of the upper mantle (olivines, pyroxenes, garnets) but must be predominantly octahedrally coordinated in the lower mantle to explain the high densities inferred from geophysical models (following Birch, 1952). This increase in coordination of silicates most probably takes place in the transition zone (350–1000 km) as the open-packed structures transform to denser phases. There is no unambiguous way of identifying any specific phase transformation with a particular seismic discontinuity. However, the 400 km discontinuity has been associated with the transformation of $(Mg, Fe)_2SiO_4$ olivine to the spinel or β-phase structures since Bernal's (1936) suggestion was adopted by Jeffreys (1937). Much discussion has recently taken place about the mineralogical nature and thermodynamic and elastic properties of the 650 km discontinuity (e.g. Jackson *et al.*, 1974; D. L. Anderson, 1976; Liebermann, Jackson and Ringwood, 1977; and the Ringwood, Chapter 1, and Liu, Chapter 6, papers in this volume). Details are still unresolved but it appears likely that the phase transformation(s) responsible for the 650 km discontinuity involves an increase in Si coordination whereas those for the 400 km discontinuity do not.

The distinction between phase transformations which involve constant or increasing cation–anion coordinations and interionic distances is an important diagnostic tool for interpreting geophysical models of the Earth's interior in terms of composition and mineralogy (Liebermann and Ringwood, 1974). We dramatize this important point by contrasting the conventional and modified interpretations of the velocity–density behavior exhibited in Fig. 4 by Model QM2 of Hart *et al.* (1977), for which absorption corrections have been incorporated.

All of the geophysical applications of elasticity systematics cited in the previous section assumed or implied that velocity (elastic modulus) depends in a unique way on density (volume) for materials of common mean atomic

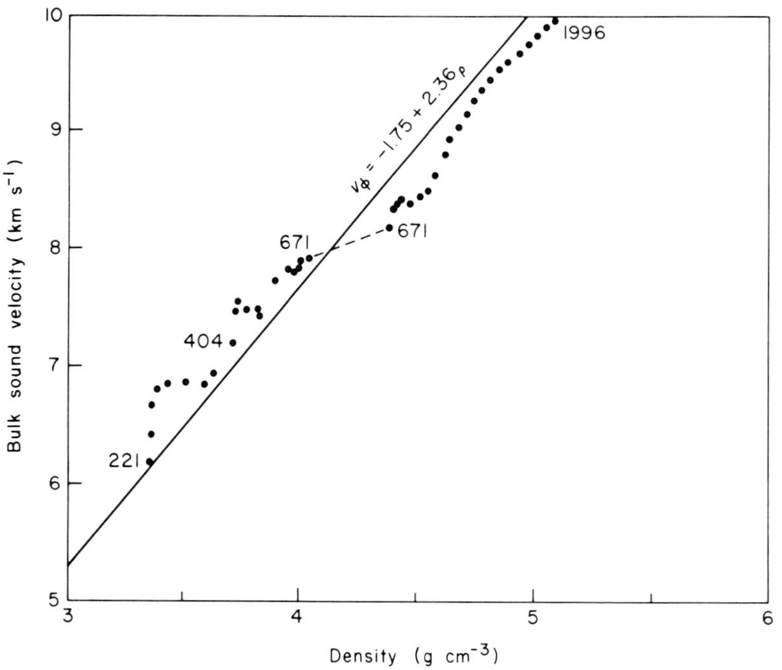

Fig. 4. Birch diagram for bulk sound velocity for Earth model QM2 of Hart *et al.* (1977) which incorporates absorption corrections. Solid line from Fig. 1.

weight $\bar{M}$, irrespective of whether the density (volume) changes are due to variations in crystallographic structure, pressure or temperature. If this assumption were true, then any deviation from such velocity–density laws in the Earth must be ascribed to changes in chemical composition (specifically, the iron content or FeO/(FeO+MgO) molecular ratio). Thus the conventional interpretation of Model QM2 in Fig. 4 suggests an increase in $\bar{M}$ across the seismic discontinuity at 671 km, thus implying that the upper and lower mantle are chemically inhomogeneous.

However, on the basis of the laboratory data summarized in Tables 1 and 2, we may expect seismic discontinuities due to phase transformations which involve no change in coordination to be characterized by high v–ρ slopes on a Birch diagram comparable to those produced by homogeneous compression and expansion of the low-pressure phase. By contrast, those discontinuities due to transformations which do involve an increase in coordination should have relatively low v–ρ slopes. In Fig. 5, we have constructed a hypothetical velocity–density trajectory for a peridotite mantle of constant chemical composition. It is clear from Fig. 5 that offsets from a linear v–ρ trend are to be *expected* in chemically homogeneous mantle and cannot, therefore, be attributed to increases in the iron content of the lower mantle relative to the upper mantle, as has been done by so many previous investigators.

3.3. The Olivine (α)–Spinel (γ) and Olivine (α)–β-phase Transformations and the 400 km Discontinuity

As mentioned above, the phase transformation of $(Mg, Fe)_2SiO_4$ olivine to the spinel or β-phase structures is widely accepted as the principal cause of the velocity and density discontinuities observed in the Earth's mantle in the neighborhood of 400 km depth. In several earlier papers, we have compared the experimental laboratory data for the α–β and α–γ transitions with several Earth models in the vicinity of 400 km depth (Liebermann, 1973, 1975). Our purpose is to test for consistency between experimental and model data and to suggest possible constraints in future inversion attempts.

We have selected for comparison six recent Earth models which are characterized either by first-order discontinuities in velocity and density near 400 km or by regions of anomalously high velocity and density gradients near this depth. In Table 5 we calculate λ_Φ for the power law form of Birch's law for the 400 km discontinuities in the Earth's models. From the phase transition data in Table 2, we had concluded that $\lambda_\Phi = 1.7$ for the α–γ transition and somewhat larger for the α–β transition. Of the Earth models, only OC-1 and 1066B are consistent with the laboratory data; note here that since the λ_Φ for these transitions are comparable with the thermodynamic λ_Φ

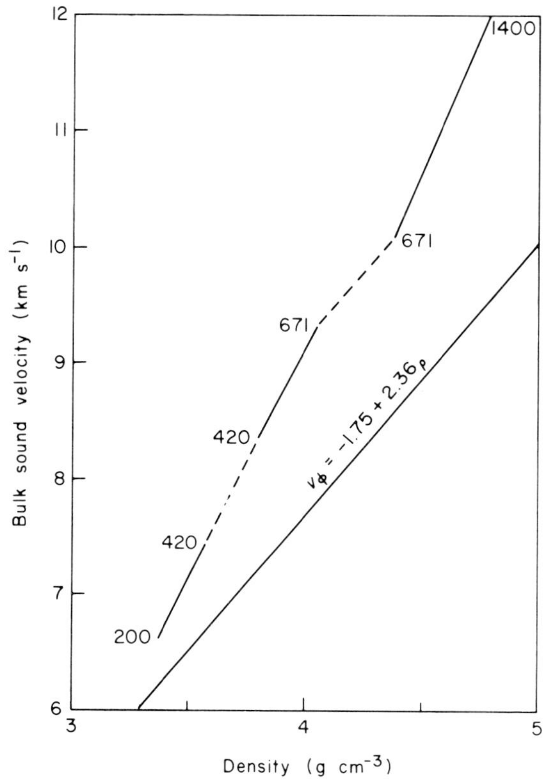

Fig. 5. Birch diagram for bulk sound velocity for chemically homogeneous peridotite mantle undergoing compression and transformation across 420 km discontinuity (no coordination change) and 671 km discontinuity (coordination increases). Heavy solid line from Fig. 1.

Table 5. Earth models: parameters of power law form of Birch's law across the 400 km discontinuity

Model	λ_ϕ	Reference
Simple Earth	1.0	Wang (1972)
OC-1	2.1	Mizutani and Abe (1972)
Bl	0.5	Jordan (1972); Jordan and Anderson (1974)
1066B	2.0	Gilbert and Dziewonski (1975)
PEM	1.2	Dziewonski *et al.* (1975)
QM2	1.3	Hart *et al.* (1977)

for the olivine structure, these comparisons remain valid if we correct the laboratory data to mantle conditions. The "simple Earth", PEM and absorption-corrected QM2 models are not compatible with the experimental data and the B1 model is completely discordant.

We conclude from this discussion that only the OC-1 (Mizutani and Abe, 1972) and 1066B (Gilbert and Dziewonski, 1975) models are consistent with the experimental data for the α–γ and α–β phase transformations in germanate and silicate compounds. Of these two models, 1066B is a more viable model as it has been constrained to satisfy a much larger body of free oscillation data. It is interesting to note that 1066B was derived by perturbing B1 to fit the new data with the requirement that the perturbation added to the starting model be as smooth as possible.

An alternative conclusion is that the experimental data should be used to provide constraints in future inversion attempts to derive geophysical models for the Earth's interior. Application of these constraints presupposes, of course, that the olivine–spinel or olivine–β-phase transformations are the cause of the velocity and density discontinuities near 400 km depth in the Earth's mantle.

3.4. The Lower Mantle

The lower mantle (depths greater than about 1000 km) is the region of the solid Earth least well-constrained by petrological considerations and most consistently described by various seismological models. Many previous authors have attempted to draw inferences about the chemical composition and crystal structure of the lower mantle by comparing the observed velocity–depth and density–depth profiles with those calculated via an equation of state.

These calculations may be performed in the forward or inverse directions and originated with Birch (1952) who extrapolated the properties of various candidate materials to mantle temperatures and pressures and compared the results with the seismic data. A balanced account of this forward approach is given by Watt and O'Connell (1978) for the region from 700–1200 km depth. Birch's method has also been used to extrapolate the observed seismic data back to low pressure and temperature conditions in order to compare with laboratory results; e.g. Butler and Anderson (1978) have tested various regions of the lower mantle for homogeneity and adiabaticity and have used both finite strain and atomistic equations of state to extract the zero-pressure and room-temperature properties of these regions.

Such forward or inverse comparisons are difficult, if not hazardous, for a variety of reasons including ambiguities in (1) defining homogeneous chemical and mineralogical regions of the mantle; (2) selecting the equation of state for the extrapolation; (3) choosing the mixing law to calculate the elastic properties of a mineral assemblage; and (4) the temperature distribution in the lower mantle.

Notwithstanding these ambiguities, we may illustrate the impact of our estimates of the elastic properties of various assemblages of high-pressure phases (Table 4) by comparing them with extrapolations from mantle conditions. D. L. Anderson (1976) has demonstrated that the bulk modulus is relatively insensitive to iron content, so that our estimates should be valid for reasonable ferromagnesian compositions. For the region from 1246 to 1546 km depth, Butler and Anderson (1978) estimate $\rho_0 = 4.06 \pm 0.02$ g cm^{-3}, $K_0 = 2.5 \pm 0.2$ mbar and $\phi_0 = 61 \pm 4$ (km^2 s^{-2}). From Table 4 we can see that the periclase + stishovite mixtures with either olivine or pyroxene stoichiometry have K_0 and ρ_0 values which are much too low to satisfy the Earth models (addition of Fe would improve the density agreement but not the K_0 contrast). It appears that only the mineral assemblages in Table 4 which contain $MgSiO_3$-perovskite have sufficiently high densities and seismic parameters to be compatible with the Earth models.

4. Summary and Future Directions

It is now evident that there is no unique dependence of elastic wave velocity v upon density ρ for materials of constant chemical composition. The changes of elasticity produced depend critically on the crystallographic details of the density changes due to compression, expansion or phase transformation. Fortunately, the latter data are now becoming available from single-crystal X-ray diffraction studies at high temperatures and pressures (e.g. Hazen, 1977). We plan to integrate these crystallographic data into our analyses of the observed elasticity–density behavior to develop more realistic interatomic force models. The experimental demonstration by Liebermann, Jackson and Ringwood (1977) that the elasticity of spinel-structure oxides depends strongly on the cation configuration makes this structure an interesting one for further theoretical investigation using approaches such as those suggested by Sammis (1970) or Weidner and Simmons (1972).

The absence of single-crystal elasticity data has been a severe limitation in theoretical approaches to describe the lattice dynamics of high-pressure phases. It is now possible to measure the single-crystal elastic moduli on specimens of 100 μm dimensions using Brillouin scattering techniques developed at Stony Brook by Weidner and Carleton (1977). This grain size limit is within reach of current laboratory expertise using solid-media high-pressure apparatus such as that employed in the hot-pressing of the polycrystals. In fact, depending on the compound, the synthesis conditions (especially time and temperature), and the phase involved, we have often produced polycrystalline specimens with grain sizes greater than 100 μm. These specimens were unsuitable for our acoustic measurements since they scattered and attenuated the high-frequency ultrasonic signals, but would have been suitable for Brillouin scattering studies. In fact, sample specimens of stishovite produced in this manner (see Liebermann *et al.*, 1976) have recently been made available for Weidner's experiments through the courtesy of A. E. Ringwood. The 100 μm value is also approximately the same as that required for crystal structure investigations using single-crystal X-ray diffraction techniques and it is highly desirable that any high-pressure phases be studied by both techniques.

The Earth's mantle is not a single crystal, of course, but a polycrystalline aggregate of a multi-component phase in terms of individual components (e.g. olivine or pyroxene and their high-pressure modifications). There exists no unambiguous method of estimating the elastic moduli of a polycrystalline aggregate of two or more crystalline phases (see Watt *et al.*, 1976 for a comprehensive and authoritative review). The Voigt and Reuss averaging schemes (arithmetic and geometric volume averages, respectively) provide theoretical upper and lower bounds, but these often fail as constraints because they are too far apart. Hashin and Shtrikman (1963) have used a variational method to narrow the theoretical bounds: some ambiguity remains but these are the best possible limits without additional knowledge or assumptions about the geometrical distribution and shape of the individual grains of the two phases in the aggregate. There is no theoretical or experimental justification for selecting any of these bounds or any average of them to be representitive of the properties of the polycrystalline aggregate. The Hashin–Shtrikan bounds are well-constrained, however, if the contrast in shear moduli of the two phases is not large. Davies (1974, 1976) has exploited this feature to calculate the elastic properties of oxide mixtures of various chemistries and stoichiometries of interest to geophysics. Additional experimental studies are necessary and desirable to clarify these important theoretical and practical problems.

One system which we will investigate to clarify the theoretical ambiguity discussed in the previous section is MgO-rocksalt and GeO_2-rutile in which the elasticity of the end members is well-constrained by ultrasonic measurements on single crystals. It is important that polycrystalline aggregates of such mixtures be carefully characterized so that such properties as shape anisotropy, crystallographic preferred orientation and inhomogeneous distribution of mineral grains can be either eliminated or fully accounted for. It may be easier to avoid such complications in systems for which the polycrystalline mixture is produced by transformation from a single phase. We have recently obtained elasticity data for mixtures of the disproportionation products of certain spinel-structure compounds which exhibit good agreement with the theoretical calculations (Liebermann, Jackson and Ringwood, 1977). Other structural systems for which this approach may be feasible and interesting include pyroxene–spinel + rutile (e.g. $FeGeO_3$, $CoGeO_3$), olivine–perovskite + rocksalt (e.g. Cd_2GeO_4, $CaMgGeO_4$, $CaMnGeO_4$), and strontium plubate–perovskite + rocksalt (e.g. Ca_2SnO_4, Cd_2SnO_4, Cd_2TiO_4).

The Earth's mantle is also not perfectly elastic and Liu *et al.*, 1976 have shown that one cannot directly compare laboratory and seismic data on the elastic properties with correcting for attenuation. Recent seismic models have incorporated these corrections when analyzing both free oscillation and body-wave data and the consequent models are appropriate for frequencies of 1 Hz (e.g. Hart *et al.*, 1977). Laboratory elastic data from ultrasonic or Brillouin scattering experiments are obtained at frequencies greater than 10^7 Hz. Until detailed attenuation measurements are available over a wide frequency range, comparisons of laboratory and seismic data must be viewed with some caution.

Acknowledgements

It is a special pleasure for me to contribute this chapter to a book in honor of Professors Jaeger and Hales, as they both played an important role in my career at the Australian National University: Professor Jaeger offered me encouragement and technical support in the critical initial phases of our program; Professor Hales provided constructive criticism and skepticism in the latter stages, amply tempered by the warmth of his personality. I am also indebted to Professor A. E. Ringwood for his generous support and advice throughout this collaborative program and to my two graduate students, Ian Jackson and Leonie Jones, for their stimulating presence. Special mention and my thanks must also go to Doug Mayson, Alan Major, the late Bill McIntyre and Bill Hibberson for their technical assistance.

This paper was written under the support of the National Science Foundation, Grant EAR 76–84066.

References

Ahrens, T. J. (1973). Petrologic properties of the upper 670 km of the Earth's mantle: geophysical implications. *Phys. Earth planet. Interiors* **7**, 167–186.

Anderson, D. L. (1967*a*). Phase changes in the upper mantle. *Science, N.Y.* **157**, 1165–1173.

Anderson, D. L. (1976*b*). A seismic equation of state. *Geophys. J. R. astronom. Soc.* **13**, 9–30.

Anderson, D. L. (1969). Bulk modulus–density systematics. *J. geophys. Res.* **74**, 3857–3864.

Anderson, D. L. (1976). The 650 km mantle discontinuity. *Geophys. Res. Lett.* **3**, 347–349.

Anderson, D. L. (1977). Composition of the mantle and core. *A. Rev. Earth planet. Sci.* **5**, 179–202.

Anderson, D. L. and Kanamori, H. (1968). Shock-wave equations of state for rocks and minerals. *J. geophys. Res.* **73**, 6477–6502.

Anderson, D. L. and Sammis, C. (1970). Partial melting in the upper mantle. *Phys. Earth planet. Interiors* **3**, 41–50.

Anderson, D. L., Sammis, C. and Jordan, T. (1971). Composition and evolution of the mantle and core. *Science, N.Y.* **171**, 1103–1112.

Anderson, O. L. (1973). Comments on the power-law representation of "Birch's law". *J. geophys. Res.* **78**, 4901–4914.

Anderson, O. L. and Nafe, J. E. (1965). The bulk modulus–volume relationship for oxide compounds and related geophysical problems. *J. geophys. Res.* **70**, 3951–3963.

Archambeau, C. B., Flinn, E. A. and Lambert, D. G. (1969). Fine structure of the upper mantle. *J. geophys. Res.* **74**, 5825–5865.

Bernal, J. D. (1936). Discussion. *Observatory*, **59**, 268.

Birch, F. (1952). Elasticity and constitution of the Earth's interior. *J. geophys. Res.* **57**, 227–286.

Birch, F. (1961*a*). The velocity of compressional waves in rocks to 10 kb., part 2. *J. geophys. Res.* **66**, 2199–2224.

Birch, F. (1961*b*). Composition of the Earth's mantle. *Geophys. J. R. astronom. Soc.* **4**, 295–311.

Birch, F. (1964). Density and composition of the mantle and the core. *J. geophys. Res.* **69**, 4377–4388.

Burdick, L. and Anderson, D. L. (1975). Interpretation of velocity profiles of the mantle. *J. geophys. Res.* **80**, 1070–1074.

Burdick, L. J. and Helmberger, D. V. (1978). The upper mantle P velocity structure of the western United States. *J. geophys. Res.* **83**, 1699–1712.

Butler, R. and Anderson, D. L. (1978). Equation of state fits to the lower mantle and outer core. Submitted for publication.

Chung, D. H. (1971). Elasticity and equations of state of olivines in the Mg_2SiO_4–Fe_2SiO_4 system. *Geophys. J. R. astronom. Soc.* **25**, 511–538.

Chung, D. H. (1972). Birch's law: why is it so good? *Science, N.Y.* **177**, 261–263.

Davies, G. F. (1974). Elasticity, crystal structure and phase transitions. *Earth planet. Sci. Lett.* **22**, 339–346.

Davies, G. F. (1976). The estimation of elastic properties from analogue compounds. *Geophys. J. R. astronom. Soc.* **44**, 625–647.

Dziewonski, A. M., Hales, A. L. and Lapwood, E. R. (1975). Parametrically simple Earth models consistent with geophysical data. *Phys. Earth planet Interiors* **10**, 12–48, 1975.

Gilbert, F. and Dziewonski, A. M. (1975). An application of normal mode theory to the retrieval of structural parameters and source mechanisms from seismic spectra. *Phil. Trans. R. Soc. A* **278**, 187–269.

Goetze, C. (1977). A brief summary of our present day understanding of the effect of volatiles and partial melt on the mechanical properties of the upper mantle. In *High-Pressure Research: Applications in Geophysics* (M. H. Manghnani and S. Akimoto, eds), Academic Press, London and New York, pp. 3–23.

Graham, E. K. (1970). Elasticity and composition of the upper mantle. *Geophys. J. R. astronom. Soc.* **20**, 285–302.

Graham, E. K. (1973). On the compression of stishovite. *Geophys. J. R. astronom. Soc.* **32**, 15–34.

Graham, E. K. and Dobrzykowski, D. (1976). Temperatures in the mantle as inferred from simple compositional models. *Am. Mineralogist* **61**, 546–559.

Green, D. H. and Liebermann, R. C. (1976). Phase equilibria and elastic properties of a pyrolite model for the oceanic upper mantle. *Tectonophysics* **32**, 61–92.

Hales, A. L., Muirhead, K. J. and Rynn, J. M. W. (1978). A compressional velocity distribution for the upper mantle. Submitted for publication.

Hart, R. S., Anderson, D. L. and Kanamori, H. (1977). The effect of attenuation on gross Earth models. *J. geophys. Res.* **82**, 1647–1654.

Hashin, Z. and Shtrikman, S. (1963). A variational approach to the theory of the elastic behavior of multiphase materials. *J. Mech. Phys. Solids* **11**, 127–140.

Hazen, R. M. (1977). Temperature, pressure and composition: structurally analogous variables. *Phys. Chem. Minerals* **1**, 83–94.

Jackson, I. N. S., Liebermann, R. C. and Ringwood, A. E. (1974). Disproportionation of spinels to mixed oxides: significance of cation configuration and implications for the mantle. *Earth planet. Sci. Lett.* **24**, 203–208.

Jeffreys, H. (1937). On the materials and density of the Earth's crust. *Mon. Not. Roy. astronom. Soc., Geophys. Suppl.* **4**, 50–61, 1937.

Jordan, T. H. (1972). Estimation of the radial variation of seismic velocities and density in the Earth. Ph.D. Thesis, California Institute of Technology, Pasadena, California.

Jordan, T. H. and Anderson, D. L. (1974). Earth structure from free oscillations and travel times. *Geophys. J. R. astronom. Soc.* **36**, 411–459.

Kumazawa, M. and Anderson, O. L. (1969). Elastic moduli, pressure derivatives, and temperature derivatives of single-crystal olivine and single-crystal forsterite. *J. geophys. Res.* **74**, 5961–5972.

Leitner, B. J. and Weidner, D. J. (1978). Elasticity of pyrope. *Trans. Am. geophys. Union* **59**, 813.

Liebermann, R. C. (1973). Elasticity of the olivine–spinl and olivine–β phase transformations and the 400-kilometer discontinuity in the mantle. *J. geophys. Res.* **78**, 7015–7017.

Liebermann, R. C. (1973). Elasticity of pyroxene–garnet and pyroxene–ilmenite phase transformations in germanates. *Phys. Earth planet. Interiors* **8**, 361–374.

Liebermann, R. C. (1975). Elasticity of olivine (α), beta (β), and spinel (γ) polymorphs of germanates and silicates. *Geophys. J. R. astronom. Soc.* **42**, 899–929.

Liebermann, R. C. (1976). Elasticity of ilmenites. *Phys. Earth planet. Interiors* **12**, P5–P10.

Liebermann, R. C. and Leitner, B. J. (1978). Elasticity of minerals *Rev. Geophys. Space Phys.* to be submitted.

Liebermann, R. C. and Ringwood, A. E. (1974). Elasticity of phase transformations and the iron content of the mantle (Abstract). *Trans. Am. geophys. Union* **55**, 416.

Liebermann, R. C. and Ringwood, A. E. (1978). Elasticity and crystallography of phase transformations. *Phys. Chem. Minerals* to be submitted.

Liebermann, R. C. and Schreiber, E. (1969). Critical thermal gradients in the mantle. *Earth planet. Sci. Lett.* **7**, 77–81.

Liebermann, R. C., Ringwood, A. E. and Major, A. (1976). Elasticity of polycrystalline stishovite. *Earth planet. Sci. Lett.* **32**, 127–140.

Liebermann, R. C., Jackson, I. and Ringwood, A. E. (1977). Elasticity and phase equilibria f spinel disproportionation reactions. *Geophys. J. R. astronom. Soc.* **50**, 553–586.

Liebermann, R. C., Jones, L. E. A. and Ringwood, A. E. (1977). Elasticity of aluminate, titanate, stannate, and germanate compounds with the perovskite structure. *Phys. Earth planet. Interiors* **14**, 165–178.

Liebermann, R. C., Ringwood, A. E., Mayson, D. J. and Major, A. (1975). Hot-pressing of polycrystalline aggregates at very high pressure for ultrasonic measurements. In *Proceedings of the 4th International Conference on High Pressure* (J. Osugi, ed), Physico-Chemical Society of Japan, Kyoto, pp. 495–502.

Liu, H. P., Anderson, D. L. and Kanamori, H. (1976). Velocity dispersion due to anelasticity: implications for seismology and mantle composition. *Geophys. J. R. astronom. Soc.* **47**, 41–58.

Liu, L. G. (1977). Mineralogy and chemistry of the Earth's mantle above 1000 km. *Geophys. J. R. astronom. Soc.*, **48**, 53–62.

McQueen, R. G., Fritz, J. N. and Marsh, S. P. (1964). On the composition of the Earth's interior. *J. geophys. Res.* **69**, 2947–2965.

Mizutani, H., and Abe, K. (1972). An Earth model consistent with free oscillation and surface wave data. *Phys. Earth planet. Interiors* **5**, 345–356.

Mizutani, H., Hamano, Y. and Akimoto, S. I. (1972). Elastic-wave velocities of polycrystalline stishovite. *J. geophys. Res.* **77**, 3744–3749.

Press, F. (1970). Regionalized Earth models. *J. geophys. Res.* **75**, 6575–6581.

Ram Datt and Muirhead, K. J. (1976). Evidence for a sharp increase in P-wave velocity at about 770 km depth. *Phys. Earth planet. Interiors* **13**, 37–46, 1976.

Ringwood, A. E. (1970). Phase transformations and the constitution of the mantle. *Phys. Earth planet. Interiors* **3**, 109–155, 1970.

Ringwood, A. E. (1975). *Composition and Petrology of the Earth's Mantle.* McGraw-Hill, New York.

Sammis, C. G. (1976). The pressure dependence of the elastic constants of cubic crystals in the NaCl and spinel structures from a lattice model. *Geophys. J. R. astronom. Soc.* **19**, 285–297.

Shankland, T. J. (1972). Velocity–density systematics: derivation from Debye theory and the effect of ionic size. *J. geophys. Res.* **77**, 3750–3758.

Shankland, T. J. and Chung, D. H. (1974). General relationships among sound speeds. II. Theory and discussion. *Phys. Earth planet. Interiors* **8**, 121–129.

Shannon, R. D. and Prewitt, C. T. (1969). Coordination and volume changes accompanying high-pressure phase transformation of oxides. *Mat. Res. Bull.* **4**, 57–62.

Simpson, D. W., Mereu, R. F. and King, D. W. (1974). An array study of P-wave velocities in the upper mantle transition zone beneath northeastern Australia. *Bull. seismol. Soc. Am.* **64**, 1757–1788.

Soga, N. (1971). Sound velocity of some germanate compounds and its relation to the law of corresponding states. *J. geophys. Res.* **76**, 3983–3989.

Spetzler, H. (1970). Equation of state of polycrystalline and single crystal MgO to 8 kilobars and 800 K. *J. geophys. Res.* **75**, 2073–2087.
Spetzler, H. and Anderson, D. L. (1968). The effect of temperature and partial melting on velocity and attenuation in a simple binary system. *J. geophys. Res.* **73**, 6051–6060.
Wang, C. Y. (1970). Density and constitution of the mantle. *J. geophys. Res.* **75**, 3264–3284.
Wang, C. Y. (1972). A simple earth model. *J. geophys. Res.* **77**, 4318–4329.
Watt, J. P. and O'Connell, R. J. (1978). Mixed-oxide and perovskite-structure model mantles from 700–1200 km. *Geophys. J. R. astronom. Soc.* in press.
Watt, J. P., Davies, G. F. and O'Connell, R. J. (1976). The elastic properties of composite materials. *Rev. Geophys. Space Phys.* **14**, 541–563.
Weidner, D. J. and Carleton, H. R. (1977). Elasticity of coesite. J. geophys. Res. **82**, 1334–1346.
Weidner, D. J. and Simmons, G. (1972). Elastic properties of alpha quartz and the alkali halides based on an interatomic force model. *J. geophys. Res.* **77**, 826–847.
Weidner, D. J., Wang, H. and Ito, J. (1978). Elasticity of enstatite. *Phys. Earth planet. Interiors* in press.
Wiggins, R. A. and Helmberger, D. V. (1973). Upper mantle structure of the western United States. *J. geophys. Res.* **78**, 1870–1880.

8. Fluid Models of Mantle Convection

D. J. STEVENSON and J. S. TURNER

Abstract

Mantle convection appears to be an inevitable consequence of the Earth's thermal evolution, but the form which this convection takes is still unclear because the relevant data have either not yet been obtained or are ambiguous. In particular, the well-established, mechanistic picture of plate tectonics explains most of what is directly observable, but this need not imply the presence of whole-mantle motions. In this review, it is shown that a variety of indirect evidence, especially postglacial uplift and gravity data, together with an increasing theoretical and laboratory understanding of model systems, suggests convection throughout the mantle, although the flow may be in the form of coupled layers rather than overturning of the whole mantle. The relationship of this convection to plate tectonics is discussed, with emphasis on subduction as a form of thermal convection in which the mantle is cooled from above. Complications such as the non-uniform rheology, irreversible chemical evolution and non-uniform distribution of radioactive heating in the mantle make it difficult to produce a detailed picture of mantle convection at this stage, but the importance of internal motions in the thermal evolution of the Earth is clearly established. Suggestions are made about the future observational, theoretical and laboratory work that is required to resolve the outstanding problems.

1. Introduction

The unifying concept of plate tectonics as an explanation for the modification of the Earth's outermost solid layer is now well-established. According to this empirical description, the Earth's surface is divided into "plates" (i.e. spherical caps) which are in relative motion. Earthquakes, volcanism and mountain building occur mostly at the boundaries of these plates, where the relative motions produce large stresses. The possibility of relative motions is an old idea (Wegener, 1912; Holmes, 1931) but it did not receive substantial support from observations until the 1950's, and it did not become almost universally accepted until the late 1960's (Cox, 1973). A detailed empirical description of plate tectonics can be found in Le Pichon *et al.* (1973).

The motions are usually attributed to some form of thermal convection driven by energy sources within the mantle. The relevance of convection in very large bodies (for which thermal conduction is ineffective in establishing equilibrium) was noted very early by Hopkins (1839) in his "Preliminary observations on the refrigeration of the globe", and mantle convection has been specifically considered by Bull (1921), Holmes (1931), Pekeris (1935), Hales (1936) and, subsequently, by many others. Despite this long history, there is still no indication of a consensus on the nature of mantle dynamics and how it is related to the relatively well-established plate tectonics. The empirical and mechanistic picture of plate tectonics explains most of what we directly observe, but it may not have direct implications for the flow at much deeper levels.

Unfortunately, most of the relevant data for constraining speculation on mantle convection are either elusive or ambiguous. Given these circumstances, we propose to discuss a very general picture of global dynamics, *before* attempting to understand the most recent 10 per cent of the Earth's history (i.e. all of geology except the Precambrian) and the outermost 5 per cent of the Earth's volume (i.e. the lithosphere). The fundamental hypothesis in this review is that thermal convection (by which we mean any motion that is primarily driven by temperature differences) plays a dominant role in the thermal and structural evolution of the Earth. We shall proceed by discussing simple models based on general theoretical considerations, guided in part by the evidence from appropriately chosen laboratory experiments. The three fundamental problems that we consider are:

1. Why is mantle convection necessary?
2. What is the nature and scale of mantle convection?
3. How is this related to plate tectonics? (i.e. what makes the plates move?)

We begin in Section 2 by demonstrating that mantle convection is inevitable for a wide range of mantle rheologies and thermal histories, and that it can persist for about 10^{10} years, once it is initiated.

In Section 3, we consider the modifications required to allow for real Earth effects: finite compressibility, viscous heating, non-linear and variable rheology, non-uniform distribution of energy sources, chemical layering and phase transitions. We conclude that convection throughout the mantle is likely, although it could be confined to layers rather than being an overturning of the whole mantle. The relationship between the idealized models and a variety of geophysical observations is also discussed.

In Section 4, we consider the forces which drive the plates. The discussion is separated from the preceeding analysis of mantle convection since many aspects of the plate motions can be understood without invoking strong coupling between the plates and deep-mantle convection. The "kinematic" modelling of plate tectonics is critically assessed, and the dynamics of subduction are discussed in some detail.

In Section 5, we conclude by outlining the ways in which the general picture of deep mantle convection may be reconciled with the observed plate tectonics. Suggestions for future observational, theoretical and laboratory work are given.

2. The Inevitability of Mantle Convection

In this section, we consider the necessary conditions for mantle convection and indicate why these conditions are almost certainly satisfied by the Earth. These conditions are (a) the ability of solid matter to behave as a viscous fluid over sufficiently long time scales at a sufficiently high (subsolidus) temperature; (b) the existence of internal heating sufficient to ensure the necessary temperature; (c) the inability of thermal conduction to eliminate this heat during the age of the Earth. The resulting finite amplitude convection is discussed, and the persistence of this convection, even when sustained solely by gradual cooling of the Earth, is established.

2.1. The Rheology of Matter

Among the mechanisms that cause the irreversible deformation (creep) of hot but solid matter subjected to stress, the two considered most relevant to the mantle are diffusion creep (Herring, 1950; Gordon, 1965; Vetter and Meissner, 1977) in which the strain rate is proportional to the stress; and dislocation glide (Weertman, 1970) in which the strain rate is proportional to the cube of the stress. Both behaviours allow arbitrarily large strain so that solids have no long-term strength. This property ensures that an "effective viscosity" can be defined for geological time scales (although this "viscosity" does depend on the average stress level, if dislocation glide is appropriate). There is a dissenting view (Jeffreys, 1976), but we shall assume that the mantle can be thought of as a viscous fluid on geological time scales. (The mechanical behaviour of solids is discussed further by Paterson, Chapter 14 in this volume.)

Regardless of the rheological details, the effective viscosity is strongly temperature dependent. For example, diffusion creep predicts that

$$\mu \sim 10^6 \exp(AT_M/T) \qquad \text{P}, \tag{1}$$

where T_M is the melting temperature and $A \sim 30$ at the pressures of interest (Turcotte and Oxburgh, 1972). For $\mu \sim 10^{22}$ P, the value inferred from postglacial uplift (Cathles, 1975; Peltier, 1976; but see Sections 3.3 and 3.4) $T \sim 0.8T_M$ and μ changes by an order of magnitude if T changes by only about 5 per cent. The values of A and T/T_M may be smaller in the lower mantle, but of most importance is the strong temperature dependence of μ, which ensures that its value is determined by the mantle convection, rather than vice versa (see Sections 2.3 and 2.4).

A viscosity of 10^{22} P has the following significance. Consider a fluid in which deviations from a neutrally stable density distribution of amplitude $\Delta\rho$ and scale λ are imposed. The time τ_b required for the deformations caused by buoyancy to reach an amplitude $\sim\lambda$ is then $\tau_b \sim \mu/g\,\Delta\rho\lambda$. For $\mu = 10^{22}$ P, $g = 10^3$ cm s^{-2}, $\lambda \sim 10^3$ km, $\Delta\rho \sim 10^{-4}$ g cm^{-3}, $\tau \sim 10^{15}$ s $\simeq 5\times10^7$ years, comparable to the tectonic time scale. (If thermal diffusion were negligible, then $\Delta\rho = \alpha\,\Delta T\rho \sim 10^{-4}$ g cm^{-3}, where α is the coefficient of thermal expansion, would correspond to temperature variations of about 10 K). The induced stresses in this example are of order 10 bar.

2.2. Alternative Processes

There are many processes besides thermal convection which might in principle produce stress and sustain motions in the Earth. These include expansion of the Earth (Carey, 1976), contraction of the Earth (Lyttleton, 1965, 1976), large-scale inhomogeneities in the mantle (Artyushkov, 1971, 1972; Sorokhtin, 1974), gradual differentiation of the core from the mantle (Urey, 1952; Runcorn, 1962), tidal stresses (Bostrom, 1971) and fossil bulges (long-lived deviations from isostasy and hydrostatic equilibrium). These should be considered before we confine ourselves to thermal convection.

Substantial expansion or contraction of the Earth is contrary to palaeomagnetic and other data (McElhinny *et al.*, 1978; and Lambeck, Chapter 2 in this volume). Primordial large-scale inohomogeneities or fossil bulges would not persist for the age of the Earth if present estimates of mantle and lithospheric rheology are even approximately correct. Differentiation of the core from the mantle was probably a catastrophic process in the early Earth (Ringwood, 1960; Tozer, 1965). Tidal stresses are inadequate to drive plate tectonics (Jordan, 1973), especially since most tidal dissipation is in the oceans (Lambeck, 1975). It is for these reasons that we concentrate on thermal convection as a source of motion. The possibility that the mantle might be unstable to convective overturn was first discussed in detail by Hales (1936), and subsequent work has confirmed his view.

2.3. The Onset of Thermal Convection

We consider mainly convection initiated by a vertical, rather than a horizontal temperature gradient. Horizontal gradients undoubtedly exist (because radioactive heating is concentrated in continental crust) and could initiate convection (Pekeris, 1935; Howard *et al.*, 1970; and see Section 4). However, the rapid apparent motion of plates containing little or no continental material (e.g. the Pacific plate) indicates that horizontal gradients cannot be the primary cause of the motions.

The stability of a fluid layer, heated from below, has become a classic problem, which is fully discussed by Chandrasekhar (1961) and (in the context of the mantle) by Turcotte and Oxburgh (1972). We omit the formal analysis and discuss only the dimensionless groups which characterize the convection. In the simplest approximation, known as the Boussinesq approximation, the fluid is assumed to be incompressible (except for thermal expansion, which provides the buoyancy) and convective onset in a non-rotating fluid is governed entirely by the Rayleigh number, Ra, which is essentially the ratio of the driving force (thermal buoyancy) to the retarding force (diffusion of heat and momentum). For a fluid layer of depth H, with constant kinematic viscosity $\nu \equiv \mu/\rho$ and thermal diffusivity κ,

$$\mathrm{Ra} \equiv \frac{g\alpha\beta H^4}{\nu\kappa}, \tag{2}$$

where g is the gravitational acceleration, α is the coefficient of thermal expansion, and β is the *difference* between the actual temperature gradient and the adiabatic temperature gradient. (An adiabatic temperature distribution is neutrally stable and only deviations from it can provide buoyancy.) If Ra exceeds a critical value $\mathrm{Ra_c} \sim 10^3$, then convection is initiated, and is characterized by a horizontal wavelength λ comparable to the vertical wavelength (i.e. an aspect ratio $\lambda/H \sim 2$). Both $\mathrm{Ra_c}$ and the aspect ratio depend somewhat on the boundary conditions (Chandrasekhar, 1961), but these details need not concern us. Similar results apply for a non-planar geometry. For internal heating, rather than an externally imposed temperature difference, the relevant Rayleigh number can still be defined as in (2), except that β is now the average superadiabatic temperature gradient that would be required for a conductive steady state (Roberts, 1967).

Since the Earth rotates, the Coriolis effect should be considered. This is conventionally parameterized by Ta, the Taylor number:

$$\mathrm{Ta} \equiv 4\Omega^2 H^4/\nu^2, \tag{3}$$

where Ω is the rotation rate. In the Earth, $\Omega \sim 10^{-4}\ \mathrm{s}^{-1}$, $H \sim 10^8$ cm, $\nu \sim 10^{22}\ \mathrm{cm}^2\ \mathrm{s}^{-1}$ implies $\mathrm{Ta} \sim 10^{-20} \ll 1$, so that the Coriolis effect should be unimportant in mantle convection. There may, however, be indirect effects. For example, rotation causes the Earth to be oblate, and this in turn modifies plate dynamics (see Section 4). Nevertheless, we can neglect rotation in our discussion of convective *onset.*

We could try to evaluate Ra in the Earth, and assess whether it exceeds $\mathrm{Ra_c}$, using the poorly known values of the physical properties. Instead, we outline below a thermal evolution model which we believe demonstrates the "inevitability" of mantle convection. The model is based on the strong temperature dependence of the viscosity, rather than on any particular value of μ inferred from present-day observations, and it indicates how the thermal evolution and convective efficiency can determine the viscosity. The point has been made before (Tozer, 1965) but has been insufficiently appreciated.

Consider an infinite horizontal layer, thickness H, of a fluid with a temperature-dependent viscosity. Suppose that it contains a uniformly distributed energy source, that the bottom ($z = H$) is bounded by a perfect insulator, and the top ($z = 0$) is fixed at $T = 0$ °C. The isothermal boundary condition is appropriate for a planet, such as the Earth, where the geothermal heat flux is much less than the insolation.

At time $t = 0$, we suppose $T = 0$ everywhere. Subsequently, but before convection occurs, the temperature distribution obeys the diffusion equation,

$$\frac{\partial T}{\partial t} = \kappa \frac{\partial^2 T}{\partial z^2} + \frac{T_{\mathrm{M}}}{\tau_{\mathrm{M}}}, \tag{4}$$

where the source term on the right has been written in terms of the melting temperature T_{M} (assumed constant) and τ_{M}, the time that the internal heat source would take to melt the material (assuming no heat loss). For $H \sim 3 \times 10^8$ cm and $\kappa \sim 10^{-2}\ \mathrm{cm}^2\ \mathrm{s}^{-1}$ (Stacey, 1977), $H^2/\kappa \sim 3 \times 10^{11}$ years, which is much longer than the age of the Earth. In other words, the conductivity is so small that the heat generated cannot escape by conduction alone during the age of the Earth. Solutions to (4), showing the increase of T with time, are shown in Fig. 1, for the (arbitrary) choice $\tau_{\mathrm{M}} = 2 \times 10^9$ years. (The Earth was probably gravitationally heated far more rapidly than this—see Section 3.)

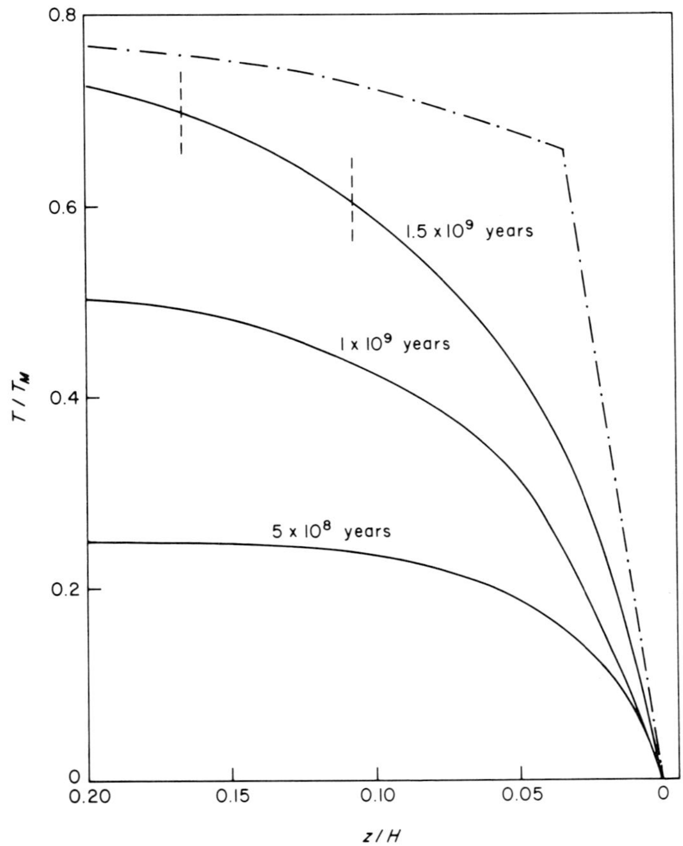

Fig. 1. Thermal diffusion in an internally heated body of depth H. The surface $z=0$ is fixed at $T=0$ °C. The three curves are for different stages of the evolution, and are labelled by the elapsed time in years since the initial isothermal state, assuming $\tau_M = 2\times 10^9$ years. The dashed vertical lines at $t=1{\cdot}5\times 10^9$ years indicate the region which first becomes convectively unstable (see text). The dot-dashed curve is a schematic representation of the final ($t\to\infty$) convective-diffusion steady state.

Suppose, now, that the effective viscosity in the layer is given by (1) with $A=30$. As T increases, μ rapidly decreases. Provided $\tau_M \lesssim H^2/\kappa$, a time is inevitably reached at which convection becomes possible. The details of the onset are complicated by the variable temperature gradient and viscosity, but we can define a Rayleigh number at each level Ra(z), based on a depth interval centred on z (in which the temperature gradient changes by a factor of two, an arbitrary choice), and the viscosity at the average temperature in the layer. (This procedure is partly justified by experiments on variable viscosity convection; Booker, 1976). It is found that the maximum of Ra(z) occurs at $z \simeq 2.0\ (\kappa t)^{1/2}$ and exceeds 10^3 at $t \sim 0.8\ \tau_M$ (see Fig. 1). In this example, the onset of convection is rather like the breakdown of a boundary layer in a fluid cooled from above (Howard, 1964), with one very important difference: the outermost layer is much more viscous than the deeper layers and does not, therefore, participate significantly in the initial flow.

The subsequent evolution is complicated, but in general one can say that all regions with the possible exception of an outermost highly viscous layer, must eventually become convective if $\tau_M \lesssim H^2/\kappa$. This follows from the fact that any non-convective region must continue to heat up, because conduction is so small, and must have a progressively lower viscosity. Convective mixing will always become possible before melting occurs, and a convective steady state ensues. The properties of this convective steady state are described below.

The relationship of the above highly idealized model to the real world is discussed in Section 3. A much more detailed evolutionary model is to be found in McKenzie and Weiss (1975).

2.4. Finite Amplitude Convection

Up to this point, our analysis has been limited to a discussion of whether or not convection is initiated. However, if $\mathrm{Ra} \sim \mathrm{Ra_c}$, then the heat flux is still primarily transported by conduction. In the model just discussed, $\tau_M \ll H^2/\kappa$, so it is clear that the eventual steady state must have a large convective heat transport and $\mathrm{Ra} \gg \mathrm{Ra_c}$.

In the finite amplitude flow, three more dimensionless groups are relevant: the Prandtl number Pr, the Reynolds number Re and the Peclet number Pe. These are defined by

$$\mathrm{Pr} \equiv \nu/\kappa, \tag{5}$$

$$\mathrm{Re} \equiv uL/\nu, \tag{6}$$

$$\mathrm{Pe} \equiv \mathrm{Pr}\,\mathrm{Re} = uL/\kappa, \tag{7}$$

where u is a characteristic convective velocity and L is some appropriately chosen length scale (usually, but not always, the depth of the layer). In the Earth's mantle, $\mathrm{Pr} \sim 10^{24}$, which means that the viscous response to a perturbation is instantaneous relative to the thermal response. For $u \sim 10^{-7}\ \mathrm{cm\ s^{-1}}$, characteristic of tectonics, and $L \sim 10^8$ cm, $\mathrm{Re} \sim 10^{-21}$ and $\mathrm{Pe} \sim 10^3$. $\mathrm{Re} \ll 1$ implies that inertial effects are negligible in the fluid flow, whilst $\mathrm{Pe} \gg 1$ implies that convective heat transport is much more important than conductive heat transport.

No analytical solution exists for $Ra \gg Ra_c$, but Turcotte and Oxburgh (1967, 1972) have developed a boundary layer theory for $Pr \to \infty$ which agrees with the numerical experiments (McKenzie *et al.*, 1974). In this theory, the fluid is assumed to be adiabatic everywhere except in two regions: the boundary layers (where the vertical convective velocity must tend to zero) and narrow plumes within which most of the vertical transport occurs. With the simplest assumptions of constant viscosity, dynamically free boundaries, two-dimensional flow, and a fluid layer that is heated from below and cooled from above, the convective pattern is shown in Fig. 2.

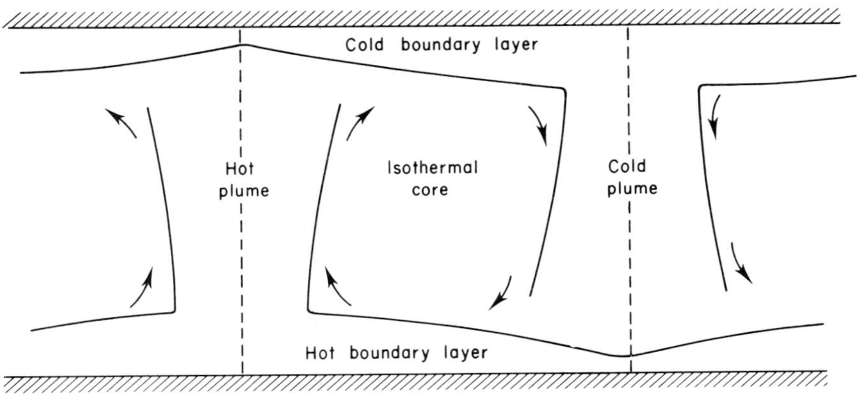

Fig. 2. Illustration of the boundary-layer model for two-dimensional thermal convection, assuming both cooling from above and heating from below. (From Turcotte and Oxburgh, 1972.)

(With internal heating alone, only the cold, descending plumes are formed.) Most of the temperature drop ΔT occurs across the boundary layers of thickness δ. In a boundary layer, the horizontal convection of heat is balanced by thermal diffusion vertically:

$$u\frac{\Delta T}{H} \sim \kappa\frac{\Delta T}{\delta^2}, \tag{8}$$

where u is the magnitude of the horizontal velocity. In the vertical plumes, the generation of vorticity is balanced by viscous dissipation:

$$g\alpha\frac{\Delta T}{\delta} \sim \nu\nabla^2\omega \sim \frac{\nu\omega}{\delta^2}, \tag{9}$$

where ω is the only non-zero component of vorticity (perpendicular to the plane of Fig. 2). Away from the plumes and boundary layers, $\nabla^2\omega = 0$, so

$u \sim v \sim \omega H$, where v is the magnitude of the vertical velocity within the plume. If we define $\mathrm{Ra} = g\alpha\, \Delta T H^3/\nu\kappa$, then it follows that

$$\delta \simeq H(\mathrm{Ra_c}/\mathrm{Ra})^{1/3},$$

$$u \sim v \simeq 0.2\ \mathrm{Ra}^{2/3}\kappa/H,$$

$$F_{\mathrm{conv}} \simeq \frac{\rho C_p v \Delta T \delta}{H} \simeq \frac{\rho C_p \kappa \Delta T}{H}\left(\frac{\mathrm{Ra}}{\mathrm{Ra_c}}\right)^{1/3}, \tag{10}$$

where F_{conv} is the convective heat flux, and the numerical factors deduced by Turcotte and Oxburgh's more refined analysis are included. A convenient measure of the heat flux is Nu, the Nussett number, defined as the ratio of the total heat flux to that which would be transported (for the same ΔT) by conduction alone. From (10).

$$\mathrm{Nu} \simeq (\mathrm{Ra}/\mathrm{Ra_c})^{1/3}. \tag{11}$$

Turcotte and Oxburgh (1967, 1972) sought to apply this analysis directly to mantle convection: downgoing plumes are identified with subduction zones and upgoing plumes with ocean ridges. This idea is attractive in its simplicity, and although the real world is undoubtedly more complicated (see Section 3), it suggests a way of crudely estimating Ra for the mantle, without any assumption for the viscosity. This estimate is obtained by noting that for $\kappa \sim 10^{-2}\ \mathrm{cm^2\ s^{-1}}$, the heat flux transported by conduction is over an order of magnitude less than the total geothermal heat flux of $3 \times 10^{20}\ \mathrm{erg\ s^{-1}}$. Since a large fraction of this is believed to originate from the mantle (Stacey, 1977), a reasonable estimate for the Nusselt number is $\mathrm{Nu} \sim 10$. From (11), this suggests $\mathrm{Ra} \sim 10^6$, and $\mu \sim 10^{22}$ or 10^{23} P, consistent with postglacial uplift (Cathles, 1975; Peltier, 1976). The viscosity is thus determined by the heat flux and the efficiency of mantle convection. However, no account has yet been taken of the variation of viscosity with depth, or the distribution of heat sources, or whether convection is throughout the mantle: these problems are considered in Section 3.

Two other properties of the convection are of importance in understanding the Earth: time variability and the dependence of the convection pattern on Ra. It is known from experiments that the convection exhibits a series of discrete transitions to more complex flows as Ra increases. For $\mathrm{Pr} \gg 1$ and $10^3 \lesssim \mathrm{Ra} \lesssim 2 \times 10^4$, convection has the form of two-dimensional "rolls" (Fig. 2) but at $\mathrm{Ra} \sim 2 \times 10^4$ there is a transition to three-dimensional bimodal flow (Krishnamurti, 1970*a*; Busse and Whitehead, 1971). At $\mathrm{Ra} \gtrsim 5 \times 10^4$ and $50 \lesssim \mathrm{Pr} \lesssim 10^4$, Krishnamurti (1970*b*, 1973) observed the onset of an oscillatory flow. Busse and Whitehead (1974) showed that the critical Rayleigh number at which the flow became oscillatory is an increasing function of Pr, and since Pr is so large in the mantle, this suggests that mantle convection may be steady at $\mathrm{Ra} \sim 10^6$ or even $\mathrm{Ra} \sim 10^9$. There are problems in interpreting these experiments (see Whitehead, 1976, for a review) and their applicability to the mantle can be questioned, but it is clear that any time-dependent behaviour in the mantle must be of thermal rather than inertial origin, since $\mathrm{Re} \ll 1$. In the boundary layer model, oscillations on a time scale similar to $H/u \sim \delta^2/\kappa$ are conceivable, although convection theory or experiment do not provide strong evidence for them. In the Earth, this time scale would be of the order of 10^8 years.

Despite the transitions in the convective pattern, the aspect ratio λ/H does not appear to be substantially modified, even as Ra changes by many orders of magnitude. Theoretical work predicts that λ/H should decrease as Ra increases (Koshchmieder, 1974) whereas the experimental situation is unclear at present (probably because the infinite Pr limit has not been approached). There is no evidence from constant viscosity experiments or theory to support the large aspect ratio that the large plate sizes on the Earth's surface would seem to suggest, if convection were confined to the upper mantle or were in layers. Possible explanations for this are considered in Sections 3 and 4.

2.5. The Persistence of Convection

It is often argued that the absence of seismicity in the lower mantle indicates that convection is confined to the upper mantle. This argument is unconvincing (see Section 3.6). If convection ever existed in the lower mantle, it could persist, driven by its own internal heat content (and also that of the liquid core), without any need for radioactive heating. In this section, it is shown that whole mantle convection might persist for the age of the Earth, driven solely by the primordial heat content.

Consider a mantle of thickness H and average temperature $\bar{T}$. In accordance with the boundary layer model, the mantle is assumed to consist of an outer conductive layer, across which is a temperature drop ΔT, and an adiabatic interior region. The heat flux through the layer is then given by equation (10). This is balanced by the gradual loss of mantle heat content:

$$\frac{\rho C_p \kappa\, \Delta T}{H}\left(\frac{\mathrm{Ra}}{\mathrm{Ra_c}}\right)^{1/3} = -\rho C_p \frac{\partial \bar{T}}{\partial t} H. \tag{12}$$

We define $\theta = \Delta T(t)/\Delta T(0)$, and measure time in units of $t' = \kappa t/H^2$. The temperature dependence of the viscosity is parameterized as $\nu \propto \exp\,(A/\theta)$, where, for example, (1) indicates $A \sim 40$. The thermal evolution is then governed by

$$\frac{\partial \theta}{\partial t'} = -\gamma \theta^{4/3} \exp\left\{-\frac{A}{3}\left(\frac{1}{\theta} - 1\right)\right\},$$

$$\gamma \equiv \frac{\Delta T}{\bar{T}}\left(\frac{\mathrm{Ra}_0}{\mathrm{Ra}_c}\right)^{1/3}, \tag{13}$$

where Ra_0 is the Rayleigh number at $t = 0$. Figure 3 shows the solutions of (13) for several choices of Ra_0 and A, assuming $H = 3000$ km and $\Delta T = \frac{1}{2}\bar{T}$. The model indicates that despite the strong temperature dependence of viscosity, gradual cooling of the Earth does not rapidly shut off convection.

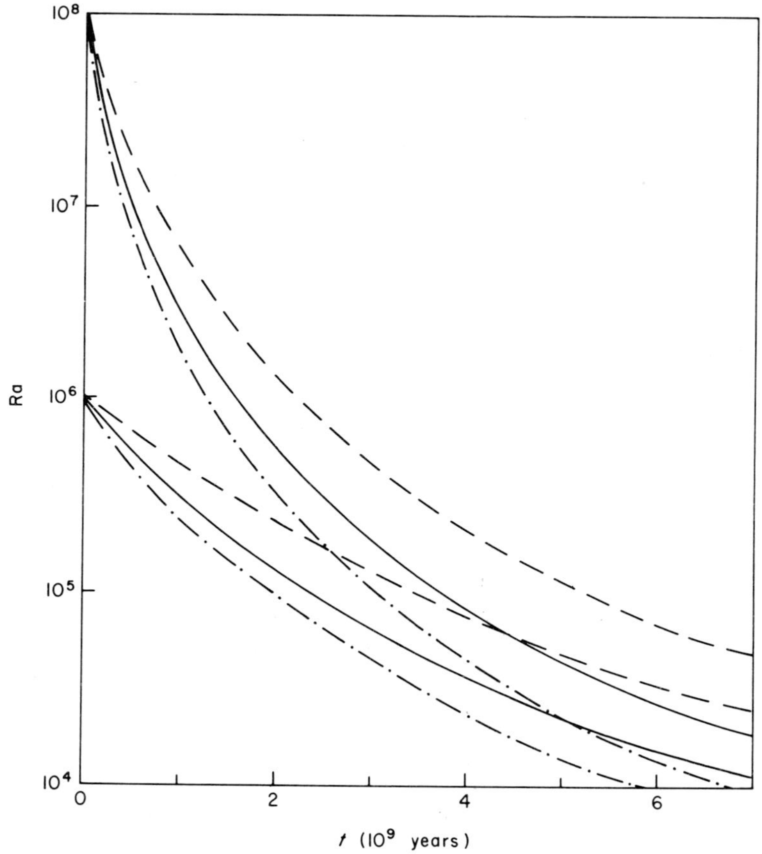

Fig. 3. Convective cooling of the mantle with no internal heat sources. Two sets of curves are shown for two choices of the initial Rayleith number Ra_0 and three choices of the viscosity exponent A (see text): ———, $A = 42$; – – – –, $A = 30$; – · – · – · –, $A = 54$.

This does *not* mean that the entire geothermal heat flux could result from cooling of the Earth, but suggests that cooling could play an important role. This is discussed further in Section 3.1.

The model eventually becomes invalid as the outer boundary layer thickness δ increases, either because this layer becomes too rigid to participate in the motions, or because conductive steady state takes longer to establish in the boundary layer than the age of the Earth. These complications probably arise at $\mathrm{Ra} \sim 10^4$, where $\delta^2/\kappa \sim 5 \times 10^9$ years.

3. The Real Earth

In this section, we consider various modifications to the above simple picture of convection which seem to be required by the actual conditions in the Earth. The emphasis is on understanding the scale and nature of the mantle flow, and we defer till the next section any detailed consideration of the relation between this and plate tectonics. The extent to which any predictions of convection theory can be supported by geophysical observations will also be discussed.

3.1. The Distribution and Nature of the Energy Sources

Only gradual cooling of the Earth and radioactive decay are important in driving mantle convection. In Section 2.5, it was shown that gradual cooling could indeed be significant. (The present total heat content of the Earth is about 1.2×10^{38} ergs, sufficient, in principle, to maintain the present geothermal heat flux for more than 10^{10} years.) This has been emphasized by many others, including Urey (1952) and Elder (1976). Although it is also possible, in principle, to account entirely for the geothermal heat flux from radioactive decay alone (Wasserburg *et al.*, 1964), this is not demanded by observations and geochemical arguments. Indeed, Ringwood (1975) requires gradual cooling to contribute significantly, since his compositional models have insufficient radioactive heating. Generation of the Earth's magnetic field also appears to require cooling (Gubbins, 1976), perhaps by about 1 °C every 10^8 years (see Lilley, Chapter 3 in this volume).

Unlike gradual cooling, which will tend to be uniform, radioactive decay is clearly non-uniform, with perhaps one-third being generated in the continental lithosphere alone (Pollack and Chapman, 1977). The distribution of radioactive elements in the mantle and core is not well understood, but there is no reason to suppose that nearly all of the energy originates very near the Earth's surface. A uniform distribution of radioactive elements in the mantle would certainly sustain convection. In summary, convection throughout the mantle is likely, and models should include heating at the bottom, as well as cooling at the top and internal heating, since the Earth's core is necessarily losing energy. The arguments for whole mantle convection are discussed further in Section 3.6.

3.2. Viscous Dissipation and Finite Compressibility

Even in the absence of phase transitions and chemical layering, the mantle would have a density gradient because of the *static* gravitational self-compression. This modifies the Boussinesq equations by introducing another length scale $H_T \equiv g\alpha/C_p$, the temperature scale height. (Equivalently, H_ρ, the density scale height, may be used.) It is apparent that finite compressibility has an important effect on the flow which depends on the magnitude of H/H_T. It is less obvious that the same parameter governs the importance of viscous dissipation. This can be seen explicitly in the boundary layer model discussed in Section 2.4. The viscous dissipation per unit layer area is $\rho\nu\omega^2H$ and it can easily be shown that this is roughly H/H_T times the heat flux. The importance of viscous dissipation is discussed in detail by Hewitt *et al.* (1975). *If* H/H_T is of order unity or larger, then they find that the emergent heat flux from the convecting layer can originate mainly from the viscous dissipation.

If the whole mantle is convecting, $H/H_T \sim 0.5$ and viscous dissipation is a significant but not dominant consideration. Of course, viscous dissipation does not alter the ultimate energy source, but it does lead to an *apparent* redistribution of energy sources and thereby modifies the flow. In general, both finite compressibility and viscous dissipation tend to decrease the flow speed for a given Ra, and act as a *stabilizing* influence, at least for constant viscosity (Turcotte *et al.*, 1974). However, the simple picture of Section 2.4 is not greatly modified. This means that laboratory experiments (where $H/H_T \ll 1$ always) can still have some relevance for mantle convection.

The possibility of non-uniqueness and oscillatory behaviour caused by the non-linear viscous dissipation has been stressed by Rice (1971) but the evidence is inconclusive.

3.3. Effect of Variable Viscosity

The strong dependence of mantle viscosity on temperature and pressure is probably the most important source of deviations from the simple picture of Section 2.3. If phase transitions and chemical layering are absent, then the depth-dependence of the viscosity most probably determines both the horizontal and vertical length scales of the convection. Since Ra is a strong function of the vertical length scale H ($\text{Ra} \propto H^4$ for a constant temperature gradient) the viscosity must increase rapidly with depth to limit the convection to near-surface regions.

As we emphasized in Section 2, it is heat flux and the convective efficiency which together determine the (average) viscosity. For example, if we suppose that the mantle consists of several convective layers, through which a constant convective heat flux is transported, then approximate equality of Ra for each layer would imply approximate equality of viscosity for each layer. A modest increase of viscosity with depth would occur if the heat sources are concentrated near the surface, or if the deeper layers are thicker. Recent work on the gravity field and on postglacial rebound indicates that the mantle has a rather uniform viscosity (Goldreich and Toomre, 1969; Cathles, 1975; Lambeck, 1975; Peltier and Andrews, 1976; Peltier, 1976). This result is *not* predicted by the theory of mantle rheology directly, but is rendered plausible by convection theory.

One region where the viscosity is certainly not uniform is in the outermost few hundred kilometres of the Earth. The outermost layer (the lithosphere) of thickness ~100 km behaves rigidly, at least with respect to perturbations of horizontal wavelengths $\lesssim$10–100 km and for time scales $\lesssim 10^7$ years (Walcott, 1970). Immediately below this layer there may be a low viscosity "channel" in which $\nu \sim 10^{20}$ cm^2 s^{-1} (Cathles, 1975). This channel is about 100 km thick and is consistent with postglacial uplift studies and the theory of diffusion creep (Turcotte and Oxburgh, 1972). It is not necessarily coincident with a region of low seismic velocities or partial melting (Auten *et al.*, 1974). In deeper regions, the viscosity may gradually increase, but the variation in viscosity throughout the mantle is probably less than two orders of magnitude. Figure 4 indicates the likely range of viscosity variation with depth. There is some evidence for a non-Newtonian rheology in the mantle (Post and Griggs, 1973), but an effective viscosity should be definable (see Section 3.4).

Numerical or laboratory experiments are required to understand the effect of variable viscosity on convection. However, one fundamental aspect of convection in a fluid with strongly varying viscosity can be understood from the properties of constant viscosity convection. Consider two layers in which the lower and upper layers have thicknesses H and d, respectively, and viscosities ν_1 and ν_2, respectively. We assume $\nu_1 \gg \nu_2$ and $H \gg d$. The two layers could be identified with the lower and upper mantle or with the mantle and low-viscosity channel. For a given superadiabatic temperature gradient β, what is the nature of the convective onset? Define

$$\begin{aligned} \mathrm{Ra}_1 &\equiv g\alpha\beta H^4/\nu_1\kappa, \\ \mathrm{Ra}_2 &\equiv g\alpha\beta d^4/\nu_2\kappa. \end{aligned} \tag{14}$$

If $H > d(\nu_1/\nu_2)^{1/4}$, then we might expect the entire fluid to participate in the convection, with horizontal and vertical length scales of order H. If $H <$

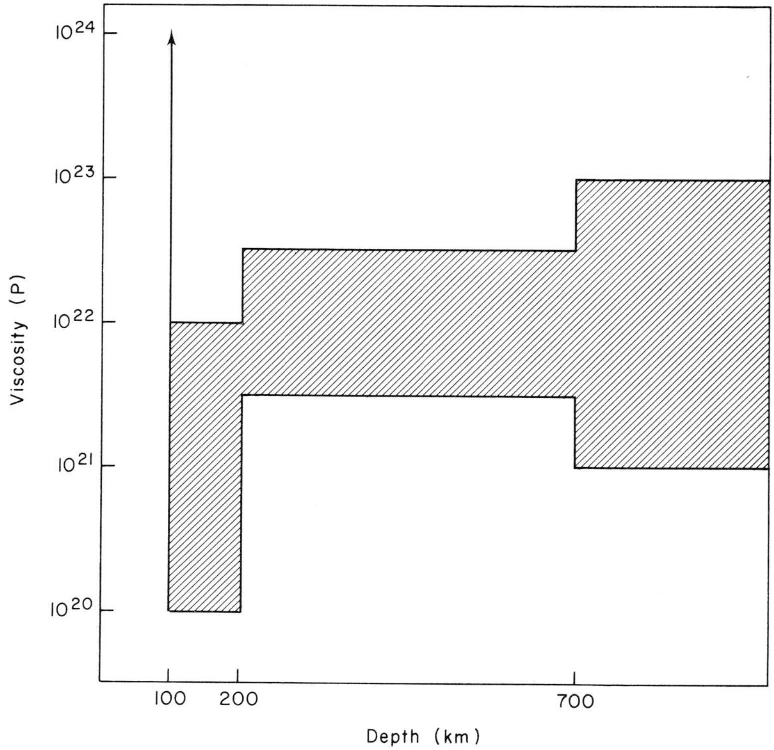

Fig. 4. Mantle viscosity as a function of depth, as inferred from uplift data and other information. The upper 100 km was treated as an elastic plate, and the shading represents the uncertainty. The data are consistent with a uniform viscosity lower mantle extending to a depth of 3000 km. (See text for references.)

$d(\nu_1/\nu_2)^{1/4}$, then $Ra_2 > Ra_1$, and most of the flow would be confined to the thin low-viscosity layer, with horizontal and vertical length scales of order $d \ll H$. A rather abrupt change in convective pattern might be expected, therefore, near $H \sim d\ (\nu_1/\nu_2)^{1/4}$. This is exactly what Davies (1977*a*) observed in his numerical experiments. He found that for $H = 4d$, the abrupt transition in convection occurred at $\nu_1/\nu_2 \sim 10^4$ or 10^5. It is evident that convection favours the largest available length scale. In the absence of phase transitions and chemical layering, this argument would suggest whole mantle convection.

We have ignored the role of the lithosphere in the above discussion. One approach is to treat it as a highly viscous fluid layer. Experiment and theory (Booker, 1976, and references therein) suggest that a thin high-viscosity boundary layer would tend to be stagnant and not participate much in the convection. If this happens, then the aspect ratio *decreases* because the convection is confined to a shallower layer. Different boundary conditions can modify the conclusion (McFadden and Smylie, 1968) and some numerical experiments (Houston and De Bremaecker, 1975; De Bremaecker, 1977) suggests that the aspect ratio could *increase*. However, numerical experiments must be carried out in a finite box, so such predictions must be viewed with caution (Torrance and Turcotte, 1971). Laboratory experiments can be similarly misleading. If the plates can be regarded as boundary layer, as the Turcotte–Oxburgh theory requires, then they participate more vigorously in mantle convection than the results of any experiment would suggest.

To summarize this section, viscosity variations are probably insufficient to limit convection to the upper mantle, and can modify the horizontal scale of the convection in a way that is not yet well understood.

3.4. Convection with a Non-Newtonian Rheology

Even if the mantle possesses long-term strength, convection may still be possible (Birger, 1976). We confine our attention, however, to non-Newtonian rheologies in which there is no long-term strength. In particular, dislocation glide is favoured by some (Post and Griggs, 1973) as the correct mantle rheology. Parmetier *et al.* (1976) have analysed finite amplitude convection for this rheology, and have found that the behaviour can be parameterized in terms of a Rayleigh number involving an effective (stress-averaged) viscosity. They conclude that the temperature and pressure dependence of the rheology is more important than its non-linearity.

This may not be true for the plates, which behave in some respects as rigid bodies. Note, however, that the plates would appear to be rigid, even if they consist of a high-viscosity fluid, because convection tends to concentrate its shearing motions in low-viscosity regions.

3.5. Phase Transitions and Chemical Layering

Seismic evidence indicates density discontinuities at the Moho (30–50 km depth under continents; 10–12 km under oceans), at 400 km and at about 650 km. The Moho is evidently a chemical boundary (Ringwood, 1975) and need not concern us here, although it may play a role in subduction dynamics (Section 4.5). Both the 400 km and 650 km transitions are believed to be polymorphic phase changes, each involving density changes of 5–10 per cent. The 650 km transition may be partly compositional, but for the moment we assume that the average molecular weight is unchanged at the transition. Flow through a phase boundary is then governed by two competing effects: latent heat and the density change. Consider, for example, the 400 km (olivine–spinel) transition, which is exothermic. Cold, downwelling material crossing the phase boundary becomes heated, because of the latent heat, and loses part of its negative buoyancy. However, the phase transition pressure is an increasing function of temperature in an exothermic reaction. This implies that cold, downwelling material locally raises the phase boundary, providing a hydrodynamic head to enhance the flow. These effects are shown schematically in Fig. 5, for both downwelling and upwelling flows.

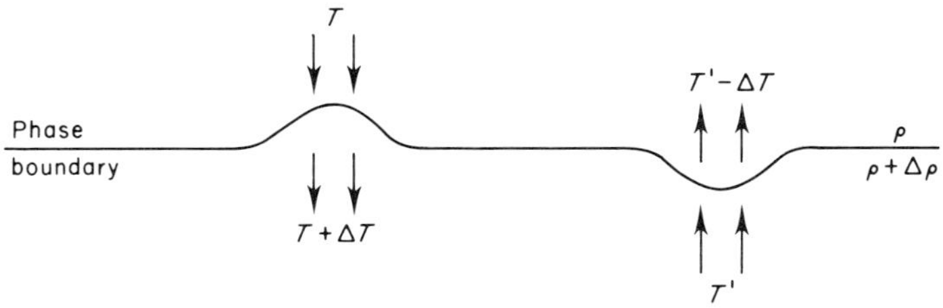

Fig. 5. The influence of an univariant phase boundary on convection. The exothermic case is shown ($\Delta T > 0$). Downgoing fluid on the left loses part of its negative buoyancy as it changes phase and heats up. This inhibits the flow. However, the cold plume elevates the phase boundary, providing a hydrodynamic head effect (because of the density difference). The net effect of these competing effects is not obvious (see text). Similar considerations apply to an upgoing plume on the right.

The net effect is not obvious, but detailed calculations (Schubert *et al.*, 1975; Sammis, 1976) indicate that convection through the phase boundary is favoured rather than convection separately above and below the transition. This conclusion, based on linear stability analysis, has been confirmed by finite amplitude calculations (Richter, 1973*a*) which indicate that flow through a phase boundary can be readily initiated. All these calculations assume instantaneous chemical equilibrium, a situation that may not be achievable if a phase can be metastable (Sung and Burns, 1976). If metastability occurs, then the density discontinuity can strongly impede vertical flows.

To illustrate this, consider the opposite extreme to a phase transition, in which the fluid consists of two chemically distinct layers which differ in density by $\Delta\rho_c$ at the same temperature. When this system is heated from below, convection can occur in the two layers separately, driven by the heat fiux across the interface, while the density difference is preserved because of the negligible solute flux across the interface (solute diffusion is much less efficient than thermal diffusion in solids). The problem has been considered explicitly for the mantle by Richter and Johnson (1974) but the related phenomenon of "double-diffusive" convection is well-documented in the heat-salt system (Turner, 1973*a*, Chap. 8). The corresponding finite amplitude stability problem has recently been discussed by Huppert and Moore (1976).

It is characteristic of double-diffusive systems that layered convection persists while the overall density gradient remains (statically) stable. Complete overturning and the breakdown of the interface occurs only when $\alpha\Delta T$ due to the temperature difference becomes comparable with $\Delta\rho_c/\rho$. For the mantle properties assumed earlier ($\nu = 10^{22}\ \mathrm{cm^2\ s^{-1}}$, $\kappa = 10^{-2}\ \mathrm{cm^2\ s^{-1}}$) we find $\Delta\rho_c/\rho \lesssim 10^{-2}$ for instability. Density discontinuities of this order, or perhaps even smaller, set up in the mantle during an early partially molten phase of the Earth's history, might persist throughout the subsequent evolution. If the spinel–postspinel transition at 650 km is partly compositional, then this interface could force the convection to be layered, with coupled convective cells above and below the chemical boundary. In the case where convection is driven by cold plumes descending from the surface and impinging an interface, downward motion could continue in the lower layer, initiated by the faster local transport of heat. On the other hand, if the buoyancy flux across the interface itself dominates the motion, a symmetrical convection pattern would be generated, as shown schematically in Fig. 6. Note the large aspect ratio of the cells in the upper layer, which take

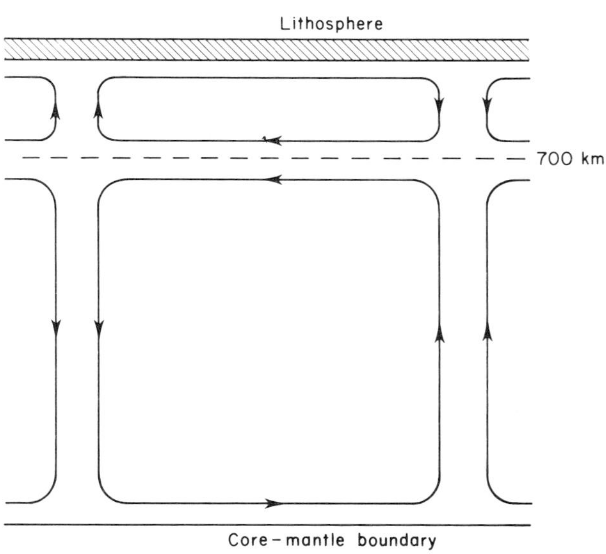

Fig. 6. Convective cells coupled across the (hypothetical) compositional boundary at ~700 km. Note the large aspect ratio of the upper mantle cells and the matching of upgoing with downgoing plumes at the interface.

Fig. 7. A laboratory experiment on double-diffusive convection in layers (P. F. Linden, private communication). Although the details of this experiment are not related to mantle convection, there is a strong net buoyancy flux across the interfaces which has become organized into coupled plumes, analogous to the layered mantle convection sketched in Fig. 6. Note the local thickening of the boundary layer as fluid is swept towards the region where the plumes break away from the interface in opposite directions.

their horizontal scale from the cells of the deeper lower layer. Figure 7 shows a laboratory experiment which is not intended as a model of the mantle (the boundary layer and plume structure have been produced in another way) but which illustrates the strong coupling of the plumes across a double-diffusive interface.

3.6. The Observational Evidence

The picture emerging from the above discussion is convection everywhere in the mantle, but possibly in layers rather than as an overturning of the whole mantle. Whole-mantle convection is not a new idea (Hales, 1936; Runcorn, 1962, 1967; Wilson, 1963; Tozer, 1965) but was in disfavour for a few years (McKenzie, 1969; Turcotte and Oxburgh, 1972). It has very recently become more favoured (McKenzie and Weiss, 1975; O'Connell, 1977; Davies, 1977*a,b*). If whole mantle overturn occurs, then the spherical geometry plays an important role in the convective pattern (Runcorn, 1962,

1967), but at present the convective pattern is not readily discerned, despite attempts to relate it to the gravity field (Runcorn, 1962) or to the geoid (Gough, 1977). At best we can indicate how the thermal convection hypothesis is compatible with current evidence.

The most obvious evidence for thermal convection is plate tectonics: the emergence of hot mantle material at ridges and the subduction of cold lithosphere at trenches. However, the relationship of this surface activity to deep mantle flow has not yet been determined. If the plates are a direct manifestation of mantle convection, then the size of the plates and their pattern of motions remains unexplained. We consider this problem in the remaining two sections.

Anomalies in the Earth's gravity field are a potential source of information about thermal convection. Unfortunately, interpretation of the gravity field is fundamentally non-unique since one cannot determine the depth of gravity anomalies from surface or near-surface observations alone. Nevertheless, lateral density anomalies exist in the mantle (Lambeck, 1976) and since these cannot be fossil effects (if the viscosity is less than about 10^{26} P), they indicate dynamic behaviour, consistent with mantle convection.

Attempts have been made to correlate gravity anomalies with other geophysical properties. McKenzie (1977) considered the correlations between residual depth anomalies (regional variations of ocean depth which do not result from contraction of plate material with age) and long-wavelength gravity anomalies. The correlation is explainable by mantle convection provided a modest (order of magnitude) increase of viscosity with depth in the mantle is allowed for.

There is much seismic evidence for inhomogeneities in the mantle (Alexander and Phinney, 1966; Julian and Sengupta, 1973; Dziewonski *et al.*, 1977), and there is some evidence for a correlation between these and the geopotential (Dziewonski *et al.*, 1977). This is consistent with whole-mantle convection.

Pollack and Chapman (1977) attempt to separate the lithospheric component of the heat flux from the total, and determine the purely mantle component. The mantle flow they find is positively correlated with the geopotential, consistent with mantle convection.

Hide and Horai (1968) propose that a major cause of gravity anomalies are "bumps" at the core–mantle boundary (for a review, see Hide, 1977). These long-wavelength bumps might be sustained by convection in the lower mantle. Jones (1977) has further proposed that intermittent breakdown of the boundary layer just above the core–mantle boundary could explain the observed long-period modulation of the Earth's magnetic field. Since there is no known core process with a characteristic time scale exceeding 10^6 years, this theory is currently the only acceptable explanation for long-period modulation, and to this extent supports the thermal convection hypothesis.

The absence of seismicity in the lower mantle is not a conclusive argument against lower mantle convection. Amongst the possible explanations are: (a) the inability of subducting slabs to penetrate the lower mantle; (b) the rise of temperature in the slab and the consequent reduction in creep strength; (c) the phase change at 650 km, causing a change in the rheological behaviour. Since the convection could be layered, whole mantle convection is not contingent on the penetration of subducting material into the lower mantle. However, Jordan (1977) presents seismic evidence for such penetration beneath the Sea of Okhotsk. The arguments for whole mantle convection are extensively reviewed by O'Connell (1977).

We conclude that there is no compelling evidence against convection throughout the mantle. The supporting evidence is encouraging, although not yet conclusive, and no competing hypothesis approaches the ability of thermal convection to explain a wide range of observations.

4. Plate Dynamics

In this section, we develop the view that plate tectonics can be understood as a separate physical system, the behaviour of which is affected (but not entirely determined) by deep mantle flows. The reconcilation of this approach with the general theory of mantle convection is left to the final section.

4.1. The Driving Forces

The forces acting on plates are of three kinds: the body forces, the forces acting on the bottom surfaces of plates, and those acting at plate boundaries. The body forces are discussed in Section 4.5, although for some purposes they can be incorporated in the boundary forces. The force acting on the bottom surface of a plate could aid plate motion, if mantle flow is a driving force, or could be a drag force, if the mantle is relatively passive. The evidence, both observational (Section 4.2) and theoretical (Section 4.3) indicate that this force is small.

Plate boundaries are of three major types: diverging, converging and shearing. The boundaries where there is a net transfer of material between lithosphere and mantle would appear to be dynamically most important: the diverging ocean ridge systems (Section 4.4) and converging subduction trench systems (Section 4.5).

4.2. Kinematic Modelling

One approach to the problem of plate motions is to hypothesize a simple set of forces on plates, guided in part by some reasoning about what types of forces are relevant. The data on sea-floor spreading and subduction, together with force or torque balances can then be used to deduce the magnitudes of the model forces. It is not obvious that this approach will work at all. For example, it is somewhat like observing only one element of a coupled oscillator system over a period less than the characteristic oscillation time. The data from such an observation might lead one to infer cause and effect, whereas this identification is irrelevant. In plate dynamics, it may be similarly irrelevant to identify cause and effect from kinematic modelling, since one is observing only a small fraction of the Earth's volume for a small fraction of the Earth's history. There is also no guarantee of uniqueness in the inversion procedure adopted in kinematic modelling.

In support of the method, the various approaches yield similar results, although this is partly explicable by the similar assumptions made. The models (Solomon and Sleep, 1974; Minster *et al.*, 1974; Harper, 1975; Kaula, 1975; Solomon *et al.*, 1975; Forsyth and Uyeda, 1975) indicate the following correlations.

(i) Plates being subducted along a major proportion of their boundary all have high apparent velocities (the prime example being the Pacific plate).
(ii) There is a negative correlation between plate velocity and the fraction of plate occupied by continent: predominantly continental plates tend to move at velocities that are only one third to one fifth the velocity of predominantly oceanic plates.
(iii) The equatorial lithosphere has a higher r.m.s. velocity than that in the two polar quarters.

It is implicit in these conclusions that "absolute" plate velocities have a physical meaning. The models indicate that the preferred "rest" frame is the frame in which "hot spots" (localized regions of high heat flux and a proposed cause of linear volcanic chains) are at rest. It is tempting to identify hot spots with narrow mantle plumes (Morgan, 1971), but the meaning of a "rest" frame is unclear.

The first conclusion above indicates that subduction plays an important role. In Section 4.5 below, we discuss subduction dynamics and the importance of the negative buoyancy of downgoing slabs. The second conclusion suggests that the continents do not provide any driving force but instead resist plate motions (presumably because of a greater viscous drag at the base of the continental lithosphere). For this reason, horizontal temperature gradients (Pekeris, 1935; Howard *et al.*, 1970) are no longer favoured as an important driving force for plate tectonics. The third conclusion suggests (but does not prove) that the Earth's rotation plays an important role in mantle dynamics, despite the smallness of the Coriolis force (see Section 2.1). One possible explanation is the oblateness of the Earth which forces plates to change their radius of curvature as they change latitude, and causes "membrane tectonics" (Turcotte and Oxburgh, 1976).

The above conclusions apply for an analysis of *present* plate motions. Application of the same modelling techniques to the tertiary era (Solomon *et al.*, 1977) does *not* support any of the above conclusions, however. This seems to imply either that the model is unphysical, or that the fraction of the Earth's history considered is too short.

4.3. Midplate Forces: Drive or Drag?

One correlation which does *not* clearly emerge in the kinematic models is that which should exist between plate area and plate velocity if the plates were driven by forces acting on their bottom surfaces. (The large Pacific plate does move rapidly, but this seems to be best explained by the length of the subducting edge.) Given the uncertain validity of kinematic modelling, it may be premature to discount deep mantle flows as the direct cause of plate motions. However, there are theoretical and laboratory indications which are consistent with the apparent unimportance of midplate forces.

First, Richter has developed a theory in which the convective flow primarily responsible for transporting heat from the deep mantle to the surface has a smaller scale than the plate size (Richter, 1973*b*; Richter and Parsons, 1975; Whitehead, 1976; Richter, 1977). Theoretical and experimental studies indicate that convective rolls tend to align with the direction of shear imposed by a horizontal moving boundary, and neither aid nor retard the boundary motion, consistent with the unimportance of midplate forces. Figure 8 shows schematically the form of mantle convection favoured by Richter and Parsons (1975). Note that participation of the lower mantle has been ignored. This model is consistent if it can be shown that the plates effectively drive themselves, for example by the negative buoyancy of downgoing slabs (see Section 4.5).

Second, the theory of lithospheric lubrication (Schubert and Turcotte, 1972; Froidevaux and Schubert, 1975; Melosh, 1976) indicates that because the viscosity is reduced as shear heating increases, the drag on a plate may decrease (or, at least, not increase) as the plate velocity increases. This may explain the apparent unimportance of drag.

4.4. Ridge Dynamics

Ridges may form either as a passive consequence of plates moving apart, or as the active uplift and injection of hot mantle material (Wilson and Burke, 1973). The elevation of the ridge can be explained by passive thermal

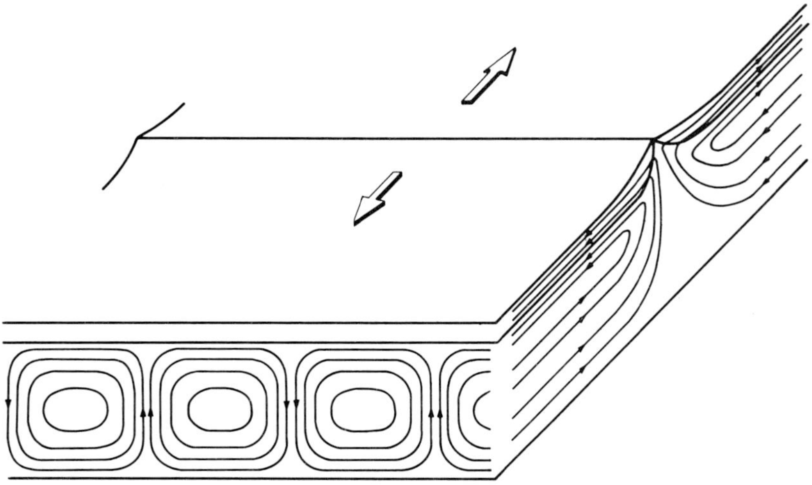

Fig. 8. Illustration of the large-scale flow in the vicinity of a ridge and the (hypothetical) superimposed small-scale longitudinal rolls. The form of the cross-section of the small-scale convection is schematic only. (From Richter and Parsons, 1975.)

expansion (McKenzie, 1967) but the elevated topography will not remain passive: the ridge will try to spread out and tend towards a lower gravitational energy. This tendency provides a "ridge push" force tending to drive plate motions (Orowan, 1964; Hales, 1969). Seismic data (Sykes and Sbar, 1973) indicate the presence of compressional intraplate stresses, which is consistent with ridge pushing, although other interpretations (such as thermal contraction) are possible. The absence of long-wavelength gravity anomalies between ridge and trench (Kaula, 1972) is evidence against ridge pushing. It may be that ridge push is most important in the *initiation* of plate motions but is subsequently unimportant, since subduction appears to require a triggering mechanism but can subsequently be self-sustaining (see below).

4.5. Subduction Dynamics

The cold downgoing slab under a trench has negative buoyancy because its density is greater than that of the surrounding mantle. If, as Elasser (1969) proposed, the lithosphere acts as a stress guide, this body force can be translated to the horizontal part of the lithosphere and help drive plate motions. Other forces acting on the slab include the viscous drag, the hydrodynamic pressure on the leading edge, and the suction (i.e. hydrodynamic pressure across the slab). These forces are indicated schematically in Fig. 9.

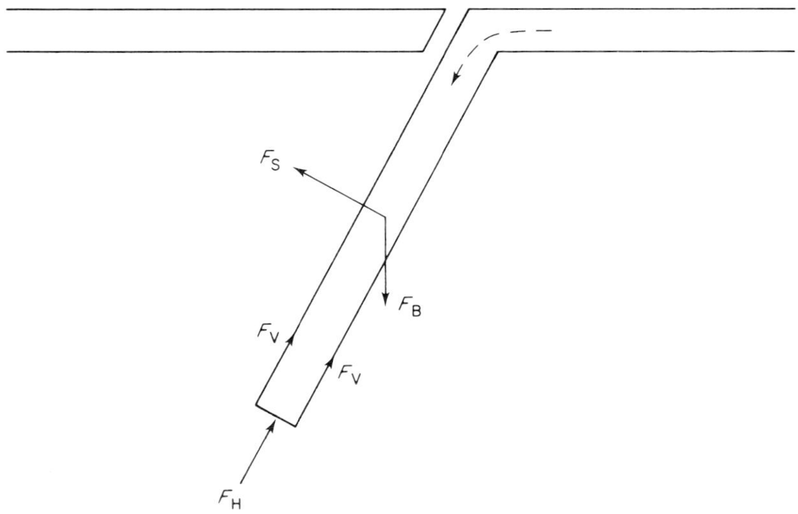

Fig. 9. Schematic representation of the forces acting on a subducting slab. F_B represents the negative buoyancy, F_H is the hydrodynamic pressure on the leading edge, F_V is the viscous drag on the sides of the slab, and F_S is the suction (i.e. the hydrodynamic pressure difference across the slab due to the induced mantle flow). The dashed arrow indicates that stresses can be transmitted through the corner region. Forces acting at the colliding boundary between the plates are not shown (they are not expected to play an essential role).

Richter (1977) has analysed the vertical forces and concludes that the primary balance is between the negative buoyancy and the hydrodynamic pressure and viscous drag deep within the mantle. This force balance is consistent with the observed focal mechanisms of slab-related earthquakes (Isacks and Molnar, 1969, 1971).

The negative buoyancy effect is essentially a form of thermal convection. As the slab descends, it slowly equilibrates thermally with the mantle, and equilibration may be almost complete at less than 100 km depth (Toksöz *et al.*, 1971). The absence of earthquakes deeper than 700 km is often interpreted as evidence that the slabs go no deeper, but alternative explanations are possible (O'Connell, 1977). Phase transitions (Turcotte and Schubert, 1971; Solomon and Kyaw Thaw Paw, 1975) and compositional differences (Oxburgh and Parmentier, 1977) may also modify subduction dynamics, but these are not dominant effects.

If subduction were merely part of a fluid convective flow, then one might expect the slab to plunge vertically into the mantle. In fact, the angle of subduction (the angle between the subducting slab and the Earth's surface) is more typically 50°. This suggests that the slab can be thought of as rigid in some respects. Stevenson and Turner (1977) suggest that the angle can be understood by considering the torque balance on the subducting slab. In the simple model considered, the balance is between the torque from the negative buoyancy (tending to make the slab go vertically) and the torque due to suction (the hydrodynamic pressure difference across the slab caused by the mantle flow induced by the subduction). It is found that no torque balance exists until the negative buoyancy exceeds a "critical" value. The speculative picture which emerges from this simple model is as follows. After subduction is initiated, the trench migrates until a steady state is reached where newly subducting material has the critical negative buoyancy and subducts at an angle of about 60°. The negative buoyancy results from the thickening and cooling of the plate as the material proceeds from ridge to trench. The observed scatter of actual values for the subduction angle (Luyendyk, 1970) results from the inevitable interaction of several ridge and trench systems, and the inability to achieve steady state. In the model, the negative buoyancy required is $\Delta\rho \sim 0.1$ g cm^{-3} density contrast between the slab and mantle, assuming a slab thickness of 100 km, velocity of 10 cm y^{-1}, slab length of 1000 km and mantle viscosity of 2×10^{22} P.

This picture explains well-developed subduction but not the *onset* of subduction. Davies (1977*b*) has analysed the mantle flow induced by a moving plate prior to the initiation of subduction, and concludes that the streamlines dip at about 30° relative to the horizontal. This he interprets as the initial angle at which subducting material is entrained. It must be emphasized, however, that if subduction *drives* plate motions then the negative buoyancy must affect the subduction angle. The onset of subduction is not understood and would appear to require a substantial triggering mechanism, since the negative buoyancy is necessarily smaller than the resisting forces in the early stages. Ringwood and Green (1966) propose that subduction is initiated by the basalt to eclogite transition at continental margins.

4.6. Laboratory Simulation

Although many experiments have been conducted on ordinary thermal convection, including some with a temperature-dependent viscosity (cf. Booker, 1976), few experiments have been conducted which attempt to incorporate the features of plate tectonics.

Elder (1967, 1971, 1977) has investigated the effect of viscosity variations, and has studied the behaviour of plastic sheets of non-uniform thickness floating on the surface of a turbulent liquid heated from below. He found that the convection pattern could propel surface layers, but his plastic sheets were compositionally different from the fluid, whereas the mantle and lithosphere are compositionally similar. Related experiments by Whitehead

(1972, 1976) used sources which were carred about by the convective motions they produced. These experiments were motivated by the hypothesis that the continents play an important role in plate dynamics. It is now increasingly evident that this is not correct, (e.g. the Pacific plate is essentially free of continental material). Horizontal temperature gradients do produce oscillatory motion (Whitehead, 1976; Busse, 1977) and may be partly responsible for the unsteadiness of mantle convection.

Turner (1973*b*) has conducted experiments using a single working fluid (glycerine), the viscosity of which varies strongly with temperature. When hot fluid is driven from below and strongly cooled from above, a thick high-viscosity surface layer is formed, which is driven away from regions of upward convection and then plunges down steeply into the less viscous region, maintaining its identity as it does so. This is shown in Fig. 10. In this

Fig. 10. Shadowgraph photograph of a plunging sheet of cold, viscous glycerine, resulting from the combination of a strong circulation produced by bubbles on the right, and surface cooling. (From Turner, 1973*b*.)

experiment, the flow is driven by bubbles released by dry ice at the bottom of the tank, and a strong almost two-dimensional circulation is produced. It would, of course, be more desirable to drive the motions purely by thermal convection, but the required velocities are difficult to achieve on a normal laboratory scale.

Another effect observed in these experiments, above the plunging "plate", was "double-diffusive" convection driven by the more rapid transfer of heat than matter into the intrinsically light sinking fluid. This is the physical process proposed by Oxburgh and Parmentier (1977) to explain behind-arc spreading in terms of convection driven by compositional, rather than temperature, differences.

The above model may simulate subduction but not seafloor spreading. An experiment that may be relevant to ridge dynamics is illustrated in Fig. 11 (Turner, 1973*b*). A line source located at the bottom of the tank heats the glycerine, producing a slow surface spreading and a more rapid closed circulation in the warmer region of low viscosity.

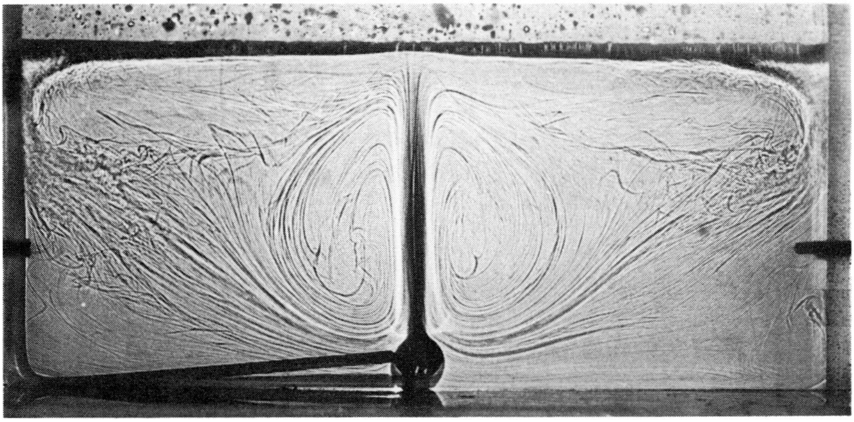

Fig. 11. The pattern of motion produced by heating glycerine at a line source located at the bottom of the tank, and cooling at the surface. There is a slow surface spreading of the cold, viscous boundary layer and a more rapid closed circulation in the warmer, less viscous fluid near the source. (From Turner, 1973*b*.)

Experiments closely related to these have been carried out by Jacoby (1976), using paraffin wax which has a different rheology from glycerine and tends to form a rigid skin. Plunging slabs can be modelled in this fluid (cf. previous experiments using flexible rubber sheets, floating on water, Jacoby, 1973) by forcing the skin to sink at one end, but irrelevant surface tension effects prevent convection alone from initiating the motion. Freezing wax has also been used to study the development of features analogous to transform faults near ridges (Oldenburg and Brune, 1972).

It has been pointed out by Duffield (1972) that the cooled surface skin on a lava lake may provide a smaller-scale natural example of convection in a fluid with temperature-dependent viscosity which merits closer study.

5. The Synthesis of Plate Tectonics with Mantle Convection

Although mantle convection and plate tectonics are terms which are used interchangeably by many, neither observation nor theory permits us to make such a simple identification yet. Whole-mantle convection, in the way we have presented it, is an inevitable consequence of the Earth's thermal evolution which manifests itself primarily by allowing a greater geothermal heat flux than conduction alone would provide. Plate tectonics, by contrast, is a mechanistic description for motions in the outermost layer of the Earth. The fundamental problem yet to be resolved in this: can the plates be thought of as a separate system which undergoes a form of thermal convection distinct from that occurring in the mantle? Or are the plate motions a direct manifestation of whole-mantle flows? An unequivocal answer would be premature at present. Instead we catalogue the future work which seems to be required to resolve this outstanding problem, and we conclude by spectulating on the type of theoretical and observational synthesis that might eventually be possible.

5.1. Directions for Future Work

Observations

The observational difficulties are caused by the great depth in the Earth at which information is required. High-resolution seismology must play a central role, and it will be particularly important to establish correlations between geophysical data sets of different kinds, incorporating a wider range of studies in geomagnetism, geothermy, geochemistry and geodesy. Palaeoscience has an important part to play, since present data are limited in time as well as space, and comparative planetology will help us further understand the thermal evolution of the Earth.

Laboratory Simulations

Historically, laboratory experiments have played an essential role in the understanding of complicated fluid dynamical systems. However, there are severe problems in scaling mantle dynamics down to the laboratory scale. Among the irrelevant features introduced into laboratory experiments are surface tension and inertial effects, which are completely negligible in the mantle. Artificial source mechanisms should also be avoided—the ideal system is an internally heated fluid.

More fundamental is the inappropriate rheology. The low viscosity immediately below the lithosphere is a consequence of the competing effects of temperature *and* pressure, whereas only temperature is readily varied in the laboratory. Shear heating also plays an important role in mantle dynamics, but cannot be reproduced on the laboratory scale.

In view of these difficulties, the most appropriate role for laboratory experiments is likely to lie in the understanding of selected parts of the system (e.g. subduction dynamics) rather than the simulation of convection for the whole Earth. Further work on fluids with temperature-dependent viscosity using somewhat larger horizontal scales than previous experiments, and on the coupling between layers of different composition (cf. Fig. 7) might be fruitful.

Theory

The greatest need in the theory is to produce a better understanding of the fundamentals rather than to elaborate the modelling of details. In particular, more work needs to be done on dynamical models of Earth evolution. Prototypes of this kind of model include those of Tozer (1972), McKenzie and Weiss (1975) and Schubert and Young (1976). In view of the inability of laboratory experiments to model whole-Earth convection adequately, numerical experiments, and especially those incorporating internal heating and variable viscosity, are likely to play an important role.

5.2. Speculations on the Nature of the Synthesis

There is little doubt that tectonic processes such as subduction constitute a form of thermal convection and induce flows within the mantle, but there are divergent views on the extent to which surface and deep-Earth phenomena are related.

Consider, first, the view that the plates can be thought of as a "separate" system, distinct from the deep mantle convection which exists but which does not determine the plate motions and geometry. The plates are driven primarily by the negative buoyancy of subducting slabs, and plate sizes are determined by the requirement that the cooling plate material must achieve a certain "critical" negative buoyancy before it can undergo steady-state subduction. The plates move readily across the underlying mantle because of shear-induced lubrication. The main result favouring this picture would be the absence of a strong large-scale correlation between plate configurations and deep Earth phenomena.

The alternative view is that the plate motions are a direct manifestation of deep mantle convection. The complicated plate configuration would be explained as a temporary effect in an unsteady convection pattern, and the lack of a coherent pattern could be attributed to the turbulent character of high Rayleigh number convection. Plate sizes would be related to the scale of overturn of the whole mantle (or, at least, to a strong coupling between convection cells in the upper and lower mantle—see Figs 6 and 7). The high viscosity in the outer layer might also favour large aspect ratios (Section 3.3). The tests of this view are theoretical or experimental confirmation that the plate sizes are achievable in thermal convection, and observations indicating a strong correlation between the plate configuration and deep Earth phenomena.

These two pictures are, of course, too simplistic, and intermediate views can be adopted. Present evidence is inconclusive, but we are confident that the outstanding problems will be solved through an interdisciplinary approach which combines fluid models and a wide range of geophysical observations.

References

Alexander, S. S. and Phinney, R. A. (1966). A study of the core–mantle boundary using P-waves diffracted by the Earth's core. *J. geophys. Res.* **71**, 5943–5975.

Artyushkov, E. V. (1971). Convective instabilities in geotectonics. *J. geophys. Res.* **76**, 1397–1415.

Artyushkov, E. V. (1972). Density differentiation of the Earth matter and processes at the core–mantle interface. *J. geophys. Res.* **77**, 6454–6458.

Auten, T. A., Gordon, R. B. and Stocker, R. L. (1974). *Q* and mantle creep. Nature, Lond. **250**, 317–318.

Birger, B. I. (1976). On the problem of the instability of a viscoelastic layer and its approximation to the problem of heat convection in the upper mantle. *Phys. Solid Earth* **12**, 163–169.

Booker, J. R. (1976). Thermal convection with strongly temperature-dependent viscosity. *J. Fluid Mech.* **76**, 741–754.

Bostrum, R. C. (1971). Westward displacement of the lithosphere. *Nature, Lond.* **234**, 536–538.

Bull, A. J. (1921). A hypothesis of mountain building. *Geol. Mag.* **58**, 364–367.

Busse, F. H. (1977). A model of time-periodic mantle flow. Preprint.

Busse, F. H. and Whitehead, J. A. (1971). Instabilities of convection rolls in a high Prandtl number fluid. *J. Fluid Mech.* **47**, 305–320.

Busse, F. H. and Whitehead, J. A. (1974). Oscillatory and collective instabilities in large Prandtl number convection. *J. Fluid Mech.* **66**, 67–79.

Carey, S. W. (1976). *The Expanding Earth*, Developments in Geotectonics Vol. 10. Elsevier, New York and Amsterdam.

Cathles, L. M. III (1975). *The Viscosity of the Earth's Mantle*, Princeton University Press, Princeton, N.J.

Chandrasekhar, S. (1961). *Hydrodynamic and Hydromagnetic Stability*, Clarendon Press, Oxford, Chapter II.

Cox, A. (ed.). (1973). *Plate Tectonics and Geomagnetic Reversals*, W. H. Freeman & Co., San Francisco.

Davies, G. F. (1977*a*). Whole mantle convection and plate tectonics. *Geophys. J. R. astronom. Soc.* **49**, 459–486.

Davies, G. F. (1977*b*). Viscous flow under moving lithospheric plates and under subduction zones. *Geophys. J. R. astronom. Soc.* **49**, 557–563.

De Bremaecker, J.-Cl. (1977). Convection in the Earth's mantle. *Tectonophysics* **41**, 195–208.

Duffield, W. A. (1972). A naturally occurring model of global plate tectonics. *J. geophys. Res.* **77**, 2543–2555.

Dziewonski, A. M., Hager, B. H. and O'Connell, R. J. (1977). Large-scale heterogeneities in the lower mantle. *J. geophys. Res.* **82**, 239–255.

Elder, J. W. (1967). Convective self-propulsion of continents. *Nature, Lond.* **214**, 657–660.

Elder, J. W. (1971). Thermal turbulence and its role in the Earth's mantle. In *Mantles of the Earth and Terrestrial Planets* (S. K. Runcorn, ed.), Wiley-Interscience, New York, p. 525.

Elder, J. W. (1976). *The Bowels of the Earth*, Oxford University Press, Oxford, Chap. 9.

Elder, J. W. (1977). Thermal convection. *J. geol. Soc. Lond.* **133**, 293–310.

Elsasser, W. M. (1969). Convection and stress propagation in the upper mantle. In *The Application of Modern Physics to Earth and Planetary Interiors.* (S. K. Runcorn, ed.), Wiley-Interscience, New York, pp. 223–246.

Forsyth, D. W. and Uyeda, S. (1975). On the relative importance of driving forces of plate motions. *Geophys. J. R. astronom. Soc.* **43**, 163–200.

Froidevaux, C. and Schubert, G. (1975). Plate motion and structure of the continental asthenosphere: a realistic model of the upper mantle. *J. geophys. Res.* **80**, 2553–2564.

Goldreich, P. and Toomre, A. (1969). Some remarks on polar wandering. *J. geophys. Res.* **74**, 2555–2567.

Gordon, R. B. (1965). Diffusion creep in the Earth's mantle. *J. geophys. Res.* **70**, 2413–2418.

Gough, D. I. (1977). The geoid and single-cell mantle convection. *Earth planet Sci. Lett.* **34**, 360–364.

Gubbins, D. (1976). Observational constraints on the generation process of the Earth's magnetic field. *Geophys. J. R. astronom. Soc.* **47**, 19–39.

Hales, A. L. (1936). Convection currents in the Earth. *Mon. Not. R. astronom. Soc., Geophys. Suppl.* **3**, 373–379.

Hales, A. L. (1969). Gravitational sliding and continental drift. *Earth planet. Sci. Lett.* **6**, 31–34.

Harper, J. R. (1975). On the driving forces of plate tectonics. *Geophys. J. R. astronom. Soc.* **40**, 465–474.

Herring, C. (1950). Diffusional viscosity of a polycrystalline solid. *J. appl. Phys.* **21**, 437–445.

Hewitt, J. M. McKenzie, D. P. and Weiss, N. O. (1975). Dissipative heating in convective flows. *J. Fluid Mech.* **68**, 721–738.

Hide, R. (1977). Towards a theory of irregular variations in the length of a day and core–mantle coupling. *Phil. Trans. R. Soc. A.* **284**, 547–554.

Hide, R. and Horai, K.-I. (1968). On the topography of the core–mantle interface. *Phys. Earth planet. Interiors* **1**, 305–307.

Holmes, A. (1931). Radioactivity and earth movements. *Trans. geol. Soc. Glasgow* **18**, 559–606.

Hopkins, W. (1839). Preliminary observations on the refrigeration of the globe. *Phil. Trans. R. Soc.* **129**, 381–385.

Houston, M. H., Jr. and De Bremaecker, J.-Cl. (1975). Numerical models of convection in the upper mantle. *J. geophys. Res.* **80**, 742–751.

Howard, L. N. (1964). Convection at high Rayleigh numbers. In *Proceedings of the 11th International Conference on Applied Mechanics*, Springer-Verlag, Berlin, Heidelberg, New York, pp. 1109–1115.

Howard, L. N., Malkus, W. V. R. and Whitehead, J. A. (1970). Self-convection of floating heat sources: a model for continental drift. *Geophys. Fluid Dynamics* **1**, 123–142.

Huppert, H. E. and Moore, D. R. (1976). Non-linear double-diffusive convection. *J. Fluid Mech.* **78**, 821–854.

Isacks, B. and Molnar, P. (1969). Mantle earthquake mechanisms and the sinking of the lithosphere. *Nature, Lond.* **223**, 1121–1124.

Isacks, B. and Molnar, P. (1971). Distribution of stresses in the descending lithosphere from a global survey of focal mechanism solutions of mantle earthquakes. *Rev. Geophys. Space Phys.* **9**, 103–174.

Jacobs, J. A. (1975). *The Earth's Core*, Academic Press, London, Chapter 2.

Jacoby, W. R. (1973). Model experiments of plate movements. *Nature, Lond.* **242**, 130–134.

Jacoby, W. R. (1976). Paraffin model experiment of plate tectonics. *Tectonophysics* **35**, 103–113.

Jeffreys, H. (1976). *The Earth*, 6th ed., Elsevier, New York and Amsterdam.

Jones, G. M. (1977). Thermal interaction of the core and the mantle and long-term behaviour of the geomagnetic field. *J. geophys. Res.* **82**, 1703–1709.

Jordan, T. H. (1973). Some comments on tidal drag as a mechanism for driving plate motions. *J. geophys. Res.* **79**, 2141–2142.

Jordan, T. H. (1977). Lithospheric slab penetration into the lower mantle beneath the Sea of Okhotsk. *J. Geophys.* **43**, 473–496.

Julian, B. R. and Sengupta, M. K. (1973). Seismic travel-time evidence for lateral inhomogeneity in the deep mantle. *Nature, Lond.* **242**, 443–447.

Kaula, W. M. (1972). Global gravity and tectonics. In *Nature of the Solid Earth* (E. G. Robertson, ed.), McGraw-Hill, New York, pp. 385–405.

Kaula, W. M. (1975). Absolute plate motions by boundary velocity minimizations. *J. geophys. Res.* **80**, 204–248.

Koschmieder, E. L. (1974). Bénard convection. *Adv. Chem. Phys.* **26**, 177–212.

Krishnamurti, R. (1970*a*). On the transition to turbulent convection. I. The transition from two- to three-dimensional flow. *J. Fluid Mech.* **42**, 295–307.

Krishnamurti, R. (1970*b*). On the transition to turbulent convection. II. The transition to time-dependent flow. *J. Fluid Mech.* **42**, 309–320.

Krishnamurti, R. (1973). Some further studies on the transition to turbulent convection. *J. Fluid Mech.* **60**, 285–303.

Lambeck, K. (1975). Effects of tidal dissipation in the oceans on the Moon's orbit and the Earth's rotation. *J. geophys. Res.* **80**, 2917–2925.

Lambeck, K. (1976). Lateral density anomalies in the upper mantle. *J. geophys. Res.* **81**, 6333–6340.

Le Pichon, X., Francheteau, J. and Bonnin, J. (1973). *Plate Tectonics*, Elsevier, New York and Amsterdam.

Luyendyk, B. P. (1970). Dips of downgoing lithospheric plates beneath island arcs. *Bull. geol. Soc. Am.* **81**, 3411–3416.

Lyttleton, R. (1965). The phase change hypothesis of the structure of the Earth. *Proc. R. Soc. A* **287**, 471–493.

Lyttleton, R. (1976). Relation of a contracting Earth to the apparent acceleration of the Sun and Moon. The Moon, **16**, 41–58.

McElhinny, M. W., Taylor, S. R. and Stevenson, D. J. (1978). Limits to the expansion of the Earth, Moon, Mars and Mercury and to variation in the gravitational constant. *Nature, Lond.* **271**, 316–321.

McFadden, C. P. and Smylie, D. E. (1968). Effect of a region of low viscosity on thermal convection in the mantle. *Nature, Lond.* **220**, 468–469.

McKenzie, D. P. (1967). Some remarks on heat flow and gravity anomalies. *J. geophys. Res.* **72**, 6261–6271.

McKenzie, D. P. (1969). Speculations on the consequences and causes of plate motions. *Geophys. J. R. astronom. Soc.* **18**, 1–32.
McKenzie, D. P. (1977). Surface deformation, gravity anomalies and convection. *Geophys. J. R. astronom. Soc.* **48**, 211–238.
McKenzie, D. P. and Weiss, N. O. (1975). Speculations on the thermal and tectonic history of the Earth. *Geophys. J. R. astronom. Soc.* **42**, 131–174.
McKenzie, D. P., Roberts, J. M. and Weiss, N. O. (1974). Convection in the Earth's mantle: towards a numerical solution. *J. Fluid Mech.* **62**, 465–538.
Melosh, H. J. (1976). Plate motion and thermal instability in the asthenosphere. *Tectonophysics* **35**, 363–390.
Minster, J. B., Jordan, T. H., Molnar, P. and Haines, E. (1974). Numerical modelling of instantaneous plate tectonics. *Geophys. J. R. astronom. Soc.* **36**, 541–576.
Morgan, W. J. (1971). Convection plumes in the lower mantle. *Nature, Lond.* **230**, 42–43.
O'Connell, R. J. (1977). On the scale of mantle convection. *Tectonophysics* **38**, 119–136.
Oldenburg, D. W. and Brune, J. N. (1972). Ridge transform fault spreading pattern in freezing wax. *Science, N.Y.* **178**, 301–304.
Orowan, E. (1964). Continental drift and the origin of mountains. *Science, N.Y.* **146**, 1003–1010.
Oxburgh, E. R. and Parmentier, E. M. (1977). Compositional and density stratification in oceanic lithosphere—causes and consequences. *J. geol. Soc. Lond.* **133**, 343–355.
Parmentier, E. M., Turcotte, D. L. and Torrance, K. E. (1976). Studies of finite-amplitude non-Newtonian thermal convection with application to convection in the Earth's mantle. *J. geophys. Res.* **81**, 1839–1846.
Pekeris, C. L. (1935). Thermal convection in the interior of the Earth. *Mon. Not. R. astronom. Soc., Geophys. Suppl.* **3**, 346–367.
Peltier, W. R. (1976). Glacial isostatic adjustment. II. The inverse problem. *Geophys. J. R. astronom. Soc.* **46**, 669–706.
Peltier, W. R. and Andrews, J. T. (1976). Glacial isostatic adjustment. I. The forward problem. *Geophys. J. R. astronom. Soc.* **46**, 605–646.
Pollack, H. N. and Chapman, D. S. (1977). Mantle heat flow. *Earth planet. Sci. Lett.* **34**, 174–184.
Post, R. L. and Griggs, D. T. (1973). The Earth's mantle: evidence of non-Newtonian flow. *Science, N.Y.* **181**, 1242–1245.
Rice, A. R. (1971). Mechanisms of dissipation in mantle convection. *J. geophys. Res.* **76**, 1450–1459.
Richter, F. M. (1973*a*). Finite amplitude convection through a phase boundary. *Geophys. Jl. R. astronom. Soc.* **35**, 265–276.
Richter, F. M. (1973*b*). Dynamical models of seafloor spreading. *Rev. Geophys. Space Phys.* **11**, 223–287.
Richter, F. M. (1977). On the driving mechanism of plate tectonics. *Tectonophysics* **38**, 61–88.
Richter, F. M. and Johnson, C. E. (1974). Stability of a chemically layered mantle. *J. geophys. Res.* **79**, 1635–1639.
Richter, F. M. and Parsons, B. (1975). On the interaction of two scales of convection in the mantle. *J. Geophys. Res.* **80**, 2529–2541.

Ringwood, A. E. (1960). On the chemical evolution and densities of the planets. *Geochim. cosmochim. acta* **15**, 257–283.
Ringwood, A. E. (1975). *Composition and Petrology of the Earth's Mantle*, McGraw-Hill, New York.
Ringwood, A. E. and Green, D. H. (1966). An experimental investigation of the gabbro-eclogite transformation and some geophysical implications. *Tectonophysics* **3**, 383–427.
Roberts, P. H. (1967). Convection in horizontal layers with internal heat generation. *J. Fluid Mech.* **30**, 33–49.
Runcorn, S. K. (1962). Palaeomagnetic evidence for continental drift and its geophysical cause. In *Continental Drift* (S. K. Runcorn, ed.), Academic Press, New York and London, pp. 1–40.
Runcorn, S. K. (1967). Flow in the mantle inferred from low degree harmonics of the geopotential. *Geophys. J. R. astronom. Soc.* **14**, 375–384.
Sammis, C. G. (1976). The effects of polymorphic phase boundaries on vertical and horizontal motions in the Earth's mantle. *Tectonophysics* **35**, 169–182.
Schubert, G. and Turcotte, D. L. (1972). One-dimensional flow of shallow-mantle convection. *J. geophys. Res.* **77**, 945–951.
Schubert, G. and Young, R. E. (1976). Cooling of the Earth by whole-mantle sub-solidus convection. *Tectonophysics* **35**, 201–214.
Schubert, G., Yuen, D. A. and Turcotte, D. L. (1975). Role of phase transitions in a dynamic mantle. *Geophys. J. R. astronom. Soc.* **42**, 705–735.
Solomon, S. C. and Kyaw Thaw Paw U (1975). Elevation of the Olivine–spinel transition in the subducted lithosphere: seismic evidence. *Phys. Earth planet interiors* **11**, 97–100.
Solomon, S. C. and Sleep, N. H. (1974). Some simple physical models for absolute plate motions. *J. geophys. Res.* **79**, 2557–2567.
Solomon, S. C., Sleep, N. H. and Jurdy, D. M. (1971). Mechanical models for absolute plate motions in the early tertiary. *J. geophys. Res.* **82**, 203–212.
Solomon, S. C., Sleep, N. H. and Richardson, R. M. (1975). On the forces driving plate tectonics: inferences from absolute plate velocities and intraplate stress. *Geophys. J. R. astronom. Soc.* **42**, 769–801.
Sorokhtin, O. G. (1974). Density convection in the Earth's mantle and the possible nature of tectonic cycles. *Akad. Nauk. SSR Phys. solid Earth* **5**, 297–305.
Stacey, F. D. (1977). *Physics of the Earth*, 2nd ed., Wiley, New York.
Stevenson, D. J. and Turner, J. S. (1977). Angle of subduction. *Nature, Lond.* **270**, 334–336.
Sung, C. M. and Burns, R. G. (1976). Kinetics of the olivine–spinel transition: implications in deep-focus earthquake genesis. *Earth planet. Sci. Lett.* **32**, 165–170.
Sykes, L. R. and Sbar, M. L. (1973). Intraplate earthquakes, lithospheric stresses and the driving mechanism of plate tectonics. *Nature, Lond.* **245**, 298–302.
Toksöz, M. N., Minear, J. W. and Julian, B. R. (1971). Temperature field and geophysical effects of a downgoing slab. *J. geophys. Res.* **76**, 1113–1138.
Torrance, K. E. and Turcotte, D. L. (1971). Thermal convection with large viscosity variations. *J. Fluid Mech.* **47**, 113–125.
Tozer, D. C. (1965). Heat transfer and convection currents. *Phil. Trans. R. Soc. A* **258**, 252–271.
Tozer, D. C. (1972). The present thermal state of the terrestrial planets. *Phys. Earth planet. Interiors* **6**, 182–197.

Turcotte, D. L. and Oxburgh, E. R. (1967). Finite amplitude convection cells and continental drift. *J. Fluid Mech.* **28**, 29–42.
Turcotte, D. L. and Oxburgh, E. R. (1972). Mantle convection and the new global tectonics. *A. Rev. Fluid Mech.* **4**, 33–68.
Turcotte, D. L. and Oxburgh, E. R. (1976). Stress accumulation in the lithosphere. *Tectonophysics* **35**, 183–199.
Turcotte, D. L. and Schubert, G. (1971). Structure of the olivine–spinel phase boundary in the descending lithosphere. *J. geophys. Res.* **76**, 7980–7987.
Turcotte, D. L., Hsui, A. T., Torrance, K. E. and Schubert, G. (1974). Influence of viscous dissipation on Bénard convection. *J. Fluid Mech.* **64**, 369–374.
Turner, J. S. (1973*a*). *Buoyancy Effects on Fluids*, Cambridge University Press, Cambridge.
Turner, J. S. (1973*b*). Convection in the mantle: a laboratory model with temperature-dependent viscosity. *Earth planet. Sci. Lett.* **17**, 369–374.
Urey, H. C. (1952). *The Planets*, Yale University Press, New Haven, Conn., p. 87.
Vetter, U. R. and Meissner, R. O. (1977). Creep in geodynamic processes. *Tectonophysics* **42**, 37–54.
Walcott, R. I. (1970). Flexural rigidity, thickness and viscosity of the lithosphere. *J. geophys. Res.* **75**, 3941–3954.
Wasserburg, G. J., MacDonald, G. J. F., Hoyle, F. and Fowler, W. A. (1964). Relative contributions of uranium, thorium and potassium to heat production in the Earth. *Science, N.Y.* **143**, 465–467.
Weertman, J. (1970). The creep strength of the Earth's Mantle. *Rev. Geophys.* **8**, 145–168.
Wegener, A. (1912). Die Entstehung der Kontinente. *Geol. Rundschau* **3**, 276–292.
Whitehead, J. A. (1972). Moving heaters as a model of continental drift. *Phys. Earth planet. Interiors* **5**, 199–212.
Whitehead, J. A. (1976). Convection models: laboratory versus mantle. *Tectonophysics* **35**, 215–228.
Wilson, J. T. (1963). Hypothesis of Earth's behaviour. *Nature, Lond.* **198**, 925–929.
Wilson, J. T. and Burke, K. (1973). Plate tectonics and plume mechanics. *Trans. Am. geophys. Union* **54**, 238–239.

9. Petrogenesis of Mid-ocean Ridge Basalts

D. H. GREEN, W. O. HIBBERSON and A. L. JAQUES

Abstract

The liquidus and near-liquidus crystallization of primitive, mid-ocean ridge basalt (DSDP3-18-7-1) has been studied experimentally from 0–15 kbar. Olivine is the liquidus phase at low pressure (<12 kbar) and is joined by plagioclase (An_{78}) and clinopyroxene at lower temperature. Calcium-rich clinopyroxene becomes the liquidus phase at ⩾12 kbar and there is no liquidus or near-liquidus field for orthopyroxene. Orthopyroxene addition experiments confirm that the liquid composition is not close to saturation with orthopyroxene. The low-pressure experiments show that DSDP3-18-7-1 could be a derivative liquid by olivine fractionation from a more olivine-rich parent. Olivine addition experiments show that a picritic composition (83 per cent DSDP3-18-7-1 + 17 per cent olivine) is saturated with olivine + orthopyroxene at ~20 kbar, 1430 °C.

An earlier model for genesis of mid-ocean ridge, high-Al_2O_3 olivine tholeiites by magma segregation from pyrolite at 9–10 kbar is retracted. The primary magma for high-Al_2O_3 mid-ocean ridge olivine tholeiites is considered to be tholeiitic picrite, segregated from residual harzburgite at ~20 kbar, 1430 °C. The primary tholeiitic picrite contains ~25 per cent normative olivine, with SiO_2 ~48 per cent, MgO ~16 per cent, 100Mg/(Mg+Fe)≈79, ~600 p.p.m. Ni and was strongly depleted in incompatible elements. The experimental studies show that there can be no simple genetic relationship between the primary tholeiitic picrite magmas which are parental to common mid-ocean ridge basalts, and the residual harzburgite or accumulate peridotite, pyroxenite, anorthosite or troctolite of ophiolite complexes. Similarly, the tholeiitic picrite magmas cannot precipitate the highly calcic plagioclase (>An_{80}) or magnesian clinopyroxene (>Mg_{88}) megacrysts which they sometimes contain. It is argued that there must exist, in mid-ocean ridge environments, a second magma type with high SiO_2 (>50 per cent?), very high CaO/Na_2O and very low TiO_2. This magma type, and its fractionation products, are assigned a complementary relation to ophiolite complexes, and it is suggested that both arise as a consequence of shallow emplacement and second-state melting of diapirs of refractory lherzolite or harzburgite.

The experimental data on DSDP3-18-7-1 are used to calculate a model source composition which differs significantly from previous estimates of the source pyrolite for mid-ocean ridge basalts. The source pyrolites for mid-ocean ridge tholeiites and "Hawaiian" or "incompatible element enriched" tholeiites are both considered to occur within the upper mantle and to characterize a chemically zoned low-velocity zone.

1. Introduction

The production and emplacement of basaltic magmas at mid-ocean ridges is volumetrically the most important step in the geochemical differentiation of the Earth's mantle. A common view is that this process yields a basaltic oceanic crustal layer about 10 km thick, overlying residual upper mantle peridotite of dunite, harzburgite or lherzolite character. The basaltic crustal layer is considered to be lithologically complex but with an idealized depth profile grading downwards from pillow lavas, through a sheeted dolerite dyke complex, to an underlying layer rich in gabbro and dolerite intrusions which individually pass downwards into layered accumulate sequences of gabbro, anorthosite, pyroxenite, troctolite and peridotite. The effects of low-grade metamorphism, particularly hydration reactions, on this idealized crustal section, produce chemical change and patchy alteration to greenschists and amphibolites.

There is a common view that the tectonically-emplaced associations of deep-sea sediments, pillow lavas, dolerites, gabbros, layered mafic-ultramafic complexes and ultramafic tectonites (collectively called the "ophiolite suite") occurring in many orogenic belts are uplifted sections through the oceanic crust. Furthermore, many authors have recognized that both ocean floor basalts and basalts of ophiolite suites commonly include lavas that have been fractionated by separation of olivine, pyroxene and plagioclase from more magnesian parental lavas. It has been inferred that the layered accumulate sequences of the oceanic crust and of ophiolite suites were formed as the complementary accumulates to the fractionation of mid-ocean ridge basalts (Bonatti *et al.*, 1971; Cann, 1974; Davies, 1971; Melson and Thompson, 1970; Moores and Vine, 1971; Hodges and Papike, 1976).

The rapid accumulation of data from the DSDP program and the FAMOUS project and the application of a variety of geochemical analytical tools to the study of ocean floor basalts have brought about a situation in which it should be possible to define more closely the processes of magma formation beneath mid-ocean ridges, the compositions of source rocks, and the nature of crystal fractionation processes and of crystalline accumulates. One essential ingredient to understanding of mid-ocean ridge magmatism is a knowledge of the phase relationships of both magmas and possible source compositions. In this paper we present part of our data gathered for this purpose and explore the relevance of the data to the general model of oceanic crust and ophiolite genesis.

2. "Primary" Magmas

Mid-ocean ridge basalts are the products of crystallization of magmas consisting of silicate liquid, variable proportions of phenocrysts and xenocrysts (Donaldson and Brown, 1977) and very minor inclusions of immiscible sulphide liquid (Mathez, 1976). Analyses of quenched glasses avoid the problem of contamination by xenocrysts and enrichment by settling of phenocrysts and are a reasonable first step towards establishing the compositions of parental, mantle-derived liquids and derivative, fractionated liquids. Since the first phases to crystallize from mid-ocean ridge basalts are magnesian olivine, magnesian clinopyroxene and calcic plagioclase, the most "primitive" liquids can be selected as those glass analyses having the highest Mg/(Mg + Fe) and/or Ca/(Ca + Na) ratios.

We use the term "primary magma" to refer to a magma which has not changed composition by crystal fractionation, contamination or volatile loss between the time of *magma segregation* from residual crystals of the source rock and the time of extrusion and quenching. Green and Ringwood (1967) discussed the concept of an "event" (time and depth) of magma segregation and showed that "primary magma" compositions are determined by crystal–liquid equilibria between residue and melt at the conditions of magma segregation. The ascent of a peridotite diapir consisting of crystalline residue and a minor liquid fraction, both changing composition with changing pressure and temperature during movement of the diapir, is considered to be the prime cause of magma generation in the upper mantle (Verhoogen, 1954; Green and Ringwood, 1967). Readjustment of equilibrium, i.e. change in composition of crystals and liquid, will occur until the liquid segregates from the solid residue (this will happen rapidly if the melt fraction exceeds 20–30 per cent, Green and Ringwood, 1967; Arndt, 1977). We refer to the liquid at this point as a *primary* liquid and the experimental petrologist, given the composition (including H_2O, CO_2 contents) of such a primary liquid, sets out to establish the *P*, *T* conditions at which it is multiply saturated with the inferred residual minerals, i.e. in the case of a peridotite source and basaltic magma, saturated with olivine ± orthopyroxene ± clinopyroxene ± garnet, spinel or plagioclase.

The above model of melting and magma segregation is the simplest one and we can devise more complex models. It is possible that a peridotite diapir at, say, 60 km depth may contain 10 per cent melt, half of which may segregate as basalt A leaving entrapped liquid among the residual crystals. We now have effectively a new source peridotite composition which may continue to rise and, as the melt fraction increases with decreasing depth, segregate a further melt fraction as basalt B derived by 20 per cent melting at ~20 km depth. If basalts A and B are sampled at the surface, both are primary magmas in the sense used here and the experimental petrologist should be able to define the depth of magma segregation and nature of residual minerals for each of basalt A and B. The source compositions for basalts A and B will, however, differ and the experimental petrologist cannot proceed further to relate these source compositions. Thus conceptual models such as the "dynamic melting model" of Langmuir *et al.* (1977) or the models of sequential batch extraction of Arndt (1977) are equally suited to establishment of "depth of magma segregation" by the experimental approach, but differ from the simple partial melting or batch melting and extraction model in that the "source composition" is not fixed but varies from an original "undepleted" mantle composition towards increasingly refractory residual compositions.

Other models of melting of mantle peridotite which include mixing of already segregated magmas or include fractional melting (i.e. a melt (A) formed at T_1 is segregated from reaction with residual crystals; at $T_1 + \Delta T$ a second melt (A′) forms from residual crystals, segregates and is added to A, etc. (Presnall, 1969)) are not amenable to the experimental approach, i.e. the compositions of the liquids so formed differ from those originally in equilibrium with residual crystals and one cannot correlate liquidus phases of the observed liquids with residual phases for the (unknown) individual increments of magma.

In a sequence of earlier related papers (see review by Green, 1976*a*), it has been profitable to explore the range of primary magmas which can be explained by the simple model of batch partial melting and magma segregation based on

(a) a constant anhydrous source peridotite composition (P and T variable);
(b) a constant source peridotite composition but with addition of variable amounts of water (P, T, per cent H_2O variable);
(c) a constant source peridotite composition but with variable H_2O and CO_2 contents (P, T, per cent CO_2 variable).

However, even from the earliest of these papers, it has been accepted that the constant source peridotite composition is an oversimplification (Green and Ringwood, 1967; Green, 1968, 1970, 1971; Frey and Green, 1974; Frey *et al.*, 1978) and that inhomogeneity, particularly in incompatible element contents, was present in the upper mantle source regions for basalts. The study of mantle derived xenoliths and of mantle-derived peridotite diapirs (Lizard, Tinaquillo, Beni Bousera, Serrania del Ronda) is one way of exploring mantle inhomogeneity and detailed geochemical studies emphasize that peridotites of these types have complex histories (Frey, 1969; Frey and Green 1974; Frey and Prinz, 1977). However, the relationship of such peridotites to basalt genesis is unclear and it is therefore necessary to examine the basalts themselves to investigate mantle inhomogeneity. If specific basalts can be identified as primary basalts and the nature of the melting conditions and residue specified, then source compositions can be estimated and compared. The use of basalts to determine regional and/or depth variations of mantle composition is still some way off, but this approach may well be important in understanding mantle and crust evolution—it is already apparent that models of mantle evolution based on trace element abundances and isotopic measurements (e.g. U/Pb and Pb isotopes; RB/Sr and Sr isotopes, Sm/Nd and Nd isotopes) need to be constrained by petrogenetic models which consider phase equilibria, major element geochemistry and well-determined element partition coefficients. The converse of this statement is equally valid.

2.1. Primary Basalts of the Ocean Floor

The arguments favouring an upper mantle composition of peridotite with 3–4 per cent CaO, 3–4 per cent Al_2O_3 and dominated (>50 per cent) by olivine (Fo_{88-90}) have been given elsewhere. If we consider equilibrium partial melting of such source peridotite to leave a more refractory residue (olivine of Mg_{88-92}) then liquids should have Mg values $(Mg/(Mg+Fe))=$ 0.68–0.78 for up to approx. 30 per cent melting (Green and Ringwood, 1967; Roeder and Emslie, 1970; Green, 1970, 1971).

Most ocean floor basalts have Mg values less than this (and contain less than 9 weight per cent MgO) and have probably undergone pre-eruptional fractional crystallization of olivine, spinel and in many cases, plagioclase and pyroxene (e.g. Kay *et al.*, 1970; Miyashiro *et al.*, 1969; Shido *et al.*, 1971; Frey *et al.*, 1974; Bryan *et al.*, 1976; Bryan and Moore, 1977). The apparent cut-off at about 9 weight per cent MgO and 200 p.p.m. Ni (Mg value $\simeq$65) led Kay *et al.* (1970) to suggest that these features typified relatively undifferentiated basalts. More magnesian lavas with Mg values of 68–72 have been reported by Frey *et al.* (1974) at DSDP 3-14, 18, by Melson *et al.* (1977) in a survey of abyssal glasses, and by Blanchard *et al.* (1976) and Bence and Taylor (1977) for DSDP 37-332B. Sampling by submersibles in the rift zone of the mid-Atlantic Ridge in the FAMOUS area has provided a large suite of highly magnesian glasses (Mg_{70-72}), e.g. Bryan and Moore (1977), Langmuir *et al.* (1977), Melson *et al.* (1977).

Examination of analyses of these relatively unfractionated basalts reveals that most have high Al_2O_3 content (>16 per cent), Al_2O_3/CaO ratios of

$\simeq 1.3$, high MgO (9–10 weight per cent), mostly low TiO_2 (<0.75 weight per cent), low Na_2O (<2.5 weight per cent), and high CaO/Na_2O ratios (>5). Fractionation by olvine, plagioclase and clinopyroxene will lead to decrease in Mg value and in Ca/Na. Silica contents are mostly in the range 48–49 weight per cent and the percentage of normative olivine ranges from about 10 to 20, averaging 12–14 per cent.

A basalt glass from the south Atlantic (Frey *et al.* 1974, Leg 3, Site 18, Sample 3-18-7-1) was chosen for experimental study. Frey *et al.* (1974) showed that this glass ($Mg_{69.5}$) which contains microphenocrysts of olivine ($Fo_{89.7-89.2}$) and of plagioclase (An_{79}) is strongly depleted in light rare-earth elements and has very low contents of large-ion lithopile (LIL) elements together with high NiO and Cr_2O_3 contents. The compositional characteristics of this basalt strongly suggests that it may be parental to other ocean floor basalts by olivine and plagioclase fractionation and, on the basis of its Mg value and olivine microphenocryst composition, it is a possible primary melt from mantle peridotite containing olivine (Mg_{88-92}). This basalt is not reported to contain megacrysts of olivine (>Mg_{90}) or calcic plagioclase ($\sim An_{85-95}$) unlike other examples of ocean floor glasses (Donaldson and Brown, 1977; Bryan *et al.*, 1976; Blanchard *et al.*, 1976).

The experimental study was directed towards establishing the liquidus phases for the chosen basalt, with the hope that some condition would be found at which olivine, orthopyroxene and possibly clinopyroxene occurred together on the liquidus, this condition then defining the probable depth of magma segregation from residual harzburgite or lherzolite for this particular magma.

3. Experimental Petrology of a Mid-ocean Ridge Basalt

The composition of the glass prepared for experimental study is listed in Table 1. Experiments at 0 kbar were carried out by placing ~20 mg sample in spec-pure iron capsules which were sealed in evacuated silica tubes and suspended in a 1-atm quenching furnace. Since previous experience, particularly on lunar compositions, had demonstrated that plagioclase nucleation and crystallization could be sluggish using glass starting material, the low-pressure (0 kbar, 5 kbar) runs used a fine-grained crystalline sample prepared by devitrifying a 250 mg sample of the glass to pyroxene + plagioclase + olivine assemblage at 0 kbar, 1050 °C, 5 h. To confirm that plagioclase observed at 0 kbar, 1210 °C (Run A520, Table 2) was stable and not relict from the starting material and to provide coarser crystals for microprobe analysis, a repeat run at 0 kbar, 1210 °C used a starting mix of (90 per cent glass + 10 per cent Run A520).

The problem of sample containers in experimental melting studies has been explored in lunar studies (O'Hara *et al.*, 1975; O'Hara and Humph-

Table. 1. Compositions used in experimental study.

	Glass of DSDP3-18-7-1 Composition	91 per cent DSDP3-18 +9 per cent olivine Mix D	83 per cent DSDP3-18 +17 per cent olivine Mix E
SiO_2	49.7	48.9	48.3
TiO_2	0.72	0.65	0.60
Al_2O_3	16.4	14.9	13.7
FeO	7.9	7.9	7.9
MnO	0.12	0.12	0.12
MgO	10.1	13.6	16.7
CaO	13.1	11.9	10.9
Na_2O	2.0	1.80	1.65
K_2O	0.01	0.01	0.01
Cr_2O_3	(0.07)†	(0.06)†	(0.06)†
NiO	(0.03)†	0.05	0.08
CIPW Norms			
Or	0.06	0.05	0.05
Ab	16.7	15.2	14.0
An	35.9	32.6	29.9
Di	23.4	21.3	19.5
Hy	10.0	9.1	8.3
Ol	12.6	20.5	27.1
Ilm	1.4	1.25	1.1
100 Mg/(Mg+Fe)	70.9	75.4	79.0

† These components were not added to the synthetic mix but the values given are those for the natural rock (Frey *et al.*, 1974)

reys, 1977; Green *et al.* 1975). Noble metal capsules (Pt, AgPd) result in Fe loss to the container and sample compositions must be monitored carefully. At low pressure (<5 kbar approx.) graphite capsules precipitate metallic iron from the charge. High-purity magnetic iron has similar problems at both low and high pressure. Spec-pure iron is suitable at low and high pressure except that Fe_2O_3 in the charge must be assumed to be reduced to FeO and the sample bulk composition after the run should be monitored to check for composition change either by addition of iron or reduction of FeO to metallic iron. Graphite is suitable as a container at pressures ≥ 10 kbar in the sense that the C–$(CO_3)^{2-}$ oxygen buffer appears to lie below the QFM oxygen buffer, but poses an uncertain risk at high pressure as (CO_3^{2-}) is soluble in natural basalts at high pressure and has been shown to affect the nature of phases on the liquidus at high pressure (Brey and Green, 1975, 1977). The best technique available for dry experiments is to use graphite capsules but to carry out confirmatory near-liquidus runs in spec-pure Fe capsules to ascertain that depression of liquidus and change of liquidus phases due to $(CO_3)^{2-}$ solubility have not occurred. The experiments in spec-pure Fe capsules must be monitored for gain of FeO.

The experimental techniques and composition used however are unsuitable for evaluation of the role of chrome spinel—trace quantity of Cr_2O_3 was not added to the DSDP 3-18 composition—and the low f_{O_2} of the experiments would have inhibited spinel crystallization. Thus in the discussion and application of our experiments we cannot comment on the presence or absence of chrome spinel as a near-liquidus phase in DSDP 3-18.

In the present study, spec-pure Fe capsules were used at 5 kbar and both graphite and spec-pure Fe were used at $\geqslant$10 kbar. The compositions of above-liquidus runs at 0, 10, 12 and 13 kbar were determined by electron microprobe analysis. This confirmed that there was no composition change in graphite capsules at 10, 12 and 13 kbar, and only very minor addition of Fe in the spec-pure Fe capsule at 0 kbar, but that the above liquidus run in spec-pure Fe at 10 kbar had absorbed iron to change its Mg value from 69.5 to 67.7 (i.e. 7.85 per cent FeO $\rightarrow$ 8.25 per cent FeO).

The composition of near-liquidus crystalline phases and of glass were determined by electron microprobe analysis and the phase relationships and compositional data are summarized in Fig. 1 and Table 2. At low pressure (0 kbar, 5 kbar) olivine (Mg_{88}) is the liquidus phases and crystallizes alone over a 10–30 °C temperature interval. Plagioclase (An_{78}) is the second phase to appear at both 0 kbar and 5 kbar. Analyses of crystal and liquid compositions show that 3 ± 0.5 per cent olivine crystallized from the melt at 0 kbar prior to the liquid reaching the plagioclase saturation surface. At 5 kbar and 10 kbar, less olivine crystallizes before reaching the plagioclase saturation surface and at 10 kbar the liquid becomes saturated with clinopyroxene and plagioclase together (within the resolution of the experimental data) and after separation of ~1 per cent olivine. The effect of pressure in the 0–10 kbar interval on the anhydrous, near-liquidus crystallization is to move the olivine–plagioclase cotectic towards more olivine-rich composition, and to expand the field of clinopyroxene crystallization towards both olivine- and plagioclase-rich compositions. The composition of plagioclase at its first appearance also becomes less anorthite-rich at higher pressure (An_{78}–0 kbar, An_{77}–5 kbar, An_{73}–10 kbar, An_{69}–12 kbar)—this may be attributed to the appearance and coprecipitation of calcium-rich clinopyroxene, but the analytical data also show a decrease in

$$\frac{(Ca/Na)_{plag}}{(Ca/Na)_{liq}}$$

with increasing pressure.

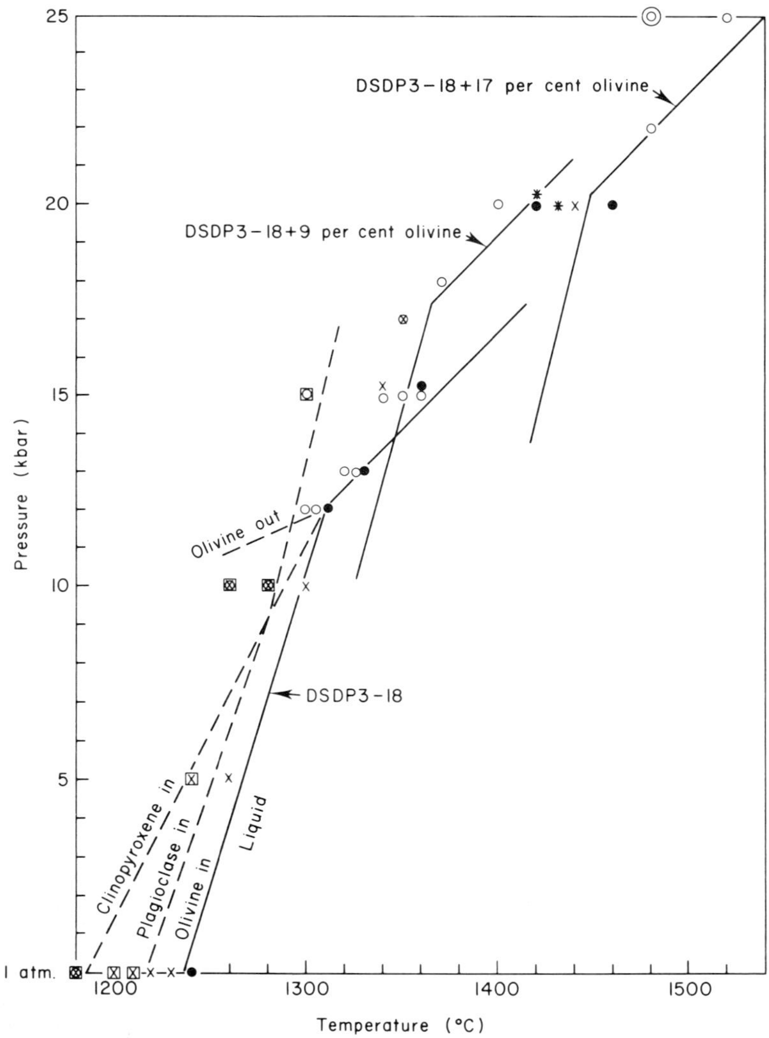

Fig. 1. Experimental determination of liquidus and near-liquidus phases for olivine tholeiite DSDP3-18-7-1 and olivine-enriched compositions (see text). ●, Above liquidus; ×, olivine; +, orthopyroxene; ○, clinopyroxene; ◯, garnet; □, plagioclase.

Table 2. Details of melting experiments on DSDP3-18 composition and derivative compositions A, B, C, D, E (see text and Table 1)

Run no.	Starting material	Capsule	Temperature (°C)	Pressure (kbar)	Time (h)	
A517	Crystallized mix	Fe	1240	0	1	Glass ($Mg_{69.2}$)
A498	Crystallized mix	Fe	1230 ± 10	0	1	Olivine (Mg_{88}) + Glass ($Mg_{68.8}$)
A493	Crystallized mix	Fe	1220	0	1.25	Olivine + glass
A520	Crystallized mix	Fe	1210	0	2	Olivine ($Mg_{86.8}$) + plagioclase (An_{78}) + glass (Mg_{67})
A521	90 per cent glass 10 per cent A520	Fe	1210	0	2	Olivine ($Mg_{86.5}$) + plagioclase (An_{78}) + glass
A519	Crystallized mix	Fe	1200	0	1.5	Olivine + plagioclase + glass
A514	Crystallized mix	Fe	1180	0	2	Olivine + plagioclase ± clinopyroxene + glass (fine grained)
5535	Glass	Fe	1260	5	1	Olivine ($Mg_{87.8}$) + glass ($Mg_{68.8}$)
5532	Glass	Fe	1240	5	1	Olivine ($Mg_{87.5}$) + plagioclase (An_{77}) + glass ($Mg_{67.4}$)
5548	Glass	C	1300	10	0.5	Glass ($Mg_{69.5}$) + very rare olivine ($Mg_{88.7}$)
5555	Glass	C	1280	10	0.5	Olivine ($Mg_{87.5}$) + clinopyroxene (Mg_{88}) + plagioclase (An_{73}) + glass ($Mg_{67.5}$)
5551	Glass	C	1260	10	0.65	Olivine ($Mg_{83.4}$) + clinopyroxene ($Mg_{85.6}$) + plagioclase (An_{71}) + glass (Mg_{61})
5537	Glass	Fe	1280	10	1	Glass ($Mg_{67.7}$)
5544	Glass	Fe	1280	10	0.5	Glass + very rare olivine
5543	Glass	Fe	1260	10	0.5	Olivine ($Mg_{86.5}$) + clinopyroxene (Mg_{87}) + plagioclase (An_{72}) + Glass (Mg_{63})
5575	Glass	C	1310	12	0.5	Glass (Mg_{70})
5590	Glass	C	1305	12	0.5	Clinopyroxene ($Mg_{88.3}$) + glass ($Mg_{68.5}$)
5578	Glass	C	1300	12	0.5	Clinopyroxene (very rare) + glass
5587	Glass	C	1290	12	0.5	Clinopyroxene ($Mg_{83.5}$) + plagioclase (An_{69}) + glass (Mg_{62})
5565	Glass	C	1330	13	0.5	Glass ($Mg_{69.5}$)
5586	Glass	C	1325	13	0.5	Glass ($Mg_{69.5}$)
5596	Glass	C	1325	13	0.5	Clinopyroxene ($Mg_{88.5}$) + Glass ($Mg_{69.5}$)
5571	Glass	C	1320	13	0.5	Clinopyroxene (Mg_{88}) + Glass ($Mg_{68.5}$)
5591	Glass	C	1320	13	0.5	Clinopyroxene ($Mg_{87.8}$) + Glass† (Mg_{67})
5569	Glass	C	1310	13	0.5	Clinopyroxene (Mg_{87}) + Glass†
5561	Glass	C	1360	15	0.5	Clinopyroxene (Mg_{88}) + Glass ($Mg_{68.5}$)
5563	Glass	C	1350	15	0.65	Clinopyroxene ($Mg_{87.5}$) + Glass ($Mg_{67.5}$)
5560	Glass	C	1340	15	0.5	Clinopyroxene ($Mg_{85.5}$) + Glass (Mg_{65})
5558	Glass	C	1300	15	0.5	Clinopyroxene (Mg_{76}) + Plagioclase (An_{66}) + Glass†
T16	Mix A	C	1360	12	1	Rare orthopyroxene (Mg_{99}–Mg_{91}) + glass
T20	Mix A	Fe	1350	12	1	Rare orthopyroxene + glass

T14	Mix A	C	1340	12	1	Orthopyroxene + calcic clinopyroxene + glass(+ relict $MgSiO_3$)
T12	Mix A	C	1320	12	1	Orthopyroxene ($Mg_{88.7}$) + clinopyroxene (Mg_{91}) + glass(+ relict $MgSiO_3$)
T15	Mix B	C	1330	12	1	Glass
T18	Mix B	Fe	1320	12	1	Orthopyroxene (Mg_{91}–Mg_{86}) + glass
T13	Mix B	C	1320	12	1	Orthopyroxene (Mg_{89}) + glass
T11	Mix B	C	1310	12	1	Relict orthopyroxene rimmed by low-Ca clinopyroxene + glass†
T19	Mix B	Fe	1300	12	1	Relict orthopyroxene rimmed by low-Ca clinopyroxene + glass†
T24	Mix C	C	1315	12	1	Glass
T22	Mix C	Fe	1310	12	1	Very rare clinopyroxene + glass
T25	Mix C	Fe	1300	12	1	Calcic clinopyroxene ($Mg_{89.3}$, 19 per cent CaO, 3.9 per cent Al_2O_3) + glass
T230	Mix D	C	1310	12	1	Olivine ($Mg_{90.5}$) + glass
T228	Mix D	C	1360	15	1	Glass
T229	Mix D	C	1340	15	1	Olivine ($Mg_{90.3}$) + clinopyroxene ($Mg_{90.3}$, 15.2 per cent CaO, 6.5 per cent Al_2O_3) + glass
T232	Mix D	C	1350	17	0.4	?Olivine + clinopyroxene (Mg_{90}, 7–12 per cent CaO, 3.6–7.1 per cent Al_2O_3) + glass†
T234	Mix D	Fe	1370	18	0.25	Clinopyroxene (Mg_{88}, 13.5–14.5 per cent CaO, 8.5 per cent Al_2O_3) + glass†
T233	Mix D	C	1370	18	0.5	Clinopyroxene + glass†
T235	Mix D	Fe	1420	20	0.5	Clinopyroxene (Mg_{89}, 12.5–13.5 per cent CaO, 8.5 per cent Al_2O_3) + glass†
T226	Mix D	C	1420	20	0.75	Glass
T227	Mix D	C	1400	20	0.75	Clinopyroxene ($Mg_{88.5}$, 14.5 per cent CaO, 9.8 per cent Al_2O_3) + glass†
T240	Mix E	C	1460	20	0.25	Glass
T245	Mix E	Fe	1450	20	0.25	Olivine (Mg_{90}) + glass (Mg_{75})
T241	Mix E	C	1440	20	0.25	?Olivine + glass
T244	Mix E	C	1430	20	0.25	Olivine ($Mg_{91.5}$) + glass ($Mg_{78.5}$)
T247	Mix E	Fe	1430	20	0.2	?Olivine + orthopyroxene ($Mg_{90.5}$, 2.6 per cent CaO, 5.5 per cent Al_2O_3) cores to low-Ca clinopyroxene + glass
T243	Mix E	C	1420	20	0.25	?Olivine + orthopyroxene (Mg_{92}, 2.7 per cent CaO, 5.3 per cent Al_2O_3) + glass†
T239	Mix E	C	1480	22	0.25	?Olivine + low calcium clinopyroxene + glass†
T238	Mix E	C	1520	25	0.25	Clinopyroxene + glass†
T237	Mix E	C	1480	25	0.35	Clinopyroxene + garnet + glass†

† Glass is accompanied by common quench clinopyroxene, analyses of glass are not reliably those of liquid in equilibrium with primary crystals.

Table 3. Compositions of near-liquidus calcic pyroxenes for DSDP3-18-7-1 and DSDP3-18+9 per cent olivine.

	DSDP3-18-7-1				91 per cent DSDP3-18 +9 per cent olivine	
P (kbar)	12	12	15		15	20
T (°C)	1305	1290	1360		1340	1420
			Low Al†	High Al†		
SiO_2	51.6	50.7	53.2	51.5	52.8	52.6
TiO_2	0.3	0.5	0.1	0.3	0.3	0.3
Al_2O_3	7.2	8.9	5.3	8.3	6.5	8.4
FeO	4.4	6.0	4.7	4.1	4.5	4.5
MnO	—	—	—	—	—	—
MgO	18.5	17.5	19.4	17.1	23.0	20.8
CaO	17.6	15.6	17.0	18.4	12.5	12.9
Na_2O	0.4	0.6	0.4	0.2	0.4	0.5
100 Mg/(Mg+Fe)	88.3	83.8	88.1	88.2	90.2	89.3

† The pyroxenes show a range of Al–Si solid solution.

Between 10 kbar and 12 kbar, olivine is replaced as the liquidus phase by clinopyroxene. The clinopyroxene (Table 3) is calcium-rich and has Fe/Mg partition coefficient with respect to the liquid

$$\frac{(\mathrm{Fe/Mg})_{\mathrm{cpx}}}{(\mathrm{Fe/Mg})_{\mathrm{liq}}} = 0.3 \pm 0.03,$$

which is similar to that for olivine (cf. Cawthorn *et al.* 1973). At 13 kbar and 15 kbar, calcium-rich clinopyroxene remains the liquidus phase and earlier studies on a variety of basaltic compositions suggest that clinopyroxene will be replaced by garnet on the liquidus phase between 20 and 30 kbar.

The experimental study confirms that the microphenocrysts present in the natural rock have compositions appropriate to low pressure liquidus and near liquidus phases of a liquid matching the glass composition, i.e. the olivine or plagioclase are not xenocrysts. (Note that the small difference ($Mg_{89.7}$ to $Mg_{88.7}$) between most Mg-rich microphenocryst and experimental liquidus olivine (10 kbar, graphite) compositions could be explained by reduction of Fe^{3+} present in the natural magma, to Fe^{2+} in the experiments.) The experimental study also confirms that liquids such as the glass from DSDP 3-18 are incapable of precipitating at any pressure either the anorthite-rich plagioclase ($>An_{80}$) or olivine of $>Mg_{90}$ composition or chrome-rich, aluminium-poor diopside ($>Mg_{88}$) which are observed as megacrysts in some ocean floor basalts, including glassy basalts similar to DSDP 3-18

(Donaldson and Brown, 1977). The experimental study also confirms that if DSDP 3-18 is not a primary liquid in the sense used herein, but has evolved by crystal fractionation from a more primitive magma, then, provided that crystal fractionation occurred at <10 kbar, olivine is the only crystalline phase involved, i.e. a more primitive melt lies along the olivine-addition fractional crystallization path from DSDP 3-18. If crystal fractionation had occurred between 10 kbar and 12 kbar, then both olivine- and calcium-rich clinopyroxene may have separated and if at pressures $\geqslant 12$ kbar, then clinopyroxene only may have separated. It should be noted however that the possibility of clinopyroxene fractionation at $P \geqslant 12$ kbar to yield DSDP 3-18 is ruled out if the constraint of an ultimate peridotite source is accepted. Thus addition of Ca-rich clinopyroxene to DSDP 3-18 liquid would produce liquids moving further away from the olivine field and further into the liquidus field of clinopyroxene. Higher pressure would accentuate this problem and the problem also exists for postulated olivine + clinopyroxene crystal fractionation, which in addition would move liquids further away from the liquidus field of orthopyroxenes (see following discussion).

3.1. Possible Derivation of DSDP 3-18 from Peridotite Source Rock

Olivine ($\sim Mg_{89}$) is the liquidus phase of DSDP 3-18 in the 0–10 kbar pressure interval and this liquid could be derived from a mantle peridotite by partial melting and magma segregation at depths to about 35 km provided the degree of melting of the peridotite was sufficient to melt out all phases other than olivine (and chrome-rich spinel probably). We return to discuss this model in a later section, but we note at this point that extremely pure dunites are rather uncommon in nature, whereas harzburgites and lherzolites in which orthopyroxene is a major phase are relatively much more common. We prefer to investigate further for a model of derivation in which olivine and one or more pyroxenes remain as residual phases.

The melting relations of two peridotite compositions (pyrolite (Ringwood, 1966) and Tinaquillo peridotite (Green 1963)) in the pressure range 0–15 kbar are currently being investigated (Jaques and Green, in preparation) and in Fig. 2 we summarize our data on the solidus and phase stability at above-solidus conditions. We have also plotted the 20, 30 and 40 per cent melting contours for both compositions. For our present purpose it is only necessary to point out that olivine and orthopyroxene are present as residual phases in both compositions over the entire P–T range defined by the liquidus of DSDP 3-18, i.e. if DSDP 3-18 was a primary melt from either peridotite, then olivine and *orthopyroxene* should occur together on its liquidus at some P, T condition. The liquidus of DSDP 3-18 overlaps the

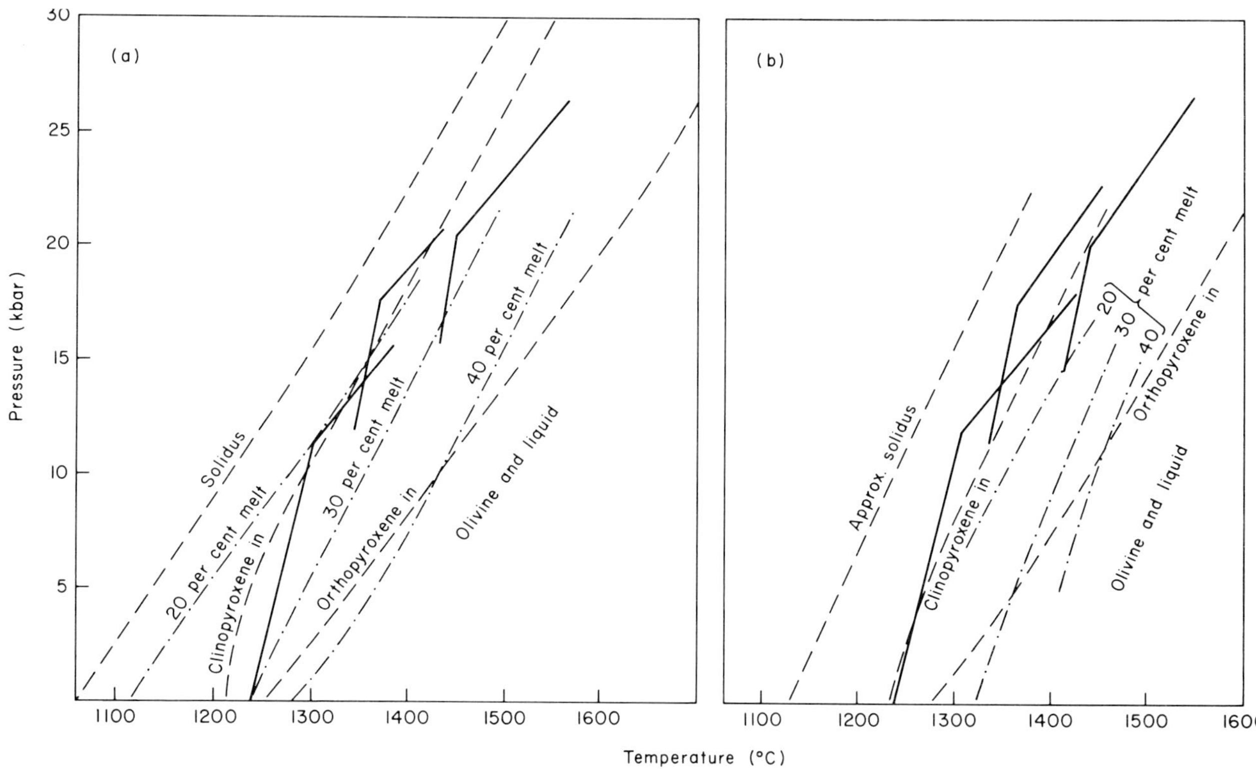

Fig. 2. (a) Summary of partial melting data for pyrolite composition showing solidus and boundaries marking the disappearance of phases above the solidus. Also shown are percentage melting contours and (solid lines) the positions of the liquidi from Fig. 1. (b) Summary of partial melting data for Tinaquillo peridotite composition showing solidus and boundaries marking the disappearance of phases above the solidus. Also shown are percentage melting contours and (solid lines) the positions of the liquidi from Fig. 1.

field for residual clinopyroxene for Tinaquillo peridotite except at low pressure and overlaps the clinopyroxene field for pyrolite only between 10 and 13 kbar. The absence of orthopyroxene and the presence of a calcium-rich pyroxene alone or with olivine is therefore prohibitive for direct derivation of DSDP 3-18 by equilibrium partial melting of peridotites with compositions similar to pyrolite or Tinaquillo peridotite.

Harzburgite and enstatite-bearing dunite are common highly refractory peridotites ($>Mg_{90}$) occurring in nature whereas wehrlite or chrome diopside-bearing dunite (enstatite-free) are very uncommon among refractory peridotites ($>Mg_{90}$). We take these observations on natural rocks as support for the general relevance of compositions close to pyrolite or Tinaquillo peridotite, i.e. peridotites with high normative hy/di ratios, and the general applicability of the olivine + orthopyroxene constraint in evaluating primary magmas and their conditions of genesis.

In earlier studies of both terrestrial and lunar basalts, it was found that orthopyroxene occurred as a liquidus phase between the P-dependent liquidus fields of olivine (lower pressure) and clinopyroxene (higher pressure) (Green and Ringwood, 1967; Ito and Kennedy, 1968; Green *et al.*, 1971, 1975). In addition it has been observed that the composition of the first calcic pyroxene to crystallize in coexistence with olivine, becomes less calcic, i.e. approaches the orthopyroxene–clinopyroxene solvus, with increasing pressure (Bultitude and Green, 1971) and clinopyroxenes coexisting with orthopyroxene on anhydrous liquidi are extremely subcalcic indeed (Green and Ringwood, 1967; Bultitude and Green, 1971; Green and Hibberson, 1970; Green *et al.* 1971, 1975). It has also been shown that orthopyroxene, present at high temperature, may react out at lower temperature in the melting interval. Finally, the careful study of liquidus pyroxenes of lunar basalts has shown that there exist experimental difficulties in the metastable nucleation and growth of subcalcic ("pigeonitic") clinopyroxenes in place of the stable coexisting orthopyroxene–clinopyroxene pair (Green, 1976*a*,*b*, and unpublished data on a wide range of subcalcic near-liquidus pyroxenes from lunar compositions studied at high pressure). It has proved very difficult to experimentally define the pyroxene miscibility gap at high pressure in the complex natural systems but our incomplete results suggest that clinopyroxene (Mg_{85}–Mg_{90}) coexisting with orthopyroxene on the liquidi of basalts at 10–15 kbar, 1300–1400 °C should contain not less than 12 weight per cent CaO and not more than 14 weight per cent CaO.

Against this background it is necessary to examine critically our data on clinopyroxene composition and conditions of crystallization to ascertain if DSDP-18 could be saturated with orthopyroxene at conditions slightly different from those tested, or is very close to saturation with orthopyroxene or is indeed saturated with orthopyroxene but crystallization of metastable clinopyroxene has obscured this. The compositions of subliquidus pyroxenes from DSDP3-18 at pressures below 12 kbar and of liquidus pyroxenes at and above 12 kbar are calcium-rich (Tables 3 and 4) and unlike compositions occurring on the solvus limb, at least with our present state of knowledge.

Table 4. Compositions of residual phases (near-liquidus at 20 kbar, 1420–1430 °C) for composition E (83 per cent DSDP3-18 + 17 per cent olivine). Also compositions of residual spinel in pyrolite at 15 kbar, 1400 °C coexisting with olivine and orthopyroxene only, and spinel inclusion within olivine megacrysts (Donaldson and Brown, 1977).

	Olivine	Orthopyroxene	Spinel (Jaques and Green, in progress)	Spinel xenocryst (Donaldson and Brown, 1977)
SiO_2	40.9	54.8	—	—
TiO_2	—	<0.1	1.2	0.4
Al_2O_3	—	5.5	19.2	24.2
FeO	8.2	4.9	17.8	15.3
MgO	50.4	32.0	17.8	15.3
CaO	0.4	2.8	—	—
Na_2O	—	—	—	—
K_2O	—	—	—	—
Cr_2O_3	—	(1.2)†	43.5	44.4
NiO	(0.4)†	(0.05)†	—	—
100 Mg/(Mg+Fe)	91.6	92.1	64	64

† Values from residual pyroxene and olivine in pyrolite at 15 kbar, 1400 °C. Note that Cr_2O_3 and NiO were not added to DSDP3-18 mix.

Orthopyroxene Addition Experiments

To test further for orthopyroxene saturation or nearness thereto, we have added enstatite to DSDP3-18. Three mixes were used:

A. 91 per cent DSDP3-18 (glass) + 9 per cent $MgSiO_3$ (Tem-Pres synthetic clinoenstatite);
B. 91 per cent DSDP3-18 (glass) + 9 per cent bronzite (Mg_{86}, 0.9 per cent Al_2O_3, 0.5 per cent CaO);
C. 95 per cent DSDP3-18 (glass) + 5 per cent bronzite (Mg_{86}, 0.9 per cent Al_2O_3, 0.5 per cent CaO).

In each case the pyroxene added was sufficiently different from anticipated liquidus orthopyroxene for the chemical distinction between relict seed cores to pyroxenes and equilibrium orthopyroxene to be readily made by electronprobe analysis. The seeds were respectively more magnesian and less magnesian than the liquidus pyroxene of DSDP3-18. The seed pyroxenes were crushed to $<10\ \mu m$ but grain size was very variable. Experiments were run at 12 kbar, 1300–1360 °C in graphite and spec-pure iron capsules. In mix C, the liquidus was at ~1310 °C and calcium-rich pyroxene ($\geqslant$19 per cent CaO, $Mg_{89.3}$) was the liquidus phase, orthopyroxene seeds disappeared and there was no low-calcium (<10 per cent CaO) pyroxene present. For mix B, the liquidus at 12 kbar was 1325 °C and the liquidus phase was orthopyroxene. The charge contained small cores of orthopyroxene with compositions approaching the seed crystals surrounded by rims of orthopyroxene with Mg_{89-91}, 2.4 per cent Al_2O_3, 2.2 per cent CaO. Similarly, for mix A, the liquidus temperature was ~1360 °C and near-liquidus runs contained rare relict seed cores (orthopyroxene up to Mg_{99}), new "equilibrium" orthopyroxene (Mg_{91}, 2.4 per cent Al_2O_3, ~3.0 per cent CaO) and metastable or quench low-calcium clinopyroxene ($<Mg_{91}$). At 12 kbar, 1320 °C and 1310 °C (i.e. just above the liquidus temperature of DSDP3-18), mix A contained relict seed orthopyroxene (Mg_{100}), orthopyroxene ($Mg_{92.3}$, 2.2 per cent Al_2O_3, 2.8 per cent CaO) and calcium-rich clinopyroxene ($Mg_{89.6}$, 2.8 per cent Al_2O_3, 10.1 per cent CaO). The experiments confirm the difficulty in nucleating coexisting orthopyroxene and calcium-rich pyroxene rather than metastable low-calcium clinopyroxene and expose the limitations of the seeding technique in that the seeds only partially react, and act as nucleii for new crystal growth. However, the experiments confirm that DSDP3-18 lies in the liquidus field of calcium-rich pyroxene at 12 kbar, and that it does not lie on the two pyroxene boundary. The orthopyroxene addition experiments demonstrate that the two-pyroxene boundary intersects the DSDP3-18 to orthopyroxene join at compositions between DSDP3-18+5 per cent orthopyroxene and DSDP3-18 +10 per cent orthopyroxene. It may be noted that this result is consistent with earlier studies in the 9–13.5 kbar interval (Green and Ringwood, 1967; T. H. Green *et al.*, 1967; Green, 1970) in which orthopyroxene accompanied or preceded clinopyroxene precipitation in compositions having higher normative hy/di or (hy+ol)/(di+an) ratios than DSDP3-18. The various data partially define the primary liquidus fields of olivine, orthopyroxene, clinopyroxene, plagioclase within the "basalt tetrahedron" (Yoder and Tilley, 1962) at 9–13 kbar, anhydrous conditions.

Olivine Addition Experiments

Previous studies have demonstrated that pressure has the effect of increasing the activity of pyroxene components at the expense of olivine in olivine-rich liquids and that a liquid which at low pressure has olivine on its liquidus, followed at lower temperature by clinopyroxene, may, at a higher pressure, have olivine replaced by orthopyroxene or orthopyroxene + clinopyroxene (Green and Ringwood, 1967; Green, 1970). The results of the experiments on DSDP3-18 demonstrated that this magma could originate by olivine extraction from a more olivine-rich parent magma. We have, on these grounds, added olivine to DSDP3-18 to evaluate whether, at higher pressure and temperature, an olivine-enriched magma could be both saturated in olivine and orthopyroxene (± calcic clinopyroxene), and thus a possible primary melt from mantle peridotite, and also parental to DSDP3-18 by extraction of olivine alone at lower pressures.

Two mixes were prepared by addition of natural olivine ($Mg_{91.7}$) to DSDP3-18 glass. The olivine was crushed to $< 10\ \mu m$ and mixed with the powdered glass. Bulk compositions of the two mixes (composition D: 91 per cent DSDP3-18 + 9 per cent olivine; composition E: 83 per cent DSDP3-18 + 17 per cent olivine) are given in Table 1. Composition D has olivine as liquidus phase ≤ 15 kbar, olivine and clinopyroxene (Mg_{90}, 7–12 per cent CaO, 3.5–7 per cent Al_{23}) on the liquidus at 17 kbar, ~1360 °C and clinopyroxene (13–14 per cent CaO, 8.5 per cent Al_2O_3) as liquidus phase at 18 kbar, 1370 °C and 20 kbar, 1400 °C. It is at present uncertain whether the clinopyroxene lies within the pyroxene solvus at 17–18 kbar, 1360–1370 °C or on the calcium-rich side of the diopside limb. The range of CaO and Al_2O_3 contents in the 17 kbar, 1360 °C run suggest metastable low-calcium clinopyroxene in place of orthopyroxene + calcic clinopyroxene.

Composition E has olivine as the liquidus phase at 20 kbar, 1450 °C, but olivine and orthopyroxene (Mg_{92}, 2.8 per cent CaO, 5.5 per cent Al_2O_3) occur together at 20 kbar, 1430 °C. The presence of near-liquidus orthopyroxenes accompanying olivine was confirmed in both graphite and spec-pure iron capsules. (Note the minor FeO addition in the latter and the barely significant difference in liquidus temperature in C and Fe capsules.) At 22 kbar, clinopyroxene (stable or metastable low-calcium clinopyroxene) and olivine occur very close to the liquidus and similar clinopyroxene occurs on the liquidus at 25 kbar, 1520 °C. Garnet joins clinopyroxene at 25 kbar, 1480 °C but does not appear in the near-liquidus runs at lower pressure.

4. A Primary Parental Magma

We infer from the olivine addition studies that a picritic magma containing ~25 per cent normative olivine and parental to DSDP3-18 by the separation of 15–17 per cent olivine (mean composition ~Mg_{91}), has appropriate liquidus phases, olivine and orthopyroxene, to be a primary magma derived at high pressure from mantle orthopyroxene-bearing peridotite. In particular, the depth of magma segregation for this primary, picritic magma is

~70 km (~20 kbar) at a temperature of ~1430 °C leaving residual olivine ($Mg_{91.5}$) and orthopyroxene (Mg_{92}, 5.5 per cent Al_2O_3, 2.8 per cent CaO). From the nickel content of DSDP3-18 and of olivine phenocrysts and xenocrysts and the similar contents of nickel in olivine (Mg_{89-92}) of mantle derived lherzolites, it may be inferred that the primary, picritic parental magma would contain ~600 p.p.m. Ni. The Ni content of source peridotite and potential primary liquids will provide an important constraint (cf. Frey *et al.*, 1978) on models of magma genesis, but uncertainties in the compositional and temperature effects on the partition coefficient for Ni between olivine and liquids are as yet not resolved (Leeman and Scheidegger, 1977; Hakli and Wright, 1967; Mysen, 1976). The predicted Ni content of ~600 p.p.m. for the DSDP3-18 parent magma is similar to that (670 p.p.m. Ni) modelled for a primitive tholeiite melting at ~17 kbar, 1400 °C leaving harzburgite residue (Frey *et al.*, 1978).

4.1. Retraction of an Earlier Model of Mid-ocean Ridge Petrogenesis

In earlier papers (Green and Ringwood, 1967; T. H. Green *et al.*, 1967; Green, 1970, 1971) the experimental data then available was used to infer that high Al_2O_3 (>14 per cent Al_2O_3) olivine tholeiites (~10 per cent normative olivine) were primary melts from mantle peridotite (pyrolite) at ~20 per cent melting, leaving residual olivine, orthopyroxene and very minor calcium-rich pyroxene, at a depth of magma segregation of ~30 km (9 kbar). Direct studies of melting of pyrolite and Tinaquillo lherzolite at 0–15 kbar (Jaques and Green, unpublished data) show that liquids derived at these conditions are indeed enriched in Al_2O_3/CaO relative to liquids at ~15 kbar but differ from liquids such as DSDP3-18 in lower Al_2O_3 contents, higher SiO_2 contents and higher alkali contents. Similarly, other studies (Green *et al.*, 1974) have confirmed a low pressure fractionation trend towards Al_2O_3 and Al_2O_3/CaO enrichment (Green and Ringwood, 1967) at low pressure (5–10 kbar), but derivative liquids are hawaiitic and mugearitic rather than high-Al_2O_3-olivine tholeiites.

The earlier model emphasized that olivine fractionation was important in magma genesis in the mid-ocean ridge situation, but envisaged primary parental magmas with ~10 per cent normative olivine (~49 per cent SiO_2, >14 per cent Al_2O_3, $\geqslant$12 per cent CaO). This model is now retracted and we support the view, earlier advocated by O'Hara (1968), that primary, parental magmas in the mid-ocean ridge situation include picrites with ~25 per cent normative olivine, ~48 per cent SiO_2, ~13–14 per cent Al_2O_3, ~11 per cent CaO. We do not infer that magmas of this composition are the only primary, parental magmas, but rather that this is one primary magma whose probable conditions of genesis can now be delineated and which will yield, by crystallization and settling out of olivine, highly magnesian liquids such as DSDP3-18 and similar compositions (Melson *et al.*, 1977).

As noted above, our conclusion that a primary parental magma for mid-ocean ridge magmatism is tholeiitic picrite rather than olivine tholeiite, is in agreement with O'Hara (1968). However, O'Hara inferred that the primary picrite was derived by melting at $\sim$30 kbar ($>\sim$100 km) whereas our data shows that this pressure is too high and the depth of magma segregation is around 70 km. Similarly, O'Hara inferred that the primary magma was formed in equilibrium with olivine + orthopyroxene + garnet $\pm$ clinopyroxene (i.e. a garnet harzburgite residue). Our data show that garnet is absent both for the near-liquidus experiments on DSDP3-18 + olivine, and for the melting studies on pyrolite and Tinaquillo peridotite at 20 kbar, 1400–1500 °C. We infer that clinopyroxene is also absent as a residual phase. The absence of garnet is a key factor in controlling rare earth fractionation in the partial melting process and is consistent with the absence of heavy rare earth element depletion in DSDP3-18 and similar mid-ocean ridge basalts (Group I of Bryan *et al.*, 1976).

Our study of the petrogenesis of one of the more primitive mid-ocean ridge basalts has provided evidence for a deep origin and a post-segregation history of crystal fractionation and cooling (from $\sim$1430 °C to $\sim$1230 °C) during movement to the Earth's surface. Other authors (including Donaldson and Brown, 1977; Flower *et al.*, 1977; Clarke and Pedersen, 1976; Esson *et al.*, 1975) have presented evidence for picritic melts in mid-ocean ridge and ?pre-ridge volcanism (western Greenland, Hebridean province). Variations in Na_2O/CaO, K_2O/Na_2O, TiO_2/CaO, or TiO_2/Al_2O_3 in such primary picritic melts may originate by:

1. *Variation in degree of melting.* Lower degrees of melting produce liquids with higher K_2O/Na_2O, TiO_2/CaO, etc., and the liquids trend towards alkali picrite.
2. *Difference in source composition.* This particularly includes earlier processes of magma extraction leaving more refractory peridotite with lower Na_2O/CaO, TiO_2/CaO ratios. There may also have been earlier melt addition (Green, 1971; Frey and Green, 1974; Frey *et al.*, 1978) enriching the source in incompatible elements with increase in CaO/Al_2O_3, TiO_2/CaO, etc.

Further work is required to evaluate whether the entire spectrum of mid-oceanic ridge basalts derive from primary *picritic melts*. Alternatively, the spectrum may include examples which are derivative from primary *olivine tholeiite* to *alkali olivine basalt* (Green and Ringwood, 1967; Green, 1971) segregated from mantle peridotite at *shallower depths*. Primary basalts formed in the latter way should have Mg-values of 75 ± 3, ~ 500 p.p.m. Ni, but have lower normative olivine and higher SiO_2 contents than the picritic magmas deduced in this study.

5. Ophiolites in Relation to Mid-ocean Ridge Petrogenesis

The experimental data obtained on DSDP3-18 and the olivine-enriched compositions define the possible crystal accumulates and residues which can be formed as complementary rocks to the melting-out of the parent magma or its low pressure fractionation. The residual minerals at ~ 20 kbar, 1430 °C would be olivine (Mg_{92}) and enstatite (+Cr-spinel)† but, although these are appropriate to harzburgite, the composition of the orthopyroxene, with >5 per cent Al_2O_3, >2.5 per cent CaO, is unlike that of orthopyroxene of harzburgite associated with ophiolites (<1.5 per cent Al_2O_3, <1 per cent CaO) (England and Davies, 1973; Moores, 1969). Residual harzburgite from extraction of DSDP3-18 + 17 per cent olivine at 20 kbar, 1430 °C would recrystallize as lherzolite (plagioclase or spinel lherzolite) at lower temperatures and would not equate chemically or mineralogically with non-cumulate harzburgite observed as "basement" to many ophiolite complexes.

If DSDP3-18 or olivine-enriched parents were to segregate from mantle peridotite at <20 kbar, then wehrlite (DSDP3-18 at 12 kbar) or dunite (DSDP3-18 at <12 kbar, or picritic parental magmas at <20 kbar) would be *residual* rock types. The common occurrence of non-accumulate harzburgite, the less common occurrence of enstatite-free dunites, the rarity of wehrlite and the accumulative nature of at least some observed wehrlites and dunites, are inconsistent with this model.

A picritic magma of DSDP3-18 + 17 per cent olivine composition can produce low-pressure accumulates of dunite (olivine ($\leqslant Mg_{92}$) + ?chrome spinel), olivine + plagioclase (more iron-rich and sodic than Fo_{87} and An_{79}, respectively), or olivine, diopside and plagioclase. There is no possibility of magmas such as DSDP3-18 or the picritic parental magma producing accumulates containing magnesian orthopyroxene ($\geqslant Mg_{87}$) or calcic plagioclase ($\geqslant An_{80}$). Similarly, it is characteristic that plagioclase precipitation should precede clinopyroxene precipitation in magmas such as

† It has been noted previously that our experiments do not allow an appraisal of the near-liquidus role of spinel for DSDP3-18 or derivative compositions. However, the experimental melting studies of pyrolite and Tinaquillo peridotite composition show that spinel ($100Cr/(Cr + Al) \approx 65$, $TiO_2 = 1.2$ per cent) would be present as a very minor residual phase at 20 kbar, 1430 °C.

DSDP3-18 or similar high-Al_2O_3 ocean floor basalts. The first clinopyroxene to appear in DSDP3-18 at low pressure has Mg-value of ~ 88 and is unlike the most magnesian clinopyroxenes occurring in rocks with cumulus texture in at least some ophiolite complexes (Mg_{92}, Mg_{94}; England and Davies, 1973). The uncommon chrome diopside megacrysts (Mg_{91}) in mid-ocean ridge basalts (Donaldson and Brown, 1977) are also inconsistent with precipitation from DSDP3-18 or picritic parental magmas.

We conclude that the primary picritic magma which we have identified as parental to DSDP3-18 and similar basalts, has no genetic relationship to ophiolite complexes having the following characteristics:

(a) basement of non-cumulate harzburgite with olivine $>Mg_{91}$, orthopyroxene (Mg_{91}, <1.5 per cent Al_2O_3, <1 per cent CaO), and lacking clinopyroxene and/or plagioclase;
(b) cumulate sequences containing plagioclase more anorthite-rich than An_{79}, clinopyroxene more Mg-rich than Mg_{88}, or orthopyroxene more Mg-rich than approx. Mg_{87}.

These characteristics are in fact typical of most ophiolite complexes and it follows that ophiolite complexes are related to some other magma type than the more primitive and common mid-ocean ridge basalts (i.e. magmas similar to DSDP3-18). Similarly, the megacrysts of very calcic plagioclase and magnesian clinopyroxene occurring in some mid-ocean ridge basalts have no genetic relation to their host magma and are more probably genetically related to the accumulates of ophiolite complexes. Curiously, highly magnesian orthopyroxene megacrysts do not occur in mid-ocean ridge basalts, although their role as accumulate phases in ophiolite complexes and residual phases in non-cumulate harzburgite is equally problematical. We suggest the need to explore several alternative and probably complementary hypotheses on the petrogenesis of ophiolite complexes. We consider that there is no single tectonic setting for development of ophiolite complexes—probable tectonic settings include:

(i) *Island arcs.* Ophiolite complexes may develop in complementary relation to genesis and fractionation of primitive highly magnesian island-arc tholeiites (boninites, high-Mg andesites of Cape Vogel, Marianas Trench, etc.; Dallwitz *et al.*, 1966; Green, 1973).
(ii) *Mid-oceanic ridge environments*, as complementary residues and accumulates to second-stage melting of refractory peridotite diapirs at shallow depths to yield highly magnesian, silica-rich olivine tholeiites, or quartz tholeiites rarely if every seen as extrusive magmas and yielding eruptive rocks only after crystal fractionation. The shallow "extrusive" diapir model of Moores (1969) may be applicable, the

bulk composition of the diapir resembling the composition of Lizard or Tinaquillo lherzolites or possibly even more refractory, i.e. low Na/Ca, low K, Ti, etc., contents, due to loss of an earlier melt fraction, perhaps identified as the "typical" mid-ocean ridge basalts.

(iii) *Back-arc basin environments*, e.g. Lau basin; possibly transitional between (i) and (ii) in that limited access of water activates melting of refractory lithosphere, residual from a previous melting episode.

The origin of megacrysts in mid-ocean ridge basalts may be found in the passage of primary picrite magmas, segregating at ~60–70 km, (or their olivine-depleted derivatives) through accumulates and incompletely crystallized magma chambers of shallow, second-stage melting, diapirs occurring at depths of 1–15 km. The two different depth regimes for origins of picrites and "ophiolite" layered sequences provides opportunity for entrainment of xenocrysts and xenoliths from the latter within the picritic magmas. This model provides opportunity for magma mixing to occur and is sufficiently complex to make testing difficult. Possibly the most rewarding area for further study is a search for primitive liquids or fractionation products of primitive liquids, which originate in the postulated shallow depth, second-stage of partial melting, i.e. liquids which leave harzburgite residue, precipitate plagioclase of An_{80-95}, clinopyroxene of Mg_{89-95}, and orthopyroxene of Mg_{89-92}, in early accumulates. Liquids with these characteristics should have high SiO_2 contents (>50 per cent), high CaO contents (>12 per cent) with low Na_2O contents (<1.5 per cent) and should be quartz normative or with small normative olivine contents. Our conclusions that high Ca, low Na liquids, unlike common mid-ocean ridge basalts such as DSDP3-18, are required to account for precipitation of megacrysts in mid-ocean ridge basalts and to produce the accumulate sequences of ophiolite complexes, are similar to those of Donaldson and Brown (1977) and Hodges and Papike (1976).

6. Composition of the Upper Mantle Source Region for Mid-ocean Ridge Basalts

Our experimental study defines two rather different possible origins for DSDP3-18 (and by inference, similar mid-ocean ridge basalts). First, DSDP3-18 may be a primary magma segregating from residual dunite at pressures less than 12 kbar, i.e. depths <~40 km. The composition of residual olivine (i.e. liquidus olivine for DSDP3-18) is best determined by the most magnesian microphenocrysts in DSDP3-18 (Frey *et al.*, 1974). These authors did not determine the NiO content of the olivine but this may be approximated as 0.4 weight per cent from partition data (Leeman and

Scheidegger, 1977) or from other analyses of similar olivine (e.g. Hodges and Papike, 1976). Similarly, the occurrence of spinel inclusions in early olivine phenocrysts, the presence of spinel megacrysts and the melting studies of pyrolite and Tinaquillo peridotite suggest that chrome-rich spinel should be a residual phase for DSDP3-18 in low-pressure magma segregation. The composition of spinel given in Table 4, column 4, is assumed to be a reasonable estimate of residual spinel—it is close to residual spinel compositions in pyrolite and Tinaquillo peridotite.

It is thus possible to calculate model source composition for DSDP3-18 by mixing calculations using DSDP3-18, olivine ($Mg_{89.7}$) and chrome–spinel (Mg_{64}). The approach used herein does not determine an estimate of the proportions of these phases. In Table 5, column A, we list source and residue compositions assuming 30 per cent melting—this figure being derived from the intersection of the liquidus for DSDP3-18 with the 30 per cent melting contour for pyrolite. This choice is not unreasonable for self-consistency as shown by comparison of key element contents (Na_2O, CaO, Al_2O_3, SiO_2, MgO) between "source peridotite" (A) and pyrolite (Ringwood, 1966) compositions (Table 5).

The second class of models for derivation of DSDP3-18 are those which derive a parental picritic magma (DSDP3-18 + 17 per cent olivine) by magma segregation at ~20 kbar (70 km) bearing residual harzburgite. Our experiments determine phase compositions (Tables 1, 3 and 4) but not the relative proportions. In Table 5 we have calculated two model sources and residue compositions (columns B, C) on the arbitrary assumption that the degree of partial melting of the source is 24 per cent, there is 1 per cent residual spinel and the residue is either an orthopyroxene-poor (column B) or orthopyroxene-rich (column C). The relative proportions were chosen to yield ~3–4 per cent CaO and 4–5 per cent Al_2O_3 in the source composition, i.e. compositions similar to previous estimates of upper mantle "pyrolite" composition (Table 5), and hence with the same support from natural occurrences of peridotite (Green and Ringwood, 1963, 1967). In calculating the model source compositions and residue compositions, it should be noted that because residual olivine contains 0.4 per cent CaO and no detectable (i.e. <0.2 per cent) Al_2O_3, and, because spinel contains high Al_2O_3 and no CaO, it is possible to choose proportions of the four phases to yield "chondritic" CaO/Al_2O_3 ratios for the source (cf. Ringwood, 1975; Sun and Nesbitt, 1977). However the TiO_2 content, TiO_2/Al_2O_3 and/or TiO_2/CaO ratios of all phases are so low that it is not possible to calculate model source composition with "chondritic" ratios for these elements—the possible source compositions are all depleted in TiO_2. Similar conclusions apply (more strongly) to the incompatible elements K_2O, P_2O_5, Rb, Sr, light rare earth elements, etc., based on the composition of DSDP3-18 and estimates

Table 5. Compositions of model source compositions (see text), i.e. model pyrolite compositions, based on: A, DSDP3-18 + dunite residue; B, picrite (from DSDP3-18) + enstatite-poor harzburgite residue; C, picrite + enstatite-rich harzburgite residue; D is the favoured model adjusted from B and C to bring the source and residue compositions into the spectrum of lherzolite compositions of Kuno and Aoki (1970).

	A		B		C		D (favoured model)		Pyrolite (Ringwood, 1966) (based on Hawaiian tholeiite)	Pyrolite (Ringwood, 1975) (based on mid-ocean ridge tholeiite)
Melt (per cent)	30		24		24		24			
Olivine (per cent)	69‡		65†		50†		57.5†			
Orthopyroxene (per cent)	—		10†		25†		18†			
Spinel (per cent)	1‡		1†		1†		0.5†			
	Source	Residue	Source	Residue	Source	Residue	Source	Residue		
SiO_2	44.3	39.9	43.5	42.2	45.6	44.7	45.0	43.9	45.2	46.1
TiO_2	0.24	—	0.16	0.02	0.17	0.03	0.17	0.04	0.71	0.2
Al_2O_3	5.7	0.34	4.0	0.97	4.8	2.1	4.4	1.4	3.5	4.3
FeO	9.6	10.0	7.9	7.9	7.4	7.2	7.6	7.5	8.6	8.2
MnO	0.12	0.11	0.12	0.12	0.11	0.11	0.11	0.11	0.14	—
MgO	37.1	48.0	40.0	47.6	37.3	43.7	38.8	45.9	37.5	37.6
CaO	4.6	0.43	3.2	0.71	3.5	1.2	3.4	0.95	3.1	3.1
Na_2O	0.66	—	0.40	—	0.40	—	0.40	—	0.57	0.4
K_2O	0.003	—	0.003	—	0.003	—	0.003	—	0.13	0.03
Cr_2O_3	0.46	0.63	0.50	0.74	0.74	0.96	0.45	0.58	0.43	—
NiO	0.29	0.40	0.28	0.35	0.23	0.28	0.26	0.32	0.20	—
100 Mg/(Mg + Fe)	87.3	89.5	90.1	91.5	90.0	91.5	90.1	88.6	89.1	

† Mineral compositions of Table 3, columns 1–3

‡ Spinel from Table 4, column 4; olivine is $Mg_{89.7}$ from Frey *et al.* (1974) (with 0.4 per cent NiO added)

of the relevant residue/liquid partition coefficients (Frey *et al.*, 1978). It is thus not reasonable to use preconceptions of "chondritic" mantle chemistry to model the proportions of phases in the source/residue/melt equilibrium. It is however reasonable to turn to natural mantle-derived peridotites to enquire whether independent evidence from this source permits choice between models such as A, B and C of Table 5.

Kuno and Aoki (1970) assembled chemical data on spinel lherzolite inclusions in basalt from many localities. It is generally accepted that these rocks are random xenoliths of the upper part of the subcrustal lithosphere and that, although they include examples of accumulates from basaltic magmas, they are predominantly residual after extraction of basaltic magmas from parental lherzolite (Kuno and Aoki, 1970; Frey and Green, 1974). Likewise, high-temperature lherzolite diapirs such as Tinaquillo, Lizard, Serrania del Ronda, Beni Bousera, etc., are interpreted as residual upper-mantle compositions, depleted by loss of a small melt fraction. Kuno and Aoki (1970) show that lherzolite compositions are variable in Mg/Fe ratios and that there is systematic change in Ca, Al, Cr, Ti and Na + K associated with Mg/Fe variation. These systematic compositional changes are consistent with melting and magma extraction. These inferences on the origin of natural lherzolites, if accepted, provide a discriminant to choose between models A, B and C, i.e. the calculated source/residue pairs should lie within the spread of natural residue/?source peridotite compositions. On this basis, model A is not appropriate as it produces a chemical trend lying athwart the natural rock trends and either source or residue compositions lie outside the compositional range of the natural rocks. Models B and C are generally compatible with the natural rock trends except that both are too high in Cr_2O_3. In detail, the plots of individual oxides against Mg/Fe (Kuno and Aoki, 1970) suggest that B is too low and C is too high in orthopyroxene and/or olivine in the residue. These "errors" are corrected in model D, in which a "favoured model" for source and residue is calculated. It is emphasized that this model source is capable of yielding a variety of residual compositions with differing olivine to orthopyroxene ratios dependent on conditions of melting and magma extraction (Fig. 2). It is also emphasized that model D is constrained only by the compositions of the phases involved in the postulated partial melting and magma extraction event and by the applicability and compositional spread of the natural peridotite compositions.

Model D may be evaluated using information on rare earth concentrations. Frey and Green (1974) showed that although light rare earth elements showed large fractionation among lherzolite inclusions and intrusive lherzolites, the heavy rare earth's were correlated with CaO-content and the degree of refractoriness of the lherzolites and, in the least refractory lherzolites (~3 per cent CaO) approach concentrations of 2× chondrites and flat, unfractionated patterns (Tb→Lu). Frey *et al.* (1978) analysed the significance of trace-element and major-element variations in a suite of tholeiitic to olivine melilite basalts from both Victoria and Hawaii to infer that the mantle source peridotites had 2.5–3.0× chondritic abundances of the heavy rare earths. Model D, using the rare earth element contents of Frey *et al.* (1974) for DSDP3-18, implies that the source region had a heavy rare earth concentration of 2.2× chondrites but showed very strong light rare earth element depletion. Thus the model deduced on phase equilibria and major element considerations is consistent with constraints implied by heavy rare earth abundances of mantle-derived basalts and mantle-derived lherzolites.

The model source composition, i.e. "pyrolite"-yielding mid-ocean ridge basalts matching DSDP3-18, is very similar in major elements to the pyrolite of Ringwood (1966) which has been so extensively studied experimentally. It is significantly lower in its orthopyroxene to olivine ratio and higher in Al_2O_3 content, so that there will be some differences in the roles of subsolidus spinel and Al_2O_3 content of pyroxenes in the spinel pyrolite field and the proportions of garnet in the garnet–pyrolite field (Green and Ringwood, 1970; Green and Liebermann, 1976). Differences in CaO and Na_2O between the two compositions have complementary effects, but the principal difference between the two model compositions is in incompatible element contents exemplified by TiO_2 (0.71 per cent and 0.17 per cent) and K_2O (0.13 per cent and 0.003 per cent). The difference in Mg/(Mg+Fe) value is a significant one and may be attributed to the rather low Mg-value of the basalt used in the original Ringwood (1966) calculation—an "error" only apparent with the benefit of hindsight! Ringwood (1975, Table 5-4) calculated a pyrolite source composition for mid-ocean ridge tholeiite on the basis of the earlier petrogenetic model of segregation at 4–10 kbar from residual harzburgite. Taking the heavy rare earth concentration of "primary" oceanic tholeiite as 12× chondrites and assuming a source concentration of ~2× chondrites, Ringwood inferred 17 per cent melting to yield a primitive mid-ocean ridge tholeiite (KD-11 of Kay, Hubbard and Gast, 1970); leaving residual orthopyroxene-rich (33 per cent enstatite, 67 per cent olivine) harzburgite. The experimental data presented herein necessitate revision of Ringwood's calculation in that KD-11 is too Fe-rich (100Mg/(Mg+Fe)=64) and the earlier petrogenetic model is now retracted in favour of a picritic parental magma. Thus, although the calculated model D source and Ringwood's (1975) model pyrolite agree (somewhat fortuitously bearing in mind the different basalts and models) in TiO_2, Al_2O_3 and Na_2O contents, the differences in CaO, Mg-value and particularly in MgO, SiO_2 and Opx/Ol ratio are significant for mantle mineralogy and petrogenesis and the source composition D is preferred as a more closely constrained estimate of upper-mantle composition for the source region of mid-ocean ridge basalts. This conclusion applies strictly only to DSDP3-18 or its derivative melts, but in so far as the petrographic and major element characteristics of DSDP3-18 are common to many mid-ocean ridge basalts (cf. Melson *et al.*, 1977) the deduced composition may have wide application.

Frey *et al.* (1978) have shown that the pyrolite compositon of Ringwood (1966) (or a slightly modified composition with 0.3–0.4 per cent TiO_2) which was based on Hawaiian olivine tholeiite composition, is a suitable source composition for a suite of south-east Australian basalts ranging from olivine tholeiite to olivine melilitite. On the basis of the recognized similarity of these continental "rift" basalt sequences to oceanic-island or light rare earth element-enriched oceanic basalts (as typified by Hawaii, Reunion, Azores, etc.) it has previously been argued that a world-wide source pyrolite of composition close to Ringwood's (1966) estimate exists or develops at the upper part of the low-velocity zone (LVZ) (Green, 1971; Green and Liebermann, 1976). This hypothesis contrasts with an alternative (e.g. Schilling 1973, 1975) view that the appropriate mantle source composition for these basalts occurs as "plumes" or "mantle blobs" rising from a *deep* mantle region underlying a more refractory peridotite (light rare earth-depleted, etc.) acting as a source for mid-ocean ridge basalts. In the context of the model for the LVZ developed by Green and Liebermann (1976), we suggest that our present study defines a composition of the lower, depleted part of the LVZ. We consider that the process of migration of incompatible elements, i.e. depletion in the deeper part of the LVZ and enrichment in the upper levels, occurs through the migration of a small melt fraction (<2 per cent melt) with composition of olivine melilitite to kimberlitic character (Green, 1971; Brey and Green, 1977; Frey *et al.*, 1978; Green and Liebermann, 1976). Although we have, by the nature of the approach used, identified two important pyrolite compositions, i.e. incompatible-element enriched or "Hawaiian type" and incompatible element depleted or "mid-ocean ridge type", the discussion above should illustrate that we envisage a continuum of such compositions and a dynamic situation in which incompatible element contents reflect both "old" events of depletion and enrichment and the presently active migration within the LVZ. Thus the model predicts minor variations and inhomogeneities in incompatible element contents in both "enriched" and "depleted" parts of the LVZ. The need for such source inhomogeneity has been noted in the detailed studies in the ridge-crest areas of the mid-Atlantic Ridge (Blanchard *et al.*, 1976; Bryan *et al.*, 1976). The time-integrated effects and overprinting of depletion and enrichment events in upper-mantle source regions for basalts are becoming apparent in the combination of trace element and isotopic studies, particularly Rb/Sr and Sm/Nd systematics (De Paolo and Wasserburg, 1976; O'Nions *et al.*, 1977). It is our view that the time scale for migration of interstitial melt and development of chemical zoning in the LVZ is short (10–20 million years) compared to the "life" of oceanic crust/lithosphere (100–200 million years) or to the "cycle" of deep mantle convection (? 1 billion years) implied by the persistence of Benioff zones to depths of ~700 km.

In the model which we propose for derivation of mid-ocean ridge basalts, ~7 km of basaltic oceanic crust implies ~21 km refractory harzburgite residue. Since the lithosphere thickens to ~90 km at distance from the mid-ocean ridge crests, it is clear that the lithosphere itself should not be regarded as refractory harzburgite but as a mix of lherzolite → harzburgite → dunite containing both country-rock to intrusive diapirs, diapirs that have lost large melt fractions (harzburgites) or small melt fractions (e.g. Lizard or Tinaquillo peridotites) and accumulates of dunite and troctolite and wehrlite resulting from fractional crystallization of picritic parent magmas *en route* to the surface. At shallower depths, both within the crust and at the crust/mantle interface, we envisage complexes of residual harzburgite, and accumulate sequences of peridotite, pyroxenite and gabbro representing emplacement, second-stage melting and crystal fractionation within diapirs of refractory lherzolite or harzburgite, these being characterized by highly calcic plagioclase ($>An_{80}$), accumulate magnesian orthopyroxene and clinopyroxene.

7. Conclusions

(1) The experimental study of one of the most magnesian glass compositions among mid-ocean ridge basalts (DSDP3-18-7-1) has shown that this composition is not a possible melt from a lherzolite source peridotite unless melting and magma segregation occurs at shallow depths (<35 km) leaving residual chrome–spinel-bearing dunite. This model is not favoured as it implies source and residue peridotite compositions which are unlike natural mantle lherzolite compositions.

(2) Olivine addition experiments (i.e. testing of the hypothesis that DSDP3-18-7-1 is not a primary magma but is derived from a primary magma by olivine fractionation at low pressure) have demonstrated that a picritic parental composition (83 per cent DSDP3-18-7-1 + 17 per cent olivine ($Mg_{91.7}$)) is a possible melt from mantle lherzolite leaving harzburgite (olivine ($Mg_{91.5}$) + enstatite (Mg_{92}, 5.5 per cent Al_2O_3) + ?spinel) as residue. The conditions of melting and magma segregation are 20 kbar (~70 km depth) and ~1430 °C.

(3) It is inferred that tholeiitic picrite magmas with ~25 per cent normative olivine are important primary magmas in mid-ocean ridge petrogenesis. An earlier hypothesis that the primary basalts of the mid-ocean ridges are high-Al_2O_3 olivine tholeiites (~10 per cent olivine) segregating from residual harzburgite or lherzolite at ~30 km depth is retracted in the light of the new experimental data in favour of the picritic model (O'Hara, 1968).

(4) Primary magmas of tholeiitic picrite or their derivatives such as DSDP3-18-7-1 cannot precipitate the highly magnesian clinopyroxene ($>Mg_{88}$) or highly calcic plagioclase ($>An_{80}$) megacrysts which occur in some mid-ocean ridge basalts, nor can they precipitate those accumulate sequences of ophiolite complexes which contain highly calcic plagioclase, magnesian orthopyroxene ($>Mg_{87}$ approx.) or highly magnesian diopside ($>Mg_{88}$ approx.). Similarly, primary magmas of this type cannot be genetically related as liquid/residue pairs to highly refractory harzburgites of ophiolite complexes in which the orthopyroxene and olivine have very low contents of Al_2O_3 (<1.5 per cent in orthopyroxene) and CaO (<0.3 per cent in olivine, <2.0 per cent in orthopyroxene) and these constituents are not redistributed in the rock on subsolidus recrystallization. To summarize, the experimental study precludes a direct genetic relationship between ophiolite complexes (with the character designated) and the type of mid-ocean ridge basalt represented by tholeiitic picrite or olivine tholeiite magmas studied in this project.

(5) We postulate that, in the mid-ocean ridge environment, there must occur primary magmas, and particularly fractionated magmas derived from them, which segregated from Na-poor, Ti-poor depleted peridotite at very shallow depths leaving harzburgite residue. These postulated primary magmas would have high SiO_2 contents, high MgO contents, very low Na_2O/CaO ratios, low TiO_2 and K_2O contents and would be quartz-normative or low in normative olivine content. These magmas and their fractionation products are considered to be complementary to the harzburgites and accumulate sequences of ophiolite complexes. The extrusive diapir model of Moores (1969) may be appropriate for the emplacement and characteristics of these complexes.

(6) We have used the information obtained on DSDP3-18-7-1 and olivine-enriched derivatives to constrain the composition of the source peridotite and to obtain a preferred model composition. This pyrolite composition differs slightly but significantly from previous estimates of upper-mantle compositions.

(7) We support a model of the lithosphere and asthenosphere (LVZ) in which the source pyrolite for mid-ocean ridge tholeiites is considered as representative of the lower, depleted part of the LVZ and the incompatible element-enriched pyrolite composition (Ringwood, 1966) previously stu-

died is considered representative of the upper, enriched part of the asthenosphere (LVZ). Both types of pyrolite source (and compositional variants between them) are considered to be of general, world-wide occurrence beneath oceanic crust and the time scale of enrichment and depletion processes is short compared with the "life" of oceanic crust/lithosphere segments. The factors determining which source region is actively involved in diapirism and magma production are the particular tectonics and dynamics of the area, i.e. a "leaky transform" fault, an oceanic island chain (migrating fracture) or a continental rift may trigger diapirs only from the upper part of the LVZ. By contrast, the continued movement apart of major lithospheric plates at a spreading centre or mid-ocean ridge involves upwelling from deeper levels and upper mantle overturn which taps diapirism from the deep, incompatible-element depleted regions of the LVZ.

Acknowledgements

This study was partly completed at the Research School of Earth Science, the Australian National University, and partly at the Department of Geology, University of Tasmania, the latter part being supported by the Australian Research Grants Committee. All authors, but particularly D. H. Green, acknowledge the stimulus and leadership of Professors J. C. Jaeger and A. L. Hales—the influence of J. C. Jaeger determined the establishment of experimental petrology at The Australian National University and the stimulus of A. L. Hales has ensured that pyrolite and the upper mantle have remained very much subjects for continuing reappraisal and review.

References

Arndt, N. (1977). Ultrabasic magmas and high-degree melting of the mantle. *Contr. Mineral Petrol.* **64**, 205–221.

Bence, A. E. and Taylor, S. R. (1977). Petrogenesis of Mid-Atlantic Ridge basalts at DSDP Leg 37 Holes 332A and 332B from major and trace element geochemistry. *Initial Repts Deep Sea Drilling Program* **37**, 705–710.

Blanchard, D. P., Rhodes, J. M., Dungan, M. A., Rodgers, K. V., Donaldson, C. H., Brannon, J. C., Jacobs, J. W. and Gibson, E. K. (1976). The chemistry and petrology of basalts from Leg 37 of the Deep Sea Drilling Project. *J. geophys. Res.* **81**, 4231–4247.

Bonatti, E., Honnorez, J. and Ferrara, G. (1971). Peridotite–gabbro-basalt complex from the equatorial mid-Atlantic ridge. *Phil. Trans. R. Soc. A* **268**, 385–402.

Brey, G. and Green, D. H. (1975). The role of CO_2 in the genesis of olivine melilitite. *Contr. Mineral. Petrol.* **49**, 43–103.

Brey, G. and Green, D. H. (1977). Systematic study of liquidus phase relations in olivine melilitite + H_2O + CO_2 at high pressures and petrogenesis of an olivine melilitite magma. *Contr. Mineral. Petrol.* **61**, 141–162.

Bryan, W. B. and Moore, J. C. (1977). Compositional variations of young basalts in the mid-Atlantic Ridge rift valley near lat. 36°49'N. *Bull. geol. Soc. Am.* **88**, 556–564.

Bryan, W. B., Thompson, G., Frey, F. A. and Dickey, J. S. (1976). Inferred geological settings and differentiation in basalts from the Deep Sea Drilling Project. *J. geophys. Res.* **81**, 4285–4304.

Bultitude, R. J. and Green, D. H. (1971). Experimental study of crystal liquid relationships at high pressures in olivine nephelinite and basanite compositions. *J. Petrol.* **12**, 121–147.

Cann, J. R. (1974). A model for oceanic crustal structure developed. *Geophys. J. R. astronom. Soc.* **39**, 169–187.

Cawthorn, R. G., Ford, C. E., Biggar, G. M., Bravo, M. S. and Clarke, D. B. (1973). Determination of liquid composition in experimental samples: discrepancies between microprobe analysis and other methods. *Earth planet. Sci. Lett.* **21**, 1–5.

Clarke, D. B. and Pedersen, A. K. (1976). Tertiary volcanic province of West Greenland. In *Geology of Greenland* (A. Escher and W. S. Watt, eds), Geological Survey of Greenland, pp. 365–385.

Dallwitz, W. B., Green, D. H. and Thompson, J. B. (1966). Clinoenstatite in a volcanic rock from Cape Vogel, Papua. *J. Petrol.* **7**, 375–403.

Davies, H. L. (1971). Peridotite-gabbro complexes in Eastern Papua: an overthrust of oceanic crust and mantle. *Bur. Miner. Resour. Australia Bull.* **128**, 1–47.

De Paolo, D. J. and Wasserburg, G. J. (1976). Nd isotopic variations and petrogenetic models. *Geophys. Res. Lett.* **3**, 249–254.

Donaldson, C. H. and Brown, R. W. (1977). Refractory megacrysts and magnesium-rich melt inclusions within spinel in oceanic tholeiites: indications of magma mixing and parental liquid composition. *Earth planet. Sci. Lett.* **37**, 81–89.

England, R. N. and Davies, H. L. (1973). Mineralogy of ultramafic cumulates and tectonites from eastern Papua. *Earth planet. Sci. Lett.* **17**, 416–425.

Esson, J., Dunham, A. C. and Thompson, R. N. (1975). Low-alkali, high-calcium olivine tholeiite laves from the Isle of Skye, Scotland. *J. Petrol.* **16**, 488–497.

Flower, M. F. J., Robinson, P. T., Schmincke, H. U. and Ohnmacht, W. (1977). Magma fractionation systems beneath the mid-Atlantic Ridge at 36–37°N. *Contr. Mineral. Petrol.* **64**, 167–195.

Frey, F. A. (1969). Rare earth abundances in a high-temperature peridotite intrusion. *Geochim. cosmochim. acta* **33**, 1429–1447.

Frey, F. A. and Green, D. H. (1974). The mineralogy, geochemistry and origin of lherzolite inclusions in Victorian basantites. *Geochim. cosmochim. acta* **38**, 1023–1059.

Frey, F. A. and Prinz, M. (1977). Ultramafic inclusions from San Carlos, Arizona: Petrologic and geochemical data bearing on their petrogenesis. *Earth planet. Sci. Lett.* **38**, 129–176.

Frey, F. A., Green, D. H. and Roy, S. D. (1978). Integrated models of basalt petrogenesis: a study of quartz tholeiites to olivine melilitites from S.E. Australia utilizing geochemical and experimental petrological data. *J. Petrol.* **19**, 463–513.

Frey, F. A. Bryan, W. B. and Thompson, G. (1974). Atlantic Ocean floor: geochemistry and petrology of basalts from Legs 2 and 3 of the Deep Sea Drilling Project. *J. geophys. Res.* **79**, 5507–5527.

Green, D. H. (1963). Alumina content of enstatite in a Venezuelan high temperature peridotite. *Bull. geol. Soc. Am.* **74**, 1397–1402.

Green, D. H. (1968). Origin of basaltic magmas. In *Basalts* (H. H. Hess and A. Poldervaart, eds), Wiley-Interscience, New York, London, Sydney, Vol. 2, pp. 835–862.

Green, D. H. (1970). The origin of basaltic and nephelinitic magmas. *Trans. Leicester lit. phil. Soc.* **64**, 28–54.

Green, D. H. (1971). Compositions of basaltic magmas as indicators of conditions of origin: application to oceanic volcanism. *Phil. Trans. R. Soc. A* **268**, 707–725.

Green, D. H. (1973). Experimental studies on a model upper mantle composition at high pressure under water-undersaturated and water-saturated conditions. *Earth planet. Sci. Lett.* **19**, 37–53.

Green, D. H. (1976*a*). Experimental petrology in Australia—a review. *Earth Sci. Rev.* **12**, 99–138.

Green, D. H. (1976*b*). Orthopyroxene in the lunar interiors and constraints on early lunar differentiation. *Proc. Seventh Lunar Sci. Conf.* **1**, 336–338.

Green, D. H. and Hibberson, W. O. (1970). Experimental duplication of conditions of precipitation of high pressure phenocrysts in a basaltic magma. *Phys. Earth planet. Interiors* **3**, 247–254.

Green, D. H. and Liebermann, R. C. (1976). Phase equilibria and elastic properties of a pyrolite model for the oceanic upper mantle. *Tectonophysics* **32**, 61–92.

Green, D. H. and Ringwood, A. E. (1963). Mineral assemblages in a model upper mantle composition. *J. geophys. Res.* **68**, 937–945.

Green, D. H. and Ringwood, A. E. (1967). The genesis of basaltic magmas. *Contr. Mineral. Petrol.* **15**, 103–190.

Green, D. H. and Ringwood, A. E. (1970). Mineralogy of peridotitic compositions under upper mantle conditions. *Phys. Earth planet. Interiors* **3**, 359–371.

Green, D. H., Ringwood, A. E., Hibberson, W. O. and Ware, N. G. (1975). Experimental petrology of Apollo 17 mare basalts. *Proc. Sixth Lunar Sci. Conf.* **1**, 871–893.

Green, D. H., Edgar, A. D., Beasley, P., Kiss, E. and Ware, N. G. (1974). Upper mantle source for some hawaiites, mugearites, and benmoreites. *Contr. Mineral. Petrol.* **48**, 33–43.

Green, D. H., Ringwood, A. E., Ware, N. G., Hibberson, W. O., Major, A. and Kiss, E. (1971). Experimental petrology and petrogenesis of Apollo 12 basalts. *Proc. Second Lunar Sci. Conf.* **1**, 601–615.

Green, T. H., Green, D. H. and Ringwood, A. E. (1967). The origin of high alumina basalts and their relationships to quartz tholeiites and alkali basalts. *Earth planet. Sci. Lett.* **2**, 41–52.

Hakli, J. A. and Wright, T. L. (1967). The fractionation of nickel between olivine and augite as a geothermometer. *Geochim. cosmochim. acta* **31**, 877–884.

Hodges, F. N. and Papike, J. J. (1976). DSDP Site 34: magmatic cumulates from oceanic layer 3. *J. geophys. Res.* **81**, 4135–4151.

Ito, K. and Kennedy, G. C. (1968). Melting and phase relations in the plane tholeiite-lherzolite-nepheline basanite to 40 kbar with geological implications. *Contr. Mineral. Petrol.* **19**, 177–211.

Kay, R., Hubbard, N. J. and Gast, P. W. (1970). Chemical characteristics and origin of oceanic ridge volcanic rocks. *J. geophys. Res.* **75**, 1585–1613.

Kuno, H. and Aoki, K. (1970). Chemistry of ultramafic nodules and their bearing on the origin of basaltic magmas. *Phys. Earth planet. Interiors* **3**, 273–301.

Langmuir, C. H., Bender, J. F., Bence, A. E. and Hanson, G. N. (1977). Petrogenesis of basalts from the Famous area: Mid-Atlantic Ridge. *Earth planet. Sci. Lett.* **36**, 133–156.

Leeman, W. P. and Scheidegger, K. F. (1977). Olivine/liquid distribution coefficients and a test for crystal/liquid equilibrium. *Earth planet. Sci. Lett.* **35**, 247–257.

Mathez, E. A. (1976). Sulphur solubility and magmatic sulphides in submarine basalt. *J. geophys. Res.* **81**, 4269–4276.

Melson, W. G. and Thompson, G. (1970). Layered basic complex in oceanic crust, Romanche Fracture equatorial Atlantic Ocean. *Science, N.Y.* **168**, 817–820.

Melson, W. G., Byerly, G. R., Nelen, J. A., O'Hearn, T., Wright, T. L. and Vallier, T. (1977). A catalog of the major element chemistry of abyssal volcanic glasses. *Mineral Sciences Investigations: Smithsonian Contributions to the Earth Sciences* **19**, 31–60.

Miyashiro, A. Shido, F. and Ewing, M. (1969). Diversity and origin of abyssal tholeiite from the Mid-Atlantic Ridge near 24° and 30° north latitude. *Contr. Mineral. Petrol.* **23**, 38–52.

Moores, E. M. (1969). Petrology and structure of the Vourinos ophiolitic complex of northern Greece. *Geol. Soc. Am. spec. Paper* **118**, 1–74.

Moores, E. M. and Vine, F. J. (1971). The Troodos Massif, Cyprus and other ophiolites as oceanic crust: evaluation and implications. *Phil. Trans. R. Soc. A* **268**, 443–466.

Mysen, B. (1976). Partitioning of samarium and nickel between olivine, orthopyroxene and liquid: preliminary data at 20 kb and 1025 °C. *Earth planet. Sci. Lett.* **31**, 1–7.

O'Hara, M. J. (1968). Are ocean-floor basalts primary magmas. *Nature Lond.* **220**, 683–686.

O'Hara, M. J. and Humphries, D. J. (1977). Problems of iron gain and loss during experimentation on natural rocks: the experimental crystallization of five lunar basalts at low pressures. *Phil. Trans. R. Soc. A* **286**, 313–330.

O'Hara, M. J., Humphries, D. J. and Waterston, S. (1975). Petrogenesis of mare basalts: implications for chemical mineralogical and thermal models for the moon. *Proc. Sixth Lunar Sci. Conf.* **1**, 1043–1055.

O'Nions, R. K., Hamilton, P. J. and Evensen, N. M. (1977). Variations in $^{143}Nd/^{144}Nd$ and $^{87}Sr/^{86}Sr$ ratio on oceanic basalts. *Earth planet. Sci. Lett.* **34**, 13–22.

Presnall, D. C. (1969). The geometrical analysis of partial fusion. *Am. J. Sci.* **267**, 1178–1194.

Roeder, P. L. and Emslie, R. F. (1970). Olivine–liquid equilibrium. *Contr. Mineral. Petrol.* **29**, 275–289.

Ringwood, A. E. (1966). The chemical composition and origin of the earth. In *Advances in Earth Science* (P. M. Hurley, ed.), M.I.T. Press, Cambridge, Mass., pp. 287–356.

Ringwood, A. E. (1975). *Composition and Petrology of the Earth's Mantle*, McGraw-Hill, New York.

Schilling, J. G. (1973). Iceland mantle plume: geochemical study of the Reykjanes Ridge. *Nature, Lond.* **242**, 565–571.

Schilling, J. G. (1975). Azores mantle blob: rare-earth evidence. *Earth planet. Sci. Lett.* **25**, 103–115.

Shido, F., Miyashiro, A. and Ewing, M. (1971). Crystallization of abyssal tholeiites. *Contr. Mineral. Petrol.* **31**, 251–266.

Sun, S. S. and Nesbitt, R. W. (1977). Chemical heterogeneity of the Archaean mantle, composition of the earth and mantle evolution. *Earth planet. Sci. Lett.* **35**, 429–488.

Verhoogen, J. (1954). Petrological evidence on temperature distribution in the mantle of the Earth. *Trans. Am. geophys. Union* **35**, 85–92.

Yoder, H. S. and Tilley, C. E. (1962). Origin of basalt magmas: an experimental study of natural and synthetic rock systems. *J. Petrol.* **3**, 342–532.

10. Thermal Regime of the Australian Continental Crust

J. H. SASS and A. H. LACHENBRUCH

Abstract

The heat-flow data for the Australian continental crust (about 90 measurements) suggest a thermal regime characteristic of stable continental blocks with the exception of a band of high heat flow about 500 km wide, stretching from the Northern Territory south-south-eastward to Tasmania. The observed variations within this small set of data of variable quality can be explained by a large range of models for the thermal and tectonic regimes of the crust. The simplest explanation consistent with pertinent geological and geophysical information involves the following interpretation. The heat flow from the mantle (or lower crust) is uniform throughout the continent (~0.65 h.f.u.) and variations in surface heat flow are caused, in the western two thirds, by lateral variations in the distribution of crustal radioactivity, and in the eastern one third, by the additional effects of magmatic heat sources in the upper crust. The western two thirds of the continent, a Precambrian craton, contains two thermal provinces distinguished by their differing characteristic depth, D, of radioactive heat production. In the Western Shield Province, (ages >2500 million years), the heat production is concentrated closer to the surface ($D \simeq 4.5$ km) than in the younger (600–2000 million years) Central Shield Province ($D \simeq 11$ km). For the same surface radioactivity, heat flow and crustal temperatures are therefore lower in the west. The Eastern Province is a Paleozoic terrane. Although it is more stable than continental regions near plate margins (e.g. western North America), it has been the site of widespread Cenozoic magmatic and tectonic activity. High heat flows have been measured in the southern part of the Eastern Province (the highest close to regions of very recent volcanism and probably tectonic extension). In contrast to central Australia, crustal radioactivity is not nearly high enough to account for the high surface heat flows, nor is it likely that they are the result of high mantle heat flow, because massive melting is precluded by seismic evidence for a 40-km crust and a thick lithosphere. The distribution of heat flow, heat production and present-day crustal stress (mainly compressional), and the systematic southward migration of the termination of basaltic volcanism during the Cenozoic Era, suggest a simple model attributing the high heat flows in eastern Australia to upper crustal intrusion, more widespread than the associated volcanism. According to the model, intrusion ceased at each latitude as the transition from crustal tension to compression swept southward; subsequently the crust cooled conductively.

1. Introduction

The Earth's surface is now viewed as a collection of mobile plates and most presently observed tectonic activity occurs at or near plate boundaries. It is possible to construct thermomechanical models that provide insight into both present tectonic activity and (because of the long time constants associated with thermal conduction), the tectonic history of much of the Earth's surface. These models must, however, be reconciled with our knowledge of the chemical composition and physical properties of rocks, the structure of the Earth's interior, and the thermal regime of the Earth as inferred from observations of heat flow near its surface.

Thermomechanical models of ocean basins are constrained chiefly by the time elapsed since the formation of the oceanic crust at spreading centers. Simple heat-conduction models, derived according to the geomagnetic time scale and the cooling and contraction of initially molten mafic material, can account for the observed heat flow and for the topography as a function of distance from spreading centers. Anomalously low heat flows observed near active ridge crests have been shown to be caused by hydrothermal convection in the permeable crustal rocks (Lister, 1972, 1974). Sclater and Crowe (1976) have demonstrated that by choosing only heat flows measured in areas of thick, impermeable sediments, the previously observed scatter in average heat flows (cf. Sclater and Francheteau, 1970) is reduced considerably, and plots of heat flow vs. distance from ridge crests conform to the simple conductive cooling models.

Near trenches, island arcs and back-arc basins, the situation is more complicated than that for the ridges and ocean basins (e.g. Watanabe *et al.*, 1977); a broader region is subject to thermal disturbance by a variety of processes, the details of which are not as well understood. Models that take into account both the heat sources and heat sinks associated with descending lithospheric plates and the extensional tectonics of back-arc basins can provide reasonable explanations for the gross features of the observed heat-flow field (see, for example, Oxburgh and Turcotte, 1971).

The analysis of heat flow from the continental crust poses additional problems (see Jaeger, 1965; Roy *et al.*, 1972; Lachenbruch and Sass, 1977), primarily those of accounting for the large and variable contribution to the surface heat flow (q_0) of crustal radioactivity and for laterally variable conditions reflecting a long and complex geologic history. The usual procedure is to define "heat-flow provinces" based on a characteristic value (or range of values) for q_r, the "reduced heat flow", which is calculated by subtracting the crustal radiogenic contribution for q_0 using empirically derived relations between heat flow and near-surface radioactivity. The patterns of reduced heat flow within continents can then be used as a constraint for thermotectonic models of the continental crust.

Several recent reviews (Cermak and Hurtig, 1977; Pollack and Chapman, 1977*a*, *b*; Jessop *et al.*, 1976; Chapman and Pollack, 1975*a*) have addressed the distribution of heat flow over the globe and its major tectonic significance. In this chapter, we shall draw on our experience from the United States and attempt to characterize the thermal regime of the Australian continental crust based on the present distribution of heat flow and some empirical relations among heat flow, tectonic age, and crustal radioactivity. We shall also examine the constraints imposed on crustal geotherms by other geological and geophysical observations with particular reference to the Australian continent.

The following symbols and units are defined:

Θ	Temperature; °C, Θ_0, surface temperature, $\Theta(x, y, z, t)$, temperature as a function of space and time;
Θ_n,	Temperature at base of crust;
K,	Thermal conductivity; 1 heat conductivity unit (h.c.u.) $= 1$ mcal cm^{-1} s^{-1} °C^{-1} $= 0.418$ W m^{-1} K^{-1};
α,	Thermal diffusivity; cm^2 s^{-1} $= 10^{-4}$ m^2 s^{-1};
$\tau(z) = z^2/4\alpha$,	Thermal conductive time constant for a layer of thickness z;
A,	Heat production; 1 heat generation unit (h.g.u.) $=$ 10^{-13} cal cm^{-3} s^{-1} $= 0.418$ μW m^{-3};
A_0,	Heat production at the Earth's surface ($z = 0$);
q,	Heat flow $\equiv$ vertical conductive heat flux defined by $q = K \cdot \partial\Theta/\partial z$, z positive downwards, 1 heat flow unit (h.f.u.) $= 10^{-6}$ cal cm^{-2} s^{-1} $=$ 41.8 mW m^{-2}.
q_0,	Conductive heat flux at the Earth's surface;
q_r,	Reduced heat flow, defined by $q_0 = q_r + DA_0$, where D is a depth characteristic of the upper crustal radioactivity;
q_n,	Conductive heat flux into base of crust;
q_a,	Conductive heat flux into base of lithosphere;
s,	Rate of increase in horizontal surface area (areal strain rate), percent per million years;
W,	Average rate of basalt supply to crust, km per million years.

2. A Heat-flow Map of Australia

2.1. Historical Background

In 1950, reliable measurements of heat flow on continents numbered only a few with scattered measurements in South Africa, Britain and North America, mainly attributable to E. C. Bullard, A. E. Benfield, Francis Birch, A. D. Misener, and their associates. No Australian data were available when J. C. Jaeger of the Australian National University initiated a geothermal studies program beginning with measurements at hydroelectric developments in Tasmania and tunnel excavations by the Snowy Mountains Hydroelectric Authority in Southern New South Wales. The first published results (Newstead and Beck, 1953; Beck, 1956) suggested heat flows that were unacceptably high in the context of world heat flow at that time (see Bullard, 1954; Birch, 1954). Subsequent studies by L. E. Howard (first reported by Jaeger and Thyer, 1960; see also Howard and Sass, 1964; Howard, 1965) corroborated earlier indications of high heat flow in eastern and central Australia and established that low-to-normal values were characteristic of the older (ages >2500 million years) Precambrian shield of Western Australia. The latter observation and that of Kraskovski (1961) were the first published statements regarding systematic differences between heat flow on Precambrian shields and younger rocks. Between 1960 and 1970, further studies confirmed the earlier observations and raised the possibility of a band of low heat flow near the coast in eastern Australia (see Lilley *et al.*, 1977 for a complete bibliography). Hyndman *et al.* (1968) related measured heat flows to near-surface radioactivity in Western Australia, and Jaeger (1970) suggested a linear heat flow–heat production relation for the Western Australian Shield.

In 1971 and 1972 a major effort was launched to fill in the Australian picture on a geographical basis, in an attempt to define heat flow–heat production provinces within the Australian continent, and to relate these results to what was then known about the gross tectonic and geologic state of Australia. These results have been published only in preliminary form (Bunker *et al.*, 1975; Munroe *et al.*, 1975; and Sass *et al.*, 1976). Subsequently, additional data have been published by Taplin (1976), Wronski (1977) and Lilley *et al.* (1977).

2.2. Present Status of Heat-flow Measurements

Although the Australian continent has been more thoroughly sampled than most with 90 or so measurements, the heat-flow coverage is still sparse and interpretations can be made only in a gross geologic or tectonic context. The present distribution suggests two regions of low heat flow separated by a broad band of high heat flow stretching from the Northern Territory south-south-eastward to Tasmania (stippled area of Fig. 1). The density of coverage is comparable to that which existed in the continental United States, a region of about the same area, in the early 1970's (see Roy *et al.*, 1972; Diment *et al.*, 1972; Sass *et al.*, 1971). At present, more than 500 heat-flow data are available from the United States (Fig. 2). The boundaries of previously discovered thermal provinces have been delineated in more detail, and additional subprovinces have been outlined. We might reasonably expect that the Australian map will evolve in a similar manner as more data become available.

For reasons to be discussed, we divide the Australian continent into three heat-flow provinces (Figs 1 and 3). The Western Shield Province comprises the old (>2000 million years) Precambrian shield contained entirely in the state of Western Australia (1, Fig. 1). The range of heat flow there is low (Figs 1 and 3(a)) and the mean of 0.93 h.f.u. (39 mW m^{-2}) is comparable to that from other shields (Rao and Jessop, 1975). In the Central Shield Province (2, Fig. 1), most basement rocks range in age from 600 to 2000 million years. "Normal" (1.1–1.4 h.f.u.) heat flows have been measured in the south-western part of this province, but the remainder of the province is characterized by high heat flow. The mean of 2 h.f.u. (83 mW m^{-2}) (Fig. 3(b)) for the Central Shield is quite anomalous when compared with other stable Precambrian platforms (Rao and Jessop, 1975). In the Eastern Province (3, Fig. 1) the basement rocks are paleozoic and heat flow is variable with a somewhat lower average (1.72 h.f.u., 72 mW m^{-2}) than the Central Province. Low-to-normal heat flows predominate in the northern part of the Eastern Province, and high heat flow is characteristic of most of the southern part.

2.3. Distribution of Heat Production

The radioactivity of crustal rocks can contribute significantly to the observed surface heat flow, and it seems reasonable to investigate whether regional variations in heat flow might be matched by corresponding variations in radioactive heat production.

K. S. Heier and I. B. Lambert (unpublished) measured and compiled heat-production data on more than 2000 specimens of granitic and metamorphic rocks collected for various purposes (age dating, physical properties measurements, etc.) by scientists of the Bureau of Mineral Resources and the Australian National University. Their compilation is summarized in a map of average heat production (in h.g.u.) of intermediate and felsic plutonic and metamorphic rocks for $4° \times 6°$ areas of the Australian

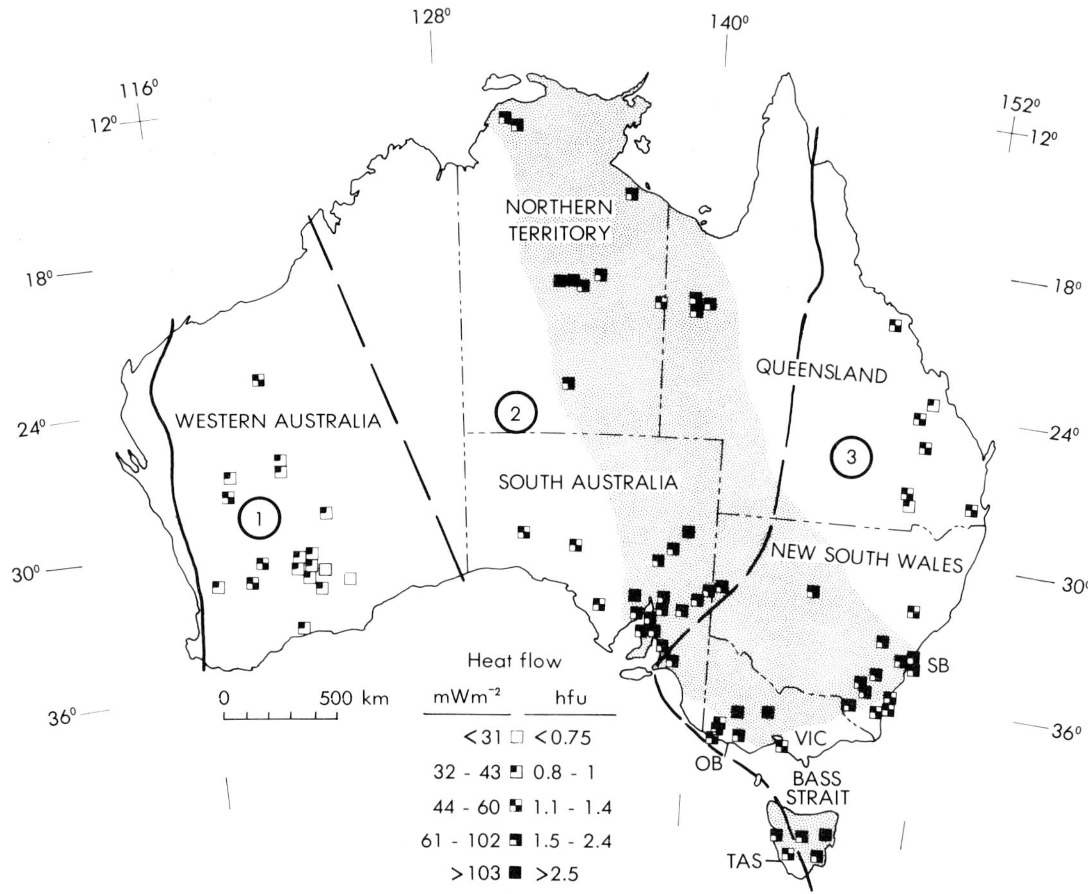

Fig. 1. Heat flow from the Australian continent. Stippled area includes all heat flows greater than 1.5 h.f.u. (61 mW m^{-2}). Heat-flow provinces are: 1, Western Shield Province; 2, Central Shield Province; 3, Eastern Province. OB is the Otway Basin and SB, the Sydney Basin.

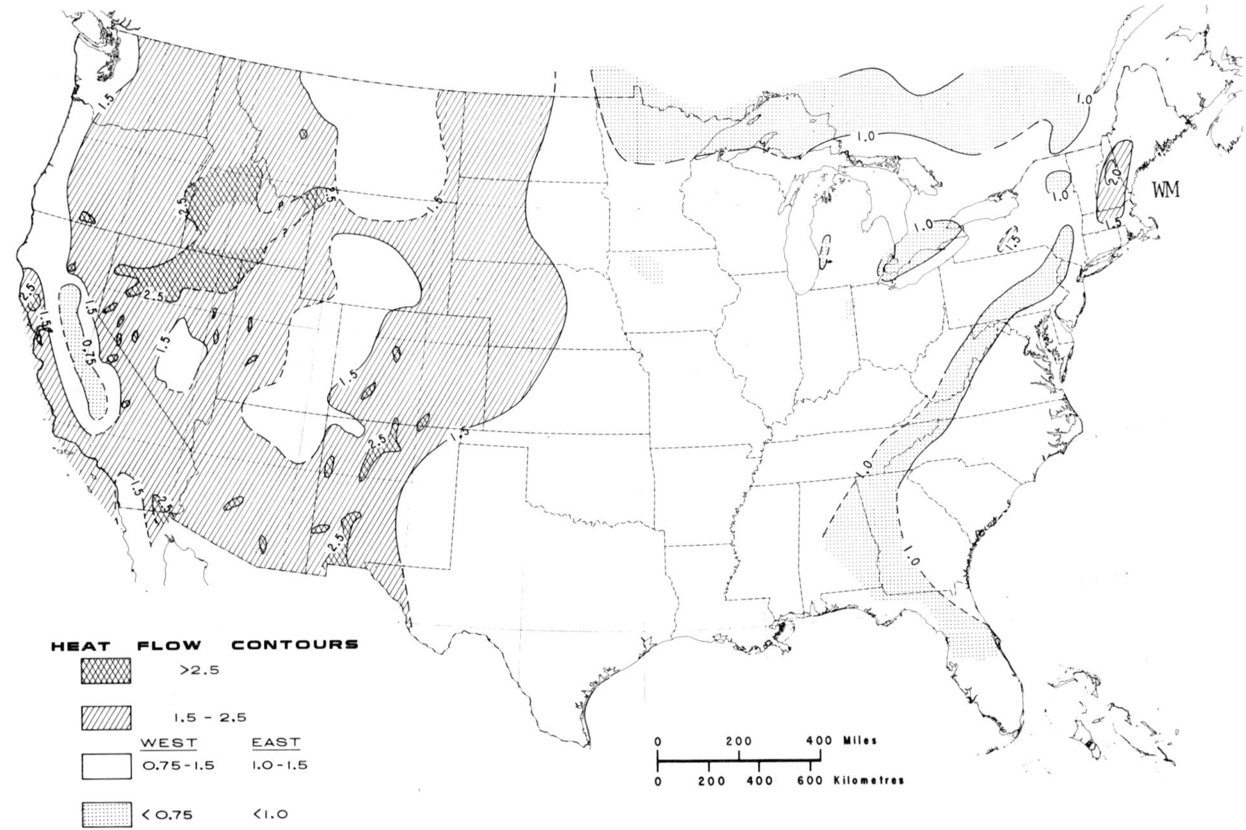

Fig. 2. A generalized heat-flow map of the conterminous United States (heat flow in h.f.u.). WM, White Mountains (Birch *et al.*, 1968).

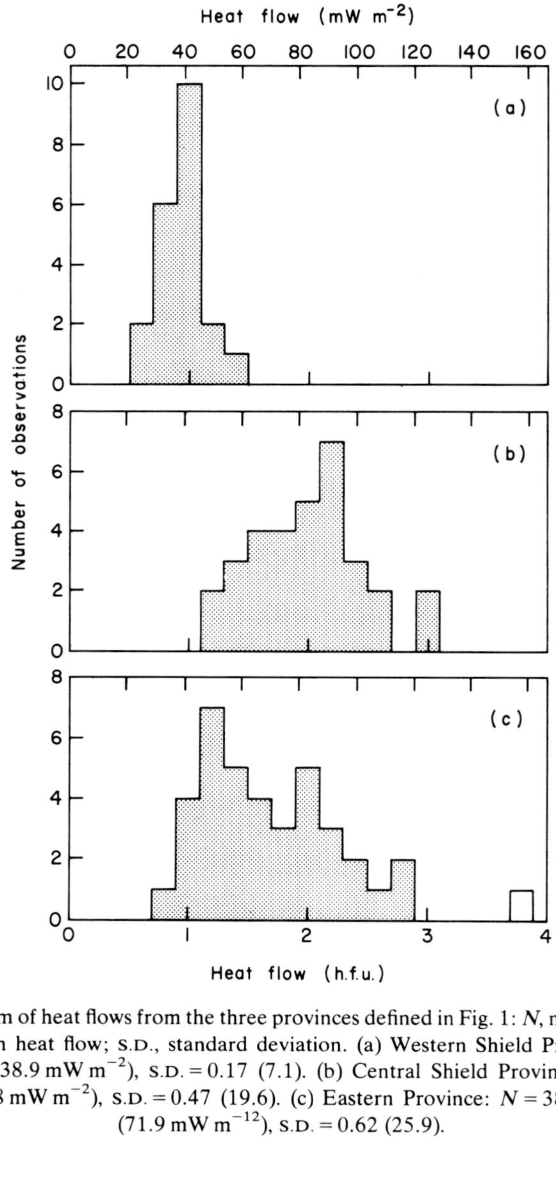

Fig. 3. Histogram of heat flows from the three provinces defined in Fig. 1: N, number of data; $\bar{q}$ arithmetic mean heat flow; S.D., standard deviation. (a) Western Shield Province: $N = 21$, $\bar{q} = 0.93$ h.f.u. (38.9 mW m^{-2}), S.D. = 0.17 (7.1). (b) Central Shield Province: $N = 32$, $\bar{q} =$ 1.98 h.f.u. (82.8 mW m^{-2}), S.D. = 0.47 (19.6). (c) Eastern Province: $N = 38$, $\bar{q} = 1.72$ h.f.u. (71.9 mW m^{-12}), S.D. = 0.62 (25.9).

continent (Fig. 4). The data suggest that Australian rocks may be more radioactive than those on other continents, even in the Western Shield Province which has low heat flow. The mean of all determinations is ~7.5 h.g.u., about 50 per cent higher than that found by Diment *et al.* (1975) for a comparable sample from the United States (see also Table 3 of

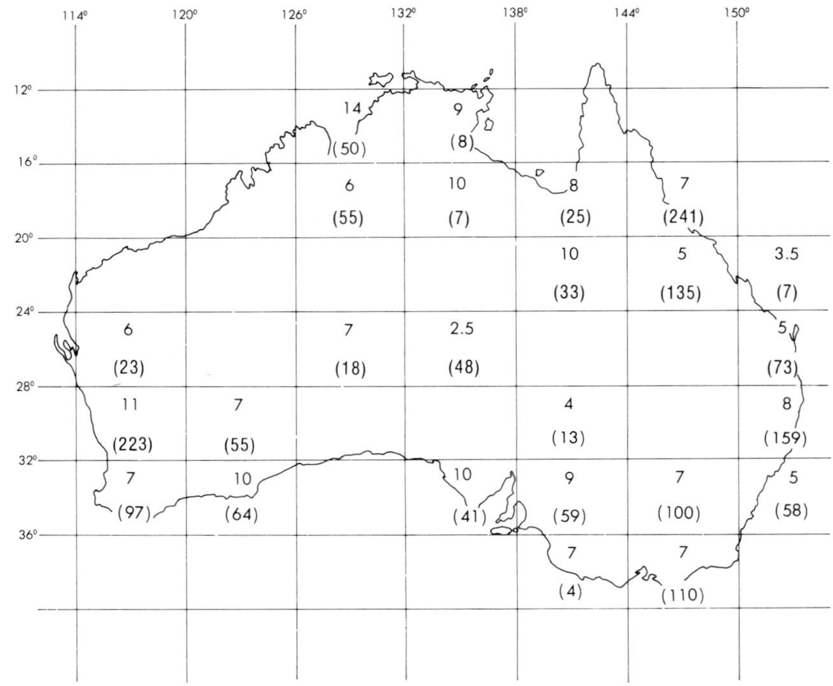

Fig. 4. Mean heat production (h.g.u.) of felsic intrusive and metamorphic rocks for 4° × 6° areas of the Australian mainland (number of determinations within each area is shown in parenthesis). (Data from K. S. Heier and I. B. Lambert, unpublished compilation.)

Rao and Jessop, 1975). As noted by Lambert and Heier (1967), there is not much difference between the mean radioelement abundances for rocks obtained from Western Australia and those from the rest of the continent, certainly not enough to account for the difference in mean heat flows. A comparison of the radioactivity distribution in the eastern and south-eastern coastal regions (Fig. 4) with the distribution of heat flow over the same area (Fig. 1) shows that the regional variations in heat flux are not matched by corresponding variations in the radioelement concentrations of the near-surface crystalline basement rocks.

3. Variability of Continental Heat Flow and some of its Causes

Except for conspicuous local anomalies associated with active hydrothermal systems, the heat conducted through the upper kilometer or so of the continental crust varies by an order of magnitude from about 0.5 to 5 h.f.u. (20–

200 mW m^{-2}). A great many factors are involved in the observed patterns of crustal heat flow, but the most important are: (1) the nature of the last major tectonic event and the time elapsed since its completion, (2) crustal radiogenic heat production, (3) convective processes in the crust, and (4) conductive flux from the mantle. In regions of contemporary tectonic activity, convective processes can be the dominant factor (e.g. Lachenbruch and Sass, 1977, 1978); in older regions, consideration of the relative contributions of all of these factors is usually required to explain the observed patterns of regional heat flow.

3.1. Heat Flow vs. Tectonic Age

In general, the *average* heat flow correlates negatively with the age of the most recent "thermotectonic event" in a region (Polyak and Smirnov, 1966, 1968; Hamza and Verma, 1969; Chapman and Pollack, 1975*a*). With the refinement of plate-tectonic models, this correlation has taken on explicit physical meaning for the major ocean basins; the last thermotectonic event is generally the creation of sea floor at a spreading axis, and the time is determined from the subsequent amount of spreading. However for the continents the problem is more difficult as evidenced by the considerable scatter about the mean heat flow for various tectonic ages (Fig. 5). There may be several types and intensities of "thermotectonic events" and we have no comprehensive general thermomechanical models of any of them (some thermotectonic events, e.g. subduction at a continental margin, can reduce the surface heat flow). Consequently, the initial thermal condition, and even the time origin from which to reckon the subsequent cooling of the continents, are largely matters of speculation. Thus, the heat flow–tectonic age relation cannot, by itself, be used to specify the thermal regime of any continental region, and individual regions must be examined in detail to identify the relative contribution of other factors to the observed patterns of heat flow. The thermal structure of the western half of the United States (Fig. 6) with well-defined regions of both very low (<0.75 h.f.u.) and very high (>2.5 h.f.u.) heat flow illustrates the Cenozoic interaction between the Pacific and North American plates involving both heat sources and heat sinks, the tectonic and thermal effects of which probably extend inland for more than a thousand kilometers (Roy *et al.*, 1972; Lachenbruch and Sass, 1977).

The Western Shield Province of Australia (Fig. 1) has a range of heat flows consistent with the empirical heat flow vs. age relation (Fig. 5); however, the Central Shield Province has an average heat flow (2 h.f.u.) significantly higher than would be predicted from Fig. 5. Heat flows of this magnitude have been measured in Precambrian rocks of the Northern and Southern

Rocky Mountains (Fig. 6) of the western United States (Birch, 1950; Blackwell, 1969; Sass *et al.*, 1971), but intense tectonic activity has occurred in the region during the Cenozoic era. High heat flows also have been measured in Precambrian rocks of the Central African Plateau by Chapman and Pollack (1975*b*, 1977) who interpret the results as indicating lithospheric thinning and incipient rifting in this region. Parts of the Central Shield Province have relatively high seismicity, but there are no surface manifestations of significant Cenozoic tectonic activity. It will be shown that this province might more readily be compared to the White Mountains of the eastern United States (WM, Fig. 2) where high heat flows are attributable to anomalously high radioactivity in the crust (Birch *et al.*, 1968).

The eastern Australian heat-flow province is a Paleozoic feature (Brown *et al.*, 1968; Oversby, 1971), but Cenozoic uplift and widespread volcanism (albeit of relatively small total volume) has occurred within it (Wellman and McDougall, 1974). The regions of most intense late Cenozoic volcanism occur in extreme northern Queensland (where there are no heat-flow data) and in southern Victoria where, with the exception of the Otway Basin (OB, Fig. 1) heat flow is high. Eastern Tasmania (Fig. 1) a region of high heat flow,

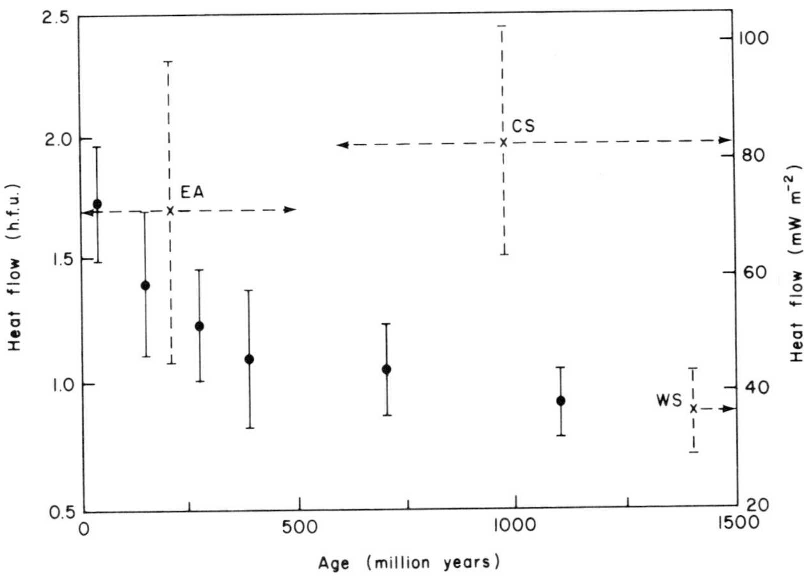

Fig. 5. Heat flow vs. tectonic age for continents (modified from Pollack and Chapman, 1977*a*). Solid dots are world-wide average heat flows for various mean ages of basement rocks based on the compilation of Jessop *et al.* (1976). Error bars represent standard deviation. ×'s represent the Australian data: WS, Western Shield; CS, Central Shield; EA, eastern Australia.

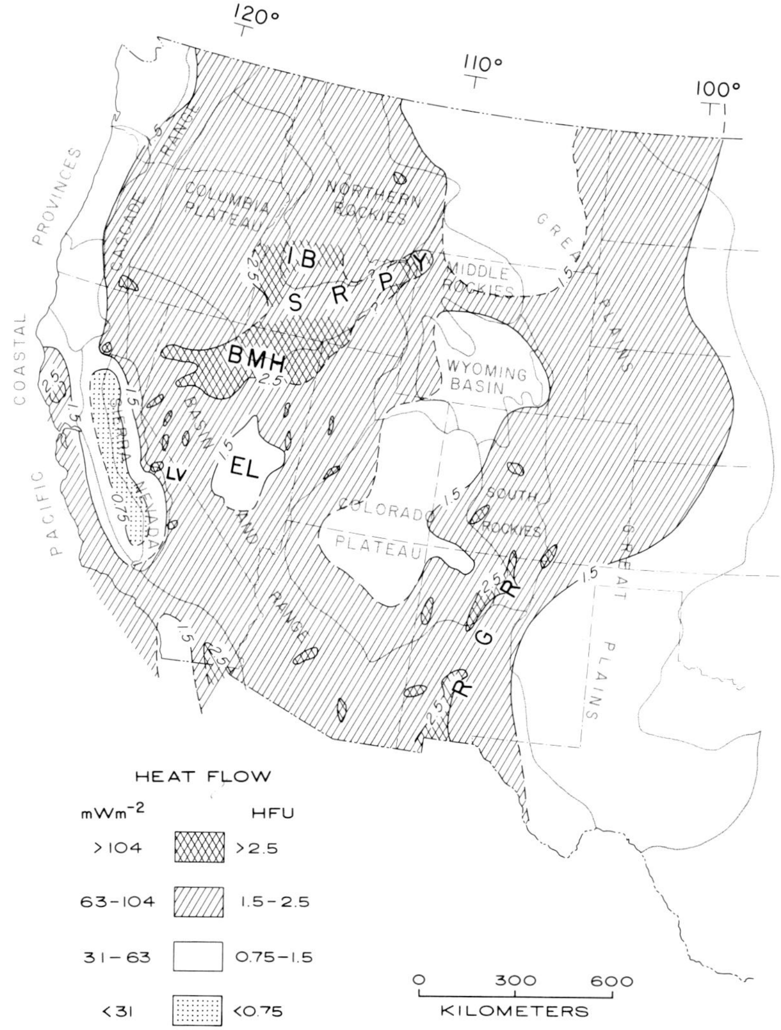

Fig. 6. Generalized map of heat flow and physiographic provinces in the western United States. Abbreviations are BMH for Battle Mountain High, SRP for eastern and Central Snake River Plain, Y for Yellowstone geothermal area, RGR for Rio Grande Rift, EL for Eureka Low, and LV for Long Valley volcanic center.

also experienced rifting and volcanism throughout most of the Cenozoic Era (Solomon, 1962).

3.2. Crustal Radioactivity

Heat Flow vs. Heat Production

The radioactivity contained in near-surface rocks could account for a substantial fraction of the observed heat flow if it were distributed uniformly through the crust. This possibility has long been considered remote, strictly on geochemical and seismological grounds, but not until the discovery by Birch *et al.* (1968) of a linear relation between heat flow and radioactive heat production for the granite rocks of the north-eastern United States could the vertical distribution of radioactivity in the continental crust be considered quantitatively. The linear relation was extended to other parts of the eastern and stable central United States by Roy *et al.* (1968) who also defined heat-flow provinces on the basis of linear relations for the Basin and Range and Sierra Nevada physiographic provinces in the western United States (Figs 7 and 8) (see also Lachenbruch, 1968). The relation between heat flow (q_0) and surface radioactivity (A_0) for various locations within the same heat-flow province may be written

$$q_0 = q_r + DA_0. \tag{1}$$

The intercept q_r is the reduced heat flow and the slope D, a "characteristic depth" for crustal radioactivity.

The simplest interpretation of D is that it represents the depth to the base of a slab of uniform radioactivity (Roy *et al.*, 1968). In another interpretation that satisfies equation (1), D is the logarithmic decrement for an exponentially decreasing concentration of radioactive elements (Lachenbruch, 1968, 1970). For this case, the relation between heat production A and depth z may be written

$$A(z) = A_0 \, e^{-z/D}. \tag{2}$$

The exponential model is reasonable on geochemical grounds and preserves the linear relation if plutons within the same heat-flow province have been eroded to different levels and the thermal transients associated with the erosion are small. Equation (1) has been used to define heat-flow provinces in many parts of the world (Rao and Jessop, 1975; Swanberg *et al.*, 1974; Rao *et al.*, 1976; Pollack and Chapman, 1977*c*; Kutas, 1977). Values of q_0, q_r and D for these provinces (including the Australian provinces to be discussed) are summarized in Table 1. Reduced heat flows range from 0.3 to 1.4 h.f.u. for the major provinces and the characteristic depth D is in the range of 7–15 km with the notable exception of the Western Shield province of Australia (Table 1). Pollack and Chapman (1977*c*) noted that there is also

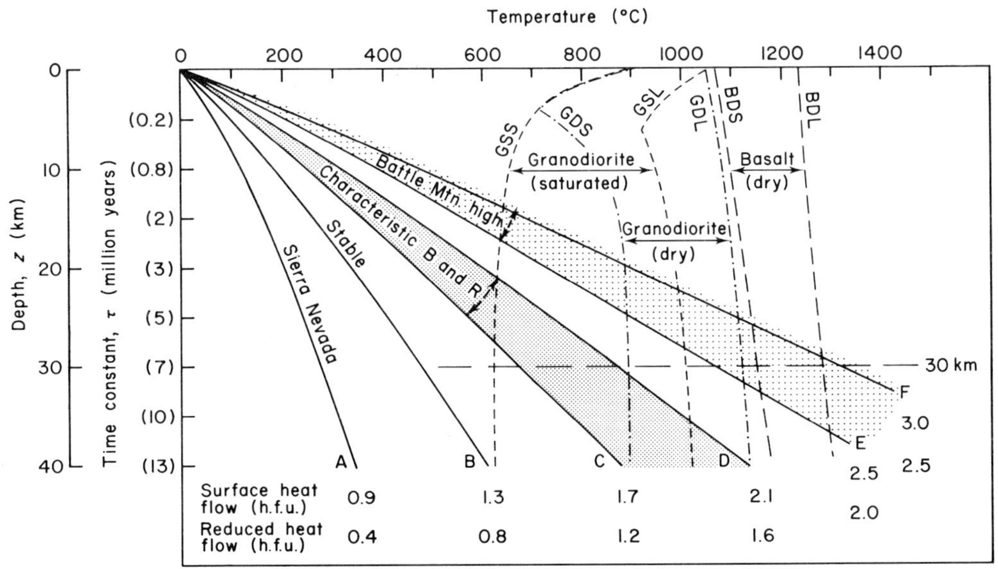

Fig. 7. Generalized conductive temperature profiles for the Sierra Nevada crust (A), a stable reference crust (B), the characteristic Basin and Range crust (C to D), and lower limiting (E) and average (F) conditions in the crust of the Battle Mountain High. All curves are drawn for a surface heat production (A_0) of 5 h.g.u., characteristic depth (D) of 10 km, and thermal conductivity (K) of 6 h.c.u. (equation (3)). Surface heat flow and reduced heat flow are shown at the bottom of each curve. Melting relations (after Wyllie, 1971) are shown for intermediate crustal rock by the curves GSS (granodiorite saturated solidus), GSL (granodiorite saturated liquidus), GDS (granodiorite "dry" solidus, Wyllie, type II), and GDL (granodiorite dry liquidus) and for basalt by BDS (basalt dry solidus), and BDL (basalt dry liquidus). In parentheses on the depth axis is shown the time constant $\tau(z) = z^2/4\alpha$ for an overburden of thickness z.

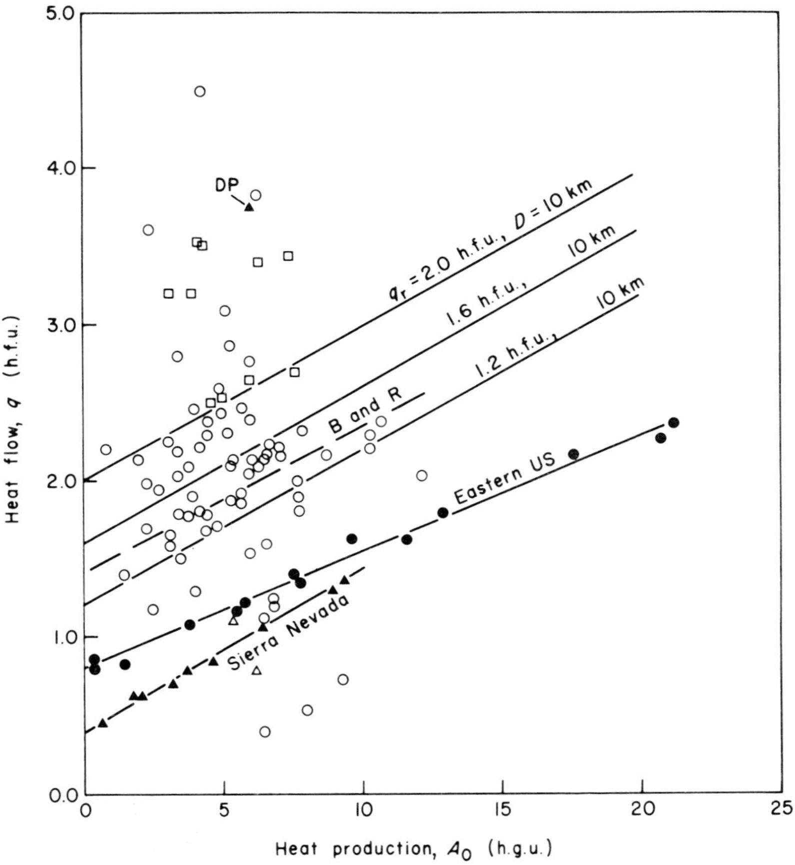

Fig. 8. Heat flow vs. heat production in crystalline rock for the heat-flow provinces in the United States (from Lachenbruch and Sass, 1978). Lines labeled B and R, Eastern US and Sierra Nevada were defined by Roy *et al.* (1968). ●, Eastern U.S.; ▲, Sierra Nevada; □, △, ○ Basin and Range (Battle Mountain High, Eureka Low, and others, respectively).

an empirical relation between q_0 and q_r, a relation that can be expressed as $q_r \simeq 0.6q_0$.

It is apparent (Table 1) that the distribution of crustal radioactivity can distort the relation between heat flow and tectonic age. For example, heat flow from Precambrian rocks in India and Central Australia is much higher than one would predict from Fig. 5 even though the reduced heat flow is consistent with that from other shields. Hamza (1976) pointed out that plotting reduced heat flow vs. tectonic age removes much of the scatter in the relation, but the effects of both age and crustal radioactivity are overwhelmed by other processes in both the Cretaceous Sierra Nevada (Fig. 6)

Table 1. Average surface heat flow (q_0), reduced heat flow (q_r) and characteristic depth (D) for several continental heat-flow provinces.†

Continent and province	q_0(h.f.u.)	q_r(h.f.u.)	K(km)
North America			
Basin and Range	2.20	1.41	9.4
Sierra Nevada	0.93	0.40	10.1
Eastern USA	1.36	0.79	7.5
Canadian Shield	0.93	0.53	12.4
		0.81	7.1
Europe			
Baltic Shield	0.86	0.53	8.5
Ukranian Shield	0.87	0.6	7.1
Africa			
Niger	0.48	0.26	8
Zambia	1.60	0.96	(11)‡
Asia			
Indian Shield	1.52	0.92	14.8
		0.79	7.5
Australia			
Western Shield	0.93	0.63	4.5
Central Shield	1.98	0.64	11.1
Eastern Australia	1.72	1.37	(11.1)‡

† Data from Pollack and Chapman, 1977*c*; Rao and Jessop, 1975; Rao *et al.*, 1976; Kutas, 1977, and present work (Tables 2, 3, 4).
‡ Value of D assumed for calculation of q_r.

and the Precambrian rocks of Zambia (Chapman and Pollack, 1975*b*, 1977). The latter anomalies probably result from the basic difficulty of a simplistic characterization of tectonic events and the difficulty of establishing when they were and were not in progress. Crough and Thompson (1976) have pointed out that these difficulties, relating to initial conditions are less important after a substantial period of quiescence.

Crustal Temperature Profiles

For any heat-flow province defined by equation (1), it is easy to construct one-dimensional steady-state, conductive crustal geotherms. If we assume an exponential decrease of heat production with depth (equation (2)), the temperature Θ is given as a function of depth (z) by (Lachenbruch, 1970)

$$\Theta(z) = \frac{q_r}{K} z + \frac{D^2}{K} A_0 \left[1 - \exp\left(-\frac{z}{D}\right)\right]. \tag{3}$$

Thus if we know q_r and D, we may draw a series of geotherms for appropriate values of A_0 or a single representative geotherm for an average value of A_0 (assuming, of course, that the thermal conductivity K can be specified within reasonable limits).

Lachenbruch and Sass (1977) reviewed the data for the continental United States and presented a number of conductive, steady-state geotherms (based on a mean thermal conductivity of 6 h.c.u. ($\sim$2.5 W m^{-1} K^{-1}), $A_0 = 5$ h.g.u. ($\sim$2 μW m^{-3}) and $D = 10$ km), representative of the three major heat-flow provinces (Fig. 7). Curves A through D present an interpretation based loosely on the parameters initially defined by Roy *et al.* (1968) (see Table 1). (Curve B, "stable," uses the reduced heat flow for the eastern United States, Table 1, but for consistency within the diagram, a D of 10 km rather than the empirically derived value of 7.5 km.) Curves E and F are the predicted steady-state conductive geotherms for values of q_0 and q_r representative of the "Battle Mountain High" (BMH, Fig. 6), a subprovince of very high heat flow in the Basin and Range. Also shown on Fig. 7 are various melting relations for possible crustal rocks after Wyllie (1971). There is no difficulty in reconciling the simple model of equation (3) with the thermal parameters of the Sierra Nevada and the stable region (Fig. 7, curves A and B), but in the hotter provinces, particularly the Battle Mountain High, steady-state conductive models imply wholesale melting in the lower crust and upper mantle, and thus are inappropriate. At some depth along the geotherms E and F (and possibly even C and D), the conductive steady-state model must give way to a more realistic one. There is an infinite number of possibilities ranging from the transient conductive decay from initial conditions resulting from one or more (convective) "thermotectonic events" in the past to a steady-state convective regime.

3.3. Convective Processes

In geologically active regions, particularly those undergoing tectonic extension, vertical convective heat transport significantly affects the thermal regime and indeed may dominate it. Effects of convective processes on the observed heat-flow patterns of the western United States have been examined by Lachenbruch and Sass (1977, 1978). In particular, we analyzed the approximately 300 heat-flow observations in the Basin and Range province (Fig. 6) and concluded that convective processes, both local and regional, involving both hydrologic and magmatic phenomena must play an important role in the observed distribution of heat flow there.

The Basin and Range province generally is characterized by high heat flow (>1.5 h.f.u.), but there are many areas of very high heat flow (>2.5 h.f.u.) and one conspicuous region (the Eureka Low, EL, Fig. 6) of low heat flow.

It is possible to explain the Eureka Low on the basis of regional groundwater flow to depths of 2 km or so with a downward vertical velocity component on the order of 1 cm y^{-1}, and there is hydrologic evidence to support this view (Lachenbruch and Sass, 1977).

The overall magnitude of convective effects (and their associated transients) can be evaluated by comparing the heat flow versus heat production relation for the Basin and Range to those in more stable tectonic environments. Figure 8 shows the available data from the Basin and Range province together with those from the Sierra Nevada and eastern United States. Points from the eastern United States are tightly grouped about a straight line originally defined by Roy *et al.* (1968), as are all but one from the Sierra Nevada. The single exception from the Sierra (DP, Fig. 8) was obtained only 3 km from the rim of the Long Valley caldera (LV, Fig. 6), an active hydrothermal system.

When only a few well-documeted q_0–A_0 pairs were available from crystalline rocks of the Basin and Range province, they were adequately described by the regression line (dashed, Fig. 8) proposed by Roy *et al.* (1968). From the subsequent work, however (Fig. 8), it seems that the Basin and Range contains several thermal subprovinces, and that it cannot be characterized by a single "intercept" value of q_r, nor can a value of D be defined. It is still possible to define the "non-radiogenic" component, q_r, for each site by rewriting equation (1) as

$$q_r \equiv q_0 - DA_0. \tag{4}$$

Using the value of D established for the adjoining Sierra Nevada province, we calculated q_r for each site in the Basin and Range province. From these data, we concluded that variations in surface heat flow caused by convective processes were greater, perhaps by a factor of 3 or 4, than those caused by lateral variations in radioactivity. Much of the scatter in q_0 (Fig. 8) is probably caused by relatively shallow hydrothermal convection whereas the higher mean value may be the effect of convective transport in the lithosphere caused by rising magma and tectonic extension (Lachenbruch and Sass, 1977).

The upward movement of mass in the lower lithosphere, whether from solid-state motion association with tectonic extension or from rising magma, will generally lead to smaller thermal gradients for a given surface heat flow, and can thereby resolve the melting paradox implied by conductive steady-state geotherms (e.g. curves D and F, Fig. 7). Three simple one-dimensional steady-state models for combined conductive and convective transfer in a uniformly extending lithosphere are illustrated in Fig. 9. The extensional strain rate is denoted by s and the conductive heat flux from the asthenosphere into the base of the lithosphere is denoted by q_a. The depth R to the

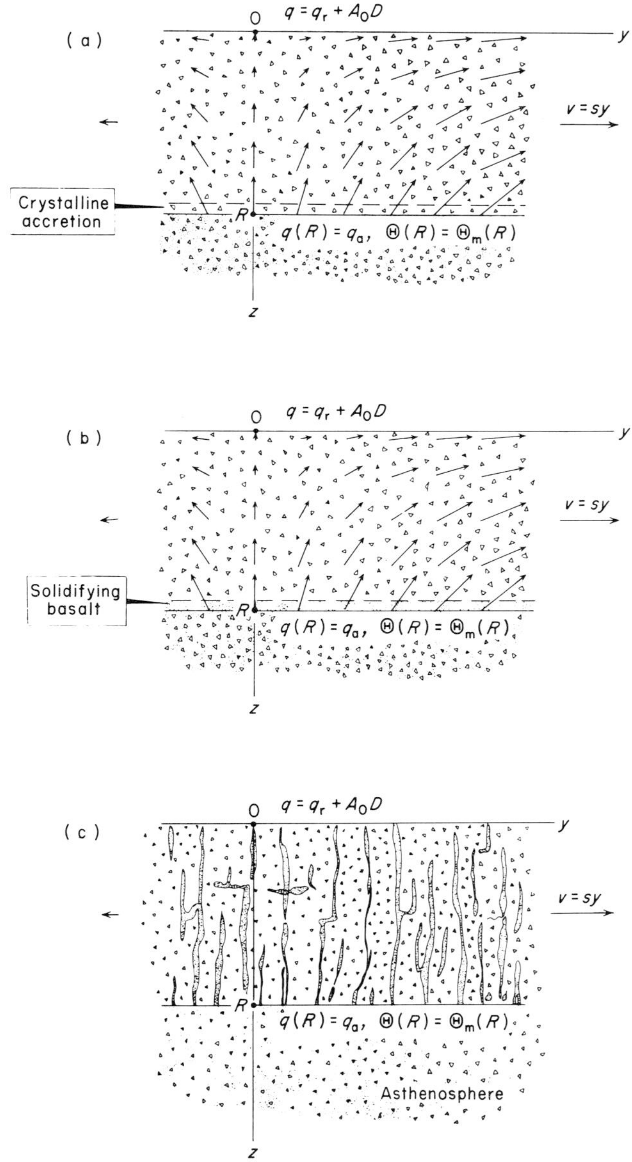

Fig. 9. Simple thermomechanical models of a steadily extending lithosphere of thickness R with base at solidus temperature, $\theta_m(R)$. (a) Stretching: homogeneous solid-state stretching with accretion of crystalline material at the base. (b) Underplating: homogeneous solid-state stretching with accretion of solidifying basalt at the base. (c) Intrusion: extension by distributed dike intrusion. Arrows in (a) and (b) are velocity vectors, and s is horizontal strain rate (assumed constant throughout the lithosphere).

base of the lithosphere is given by the intersection of the geotherm and the basalt dry solidus $\Theta_m(z)$ (cf. BDS, Fig. 7). Analytical results and further discussion of these models are presented elsewhere (Lachenbruch and Sass, 1978; Lachenbruch, 1978); models similar to Fig. 9(c) have been discussed by Bodvarson (1954) and Palmason (1973).

The steady-state effects on surface heat flow of solid-state convection, and magmatic heat are shown approximately in Figs 10 and 11, respectively. Thus, solid-state convection across a 40-km layer extending in the stretching mode (Fig. 9(a)) at 2.7 per cent per million years will cause the heat flux from its upper surface (q_r) to be twice that conducted into its base (q_a) (see Fig. 10). The magmatic contribution to q_r depends upon the rate at which basalt is supplied to the base of the lithosphere (W, Fig. 11) and the average ambient temperature Θ' at which the magma gives up its heat. Thus in the underplating mode (Fig. 9(b)), basalt solidifies at temperatures near its solidus ($\Theta' \sim 1200\,°C$, Fig. 11) and only the latent heat is given up. For $W \sim 1$ km per million years, this would add about 0.65 h.f.u. to q_a (see Fig. 11), and the sum $q_a + 0.65$ is then enhanced by solid-state convection as illustrated in Fig. 10. The dike-intrusion mode (Fig. 9(c)) applied to the lithosphere as a whole corresponds approximately to $\Theta' \simeq 700°C$ (Lachenbruch and Sass, 1978) and for $W \sim 1$ km per million years, q_r would be

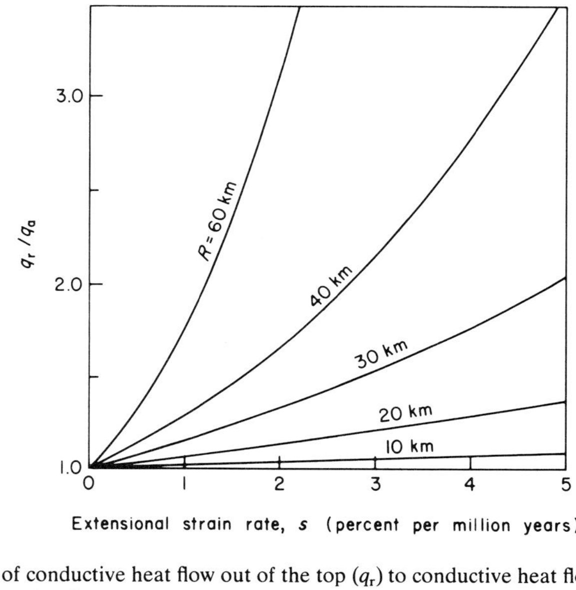

Fig. 10. Ratio of conductive heat flow out of the top (q_r) to conductive heat flow into the base (q_a) for a (lithosphere) layer of thickness R extending in the stretching mode (Fig. 9(a)) at uniform horizontal strain rate s.

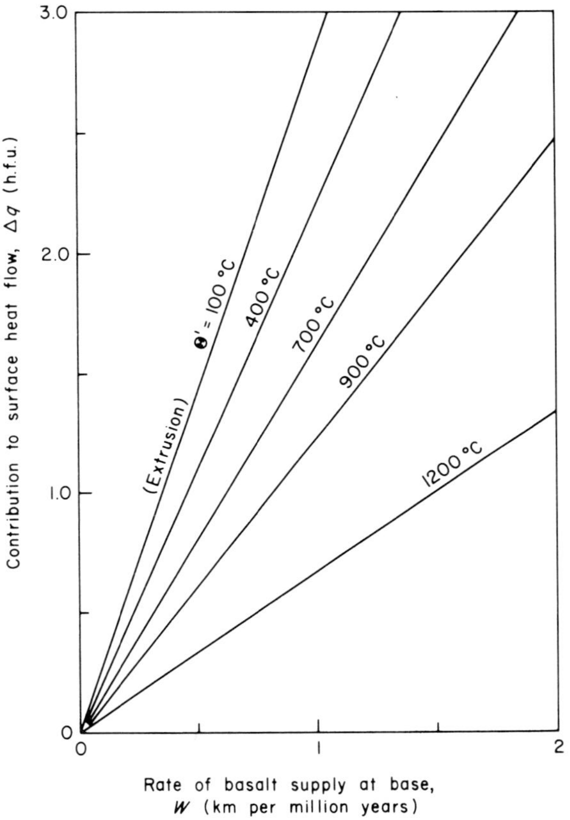

Fig. 11. Steady-state contribution (Δq) to surface heat flow from a layer receiving basalt (at 1200 °C) through its base at the rate W; the basalt's heat is lost to the layer at the average ambient temperature Θ'. The magmatic contribution for underplating (Fig. 9 (b)) and the total contribution for dike intrusion (Fig. 9(c)) are given approximately by $\Theta' = 1200$ °C and $\Theta' = 700$ °C, respectively, where $W = Rs$.

enhanced by about 1.6 h.f.u. (Fig. 11); in this mode, however, there is no additional enhancement by solid-state convection.

In the two modes involving a magmatic phase (Fig. 9(b) and (c)), we require that the layer thickness R remain constant and that the rate of magma supply just compensate for the volume increase caused by extension, i.e. $W = Rs$. However, since horizontal extension is not required for sill-like intrusion (Fig. 9(b) is only a special case), Fig. 11 may be applied more generally to estimate the steady magmatic contribution to heat flow whether extension is occurring or nor. If no extension occurs above the intruded layer, the top of the block beneath it must move downward at the rate W.

This could be accommodated by rigid downward displacement of the lower block or by stretching and thinning if the lower block were decoupled by the intruded layer. The associated downward solid-state convection introduces a thermal sink in the lower crust which can subtract appreciably from the magmatic contribution to surface heat flow. For a given rate of supply, the magmatic contribution to surface heat flow is generally greater for upper crustal intrusion (e.g. $\Theta' \sim 400$ °C, Fig. 11) than for lower crustal intrusion (e.g. $\Theta' \sim 700$ °C, Fig. 11). Intermittent sill intrusion with period small relative to the time constant of the overburden (τ, Fig. 7) will have the same effect on surface heat flow as continuous intrusion at a uniform (time-averaged) rate. Under conditions favorable for intrusion (a large basalt supply at depth and an extensional stress regime in the lithosphere), the simple sill-injection mode is not mechanically unreasonable; basalt might rise through the fractures in the lower lithosphere and spread horizontally where it loses its buoyancy in low density upper crustal rocks. In this mode the contribution of magmatic heat is, of course, quite independent of horizontal extensional strain.

Although these simple one-dimensional steady-state considerations will give a poor account of conditions deep in the lithosphere for local or short-lived tectonic and magmatic events, they describe useful limiting conditions for convective transfer in the lithosphere and provide insight from simple analytical results. Figure 12 shows an application to the entire lithosphere (from Lachenbruch and Sass, 1978) in which results for the intrusion (Fig. 9(c)) and stretching (Fig. 9(a)) modes were combined by matching boundary conditions at their interface. The dashed curves labeled C through F are the steady-state conductive geotherms ($s = 0$) shown in Fig. 7. The corresponding solid curves represent geotherms in the extending lithosphere for the same reduced heat flows. Numbers on the solid curves are the required extension rate (s) in percent per million years. For the average Battle Mountain High condition (curves F, Fig. 12) this amount of convection in the extending lithosphere almost doubles the estimate of lithosphere thickness and decreases estimates of temperature in the lower crust more than 300 °C. The solid curves of Fig. 12 represent only one of many models that can be applied to regions with high heat flow, extensional tectonics and basaltic magmatism. Constraints on such heat-flow models depend upon information from other sources regarding the rates and history of extension and the thickness and composition of crustal layers and of the lithosphere mantle.

Many of the concepts introduced thus far have been developed with particular reference to the continental United States, which has been studied more intensively than most other continental blocks of comparable size. Data from some other continental regions are very sparse (Pollack and

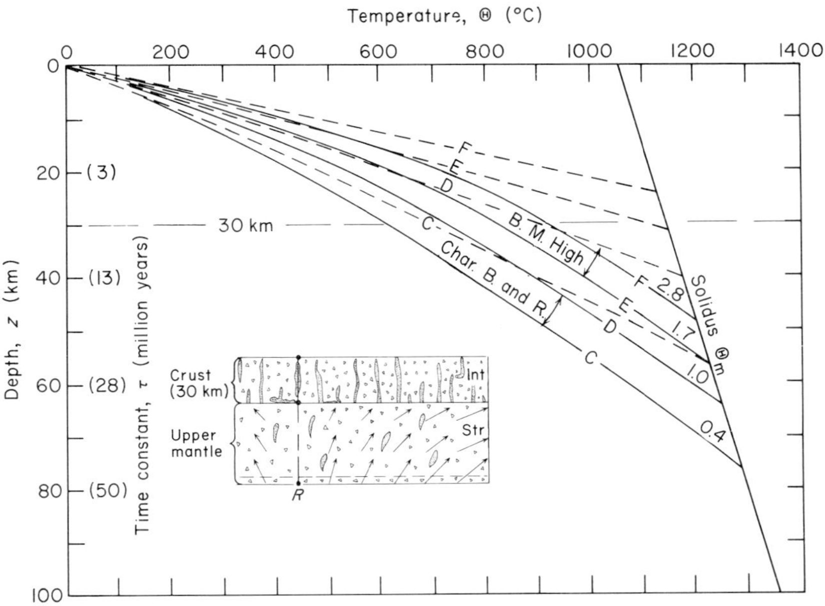

Fig. 12. Solid curves show geotherms for a lithosphere in which the 30-km crust extends in the intrusive mode and the upper mantle extends in the stretching mode ($g_a = 0.8$) (see inset and text). Dashed curves are corresponding geotherms for static case $s = 0$, $q_a \equiv q_r$. Numbers on solid curves are extension rates in percent per million years. Curves labeled C, D, E and F are for reduced heat flows of 1.2, 1.6, 2.0 and 2.5 h.f.u., respectively.

Chapman, 1977*a*; Jessop *et al.*, 1976). In Europe and the Mediterranean, many new data have recently been published (Cermak and Hurtig, 1977) and a considerable effort is being made by investigators familiar with that region to reconcile the complicated thermal and tectonic patterns observed there.

In the remainder of this discussion, we shall focus on the 90 or so heat-flow determinations presently available for the Australian continent (Fig. 1) in an attempt to test (and possibly to extend) the generality of some concepts developed from observations over a fairly small fraction of the Earth's continental area, and to characterize the thermal regime of the Australian continental crust to the extent permitted by the existing data.

4. Heat-flow Provinces in Australia

For each of the three heat-flow provinces already outlined (Fig. 1), we shall examine the relation between heat flow and radioactivity, construct

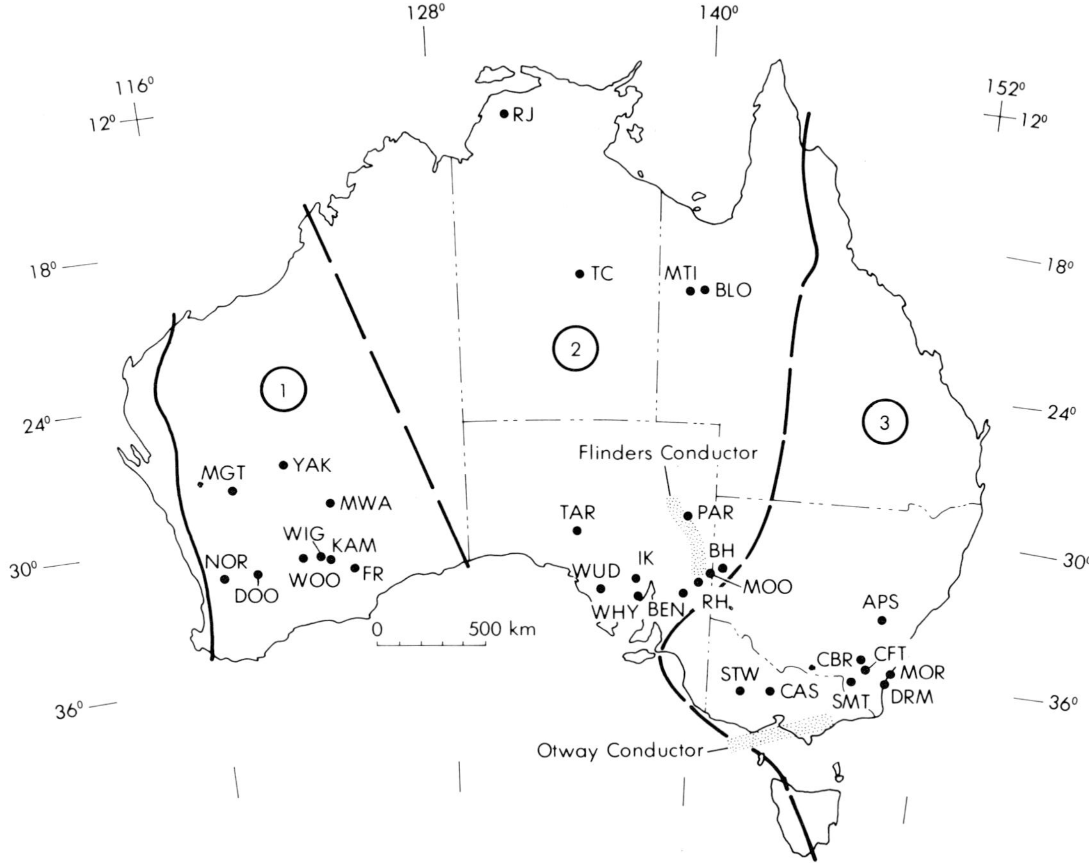

Fig. 13. Locations of heat flow–heat production pairs within the Australian continent (Tables 2, 3, 4). Heat-flow provinces are: 1, Western Shield Province; 2, Central Shield Province, 3, Eastern Province. Geoelectric anomalies (Lilley, 1975) are shown as patterns.

geotherms based on a one-dimensional steady-state model (equation (3)) with $K = 6$ h.c.u. (2.5 W m^{-1} K^{-1}), and examine other geological and geophysical observations that might constrain our interpretation of heat flow. The sites for which we have both heat-flow and heat-production data are shown in Fig. 13.

4.1. Western Shield Province

Heat Flow vs. Heat Production

Heat flows and heat productions for the Precambrian shield of Western Australia are listed in Table 2 and are plotted in Fig. 14. With nine points, the relation defined by Jaeger (1970) (solid line, Fig. 14) on the basis of three points is strengthened considerably. (A new least-squares relation was not calculated because only one of the new points, KAM, Table 2 and Fig. 13, was sufficiently well documented to be of equal status with Jaeger's points, and it corresponds exactly to his least-squares parameters.) The relation clearly is different from that defined by Roy *et al.* (1968) for the stable eastern United States (dashed line, Fig. 14).

The calculation of q_r from the parameters of the q_0–A_0 line (equation (4)) provides a convenient method of evaluating individual sites within the province. Except for DOO, individual values of q_r (Table 2) are represented

Table 2. Heat flow–heat production pairs for Precambrian granites and gneisses of the Western Shield Province.

Locality	N†	A_0 (h.g.u.)	Rock	q_0 (h.f.u.)	q_r‡ (h.f.u.)
Doodlakine (DOO)§	36	21.3 ± 1.1	Granitic gneiss	1.28	0.32
Fraser Range (FR)	60	1.3 ± 0.1	Pyroxene granulite	0.69	0.63
Kambalda (KAM)	13	3.09 ± 0.05	Granite	0.75	0.61
Mount Magnet (MGT)	7	16.3 ± 2.0	Granite	1.3	0.57
Mount Windarra (MWA)	5	2.8 ± 0.2	Gneissic granite	0.96	0.83
Northam (NOR)§	84	5.1 ± 0.7	Gneiss	0.85	0.62
Widgiemooltha (WIG)	4	2.75 ± 0.09	Granite	0.79	0.67
Woolgangie (WOO)§	54	7.6 ± 0.9	Granite and gneissic granite	0.97	0.63
Yakabindie (YAK)	6	4.6 ± 0.2	Granodiorite	0.82	0.61

† Number of samples.
‡ $q_r = q_0 - 0.45\,A_0$ (Jaeger, 1970).
§ From Jaeger (1970).

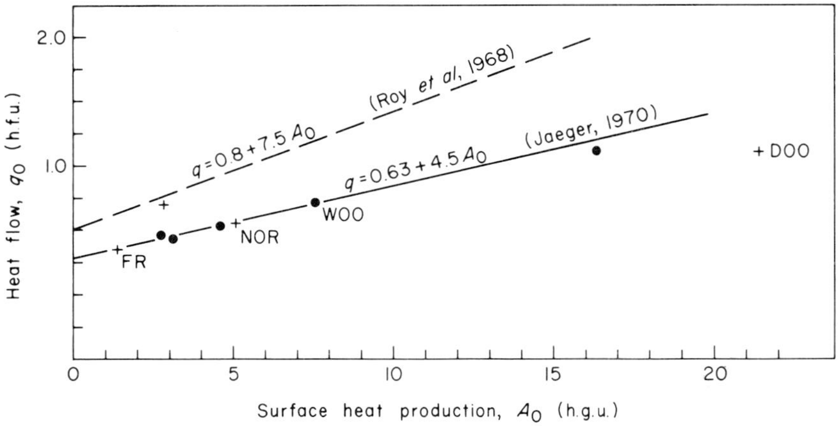

Fig. 14. Heat flow vs. heat production in crystalline rocks of the Western Shield Province, 1, Figs. 1 and 13. Labeled points are those published by Jaeger (1970). Plus signs, Gneisses, dots, granites. Dashed line is q_0–A_0 line for the Eastern United States. Solid line is least-squares line fitted by Jaeger (1970) to FR, NOR and WOO.

reasonably well by Jaeger's intercept value of 0.63 h.f.u. Two outcrops within 15 km of DOO have radioelement concentrations comparable to those measured in the hole but within a radius of 50 km, the mean A_0 is about 10 h.g.u. (Jaeger, 1970; Hyndman *et al.*, 1968). Jaeger (1970) suggested that the apparent anomaly might be caused by a thin veneer (<2 km) of the high radioactivity material overlying rocks with the characteristic regional heat production.

The low value of D may have important implications. If we assume that D simply represents the thickness of a slab of (vertically) uniform radioactivity in the upper crust (Roy *et al.*, 1968), then we should conclude that these ancient rocks have never been thicker than 4.5 km or (more plausibly) that they have subsequently been eroded to this thickness. However, since D from other shields of comparable age is higher (Table 1), we must either reject this interpretation or search for evidence that the Western Australian shield has been more highly differentiated or more deeply eroded than others. If, on the other hand, we assume that D represents the logarithmic decrement for a geochemically more realistic vertical distribution of radioactivity that decays exponentially with depth (equation (2)), then we still must explain why the radioelements of this province are more strongly differentiated upward than those of other shields. More values of both heat flow and heat production, particularly in the rocks of higher radioactivity, are required to establish whether or not this region is geochemically unique as the present results suggest.

Other Observations
This province is similar to other Precambrian Shields of comparable age. The crust is about 40 km thick, and seismic velocities are high (Cleary, 1973). There is some seismic activity, but it is localized along ancient pre-existing faults (Fitch *et al.*, 1973), and there is no reason to infer a thermal origin for the activity.

Crustal Thermal Regime
To calculate crustal geotherms, we use equation (3) with the parameters (q_r, D) from the linear relation defined by Jaeger (1970) and confirmed by the present work (Fig. 14). Four curves are shown (Fig. 15) for values of 0, 7, 14 and 20 h.g.u. which covers the range of observed radioactivity. The steady-state temperature profiles predict a narrow range of temperatures (~500 °C) for Θ_n, the temperature at the base of the crust (cf. Hyndman *et al.*, 1968). The geotherms would preclude melting in the upper mantle, which is consistent with models of the velocity structure for western Australia (Goncz and Cleary, 1976).

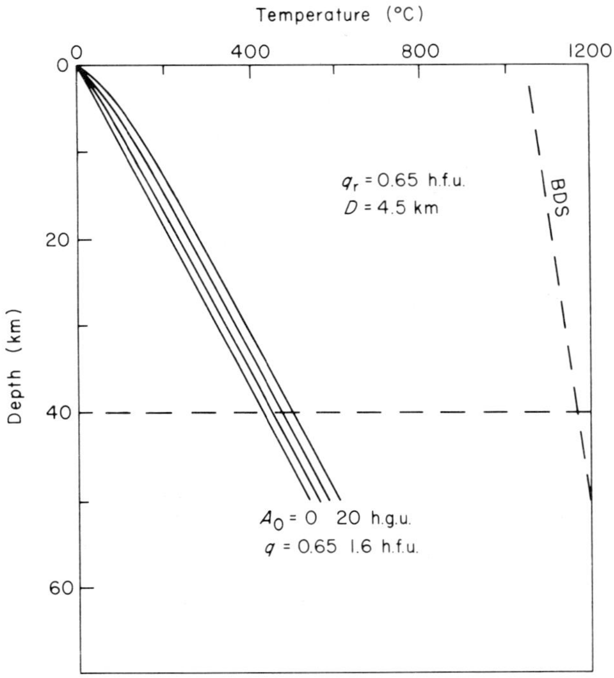

Fig. 15. Theoretical steady-state geotherms (equation (3)) for the Western Shield Province. Thermal conductivity K assumed constant = 6 h.c.u.; values of A_0 are 0, 7, 14 and 20 h.g.u; BDS is the basalt dry solidus.

4.2. Central Shield Province

Heat Flow vs. Heat Production

Table 3 and Fig. 16 show heat flow–heat production pairs for the younger Precambrian rocks of central Australia. There is considerable scatter; the coefficient of correlation is only 0.8. The least-squares line has about the same intercept q_r as that defined for the Western Shield Province by Jaeger (1970), but the slope, D, of 11.1 km is more than double that from western Australia. In this province, the parameters for the eastern United States apply almost as well (dashed line, Fig. 16) as those we calculated. If the less reliable q_0–A_0 pairs (see footnote § in Table 3, and open circles, Fig. 16) are included in the least-squares fit, q_r and D are 0.97 h.f.u. and 8.3 km, respectively. Thus, caution must be exercise in using the parameters of the q_0–A_0 line. We believe that the q_0–A_0 relationship in central Australia is consistent with that from other cratons, but the exact agreement of q_r between this province and the Western Shield Province (Fig. 14) is certainly fortuitous. In view of the observed scatter in the q_0–A_0 relation, it is useful to examine values of q_r calculated according to equation (4) at individual sites. Low values of q_r for TAR, WUD, WHY and RJ (Table 3 and Fig. 13) are consistent with a crustal regime with D intermediate between that defined for the Western Shield Province (dotted line, Fig. 16) and the least-squares value. Values of q_r at three of the sites in the northern part of the province

Table 3. Heat flow–heat production pairs for Precambrian granites and gneisses of the Central Shield Province.

Locality	N†	A_0(h.g.u.)	Rock	q_0(h.f.u.)	r_r‡(h.f.u.)
Bendigo Sta. (BEN)	12	9.09 ± 0.29	Granite	1.52	0.51
Blockade (BLO)§	2	7.5 ± 2	Granite	1.8	0.97
Broken Hill (BH)	56	11.0	Gneiss	1.93	0.71
Iron Knob (IK)§	6	18 ± 5	Granite	2.6	0.60
Mootooroo (MOO)	11	7.4 ± 0.4	Gneiss	1.61	0.79
Mount Isa (MTI)	8	9.3 ± 0.4	Granite	2.0	0.97
Parabarana (PAR)	4	18.8 ± 0.1	Granite	3.0	0.91
Radium Hill (RH)	4	9.1 ± 1.4	Paragneiss	1.8	0.79
Rum Jungle (RJ)	33	15.0	Granite	2.0	0.34
Tarcoola (TAR)	20	6.35 ± 0.11	Granite	1.20	0.50
Tennant Creek (TC)	7	9.3 ± 0.4	Granite	2.0	0.97
Whyalla (WHY)§	2	20 ± 5	Granite	2.2	0
Wuddina (WUD)	17	11.8 ± 0.6	Granite	1.38	0.07

† Number of samples.

‡ $q_r = q_0 - 11.1\,A_0$.

§ Excluded from least-squares calculation because of small sample and/or large scatter.

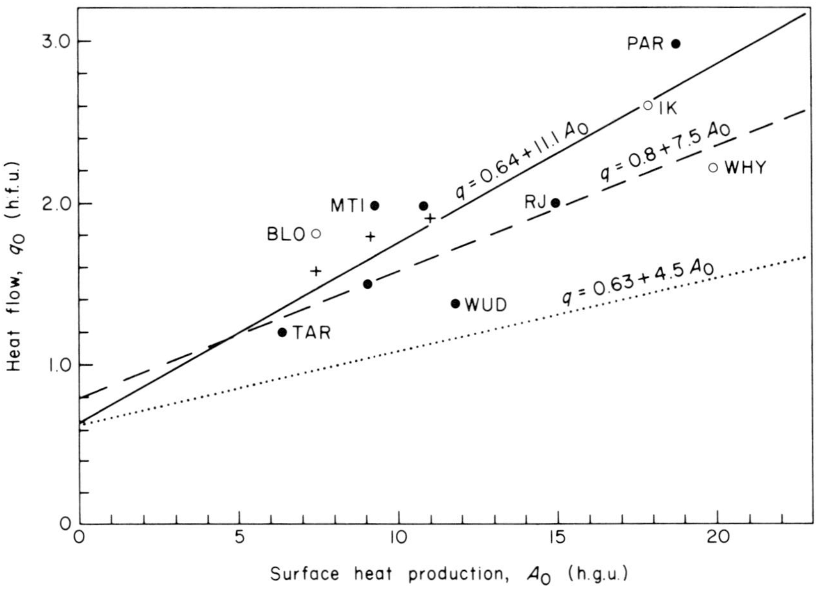

Fig. 16. Heat flow vs. heat production for the Central Shield Province, 2, Figs. 1 and 13. Solid line represents least-squares fit to observed data (excluding open circles). Dashed line is that for the eastern United States, dotted line, for the Western Shield Province (Fig. 14). Plus signs, gneisses; dots and open circles, granites.

(TC, MTI and BLO, Fig. 13 and Table 3) are about 1 h.f.u., comparable to those found in Zambia by Chapman and Pollack (1977). Thus, the possibility that renewed tectonic activity beneath the northcentral part of the Central Shield Province is contributing to the observed heat flow cannot be excluded.

Other Observations

The Flinders Conductor. In 1970, Gough *et al.* (1974) deployed a large (1200 km × 300 km) array of recording magnetometers in South Australia and western New South Wales. From the differences in response to magnetic storms among the various magnetometers, they inferred that a section of the lithosphere with approximately the shape and extent shown in Fig. 13 has a higher electrical conductivity than the surrounding region. The conductor lies within a zone of relatively high seismicity and possible crustal thinning (Cleary and Simpson, 1971; Muirhead *et al.*, 1977). Magnetotelluric studies made simultaneously with the array study by Tammemagi and Lilley (1973) suggest that the conductor is a crustal feature possibly caused by brines in porous sedimentary rocks or highly fractured basement rocks at depths of between 5 and 15 km, and that it need not have a thermal origin.

Seismicity. Seismic activity in the Central Shield Province is largely confined to the southern half. Focal depths are shallow (<15 km) and focal mechanism studies indicate that most of the activity is the result of strike-slip motion along old fracture zones with some normal faulting associated with relative motion between parallel fracture zones (Stewart and Mount, 1972).

Seismic Attenuation. Mereu *et al.* (1974) postulated layers of high seismic attenuation (low Q) and consequently of high temperature at depths between 85 and 650 km in the mantle beneath northern Australia, based on observations at the Warramunga seismic array near Tennant Creek (TC, Fig. 13) of earthquakes originating in New Guinea, New Britain and the Solomon Islands. They suggested that the lowermost ~100 km of the low Q region might be associated with the return path of material from a subduction zone. In view of the time constants involved for thermal events originating deeper than 80 km (τ, Fig. 12), it is possible to have a partially molten zone presently in the deep mantle with no surface thermal expression. P-arrivals from teleseismic events are early at Warramunga (Cleary, 1967), and there is no evidence for low P_n velocities (Cleary, 1973), suggesting that high temperatures are not characteristic of the crust and upper mantle near TC.

Crustal Thermal Regime

The least-squares parameters (q_r, D) from Fig. 16 were used in equation (3) to calculate steady-state crustal geotherms for the Central Province (Fig. 17). The effect of D is seen by comparing Figs 15 and 17, where D is the only parameter changed. For very low values of A_0, the geotherms are the same, but for $A_0 = 20$ equation (3) predicts values of Θ_n greater than 800 °C, some 300 °C higher than the corresponding geotherm for the Western Shield Province. At these temperatures, we might even expect partial melting of crustal material if a volatile phase were present (see melting relationships for granodiorite, Fig. 7).

The geotherms (Fig. 17) represent our best estimates of the *average* crustal temperatures for the entire Central Shield Province. They are consistent with surface geological observations and other geophysical data which indicate a stable continental crust with high average radioactivity and little, if any, contemporary tectonic activity of thermal origin. Our simple interpretation is, however, based on a limited number of data of variable quality. Further studies may well outline several thermal subprovinces and consequently a more complicated crustal thermal regime than that represented by Fig. 17.

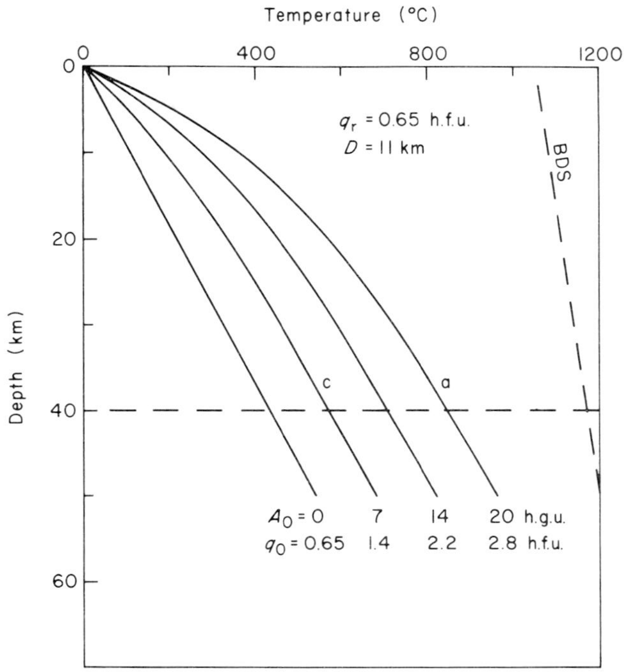

Fig. 17. Theoretical steady-state geotherms (equation (3)) for the Central Shield Province. Curves labeled a and c are compared with the corresponding curves in Figs 21 and 23 (see text). Thermal conductivity K assumed constant = 6 h.c.u. BDS is the basalt dry solidus.

4.3. Eastern Province

Heat Flow vs. Heat Production

The eight q_0–A_0 pairs (Table 4, Figs 13 and 18) are all from the southern part of the Eastern Province which is characterized by higher-than-normal heat flow (Fig. 1). A formal analysis of the q_0–A_0 relation results in a slightly higher correlation coefficient (0.86) than that obtained for the Central Shield Province. The least-squares line (dotted, Fig. 18) has an intercept q_r of 0.19 h.f.u. and a slope D of 31.3 km. These parameters are very different from those for any of the other thermal provinces so far defined (cf. Table 1) and it is difficult to reconcile either the low q_r or the high D with plausible physical and chemical models for the crust and upper mantle. Thus, we consider an alternative explanation whereby the points lying above the reference lines for the Central Shield (solid line) and the Eastern United States (dashed, Fig. 18) represent thermal disturbances superimposed on a crustal regime characteristic of cratons. For convenience, we calculate the "non-radiogenic" heat flow q_r (equation (4)) based on the value of D from

Table 4. Observed heat flow, q_0, radioactivity, A_0, and reduced heat flow, q_r, (assuming $D = 11.1$ km) for granites from the Eastern Province

Locality	N†	A_0(h.g.u.)	q_0(h.f.u.)	q_r‡(h.f.u.)
Apsley (APS)	3	5.9±0.2	1.5	0.84
Canberra (CBR)	14	6.5±0.6	2.1	1.38
Captain's Flat (CFT)	5	5.2±0.7	2.4	1.82
Dromedary (DRM)	3	2.7±0.3	1.07	0.77
Moruya (MOR)	12	3.47±0.2	1.28	0.90
Snowy Mountains (SMT)				
North (SMN)	84	6.3±0.2	1.8	1.10
South (SMS)	67	6.25±0.2	2.1	1.41
Castlemaine (CAS)	1§	8.2	2.9	1.99
Stawell (STW)	1§	7.5	2.8	1.97

† Number of samples.
‡ $q_r = q_0 - 11.1\,A_0$.
§ Single fresh specimens of granite obtained by blasting for K–Ar dating.

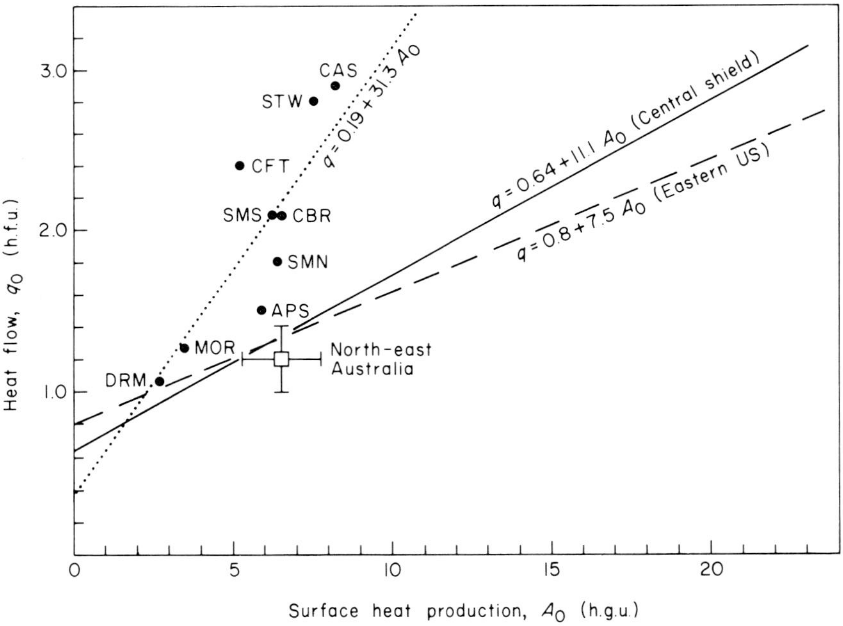

Fig. 18. Heat flow vs. heat production for intrusive rocks of the Eastern Province, 3, Figs 1 and 13. Dotted line is least-squares fit to observed data. Solid and dashed reference lines are for the Central Shield Province (Fig. 16) and the eastern United States, respectively. Square is mean heat flow and mean heat production (±standard deviation, error bars) for north-eastern Australia between latitudes 20° and 32°S (see text).

the adjoining Central Shield Province. With due consideration of the uncertainties involved in D and in the individual values of q_0 and A_0, values of q_r for APS, MOR and DRM (Figs 13 and 18, Table 4) are consistent with those from the eastern United States or the Central Shield. The mean of the eight heat flows (Fig. 1) and the weighted mean of the approximately 600 radioactivity data (Fig. 4) from the northern part of the Eastern Province (with error bars indicating standard deviations) are also shown in Fig. 18. If we assume that the regional average A_0 of about 6.5 h.g.u. is representative of the radioactivity of basement rocks beneath the heat-flow sites, then the q_0–A_0 point representing average conditions for eastern Australia between latitudes 20° and 32°S also suggests a thermal regime characteristic of stable continental blocks. According to this interpretation there remain six anomalous values of q_r inland from the coast in south-eastern Australia between CFT and STW (Table 4, Figs 13 and 18). Based on well-documented values of high unreduced heat flow q_0, this region might extend north-eastward as far as the Sydney Basin (SB, Fig. 1).

Other Observations

Seismicity. The entire eastern coastal region is seismically active, but more earthquakes have been observed in the southern part of the Eastern Province, where heat flow generally is high (Fig. 1).

Crustal Structure. Cleary (1973) reviewed all of the available data and concluded that the average crustal thickness (inland from the coast) is about 40 ± 5 km with no clear relation between crustal thickness and topography. More recently, Muirhead *et al.* (1977) have examined south-eastern Australia in detail. They suggested that the crust thins rapidly by several kilometers west of the Snowy Mountains (SMT, Fig. 13) probably resulting from isostatic adjustment as the mean elevation drops. They found evidence for crustal thinning of "a few kilometers" at a station near STW (Fig. 13).

Seismic Velocities. There is general agreement that the velocity structure beneath south-eastern Australia is markedly different from that beneath both the Western Shield Province and the Central Shield Province (Muirhead *et al.*, 1977). Little evidence exists for a low-velocity layer at depths above 400 km in the Precambrian provinces. Goncz and Cleary (1976) suggested (from studies of Rayleigh waves) a low velocity zone for shear waves extending from about 90 to 180 km beneath south-eastern Australia. A more recent study by Mills and Fitch (1977) concluded that if such a layer exists, it is deeper than 140 km. This interpretation is consistent with the models of Muirhead *et al.* (1977) which suggest a P-wave low velocity layer between 150 and 200 km in the region. Further evidence supporting the differences in velocity structure between the eastern province and the

Precambrian provinces was provided by an analysis of P-wave travel times from nuclear explosions occurring at distances of from 35° to 96° from the stations (Cleary, 1967). The Precambrian provinces are characterized by early arrivals and the southern part of the eastern province, by late arrivals. Cleary (1973) also noted that there is evidence for low P_n velocities in south-eastern Australia, but none for the two Precambrian provinces.

Plate-tectonic Setting. The Australian land mass is the most conspicuous physiographic feature within the Indian Plate, but no part of it is close to a currently active plate boundary. Between 60 and 80 million years ago, sea-floor spreading resulted in the creation of the Tasman Sea, separating Australia from New Zealand and the Lord Howe Rise (Hayes and Ringis, 1973; Weissel and Hayes, 1977). Beginning about 50 million years ago, spreading along the Antarctic Ridge has resulted in the separation of Australia and Antarctica at the rate of about 7 cm y^{-1} (Jones, 1971; Weissel and Hayes, 1971).

Cenozoic Volcanism. Within the eastern Australian thermal province, Wellman and McDougall (1974) have defined a number of Cenozoic volcanic "provinces" primarily of two types: (1) central volcano provinces; mainly basaltic lavas but with some felsic flows and intrusions, (2) lava field provinces; formed exclusively of basaltic flows with associated mafic intrusions. Both types of volcanism occur within the same physiographic region, the "Eastern Highlands" (Fig. 19), but their distribution with time shows contrasting patterns (Figure 20). Wellman and McDougall have made the remarkable observation that the termination of volcanism at each latitude was marked by the occurrence of a unique episode producing central volcanos, and that the southward sweep of the volcanos occurred at the same rate as Australia moved northward over the asthenosphere in its migration from Antarctica (~7 cm y^{-1}); that is, the central volcanos seem to occur over a location fixed in the asthenosphere, and an analysis by Vogt and Conolly (1971) suggests that a parallel chain of guyots in the Tasman Sea might have a similar origin. Wellman and McDougall (1974) concluded that "volcanism [probably] only occurred in areas of tension and that such areas at any one time were restricted to a relatively narrow band parallel to the axis of the eastern highlands," and further "that the rate of crustal extension controls the rate of dike and plug intrusion." They estimate that the volume of intrusive dikes and plugs exceeds that of the associated overlying volcanic rocks by a factor of two. Recent studies of earthquake mechanisms (Mills and Fitch, 1977) imply a present-day stress field dominated by east–west horizontal compression in eastern Australia with the possible exceptions of northern Queensland and of western Victoria, where Hills (1959) found

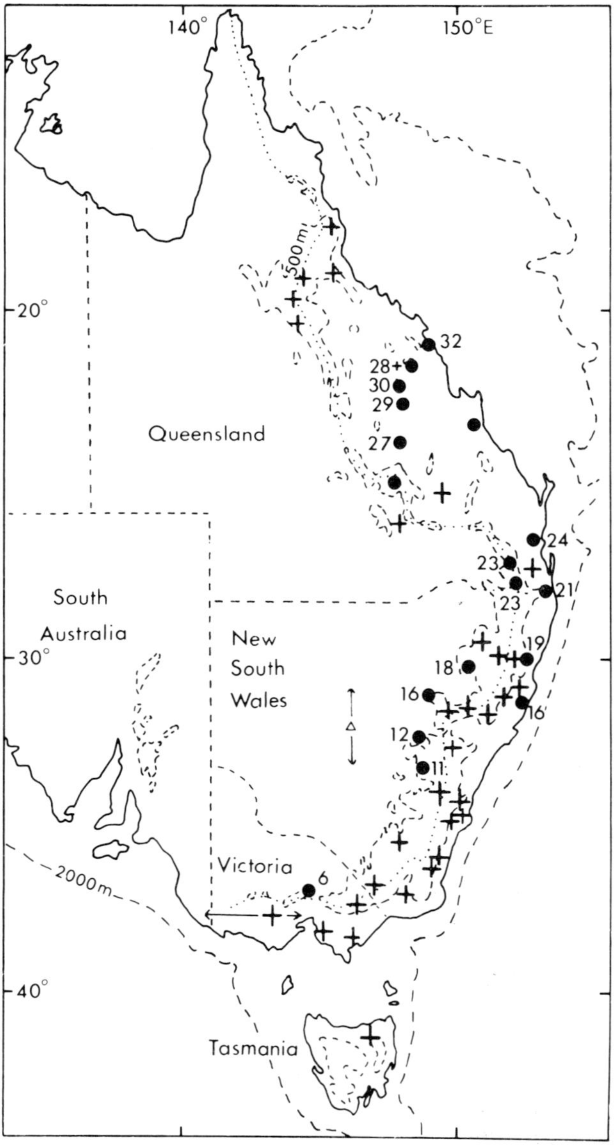

Fig. 19. Location of Cenozoic volcanic provinces in eastern Australia (reproduced from Wellman and McDougall, 1974). Plus signs, lava field provinces; dots, central volcano provinces with ages in million years. The eastern highlands are outlined by the 500-m contour and the drainage divide is indicated by the dotted line.

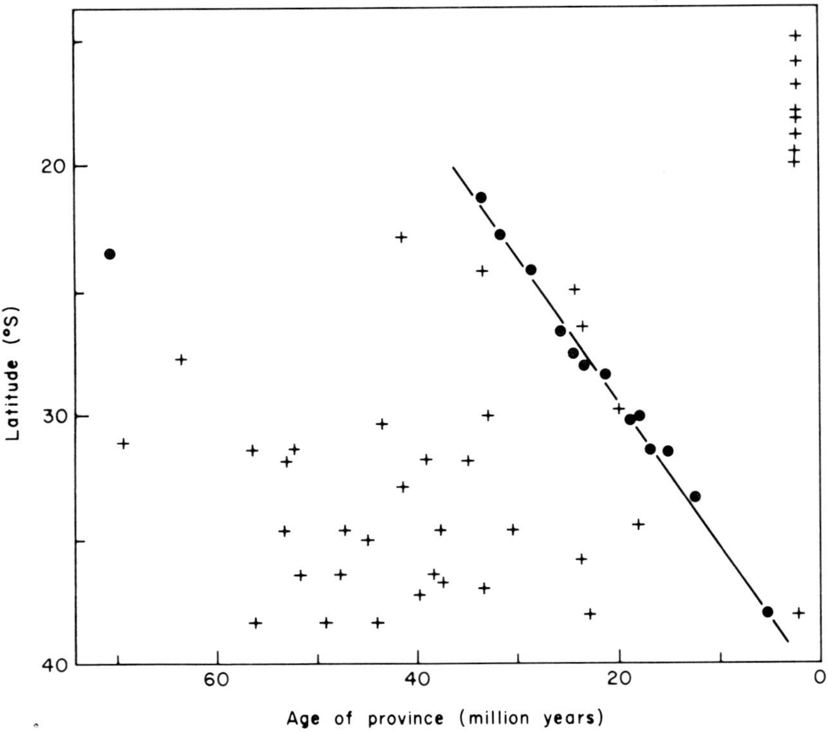

Fig. 20. Radiometric ages of Australian Cenozoic volcanic provinces as a function of latitude (reproduced from Wellman and McDougall, 1974). Plus signs, lava field province; dots, central volcano provinces. Solid line emphasizes the relation between latitude and the age of central volcano provinces.

evidence for normal faulting and crustal extension corresponding to the observed quaternary volcanism.

The Otway Conductor. A series of magnetometer-array studies by Bennett and Lilley (1974) and Lilley (1976) resulted in the discovery and delineation of the Otway Conductor near the south coast of Victoria (Fig. 13). This anomaly is associated with relatively high seismicity (Lilley, 1975) and with the "newer basalts," a series of Pliocene through Holocene extrusive rocks (McDougall *et al.*, 1966; Aziz-Ur-Rahman and McDougall, 1972; Irving and Green, 1976). In contrast to the Flinders Conductor, Lilley (1976) presented evidence that the top of the Otway Conductor is deep, in the lower crust or upper mantle.

Heat Flow from the Otway Basin. Low-to-normal heat flows from the Otway Basin in south-western Victoria (OB, Fig. 1) are puzzling in view of the high regional heat flow, the presence of a deep electrical conductor, high seismicity and very young volcanic rocks. Thermal disequilibrium associated with the late stages of the evolution of this basin might account for some reduction in surface heat flow, but to reconcile the observed heat flow with the geological and geophysical evidence for high temperatures in the lithosphere, we must invoke a regional hydrologic heat sink similar to that proposed for the "Eureka Low" (EL, Fig. 6) in Nevada by Lachenbruch and Sass (1977). Taplin (1976) presented evidence for systematically decreasing temperature gradients from north to south based on bottom-hole temperatures obtained during the drilling of exploratory wells in the basin. Aquifers in the basin are confined with water movement southward along the topographic gradient and natural discharge is in lakes, streams and submarine springs (Lawrence, 1976). The apparent fresh-water flushing of hydrocarbons from Tertiary sedimentary formations and the persistence of fresh waters offshore also provide evidence for lateral water movement with a downward vertical component within the Otway Basin (Wopfner *et al.*, 1971). This fragmentary evidence, while suggestive of regional hydrologic disturbances to the thermal regime, does not rule out a subprovince of normal heat flow in the Otway Basin.

Crustal Thermal Regime

A literal interpretation of the least-squares line for data from eastern Australia (Fig. 18) in terms of equations (1) and (2) implies an extremely anomalous vertical distribution of crustal radioactivity ($D \simeq 30$ km). (It also ignores the likely association of high heat flows with convective processes accompanying Cenozoic volcanic activity.) As the lower heat flows are fairly well accounted for by the line for the Central Shield Province it seems reasonable to assume a more normal radioactive distribution ($D \simeq 11$ km) and interpret the high heat flows of Fig. 18 as representing anomalous values of q_r. Figure 21 shows the steady-state conductive geotherms resulting from this interpretation for an assumed surface heat production, A_0, of 7 h.g.u. and reduced heat flows q_r bracketing those calculated in Table 4.

A comparison of Figs 17 and 21 illustrates the effect of increasing the deep crustal and mantle contribution to the heat flow. The curve labeled c in Fig. 17 is the same as curve c in Fig. 21; it represents a deep crustal heat flow typical of the western two thirds of the continent. The curve labeled a in Fig. 21 corresponds to conditions at STW and CAS in Victoria (Fig. 13, Table 4); although it represents the same surface heat flow (2.8 h.f.u.) as the warmest curve (a) in Fig. 17, it yields a subcrustal temperature (Θ_n) greater by 600 °C, and confronts us with the "melting paradox" discussed above with reference to the western United States (Fig. 7). Lower crustal temperatures are well above the melting temperature (BDS, Fig. 21) for dry rock of basaltic

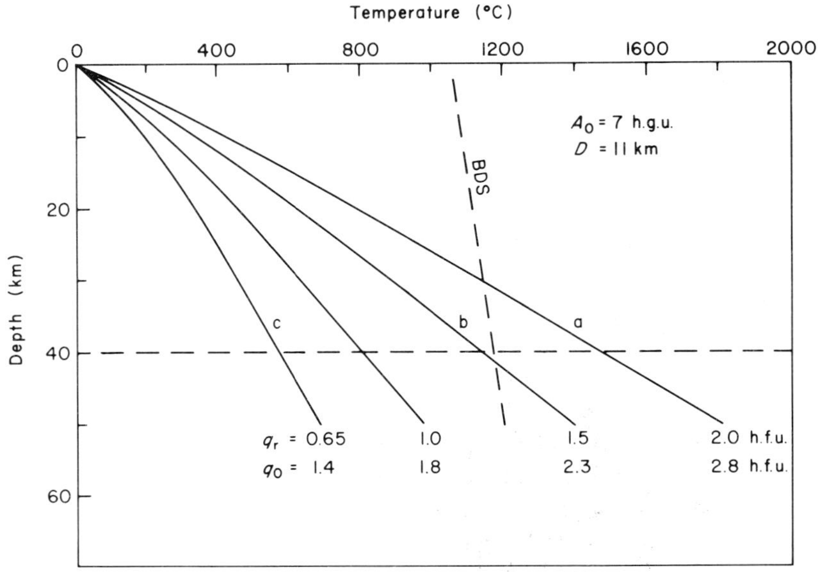

Fig. 21. Theoretical steady-state geotherms (equation 3) for the Eastern Province. Thermal conductivity K assumed constant = 6 h.c.u.; BDS is the basalt dry solidus. Curves a, b and c are compared to the corresponding curves in Figs 17 and 23 (see text).

composition which contradicts seismic evidence for an essentially rigid 40-km crust. Similarly, the high mantle heat flow implied by curve b, Fig. 21 would require steady conductive gradients in the uppermost mantle greatly exceeding the local melting gradient, contradicting seismic evidence for a lithosphere more than 100-km thick. These examples illustrate the difficulty of accounting for high heat flow in regions of moderate crustal radioactivity by invoking high mantle heat flow (q_n), which for the conductive model differs little from q_r. To explain high q_r with a moderate q_n it is necessary to call upon additional sources of heat in the crust; their effects could be transient or stationary. Possibilities include anomalous lower crustal radioactivity, frictional heating, chemical reactions, phase changes, or vertical convective transport within the crust. Our knowledge of conditions at each of the 8 or 10 high-heat-flow sites in eastern Australia is quite inadequate for a selection from among these possibilities. However, the Wellman–McDougall relation, represented by the straight line of Fig. 20, suggests a simple speculation, namely that the complex eastern thermal province would be part of the central thermal province ($q_r \sim 0.65$ h.f.u., $D \sim 11$ km) were it not for the convective disturbance caused by Cenozoic magmatism and the associated motion of framework rocks.

This simple view is supported by the observation that in the northern part of the province south of latitude 20°, where volcanism terminated 15–30 million years ago, and tectonic stress is probably compressional, the heat flow and regional radioactivity seem consistent with values measured in the Central Province. On the other hand, our highest reduced heat flows (CAS and STW) occur near the youngest volcanic rocks, the "newer volcanics" in Victoria where tectonic extension may still be in progress. Figure 20 suggests that volcanism terminates within a few million years of the central volcano event at latitudes where such volcanos occurred. If, as suggested by Wellman and McDougall, volcanism probably occurs only in regions of tension, we might generalize the information in Fig. 20, and suppose that the straight line represents the start of a transition from extensional tectonics (on the left) to compressional tectonics (on the right). Figure 22(a) shows that the high and variable heat flows occur south of latitude 33°S where the hypothetical transition started about 12 million years ago. (The lower heat flows from the Otway Basin are not included in Fig. 22 because of conflicting evidence from resistivity studies and likely systematic hydrologic effects discussed above.) By this view, Fig. 22(b) suggests that the heat-flow anomalies associated with magmatism and extension might decay in about 10 million years.

But why should high and variable heat flow on a regional scale be associated with episodes of local volcanic activity? The simplest answer seems to be that the extensional stress, presumed to accompany the magmatic episodes, facilitates crustal intrusion over an area considerably larger than that occupied by volcanic rocks; where there was no extrusive association, any extension that might have occurred in the uppermost layers was accommodated by normal faulting. We shall assume that this is so, and that with the onset of tectonic compression at each latitude, convection stopped and the crust cooled conductively.

It has been pointed out that the relations among steady-state heat flow, extension rate, and crustal evolution depends upon the mechanical mode of extension and magmatism. Any combination of the three modes of Fig. 9 will result in enhancement of surface heat flow (relative to mantle heat flow) from both magmatic heat and solid-state convection, and the steady-state heat-flow anomaly will be a function of extensional strain rate (s). For a contribution by the fourth mode, simple sill injection, there will be no associated extensional strain and part of the magmatic contribution will be consumed by downward solid-state convection. Thus even for steady-state one-dimensional conditions, we have a wide range of options for relating high heat flow and magmatism, each with specific implications for extensional strain and changes in composition and thickness of crustal layers. For analytical convenience, we select examples from the modes of Fig. 9;

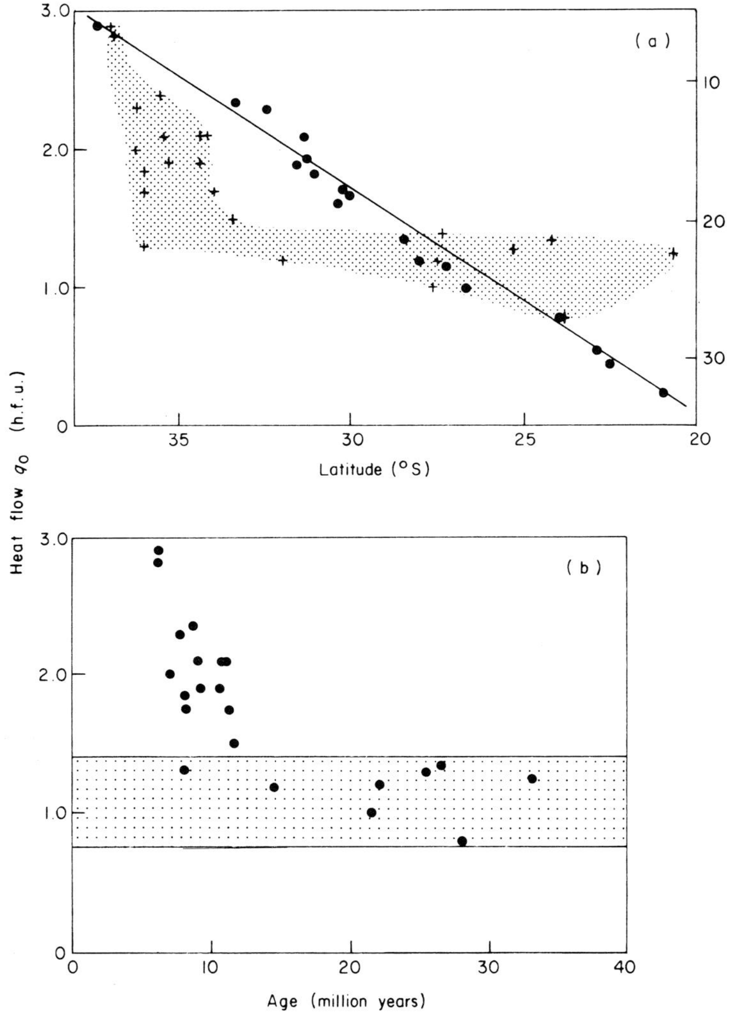

Fig. 22. (a) Ages of central volcano provinces (dots) and heat flow in eastern Australia (plus signs) as a function of latitude. Shaded area encloses all heat-flow values. Solid line defines latitude–age relation for central volcano provinces. (b) Heat flow in the Eastern Province as a function of age of central volcano provinces as inferred from the age-latitude relation. Shaded area indicates range of "normal" heat flow in Mesozoic and older terranes (cf. Fig. 5).

the discussion can be generalized somewhat with the aid of Figs 10 and 11.

Figure 23 illustrates how we might reconcile the high heat flow with a general subsolidus condition in the crust with these simple models. In model

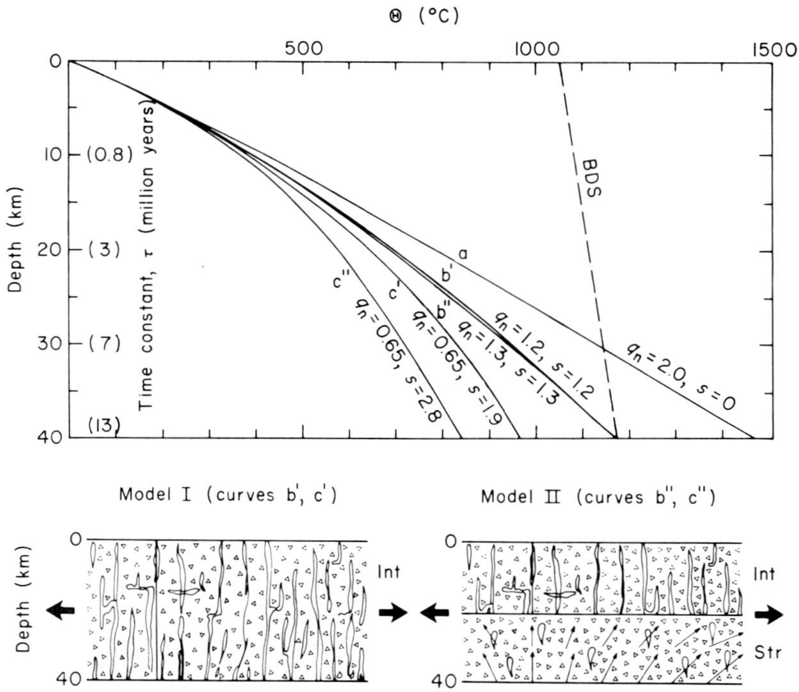

Fig. 23. Theoretical steady-state crustal geotherms for a reduced heat flow q_r of 2.0.h.f.u., radioactive heat production A_0 of 7 h.g.u., thermal conductivity K, of 6 h.c.u. and $D = 11$ km. Curve a is the steady-state conductive geotherm calculated from equation 3 (curve a, Fig. 21), curves b′ and c′ refer to model I and curves b″ and c″ to model II. Values of mantle heat flow q_n (h.f.u.) and horizontal strain rate s (per cent per million years) are shown on the appropriate curves.

I extension of the crust at uniform strain rate s is accommodated (without crustal thinning) by intermittent distributed dike-like intrusions as suggested by Wellman and McDougall (1974). Heat is convected through the crust by the rising magma and an average thermal stationary state obtains. Convective transport in the crust enhances the surface heat flow in the amount $q_r - q_n$; for the static conductive case ($s = 0$) illustrated by curve a, Figs 21 and 23, $q_r \simeq q_n$. Curves b′ and c′, Fig. 23, illustrate how conduction-theory estimates of the geotherm (curve a) can be modified by small rates of crustal extension in the dike-intrusion mode (model I, Fig. 23). For steady extension at a little more than 1 percent per million years, θ_n is revised from 1470° to an average value of 1170° (curve b′, Fig. 23) and the crustal geotherm would lie below BDS. For extension at rates less than 2 percent per million years (~6×10^{-16} s^{-1}, or 2 mm y^{-1} for a band 100 km wide (q_n

would be only 0.65 h.f.u. (curve c′, Fig. 23), the value characteristic of the rest of the continent, although the average surface heat flow would still be 2.8 h.f.u.

From the transient solution for a conducting slab with uniformly distributed sources (Carslaw and Jaeger, 1959, p. 130, equation (7)), it can be estimated that the build-up or decay of the steady-state condition for model I would be almost complete in about one crustal time constant if Θ_n were held constant, or in about three time constants if q_n were held constant. (More detailed considerations of build-up and decay are not warranted here as the actual time constant for the system can vary over a broad range depending upon unknown details of heat and mass transport in the upper mantle.) The time constant for a 40-km crust is about 13 million years and for the 20-km upper crust, it is about 3 million years (Fig. 23). Thus if the entire crust extended according to model I while the mantle flux, q_n, remained at 0.65 h.f.u. it might take about 40 million years for the Central Province geotherm c (Figs 17 and 21) to approach the corresponding Eastern Province geotherm c′ (Fig. 23), and a comparable period for restoration of equilibrium after cessation of extension and magmatism. If equilibrium were restored with Θ_n at BDS (1170 °C), curve b′ (Fig. 23) might approach curve b (Fig. 21) after 10 million years or so, but the surface heat flow would still be 2.3 h.f.u. These results are inconsistent with our interpretation of Fig. 22 which suggests that where high heat flow is associated with magmatism and extension, it decays to normal values in times of perhaps 10 million years or less.

Although other options are possible (e.g. short-lived episodes of deep crustal intrusion at rapid rates), these transient considerations suggest a refinement (Model II, Fig. 23) similar to that shown in the inset of Fig. 12, in which the sources associated with intrusion are confined primarily to the upper crust. Magma from a subcrustal source rises through the lower crust at temperatures near its solidus; there the magma gives up a small fraction of its sensible heat but it ultimately comes to rest in dike-like bodies in the upper crust, thereby accommodating volume extension there. Solid-state flow ("stretching") accommodates extension in the lower crust causing it to thin in the process (see Lachenbruch and Sass, 1978). Average geotherms for $\Theta_n = 1170$ °C (BDS) and for $q_n = 0.65$ h.f.u. are shown respectively by curves b″ and c″, Fig. 23. In this model, 90 per cent of the convective contribution $(q_r - q_n)$ to surface heat flow originates in the 20-km upper crust; after tectonic activity ceased, this part of the anomaly would decay rapidly, (governed by the 3 million year upper crustal time constant), consistent with our interpretation of the data in Fig. 22.

For the case represented by c″, Fig. 23, the non-conductive contribution $(\Delta q = q_r - q_n)$ is 1.35 h.f.u., and only about 0.1 h.f.u. is supplied by solid-state stretching of the 20-km lower crust (see Fig. 10, $R = 20$ km, $s = 2.8$ percent per million years, $q_a = 0.65$ h.f.u.). Steady dike intrusion of the upper crust, supplied with basalt at the rate W of about 0.55 km per million years (2.8 percent × 20 km) contribute the remaining 1.25 h.f.u., and according to Fig. 11 this implies that the magma loses its heat at an average ambient temperature of about 400 °C. Distributed sill injection of the upper crustal region at similar ambient temperatures would generate a similar geotherm and heat-flow anomaly without crustal extension, but a somewhat greater rate of basalt supply would be required (perhaps as much as 50–100 percent) depending upon details of solid-state convection associated with downward displacement in the thickening crust.

The foregoing analysis suggests that in the Eastern Province, even where the reduced heat flow is (or was) 2 h.f.u., the mantle heat flow might not have been much greater than the value characteristic of the rest of the continent, for otherwise the heat flow would decay too slowly after the cessation of extension and magmatism. The model suggests that intrusion is confined to the upper crust and that corresponding increases in upper crustal volume at rates in the neighborhood of 3 percent per million years, persisting for several million years, may be required to generate the higher heat flows; events of shorter duration would require even greater rates of magma supply. (Note that the geotherm c″, Fig. 23, which represents a heat flow of 2.8 in the Eastern Province is similar to the geotherm for the same heat flow (curve a, Fig. 17) in the Central Province; both have the same mantle flux, but upper crustal magmatic sources in one, proxy for upper crustal radioactivity in the other.) The upper crustal volume increase, which would be accompanied by basalt supply to the crust at rates of 1/2 km per million years or more, could be accommodated by any combination of crustal extension and thickening (the latter requiring larger rates of intrusion). At least a small component of extension is likely, however, to accommodate formation of dikes associated with volcanism and to facilitate crustal intrusion.

The eastern volcanic highlands province (Fig. 19) is about 3000 km long and 300 km wide. If a reduced heat flow of 2.h.f.u. were a typical regional value throughout its 60 million year volcanic history, then extension at the rate of 3 percent per million years would increase its area several fold, or sill intrusion at more than 0.5 km per million years without extension would double the crustal thickness; these are unacceptable alternatives. At the other extreme, if the only upper crustal intrusion had been in dikes associated with the volcanic provinces (about 40 000 km^3 according to the estimate of Wellman and McDougall, 1974) and the volume had been distributed throughout the 20-km upper crust, the area of the province would have increased only 0.2 percent during the 60 million year volcanic history. This would correspond to a region 10^3 km^2 extending at the rate of 3 percent per million years at any time; elevated heat flow would occur only localty at the volcanic centers. There seems to be room for reasonable intermediate alternatives. For example, a region on the order of 10^4 km^2 extending at 3 percent per million years at any time, would lead to a 2 percent increase in the area of the province over the volcanic history and the region of elevated heat flow would be an order of magnitude larger than the region of volcanic rocks. A similar heat flow, crustal geotherm, and total areal strain (2 percent), would result from a 10^5 km^2 region extending at 0.3 percent per million years at any time with combined upper crustal sill and dike intrusion at the rate of perhaps 0.8 km per million years; the upper crust would thicken by about 7 km in 10 million years and the region of elevated heat flow would be two orders of magnitude larger than the area of volcanic centers. Conditions such as these could each generate heat flows comparable to those observed at STW and CAS in western Victoria today. If they persisted for 10 million years, about one fourth of the upper crust would locally become basaltic; much longer durations of intrusion at such rates might lead to an unacceptably mafic upper crust. Reasonable limits to Cenozoic crustal extension and to permissible changes in composition and thickness of the upper and lower crust (best established by specialists in these matters) are needed for verification of thermal models of this kind.

The sparse thermal data seem consistent with the speculation that major differences between the Eastern thermal province and the Central Shield Province might be the effects of Cenozoic magmatism and tectonic extension, essentially as they have been interpreted by Wellman and McDougall (1974). By this view, in early Cenozoic time (following the opening of the Tasman Sea) eastern Australia was undergoing tectonic extension, upper crustal intrusion, and more locally, basaltic volcanism. Starting in the north (~latitude 20°S) in Oligocene time the tectonic mode changed to compression and the boundary between the modes swept southward relative to Australia over a point fixed in the asthenosphere, as Australia drifted northward relative to Antarctica. As the compression swept southward, it might have been instrumental in generating the succession of central volcanos, pumping residual silicic melt from the crust to provide the associated felsic magma; crustal extension and intrusion stopped, and those portions of the crust that were intruded began to cool conductively. The rapid decay of the heat-flow anomaly, the improbability of massive melting in the thick 40-km crust, and the relatively thick lithosphere suggest that the mantle heat flow during these events might not have been substantially greater than that believed to be characteristic of the rest of the continent today. Thus, from the thermal point of view, there seems to be no need to invoke a mantle "hot spot" but simply a migrating stress field and a large magma supply that could rise where lithosphere extension prevailed. Average crustal geotherms during these events might not have been much higher than those that we have suggested for similar surface heat flows in the Central Shield Province (Fig. 17).

5. Summary

With several thousand data available, the heat flowing from much of the continental crust can be discussed in terms of useful generalizations which impose significant constraints on models of thermal, tectonic, and magmatic processes in the crust. The broadest of these generalizations is the inverse correlation between the average heat flow in a geologic province and the age of its crystalline basement. The relation suggests that no anomalous sources of heat existed beneath the region since termination of the thermotectonic event responsible for the formation of the crystalline basement. In the many regions throughout the world where this generalization is violated, anomalous radioactivity or subsequent thermotectonic events (generally involving convective transfer in the lithosphere associated with subduction, extension, or magmatism) are indicated.

Much of the observed variability in heat flow within large continental regions can be accounted for by variations in crustal radioactivity. For many such regions, a linear relation between heat flow and near-surface radiogenic heat production has been observed for granitic and some types of metamorphic rocks. Heat-flow provinces have been defined on the basis of the slope and intercept of the heat-flow vs. heat-production line (equation (1)). The slope D can be interpreted as the average thickness of the plutons in a given province or as a measure of the degree of upward differentiation of radioelements (low D reflecting a higher degree of differentiation), depending on the model used. The intercept or reduced heat flow q_r usually is interpreted as the heat flowing from some depth which can range from D to three or four times D, again depending on the model used. As lower crustal radioactivity is not expected to be large, for many purposes q_r can be identified with q_m, the heat flowing from the mantle. Values of q_r for various provinces range from about 0.4 to 1.5.h.f.u. (17–60 mW m^{-2}) and for D, from 4.5 to 15 km (Table 1). For a given heat-flow province, the empirically determined values of q_r and D can be combined with suitable values for thermal conductivity to construct crustal geotherms over the range of radioactivity found in the province. These steady-state, conductive models predict unacceptably high temperatures in the lower crust for regions of very high heat flow ($\gtrsim$2.5 h.f.u., $\gtrsim$100 mW m^{-2}) and moderate crustal radioactivity. In these regions, non-conductive processes must play (or have played) some role, and thus must be incorporated into realistic thermal models.

Convection in the crust associated with the motion of solid rock, magma and/or water is inevitable in regions of tectonic extension and volcanism. In such regions, the convective contribution to crustal heat flow (positive or negative) can be several times larger than the radiogenic contribution. Since convection is inherently a more efficient mode of heat transport than conduction, the net effect is for the average temperature in the lithosphere to be lower than that predicted by conduction models for a given surface heat loss. A large number of possible convective models can be constructed to reconcile the observed heat flow and radioactivity with geological and geophysical observations in regions of Cenozoic basaltic volcanism like the Basin and Range province of the western United States or eastern Australia. The range of possible models can be narrowed by a detailed study of such quantities as rate of extension, volumes of extrusive rocks, tectonic history, and the structure and composition of the lithosphere.

Constrained by fewer than 100 heat-flow data, the gross features of the thermal regime of the Australian continental crust admit the following simple interpretation. The heat flow from the mantle (or lower crust) is uniform throughout the continent (~0.65 h.f.u.) and variations in surface heat flow are caused, in the western two thirds, by variations in the distribution of radioactivity, and in the eastern one third, by the additional effects of magmatic heat sources in the upper crust. The western two thirds of the continent, a Precambrian craton, contains two thermal provinces distinguished by their differing chracteristic depth, D, of radioactive heat production. In the western third, the heat production is concentrated closer to the surface ($D \simeq 4.5$ km) than in the central third ($D \simeq 11$ km); for the same surface radioactivity, heat flow and crustal temperatures are therefore lower in the west.

The eastern third of the continent is a Paleozoic terrane. Although it is more stable than continental regions near plate margins (like the western US), it has been the site of widespread Cenozoic magmatic and tectonic activity. High heat flows have been measured in the southern part of the province (the highest close to regions of very recent volcanism and probably tectonic extension). Crustal radioactivity is not nearly high enough to account for these high surface heat flows, nor is it likely that they are the result of high mantle heat flow, because massive melting is precluded by seismic evidence for a 40-km crust and a thick lithosphere. The distribution of heat flow, heat production and present-day crustal stress, and the systematic southward migration of the termination of basaltic volcanism, suggest a speculative model attributing the high heat flows to upper crustal intrusion, more widespread than the associated volcanism. According to the model, intrusion ceased at each latitude as the transition from crustal tension to compression swept southward and subsequently the crust cooled conductively.

There are, of course, other alternatives for accommodating the heat-flow data with models of the thermal regime of the Australian continental crust, particularly in eastern Australia. However, the models presented here are among the simplest consistent with the information presently available.

Acknowledgements

We thank our colleague, Robert Munroe, for his help in compilation of the Australian heat-flow map. David Chapman, James Cull, William Diment, Marvin Lanphere, Lawrence Lawver, Ted Lilley, Ian McDougall and Peter Wellman offered valuable comments and criticisms on the manuscript.

References

Aziz-Ur-Rahman and McDougall, I. (1972). Potassium–argon ages on the Newer volcanics of Victoria. *Proc. R. Soc. Victoria* **85**, 61–69.

Beck, A. E. (1956). The measurement of the flow of heat through the crust and through Rocks. Ph.D. Thesis, Australian National University, Canberra.

Bennett, D. J. and Lilley, F. E. M. (1974). Electrical conductivity structure in the south-east Australian region. *Geophys. J. R. astronom. Soc.* **37**, 191–206.

Birch, F. (1950). Flow of heat in the Front Range, Colorado. *Bull. geol. Soc. Am.* **61**, 567–630.

Birch, F. (1954). The present state of geothermal investigations. *Geophysics* **19**, 649–659.

Birch, F., Roy, R. F. and Decker, E. R. (1968). Heat flow and thermal history in New England and New York. In *Studies of Appalachian Geology: Northern and Maritime* (E-an Zen, W. S. White and J. B. Hadley, eds), Interscience, New York, pp. 437–451.

Blackwell, D. D. (1969). Heat-flow determinations in the northwestern United States. *J. geophys. Res.* **74**, 992–1007.

Bodvarsson, G. (1954). Terrestrial heat balance in Iceland. *Timarit Verkfraedingafelags Islands*, 69–76.

Brown, D. A., Campbell, K. S. W. and Crook, K. A. W. (1968). *The Geological Evolution of Australia and New Zealand*, Pergamon Press, London.

Bullard, E. C. (1954). The interior of the earth. In *The Earth as a Planet* (G. P. Kuiper, ed.), University of Chicago Press, Chicago, pp. 57–137.

Bunker, C. M. Bush, C. A. Munroe, R. J. and Sass, J. H. (1975). Abundances of uranium, thorium, and potassium for some Australian crystalline rocks. Open-File Rept 75–393, U.S. Geological Survey, Menlo Park, California.

Carslaw, H. S. and Jaeger, J. C. (1959). *Conduction of Heat in Solids*, Oxford University Press, London and New York.

Cermak, V. and Hurtig, E. (1977). Preliminary Heat Flow Map of Europe, 1:5,000,000. Geophysics Institute, Prague, and Central Earth Physics Institute, Potsdam.

Chapman, D. S. and Pollack, H. N. (1975a). Global heat flow: a new look. *Earth planet. Sci. Lett.* **28**, 23–32.

Chapman, D. S. and Pollack, H. N. (1975*b*). Heat flow and incipient rifting in central Africa. *Nature, Lond.* **256**, 28–30.

Chapman, D. S. and Pollack, H. N. (1977). Heat flow and heat production in Zambia: Evidence for lithospheric thinning in central Africa. *Tectonophysics* **41**, 79–100.

Cleary, J. (1967). P times to Australian stations from nuclear explosions. *Bull. seismol. Soc. Am.* **57**, 773–781.

Cleary, J. (1973). Australian crustal structure. *Tectonophysics* **20**, 241–248.

Cleary, J. R. and Simpson, D. W. (1971). Seismotectonics of the Australian continent. *Nature, Lond.* **230**, 239–241.

Crough, S. T. and Thompson, G. A. (1976). Thermal model of continental lithosphere. *J. geophys. Res.* **81**, 4857–4862.

Diment, W. H., Urban, T. C. and Revetta, F. A. (1972). Some geophysical anomalies in the eastern United States. In *The Nature of the Solid Earth* (E. C. Robertson, ed.), McGraw-Hill, New York, pp. 544–572.

Diment, W. H., Urban, T. C., Sass, J. H., Marshall, B. V., Munroe, R. J. and Lachenbruch, A. H. (1975). Temperatures and heat contents based on conductive transport of heat. *U.S. geol. Survey Circ.* **726**, 84–103.

Fitch, T. J., Worthington, M. H. and Everingham, I. B. (1973). Mechanisms of Australian earthquakes and contemporary stress in the Indian Ocean Plate. *Earth planet. Sci. Lett.* **18**, 345–356.

Goncz, J. H. and Cleary, J. R. (1976). Variations in the structure of the upper mantle beneath Australian from Rayleigh wave observations. *Geophys. J. R. astronom. Soc.* **44**, 507–516.

Gough, D. I., McElhinny, M. W. and Lilley, F. E. M. (1974). A magnetometer array study in southern Australia. *Geophys. J. R. astronom. Soc.* **36**, 345–362.

Hamza, V. M. (1976). Possible extension of oceanic heat flow age relation to continental regions and the thermal structure of continental margins. *An. Acad. Sci. Brazil* **48**, 121–129.

Hamza, V. M. and Verma, R. K. (1969). The relationship of heat flow with age of basement rocks. *Bull. Volcanol.* **33**, 123–152.

Hayes, D. E. and Ringis, J. (1973). Sea floor spreading in the Tasman Sea. *Nature, Lond.* **243**, 454–485.

Hills, E. S. (1959). *The Physiography of Victoria*, Whitcombe and Tombs Pty. Ltd., Melbourne.

Howard, L. E. (1965). Heat flow in Australia. Ph.D. Thesis, Australian National University, Canberra.

Howard, L. E. and Sass, J. H. (1964). Terrestrial heat flow in Australia. *J. geophys. Res.* **72**, 527–539.

Hyndman, R. D., Lambert, I. B., Heier, K. S., Jaeger, J. C. and Ringwood, A. E. (1968). Heat flow and surface radioactivity measurements in the Precambrian Shield of Western Australia. *Phys. Earth planet. Interiors* **1**, 129–135.

Irving, A. J. and Green, D. H. (1976). Geochemistry and petrogenesis of the Newer basalts of Victoria and South Australia. *J. geol. Soc. Australia* **23**, pt. 1, 45–66.

Jaeger, J. C. (1965). Application of the theory of heat conduction to geothermal measurements. In *Terrestrial Heat Flow* (W. H. K. Lee, ed.), Geophysics Monograph Series, vol. 8, American Geophysical Union, Washington, D.C., pp. 7–23.

Jaeger, J. C. (1970). Heat flow and radioactivity in Australia. *Earth planet. Sci. Lett.* **8**, 285–292.

Jaeger, J. C. and Thyer, R. F. (1960). Report on progress in geophysics, geophysics in Australia. *Geophys. J. R. Astronom. Soc.* **3**, 450–461.

Jessop, A. M., Hobart, M. A. and Sclater, J. G. (1976). *The World Heat Flow Data Collection—1975*, Geothermal Service of Canada, Ottawa.

Jones, J. G. (1971). Australia's Caenozoic drift. *Nature, Lond.* **230**, 237–239.

Kraskovski, S. A. (1961). On the thermal field in old shields. *Izv. Akad. Nauk SSSR Ser. Geofiz.* **3**, 247–250.

Kutas, R. I. (1977). Investigation of heat flow in the territory of the Ukraine. *Tectonophysics* **41**, 139–145.

Lachenbruch, A. H. (1968). Preliminary geothermal model of the Sierra Nevada. *J. geophys. Res.* **73**, 6977–6989.

Lachenbruch, A. H. (1970). Crustal temperature and heat production: Implications of the linear heat-flow relation. *J. geophys. Res.* **75**, 3291–3300.

Lachenbruch, A. H. (1978). Heat flow in the Basin and Range province and thermal effects of tectonic extension. *Pure appl. Geophys.* **117**, 34–50.

Lachenbruch, A. H. and Sass, J. H. (1977). Heat flow in the United States and the thermal regime of the crust. In *The Earth's Crust* (J. G. Heacock, ed.), Geophysics Monograph Series, vol. 20, American Geophysical Union, Washington, D.C., pp. 626–675.

Lachenbruch, A. H. and Sass, J. H. (1978). Models of an extending lithosphere and heat flow in the Basin and Range province. In *Cenozoic Tectonics and Regional Geophysics of the Western Cordillera* (R. B. Smith and G. P. Eaton, eds), Geological Society of America Memoir 152. Geological Society of America, Boulder, Colorado, pp. 209–250.

Lambert, I. B. and Heier, K. S. (1967). The vertical distribution of uranium, thorium, and potassium in the continental crust. *Geochim. cosmochim. acta* **31**, 377–390.

Lawrence, C. R. (1976). Groundwater. In *Geology of Victoria* (J. G. Douglas and J. A. Ferguson, eds), Geological Society of Australia Special Publication 5, pp. 411–417.

Lilley, F. E. M. (1975). Electrical conductivity anomalies and continental seismicity in Australia. *Nature, Lond.* **257**, 381–382.

Lilley, F. E. M. (1976). A magnetometer array study across southern Victoria and the Bass Strait area, Australia. *Geophys. J. R. astronom. Soc.* **46**, 165–184.

Lilley, F. E. M., Sloane, M. N. and Sass, J. H. (1977). A compilation of Australian heat-flow measurements. *J. geol. Soc. Australia*, **24**, 439–445.

Lister, C. R. B. (1972). On the thermal balance of a mid-ocean ridge. *Geophys. J. R. astronom. Soc.* **26**, 515–535.

Lister, C. R. B. (1974). On the penetration of water into hot rock. *Geophys. J. R. astronom. Soc.* **39**, 465–509.

McDougall, I. Alsopp, H. L. and Chamalaun, F. H. (1966). Isotopic dating of the newer volcanics of Victoria, Australia, and geomagmatic polarity epochs. *J. geophys. Res.* **71**, 6107–6118.

Mereu, R. F., Simpson, D. W. and King, D. W. (1974). *Q* and its effect on the observation of upper mantle travel-time branches. *Earth planet. Sci. Lett.* **21**, 439–447.

Mills, J. M. and Fitch, T. J. (1977). Thrust faulting and crust-upper mantle structure in east Australia. *Geophys. J. R. astronom. Soc.* **48**, 351–384.

Muirhead, K. J., Cleary, J. R. and Finlayson, D. M. (1977). A long-range seismic profile in south-eastern Australia. *Geophys. J. R. astronom. Soc.* **48**, 509–519.

Munroe, R. J., Sass, J. H., Milburn, G. T., Jaeger, J. C. and Tammemagi, H. Y. (1975). Basic data for some recent Australian heat-flow measurements. Open-File Rept 75–567, U.S. Geological Survey, Menlo Park, California.

Newstead, G. and Beck, A. (1953). Borehole temperature measuring equipment and the geothermal flux in Tasmania. *Australian J. Phys.* **6**, 480–489.

Oversby, B. (1971). Paleozoic plate tectonics in the Southern Tasman Geosyncline. *Nature, Lond.* **234**, 45–47.

Oxburgh, E. R. and Turcotte, D. L. (1971). Origin of paired metamorphic belts and crustal dilation in island arc regions. *J. geophys. Res.* **76**, 1315–1327.

Palmason, G. (1973). Kinematics and heat flow in a volcanic rift zone, with application to Iceland. *Geophys. J. R. astronom. Soc.* **33**, 451–481.

Pollack, H. N. and Chapman, D. S. (1977*a*). The flow of heat from the earth's interior. *Scient. Am.* **237** (2), 60–76.

Pollack, H. N. and Chapman, D. S. (1977*b*). On the regional variation of heat flow, geotherms and lithospheric thickness. *Tectonophysics* **38**, 279–296.

Pollack, H. N. and Chapman, D. S. (1977*c*). Mantle heat flow. *Earth planet. Sci. Lett.* **34**, 174–184.

Polyak, B. G. and Smirnov, Y. B. (1966). Heat flow on continents. *Dokl. Akad. Nauk SSSR* **168**, 26–29.

Polyak, B. G. and Smirnov, Y. B. (1968). Relationship between terrestrial heat flow and the tectonics of continents. *Geotectonics* **4**, 205–213.

Rao, R. U. M. and Jessop, A. M. (1975). A comparison of the thermal characters of shields. *Can. J. Earth Sci.* **12**, 347–360.

Rao, R. U. M., Rao, G. V. and Narain, H. (1976). Radioactive heat generation and heat flow in the Indian Shield. *Earth planet. Sci. Lett.* **30**, 57–64.

Roy, R. F., Blackwell, D. D. and Birch, F. (1968). Heat generation of plutonic rocks and continental heat-flow provinces. *Earth planet. Sci. Lett.* **5**, 1–12.

Roy, R. F., Blackwell, D. D. and Decker, E. R. (1972). Continental heat flow. In *The Nature of the Solid Earth* (E. C. Robertson, ed.), McGraw-Hill, New York, pp. 506–544.

Sass, J. H., Jaeger, J. C. and Munroe, R. J. (1976). Heat flow and near-surface radioactivity in the Australian continental crust. Open-File Rept 76–250, U.S. Geological Survey, Menlo Park, California.

Sass, J. H., Lachenbruch, A. H., Munroe, R. J., Greene, G. W. and Moses, T. H., Jr. (1971). Heat flow in the western United States. *J. geophys. Res.* **76**, 6376–6413.

Sclatter, J. G. and Crowe, J. (1976). On the reliability of oceanic heat flow averages. *J. geophys. Res.* **81**, 2997–3006.

Sclater, J. G. and Francheteau, J. (1970). The implications of terrestrial heat flow observations on current tectonic and geochemical models of the crust and upper mantle of the earth. *Geophys. J. R. astronom. Soc.* **20**, 509–542.

Solomon, M. (1962). The tectonic history of Tasmania. *J. geol. Soc. Australia* **9**, pt. 2, 311–339.

Stewart, I. C. F. and Mount, T. J. (1972). Earthquake mechanisms in South Australia in relation to plate tectonics. *J. geol. Soc. Australia* **19**, pt. 1, 41–52.

Swanberg, C. A., Chessman, M. D., Simmons, G., Smithson, S. B., Gronlie, G. and Heier, K. S. (1974). Heat flow–heat generation studies in Norway. *Tectonophysics* **23**, 31–48.

Tammemagi, H. Y. and Lilley, F. E. M. (1973). A magnetotelluric traverse in southern Australia. *Geophys. J. R. astronom. Soc.* **31**, 433–445.
Taplin, R. E. (1976). Geothermal aspects of the Otway Basin. B.A. (Hons.) Thesis, Macquarie University, Sydney.
Vogt, P. R. and Conolly, J. R. (1971). Tasmantid Guyots, the age of the Tasman Basin and motion between the Australian Plate and the mantle. *Bull. geol. Soc. Am.* **82**, 2577–2584.
Watanabe, T., Langseth, M. G. and Anderson, R. N. (1977). Heat flow in back-arc basins of the western Pacific. In *Island Arcs, Deep Sea Trenches and Back-Arc Basins* (M. Talwani and W. C. Pitman, III, eds), Maurice Ewing Series 1. American Geophysical Union, Washington, D.C., pp. 137–161.
Weissel, J. K. and Hayes, D. E. (1971). Asymmetric sea floor spreading south of Australia. *Nature, Lond.* **231**, 518–521.
Weissel, J. K. and Hayes, D. E. (1977). Evolution of the Tasman Sea reappraised. *Earth planet. Sci. Lett.* **36**, 77–84.
Wellman, P. and McDougall, I. (1974). Cainozoic igneous activity in eastern Australia. *Tectonophysics* **23**, 49–65.
Wopfner, H., Kenby, P. R. and Thornton, R. C. N. (1971). Hydrocarbon occurrences and potential of the Otway Basin. In *The Otway Basin of Southeastern Australia* (H. Wopfner and J. G. Douglas, eds), Geological Surveys, South Australia and Victoria, Special Bulletin, pp. 385–435.
Wronski, E. B. (1977). Two heat flow values for Tasmania. *Geophys. J. R. astronom. Soc.* **48**, 131–133.
Wyllie, P. J. (1971). Experimental limits for melting in the earth's crust and upper mantle. In *The Structure and Physical Properties of the Earth's Crust* (J. G. Heacock, ed.), Geophysics Monograph Series, vol. 14, American Geophysical Union, Washington, D.C., pp. 279–301.

11. Chemical Composition and Evolution of the Continental Crust: the Rare Earth Element Evidence

S. R. TAYLOR

Abstract

The uniformity of rare earth element patterns in sedimentary rocks is used to provide estimates for the composition of the Post-Archean upper crust. This approximates to granodiorite. The island-arc model for continental growth is used to provide a total crustal composition, from which the upper crust is derived by partial melting, leaving a depleted lower crust.

The Archean sedimentary rock patterns reveal a different composition for the upper crust in that era. They closely resemble those of present-day island-arc suites, which the composition of the Archean crust is inferred to resemble (except for higher Ni and Cr contents). The Archean crust is considered to be less fractionated, with much less development of high level granodiorites.

Continental evolution appears to be episodic, rather than uniform, with a major period of production of differentiated upper crust at about 2.5 aeons, long recognized as the Archean–Proterozoic boundary. The Archean crust was probably as thick as present-day crust. Conventional plate tectonic processes do not appear to have operated in the Archean, although volcanic rocks similar in chemistry to modern island-arc suites were produced. There is no evidence for a widespread primordial sialic crust, nor for an anorthositic crust, analogous to the lunar highlands. The Earth has not undergone any significant change (>50 km) in radius throughout geological time.

1. The Continental Crust

The concentration of many elements (e.g. Ba, Th, U, K) in the continental crust is a significant fraction of their abundance in the whole Earth. This fact gives the crust a geochemical significance out of proportion to its volume, and theories for the origin of the crust must account for this efficient elemental fractionation.

The first step is to establish the composition of the presently observable crust. Is it uniform with depth? Has it changed with time? Was the crust formed early in the history of the Earth, or is it the product of continuous or episodic addition from the mantle? Was there an early crust of different composition, either anorthositic, from analogy with the moon, or ultrabasic? If island-arc volcanism is now contributing to continental growth, did it do so in the Archean? What is the relation between the fractionated crustal composition, and current plate tectonic activity?

In this chapter, these questions will be addressed. First, an estimate of the composition of the observable post-Archean upper crust is presented. Models for the origin of this highly fractionated chemistry are discussed, and the composition of the whole crust, and of the lower crust, is inferred. The question of the nature and composition of the Archean crust is next addressed, followed by speculations as to its origin. The overall evolution of the crust through time is assessed, with particular reference to the break between the Archean and later eras. Broader questions relating to episodic or uniform growth, mantle convection, and to the possible occurrence of primitive crusts of differing composition are considered briefly.

The base of the continental crust is taken as the Mohorovičić discontinuity, typically about 40 km, which forms a compositional boundary. The "lithosphere", which forms part of the upper mantle, is not considered, except to note that compositional distinctions may exist between subcontinental and suboceanic mantle. As defined here, the crust constitutes 0.4 per cent of the volume of the Earth. It is divided into the upper crust, defined as that thickness exposed to weathering and erosion, and the lower crust. The division between these two units is made at about 15 km, so that the upper crust constitutes about one third of total crustal volume.

This division has some important geochemical consequences, as will become evident. It is recognized that the lower crust is likely, in detail, to be at least as heterogeneous and diversified as the upper crust, but that broad differences in composition, particularly in the abundances of the heat-producing elements, persist. The chemical evidence summarized in this paper indicates a difference in composition between the upper crust in Archean and post-Archean times.

2. The Post-Archean Crust

A primary purpose of geochemists has been to establish the composition of this crust. Summaries of earlier attempts have been given by Parker (1967) and Taylor (1964*a*,*b*). Three basic approaches have been used. One is to estimate the areal or volumetric proportions of the principal rock types (so far as these can be identified) and to calculate an average composition from these values (e.g. Poldervaart, 1955; Tan and Chi-Lung, 1970; Galdin, 1974; Ronov and Yaroshevski, 1976). A second method is to carry out wide-scale sampling, with analysis of individual or composite samples, representative of thousands of square kilometres. Recent work of this type has been carried out by Eade and Fahrig (1971, 1973), Holland and Lambert (1972), Ronov *et al.* (1974) and Shaw *et al.* (1976). The third approach is to utilize geochemical regularities in element distribution, which enables widespread compositional information to be obtained from a relatively few samples. Examples of the latter approach include the use of compositional data from glacially derived clays (Goldschmidt, 1954) and of rare earth patterns (Taylor, 1964*a*).

All these methods provide useful information. Methods which employ elemental or isotopic parameters possess the additional advantage of tracing changes in composition with time.

2.1. Upper Crustal Rare Earth Element (REE) Patterns

The general uniformity of REE patterns in sedimentary rocks is well established (Haskin and Haskin, 1966; Haskin *et al.* 1968; Nance and Taylor, 1976). This regularity is illustrated in Fig. 1, which shows REE patterns in Post-Archean Australian shales (PAAS), normalized, element by element, to the abundances in chondritic meteorites (Taylor and Gorton, 1977). This diagram shows three differences from the chondritic pattern, which is our best approximation to primordial solar nebula REE patterns. First, there is an overall enrichment of REE; second, a relative enrichment of the light rare earth elements (LREE = La–Sm), and third, a relatively depletion in europium (Eu). Data from many sources shows that the REE patterns in diverse sedimentary rocks are parallel to one another. For the example shown, the La/Yb ratio is 13.6 ± 2 ($La_C/Yb_C = 8.6$ where C = chondrite normalized). The Eu depletion is constant ($Eu/Eu^* = 0.67 \pm 0.05$, where Eu^* is the theoretical value for no Eu anomaly).

The uniformity of these patterns is not dependent on lithology. Shales, subgreywackes, limestones and orthoquartzites all give similar patterns (Nance and Taylor, 1976). These results are not restricted to one continent. The average REE patterns for the Post-Archean Australian shale (PAAS) is identical to that for the North American Shale Composite (NASC) of 240 Palaeozoic and Mesozoic shales (Haskin *et al.*, 1968).

How does this uniformity arise? Igneous rocks show a wide variety of REE patterns. Mixing processes during sedimentation must be remarkably efficient in averaging these differences, and in providing a key to the nature

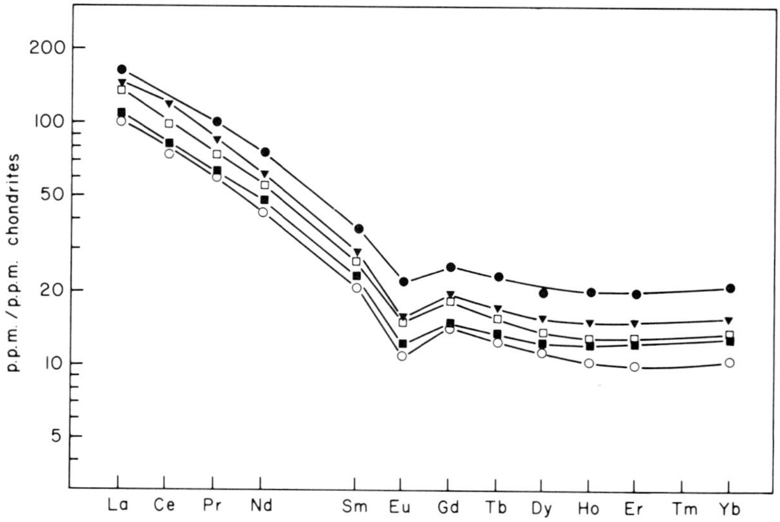

Fig. 1. REE patterns, normalized to chondrites, for averages of Post-Archean Shale samples. Data from Nance and Taylor (1976). Note uniformity of slope of patterns and of Eu depletion. There is an increase in total REE in the younger samples. (Reproduced with permission of Pergamon Press from *Geochemica et cosmochimia acta*, **40**, 1543, 1976) ●, Triassic, PW 4, 5, 7; ▼, Carboniferous, PL 6, 7; □, Silurian, SC 1, 2, 3, 4, 5, 6, 7, 8; ■, Upper Proterozoic (850 million years), AO 7, 8, 9, 10, 11, 12; ○, Mid-Proterozoic (1500 million years), M 11, 12, 14, 15.

of the upper crust. The uniform nature and widespread occurrence of these sedimentary REE patterns is interpreted here as indicative of *the average REE pattern of the upper crust.*

How far back in time does this uniformity hold? Current studies (Nance and Taylor, 1976) show that it extends back for at least 1500 million years (Mt. Isa sediments). In Canada, the oldest sedimentary rocks which show these patterns occur toward the top of the Huronian succession (Gordon Lake Formation), probably older than 2.2 billion years (McLennan *et al.*, 1978). Thus these patterns extend back to Lower Proterozoic time. Work is currently in progress to refine these estimates.

Evidence from studies of the Sm and Nd isotopic systematics (McCulloch and Wasserburg, 1978) show that selective fractionation of Sm and Nd does not occur during the processes producing sedimentary (and metamorphic) rocks. This finding is consistent with the short residence times (<400 years) of REE in seawater (Piper, 1974).

A final comment on sedimentary rock REE patterns is that they are identical to the patterns observed in tektites (Taylor, 1973), a piece of evidence consistent with the terrestrial origin of these glassy objects.

2.2. The Composition of the Upper Crust

Estimates of crustal composition have not always clearly distinguished between the upper crust and total crust (down to the Mohorovičić Discontinuity). The abundance tables of Vinogradov (1962) and Taylor (1964*a*) refer to the whole crust, whereas those of Eade and Fahrig (1971, 1973) and Shaw *et al.* (1976) refer to the upper crust, exposed to weathering and erosion.

An estimate of the composition of the upper crust is given in Table 1. This estimate is based on a synthesis of the data available from many sources, and the several different approaches adopted. There is a consensus with respect to the major element compositions. Those adopted here are from the widescale sampling of Eade and Fahrig (1971). This composition approximates to that of granodiorite. The REE average values from shales provide the relative REE pattern for the upper crust. This pattern and the absolute values are consistent with a granodioritic composition. Data for most of the other trace elements come from Shaw *et al.* (1976). The absolute values for the REE are consistent, for example, with those of Th, U, Ba, Rb, Zr, Hf and Nb, and indicate about comparable levels of enrichment of these incompatible elements.

It is well known that this composition, at least for the heat-producing elements, cannot extend to depths below about 15 km. This strong upward concentration of the incompatible elements tends to bias estimates of near-surface abundances of trace elements toward high values, and revision downward of the estimates has been an historical trend. The composition of the lower crust is unknown, but may be inferred. Before proceeding to this model-dependent step, it is necessary to consider processes through which the composition of the upper crust might be produced.

2.3. Models for Post-Archean Crustal Evolution

The observation that the heat-producing elements are concentrated near the continental surface has led to the notion that the crust is vertically stratified as a consequence of intra-crustal metamorphic or igneous processes (e.g. Fyfe, 1973*a,b*; Heier, 1973). The REE data provide significant information in this context. There is mounting evidence, both from the study of deep-sea basalts (Langmuir *et al.*, 1977) and from Sm–Nd systematics (De Paolo and Wasserburg 1976*a,b*, 1977) that the primitive REE pattern in the Earth is chondritic.

Upper mantle patterns are either chondritic or LREE depleted. There is no evidence that the upper mantle has an inbuilt Eu anomaly. Accordingly, the well-established Eu depletion in the upper crust must originate within

Table 1. Average compositions for the Post-Archean and Archean continental crust. The Post-Archean composition lists values for the upper crust (15 km), total crust (to 40 km) and the lower crust.

	Post-Archean continental crust			Archean crust (%)
	Upper (%)	Total (%)	Lower (%)	
SiO_2	66.0	58.0	54.0	57.4
TiO_2	0.6	0.8	0.9	0.9
Al_2O_3	16.0	18.0	19.0	15.6
FeO†	4.5	7.5	9.0	9.5
MgO	2.3	3.5	4.1	5.2
CaO	3.5	7.5	9.5	7.3
Na_2O	3.8	3.5	3.4	3.1
K_2O	3.3	1.5	0.6	0.9
Σ	100.0	100.3	100.5	99.9

	Post-Archean continental crust			Archean crust (p.p.m.)
	Upper (p.p.m.)	Total (p.p.m.)	Lower (p.p.m.)	
Rb	110	50	20	‡
Pb	15	7	3	‡
Ba	700	350	175	‡
Sr	350	400	425	‡
La	38	19	9.5	‡
Ce	80	38	17	‡
Pr	8.9	4.3	2.0	‡
Nd	32	16	8.0	‡
Sm	5.6	3.7	2.8	‡
Eu	1.1	1.1	1.1	‡
Gd	4.7	3.6	3.1	‡
Tb	0.77	0.64	0.58	‡
Dy	4.4	3.7	3.4	‡
Ho	1.0	0.82	0.73	‡
Er	2.9	2.3	2.0	‡
Tm	0.5	0.4	0.35	‡
Yb	2.8	2.2	1.9	‡
Lu	0.4	0.3	0.25	‡
ΣREE	183	97	54	‡
Eu/Eu*	0.67	1.0	1.1	1.0
Y	27	22	20	‡
Th	10.5	2.5	—	‡
U	2.5	1.0	0.25	‡
Zr	240	100	30	‡

Table 1—*continued*

	Post-Archean continental crust			Archean crust (p.p.m.)
	Upper (p.p.m.)	Total (p.p.m.)	Lower (p.p.m.)	
Hf	5.8	2.2	0.4	‡
Nb	25	11	4	‡
Cr	35	55	65	200
V	60	175	230	‡
Sc	10	30	40	‡
Ni	13	20	23	100
Co	10	25	33	‡
Cu	25	60	78	‡
Zn	52	—	—	—

Data for upper crustal composition adapted from the following sources: Major elements, Eade and Fahrig (1971) Canadian Shield data; REE from Nance and Taylor (1976) average post-Archean Australian sedimentary rocks. Most other trace element data from Shaw *et al.* (1976) values for Canadian Shield rocks, with some values from average granodiorite of south-eastern Australia from Kolbe and Taylor (1966). Data for overall continental crust composition is based on the assumption that the crust is derived from island-arc type volcanism. The overall composition is derived from the estimates by Taylor (1968) for the average andesite, modified by later estimates of Jakeš and White (1971), Jakeš (1973) and Whitford (1975).

The lower crustal composition is the residual remaining following extraction of the upper crustal values from the overall crustal composition.

The major elements for the Archean crust are based on the average of 1629 analyses of Archean volcanic rocks from the Superior Province of the Canadian Shield (Goodwin, 1977)

† All Fe expressed as FeO.

‡ The trace element data are taken as equivalent to those estimated for the total post-Archean crust, with the exception of Ni and Cr which appear to be significantly higher. Data for Ni and Cr from Goodwin (1977), Condie (1976), Ahmad (1977) and Taylor and Hallberg (1977).

the crust. Thus the origin of the upper crustal composition demands at least two stages of chemical fractionation from the mantle.

In the first stage, material is added from the mantle, followed by intracrustal melting to provide the observed upper crustal material of granodioritic composition. Since the production of the Eu depletion requires liquid–crystal equilibria, partial melting rather than metamorphism appears to be the dominant process.

Whether either or both of these stages occurred very early in the history of the Earth or whether the crust has grown, either continuously or episodically with time, has been frequently debated. The evidence from Sr isotopes (Moorbath, 1975, 1977) and Sm–Nd studies (McCulloch and Wasserburg, 1978) appears to be decisively in favour of growth rather than recycling, and this conclusion is adopted here.

Of the possible processes which provide chemically fractionated material from the mantle, only island-arc-type volcanism possesses the twin requirements of adequate volume and appropriate chemistry to provide a realistic source for continental growth (Taylor, 1967). Intra-crustal partial melting can produce, from this composition, granodioritic magmas which invade and form the upper crust, leaving a lower crust relatively depleted in the lithophile elements.

The model advocated here for the production of the upper crustal composition, with its well-developed Eu depletion, does not preclude the occurrence of intrusive bodies of tonalitic (=andesitic) composition which may form during the island-arc stage (cf. Brown, 1977). The production of large bodies of granodioritic composition is, however, attributed to intra-crustal melting, in accord with much experimental evidence (Tuttle and Bowen, 1958). Depletion of high-grade granulites in Th and U has been frequently noted, and ascribed to high-grade metamorphism (Heier, 1973) although the possibility that such depletion is inherited from the source materials has been less frequently examined (Holland and Lambert, 1972).

2.4. Composition of the Whole Crust

At present, this composition is model dependent. Based on the model developed in the previous section, the composition of the whole crust is equivalent to the total product of volcanic and intrusive activity in the island arc environment. The composition given in Table 1 is based on estimates by Taylor (1968) Jakeš and White (1972); Jakeš (1973) and Whitford (1975).

A consensus about the overall composition of volcanic activity in island arcs is emerging. Ewart (1976) has provided the most recent reliable statistical treatment. His estimates for major elements are compared in Table 2 with the values adopted here, and show close agreement.

2.5. Lower Crustal Compositions

Early simplistic views of the nature of the lower crust have given way to the recognition that it is probably as complex in detail as the upper crust (Smithson and Brown, 1977). Thus the question of establishing an "average" composition becomes much more difficult than for the observable upper

Table 2. Comparison of average island arc volcanic rocks (this paper) with values from Ewart (1977, Table 3, p. 424). All values in weight per cent.

	1	2	3	4
SiO_2	58	57.7	56.0	56.8
TiO_2	0.8	0.80	0.80	1.1
Al_2O_3	18	17.2	17.6	17.5
FeO	7.5	7.6	8.6	7.4
MgO	3.5	4.5	4.2	4.5
CaO	7.5	7.8	8.9	7.6
Na_2O	3.5	3.1	2.9	3.7
K_2O	1.5	1.3	1.0	1.4
Σ	100.3	100.0	100.0	100.0

Column 1, continental crust, based on island-arc model (this paper). Column 2, average of 195 "continental" island arcs (Ewart, 1977). Column 3, average of 153 "intra-oceanic" island arcs (Ewart, 1977). Column 4, average of 145 western American orogenic lavas (Ewart, 1977).

crust. The problem will be further exacerbated in detail by localized partial melting or high-grade metamorphism. The random nature of such samples as can be obtained precludes much direct evaluation of the composition.

An estimate of the bulk composition of the lower crust can be obtained from the model presented in this chapter. The composition of the lower crust given in Table 1 is derived from the whole crustal composition, following the extraction of the upper crustal granodioritic composition. A CIPW norm is given in Table 3. It should be emphasized that this is an "average" composition. It contains a substantial plagioclase component. The model that the production of the upper crustal Eu depletion is intra-crustal requires the presence of a stable residual Eu-bearing phase in the lower crust. The reason for the separation of Eu from the other REE is due to the formation

Table 3. CIPW norm of proposed average lower crustal composition.

Quartz	1.5
Orthoclase	3.5
Albite	28
Anorthite	35
Diopside	10
Hypersthene	20
Ilmenite	1.7

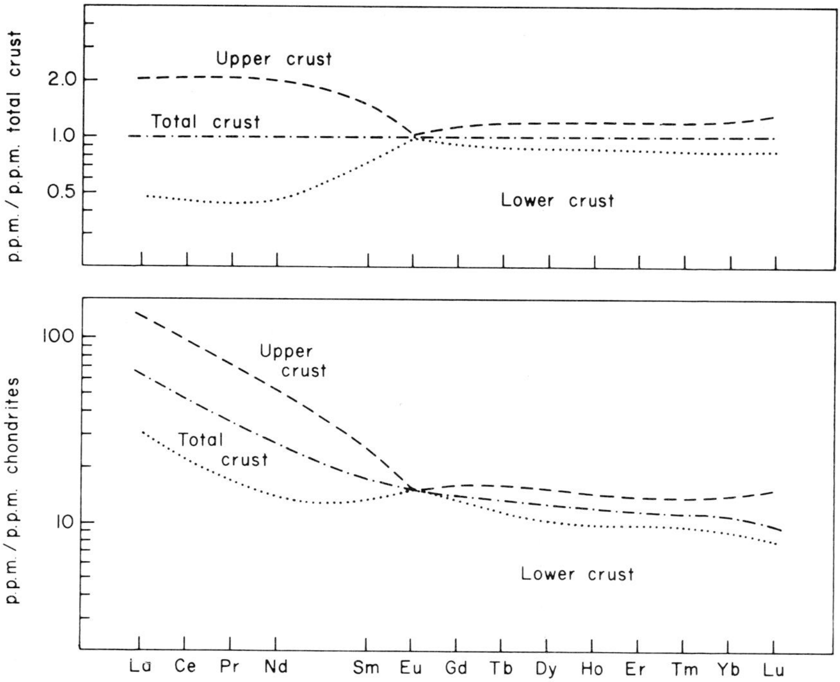

Fig. 2. Proposed REE patterns for the whole continental crust, compared with the REE patterns for the upper crust and the lower crust. The data for the whole crust, from Table 1, are based on the assumption that the continental crust has the composition of average island arc volcanic rocks. The values for the upper crust are assumed to be represented by the average sedimentary rock pattern (data from Table 1). The lower crustal composition is calculated from these data, assuming that it comprises two thirds of the total crust.

of Eu^{2+} ($r = 1.25$ Å), significantly larger than Eu^{3+} ($r = 1.004$ Å, Shannon, 1976). Eu^{2+} is nearly the same size as Sr^{2+} (1.26 Å) and enters similar lattice sites.

A comparison of REE patterns in the whole crust, and in the upper and lower crusts, is given in Fig. 2. Very recently, some evidence in support of a positive europium anomaly in lower crustal rocks has appeared. N. W. Rogers (1977) found such Eu enrichment in granulite xenoliths from Lesotho kimberlites. The granulites have a mineralogy consistent with an origin near the base of the crust ($T = 660–830$ °C; $P = 9–13$ kbar). These basic granulites do not appear to be "fragments of meta-gabbros or cumulates but are most likely derived from the metamorphosed solid residuum from the partial fusion of an amphibolite facies assemblage" (N. W. Rogers, 1977, p. 684).

3. The Archean Crust

The Archean crust contains many features which distinguish it from the Post-Archean crust. Differing tectonic styles, the presence of greenstone belts, komatiites and abundant trondjhemites have all been interpreted as indicating that distinctive processes operated during the formation of the Archean crust. These features are world wide. However, other features such as the presence of granodiorites, indistinguishable in composition from later examples (e.g. Tarney, 1976) raise the possibility that uniformitarianism has indeed been operating. A partial list of recent papers illustrative of these features includes the following (Annhaeusser, 1973; Arth and Hanson, 1975; Glikson and Lambert, 1973; Naqvi *et al.*, 1974; Radhakrishna and Vasudev, 1977; Tarney *et al.* 1972; Wilson *et al.*, 1978; Windley, 1976).

A basic question turns on the thickness of the Archean crust. Wide variations in the estimates exist (Condie, 1973; Fyfe, 1973*a*,*b*; Bridgwater *et al.* 1974; Hargraves, 1976; Windley, 1977). Experimental evidence cited by Newton (1977) appears to be decisive in requiring that pressures equivalent to those now typical of the base of the continental crust were also operative during the Archean. Temperatures in the range 700–900 °C and pressures from 8–13 kbar (Newton, 1977) indicate that the Archean crust was equivalent in thickness to the present continental crust (Windley, 1977).

3.1. Archean sedimentary REE patterns

REE patterns in Archean sedimentary rocks show distinctive differences from those of Post-Archean times. Figure 3 shows REE patterns in sedimentary rocks from Kambalda, Western Australia (Bavinton and Taylor,

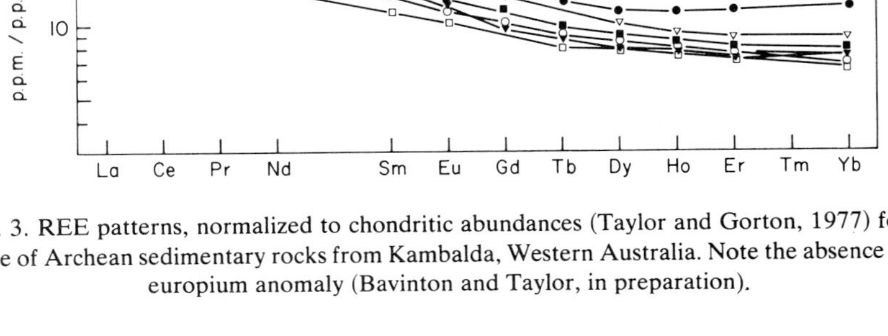

Fig. 3. REE patterns, normalized to chondritic abundances (Taylor and Gorton, 1977) for a suite of Archean sedimentary rocks from Kambalda, Western Australia. Note the absence of a europium anomaly (Bavinton and Taylor, in preparation).

in press) which illustrate the general characteristics. Compared with Post-Archean sedimentary patterns, the Archean patterns show lower overall abundances, and less enrichment of the LREE (La–Sm). The most distinctive feature, however, is that there is no Eu depletion, relative to chondrites.

These features are characteristic of an extensive suite of Archean sedimentary rocks (Fig. 4). This figure shows REE patterns from the Kalgoorlie Black Flag shales, (2600 million years; Nance and Taylor, 1977), the Wind River greywackes (2700 million years) and the Fig Tree shale (3400 million years; Wildeman and Condie, 1973; Wildeman and Haskin, 1973) and from pre-3700 million year sedimentary enclaves in the Amitsoq gneisses (McGregor and Mason, 1977; and unpublished data, this laboratory). The occasional positive Eu anomalies observed appear to be due to local provenance effects such as accumulation of feldspar, than to any more fundamental cause (Nance and Taylor, 1977; Taylor, 1977).

The occurrence of similar REE patterns in Archean sedimentary rocks from Australia, South Africa, Wyoming and West Greenland, spanning a time interval of over 1000 million years, is interpreted here to indicate that sedimentary processes during Archean time were carrying out an effective sampling of the Archean crust, exposed to weathering and erosion. The REE patterns are thus providing evidence for the composition of the Archean crust.

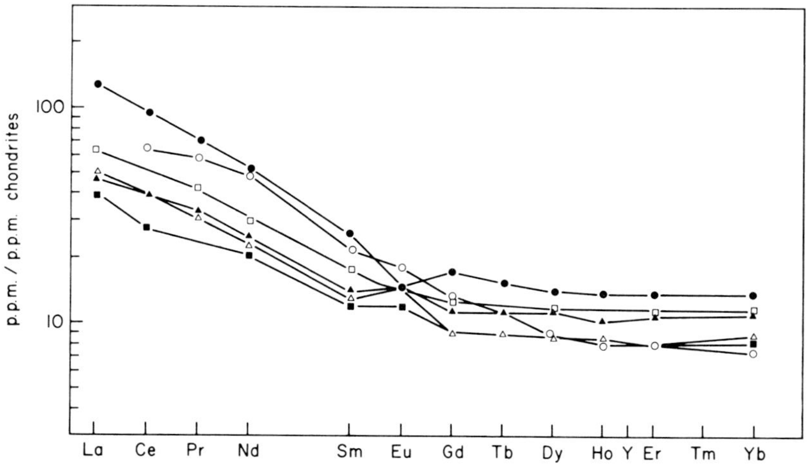

Fig. 4. Rare earth element data from a diverse suite of Archean sedimentary rocks (see text), compared with the Post-Archean average shale. ●, Post-Archean average Australian sedimentary rock; ○, Wind River Greywacke; △, Kalgoorlie Archean; ▲, West Greenland 429; ■, Fig Tree shale; □, Akilia.

There are examples of very ancient gneisses of granodioritic composition, with typical REE patterns including negative Eu anomalies (Amitsoq gneiss; Mason, 1975). Other examples include the Lewisian "Grey Gneisses" (Tarney, 1976) for which REE patterns are at present lacking. They may be predicted to show typical granodioritic patterns. The Archean sedimentary REE patterns do not record a significant contribution from these rocks, and they presumably did not constitute a significant fraction of the exposed upper crust in Archean times. This could be consistent with the proposal by Hargraves (1976) that the Archean crust was only rarely above sea level.

3.2. The Composition of the Archean Crust

The average composition of the Archean crust is inherently more difficult to establish than that of the Post-Archean crust. The restricted areal occurrences and uncertainty regarding the extent of the original Archean crust impose a non-statistical bias to sampling. Thus processes which provide wide-scale sampling of the crust acquire extra significance. The uniform nature of the REE patterns in Archean sediments is thus a key piece of evidence, and enables the overall crustal composition to be inferred.

An average REE pattern for the Archean sedimentary rocks is given in Fig. 5, normalized to chondritic abundances, and in Fig. 6, normalized to Post-Archean sedimentary rocks. In Fig. 5, the Archean sedimentary rock

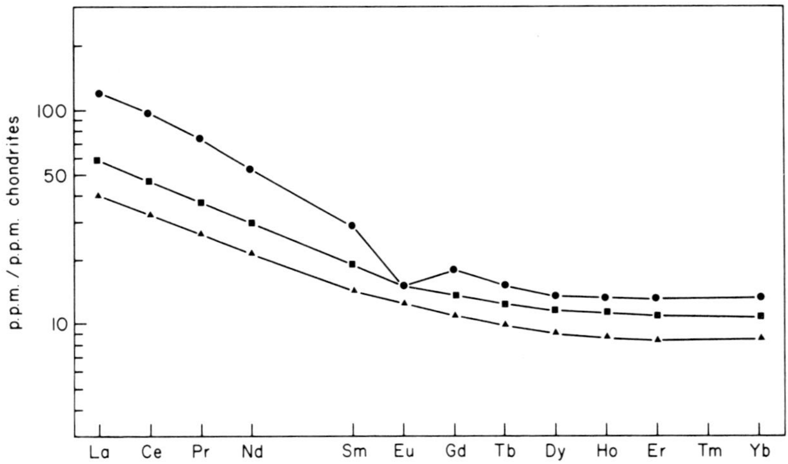

Fig. 5. A comparison between REE chondrite-normalized patterns of Post-Archean (●——●) and Archean sedimentary (▲——▲) rocks, and modern island-arc volcanic suites (■——■).

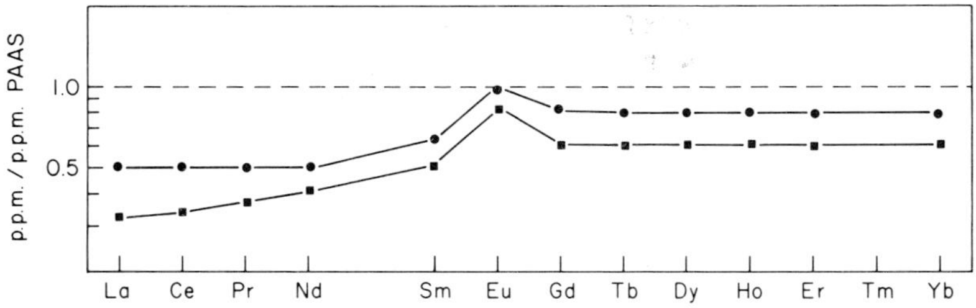

Fig. 6. Average Archean sedimentary REE patterns (●——●), and those of modern island arc suites (■——■) normalized to the post-Archean average shale (PAAS).

patterns are compared with the average Post-Archean patterns. The Archean REE patterns closely resemble those of island-arc volcanic rocks, with respect to relative abundance, slope, and lack of Eu anomalies, relative to chondrites (Figs 5 and 6). Both patterns, of course, possess a positive Eu anomaly relative to the Post-Archean average shale (PAAS) pattern (Fig. 6).

The similarity between the island-arc and Archean sedimentary patterns allows the deduction that the overall composition of the Archean crust, serving as the source of sediments, had an "andesitic" or "tonalitic" composition similar to the average composition of present day island arc volcanic rocks. A similar conclusion has been reached by J. J. W. Rogers (1977). This composition is similar to that proposed for the present day whole crust (Table 1).

However, the composition of Archean volcanic rocks of intermediate composition are generally higher in Ni and Cr than their more modern equivalents (Ahmad *et al.* 1977; Condie, 1976; Goodwin, 1977; Hawkesworth and O'Nions, 1977; Taylor and Hallberg, 1977). The composition given in Table 1 for the Archean crust is thus broadly similar to that for the total Post-Archean crust, a conclusion agreeable to devotees of uniformitarianism. Note, however, Al_2O_3 is lower, and FeO and MgO are higher, as are Ni and Cr. The present state of the art does not allow significant distinctions to be made for the other trace elements. To a first approximation, their abundances are taken as equal to those proposed for the present total crust. This assumption is probably valid within a factor of two.

3.3. Models for the Evolution of the Archean Crust

In the previous section, the assumption was made that the Archean sedimentary rock patterns provide an adequate sampling, and hence a key to the composition of the Archean crust. This composition is equivalent to tonalite or andesite.

How might such a composition arise? An immediate inference, based on analogy with present day volcanogenic sediments or older examples (e.g. Baldwin greywackes, Chappell, 1968; Nance and Taylor, 1977), is that the source of the Archean sediments was island-arc type volcanic rocks. Jakeš and Taylor (1974) proposed such a model, implying that the Archean crust differed from the Post-Archean crust in being composed dominantly of unfractionated island-arc type volcanic rocks. Intra-crustal melting and upward transfer of elements occurred only to a limited extent, although granodiorites similar to more recent examples were produced. A "critical" question for this model is whether island-arc type rocks were produced in quantity from the mantle during the Archean. Goodwin ((1977), Hallberg *et al.* (1976) and Taylor and Hallberg (1977) have described examples of calc-alkaline rocks in Archean terrains whose chemistry is similar to that of present-day island-arc volcanic rocks. As noted, higher Ni and Cr and lower Al contents comprise the main differences. Nevertheless, these rocks are not widespread, although a possible cause for this paucity includes the fact that such rocks are easily eroded, particularly in an environment free of plant cover. Volcanogenic greywackes are common in the Archean (e.g. Vermillion greywacke; Ojakangas, 1972), However, a further problem is that the production of present-day island-arc rocks appears to be closely linked with subduction processes, associated with plate tectonics. The evidence for plate tectonic activity in the Archean is not strong. Goodwin (1973) argues for plate tectonic models in the Archean, although Burke and Dewey (1973) conclude that such mechanisms operated only in Post-Archean time. More recent evidence (McElhinny and McWilliams, 1977; Kröner, 1977) appears to be decisively against the operation of conventional plate tectonics until much later in geological time.

Barker and Peterman (1974) and Arth and Hanson (1975) have argued that a bimodal tholeiitic–felsic volcanic suite dominated the Archean. An appropriate average of these rock types would produce the "tonalitic" composition. Decisive tests between these alternatives are lacking at present. Both models predict somewhat higher Ni and Cr contents in the Archean crust, but such effects have not yet been observed in the few data available for the sediments (Taylor, 1977). One exception is the Fig Tree shale (Danchin, 1967), but the high Cr and Ni contents of that formation are probably of local derivation (Condie *et al.*, 1970). The presence of Ni-rich komatiites (Grant-Cawthorn and McIver, 1977) will also be expected to provide local high concentrations. Such occurrences do not, however, constitute evidence for an early ultramafic crust.

An Early Anorthositic Crust?

The lunar highland crust is dominated by anorthositic material (Taylor, 1975). It has sometimes been suggested, by analogy, that an early anorthositic crust formed on the Earth. The lunar anorthositic crust is derived by crystallization, at or before 4.4 aeons, from a molten zone including at least the outer few hundred kilometres of the moon. The crystallization of the 60–80 km thick plagioclase rich crust removed Eu from the interior, so the later basalts, which flooded the lunar maria, and which were derived by partial melting from the lunar interior, exhibit a characteristic Eu depletion. No similar depletion appears in terrestrial basalts, which are derived from source regions either with chondritic or LREE depleted patterns (De Paolo and Wasserburg, 1976*a*,*b*). Such minor Eu anomalies as are observed in mantle derived basalts are due to near surface addition or subtraction of plagioclase.

An early anorthositic crust would likewise incorporate much Sr, with accordingly low Rb/Sr ratios. Thus large amounts of Sr with primitive $^{87}Sr/^{86}Sr$ ratios would be expected to be isolated. Evidence for such a reservoir of primitive Sr appears to be lacking. The high Ca and Al contents required to produce an anorthositic crust should lead to the extensive occurrence of garnet in the upper mantle of the Earth. The appropriate REE signatures indicative of this appear to be absent.

Impact Events in the Archean

It has been suggested occasionally, from analogy with early lunar history, that large-scale impacts were in part responsible for the development of Archean greenstone belts (Green, 1972; Glikson, 1976; Wilson *et al.*, 1974). The basin-producing stage of the lunar bombardment extends to 3.9 aeons, the last events being the production of the Imbrium and Orientale ringed basins (Taylor, 1975). The oldest terrestrial rocks are tantalizingly close to, but younger than this date.

The notion that such impact events triggered large-scale melting and basaltic volcanism (e.g. Wilson *et al.*, 1974) is based on a misconception. On the Moon, the large basin-forming impacts produce only localized melting. They are not the cause of the vast floods of basaltic lava which fill many, but not all, of the giant basins. Note that the basalt fill in Mare Orientale is very localized, and many farside basins have no associated basalt. The returned lunar basalt samples provide evidence for filling of the ringed basins up to 700 million years after the last basin forming collisions.

Whether large impacts occurred on the Earth at or before 3.9 aeons remains speculative. The large quantities of breccia produced by such events might not survive for very long, and the remnants of any primordial crust so destroyed would be more difficult to identify on the Earth than on the Moon.

Frey (1977), following earlier suggestions by Gilvarry (1961), proposes that the original ocean basins on the Earth were produced by impacts analogous to those forming the large ringed basins on the Moon. According to Frey (1977) these basins were punched in an original global "andesite" crust. The lack of evidence for recycling of ancient continental crust appears to preclude such primordial crust hypotheses.

4. Crustal Evolution with Time

The difference in composition between the upper crust in Archean and younger eras has been considered in previous sections. It has been ascribed to a major change in crustal evolution, resulting in the large-scale upward transfer, by intra-crustal partial melting processes, of material of granodioritic composition. This is a continuing process, and occurred to a limited extent in the Archean. Nevertheless, the evidence from the REE data supports a massive change roughly coincident with the Archean–Proterozoic boundary, at about 2500 million years Ago (Fig. 7). A major change in continental crustal evolution has been widely recognized at about this time by students of the early history of the Earth (e.g. papers in Windley, 1976).

Can the estimate be refined? In a study of REE patterns in the Huronian succession in Canada, McLennan *et al.* (1978) found that sediments at the base of the Huronian sequence (McKim Fm) had "Archean" REE patterns. Rapid change in the patterns occurs as the sequence becomes younger, with patterns at the top (Gordon Lake Formation) resembling typical Post-Archean sedimentary rock patterns. This lower Proterozoic sequence (Aphebian) lies unconformably on the Archean basement and was laid down, probably over a rather short time interval, some time between 2.6 and 2.16 aeons (McLennan *et al.*, 1978).

The change in crustal development is not necessarily synchronous on a world-wide scale. Some delay will also occur between the production of a differentiated upper crust and the record of this event in the sedimentary rocks. Parallel changes are observed in the Sr isotopic composition of seawater. Veizer (1973) notes a linear increase in $^{87}Sr/^{86}Sr$ from 0.701 to 0.702 in the Archean, followed by a sudden rise to 0.705 between 2500 and 2100 million years ago, followed by an increase to 0.709 for recent seawater. This is consistent with the REE evidence for an increase in large ion lithophile elements (e.g. Rb) in the upper crust, accessible to weathering.

Recycling vs. Continental Growth

As noted earlier, the consensus emerges from both Sr isotope studies (Moorbath, 1975, 1977) and Sm–Nd work (McCulloch and Wasserburg,

1978) that the continental crust grows throughout time by addition of new material from the mantle, with a rather short period during which intra-crustal melting takes place. This evidence rules out theories of primordial sialic crusts, models which invoke the early (pre 4 aeon) addition of continental material from the solar nebula, and impact theories which form continents as rim debris from large basin-forming impacts. The REE and isotopic evidence suggest episodic, rather than uniform growth rates for the addition of new continental material.

Recycling of Sedimentary Rocks
A separate question involves the recycling of sedimentary rocks themselves (Garrels and MacKenzie, 1969; Ronov and Yaroshevski, 1969; Veizer, 1973). Two aspects of the REE patterns in sedimentary rocks could be interpreted in support of this concept. In the Post-Archean shales, of similar major element chemistry, the REE patterns are strictly parallel, but the absolute abundances increase with time (Fig. 7, Nance and Taylor, 1976). This effect would be consistent with some recycling. Other effects are predictable. For example, the Th/U ratio in sedimentary rocks is usually greater than 4. This is attributed to the oxidation of U^{4+} to U^{6+}, and its removal as soluble $(UO_2)^{2-}$ during weathering. Thus sediments have higher Th/U ratios than the parental igneous rocks. If sediments are subjected to repeated recycling, a secular increase in Th/U ratios with time should develop. The data at present are not sufficient to demonstrate this effect.

4.1. Episodic or uniform crustal growth?

This is a long-standing question (e.g. Umbgrove, 1947). The REE evidence supports a rather sudden change from a "tonalitic" to a "granodioritic" upper crust at about 2.5 aeons. This is consistent with the change recorded in the Sr isotope ratios (Veizer, 1973). The Sm–Nd systematics indicate that the "Superior, Slave and Churchill Structural Provinces were all formed within the period 2.5–2.7 aeons—and that some truly major periods of continental growth are sharply episodic" (McCulloch and Wasserburg, 1978). The REE data from the sedimentary rocks records this as the major period of change in upper crustal composition. Figure 7 illustrates this change as shown by the variation in the Eu/Eu^* ratio from a chondritic ratio of about one in the Archean to a Post-Archean ratio of 0.67. Changes also occur in the slope of the REE patterns, as shown by the La/Yb and $\Sigma LREE/\Sigma HREE$ ratios (Fig. 7).

Such evidence for episodic growth raises larger issues. Are such episodes related to large-scale mantle convection? Although perhaps half of the large-ion lithophile elements (Rb, Th, U, Ba, LREE, etc.) have been

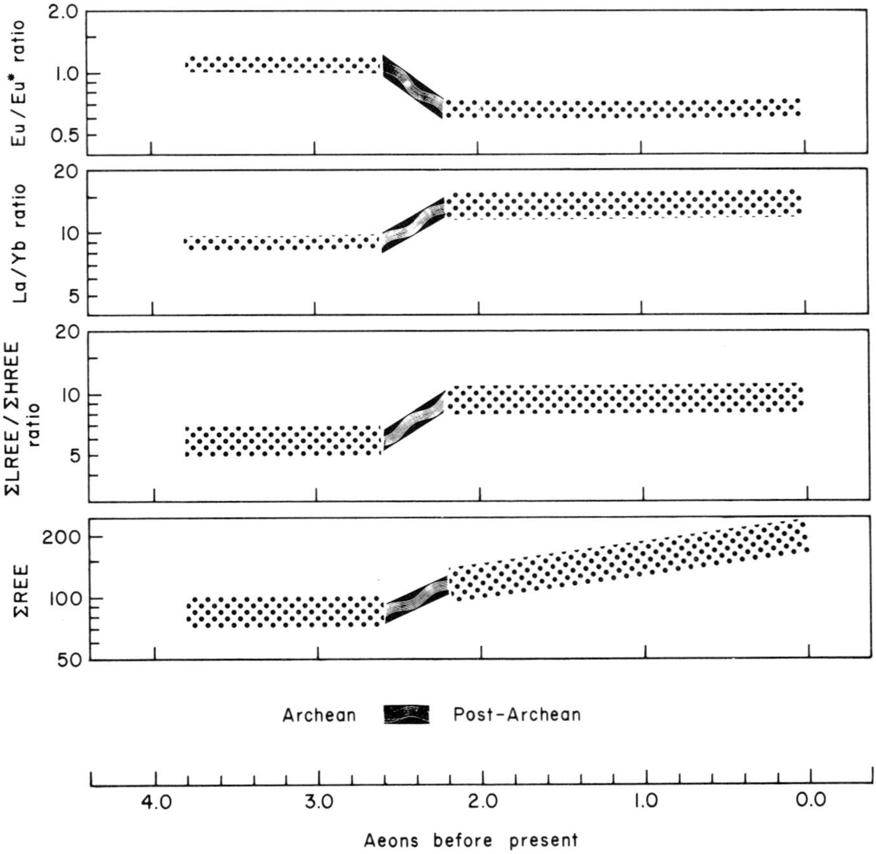

Fig. 7. The variation with time of REE in sedimentary rocks. Note that changes in the Eu anomaly are paralleled by those in La/Yb and ΣLREE/ΣHREE ratios. Data are for shales approximately equivalent in major element chemistry (Nance and Taylor, 1976, 1977; McLennan *et al.*, 1978).

extracted from the mantle into the continental crust, mantle with chondritic REE patterns is still being sampled near the surface (e.g. Columbia River Basalts, De Paolo and Wasserburg, 1976*a*,*b*, 1977). Whether this material represents primitive mantle trapped beneath the continents, implying lateral heterogeneity in the upper mantle (Jordan, 1975) or the operation of large-scale mantle convection remains as a problem for the future.

An Expanding Earth?

A variation on the primitive crustal models for continental origins proposes that, if the Earth was formerly about half its present radius, then the present continental masses form a world-encircling sialic shell. The continents have then reached their present positions principally through expansion of the Earth over geological time, to its present radius.

This view is in conflict with most of the chemical and isotopic evidence summarized in this chapter. Evidence that the radius of the Earth has remained essentially at its present value for the past 400 million years comes from palaeomagnetic data (McElhinny *et al.*, 1978). The Moon and Mercury show no evidence of expansion over periods exceeding 4 aeons. This evidence, in conjunction with the paleomagnetic calculations of the paleoradius of the Earth rule out any current theory of Earth expansion (McElhinny *et al.*, 1978).

Acknowledgements

I am grateful to M. T. McCulloch, S. M. McLennan, G. J. Wasserburg and to my colleagues at the Archean Geochemistry Conference at Hyderabad, India, November, 1977, for enlightening discussions.

References

Ahmad, S. M. *et al.* (1977). Archean geochemistry *Geophys. Res. Bull. natn. Geophys. Res. Inst., Hyderabad,* **15**.

Annhaeusser, C. R. (1973). The evolution of the early Precambrian crust of Southern Africa. *Phil. Trans. R. Soc. A***273**, 359–388.

Arth, J. G. and Hanson, G. N. (1975). Geochemistry and origin of the early Precambrian crust of northeastern Minnesota. *Geochim. cosmochim. acta* **39**, 325–362.

Barker, F. and Peterman, Z. E. (1974). Bimodal tholeiitic-dacitic magmatism and the early Precambrian crust. *Precambrian Res.* **1**, 1–12.

Bavinton, O. A. and Taylor, S. R. (1978). Rare earth elements patterns and crustal evolution. III. Archean sedimentary rocks from Kambalda, Western Australia. *Geochim. cosmochim. acta* (in press).

Bridgwater, D., McGregor, V. R. and Myers, J. S. (1974). A horizontal tectonic regime in the Archean of Greenland, and its implications for early crustal thickening. *Precambrian Res.* **1**, 179–197.

Brown, G. C. (1977). Mantle origin of Cordilleran granites. *Nature, Lond.* **265**, 21–24.

Burke, K. and Dewey, J. F. (1973). An outline of Precambrian plate development. In *Implications of Continental Drift to the Earth Sciences* (D. H. Tarling and S. K. Runcorn, eds), Academic Press, London and New York, pp. 1035–1045.

Carey, S. W. (1977). *The Expanding Earth*, Elsevier, New York and Amsterdam.

Chappell, B. W. (1968). Volcanic greywackes from the Upper Devonian Baldwin Formation, Tamworth-Barraba District, New South Wales, *J. geol. Soc. Australia* **15**, 87–102.

Condie, K. C., Macke, J. E. and Reimer, T. O. (1970). Petrology and geochemistry of Early Precambrian greywackes from the Fig Tree Group, South Africa. *Bull. geol. Soc. Am.* **81**, 2759–2776.

Condie, K. C. (1973). Archean magmatism and crustal thickening. *Bull. geol. Soc. Am.* **84**, 2981–2992.

Condie, K. C. (1976). Trace element geochemistry of Archean greenstone belts. *Earth Sci. Rev.* **12**, 393–417.

Danchin, R. V. (1967). Chromium and nickel in the Fig Tree shale from South Africa. *Science, N.Y.* **158**, 261–262.

De Paolo, D. J. and Wasserburg, G. J. (1976*a*). Nd isotopic variations and petrogenetic models. *Geophys. Res. Lett.* **3**, 249–252.

De Paolo, D. J. and Wasserburg, G. J. (1976*b*). Inferences about magma sources and mantle structure from variations in $^{143}Nd/^{144}Nd$. *Geophys. Res. Lett.* **3**, 743–746.

De Paolo, D. J. and Wasserburg, G. J. (1977). The sources of island arcs as indicated by Sm and Nd studies. *Geophys. Res. Lett.* **4**, 465–468.

Eade, K. E. and Fahrig, W. F. (1971). Chemical evolutionary trends of continental plates—a preliminary study of the Canadian shield. *Bull. geol. Survey Can.* **179**.

Eade, K. E. and Fahrig, W. F. (1973). Regional lithological and temporal variation in the abundances of some trace elements in the Canadian shield. *Geol. Survey Can. Paper* 72–46.

Ewart, A. (1976). Mineralogy and chemistry of modern orogenic lavas—some statistics and implications. *Earth planet. Sci. Lett.* **31**, 417–432.

Frey, H. (1977). Origin of Earth's ocean basins. *Icarus* **32**, 235–250.

Fyfe, W. S. (1973*a*). The granulite facies, partial melting and the Archean crust. *Phil. Trans. R. Soc.* **A273**, 457–462.

Fyfe, W. S. (1973*b*). The generation of batholiths. *Tectonophysics* **17**, 273–83.

Galdin, N. E. (1974). Composition and properties of crust in the ancient shields. *Geochem. Int.* **1974**, 270–281.

Garrels, R. M. and MacKenzie, F. T. (1969). Sedimentary rock types: relative proportions as a function of geological time. *Science, N.Y.* **163**, 570–571.

Gilvarry, J. J. (1961). The origin of ocean basins and continents *Nature, Lond.* **190**, 1048–1053.

Glikson, A. Y. (1976). Early Precambrian ultramafic–mafic volcanic rocks: ancient oceanic crusts or relic terrestrial maria? *Geology* **4**, 202–205.

Glikson, A. Y. and Lambert, I. B. (1973). Relations in space and time between major Precambrian shield units: an interpretation of Western Australian data. *Earth planet Sci. Lett.* **20**, 395–403.

Goldschmidt, V. M. (1954). *Geochemistry*, Oxford University Press, Oxford.

Goodwin, A. M. (1973). Plate tectonics and evolution of Precambrian crust. In *Implications of Continental Drift to the Earth Sciences* (D. H. Tarling and S. K. Runcorn, eds), Academic Press, London and New York, pp. 1047–1069.

Goodwin, A. (1977). Archean volcanism in Superior Province, Canadian shield. *Geol. Assoc. Canada spec. Paper* **16**, 205–241.

Grant-Crawthorn, R. and McIver, J. R. (1977). Nickel in komatiites. *Nature, Lond.* **266**, 716–718.

Green, D. H. (1972). Archean greenstone belts may include equivalents of lunar maria. *Earth planet. Sci. Lett.* **15**, 263–270.

Hallberg, J. A., Johnson, C. and Bye, S. M. (1976). The Archean Marda igneous complex, Western Australia. *Precambrian Res.* **3**, 111–136.

Hargraves, R. B. (1976). Precambrian geologic history. *Science, N.Y.* **193**, 363–371.

Haskin, L. A. and Haskin, M. A. (1966). Rare earths in European shales: a redetermination. *Science, N.Y.* **154**, 507–509.

Haskin, L. A., Haskin, M. A., Frey, F. A. and Wildeman, T. R. (1968). Relative and absolute terrestrial abundances of the rare earths. In *Origin and Distribution of the Elements.* (L. H. Ahrens, ed.), Pergamon Press, Oxford, pp. 889–912.

Hawkesworth, C. J. and O'Nions, R. K. (1977). The petrogenesis of some Archean volcanic rocks from southern Africa. *J. Petrol.* **18**, 487–520.

Heier, K. S. (1973). A model for the composition of the deep continental crust. *Fortschr. Miner.* **50**, 174–187.

Holland, J. G. and Lambert, R.St.J. (1972). Major element chemical composition of shields and the continental crust. *Geochim cosmochim. acta* **36**, 673–683.

Jakeš, P. (1973). Geochemistry of continental growth. In *Implications of Continental Drift to the Earth Sciences,* (D. H. Tarling and S. K. Runcord, eds), Academic Press, London and New York, Vol. 2, pp. 991–1001.

Jakeš, P. and Taylor, S. R. (1974). Excess europium content in Precambrian sedimentary rocks and continental evolution. *Geochim. cosmochim. acta* **38**, 739–745.

Jakeš, P. and White, A. J. R. (1972). Major and trace element abundances in volcanic rocks of organic areas. *Bull. geol. Soc. Amer.* **83**, 29–40.

Jordan, T. H. (1975). Lateral heterogeneity and mantle dynamics. *Nature, Lond.* **257**, 745–750.

Kröner, A. (1977). The Precambrian geotectonic evolution of Africa: plate accretion versus plate destruction. *Precambrian Res.* **4**, 163–213.

Langmuir, C. H., Bender, J. F., Bence, A. E. Hanson, G. N. and Taylor, S. R. (1977). Petrogenesis of basalts from the Famous area: Mid-Atlantic Ridge. *Earth planet. Sci. Lett.* **36**, 133–156.

Mason, B. H. (1975). Mineralogy and geochemistry of two Amitsoq gneisses from the Godthåb region, West Greenland. *Gron. Geol. Under Rep.* **71**, 1–11.

McCulloch, M. T. and Wasserburg, G. J. (1978). Sm–Nd and Rb–Sr chronology of continental crust formation. *Science, N.Y.* **200**, 1003–1011.

McElhinny, M. W. and McWilliams, M. O. (1977). Precambrian geodynamics—a palaeomagnetic view. *Tectonophysics* **40**, 137–159.

McElhinny, M. W., Taylor, S. R. and Stevenson, D. J. (1978). Limits to the expansion of Earth, Moon, Mars and Mercury and to changes in the gravitational constant. *Nature, Lond.* **271**, 316–321.

McGregor, V. R. (1973). The early Precambrian gneisses of the Godthåb district, West Greenland. *Phil. Trans. R. Soc. A* **273**, 343–358.

McGregor, V. R. and Mason, B. (1977). Petrogenesis and geochemistry of metabasaltic and metasedimentary enclaves in the Amitsoq gneisses, West Greenland. *Am. Mineral.* **62**, 887–904.

McLennan, S. M., Fryer, B. J. and Young, G. M. (1978). Rare earth elements in Huronian sedimentary rocks: Composition and evolution of the Post-Kenoran upper crust. *Geochim. cosmochim. acta* (in press).

Moorbath, S. (1975). The geological significance of Early Precambrian rocks. *Proc. geol. Assoc.* **86**, 259–279.

Moorbath, S. (1977). Ages, isotopes and evolution of the Precambrian continental crust. *Chem. Geol.* **20**, 151–187.

Nance, W. B. and Taylor, S. R. (1976). Rare earth element patterns and crustal evolution. I. Australian post-Archean sedimentary rocks. *Geochim. cosmochim. acta* **40**, 1539–1551.
Nance, W. B. and Taylor, S. R. (1977). Rare earth element patterns and crustal evolution. II. Archean sedimentary rocks from Kalgoorlie, Australia. *Geochim. cosmochim. acta* **41**, 225–231.
Naqvi, S. M., Rao, V. D. and Narain, H. (1974). The protocontinental growth of the Indian shield and the antiquity of its rift valleys. *Precambrian Res.* **1**, 345–398.
Newton, R. C. (1977). Experimental evidence for the operation of very high pressures in Archean granulite terrains. *Abstracts IUGS-IGCP Archean Geochemistry Conference, Hyderabad, India*, 73–75.
Ojakangas, R. W. (1972). Archean volcanogenic greywackes of the Vermillion District, Northeastern Minnesota. *Bull. geol. Soc. Am.* **83**, 429–442.
Parker, R. D. (1967). Composition of the Earth's crust. *U.S. geol. Survey prof. Paper* 440-D.
Piper, D. Z. (1974). Rare earth elements in the sedimentary cycle: a summary. *Chem. Geol.* **14**, 285–304.
Poldervaart, A. (1955). Chemistry of the Earth's crust. In *The Crust of the Earth* (A. Poldervaart, ed.). *Geol. Soc. Am. spec. Paper* 62, 119–144.
Radhakrishna, B. P. and Vasudev, V. N. (1977). The early Precambrian of southern India. *J. Geol. Soc. India* **18**, 525–541.
Rogers, N. W. (1977). Granulite xenoliths from Lesotho kimberlites, and the lower continental crust. *Nature, Lond.* **270**, 681–684.
Rogers, J. J. W. (1977). Inferred composition of early Archean crust. *Abstracts, IUGS-IGCP Archean Geochemistry Conference, Hyderabad, India*, 126–128.
Ronov, A. B., Balashov, Yu. A., Girin, Yu. P., Bratishko, R. and Kazakov, G. A. (1974). Regularities of rare-earth element distribution in the sedimentary shell and in the crust of the Earth. *Sedimentology* **21**, 171–193.
Ronov, A. B. and Yaroshevski, A. A. (1976). A new model for the geochemical structure of the Earth's crust. *Geochem. Int.* **13**, 89–121.
Ronov, A. B. and Yaroshewski, A. A. (1969). Chemical composition of the Earth's crust. In *The Earth's Crust and Upper Mantle*, American Geophysical Union, Washington, D.C., pp. 37–57.
Shannon, R. D. (1976). Revised effective ionic radii and systematic studies of interatomic distances in halides and chalcogenides. *Acta crystallogr. A* **32**, 751–757.
Shaw, D. M., Dostal, J. and Keays, R. R. (1976). Additional estimates of continental surface Precambrian shield composition in Canada. *Geochim. cosmochim. acta* **40**, 73–83.
Smithson, S. B. and Brown, S. K. (1977). A model for the lower continental crust. *Earth planet. Sci. Lett.* **35**, 134–144.
Tan, L. and Chi-Lung, Y. (1970). Abundance of chemical elements in the earth's crust and its major tectonic units. *Int. geol. Rev.* **12**, 778–786.
Tarney, J. (1976). Geochemistry of Archean high-grade gneisses, with implications as to the origin and evolution of the Precambrian crust. In *The Early History of the Earth* (B. F. Windley, ed.), Wiley, New York and Chichester, pp. 405–424.
Tarney, J., Skinner, A. C. and Sheraton, J. W. (1972). A geochemical comparison of major Archean gneiss units from Northwest Scotland and East Greenland. *Proc. 24th int. Geol. Cong.* **1**, 162–174.

Taylor, S. R. (1964*a*). Abundances of chemical elements in the continental crust: a new table. *Geochim. cosmochim. acta* **28**, 1273–1285.

Taylor, S. R. (1964*b*). Trace element abundances and the chondritic earth model. *Geochim. cosmochim. acta* **28**, 1989–1998.

Taylor, S. R. (1967). The origin and growth of continents. *Tectonophysics* **4**, 17–34.

Taylor, S. R. (1968). Geochemistry of andesites. In *Origin and Distribution of the Elements* (L. H. Ahrens, ed.), Pergamon Press, Oxford, pp. 559–585.

Taylor, S. R. (1973). Tektites: a post-Apollo view. *Earth Sci. Rev.* **9**, 101–123.

Taylor, S. R. (1975). *Lunar Science: A Post-Apollo View*, Pergamon, Oxford.

Taylor, S. R. (1977). Island-arc models and the composition of the continental crust. In Ewing Series Vol. 1, American Geophysical Union, Washington, D.C., pp. 325–335.

Taylor, S. R. and Gorton, M. P. (1977). Geochemical application of spark source mass spectrography. III. Element sensitivity, precision and accuracy. *Geochim. cosmochim. acta* **41**, 1375–1380.

Taylor, S. R. and Hallberg, J. A. (1977). Rare-earth elements in the Marda calc-alkaline suite: an Archean geochemical analogue of Archean-type volcanism. *Geochim. cosmochim. acta* **41**, 1125–1129.

Tuttle, O. F. and Bowen, N. L. (1958). Origin of granite in the light of experimental studies in the system $NaAlSiO_3O_8$–$KAlSi_3O_8$–SiO_2–H_2O. *Geol. Soc. Am. Mem.* 74.

Umbgrove, J. H. F. (1947). *The Pulse of the Earth*, Martinus Nijhoff, The Hague.

Veizer, J. (1973). Sedimentation in geologic history: Recycling versus evolution or recycling with evolution. *Contr. Min. Petrol.* **38**, 261–278.

Vinogradov, A. P. (1962). Average contents of chemical elements in the principal types of igneous rocks of the Earth's crust. *Geochemistry* 641–664 (English trans.).

Whitford, D. J. (1975). Geochemistry and petrology of volcanic rocks from the Sunda Arc, Indonesia. Ph.D. Thesis. Australian National University.

Wildeman, T. R. and Condie, K. R. (1973). Rare earths in Archean graywackes from Wyoming and from the Fig Tree Group, South Africa. *Geochim. cosmochim. acta* **37**, 439–453.

Wildeman, T. R. and Haskin, L. A. (1973). Rare earths in Precambrian sediments. *Geochim. cosmochim. acta* **27**, 419–438.

Wilson, H. D. B., Morrice, M. G. and Ziehlke, D. V. (1974). Archean Continents. *Geosci. Can.* **1**, 12–20.

Wilson, J. F., Bickle, M. J., Hawkesworth, C. J. and Martin, A. (1978). Granite-greenstone terrains of the Rhodesian Archean Craton. *Nature, Lond.* **271**, 23–27.

Windley, B. F. (1976). *The Early History of the Earth*, Wiley, New York and Chichester.

Windley, B. F. (1977). Timing of continental growth and emergence. *Nature, Lond.* **270**, 426–427.

12. Sr-Isotope Evolution of Granitoid Source Rocks

W. Compston and B. W. Chappell

Abstract

The hypothesis of restite control for the geochemical variation in granitoids (Chappell, 1966) in which linear variation lines are interpreted as mixing lines between a melt end-member and refractory source residue termed "restite", has been developed further to estimate Rb/Sr for the individual granitoid source regions. The chemistry of (model) restite is identified when normative quartz disappears as the variation lines of the major elements are extrapolated towards lower SiO_2. The melt end-members have known SiO_2 contents in the range 70–76 per cent. The Rb and Sr contents of the restites and melts are determined by combining known Rb, Sr distribution coefficients for the restite minerals with their observed abundances in the granitoids. The source Rb, Sr is then expressed as the fraction F of melt component plus (1-F) of the restite, and Rb/Sr evaluated for each granitoid source as a function of F. Adapting an analysis of melt extraction by Turcotte and Ahern (1978), F may be constrained as greater than $\sim$15 per cent, and by an argument based on the very low mechanical strength of rocks containing 25 per cent or more liquid, as less than 25 per cent for non-minimum melts. For granitoids containing minimum melts, F may be 25 per cent or greater and a maximum constraint is available only by modelling the composition of the source rocks. Regional averages of modern island-arc volcanics remote from continental areas indicate that SiO_2 would not exceed $\sim$65 per cent over the source volumes of most granitoids, which is equivalent to 61 per cent for F.

The source Rb/Sr thus calculated for nine igneous-derived granitoids of the non-minimum melt type correlates strongly and positively with their initial $^{87}Sr/^{86}Sr$, thereby defining a "source-rock isochron". The slope of this isochron is $\sim 1.1 \times 10^9$ years, which denotes the common time when the various source rocks were first separated from the mantle with $\sim$equal and low $^{87}Sr/^{86}Sr$. The sources for eight minimum melt granitoids do not define a source isochron for any approximately equal values of F, indicating that large differences in F existed between sources and/or that the source rocks were different in age. In addition, as their initial $^{87}Sr/^{86}Sr$ are lower than the non-minimum melt granitoids despite their higher F, the average age of their source rocks must be substantially less than 1.1×10^9 years. An isochron of $\sim0.75\times10^9$ years can be fitted to all the minimum-melt sources within the 25–60 per cent limits of *F*.

The results show that Precambrian continental material was present in the lower crust in south east Australia during the early Palaeozoic, so that the crust in this region did not originate *solely* as new oceanic crust during that period. Instead, sizeable continental blocks remained, containing both middle and late Precambrian metaigneous rocks and presumably also contemporary metasediments. Precambrian material at least 800 million years old has been independently detected as inherited zircons in one of the minimum-melt granitoids by Williams (1977), who has also found still older zircons in a number of sedimentary-derived granitoids not considered here.

The derivation of specific values of source Rb/Sr for particular granitoids in this manner greatly strengthens the use of initial $^{87}Sr/^{86}Sr$ in tracing the origin of igneous rocks.

1. Introduction

No rocks older than Cambro-Ordovician are known to be exposed in south-eastern Australia. The crust as sampled by the present land surface was fully formed by the early Devonian. The question arises: upon what kind of previous crust were these Palaeozoic sediments deposited? Two fundamentally different models must be considered. The first is that the previous crust was oceanic and was newly made in the early Palaeozoic during the rifting of plates; this view has been summarized by Harrington (1974). The second is that there existed a previous continental crust which is now totally covered by younger rocks and was modified in contributing to production of the younger crust. The age of any such earlier crust is unknown. It might be as young as earliest Palaeozoic or as old, at least in part, as the nearest exposed crystalline basement to the west, the Willyama Block at $\sim$1800 million years. The latter possibility has been discussed by Rutland (1976).

Approximately 30 per cent of the presently-exposed crust in the region is composed of Palaeozoic granitoids (White *et al.*, 1974), produced by partial melting of older rocks at depth, at least some of which probably constituted the earlier crust. The use of the initial $^{87}Sr/^{86}Sr$ in igneous rocks in tracing their source materials is well known. High values for initial $^{87}Sr/^{86}Sr$ would be expected in the granitoids if they were derived from an earlier crust that was substantially older than Palaeozoic and if it had a "crustal" value for its average Rb/Sr. Thus a Middle Precambrian sialic crust should easily be detected by this method. This argument is not conclusive however, because the crust is heterogeneous in its contents of Rb and Sr so that high initial $^{87}Sr/^{86}Sr$ can also be produced from a younger source rock with an unusually high Rb/Sr. This ambiguity can only be avoided if the Rb/Sr in the *particular* source rocks that were melted to form a given pluton can be estimated.

It has now become possible to estimate source Rb/Sr for certain types of granitoid plutons, using the restite model of Chappell (1966) and White and Chappell (1977). This model was developed to account for the petrographic character and the patterns of chemical variation in many eastern Australian granitoids. It ascribes the variation in granitoid plutons to progressive unmixing of material residual from partial melting (restite) from the more felsic melt. A consequence of this model is that element abundances in the source material are constrained to lie somewhere on the trends established by variation in cogenetic granitoids. The possibility of thus establishing more definite values for source Rb/Sr of even of a few plutons would enormously strengthen the power of the initial $^{87}Sr/^{86}Sr$ argument. We will be concerned here, therefore, to apply the restite model for this purpose to as many granitoids in south-east Australia for which appropriate geochemical and isotopic data are available. Detailed accounts of the geology and geochemistry (Chappell, 1978) and the Sr-isotope geochemistry and ages (Compston *et al.*, 1979) for the southern part of the Late Palaeozoic New England Batholith are in preparation and will not be repeated here. However, we will employ some of the New England data to illustrate the geochemical model and cite some of its important isotopic conclusions to support the somewhat less well-documented interpretations for other granitoids from the Lachlan Fold Belt.

The ultimate goal of this paper is to establish the age, or ages, of the pre-Ordovician crust by constructing an Rb–Sr isochron for the granitoid source rocks. We see this as a new contribution towards answering the long-standing question concerning the easterly extent of the Australian Precambrian Shield, and as a further development of the $^{87}Sr/^{86}Sr$ method for tracing the origin of igneous rocks. The use of Sr isotopes is applicable to all restite-controlled granitoids and study by this method of Archaean tonalites, for example, would be especially valuable in tracing the nature or age of the oldest parts of the terrestrial crust. We consider that the history of the continental crust in south-eastern Australia is a model for the history of the crust in general, and that granitoid production as described by the restite model is the principal mechanism for the upward geochemical fractionation of elements such as Rb and U that are partitioned into felsic melts during partial melting events.

For the Rb–Sr ages reported here, we have used $1.42 \times 10^{-11}\ y^{-1}$ for λ, the half-life of ^{87}Rb.

2. Distribution of Rb and Sr in Granitoids

Accurate major and trace element analyses by X-ray spectrometry, using many samples from a single pluton, show that plutons in a given batholith either have their own individual chemical characteristics or can be grouped into suites of plutons that have an individual chemical coherence. This can be illustrated for some of the New England granitoids by the elements Mg and Sr, for example, plotted against SiO_2 (Fig. 1). The data for the Inlet

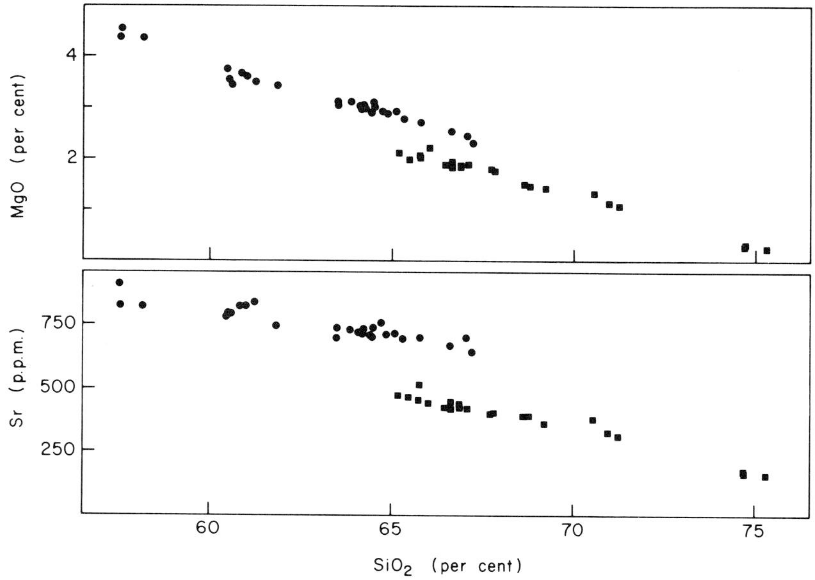

Fig. 1. MgO and Sr vs. SiO_2 for the Inlet Quartz Monzonite (●) and the Bendemeer Suite (■) of the Moonbi district, eastern Australia. The well fitted linear variation lines signify the unmixing of a felsic melt from a mafic refractory residue after melting, termed the restite. The chemistries of the melt, restite and source region are unique to each magma.

Quartz Monzonite are clearly separated from the combined data for the Moonbi and Bendemeer Adamellites (Bendemeer Suite). The striking linearity of the variation lines is more suggestive of mixing, or unmixing, between two specific end-members than of the effects of crystal fractionation. Furthermore, the discrete mafic xenoliths which these plutons contain themselves exhibit many of the distinctive chemical characteristics of their host granitoids. Xenoliths can also be recognized in all stages of disaggregation within the granitic matrix. Chappell (1966) interpreted the above observations by using the hypothesis that the variation in chemistry within an individual granitoid reflects the variable unmixing of a felsic end-member, which is usually the granitic minimum melt at about 76 per cent SiO_2, from a mafic end-member termed the restite, which is the refractory residue of the source that is retained within the magma as it moves away from the site of melting. Most xenoliths are interpreted as restite fragments, but with their mineralogy and consequently part of their chemistry modified by continued reaction with the melt as the load pressure and H_2O activity change during the upward movement of the magma.

Subsequently, Chappell and White (1974) recognized a first-order difference in the chemistry and mineralogy of granitoids and their inclusions, which reflected their origin from the partial melting of sedimentary rocks (S-type) or of igneous rocks (I-type). The great majority of the I- and S-type granitoids in south-eastern Australia show chemical variations largely controlled by mixing or unmixing processes involving melt and restite. However, there are also plutons that were emplaced as nearly complete liquids which have differentiated dominantly by crystal fractionation. There are petrographic and other criteria by which such rocks can be recognized. We show an example here of their chemistry to emphasize the difference in comparison to the restite-controlled I-type plutons. Figure 2 illustrates the

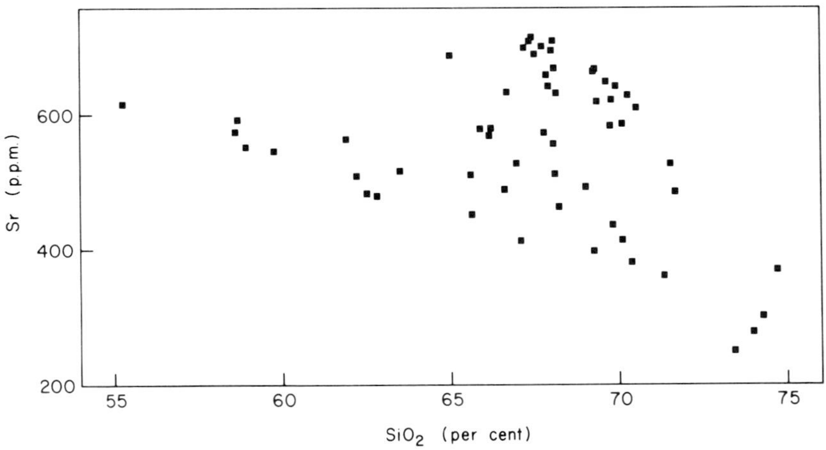

Fig. 2. The variation of Sr with SiO_2 for the Tuolumne Series, Sierra Nevada batholith. The poorly fitted, curved variation results from crystal fractionation during cooling of a totally molten magma, and contrasts greatly with Fig. 1 for the restite-bearing magma.

widely scattered Sr variation with silica for the Tuolumne Series of the Sierra Nevada batholith (Bateman and Chappell, 1978) which is indicative of mineral phases precipitating at varying rates during progressive solidification, contrasting with the good linear fit expected for restite control (Fig. 1). It must be understood that the subsequent discussion in this paper refers only to the well-fitted restite-controlled patterns, although it is likely that crystal fractionation from non-minimum melts is responsible for some minor perturbations in the variation lines.

I-type granitoids usually show better-fitted variation lines than S-types. Apparently, this is the result of their derivation from a more homogeneous source and only the I-types will be discussed here. Figure 1 shows the typical inverse correlation of Sr with SiO_2, which is consistent with the expectation that the melt end-member should have a lower Sr content than the restite which contains plagioclase ± K-feldspar. Rb and K are always positively correlated with SiO_2 but, compared with Sr, the correlation is less well-defined, indicative of the mobility of K and Rb in the melt. The positive slope of the variation line is consistent with an expected low restite content for these elements compared to the melt. Examples of Rb variation will be given later.

We have observed a range of initial $^{87}Sr/^{86}Sr$ of ~0.704–0.710 in I-type granitoids of the Lachlan Fold Belt. That this range might be ascribed to variation in source rock compositions, so that, with time, those sources with higher Rb/Sr ratios evolve to a higher ^{87}Sr-content, can be seen qualitatively by reference to Fig. 3. The source rock Sr-contents of the three suites in Fig. 3 must lie on the trends defined by the granitoids of each suite. It is clear

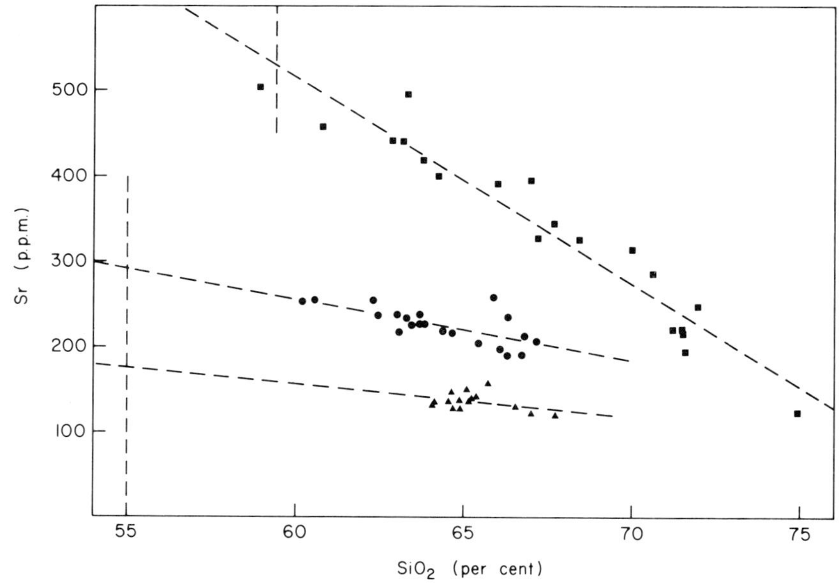

Fig. 3. Variation of Sr with SiO_2 for three restite-controlled igneous suites from south-eastern Australia. The approximate SiO_2 contents of the source regions are indicated by the broken vertical lines. For all estimates of source SiO_2, the diagram shows that Moruya Suite source Sr > Jindabyne Suite > Finister Suite, which correlates inversely with the order observed in their initial $^{87}Sr/^{86}Sr$ (denoted R_i) and is consistent with ageing of Rb and Sr in the source regions. ■, Moruya ($R_i = 0.7045$); ●, Jindabyne ($R_i = 0.707$); ▲, Finister ($R_i = 0.710$).

that the relative Sr-contents of the three source materials must be Moruya > Jindabyne > Finister. A similar argument shows that for Rb the reverse is the case, with Moruya < Jindabyne < Finister. Consequently, for Rb/Sr of the source rocks, the sequence is Finister > Jindabyne > Moruya. There is a general correlation with decreasing initial $^{87}Sr/^{86}Sr$ values which are ~0.710 (Finister), ~0.707 (Jindabyne) and ~0.704–0.705 (Moruya). These data are consistent with isotopic evolution of source materials with the same Sr-isotopic composition but differing Rb/Sr ratios, between the time of their first formation with low $^{87}Sr/^{86}Sr$ and the partial melting event that produced the granitoids. The purpose of this study is to attempt to quantify these relationships and thus arrive at an estimate of the time at which the source materials were derived from the mantle.

3. Calculation of Granitoid Source Rb/Sr

3.1. Composition of Granitoid Melts

Compositions of the melt component of some granitoid suites can be estimated with precision since the most felsic rocks in the suite have the major element chemistry of a minimum melt. The SiO_2 content of such rocks is about 76 per cent, and the concentration of certain elements such as P, Ni and Mg when extended towards higher SiO_2 along their variation lines do in fact vanish near this value. For such plutons or suites, therefore, the extrapolation of elemental concentrations along a well-defined variation line to a SiO_2 content of about 76 per cent gives the chemistry of the particular melt end-member for that suite. On the other hand, not all melt end-members are minimum melts. For example, Mg and other critical elements for the Inlet Quartz Monzonite in Fig. 1 do not extrapolate to zero until at least 80 per cent SiO_2, and the same is true in certain other granitoids to be discussed later. The restite model can still be applied to such bodies, but knowledge of the SiO_2 content of their melt end-members is less secure. We have observed no samples from this type of pluton that have SiO_2 contents greater than 70 per cent. This is analogous to the absence of samples having SiO_2 greater than 76 per cent in minimum melts, and we consider that 70 per cent can be taken as an approximate estimate for the SiO_2 content of the non-minimum melts under discussion.

3.2. Composition of Restite

An approach analogous to that used in determining melt compositions cannot be applied to restite. Primary restite that coexisted in equilibrium with the melt during the partial melting event is subsequently modified by reaction with that melt and its chemical and mineralogical characteristics are changed. Also, only a few elements normally show positive correlations with SiO_2 and none of these elements in general can be expected to vanish in the pure restite end-member. The constraints that can be placed on the primary restite chemical composition are that the composition must lie on the trends established for the granitoids and it must contain quartz, K-feldspar and plagioclase to coexist in equilibrium with a minimum melt. The limiting case is provided by the disappearance of one of those three minerals as possible compositions are extrapolated to lower SiO_2 values. Precise definition is difficult since the abundance of quartz and K-feldspar depends on the nature of the coexisting mafic minerals.

We are concerned here with the partial melting of igneous rocks to produce I-type granitoids. Such melting involves production of a hydrous melt, the water being derived from breakdown of biotite and hornblende in an original amphibolite facies assemblage. As Brown and Fyfe (1970) have pointed out, the bulk of melt production in nature is probably coupled with the destruction of biotite and hornblende, and that as hydrated phases react to provide water for melting, refractory residues remain. Complete breakdown of the hydrous minerals would result in anhydrous residual phases of granulite facies mineralogy in equilibrium with the felsic hydrous melt. The two-pyroxene mineralogy of the primary restite is generally largely or completely modified to a biotite–hornblende assemblage during cooling of the melt+restite magma at higher levels in the crust. Occasional pyroxene cores in hornblende crystals in the granitoids are the only direct evidence of their prior existence. Under favourable conditions, a large proportion of the primary restite mineralogy may be retained, for example, in the pyroxene-bearing rhyodacite associated with rocks of the northern part of the New England Batholith (Wilkinson *et al.*, 1964; Flood *et al.*, 1977). The rhyodacite contains megacrysts of clinopyroxene and rare orthopyroxene. Some of the abundant small xenoliths consist of coarse-grained clinopyroxene, orthopyroxene and plagioclase with partial to extensive replacement of pyroxene by hornblende and/or biotite. Quenching of the rhyodacite on eruption has preserved a significant proportion of the primary restite.

We define *model restite* as the composition at which normative quartz (or K-feldspar) disappears as compositions established by the granitoid trends are extrapolated to more mafic values. Two limiting values for the composition may be calculated. One is given by the usual CIPW norm in which ferric iron and titanium are assigned to magnetite and ilmenite, and this produces a lower limit for the restite SiO_2 because much of the Fe^{3+} and Ti probably occurred in solid solution in silicates. The other limiting value is obtained by including all the Fe^{3+} and Ti with Fe, Mg and Mn in the production of normative pyroxenes, which gives an upper limit to the restite SiO_2, as it assumes zero normative opaques. We have chosen the mean of the above limits as a working value for the SiO_2 content of the model restite. Model restite otherwise corresponds to the lowest SiO_2 content of a two-pyroxene assemblage coexisting with a felsic melt. A lower degree of partial melting will result in a more SiO_2-rich restite, whereas non-minimum melting will involve melting of the pyroxene+plagioclase components of the model restite.

3.3. Reconstruction of Source Rb/Sr

The linear variation lines for each element within a single pluton and between the cogenetic plutons of the one suite, are generated by a mixing relationship of the following form:

$$x\mu_M + (1-x)\mu_R = \mu_x, \tag{1}$$

where x is the weight fraction of melt in the particular sample and μ_M, μ_R and μ_x are the concentrations of any element in the melt, the restite and the observed sample, respectively. The essential point in reconstructing the source chemistry is the material balance requirement that the average source material should also lie on the mixing line between model restite and melt. If F is the maximum fraction of minimum melt that can be extracted from a given source, then

$$F\mu_M + (1-F)\mu_{MR} = \mu_S, \tag{2}$$

where μ_M, μ_{MR} and μ_S are the elemental concentrations in the melt, model restite and source, respectively. If f is the fractional mass of melt actually produced by a particular melting event, then for minimum melts, $f \leqslant F$ and for non-minimum melts, $f > F$. We will now discuss the constraints that can be placed on the values of f and F.

Limits of f

Certain types of basalt have evidently formed by the extraction of very small fractions of melt, say 5 per cent, from the mantle. We will now consider whether similar small fractions of granitic minimum melt can be extracted from rocks in the lower crust. In both cases, we may visualize the same physical situation: the movement of a melting isotherm through the source rocks which transforms them into a saturated permeable medium by the formation of melt. The complete fraction f of low melting component is formed as a melt, but only a smaller fraction f' remains as liquid in the restite. The remainder flows upwards away from the restite because of the lower density of the liquid.

The extraction of basaltic melts in this way has been modelled by Turcotte and Ahern (1978). Although their treatment may not be appropriate for granitoids in detail, we will employ their equations to illustrate the extreme dependence of melt extraction upon the viscosity of the melt. We have chosen their case corresponding to slow ascent of a diapir for which pressure-release melting occurs at a single depth, and we have plotted the ratio f'/f as a function of viscosity (Fig. 4). It is plain from Fig. 4 that virtually no extraction of melt from restite can occur for f as low as 5 per cent if the viscosity of the melt is greater than $\sim 10^6$ poise. No siliceous melts at ~700 °C have viscosities less than 10^6 poise (e.g. Shaw, 1963) and therefore no such melts at 5 per cent concentration would be extracted. In contrast to this, virtually complete extraction (≥90 per cent) will occur for the same f if the viscosity is $\sim 10^3$ poise, which is the value observed for most basalts. Very hot (900 °C), water-rich granitic melts have lower viscosities and could probably be extracted with moderate efficiency (~75 per cent), but such melts are probably uncommon. Figure 4 also shows that extraction efficiency

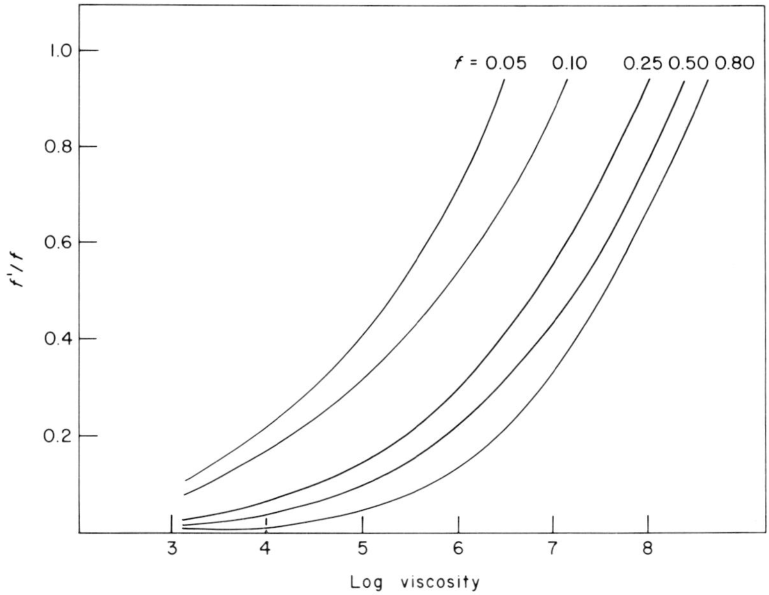

Fig. 4. The buoyant separation of melts from the source region as a function of melt viscosity, adapted from Turcotte and Ahern (1978). The term f denotes the total minimum melt fraction, which is dependent on source composition, and f' denotes the fraction of melt retained in the source. Melts having viscosities around 10^6 poise, such as granitic minimum melts at 700 °C, are effectively completely retained by the source ($f'/f \geq 0.9$) if f is 5 per cent or less. By contrast, melts having viscosities around 10^3 poise, such as basalts, are effectively completely extracted ($f'/f \leq 0.1$).

increases with the value for f. For f at ~ 20 per cent, granitic melts of moderate water content and temperature can be partially separated from restite, but with nothing like the speed and efficiency with which tholeiitic melts move after ~ 20 per cent melting of the mantle. We therefore choose 0.15 as the lowest reasonable value of f to produce a granitoid magma.

For values of f exceeding ~ 25 per cent, the model of laminar flow of melt under buoyancy through a permeable restite is inappropriate, as the mineral components of the restite no longer represent a continuous rock framework. This has been discussed by Arndt (1977). In addition, Van der Molen and Paterson (1978) have shown that flow-stress decreases by at least an order of magnitude in granitic partial melts as the melt fraction increases from 20 to 25 per cent. Such magmas (melt+restite) should therefore easily deform and may rise as a whole with or without much separation of melt to form a granitoid pluton. We have selected 0.25 as the probable highest value of f, since magma (melt + restite) would become highly plastic at this point and ascend away from the heat source as a buoyant pluton.

Limits of F for Non-minimum Melts

For non-minimum melts, the value of F is too low to produce sufficient minimum-melt to provide a mobile magma. Consequently, the magma is retained at depth until its temperature has risen and part of its (model) restite melted to dilute the minimum melt. Conversion of a minimum melt with ~ 76 per cent SiO_2 to an non-minimum melt with ~ 70 per cent SiO_2 to give 0.25 for f, by the addition of restite with ~ 50 per cent SiO_2, implies a value of $F \sim 0.19$. In this discussion we will model the source compositions of non-minimum melt granitoids by taking F as 0.19 in (2). We will then discuss the effect of variations in F on these model compositions and the deduced Rb–Sr systematics of the source rocks.

Limits to F for Minimum Melts

For minimum melts, we have $f \leqslant F < 1$. High values of F correspond to source rocks composed of highly felsic igneous bodies having SiO_2 contents of $\leqslant 76$ per cent. However, such material cannot be produced in large amounts directly from the mantle (Green and Ringwood, 1967), and we consider that its occurrence deep in the crust would be rare. Rhyolitic or at least dacitic flows occur in such island arcs as the Marianas, where there is no question of recycling of siliceous continental material. However, the upper limit for F must be assessed in terms of the maximum SiO_2 content of large volumes of rock, rather than for individual flows. For example, the Buckleys Lake Adamellite, the largest lithological unit in the Berridale Batholith, is exposed over an area of 470 km^2 and requires a huge source region even for F taken as close to unity.

We cannot *accurately* assess the random fluctuations in the SiO_2 contents of regional-scale samples of the source rocks. We have made an approximate estimate by assuming first that the Palaeozoic granitoids under discussion were formed by partial melting of an unseen older island-arc rock assemblage, so that regionally-averaged chemical analyses of modern island-arc rocks become relevant. We have employed the published data for the two main Fijian Islands, the Lau Group Islands, and the Tongan Islands as a northern and southern group, to generate a population of five regional averages in observed SiO_2 content (Table 1). We have excluded samples having less than 53 per cent SiO_2 on the grounds that they contain too little minimum melt component ($F \leqslant 0.19$) to be extracted. The mean of the above population is 60.4 per cent SiO_2, so the highest expected SiO_2 on the

Table 1. The SiO_2 content of island-arc volcanics on a regionally averaged basis. The uncertainty shown for each region is the standard error of the mean SiO_2 content.

		Number of samples	Mean SiO_2 (weight per cent)	Mean $^{87}Sr/^{86}Sr$
Viti Levu, Fiji†		11	61.3 ± 2.3	0.7040
Vanua Levu, Fiji‡		34	63.8 ± 1.3	0.7042
Lau Group§		20	58.2 ± 0.8	0.7032
"northern" Tongan Is.¶ (Metis, Late, Fonualei)		23	60.4 ± 1.1	0.7040
"southern" Tongan Is.¶ ‖ (Hunga Ha'apai, Eua)		6	58.2 ± 1.7	0.7037
		Mean	60.4 ± 2.3	
Northern volcanic zone, south-east Papua††		10	63.0 ± 1.9	0.7040
Sunda arc, Indonesia‡‡	# 1	6	59.9 ± 1.9	0.7046
	# 2	6	58.5 ± 2.0	0.7051
	# 3	14	56.4 ± 0.5	0.7049
	# 4	11	57.3 ± 0.6	0.7044
	# 5	7	57.3 ± 1.6	0.7043
		Mean of eleven groups	59.5 ± 2.4 (σ pop)	

† Gill (1970).
‡ Hindle and Gill, in Gill and Gorton (1973).
§ Gill (1976).
¶ Ewart *et al.* (1973).
‖ Ewart and Bryan (1972).
†† Smith and Compston (1979).
‡‡ Whitford (1975).

basis of the standard error of 2.3 per cent for the population will be 65.0 per cent, which corresponds to ~0.58 for F. A very slightly lower maximum is obtained by including regional samples from the Sunda arc (Table 1) divided into five loose geographic zones, and from south-east Papua. The Sunda arc samples could be disregarded as they have significantly higher and more variable $^{87}Sr/^{86}Sr$ than the Fijian and Tongan material, which might indicate contamination by older crustal rocks. A higher SiO_2 limit would be obtained if rhyolites from the Taupo volcanic zone of New Zealand were included, but as they were evidently formed by the fusion of greywackes and argillites (e.g. Ewart and Stipp, 1968), the Taupo rhyolites cannot be considered as first-generation island-arc material.

In summary, our preferred estimate for the *maximum* SiO_2 content of large volumes of island-arc material is 65 per cent, equivalent to ~0.58 for F. We consider that this estimate cannot reasonably be much increased, although it can certainly be improved by wider compilations of data. For large granitoids, we will therefore select 0.60 as the upper limit for F. For small granitoids, F may possibly be larger in extreme cases. However, we observed that the Tongan island of Fonaulei, which is predominantly dacitic in composition (Ewart *et al.*, 1973) and appears to be the most felsic island of any size in the group, averages only 64.2±0.5 per cent SiO_2, which is less than our estimated upper limit. For non-minimum melts having SiO_2 contents of 70 per cent the maximum value for F corresponding to a source SiO_2 of 65 per cent would be 0.58 also, but, as already discussed, we consider such a situation as unrealistic, on the grounds that such a felsic parental composition would have produced a minimum melt only. In the discussion of individual granitoids that follow, it will be convenient to employ an average value for F to compare estimates of the source Rb/Sr. For minimum melt-type granitoids, we will choose 0.36 equivalent to the average SiO_2 of 59.5 per cent for all eleven regional samples in Table 1. For the non-minimum melt types, it is more appropriate to employ 0.19 for F, according to the argument given in the preceding subsection, "Limits of F for non-minimum melt", which is equivalent to 25 per cent partial melting for a melt having SiO_2 of 70 per cent.

3.4. The Equilibrium Partitioning of Rb and Sr between Melt and Restite

For dilute solutions of the same element E in a melt M and in a mineral phase C in equilibrium with the melt, we expect from Henry's law that at fixed temperature

$$\mu_C^E/\mu_M^E = \text{constant} \equiv D_C^E,$$

where D_C^E, known as the distribution coefficient, is specific for the given element and mineral phase. For a restite composed of plagioclase (PL), K-feldspar (KF) and pyroxenes (PX), we have:

$$\mu_R^E/\mu_M^E = D_R^E = kD_{KF}^E + pD_{PL}^E + cD_{PX}^E, \tag{3}$$

where k, p, c are the weight fractions of K-feldspar, plagioclase and pyroxene in the restite. Estimates for D_{KF}, D_{PL} and D_{PX} in natural systems have been given for various elements, by Philpotts and Schnetzler (1970) for example, from which it is clear that for Rb and Sr, the term cD_{PX}^E in (3) is negligible compared with the other two terms for the granitoid melts considered here. In addition, the term kD_{KF}^{Sr} is usually small compared with pD_{PL}^{Sr}, so that D_R^{Sr} depends almost solely upon the plagioclase content of the restite.

From simple geometry, the intercept S_0 of any variation line with the SiO_2 axis is:

$$S_0 = \frac{D_R S_M - S_R}{D_R - 1}, \tag{4}$$

where S_M, S_R are the SiO_2 contents of the melt and restite. Thus if D_R can be estimated independently, we can determine S_0 by (4) and then construct the elemental variation line through the data point μ_i^E, S_i for a single analysed sample. We therefore have a second method, besides the use of empirically observed variation lines, for extrapolating from observed analytical data to the Rb and Sr contents of the restite. The second method is essential for those plutons which do not have sufficient internal dispersion of Rb and Sr to accurately define variation lines.

Sun *et al.* (1974) and Drake and Weill (1975) present experimental observations of D_{PL}^{Sr} for mafic melts at 1150 °C and higher temperatures and show that there is a marked increase in D_{PL}^{Sr} with decreasing temperature. The strong temperature dependence complicates the application of their results to plagioclase in granitoid minimum melts, which we assume may form at temperatures as low as 700 °C. In addition, the results for D_{PL}^{Sr} at the same temperatures by the above authors are systematically different, which they attribute to compositional differences between the melts employed. In the absence of experimental data for granitic melts, we have constructed a compromise temperature curve from the mean observed value for ln D_{PL}^{Sr} at 1150 °C and the slope observed by Drake and Weill (1975), which was based on a much wider temperature range than that of Sun *et al.* (1974). This curve is

$$\ln D_{PL}^{Sr} = \frac{9050}{T} - 5.44, \tag{5}$$

where the temperature T is given in kelvins, and it appears to give reasonable *comparative* estimates of temperatures for those of our granitoids for which D_{PL}^{Sr} can be estimated. Details of our calculations will be given later for individual plutons. Here we wish to point out that there is a good correlation between our estimated values for D_{PL}^{Sr} for the various plutons and their relative temperatures. The Inlet Quartz Monzonite, Iona Granodiorite and the Jindabyne Tonalite, which are non-minimum melts, are indicated as hotter than the others, which are minimum melts. This gives some confidence that we can employ the experimentally-determined values for D_{PL}^{Sr} to estimate D_{R}^{Sr} for the several other plutons which show low internal dispersion in Sr. On the other hand, as their apparent temperatures range from 800 °C to 950 °C, it is plain that we cannot assign the *same* value of D_{PL}^{Sr} to all minimum melts, so the method is not precise.

Rubidium distribution coefficients for K-feldspar have been determined by Long (1978) in granitic melts between 700 °C and 750 °C. We have no information concerning the temperature coefficient and we have employed Long's result, $D_{KB}^{Rb} = 0.95$, in all cases. In contrast to Sr, Rb in the restite may be distributed in comparable amounts between K-feldspar and plagioclase because the weight fraction of the latter is usually large (~ 0.6). Rather than rely upon the observed variation of Rb and K_2O with silica, which can lead to large uncertainties in Rb and K_2O when extrapolated to the restite silica content, we have determined values for the weight fraction of K-feldspar in the restite k through use of the following relationships:

(a) $\mu_R^{K_2O} = 16.9\,k + 0.1\,p$, from K_2O balance, which assumes 0.1 per cent K_2O in all plagioclases;
(b) $\mu_R^{Rb} = A + B\mu_R^{K_2O}$, where A and B are the regression parameters for the observed Rb vs. K_2O correlation of the granitoid, which is always much better defined than the alkali vs. SiO_2 correlations;
(c) $D_R^{Rb} = 0.95k + 0.017p$, where 0.017 is the value of D_{PL}^{Rb} measured by McKay and Weill (1976) for low K lunar basalts;
(d) $\mu_M^{Rb} = \mu_R^{Rb}/D_R^{Rb}$ and

$$\frac{\mu_M^{Rb} - \mu_R^{Rb}}{S_M - S_R} = \frac{\mu_C^{Rb} - \mu_R^{Rb}}{S_c - S_R},$$

which is the equation of the Rb vs. SiO_2 variation line fitted to the centroid C of the observed data.

The above relationships give a quadratic equation in k. The resulting value for $\mu_R^{K_2O}$ may be used to improve the K_2O vs. SiO_2 variation and, if necessary, fed back into the calculation of the norms and of the plagioclase content of the restite. A single iteration usually stabilizes all values.

3.5. Sr Isotopes within Individual Granitoids

The restite model for granitoid chemistry immediately raises a question concerning the interpretation of Rb–Sr total-rock isochrons. In cases in which the distribution of Rb and Sr within a granitoid is so plainly controlled by physical mixing processes, it cannot necessarily be assumed that the Sr isotopic composition of the melt component has completely equilibrated with that of the restite. Any such lack of equilibrium would result in the melt having the higher $^{87}Sr/^{86}Sr$ as it would selectively incorporate minerals rich in Rb and, hence, rich in radiogenic ^{87}Sr.

The question was examined for the New England plutons by closely comparing Rb–Sr mineral ages and precise total-rock isochron ages (Compston *et al.*, 1979). For the Inlet Quartz Monzonite and the Bendemeer Suite, the two types of Rb–Sr ages were identical at sensitivity of ~1 per cent; likewise, no significant differences could be detected in the several other plutons examined. Hofman (1974) and Hofman and Magaritz (1976) have shown that volume diffusion in melts is ineffective as a mechanism for isotopic homogenization. We therefore conclude that for the New England magmas either the source rocks were homogeneous in Rb/Sr or the melting events sampled the source rocks in a highly random manner and/or that physical stirring of the resulting mixtures of melt and restite was extensive enough to randomize any initial correlation between $^{87}Sr/^{86}Sr$ and Rb/Sr. On the other hand, the total-rock data for the Bendemeer Suite do show a small amount of random excess scatter about the regression line, which demonstrates that isotopic homogenization within the plutons was not complete.

Analyses of separated mafic xenoliths from the New England plutons show that the mean initial $^{87}Sr/^{86}Sr$ of the xenoliths agrees with that of the individual granitoids. This is consistent with the restite hypothesis that the xenoliths are cognate with their granitic host and constitute part of the restite. Xenoliths fit the total-rock isochrons to within the same precision as granitoid samples.

4. Restite-controlled I-type Granitoids

4.1. The Moruya Batholith

This is the most easterly of the batholiths of the region under consideration and has been described by Griffin *et al.* (1978). It consists of a suite of nine distinct plutons emplaced in folded sediments of which some are certainly Ordovician and others may be Cambrian (e.g. Brown *et al.*, 1968). Stratigraphic constraints do not furnish a precise younger limit to the ages of the plutons. K–Ar ages cited by Griffin *et al.* (1978) and new K–Ar ages by Richards (personal communication) range between 370 and 395 million years, and new Rb–Sr mineral isochrons (Fig. 5) give 389 million years (Moruya Tonalite) and 395 million years (Nelligen Granodiorite). Thus the minimum age of emplacement is 395 million years, i.e. Late Devonian. The Rb–Sr total-rock data (Table 2) collectively give an approximate maximum age for emplacement (Fig. 5). It is 425±5 million years, on the assumption that all plutons have exactly the same age and the same average initial $^{87}Sr/^{86}Sr$, which is significantly greater than the mineral ages. Slow cooling or later metamorphism are commonly proposed to explain such an age difference. However Griffin *et al.* (1978), while advocating a very similar source material for all the plutons because of their indistinguishable trends in major element chemistry, point out the presence of small differences in their contents of Rb, Sr and other trace elements that imply differences in Rb/Sr between their individual sources. Under these circumstances it is likely that systematic differences in initial $^{87}Sr/^{86}Sr$ exist between plutons, which would limit the usefulness of the collective total-rock data in constructing an Rb–Sr isochron. We will assume here that the true age of the

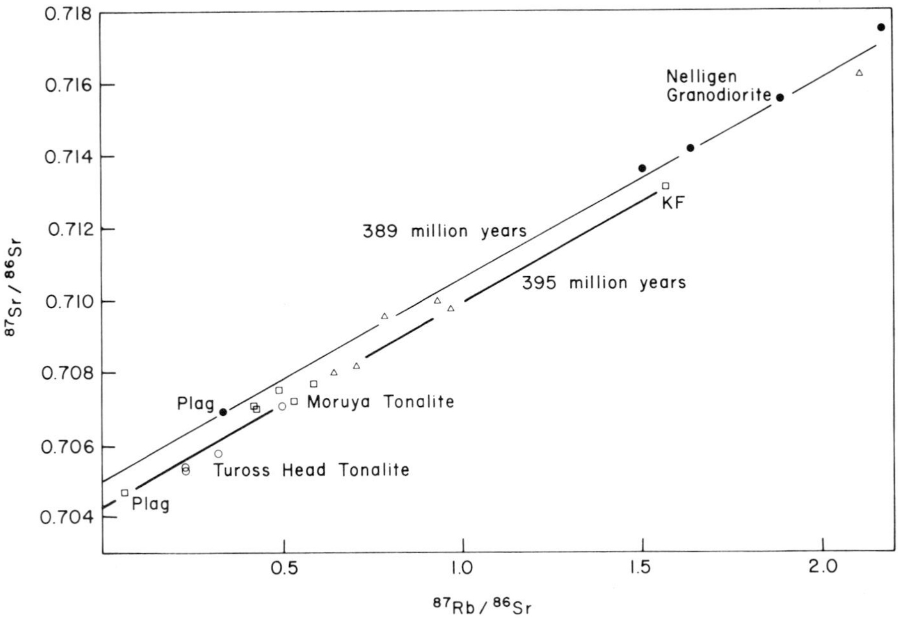

Fig. 5. Rb–Sr evolution diagram for rocks and minerals of the Moruya Suite. The emplacement of all plutons probably occurred during a short period around 390 million years, and the initial $^{87}Sr/^{86}Sr$ of the Nelligen Granodiorite is systematically greater than that of the Moruya and Tuross Head Tonalites and most of the other plutons (denoted △).

Table 2A. Analytical data for Moruya Suite rocks.

ANU #	SiO_2	K_2O	Rb	Sr	$^{87}Sr/^{86}Sr$
Nelligen Granodiorite					
34718	71.19	3.20	125	220	0.714 11
34719	71.47	3.40	115	220	0.713 56
27239	71.50	3.70	141	215	0.715 50
34720	71.57	3.73	146	194	0.717 42
Moruya Tonalite					
34723	62.84	2.01	65	442	0.707 06
34725	63.76	1.92	68	419	0.706 96
34726	64.21	1.94	68	400	0.707 47
34727	66.97	2.37	81	395	0.707 68
34728	67.18	2.21	61	327	0.707 17
Tuross Head Tonalite					
34730	58.91	1.55	41	504	0.705 30
34731	60.78	1.75	51	457	0.705 76
27249	63.30	1.42	40	495	0.705 36
34732	65.98	2.11	68	391	0.707 06
Other plutons					
34721	67.66	2.66	94	344	0.709 51
27238	71.93	2.71	83	247	0.709 77
34722	70.60	2.73	93	286	0.709 87
27236	69.98	2.35	77	314	0.708 16
34729	68.41	2.19	73	325	0.708 01
34733	74.91	3.50	90	123	0.716 18

Table 2B. Analytical data for Moruya Suite minerals.

RSES No.	Rb (p.p.m.)	Sr (p.p.m.)	$^{87}Rb/^{86}Sr$	$^{87}Sr/^{86}Sr$
76–39 biotite (Nelligen Granodiorite)	595.5	4.88	436.4	3.1562
76–39 plagioclase	42.7	365.0	0.338	0.706 92
76–39 K-feldspar	407.9	145.8	8.110	0.749 90
76–6 K-feldspar (Moruya Tonalite)	159.9	294.1	1.5718	0.713 11
76–6 plagioclase	9.18	411.4	0.0644	0.704 63

Moruya Batholith is 395 million years, as given by the mineral ages. On this basis, we have calculated the populations of initial $^{87}Sr/^{86}Sr$ for each pluton (Table 3), and small but apparently real differences between plutons are detected. In particular, the Nelligen Granodiorite has a higher initial $^{87}Sr/^{86}Sr$ than the Tuross Head Tonalite, which we will now show closely matches the differences in Rb/Sr deduced for their source rocks.

Table 3. Restite, melt and estimated source Rb–Sr parameters for south-east Australian granitoids.

	Rubidium (p.p.m.)			Strontium (p.p.m.)			Source Rb/Sr	Initial	Age million	Source $^{87}Sr/^{86}Sr$	
	(a)	(b)	(c)	(a)	(b)	(c)	(c)	$^{87}Sr/^{86}Sr$	(years)	Present-day[(d)]	420 million years ago
1. Minimum melts											
Moruya Tonalite	4.28	124	47.4	696	153	501	0.095	0.70445 ± 7	389	0.705 97	0.704 33
Tuross Head Tonalite	3.84	111	42.4	696	153	501	0.085	0.70405 ± 8	389	0.705 41	0.703 94
Nelligen Granodiorite	5.49	159	60.8	578	128	416	0.146	0.70514 ± 6	395	0.707 51	0.704 99
Buckleys Lake Adamellite	5.9	225	84.8	797	66	534	0.159	0.70665 ± 14	417	0.709 38	0.706 63
Delegate Adamellite	4.6	176	66.3	942	78	631	0.105	0.70560 ± 8	450	0.707 55	0.705 73
Maffra Adamellite	5.2	197	74.2	1141	95	764	0.097	0.70488 ± 5	413	0.706 53	0.704 85
Wullwye Adamellite	4.16	130	49.5	981	149	681	0.073	0.70543 ± 5	415	0.706 68	0.705 42
2. Non-minimum melts											
Tara Granodiorite	6.4	99	29.6	557	205	469	0.063	0.70620 ± 3	411	0.707 27	0.706 18
Currowong Suite	9.8	148	44.5	424	199	368	0.121	0.70762 ± 48	415	0.709 69	0.707 60
Iona Granodiorite	3.5	142	38.1	419	198	364	0.105	0.70678 ± 10	415	0.708 57	0.706 75
Finister Granodiorite	2.95	157	41.5	206	111	182	0.228	0.71023 ± 15	450	0.714 45	0.710 51
Merumbago Granodiorite	3.05	162	42.9	228	122	202	0.212	0.70970 ± 17	450	0.713 63	0.709 97
Round Flat Tonalite	2.10	107	28.4	369	165	318	0.089	0.7066 ± 4	420	0.7081	0.7066
Gaden Tonalite	2.39	122	32.3	385	172	332	0.097	0.7067 ± 3	420	0.7084	0.7067
Jindabyne Tonalite	2.57	131	34.8	374	167	322	0.108	0.7069 ± 2	420	0.7088	0.7069
Pendergast Tonalite	2.56	131	34.6	454	203	391	0.088	0.7069 ± 1	420	0.7084	0.7069

(a) Restite composition.
(b) Melt composition.
(c) Calculated source parameters, assuming F is 0.36 for the minimum-melts and 0.19 for the non-minimum melts.
(d) "Present-day source $^{87}Sr/^{86}Sr$" is calculated by adding the quantity $(e^{\lambda\tau} - 1)(^{87}Rb/^{86}Sr)_{source}$, where τ is the age taken for the granitoid, to the initial $^{87}Sr/^{86}Sr$ of the granitoid at τ years ago.

The chemistry of the melt end-member may be estimated by extrapolating the variation line of each element to 76 per cent SiO_2, the approximate silica content of the minimum melt. Similarly, we may extrapolate the variation lines towards low SiO_2 values and calculate model restite compositions by the normative method described previously, which gives 51.6 per cent for the SiO_2 content of the restite, this being the means of the limiting values 53.6 and 49.5 per cent.

We may now calculate Sr and Rb for the end-members of the Moruya Suite at 76 per cent SiO_2 for the minimum melt and at 51.6 per cent SiO_2 for the restite. The variation of Sr with SiO_2 (Fig. 3) as defined by all members of the suite is well-fitted and linear, typical of restite-controlled unmixing. Regression analysis gives 153 p.p.m. and 696 p.p.m. Sr for melt and restite, respectively, with the Nelligen Granodiorite excluded as this pluton is marked as different by its higher K, Rb. An apparent equilibration temperature of the magma can be estimated by calculation of D_{PL}^{Sr} and use of the temperature curve. Thus,

$$D_{R}^{Sr} = kD_{KF}^{Sr} + pD_{PL}^{Sr} = \frac{696}{153} = 4.55.$$

At 51.6 per cent SiO_2, the restite contains 64 per cent normative plagioclase and up to 4.0 per cent K-feldspar. The latter contributes an amount of only $0.04 \times 2.7 \simeq 0.1$ to D_{R}^{Sr} so that uncertainty in the value of this term due to error in k and in D_{KF}^{Sr} is not important. D_{PL}^{Sr} may be evaluated at 6.9 and substituted into (5) to obtain 950 °C for T. The absolute value for the temperature exceeds our prior expectations for a minimum melt, and it probably reflects the inadequacy of the compromise temperature curve for granitic melts in determining *absolute* temperatures. On the other hand, such high temperatures would simplify the production of granitoid magmas and the separation of melt from restite by lowering the melt's viscosity, which, as discussed previously, is a critical limiting property.

We now wish to estimate the average Sr content of the Moruya Suite source rocks. This requires knowledge of the term F in (2), the fractional mass of minimum melt that the source rocks would yield to produce the model restite. As already discussed, the value for F can be limited between the extremes of 0.25 and 0.6, and, for the present, we will assign it the value 0.36 to obtain comparative values for the source Sr contents of each magma. Thus, the source for the Moruya Suite has a comparative Sr content of 501 p.p.m. For the Nelligen Granodiorite, we may assume the same equilibration temperature and the same model restite as the Moruya Suite proper, which is equivalent to assuming the same D_{R}^{Sr}, 4.55. The relationships (d) on p. 391, may be rearranged to evaluate the slope m of a variation line through the centroid point (μ_C, S_C) and having known values for D_R, S_M and S_R. This is

$$m = \frac{\mu_C(D-1)}{S_R - S_C - D(S_M - S_C)}. \tag{6}$$

Thus we can obtain a separate Sr variation line through the Nelligen Granodiorite data, and its μ_R^{Sr} and μ_M^{Sr} may be determined (Table 3). For F taken as 0.36, we obtain a comparative Sr content for the Nelligen source of 416 p.p.m., lower than the Moruya source.

The Rb variation line for the Moruya Suite (Griffin *et al.*, 1977) is less well-defined than the Sr line, both because the Nelligen Granodiorite points lie systematically above the regression line fitted to all data and also because of appreciable internal scatter within each pluton. The same is true for the K_2O vs. SiO_2 variation. On the other hand, the K_2O vs. Rb variation is comparatively good (Fig. 6), presumably reflecting both the strong coherence between those elements and the location of the model restite on the variation line. We will therefore combine this relationship with known values of Rb distribution coefficients to estimate the Rb contents of the melt and model restite. This procedure will avoid direct use of the imperfect Rb, K_2O vs. SiO_2 variation lines and has been outlined already on p. 391.

As Rb and K in the model restite are located almost entirely in K-feldspar and plagioclase, we have

$$D_R^{Rb} = kD_{KF}^{Rb} + pD_{PL}^{Rb} = 0.95\,k + 0.0109,$$

where k is the (unknown) weight fraction of K-feldspar in the restite, and p is 0.64 from the normative plagioclase content at 51.6 per cent SiO_2. We will define the Fig. 6 variation line by the regression of Rb against K_2O, viz.

$$\mu_R^{Rb} = 39.37\,\mu_R^{K_2O} - 14.96.$$

The restite K_2O may be written

$$\mu_R^{K_2O} = 16.9\,k + 0.064$$

and the Rb variation may be fitted to the centroid of the Nelligen Granodiorite data (131.8, 71.43) using (6). The resultant equations may be solved to find 5.5 p.p.m. Rb and 0.52 per cent K_2O for the restite, and 60.8 p.p.m. Rb for the comparative content of the Nelligen source at $F = 0.36$ (Table 3). If the same D_R^{Rb} is applied to the other plutons, slightly smaller values for the restite Rb ($\sim$4.2 p.p.m.), restite K_2O (0.49 per cent), and melt Rb contents (116 p.p.m.) are obtained which give 44.9 p.p.m. Rb for the source rocks. The K_2O vs. SiO_2 regression for all data except the Nelligen Granodiorite shows 0.59 per cent K_2O for 51.6 per cent SiO_2, in good agreement with the result of the above treatment. If one "outlier" in each of

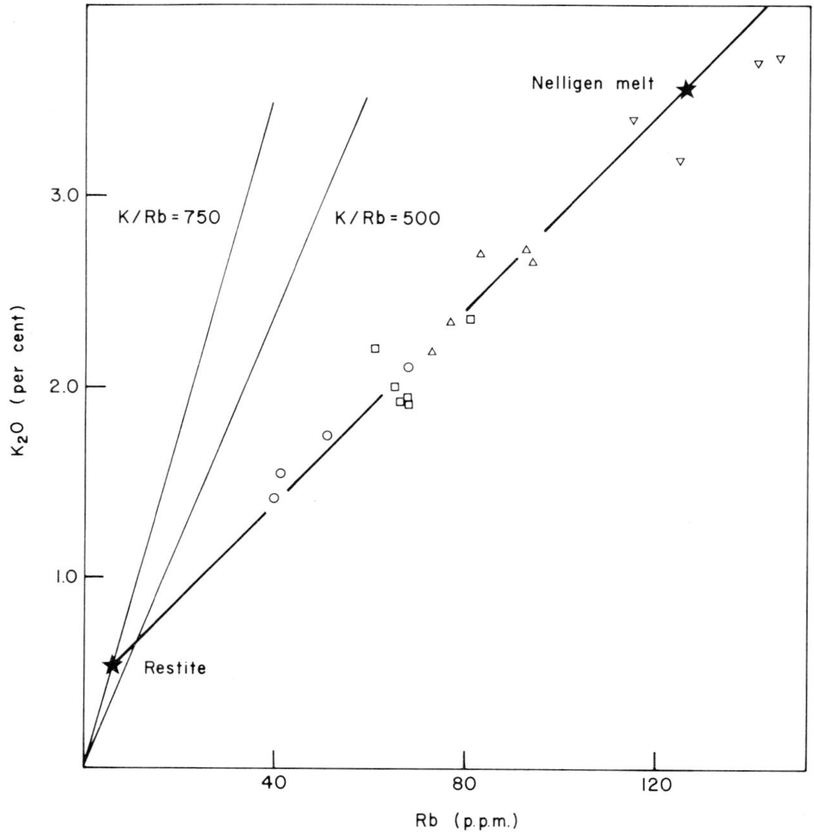

Fig. 6. K_2O vs. Rb variation for Moruya Suite rocks. The correlation line is an unmixing line between the melt end-member having low K_2O/Rb and the restite having high K_2O/Rb. ▽, Nelligen Granodiorite; □, Moruya Tonalite; ○, Tuross Head Tonalite; △, other plutons

the Tuross Head and Moruya Tonalite analyses is discarded, a small difference is marginally detected in the Rb variation of the two plutons, which may indicate slightly difference Rb contents in their source. We reject the Rb vs. SiO_2 regression as a valid representation of the original Rb variation, as it shows far too much Rb at 51.6 per cent SiO_2 to be accommodated in a biotite-free model restite. We consider that Rb has been disturbed relative to SiO_2 during the continued reaction between restite xenoliths and granitoid matrix in the time interval between melting at depth and final emplacement in the upper crust.

Combining the above Rb and Sr data, we estimate that the source rocks for the Nelligen Granodiorite had a mean Rb/Sr of 0.146, whereas those for the remainder of the Moruya Suite were some 60 per cent lower at ~0.09 (Table 3). Consequently, we would expect in general that the Nelligen source rocks had a higher mean $^{87}Sr/^{86}Sr$ than the others, and therefore that the initial $^{87}Sr/^{86}Sr$ of the Nelligen Granodiorite would be greater than that of the other plutons. This is exactly the situation displayed in Fig. 5 with the Nelligen Granodiorite showing 0.705 14±0.000 07 for the initial $^{87}Sr/^{86}Sr$ as compared with 0.704 25±0.000 07 for the others. For this situation, it would be incorrect to construct a total-rock isochron by combining the Nelligen Granodiorite data with the others, and we consider that the 425±5 million year total-rock age derived in this way earlier should be disregarded. It is possible that other systematic differences in initial $^{87}Sr/^{86}Sr$ exist between plutons within the Moruya Suite. For example, the Tuross Head Tonalite data in Fig. 6 appear slightly lower on average than that of the Moruya Tonalite, and Table 3 indicates the possibility that it might be slightly lower in source Rb/Sr also. More intensive sampling would clarify these indications. Fig. 7 shows the initial $^{87}Sr/^{86}Sr$ as measured on each individual sample plotted against the comparative value for the source $^{87}Rb/^{86}Sr$ calculated for the same sample. The distinct positive correlation could signify that the separate source regions of the different granitoids may have acquired their different values of Rb/Sr as early as ~350 million years prior to the production of the granitoids at ~390 million years ago. The implications of this interpretation and some variations of detail will be discussed when the source parameters for the other south-east Australian granitoids have been presented.

4.2. The Berridale Batholith

The geology and geochemistry of the I- and S-type plutons that constitute this batholith have been described by White *et al.* (1977). Mica ages by Rb–Sr have been reported by Williams *et al.* (1976) and Williams (1977) has made a detailed study of the geochronology using zircon and monazite U–Pb. Here we will report the Rb–Sr systematics of eleven I-type plutons for the purpose of reconstructing the Rb–Sr parameters of their source regions. The primary data for these bodies are given in Tables 4A and 4B, but, for clarity, the results will be discussed using particular groups of plutons that have similar major element chemistry.

Buckleys Lake, Maffra and Delegate Adamellites

Although small differences can be seen in the chemistry of these three bodies, the dominant feature is their similarity compared with other plutons within the batholith. All are highly felsic and only the Buckleys Lake Adamellite has sufficient dispersion in SiO_2 to define regression lines for the

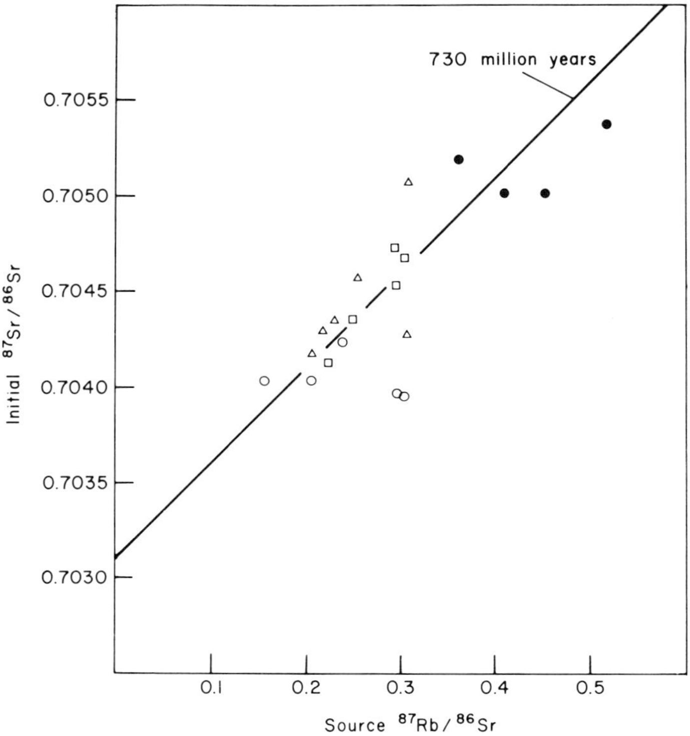

Fig. 7. Source-rock isochron for the Moruya Suite, in which the Rb, Sr and $^{87}Sr/^{86}Sr$ data for each sample are employed to estimate the source $^{87}Sr/^{86}Sr$ at 390 million years ago and the source $^{87}Sr/^{86}Sr$. The source rocks are taken as a 36:64 mixture of melt and restite, corresponding to the mean SiO_2 of regionally-averaged island arc volcanics. The indicated "age" refers to 730 million years before the present. ●, Nelligen Granodiorite; □, Moruya Tonalite; ○, Tuross Head Tonalite; △, other plutons.

major elements of reasonable accuracy. For zero opaques, normative quartz vanishes at 49.7 per cent SiO_2 and normative orthoclase vanishes at 46.5 per cent SiO_2 in the presence of 11 per cent normative opaques. We have taken the mean value, 48.1 per cent SiO_2, for the model restite, and assumed that the same value is applicable for the restite of the Delegate and Maffra Adamellites.

There are obvious differences in the Rb and K_2O variation lines (Fig. 8) but Rb and K_2O for the combined granitoids are well correlated, and we have combined the latter with material balance in restite K_2O and D_R^{Rb} to deduce the Rb and K_2O contents of the model restite. We employed the Rb vs. K_2O regression to predict the Rb content of the restite, excluding the highest of the Buckleys Lake Rb values, which appears to be an outlier, and fitted the variation line to the centroid (203.7 p.p.m. Rb, 73.35 per cent SiO_2) of the Buckleys Lake data. This procedure gives 5.9 p.p.m. Rb, 0.337 per cent K_2O and 1.6 per cent K-feldspar for the restite, 225 p.p.m. Rb for the melt, and 84.8 p.p.m. Rb for the source of the Buckleys Lake Adamellite assuming 36 per cent for F (Table 3), and the corresponding Rb variation lines are shown in Fig. 8. The same relationships have been employed to calculate the separate restite parameters for the Maffra and Delegate Adamellites. To a good approximation, the value for D_R^{Rb} is the same for all three granitoids at 0.0262. It is seen from Fig. 8 that any regression of Rb against SiO_2 gives only a very poor representation of the true Rb variation line. In contrast to this, the restite K_2O content is well predicted by the Buckleys Lake K_2O vs. SiO_2 regression at 0.21 per cent K_2O.

The Sr variation for the Buckleys Lake Adamellite is distinctive and well-fitted (Fig. 9), much steeper and therefore probably colder than any other granitoid discussed so far. The Sr contents of the melt (66.1 p.p.m.) and model restite at 48.1 per cent SiO_2 (797 p.p.m.) are adequately defined despite the long extrapolation, and lead to the comparative value of 534 p.p.m. for the Buckleys Lake source (Table 3). If we assume that the other two granitoids equilibrated with the same model restite at the same temperature, then the two separate Sr variation lines shown in Fig. 9 are defined. They indicate that the Sr contents of the source rocks for both plutons were different to each other and that both were higher than the source of the Buckleys Lake (Table 3). The indicated temperature using (5) is approximately 800 °C, apparently much lower than the Moruya Suite.

The Buckleys Lake biotite age is 417 million years (Williams *et al.*, 1976) but the total-rock age is model dependent. If all the scatter in the isochron is assigned to random variation in initial $^{87}Sr/^{86}Sr$ we obtain an age of 422 ± 10 million years (2σ) and mean initial $^{87}Sr/^{86}Sr$ of 0.7068 ± 0.0001 (2σ), but if the scatter is taken as proportional to Rb/Sr, the age becomes 433 ± 14 million years. The values for initial $^{87}Sr/^{86}Sr$ in Table 3 are calculated for 422 million years. The Delegate total-rock age, 452 ± 13 million years, is less model dependent, but definitely exceeds its biotite age of 415 million years. Monazite also gives older ages than the biotites, and these discrepancies and the question of the true age of emplacement of the Delegate Adamellite have been discussed elsewhere (Williams, 1977). Whether the 452 million year age refers to emplacement or to a preceding melting event, we consider that the total-rock alignment correctly gives the values for "initial" $^{87}Sr/^{86}Sr$ required to specify the source at 452 million years. The Maffra Adamellite is a rare instance in which mineral ages by several different methods and a two-point total-rock age are concordant at 413 million years.

Table 4A. Analytical data for Berridale Batholith rocks. Xenoliths are denoted ×.

ANU #		SiO_2	K_2O	Rb	Sr	$^{87}Sr/^{86}Sr$
Tara Granodiorite						
27544		66.76	2.25	86	246	0.712 05
27474		66.78	2.25	88	256	0.712 14
27531		67.14	2.21	86	281	0.711 27
16251		66.80	2.15	84	246	0.712 06
27542		67.32	2.24	86	252	0.712 03
27543		67.67	2.31	84	278	0.711 29
27541		67.23	2.25	85	245	0.712 09
27487		67.44	2.29	89	260	0.712 08
27514		69.11	2.71	115	207	0.715 62
30170	×	54.45	2.91	125	221	0.715 45
30173	×	56.16	1.49	61	232	0.710 55
30171	×	55.03	1.88	83	248	0.711 31
30169	×	54.68	1.74	73	248	0.711 31
30165	×	53.78	1.51	75	260	0.711 20
30164	×	54.35	1.79	82	254	0.711 44
30168	×	55.14	2.30	95	251	0.712 54
30163	×	54.21	1.64	68	262	0.710 72
30167	×	55.63	2.25	96	256	0.712 31
30174	×	56.33	1.39	56	245	0.709 79
30172	×	56.61	2.24	91	268	0.711 90
30176	×	57.08	1.20	54	251	0.709 64
30175	×	57.41	1.50	70	276	0.709 77
30166	×	57.93	1.96	77	259	0.711 28
Wullwye Granodiorite						
27505		74.86	3.25	119	189	0.716 17
27549		74.77	3.16	130	186	0.717 27
27499		74.87	3.20	124	184	0.717 01
Maffra Adamellite						
27497		75.16	3.81	177	121	0.729 86
27489		76.56	4.16	214	99	0.741 73
Buckleys Lake Adamellite						
27565		71.89	4.19	190	166	0.726 30
27482		71.15	3.98	190	191	0.723 52
27566		72.13	3.76	195	175	0.725 55
27479		72.96	4.32	212	152	0.730 44
27524		73.74	4.17	207	125	0.735 51
27498		73.86	4.05	204	128	0.734 64
27480		76.37	4.73	268	59	0.785 05
27496		74.68	4.71	209	89	0.748 33
27527	×	60.78	2.06	223	98	0.743 72
30202	×	61.87	1.58	148	161	0.720 99
30204	×	62.90	1.57	118	185	0.716 71
30203	×	67.55	1.42	151	160	0.722 77
30201	×	66.75	2.36	181	153	0.723 84

Table 4A—*continued*

ANU #		SiO_2	K_2O	Rb	Sr	$^{87}Sr/^{86}Sr$
Delegate Adamellite						
27530		73.13	3.81	176	144	0.728 27
27559		73.73	3.41	159	147	0.725 47
18679		74.07	3.51	164	140	0.727 07
27560		73.89	3.59	173	131	0.729 06
18678		74.55	3.50	155	137	0.726 38
27561		74.98	3.53	157	130	0.728 33
30209	×	57.56	1.38	73	274	0.710 19
30207	×	62.08	1.63	160	157	0.724 03
30208	×	65.86	1.90	143	161	0.722 55
30205	×	63.65	1.78	103	275	0.712 45
30206	×	66.39	1.92	189	172	0.725 54
Finister Granodiorite						
27519		64.11	2.22	120	135	0.726 05
27518		64.06	2.31	114	131	0.725 63
30214		64.89	2.55	126	127	0.729 29
27556		64.54	2.55	132	135	0.728 51
27488		65.09	1.90	81	149	0.719 60
30215		64.68	2.59	127	127	0.729 10
30212		64.87	2.61	123	137	0.727 10
30213		66.53	2.71	137	129	0.730 07
30210		67.00	2.59	127	121	0.730 44
27477		67.72	2.92	142	119	0.732 35
30221	×	64.43	2.33	113	133	0.725 50
Merumbago Granodiorite						
30218		64.63	2.63	123	146	0.725 41
27504		65.22	2.57	130	139	0.726 91
30217		65.14	2.78	136	135	0.727 90
30219		65.36	2.64	129	141	0.726 58
30220		65.72	2.75	135	156	0.726 34
Currowong Granodiorite						
27562		63.76	2.42	98	258	0.712 06
18683		64.71	2.60	119	247	0.715 18
18682		66.61	2.77	126	231	0.717 11
18685		67.09	2.80	120	237	0.716 88
18684	×	57.39	1.55	70	261	0.711 64
Bimbimbie Granodiorite						
27517		66.35	2.75	126	242	0.717 63
18687		68.05	2.98	130	236	0.717 89
Iona Granodiorite						
27557		63.06	2.24	95	277	0.712 42
27529		62.67	2.16	86	271	0.712 05
19012		64.22	2.45	104	264	0.713 57
27558		70.03	3.44	148	197	0.719 87

Table 4B. Mineral data for ANU sample 30215, Finister Granodiorite.

	Rb (p.p.m.)	Sr (p.p.m.)	$^{87}Rb/^{86}Sr$	$^{87}Sr/^{86}Sr$
K-feldspar	267.0	163.7	4.726	0.739 79
Hornblende	9.65	19.90	1.402	0.720 35
Plagioclase 1	53.9	227.1	0.686	0.716 03
Plagioclase 2	104.8	280.9	1.079	0.718 14

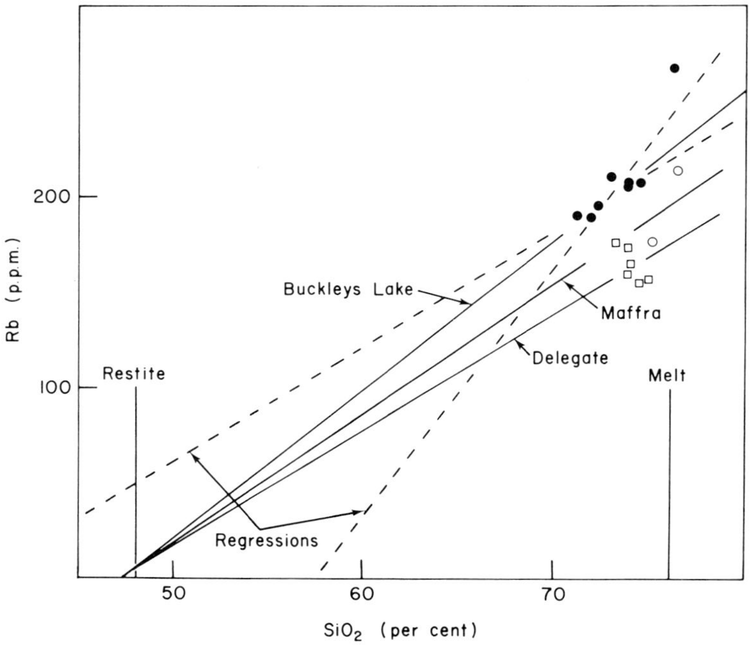

Fig. 8. Variation of Rb with SiO_2 for the Buckleys Lake(●), Maffra(○) and Delegate(□) Adamellites of the Berridale Batholith, based on the use of distribution coefficients to calculate the restite Rb content. The two broken lines represent regression lines for the Buckleys Lake data alone, with and without the highest Rb sample, and both are poor representations of the true Rb variations. The source SiO_2 for the three plutons is probably ~60 per cent, and it is clear that the Rb content of the Buckleys Lake source is greater than that of Maffra and Delegate.

The values for source Rb/Sr in Table 3 show an approximate positive correlation with the initial $^{87}Sr/^{86}Sr$, resembling that of the Moruya Batholith. The Maffra source point cannot be distinguished from that of the Moruya Suite but the other two are above the 730 million year reference line of Fig. 7. However, before close comparisons can be made, it is necessary to

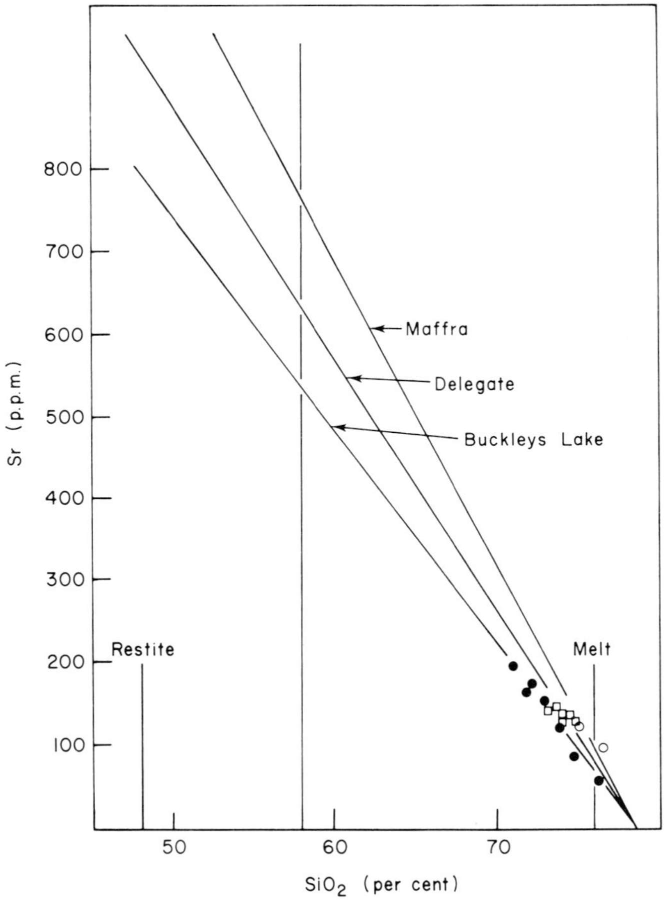

Fig. 9. Variation of Sr with SiO_2 for the Buckleys Lake, Maffra and Delegate Adamellites, based on the well fitted regression line for the Buckleys Lake data alone and assuming the same Sr distribution coefficient between melt and restite for the other two plutons. The Sr content of the Buckleys Lake source region is probably less than that of the Delegate and Maffra.

allow for the variation in age of the different plutons. This adjustment and further discussion will be reserved until the complete body of data has been presented.

Wullwye Granodiorite

This small, highly felsic body intrudes the Buckleys Lake Adamellite and its biotite age of 415 million years probably measures the time of emplacement very closely. The initial $^{87}Sr/^{86}Sr$ is well determined at 0.705 43 from this age, using three separate samples.

The Granodiorite is uniquely high in Na_2O/K_2O in comparison with other granitoids in the Berridale Batholith and is characteristically different in its content of a number of trace elements, signifying an origin from a separate source region. There is no detectable dispersion in its chemistry, so that Sr and Rb variation lines for Wullwye cannot be defined empirically. However, we can estimate the melt and restite Rb by assuming that its Rb vs. K_2O variation resembles that of the Moruya Suite (on which the observed point lies), and that its model restite contains 50 per cent SiO_2 and 65 per cent plagioclase, as observed for most other granitoids. The relationships (a)–(d) of Section 3 may then be combined with the observed chemical composition to estimate 130 p.p.m. Rb in the melt and 4.16 p.p.m. Rb in the restite (Table 3). Similarly, we may assume that the equilibration temperature of the Wullwye magma was the same as, for example, the Moruya Suite, so that we obtain 4.61 for D_R^{Sr} from (5), from which the Sr variation line can be determined using the relationship (d). We emphasize that the slope of the variation line, and hence the restite and melt Sr contents, are strongly dependent upon the assigned temperature. It is likely that the Moruya Suite temperature ("950 °C") is a maximum value and the Buckleys Lake ("800 °C") a minimum. The figures given in Table 3 for the Wullwye restite and melt Sr correspond to a compromise temperature of "895 °C".

Currowong, Bimbimbie and Tara Granodiorites

The Currowong and Bimbimbie Granodiorites cannot be distinguished from each other geochemically and are grouped as the Currowong Suite. The Tara Granodiorite shares many chemical features, e.g. high Ti, P, but is distinctive because of its high Na and lower K, Rb, Th and Cr. None of them separately has much dispersion in SiO_2 content, and, for Tara, the range comprises a tight cluster of eight samples and a single sample which is 2.0 weight per cent higher. Regression through this single sample causes K_2O to decrease so rapidly with SiO_2 that normative orthoclase disappears before normative quartz. This does not occur for other granitoids in which the regression lines are well defined, and we will assume therefore that it is incorrect. Instead, we have estimated the restite K_2O for Tara by combining the data for all three granitoids to define the Rb vs. K_2O variation and applying the same material balance relationships used already for other granitoids. We employed 51.5 per cent SiO_2 for the Tara restite, this being the mean between limiting values of 53.2 and 49.9 per cent set by the normative quartz argument, and fitted the Rb variation line to the strong cluster of Tara analyses at 88 p.p.m. Rb, 67.2 per cent SiO_2 (Fig. 10). The melt SiO_2 was taken as 70 per cent, as the Grandiorites are non-minimum melts. The results for the restite and melt are shown in Table 3. We view the single isolated Tara point that controls the K_2O and Rb vs. SiO_2 regressions as an outlier, at least for K_2O and Rb.

In contrast to the Tara Granodiorite, the data for the Currowong Suite are evenly spread over their range of SiO_2 and the regressions of K_2O, Rb vs. SiO_2 are fairly well fitted. Normative quartz vanishes between SiO_2 limits of 52.3 and 49.6 per cent, giving 51.0 per cent SiO_2 for the model restite, at which, according to the regression lines, K_2O is 0.93 per cent and Rb 27 p.p.m. Calculating values for the model restite in the usual manner, we find 1.07 per cent K_2O, which is good agreement, and 9.8 p.p.m. Rb (Table 3), which suggests that the Rb vs. SiO_2 regression is again unreliable. Alternatively, it may indicate that the value selected for the restite SiO_2 content is too high and that the lower limit, 49.6 per cent SiO_2, for which the regression line indicates 18.4 p.p.m. Rb, is more appropriate for this granitoid.

The Sr variation with SiO_2 for the three granitoids is not well defined (Fig. 10) and in order to estimate the source Sr, we have chosen to model the Sr variation on assumed values for the equilibration temperature. In view of the apparent temperature of 950 °C for the Moruya Suite discussed previously, which must be less than that of the Tara as the Moruya Suite is a minimum melt, we will assume a comparative temperature of "1020 °C". From this, the temperature curve (5) gives 4.76 for D_{PL}^{Sr}, which leads to 2.69 for D_R^{Sr} and thus to the variation line (Fig. 10) fitted to the centroid of the measured data. For the Currowong Suite, we have assumed equilibration at "1070 °C" and used the same procedure, which gives a good fit to the observed data.

The total-rock Rb–Sr data for the Tara Granodiorite, including the mafic xenoliths, is comparatively well fitted (although not quite to within experimental error), and gives an age of 418 ± 24 million years, in agreement with Rb–Sr biotite ages close to 411 million years (Williams *et al.*, 1976). Only biotite ages of 415 million years are available for the Currowong and Bimbimbie Granodiorites but values for their initial $^{87}Sr/^{86}Sr$ calculated using the biotite age (Table 3) are greater than that of the Tara Granodiorite, in accord with their higher source Rb/Sr. We do not know why the initial $^{87}Sr/^{86}Sr$ is apparently rather widely scattered for the Currowong Suite and further analytical work is necessary here.

Iona Granodiorite

This is an isolated pluton a little south of the others emplaced in Eastonian sediments ($\leq$450 million years) and having a well fitted though sparsely defined total-rock isochron of 438 ± 22 million years. The corresponding initial $^{87}Sr/^{86}Sr$, 0.7063 ± 2, is not distinguishable from that of the Tara Granodiorite, but this is a consequence of the greater age used for the Granodiorite. The slope of the Iona isochron is controlled principally by a single sample, and if the internal scatter within the Iona Granodiorite is comparable with that observed in many other granitoids, the 438 million year result may be spuriously high. If the same age of 415 million years is

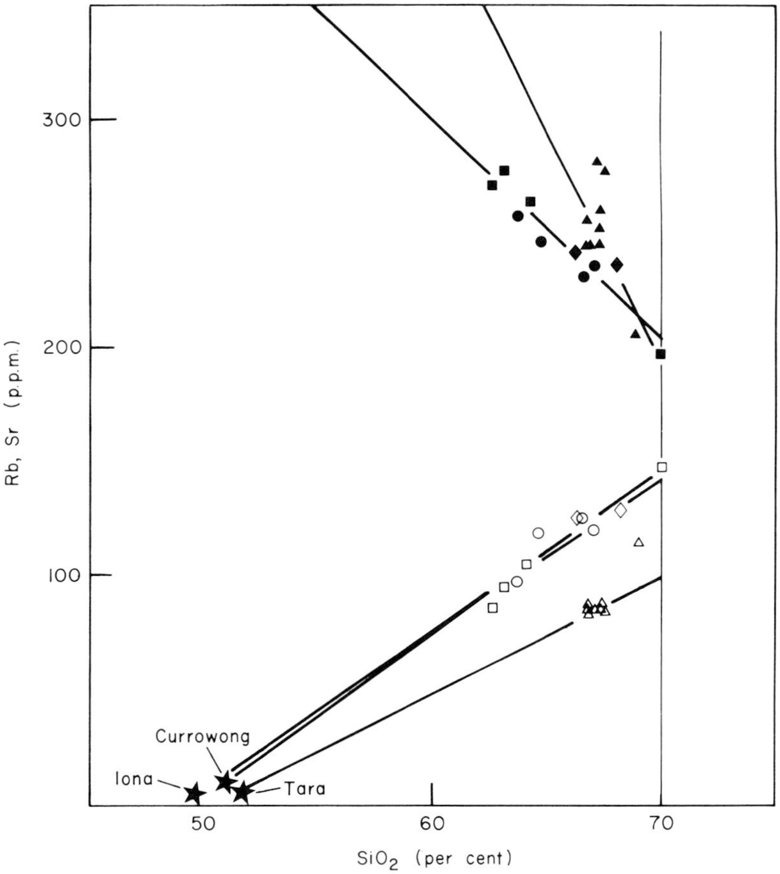

Fig. 10. Variation of Rb, Sr and SiO_2 for the Tara, Iona, Currowong and Bimbimbie Granodiorites of the Berridale Batholith. The Rb variation lines use restite Rb contents calculated from Rb distribution coefficients and the K_2O vs. Rb variation line. It is plain that the Tara source has a much lower Rb content than the others. The Sr variation is based on the Sr regression line for the Iona Granodiorite, and on the assumption that the temperatures of the fusion events were comparable for the other three plutons. ▲, Tara Sr; △ Tara Rb; ■, Iona Sr; □, Iona Rb; ●, Currowong Sr; ○, Currowong Rb; ◆, Bimbimbie, Sr; ◇, Bimbimbie, Rb.

used for all four plutons, the initial $^{87}Sr/^{86}Sr$ for the Iona Granodiorite will be intermediate between that of the Tara and Currowong Granodiorites at 0.7068 ± 0.0001.

K_2O and Rb are well correlated with SiO_2, but both variation lines are too steep to be internally consistent with material balance in the restite. Using the relationships described previously and taking 49.5 per cent SiO_2 for the model restite, we obtain 0.31 per cent K_2O and 3.48 p.p.m. Rb for the model restite, and 3.30 per cent K_2O and 142.1 p.p.m. Rb for the melt. In comparison, the regression fits appear to be "too good", with the highest point having slightly too much K_2O and Rb and the lowest, slightly too little.

The Sr variation for the Iona Granodiorite is similar to the assumed slope for the Currowong Suite and gives an apparent temperature of "1085 °C". This is exactly in the range expected relative to the other granitoids, which gives confidence to our use of temperature in estimating Sr variation lines for limited data. Table 3 lists the restite and melt parameters with the initial $^{87}Sr/^{86}Sr$.

Finister Suite

This comprises two bodies, the Finister Granodiorite and the Merumbago Granodiorite, which are indistinguishable from each other chemically and petrographically but are grossly different from all other plutons in the Berridale Batholith. More specifically, relative to the other I-type plutons, they are low in Al, Na, P, Sr and Y and high in Fe, Mg, Rb, Sc, V, Cr, Co, Ni and Cu, and both must have formed from at least a very similar source region. In addition, although many samples have been analyzed, the dispersion in SiO_2 content is so small that the precise empirical definition of Rb and Sr variation lines is not possible using the observed data alone. It is necessary instead to model the Sr variation on an assumed equilibration temperature and to set limits for the Rb and K_2O variations by estimating a model restite.

We will assume "1070 °C" for the apparent temperature of the (non-minimum) melt which is consistent with the apparent equilibration temperatures of other non-minimum melts. We will employ the (uncertain) extrapolations of the major elements towards low SiO_2 to define the restite chemistry by the disappearance of normative quartz and orthoclase. This occurs at 46.9 per cent SiO_2 (quartz zero, no opaques) and at 45.1 per cent SiO_2 (orthoclase zero, 8.1 per cent opaques), and we will select the mean at 46.0 per cent SiO_2. At this point, the normative plagioclase is 51.1 per cent, which can be combined with 3.59 for D_{PL}^{Sr} from the temperature curve. Assuming a K-feldspar content for the restite of 1 per cent (see later), D_R^{Sr} can be evaluated as 1.86. This can be combined with the means of the observed data, 129 p.p.m. Sr for the Finister Granodiorite and 143 p.p.m. Sr for the Merumbago at 65 per cent SiO_2, to determine the restite and melt Sr contents (Fig. 11, Table 3).

The K_2O vs. Rb correlation, unusually, is not very well fitted and we have been obliged to employ a sample which is an outlier on the K_2O and Rb vs. SiO_2 diagrams to obtain even moderate definition of the line. Nevertheless, we have combined the Rb vs. K_2O regression line in the usual way with material balance for the restite in K_2O and in Rb, via D_R^{Rb}. The calculations

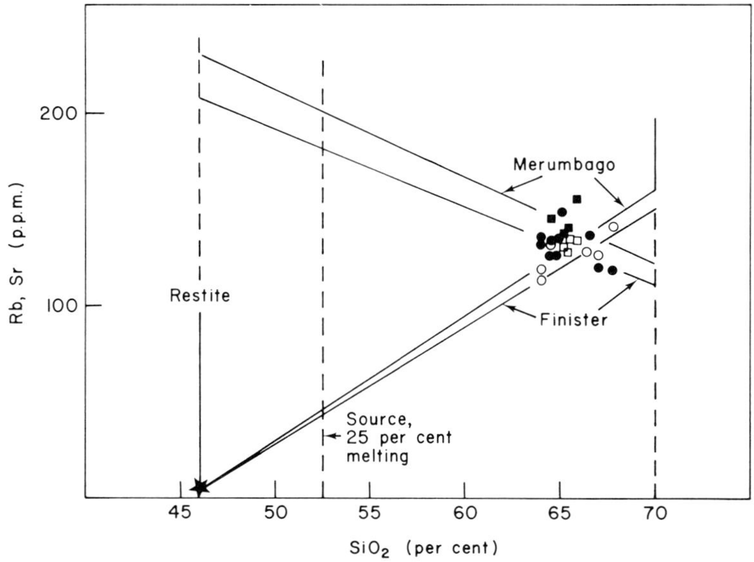

Fig. 11. The variation of Rb, Sr and SiO_2 for the Finister and Merumbago Granodiorites, based on restite Rb, Sr contents calculated using Rb and Sr distribution coefficients. The small difference predicted between the source Rb/Sr of these plutons correlates with a difference in their initial $^{87}Sr/^{86}Sr$, ●, Finister Sr; ○, Finister Rb; ■, Merumbago Sr; □, Merumbago Rb.

give 1.1 per cent K-feldspar, 0.23 per cent K_2O and 3.0 p.p.m. Rb for the Finister restite and 158 p.p.m. Rb for the melt. For the Merumbago Granodiorite, we obtain slightly greater Rb in the source rocks (Table 3).

The total-rock Rb–Sr data for Finister alone, and for Finister combined with Merumbago, define an apparent isochron at 497 ± 37 million years that exceeds the stratigraphic age limit of ~ 450 million years. It also exceeds the biotite ages at 418 million years (Williams *et al.*, 1976) and a feldspar–hornblende internal isochron of 413 ± 9 million years. Williams (1977) has shown that the Finister Granodiorite is at least 440 million years old, from the monazite U–Pb age for a younger pluton. These discordant results probably indicate that the minerals and at least some total-rock samples have been open to Rb and/or Sr movement since crystallization, which may be the reason for the scatter in the alkali variation diagrams, and we consider that the initial $^{87}Sr/^{86}Sr$ is best calculated using the 450 million year maximum age limit. Table 3 shows a slight difference in mean initial $^{87}Sr/^{86}Sr$ between Finister and Merumbago which correlates positively with the small difference in their source Rb/Sr. The latter is substantially higher than the source Rb/Sr for any of the other granitoids discussed, which makes the Finister–Merumbago data particularly valuable in the interpretations to follow. The higher Rb/Sr is not immediately matched by a proportionately higher initial $^{87}Sr/^{86}Sr$, but comparisons have been obscured by the different ages taken for the melting events, as we will shortly clarify.

4.3. The Kosciusko Batholith

The I-type granitoids of the eastern Kosciusko Batholith have been described by Hine *et al.* (1978), who recognized the petrographic and geochemical affinity of several of the plutons, ranging from Al-rich gabbro to granodiorite, which they termed the Jindabyne Suite. Rb–Sr biotite ages by Williams (1973) give 412 million years as a minimum estimate for the age of emplacement of one member of the Suite (the Round Flat Tonalite), and a monazite age of 431 million years from an S-type pluton intruded by members of the Jindabyne Suite can be taken as a maximum estimate for

Table 5. Analytical data for Jindabyne suite rocks.

ANU #	SiO_2	K_2O	Rb	Sr	$^{87}Sr/^{86}Sr$
Round Flat Tonalite					
31863	60.18	1.66	70	253	0.710 70
31862	60.56	1.55	62	254	0.710 19
31861	66.07	2.00	86	197	0.714 62
31860	66.71	2.09	90	190	0.715 56
Gaden Tonalite					
31864	63.00	2.00	85	237	0.712 87
31866	63.28	2.10	89	233	0.712 78
31859	66.28	2.34	106	190	0.715 99
34376	66.79	2.21	102	212	0.715 27
31865	67.15	2.39	106	206	0.716 30
31865	67.15	2.39	106	206	0.716 30
Jindabyne Tonalite					
34368	62.29	2.01	89	254	0.713 62
34370	63.43	2.22	97	225	0.714 44
34367	63.06	2.09	93	217	0.713 78
34375	64.35	2.38	102	218	0.715 13
34364	64.64	1.99	193	216	0.714 58
34366	65.43	2.52	108	204	0.715 70
Pendergast Tonalite					
34380	65.87	1.88	105	258	0.713 92
34381	66.31	1.96	113	234	
Other plutons					
34386	63.66	2.17	92	237	
31874	63.67	2.18	91	226	
34371	63.81	2.01	91	226	
34372	62.44	1.90	85	236	

their age of emplacement (Williams, 1977). Stratigraphically, all plutons are younger than Eastonian, i.e. about 450 million years. We have employed an age of 420 million years to calculate the initial $^{87}Sr/^{86}Sr$. Table 5 lists the samples for which Sr isotopic analyses are available.

The K_2O vs. Rb plot (Fig. 12) shows the strong coherence of the Jindabyne Suite proper and separates the Pendergast Tonalite from it, as

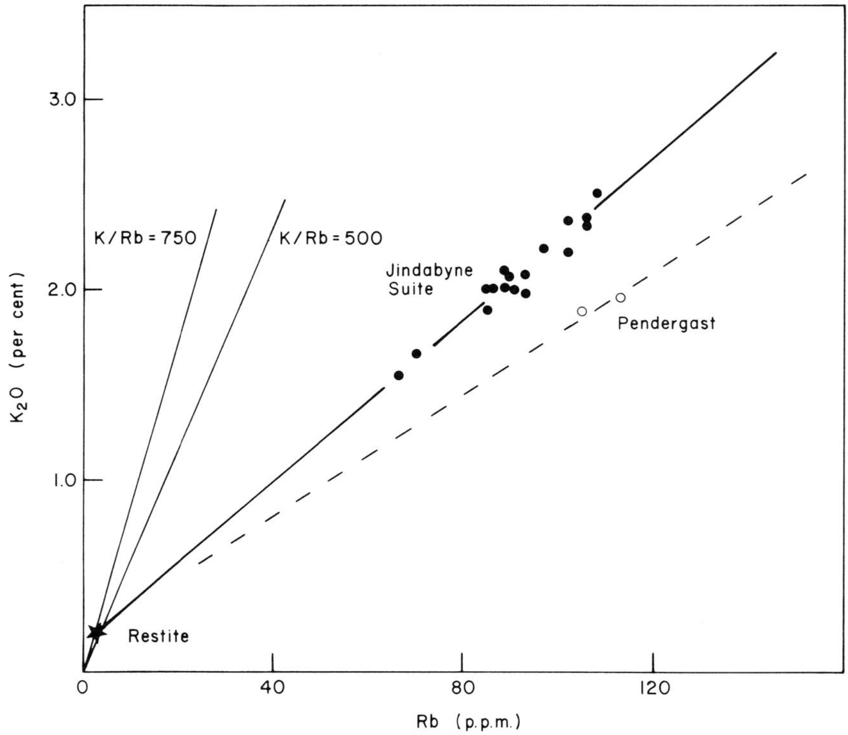

Fig. 12. The well defined K_2O vs. Rb correlation of the Jindabyne Suite, which separates the Pendergast Tonalite as being generated from a different source, and which is employed in the calculation of restite composition for the Rb variation lines of individual plutons shown in Fig. 13.

does the Sr vs. SiO_2 plot. From the major element regressions, the limits for the SiO_2 content of the model restite are 50.76 per cent (normative quartz and opaques zero) and 47.06 per cent (normative quartz zero with 7.7 per cent opaques). In this instance, we have selected the lower SiO_2 limit in order to obtain the best agreement between the calculated and observed K_2O variation lines. Table 3 lists the parameters for the model restite and Fig. 13 shows the Rb variation lines. We have assumed the same value of

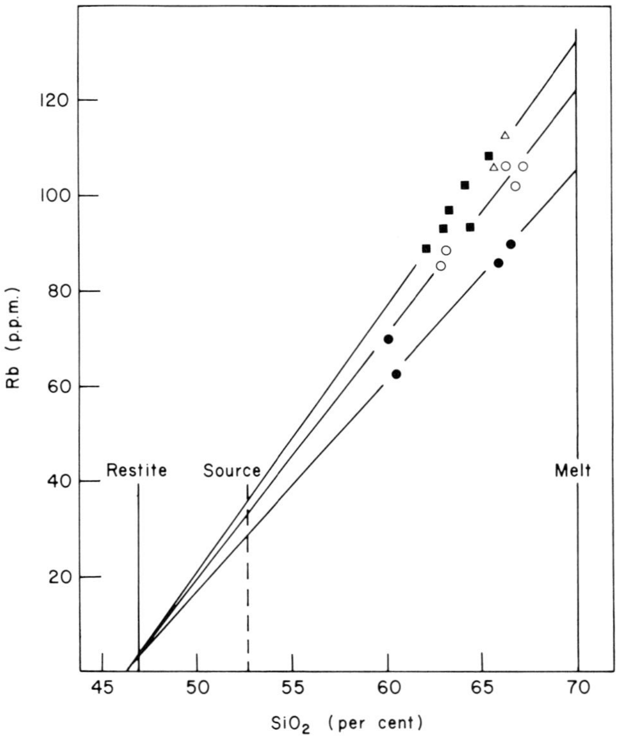

Fig. 13. The variation of Rb and SiO_2 for I-type plutons within the Kosciusko Batholith, based on restite Rb contents calculated using Rb distribution coefficients. Regression lines fitted to all data or to those of each plution separately are considered to be only poor estimates of the true Rb variation. △, Pendergast, ○, Gaden; ●, Round Flat; ■, Jindabyne.

D_R^{Rb} for each individual pluton. There are obviously systematic differences in Rb between plutons, so that a single Rb vs. SiO_2 regression fitted to all the data will be a poor measure only of the true variation lines. To a lesser degree, the same is true for the K_2O vs. SiO_2 regression, which indicates 0.62 per cent K_2O for the model restite as compared with 0.26 per cent calculated. The calculated Rb variation lines are well fitted to the measured data, with the exception of one sample from the Round Flat Tonalite.

The Sr variation with SiO_2 is not precise, but it is plain from Fig. 3 that there are only small differences between plutons except for the Pendergast Tonalite. Regression of all the data gives 2.24 for D_R^{Sr} and leads to 3.64 for D_{PL}^{Sr}, for 1.2 per cent K-feldspar and 60.6 per cent plagioclase in the model restite. From the temperature curve (5), this gives "1070 °C" for the equilibration temperature, which is similar to the values found for other non-minimum melts in which the Sr variation is better defined. Each pluton has been assigned the same D_R^{Sr} and variation lines fitted to their separate centroids to obtain the data in Table 3.

Within the Jindabyne Suite proper, the comparative values for source Rb/Sr correlate with the initial $^{87}Sr/^{86}Sr$ (Table 3) and both are comparable with results from granitoids previously discussed. In detail, the Pendergast source is slightly different.

5. Source Rock Rb–Sr Isochron

Let us suppose that the source rocks that produced the various plutons discussed above still remain in the lower crust at the present day. We may calculate the (mean) present-day $^{87}Sr/^{86}Sr$, R_p, for each source rock from the (mean) initial $^{87}Sr/^{86}Sr$ of the pluton, R_τ, and the (known) time τ either of melting or emplacement, which was used to determine the initial $^{87}Sr/^{86}Sr$.

Thus,

$$R_P = R_\tau + (e^{\lambda\tau} - 1)(^{87}Rb/^{86}Sr)_S\,, \qquad (7)$$

where $(^{87}Rb/^{86}Sr)_S$ denotes the present-day source $^{87}Rb/^{86}Sr$. The same term R_p is related to the mean initial $^{87}Sr/^{86}Sr$, R_{τ_0}, and the age, τ_0 of the *source rock*, both of which are unknown, by the standard isochron equation:

$$R_P = R_{\tau_0} + (e^{\lambda\tau_0} - 1)(^{87}Rb/^{86}Sr)_S\,. \qquad (8)$$

If now, two or more of the source rocks in the lower crust had closely the *same* age and initial $^{87}Sr/^{86}Sr$, we may estimate both of these unknowns through the definition of a *source isochron*. Table 3 lists the comparative values for source $^{87}Rb/^{86}Sr$ determined at the selected values of F for the 17 plutons in the preceding section, and corresponding values of R_P as calculated by equation (7) above. There is a distinct linear correlation between the two, particularly for the non-minimum melts, which implies that a source isochron as expressed by equation (8) may indeed exist. However, before considering the correlation in detail, we shall first remove an unwanted covariance that arises from the Palaeozoic age of crystallization of all granitoids. The ~400 million years of growth in ^{87}Sr in the source regions between the Palaeozoic and the present day has injected a correlation that may obscure the essential observation of whether or not there was a significant correlation between initial $^{87}Sr/^{86}Sr$ and source $^{87}Rb/^{86}Sr$ *at the time of granitoid emplacement.* Consequently, we have subtracted the radiogenic ^{87}Sr accumulated in the various source regions for the past 420 million years (this being the most frequent age value used for calculating an initial $^{87}Sr/^{86}Sr$), which reduces the unwanted covariance to a small quantity. The result, which is the source isochron as observed at 420 million years ago, is shown in Fig. 14. We shall discuss the implications of the diagram for each type separately. To simplify the discussion, we will first model the source rocks themselves upon modern island-arc volcanics.

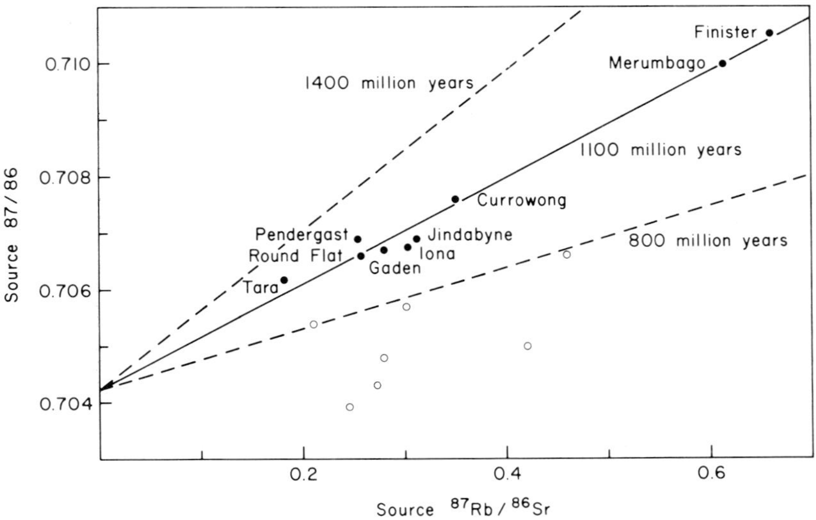

Fig. 14. Source isochron for the non-minimum melt bearing granitoids (●), as registered 420 million years ago. The isochron slope is $e^{\lambda\tau_0}-e^{\lambda\tau}$, where τ_0 is 1100 million years and τ is 420 million years. The source $^{87}Rb/^{86}Sr$ values are calculated for all sources as 25:75 mixtures of non-minimum melt and restite, which is equivalent to ~19 per cent for the minimum melt component *F*, and employ the Rb, Sr variation lines discussed previously. The upper and lower age limits correspond to the limiting constraints on source composition that arise from melt viscosity (Fig. 4) and magma strength, which are equivalent to 15 and 25 per cent for *F*. For comparison, the sources for the minimum melt-bearing granitoids (○) are also shown, calculated for an average *F* of 36 per cent according to the island-arc model. Some, but not all, of the latter could fit the 1100 million year alignment within their lower constraint of 25 per cent for *F*.

5.1. Isotopic Characteristics of Island-arc Volcanics

We have already discussed the SiO_2 contents of island-arc volcanics with respect to their potential content, *F*, of granitic minimum melt. Here we will emphasize that, regardless of their SiO_2 contents, the Sr isotopic compositions of island-arc volcanics are usually uniform and low in comparison with continental rocks. In addition, there is usually no clear correlation between their $^{87}Sr/^{86}Sr$ and Rb/Sr, although such correlations have been suggested locally (e.g. Gill and Compston, 1973). These generalizations are illustrated in Table 1 and Fig. 15(a), for samples from the Fiji–Tonga arc plus others

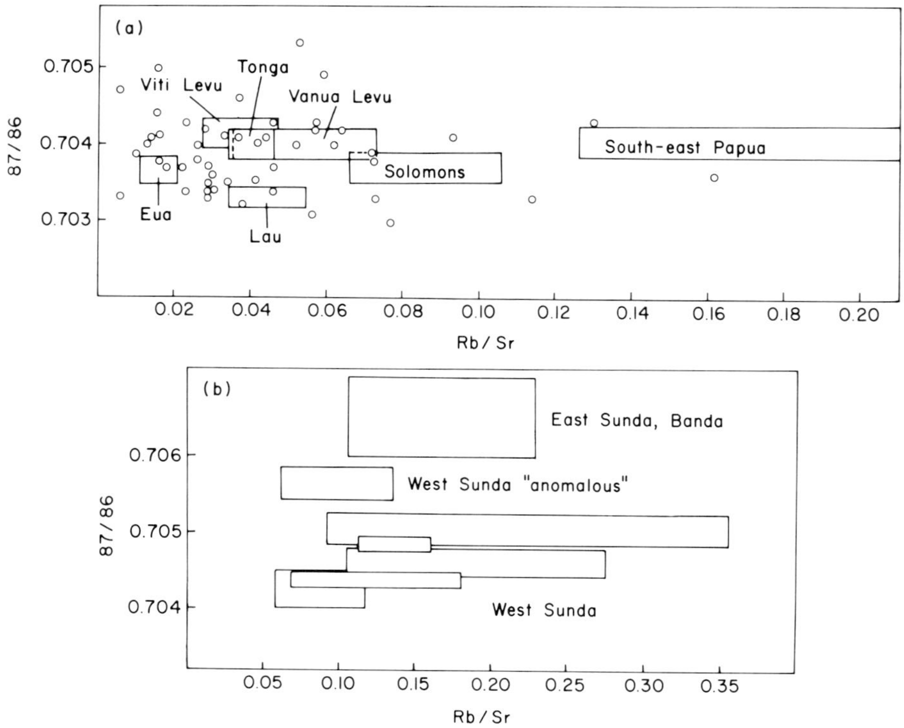

Fig. 15(a). $^{87}Sr/^{86}Sr$ and Rb/Sr for "normal" island-arc volcanics. Individual samples form the Fiji, Lau and Tongan Islands are shown as open circles(○). Regional averages are shown by boxes two standard errors in size. (b) $^{87}Sr/^{86}Sr$ and Rb/Sr for volcanics from the Sunda and Banda arcs which have been increased by continental "contamination".

from Papua and the Solomon Islands. For the seven "regional average" type samples shown in Fig. 15(a), there is a random variation in $^{87}Sr/^{86}Sr$ of ~0.000 35 and zero slope on the $^{87}Sr/^{86}Sr$ vs. Rb/Sr diagram. We regard this as the normal condition for island-arc volcanics that are free of contamination effects from pre-existing older continental material. Similar values for $^{87}Sr/^{86}Sr$ are found in the Marianas (Meijer, 1976) and South Sandwich (Hawkesworth *et al.*, 1977) arcs, to single out two that are particularly remote from continents.

In contrast to the above, we also show regional averages for the Sunda and Banda arcs (Fig. 15(b)) which we consider to be exceptional because of their high mean $^{87}Sr/^{86}Sr$ at 0.7047, cf. 0.7038 in "normal" arcs. For the west Sunda arc alone, the random variation in $^{87}Sr/^{86}Sr$ between regional samples is comparable with "normal" arcs and any correlation with Rb/Sr is at best, poor. However, the addition of the east Sunda and "anomalous" west Sunda groups greatly increases the variability in regional $^{87}Sr/^{86}Sr$ and leads to their explanation as due to some form of contamination by or association with the adjacent Australian continental crust (Whitford *et al.*, 1977). Any correlation with Rb/Sr remains poor.

We shall take the "normal" island-arc volcanics as our model for the granitoid source rocks. Thus we expect that their initial $^{87}Sr/^{86}Sr$ will average about 0.7035 (this being reduced slightly from 0.7038 because the age of the source is at least Palaeozoic), and have a regional variability from one part of the source to another of the order 0.0003. We also expect that there will be little or no correlation between source $^{87}Sr/^{86}Sr$ and mean source Rb/Sr, i.e. that the initial slope of the source isochron is zero. Thus the slope observed for the source isochron will be attributed wholly to radiogenic growth of ^{87}Sr in the source regions after they were first formed, and can be used to calculate the age of the original island arc in the usual way.

5.2. Source Isochron for the Non-minimum Melts

Figure 14 represents the source isochron for granitoids containing non-minimum melts, based on the use of an equal value for the content of minimum melt component F taken as 0.19 and equivalent to 0.25 for f, to calculate the source Rb/Sr for each magma. The good alignment of the points is due not only to the use of equal values for F, but also to the facts (a) that the relative values for calculated source Rb/Sr remain in the same sequential order for a very wide range in F, and (b) that the initial $^{87}Sr/^{86}Sr$ of the various granitoids correlates with this order. The variation in calculated source Rb/Sr with F is shown in Fig. 16. For all equal F, the source Rb/Sr for the Finister Suite is substantially higher than the others, that of the Tara Granodiorite is the lowest, and that of the Currowong Suite is the highest of the remainder for F up to ~ 0.6. The same ordering is shown by the initial $^{87}Sr/^{86}Sr$, and it is this that gives the correlation shown in Fig. 14. The interpretation using the island-arc model is that the age of the original island arc was approximately 1200 million years for a mean original $^{87}Sr/^{86}Sr$ of 0.7035, and that the small scatter of points about this isochron can be attributed both to small variations in F and to the expected variation in regional $^{87}Sr/^{86}Sr$.

5.3. Source Isochron for the Minimum Melts

It is obvious from Fig. 14 that the concepts of equal F and/or equal age do not apply to the source rocks of the minimum melt-bearing granitoids. Equal values for F should not be expected, as we are dealing with source rocks that

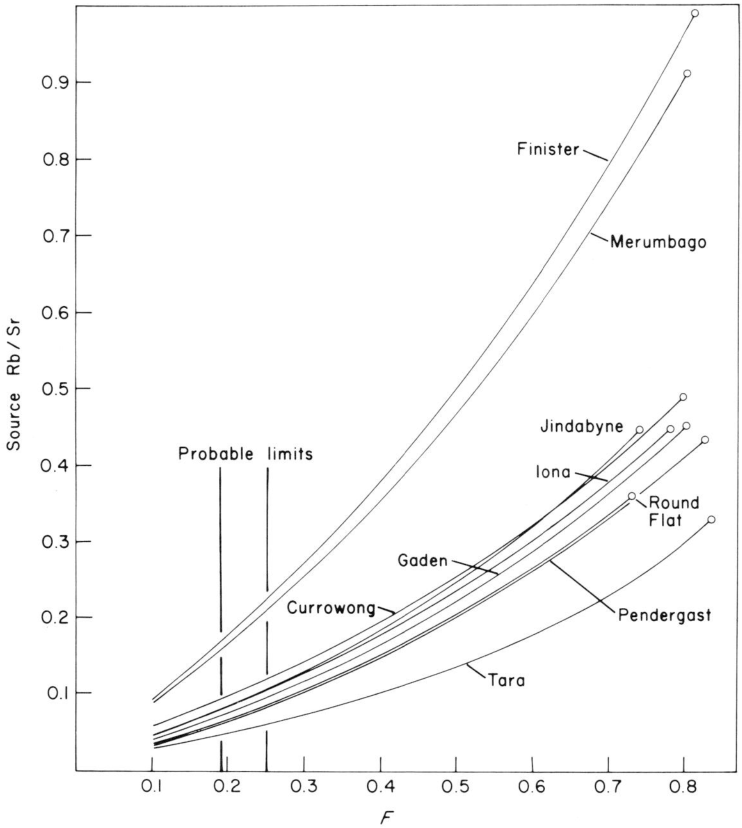

Fig. 16. Source Rb/Sr for the non-minimum melt bearing granitoids as a function of *F*, the fraction of minimum melt component in the source. The limits shown for *F* arise from considerations of melt viscosity and magma strength, and, to first order, *F* may be taken as approximately equal for all granitoids. The diagram shows that for all equal *F*, the order of source Rb/Sr is Finister > Currowong > Jindabyne > Tara, which correlates closely with the order found for the initial $^{87}Sr/^{86}Sr$.

The termination of each line at large *F* in this and in Fig. 17 represents the mean Rb/Sr observed for an individual granitoid.

can vary in *F* from 0.25 to 0.60, according to our previous analysis of the island-arc model. We may explore whether *F* may be reduced for these granitoids to move the points in Fig. 14 nearer to the non-minimum melt alignment. For *F* taken as 0.25, the points for the Buckleys Lake, Delegate and Wullwye Adamellites do in fact fall close to the 1100 million year reference line, but not the Maffra, Nelligen and Moruya Suite points. However, there is a basic inconsistency to this approach. We expect the source regions of the minimum melt-type granitoids to be more siliceous, i.e. to contain higher contents of extractable minimum melt than those giving the non-minimum melts. Such sources would have higher average values for Rb/Sr also, and therefore, if they had the same age, they should have higher average $^{87}Sr/^{86}Sr$. Figure 14 shows that the source $^{87}Sr/^{86}Sr$ for the minimum melt-type granitoids is substantially *lower* on the average than that of the other, contrary to what is expected for an equal age. We conclude from this relative order of the source $^{87}Sr/^{86}Sr$ that probably all the minimum melt source rocks are appreciably younger than the 1200 million year estimate for the other type.

There are important regularities in the relative values of calculated source Rb/Sr as F changes. Figure 17 shows that the source of the Buckleys Lake Adamellite has the highest Rb/Sr for all (equal) F, and, consistent with this, Table 3 shows that the initial $^{87}Sr/^{86}Sr$ of this granitoid is the highest in the minimum melt group. In contrast to this, the Wullwye source has the lowest Rb/Sr for equal F, but its source $^{87}Sr/^{86}Sr$ is intermediate in value rather than the lowest. Figure 18 shows one reconstruction for the minimum melt group of sources that assigns the same younger age to all the source rocks, and which is consistent with the constraints imposed on the relative values for F set by Fig. 17. Essentially the reconstruction gives the maximum probable value for F, 0.60, to the Wullwye source and makes a reduction in F for the Tuross Head Tonalite to about 0.25, so that the source isochron has a maximized common initial $^{87}Sr/^{86}Sr$ of 0.7031. The sources for all other granitoids are then moved on to the Wullwye–Tuross Head tie-line by appropriate changes in F. The Nelligen and Moruya sources hardly change so that the same source isochron as indicated in the earlier Fig. 7 is retained. The Buckleys Lake and Delegate values for F are considerably increased, although they remain well below their maximum values, and the Maffra F is also increased.

6. Concluding Remarks

6.1. Age of the Lower Crust

The principal conclusion of this work concerns the age of the lower crust in south-eastern Australia: it is probably Late and Middle Precambrian, rather than early Palaeozoic. We qualify this as "probable" because the conclusion has been reached through a number of intermediate hypotheses. First, we have accepted the model of restite control for the geochemistry of granitoids and disregarded any crystal fractionation effects. Second, we have sought to identify the chemistry of the restite of the various magmas by a geochemical argument, the disappearance of normative quartz or K-feldspar, and we

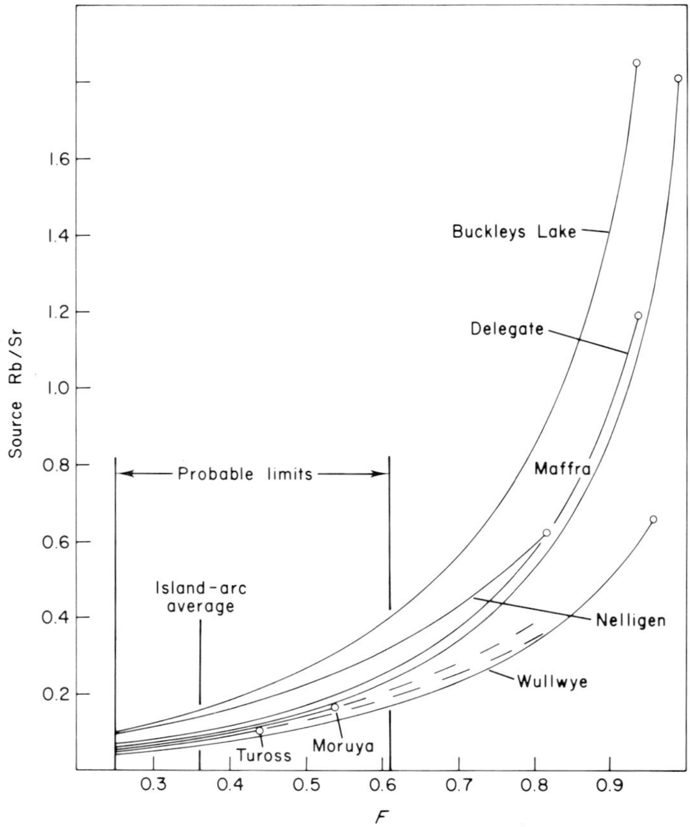

Fig. 17. Source Rb/Sr for the minimum melt-bearing granitoids as a function of *F*. There are no constraint on *F* within the limits shown, so that the order in source Rb/Sr shown at equal *F* need not represent the actual order, but it is clear that Rb/Sr for the Buckleys Lake source will be greater than any other despite the latitude in *F*. This is corroborated by the initial $^{87}Sr/^{86}Sr$ observed from Buckleys Lake, which has the highest value within the minimum melt granitoids. A comparatively small separation of melt from restite is evident for the Tuross Head and Moruya Tonalites.

have considered a specific mineralogy for the model restite. Then we have employed the "known" distribution coefficients for Rb and Sr to estimate the Rb, Sr contents of the restite and thus to define the variation of these elements with SiO_2 for granitoids that contain very little chemical dispersion. Fourth, having identified the end-members of the "unmixing" lines, we have been obliged to employ still further arguments to constrain the proportions of melt and restite which must be known to reconstruct the source chemistry. These are based on the viscosities of melts and magmas,

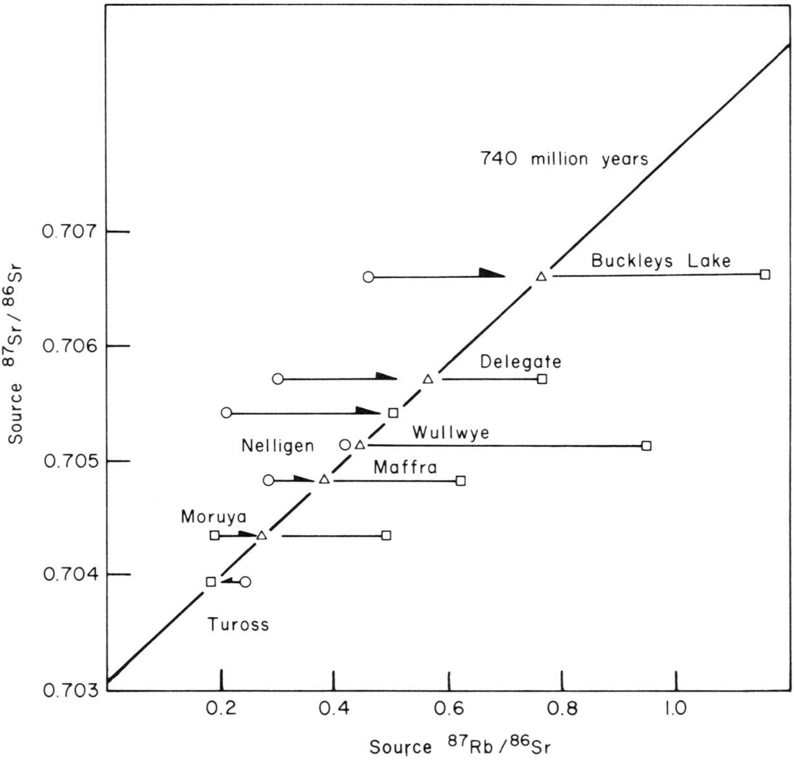

Fig. 18. A possible isochron (△——△) for source regions of the minimum melt-bearing granitoids. ○, $^{87}Rb/^{86}Sr$ calculated for F equal at 0.36 for all sources; □, the limiting $^{87}Rb/^{86}Sr$ for F at 0.60 or 0.25. All source rocks are assumed to separate from the mantle at approximately the same time and with the same low $^{87}Sr/^{86}Sr$. The indicated age of 740 million years refers to before present.

and on a model for the type of source rocks themselves, for which we have selected "normal" island-arc volcanics. In detail, each one of the above steps may be contentious, and each bears uncertainties which are difficult to quantify but which may accumulate as an implicit error in age additional to the uncertainty that we have cited ($\pm .3 \times 10^9$ years). Despite these intermediate hypotheses, we are nevertheless impressed that the slope of the isochron for the non-minimum melt sources in Fig. 14 is so large. It is a factor of five greater than the slope expected for an early Palaeozoic source, and this fact cannot be ignored.

We do not deny that at least one other interpretation of Fig. 14 is possible. This is to model the source rocks as containing "abnormal" island-arc volcanics, and to suppose that the Finister Suite source resembled the Banda arc volcanics in having high $^{87}Sr/^{86}Sr$ when it was first formed. We could not then attach a direct time significance to the apparent source isochron. Instead it would be a mixing line between unrelated Sr sources, and the source rocks may have formed as late as the Ordovician. However, this type of interpretation still implies the presence of abundant Precambrian material in the lower crust, as the latter is necessary to produce the proposed "contaminated" volcanics having high $^{87}Sr/^{86}Sr$. Thus, our principal conclusion, that Precambrian metaigneous rocks were part of the source region of the granitoids, is unaffected.

Our proposed 1.1 ± 0.3 billion year "age" for the lower crust in south-east Australia should be seen as a model capable of independent experimental testing. Part of this is already done. Williams (1977) has shown that some of the zircons from the Finister Granodiorite crystallized at least 540 million years and possibly as early as 1.1 ± 0.4 billion years ago, and that some of the zircons in the Tara Granodiorite crystallized at least 0.8 billion years ago. Still older zicrons were detected in some of the associated S-type granitoids, although the meaning of these is necessarily somewhat ambiguous. The zircon ages from the I-type granitoids are consistent with our Precambrian interpretation of the Rb–Sr data, and very difficult to reconcile with a primary origin of the crust in the early Palaeozoic.

An alternative U–Pb method of testing the model is available in principle from the isotopic composition of common Pb in the granitoids. For consistency with the proposed 1.1 billion year age of the non-minimum melt sources, such Pb should be approximately constant in each granitoid but vary between granitoids along a secondary Pb–Pb isochron. Williams (1977) again presents relevant data in the form of K-feldspar Pb isotopic compositions from the Finister and Tara Granodiorites, which imprecisely define a two-point secondary isochron of slope ~ 0.09. This is very close indeed to the value required for primary U/Pb differentiation between the granitoid source rocks at 1.1 billion years. Clearly, it is desirable to make a comprehensive study of the common Pb to see whether a secondary isochron can be convincingly defined, and to compare its slope with the expected value.

6.2. The Upwards Concentration of K, Rb, U in the Crust

We regard the production of felsic melts by partial fusion of lower crust as the principal mechanism for producing chemical fractionation within the crust. Progressive fractionation of melt from restite will produce chemically fractionated granitic and rhyolitic rocks that will contribute to more felsic upper crust compositions. The lower crust will become correspondingly more mafic. The process is illustrated in terms of Rb/Sr by Figs 16 and 17, which has been increased in the granitoids emplaced in the upper crust by a factor of five or more relative to their original source rocks. Heier (1973) has quoted several authors to support the view that granitoids are rare in medium- to high-pressure granulite terrains. We contend that such areas have lost their granitoid component by partial melting and its ascent to higher crustal levels.

Heier (1973) points out that a consistent feature of granulite facies terrains is a depletion of Th and U compared with rocks of lower metamorphic grade. Rb is likewise often depleted so that K/Rb ratios are generally higher than for upper crust rocks. We would ascribe such chemical differences to the behaviour of elements during their partitioning between anhydrous restite and hydrous melt. These two components, for example, will have markedly different K/Rb ratios. This is a consequence of a decrease in the biotite (low K/Rb): K-feldspar (high K/Rb) ratio during partial melting. Our calculations show that primary restite will contain vanishingly low amounts of biotite. Thus the line joining the minimum melt to the restite on a K_2O vs. Rb plot (see Fig. 6) lies at an angle to, and intersects, lines of constant K/Rb. Heier and Thoresen (1971) have noted that rocks of amphibolite and granulite facies grade with the same major element composition can have significant differences in trace element chemistry. However, Nesbitt (1978) has shown that, at least in some cases, the transition from amphibolite to granulite facies is associated with changes in major element chemistry. He has shown that these changes are consistent with the removal of a felsic melt.

Our main concern has been to use the restite model of granitoid genesis in an attempt to quantify "initial" Sr isotopic compositions in a way that leads to a model for the age of the source rocks. It is clear that the model is also capable of being applied to more general questions relating to the geochemical and isotopic evolution of the crust.

Acknowledgements

Many of the isotopic analyses reported here were made some years ago by T. Shirahase when visiting from the Geological Survey of Japan, and by I. S. Williams in the course of his doctoral thesis. Their applications to the emplacement ages of plutons will be fully described elsewhere. In writing the present text, we have benefitted by discussions with D. H. Green, I. Van der Molen and M. S. Paterson, and critical reading by I. S. Williams.

Professor Jaeger was responsible for the original establishment of Rb–Sr and K–Ar dating at the Australian National University and supported the continued development of research programmes and laboratory facilities such as the lunar project. Professor Hales, also a long-term friend of the mass spectrometer, sponsored our entry into U–Pb dating using the modern zircon technology, besides other more ambitious programmes. We are privileged to dedicate this paper to them both.

References

Arndt, N. T. (1977). Ultrabasic magmas and high-degree melting of the mantle. *Contr. Mineral. Petrol.* **64**, 205–221.

Bateman, P. and Chappell. B. W. (1978). Crystallization, fractionation, and solidification of the Tuolumne Intrusive Series, Yosemite National Park. *Bull. geol. Soc. Am.* (in press).

Brown, D. A., Campbell, K. S. W. and Crook, K. A. W. (1968). *The Geological Evolution of Australia and New Zealand*, Pergamon, Oxford.

Brown, G. C. and Fyfe, W. S. (1970). The production of granitic melts during ultrametamorphism. *Contr. Mineral. Petrol.* **28**, 310–318.

Chappell. B. W. (1966). Petrogenesis of the granites at Moonbi, New South Wales. Ph. D. Thesis, Australian National University (unpublished).

Chappell, B. W. (1978). Granitoids from the Moonbi district, New England Batholith, eastern Australia, *J. geol. Soc. Australia* **25**, 267–283.

Chappell, B. W. and White, A. J. R. (1974). Two contrasting granite types. *Pacific Geol.* **8**, 173–174.

Compston, W., Chappell, B. W. and Shirahase, T. (1979). Sr isotope evolution of granitoids and their source rocks in the Moonbi district, eastern Australia (in preparation).

Drake, M. J. and Weill, D. F. (1975). Partition of Sr, Ba, Ca, Y, Eu^{2+}, Eu^{3+} and other REE between plagioclase feldspar and magmatic liquid: an experimental study. *Geochim. cosmochim. acta* **39**, 689–712.

Ewart, A. and Bryan, W. B. (1972). Petrography and Geochemistry of the Igneous Rocks from Eua, Tongan Islands. *Bull. geol. Soc. Am.* **83**, 3281–3298.

Ewart, A., Bryan, W. B. and Gill, J. B. (1973). Mineralogy and Geochemistry of the Younger Volcanic Islands of Tonga, S. W. Pacific. *J. Petrol.* **14**, 429–465.

Ewart, A. and Stipp, J. J. (1968). Petrogenesis of the volcanic rocks of the Central North Island, New Zealand, as indicated by Sr^{87}/Sr^{86} ratios, and Sr, Rb, K, U and Th abundances. *Geochim. cosmochim. acta* **32**, 699–736.

Flood, R. H., Vernon, R. H., Shaw, S. E. and Chappell, B. W. (1977). Origin of pyroxene–plagioclase aggregates in a rhyodacite. *Contr. Mineral. Petrol.* **60**, 299–309.

Gill, J. B. (1970). Geochemistry of Viti Levu, Fiji, and its evolution as an Island Arc. *Contr. Mineral. Petrol.* **27**, 179–203.

Gill, J. B. (1976). Composition and age of Lau Basin and Ridge volcanic rocks: Implications for evolution of an interarc basin and remnant arc. *Geol. Soc. Am. Bull.* **87**, 1384–1395.

Gill, J. B. and Compston, W. (1973). Strontium isotopes in island arc volcanic rocks. In *The Western Pacific: Island Arcs, Marginal Seas, Geochemistry* (P. J. Coleman, ed.), University of Western Australia Press, Perth, pp. 483–496.

Gill, J. B. and Gorton, M. (1973). A proposed geological and geochemical history of Eastern Melanesia. In *The Western Pacific: Island Arcs, Marginal Seas, Geochemistry* (P. J. Coleman, ed.), University of Western Australia Press, Perth.

Green, D. H. and Ringwood, A. E. (1967). The genesis of basaltic magmas. *Contr. Mineral. Petrol.* **15**, 103–190.

Griffin, T. J., White, A. J. R. and Chappell, B. W. (1978). The Moruya Batholith and geochemical contrasts between the Moruya and Jindabyne suites. *J. geol. Soc. Australia* **25**, 235–247.

Harrington, H. J. (1974). The Tasman geosyncline in Australia. In *The Tasman Geosyncline* (A. K. Denmead, G. W. Tweedale and A. J. Wilson, eds), Geological Society of Australia, Queensland Division, Brisbane, Qld, pp. 383–407.

Hawkesworth, C. J., O'Nions, R. K., Pankhurst, R. J., Hamilton, P. J. and Evensen, N. M. (1977). A geochemical study of island-arc and back-arc tholeiites from the Scotia Sea. *Earth. planet. Sci. Lett.* **36**, 253–262.

Heier, K. S. (1973). Geochemistry of granulite facies rocks and problems of their origin. *Phil. Trans. R. Soc.* **273**, 429–442.

Heier, K. S. and Thoresen, K. (1971). Geochemistry of high grade metamorphic rocks, Lofoten-Vesterålen, North Norway. *Geochim. cosmochim. acta* **35**, 89–99.

Hine, R., Williams, I. S., Chappell, B. W. and White, A. J. R. (1978). Geochemical contrasts between I- and S-type granitoids of the Kosciusko Batholith. *J. geol. Soc. Australia* **25**, 219–234.

Hofmann, A. W. (1974). Strontium diffusion in a basalt melt and implications for Sr isotope geochemistry and geochronology. *Carnegie Inst. Washington Yearbook* **73**, 935–941.

Hofmann, A. W. and Magaritz, M. (1976). Diffusion in silicate melts and glasses: *Carnegie Inst. Washington Yearbook* **75**, 249–259.

Long, P. E. (1978). Experimental determination of partition coefficients for Rb, Sr, and Ba between alkali feldspar and silicate liquid. (Abstract). In *Papers presented to the International Conference on Experimental Trace Element Geochemistry*, pp. 78–79.

McKay, G. A. and Weill, D. F. (1976). Petrogenesis of KREEP. *Proc. 7th Lunar Sci. Conf.* 2427–2447.

Meijer, A. (1976). Pb and Sr isotopic data bearing on the origin of volcanic rocks from the Mariana island-arc system. *Bull. geol. Soc. Am.* **87**, 1358–1369.

Nesbitt, H. W. (1979). Chemical equilibrium and material balance calculations applied to the origin of granulites and granitoids (in preparation).

Philpotts, J. A. and Schnetzler, C. C. (1970). Phenocryst-matrix partition coefficients for K, Rb, Sr and Ba, with applications to anorthosite and basalt genesis. *Geochim. cosmochim. acta* **34**, 307–322.

Rutland, R. W. (1976). Orogenic evolution of Australia. *Earth Sci. Rev.* **12**, 161–196.

Shaw, H. R. (1963). Obsidian–H_2O viscosities at 1000 and 2000 bars in the temperature range 700° to 900 °C. *J. geophys. Res.* **68**, 6337–6343.

Smith, I. E. M. and Compston, W. (1979). Strontium isotopes in Cenozoic volcanics from south-eastern Papua New Guinea (in preparation).

Sun, C.-O., Williams, R. J. and Sun S.-S. (1974). Distribution coefficients of Eu and Sr for plagioclase–liquid and clinopyroxene–liquid equilibria in oceanic ridge basalt: an experimental study. *Geochim. cosmochim acta* **38**, 1415–1433.

Turcotte, D. L. and Ahern, J. L. (1978). A porous flow model for magma migration in the Asthenosphere. *J. geophys. Res.* **83**, 767–772.

Van der Molen, I. and Paterson, M. S. (1979). Experimental deformation of partially-melted granite. (in preparation)

White, A. J. R. and Chappell, B. W. (1977). Ultrametamorphism and granitoid genesis. *Tectonophysics* **43**, 7–22.

White, A. J. R., Chappell, B. W. and Cleary, J. R. (1974). Geologic setting and emplacement of some Australian Palaeozoic batholiths and implications for intrusive mechanisms. *Pacific Geol.* **8**, 159–171.

White, A. J. R., Williams, I. S. and Chappell, B. W. (1977). The geology of the Berridale 1:100000 sheet 8625. *Geol. Survey New South Wales, Sydney* 138.

Whitford, D. J. (1975). Strontium isotopic studies of the volcanic rocks of the Sunda arc, Indonesia, and their petrogenetic implications. *Geochim. cosmochim. acta* **39**, 1287–1302.

Whitford, D. J., Compston, W. and Nicholls, I. A. (1977). Geochemistry of late Cenozoic lavas from eastern Indonesia: role of subducted sediments in petrogenesis. *Geology* **5**, 571–575.

Wilkinson, J. F. G., Vernon, R. H. and Shaw, S. E. (1964). The petrology of an adamellite–porphyrite from the New England Batholith (New South Wales). *J. Petrol.* **5**, 461–488.

Williams, I. S. (1973). Granites north of Jindabyne, N.S.W. B.Sc. (Hons) Thesis, Australian National University (unpublished)

Williams, I. S. (1977). The Berridale Batholith: A lead and strontium isotopic study of its origin. Ph.D. Thesis, Australian National University (unpublished)

Williams, I. S., Compston, W., Chappell, B. W. and Shirahase T. (1976). Rb–Sr age determinations on micas from a geologically controlled, composite batholith. *J. geol. Soc. Australia* **22**, 497–505.

13. Porphyry Copper Deposits and Calc–Alkaline Volcanism

L. B. Gustafson

Abstract

Porphyry copper deposits are a major source of the world's copper and of information on processes by which igneous melts and attendant aqueous fluids have affected distribution of elements in the Earth's crust. The major and minor element character of intrusive rocks from El Salvador, Chile, typical of deposits in continental margin settings, is compared with that of intrusives from Panguna, Bougainville, typical of deposits in island-arc settings. Both are compared with local volcanic rocks to determine whether there are any distinctive petrologic characteristics of the mineralized intrusives.

Intrusives from both El Salvador and Panguna show wide ranges of major and minor element contents and general similarities with andesitic volcanics in their respective tectonic environments. Strong fractionation of rare earths and heavy rare earth depletion suggest garnet and/or amphibole involvement in the formation of El Salvador intrusive melts and some Chilean andesites, but not of Panguna intrusives or Bougainville andesite. Low Cr/Ni ratios at El Salvador contrast with very high ratios at Panguna and support the rare earth data in indicating clinopyroxene and/or amphibole and garnet dominance in partitioning relative to melts at El Salvador vs. olivene dominance at Panguna. Different crystallization histories and source regions, in terms of mineralogy and probably also composition, pressure and temperature, are indicated for intrusive melts in these two deposits. Sparse data from other deposits confirm that extension of sampling to many more deposits would simply increase the petrologic variability of porphyry copper intrusives. Only sulphur and the chalcophile elements which are concentrated in the deposits themselves distinguish the bulk chemistry of these intrusives within a broad range of orogenic calc–alkaline volcanic–intrusive rocks.

Recent experiments in the system andesite–H_2O–CO_2 and analogous system albite–H_2O–CO_2 have led to the conclusion that felsic melts will commonly be saturated with a CO_2-rich vapour phase throughout their rise through the crust. The H_2O content of this phase is now known, and it is postulated that volatiles such as S, Cl and F which control metal transport partition into this vapour phase, which is lost at depth from most orogenic melts. The small proportion of orogenic melts which start with low CO_2 contents would be the only ones which arrive at shallow reservoirs capable of forming porphyry copper deposits by volatile concentration in "cupolas".

1. Introduction

1.1. Porphyry Copper Deposits

Porphyry copper deposits represent a large and diverse class of ore deposits. Their economic importance is measured by the fact that they are the source of more than half of the world's copper production, from nearly a hundred mines. Consequently, they have been extensively studied both by exploration geologists and by academic geologists and geochemists. They are scientifically important in that they provide some of the most direct and well documented evidence of the processes by which igneous melts and attendant aqueous fluids have concentrated much of mankind's mineral wealth. They are characterized by large tonnages, measured in hundreds of millions of tons, and are relatively low grade, commonly less than 1 per cent Cu. Copper-bearing sulphur minerals occur as disseminated grains and veinlets throughout large volumes of rock. Minor amounts of associated metals include molybdenum, gold, lead, zinc and silver. Exploitation is by highly mechanized large-tonnage operations in open-pit and underground mines.

Their name derives from the fact that the host rocks of most deposits first mined in the western United States were porphyritic grandioritic to quartz monzonitic intrusives. Even in the many examples of deposits hosted primarily by volcanic, sedimentary, metamorphic and plutonic rocks, porphyritic intrusions with a common set of textures are usually seen. The close genetic association between mineralization with associated alteration and intrusions of porphyritic magmas was recognized long ago (Emmons, 1927; Lingren, 1933; Sales, 1954). Many more recent studies have documented this association and have led to recognition of a variety of mineral associations and space–time patterns in these deposits. Major areas of ignorance and disagreement among workers in this field relate to details of the mechanisms of concentration of mineralizing fluids in the shallow magma reservoir, solubility and reaction mechanisms which control transport and deposition of the metals, and the relative importance of meteoric ground water in the mineralizing fluids. General genetic aspects of an evolutionary model of porphyry copper formation, after Gustafson and Hunt (1975), are presented in Fig. 1.

Processes which have lead to the emplacement in a shallow reservoir of magma containing enough water, sulphur, metals and other volatile elements to allow concentration and formation of an economic mineral deposit are even less amenable to direct study, and are more speculative than the intrusion-mineralization processes themselves. Some deep source for the igneous melts is required, and Fig. 2 gives a schematic representation of the chain of events required to produce porphyry copper deposits. The

1. General association with andesitic to dacitic volcanism in orogenic belts, usually associated with uplift and tensional tectonics.
2. Shallow (1–3 km) emplacement of a usually complex series of porphyritic stocks and dikes in and above the cupola zone (commonly multiple zones) of an underlying plutonic body.
3. Metasomatic introduction of sulphur, copper and other metals, alkalis and hydrogen ions from solidifying melt into both porphyries and country rock, usually during only part of the intrusive process.
4. Interaction of ground water with the cooling mineralized centre.
5. Evolutionary trends of decreasing pressure and temperature, transition from alkali to hydrogen ion metasomatism, and increasing sulphide-ion activity in mineralizing fluids.

Fig. 1. Summary of evolutionary model of porphyry copper deposit genesis.

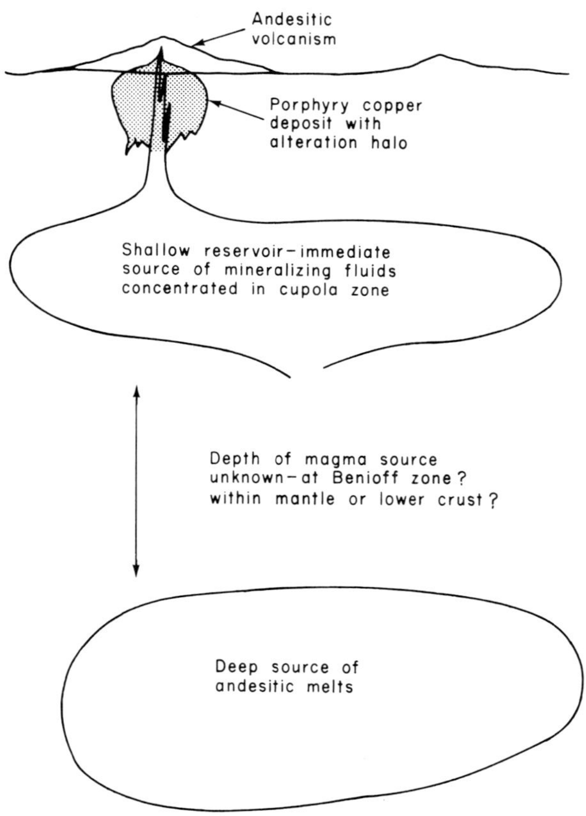

Fig. 2. Schematic illustration of porphyry copper formation in the cupola zone of a small, shallow batholith. The magma is labelled "andesitic" but represents a wide range of composition from quartz diorite to quartz monzonite and, rarely, granite or syenite.

conclusion that the bulk of the copper and sulphur at El Salvador, Chile, was derived directly from the magmatic source of the contemporaneous porphyritic intrusions (Gustafson and Hunt, 1975, summarized below) is taken as representing the general case in porphyry copper deposits, even though in some deposits an important contribution from meteoric waters may be argued. Only indirect evidence can be brought to bear on the problem of whether water, sulphur and metals are derived from the same deep source as the igneous melts or acquired by the melts during the rise through the crust and upper mantle. These problems are discussed more fully subsequently in this paper, but the reader is referred to the extensive literature on porphyry copper, especially the recent review by R. L. Nielsen (1976), for more detail on the nature of the deposits and the status of thinking about their origin.

1.2. Tectonic Setting

Porphyry copper deposits have long been recognized as being associated with igneous activity of intermediate composition in tectonic belts. Intensive exploration in the past 30 years has resulted in the discovery of hundreds of deposits and the further documentation of this association. Deposits tend to occur in clusters within mineral districts, and to form elongate belts generally coincident with andesitic volcanism, folding and uplift (Fig. 3). With the advent of plate tectonic theory, such belts were recognized as forming at convergent margins of crustal plates (e.g. Isacks *et al.*, 1968; Dewey and Bird, 1970). The correlation with plate subduction is most appealing in the western Pacific and the Andes of South America, where relatively young deposits (less than 45 million years old) are found in tectonic–volcanic belts presently at active plate margins. Several plate tectonic models of porphyry copper genesis have been published (Guilbert and Sumner, 1968; Guild, 1971; Mitchell and Garson, 1972; Sawkins, 1972; Sillitoe, 1972). The general idea is that oceanic crust is subducted beneath continental margins or island arcs, partially melts and rises to form porphyry copper deposits. However, the application of a subduction model to porphyry copper occurrences in the south-western United States, the most important single province in the world, has been challenged on the grounds that the apparent ages of the deposits and reconstructions of the tectonic events do not fit a simple subduction model (Livingston, 1973; Lowell, 1974). Particularly hard to reconcile is the occurrence of the major deposit at Bingham, Utah, far inland from any probable contemporaneous subduction at the time of formation of the deposit 35 million years ago. Part of the problem arises from the fact that even the youngest known porphyry copper deposits are 1–5 million years old. A relatively rapid rate of plate movement, measured in centimetres per year, can produce very appreciable

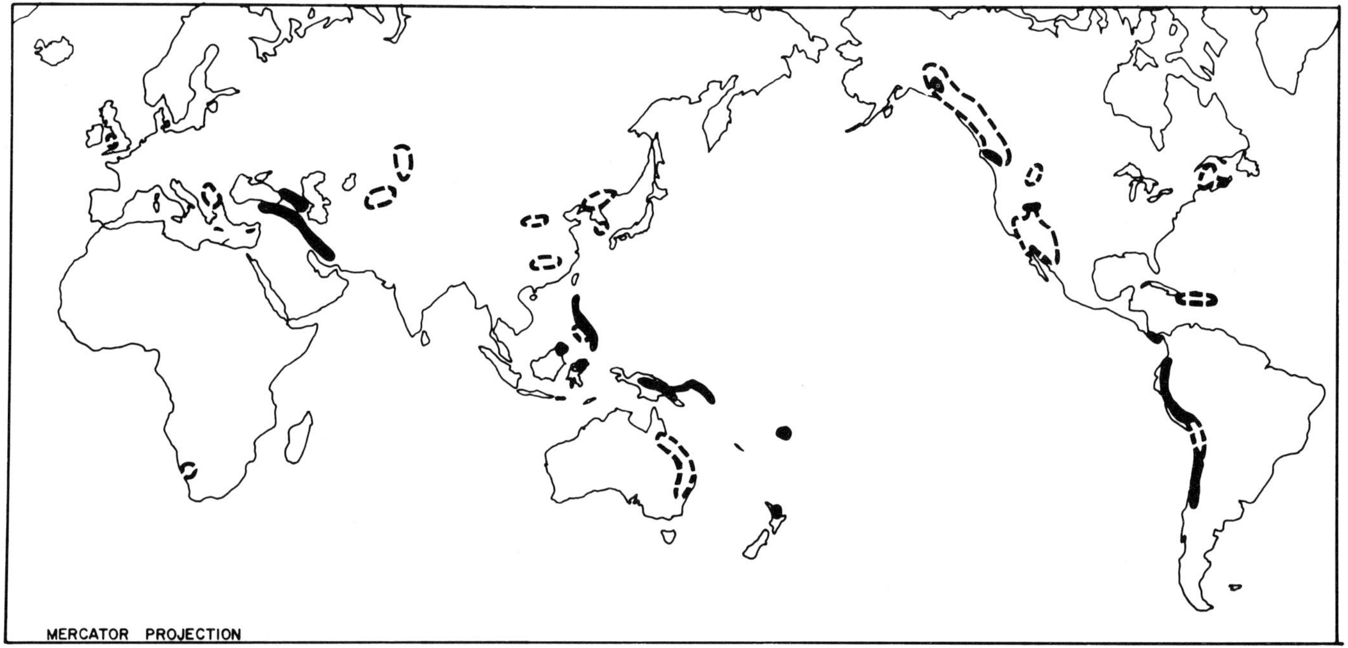

Fig. 3. Distribution of belts of porphyry copper deposits. Young deposits, less than 45 million years old (indicated by solid black areas) are distinguished from older deposits (encircled by dashed lines). Note the distribution of young deposits along orogenic belts formed at convergent plate boundaries; compare Fig. 1, p. 492 in this volume.

shifts in the configuration of subduction zones over times shorter than a million years. This is particularly well illustrated by the work of Barazangi and Isacks (1976, Fig. 6) in the southern Andes. Therefore, it is hazardous to extrapolate present day Benioff zone activity backwards in time.

Calc–alkaline volcanism has been taken as indicative of plate convergence above a subduction zone (e.g. Sillitoe, 1972), but this interpretation is also questionable. Late Cenozoic andesitic volcanism in the highlands of Papua New Guinea does not appear to be related to active subduction (Johnson *et al.*, 1978). Hence, very young porphyry copper mineralization in Papua New Guinea, including the 1.2 million year old deposit at Ok Tedi, may also not have all been related to active subduction. Clearly a closer look at the tectonic–petrologic setting of porphyry copper deposits is required.

1.3. An Exploration Problem

Although we recognize the occurrence of porphyry copper deposits along many tectonic belts, we do not understand why only some volcanic-intrusive complexes within the belt contain mineralization, and we do not really know how to recognize a potentially productive but partially covered intrusive complex if there is no mineralization and alteration exposed. It would be useful if we could recognize a distinctive petrologic signature in such apparently fresh, unmineralized intrusives or volcanics, which would indicate a favourable source region or magmatic history, and suggest shallowly concealed mineralization. What is it that makes the small proportion of productive shallow intrusives within a volcanic belt so special? The recognition of chloride complexes as characteristic of metalliferous hydrothermal fluids evolved from igneous melts (e.g. Holland, 1972) has led several investigators to examine the halogen content of intrusives associated with porphyry copper mineralization in the hope of finding a characteristic halo or "fingerprint" of favourable intrusives. Unfortunately, this approach has proved to be of little help outside zones of alteration-mineralization themselves.

The approach taken in the study reported here was to investigate the major and minor element petrology of the intrusive suite from the porphyry copper deposit at El Salvador, Chile, and compare this with a reconnaissance sampling of a few other deposits in contrasting tectonic settings.

2. El Salvador, Chile

2.1. Geological Summary

The deposit at El Salvador, in northern Chile, is a typical porphyry copper deposit which has been the subject of intensive geological investigation

(Gustafson and Hunt, 1975; Field and Gustafson, 1976; Sheppard and Gustafson, 1976). The original secondarily enriched ore body contained about 300 million tons of ore with a grade of 1.6 per cent Cu. There is a minor recoverable molybdenum content (0.02 per cent Mo) but very low gold (0.1 $g\,t^{-1}$ Au). The mineralization was formed about 41 million years ago at the culmination of about 9 million years of volcanic history within a terrestrial volcanic terrain late in the evolution of the Andean tectonic cycle (Jurassic to present). This occurred at a continental margin underlain by a thick and old continental crust. Detailed geological mapping and supporting mineralogical, chemical and isotopic studies clearly reveal that the bulk of primary mineralization and alteration accompanied emplacement of a granodioritic porphyry complex, and formed adjacent to, and simultaneously with, recurrent intrusions of porphyry magma. Hydrothermal fluids responsible for the mineralization were introduced at near magmatic temperature and pressure (well above hydrostatic). These fluids carrying the bulk of the emplaced metals and sulphur must have been derived from the underlying magmatic reservoir, though subsequent incursion of groundwater convecting under hydrostatic conditions accomplished significant redistribution and possibly some addition of these elements.

Figure 4 is a simplified geological map of the area. Upper Cretaceous andesitic flows with associated sediments were folded and truncated by an unconformity on which gently dipping Miocene rhyolitic pyroclastics were deposited. The first intrusive event in the district, about 50 million years ago, was the emplacement of a series of rhyolitic domes. These domes of Indio Muerto rhyolite were roughly contemporaneous with, and probably feeders for, pyroclastics in the upper part of the Miocene volcanic pile. A second intrusive event about 46 million years ago resulted in emplacement of quartz rhyolite as a volcanic neck, and of quartz porphyry as an apparently contemporaneous irregular sequence of dikes and sills. The main porphyry copper event occurred at about 41 million years ago, accompanying emplacement of the granodioritic porphyry complex. It is this complex of porphyritic intrusions which was so intimately associated with copper mineralization and alteration described above.

Later hydrothermal events tend to overprint early magmatic events. In order to study the magmatic evolution, it is necessary to find samples of igneous rocks which have not been affected chemically by hydrothermal modification. These are found in areas within rhyolites and quartz porphyries at the periphery of the main porphyry copper pattern, and within the youngest intrusives which were intruded into the main porphyry complex late enough to have avoided most of the hydrothermal activity.

It is not appropriate to deal here with the details of the chemical effects of hydrothermal alteration. These will be reported elsewhere but, in summary,

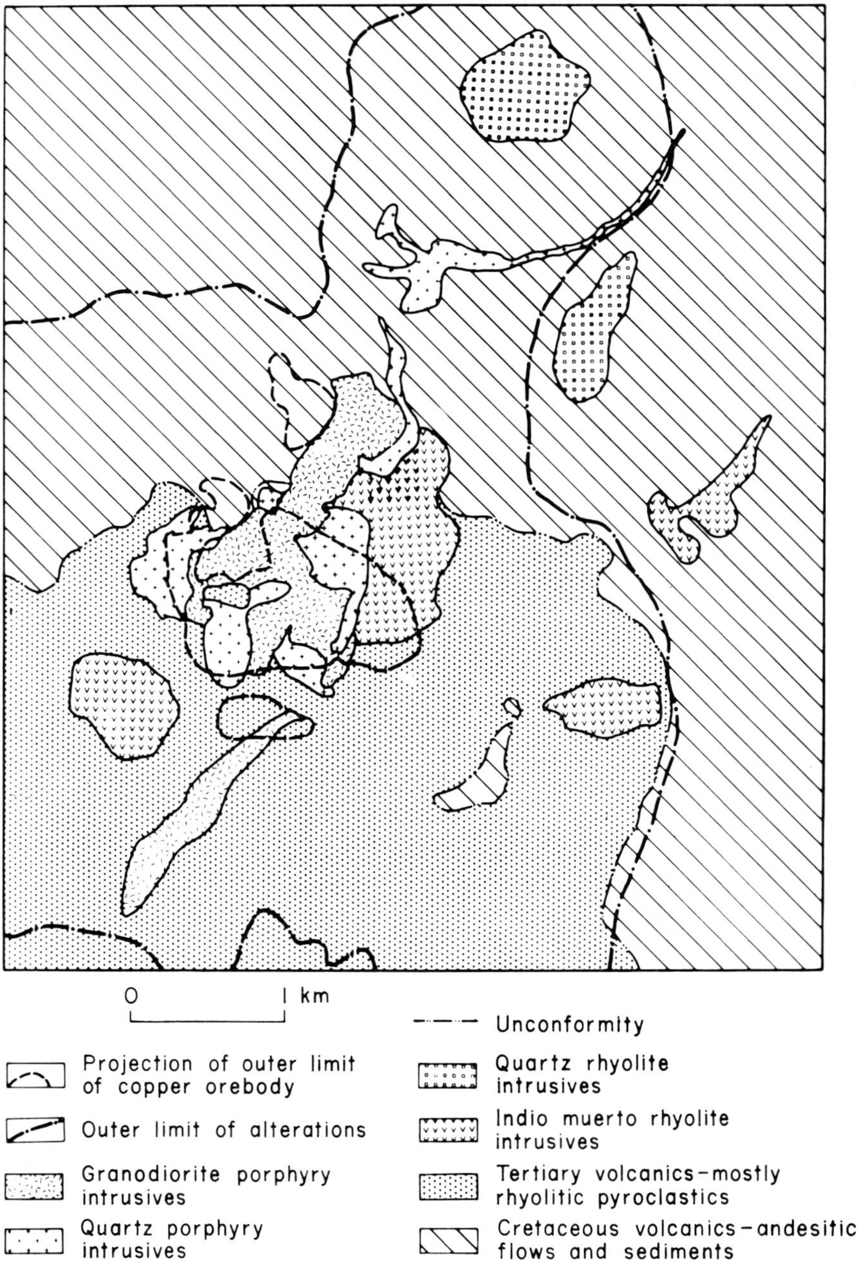

Fig. 4. Simplified geological map of the El Salvador (Indio Muerto) district, Chile, showing distribution of principal rock types analysed in this report, the outer limit of hydrothermal alteration, and the projection of the main ore zone. After Gustafson and Hunt (1975), Fig. 3.

it has been found that even moderate degrees of alteration to assemblages characterized by K-feldspar–biotite or sericite–chlorite do not have a large effect on the distribution of rare earth elements and most other trace elements. However, even minor alteration produces an appreciable depletion of MgO and CaO with increases in K_2O, sulphur and H_2O. Correspondingly, strontium drops and there is a small irregular increase in rubidium. As magnetite is removed by alteration, vanadium decreases significantly. Other trace elements are not markedly affected until alteration produces pervasive sericite or advanced argillic assemblages. Such alteration can produce strong depletion in the rare earths as well as most other trace elements. The interpretation presented below is based on analyses for fresh rocks only (Tables 1 and 2). An exception is the "X" porphyry which, even in its freshest exposures, has moderate K-feldspar–biotite alteration. For comparison, average analyses of a group of modern andesites from south-central Chile are also presented. In the following discussion, "andesite" is used in a broad sense, without distinction as to low-silica or high-K varieties (Taylor, 1969). Two andesites from the Tupungato and Marmolejo volcanoes (Lopez-Escobar *et al.*, 1977), are closely comparable with the El Salvador suite and have been listed separately in Tables 1 and 2.

2.2. Major Elements

The El Salvador intrusives fall naturally into two groups, early rhyolites of the first two volcanic episodes represented by a single analysis each, and the main porphyry sequence of quartz dioritic to granodioritic composition represented by five analyses. The quartz porphyry analysis represents an intrusive event roughly contemporaneous with the second volcanic episode, but of a composition closer to that of the main porphyry series. There is considerable textural variation within both the quartz rhyolite and the quartz porphyry at El Salvador, chemically complex units inadequately characterized by two analyses.

El Salvador rhyolites appear to be similar to younger rhyolitic ignimbrites in northern Chile in terms of major elements. The El Salvador sampling is too sparse to be sure that the low iron and low magnesium contents are significantly different. The rhyolitic ignimbrites for which the Indio Muerto rhyolite domes appear to have been feeders, are much less extensive than northern Chilean ignimbrites of the "rhyolite formation".

Among the group of porphyries, there is as much variation within the single "L" porphyry intrusive unit as across the entire intrusive sequence. Though "L" porphyry is the only unit fresh enough to have allowed determination of the extent of variation of the original magmatic composition, there do not appear to be clear compositional trends with the time across the intrusive sequence.

For comparison purposes, major element data have been plotted on standard AMF and K_2O vs. SiO_2 diagrams (Fig. 5(a) and (b)), along with the field for andesitic to rhyolitic volcanics in Chile (Pichler and Zeil, 1972; Lopez-Escobar *et al.*, 1976, 1977) (Fig. 5). The analyses of intrusive rocks show a large scatter and do not fall entirely within the broad volcanic field. Variation tends towards low K_2O, though total K_2O+Na_2O values (not plotted) fall well within the volcanic field. Low FeO, or FeO/MgO, relative to the volcanic field characterizes half the El Salvador porphyry samples in Fig. 5(a). TiO_2 and P_2O_5 contents appear to be comparable to Andean

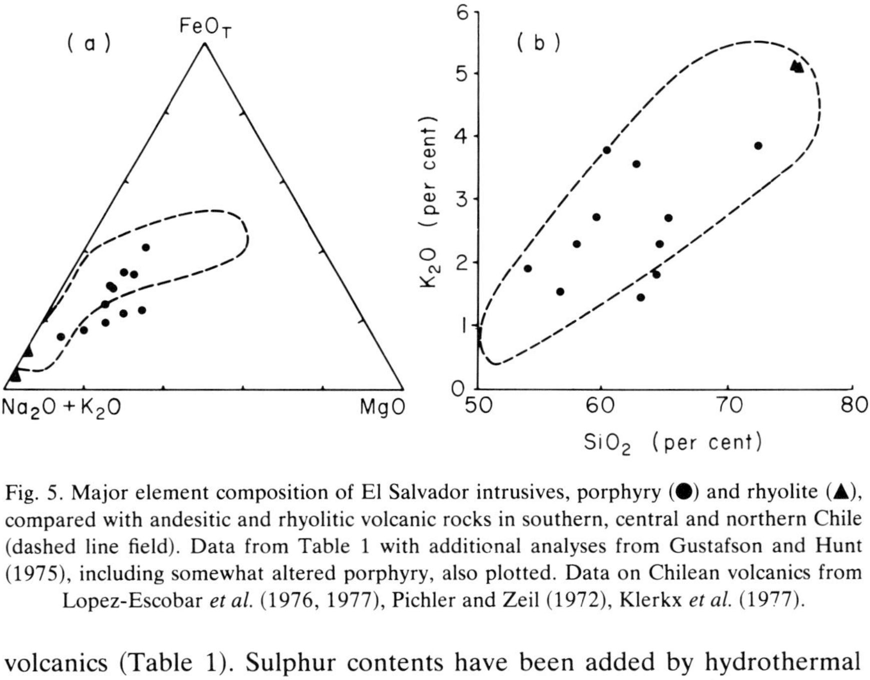

Fig. 5. Major element composition of El Salvador intrusives, porphyry (●) and rhyolite (▲), compared with andesitic and rhyolitic volcanic rocks in southern, central and northern Chile (dashed line field). Data from Table 1 with additional analyses from Gustafson and Hunt (1975), including somewhat altered porphyry, also plotted. Data on Chilean volcanics from Lopez-Escobar *et al.* (1976, 1977), Pichler and Zeil (1972), Klerkx *et al.* (1977).

volcanics (Table 1). Sulphur contents have been added by hydrothermal solutions following consolidation of the rocks, as anhydrite, pyrite and chalcopyrite, largely in small veinlets. Only in the "X" porphyry samples is there abundant disseminated mineralization and significant alteration, and even here effects other than the addition of sulphur have been minor.

2.3. Minor Elements

The El Salvador deposit has been so thoroughly exposed and studied only because of the addition of nearly 10^7 tons of the minor element copper which

Table 1. Major element content of El Salvador intrusives and south-central Chile andesites.

	El Salvador								South-central Chile	
	Indio Muerto rhyolite	Quartz rhyolite	Quartz porphyry	"X" porphyry	"L" porphyry	"L" porphyry	"L" porphyry	Latite	Andesites: average of of ten	Marmolejo and Tupungato: average of two
SiO_2	75.86	75.74	72.86	57.91	65.59	64.57	56.88	60.61	58.44	61.03
Al_2O_3	12.87	14.04	15.51	16.49	15.14	16.36	17.50	15.95	16.86	17.33
Fe_2O_3	0.44	—	—	0.59	2.07	2.64	3.46	3.17	2.90	2.27
FeO	0.75	0.32†	1.56†	2.23	1.28	1.78	2.73	1.75	3.08	2.78
MgO	0.05	0.10	0.63	2.59	1.32	1.61	2.16	2.18	3.69	2.55
CaO	0.23	0.25	1.55	6.41	3.90	4.36	6.17	5.54	5.95	4.84
Na_2O	3.44	4.20	3.82	4.05	3.59	4.81	4.67	4.41	3.91	4.66
K_2O	5.13	4.34	3.86	2.29	2.70	1.80	1.58	2.79	1.69	2.28
$H_2O(+)$	0.39	0.45	1.11	1.04	2.49	0.98	1.19	1.35	2.67‡	0.66‡
P_2O_5	0.00	0.00	0.02	0.85	0.22	0.32	0.48	0.30	0.20	0.27
TiO_2	0.24	0.10	0.34	0.99	0.48	0.71	0.73	0.98	0.80	0.72
MnO	0.00	0.08	0.02	0.06	0.02	0.03	0.03	0.05	0.09	0.10
SO_3	0.00	0.05§	0.02§	3.80	1.84	0.55	2.25	Trace	—	—
S	0.02	—	—	0.20	0.57	0.08	0.22	Trace	—	—
CO_2	0.17	0.13	0.20	0.08	0.35	0.24	0.33	1.59	—	—
Cu	0.00	0.00	0.23	0.50	0.13	0.03	0.06	0.01	—	—
Total	99.60	99.80	101.73	100.08	101.70	100.87	100.46	100.69	100.29	99.51
Less S equivalent to O	0.01	0.03	0.01	0.10	0.29	0.04	0.11	—	—	—
Total	99.59	99.77	101.72	99.98	101.42	100.83	100.35	100.69	—	—
$MgO/(MgO+FeO_T)$	0.04	0.23	0.29	0.48	0.30	0.28	0.27	0.32	0.39	0.35
K_2O/Na_2O	1.49	1.03	1.01	0.57	0.75	0.37	0.34	0.63	0.43	0.49

† Total iron as FeO. ‡ CO_2+H_2O. § Total sulphur as SO_3.

El Salvador analyses from Gustafson and Hunt (1975), except quartz rhyolite and quartz porphyry; average of ten south-central Chile andesites from Lopez-Escobar *et al.* (1976, 1977); average of two andesites, Marmolejo and Tupungato, from Lopez-Escobar *et al.* (1977).

Table 2. Minor element content of El Salvador intrusives and south-central Chile andesites.

	El Salvador								South-central Chile	
	Indio Muerto rhyolite	Quartz rhyolite	Quartz porphyry	"X" porphyry	"L" porphyry	"L" porphyry	"L" porphyry	Latite	Andesites: average of ten	Marmolejo and Tupungato: average of two
U	3.8	2.2	1.5	0.4	0.6	0.7	0.7	0.8	1.8	1.8
Th	17.0	3.0	5.0	1.0	2.2	1.4	1.6	1.8	4.3	5.0
Ba	400	250	1020	375	635	830	450	630	355	475
Rb	76	39	44	23	30	48	27	11.7	74	68
Sr	137	49	329	1015	651	659	1030	910	316	630
La	43.9	2.9	11.6	13.5	16.1	20.5	19.3	17.7	15.8	22.8
Ce	81.0	8.5	24.2	35.4	37	44.5	42.3	37.1	33.7	53.0
Pr	10.3	1.20	3.0	4.43	4.7	4.97	5.37	4.78	—	—
Nd	2.21	4.54	12.8	21.6	18.4	19.4	21.9	20.4	18.2	22.4
Sm	7.34	1.55	2.4	4.22	2.95	3.22	0.39	3.62	4.1	4.1
Eu	0.69	0.11	0.60	1.28	0.88	0.89	1.24	1.02	1.06	1.1
Gd	6.82	1.43	(2.2)	3.76	2.7	2.10	2.95	2.49	3.86	3.7
Tb	0.91	0.22	0.27	0.53	0.28	0.26	0.39	0.32	—	—
Dy	5.03	1.53	1.57	2.77	1.50	1.17	1.92	1.45	—	—
Ho	0.93	0.39	0.38	0.47	0.22	0.20	0.28	0.23	0.87	0.45

Er	2.45	1.15	1.05	1.19	0.46	0.41	0.63	0.47	—	—
Yb	2.62	1.1	1.01	1.16	0.39	0.44	0.53	0.40	2.14	1.0
REE†	165.0	24.95	61.39	90.66	85.70	98.19	97.36	90.12	—	—
Hf	5.35	1.70	2.69	4.44	2.49	2.42	6.30	3.56	4.2	4.8
V	2.8	≈0.06	42	113	68	81	119	94	—	—
Co	3.6	<0.08	8.4	16.6	13.5	7.7	13.5	9.5	20.6	14.2
Cr	<3	4	11	2	9	7	15	15	129	60
Ni	4	10	4	14	8	10	10	18	42	27
Pb	5.7	10.0	11.6	4.6	5.3	3.7	6.4	9.0	—	—
Zn	41	26	86	32	42	53	46	74	—	—
Cu	24	11	2315	4990	1350	260	616	53	—	—
K/Rb	560	923	728	826	747	311	486	1923	189	278
Rb/Sr	0.56	0.80	0.13	0.02	0.05	0.07	0.03	0.01	0.23	0.11
Ba/Rb	5.2	6.4	23.2	16.3	21.0	17.4	16.7	53.9	4.8	7.0
Th/U	4.5	1.4	3.4	2.5	3.7	2.0	2.3	2.3	2.4	2.8
Ni/Co	1.1	>125	0.5	0.8	0.6	1.3	0.7	1.9	2.0	1.9
Cr/Ni	<0.75	0.4	2.8	0.14	1.13	0.7	1.5	0.8	3.1	2.2
V/Ni	0.7	0.01	10.4	8.1	8.5	8.1	11.9	5.2	—	—
La/Yb	16.8	2.6	11.5	11.6	41.3	46.6	36.4	44.3	7.4	22.8
Eu*/Eu	0.30	0.21	(1)	0.99	0.96	1.05	1.07	1.0	0.8	0.87

† Includes estimate of Lu and Tm (both are minor)

Volcanic analyses from Lopez-Escobar *et al.* (1976, 1977); average of two andesites from Marmolejo and Tupungato. El Salvador analyses: Rb, Sr, Ni by X-ray fluorescence; Co, Cr, Pb, Zn, Cu by atomic absorption; V by optical spectrometry; Th, U, Ba, Hf and REE by spark source mass spectrometry using method described by Taylor and Gorton (1977).

accompanied nearly 10^9 tons of sulphur in the formation of the deposit. The distribution of copper, as well as minor amounts of molybdenum, lead, zinc, silver and gold, has been well documented. It is primarily related not to the rock forming igneous processes of interest here but to superimposed hydrothermal events. Therefore, the abundance of chalcophile elements, particularly in a small number of individual samples, cannot be interpreted in terms of the petrology of the rocks themselves.

At the time of writing, most of the data available on trace element abundances in Andean volcanics were obtained in samples from south-central Chile (Lopez-Escobar *et al.*, 1976, 1977). However, rubidium,

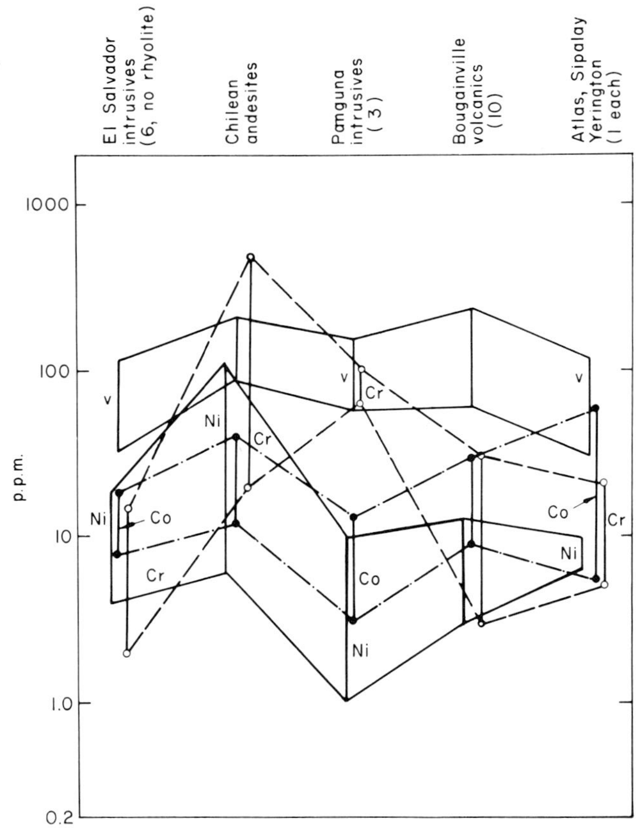

Fig. 6. Minor element content of porphyry copper intrusives and regional volcanics—compatible or ferromagnesian elements. Data from Tables 2 and 4 for intrusive porphyries; El Salvador rhyolites not included; additional data on Panguna intrusives included from Mason (1975); data on Chilean andesites (no rhyolites) from Lopez-Escobar *et al.* (1976, 1977), Pichler and Zeil (1972), Klerkx *et al.* (1977); data on Bougainville volcanics from Taylor *et al.* (1969*a*).

strontium and strontium isotope data from northern Chile are also available for comparison (Klerkx *et al.*, 1977) (Table 2 and Figs 6 and 7). The small suite of analyses of El Salvador porphyries shows a wide variation in trace element abundance which is generally typical in both values and variability of andesites in Chile and elsewhere (Taylor, 1969). The "compatible" or "ferromagnesian" elements (cobalt, chromium and nickel) are markedly depleted in El Salvador porphyries relative to southern Chilean andesites, and are more similar to andesitic lavas from Tupungato and Marmolejo. El Salvador rhyolites (tabulated but not included in Figs. 6 and 7) are even more depleted in nickel, chromium and cobalt.

The barium contents of the porphyries are high compared to south-central Chilean andesites, and are also variable. Strontium is somewhat enriched

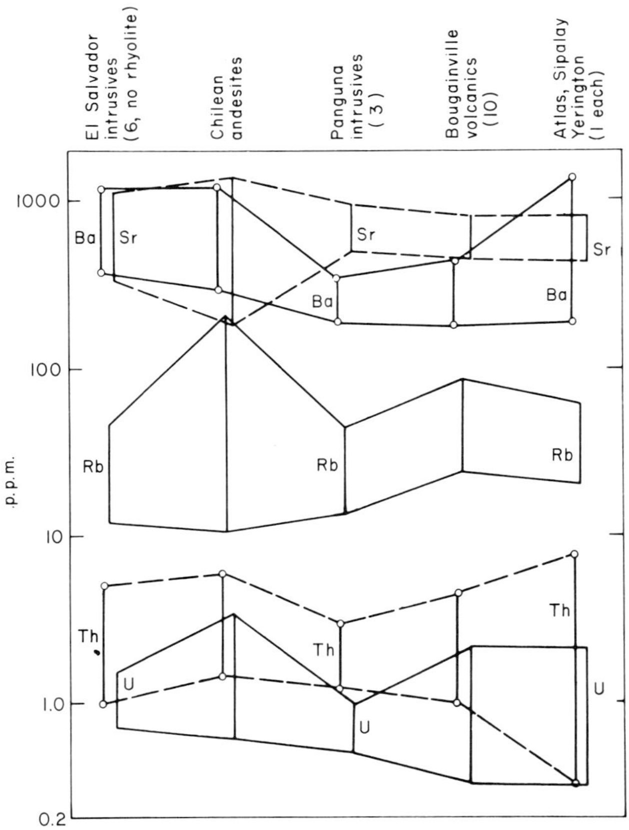

Fig. 7. Minor element content of porphyry copper intrusives and regional volcanics—large ion lithophile (LIL) elements. Same samples and sources as Fig. 6.

compared to the andesites, being comparable to, or higher than, northern Chilean andesites (Klerkx *et al.*, 1977). The porphyries are in the low end of the range of rubidium values for northern Chilean andesites, and even lower compared to south-central Chilean andesites. Consequently, Rb/Sr values are slightly lower than the northern Chilean andesites, and substantially lower than south-central Chilean andesites. El Salvador rhyolites have lower strontium but only slightly lower rubidium contents relative to north Chilean ignimbrites (Klerkx *et al.*, 1977), resulting in slightly higher Rb/Sr ratios. El Salvador thorium and uranium contents are somewhat low compared to south-central Chilean andesites, and, like most andesites, are markedly depleted relative to upper crustal values (Taylor, 1969). The Th/U ratios are typical of andesites.

Rare earth element (REE) values are tabulated in Table 2 and normalized abundances are plotted in Fig. 8. The main porphyry sequence is a fairly coherent grouping characterized by high REE content and strong fractionation ($\Sigma REE = 61–98$ p.p.m., La = 40–70 times chondritic values, and

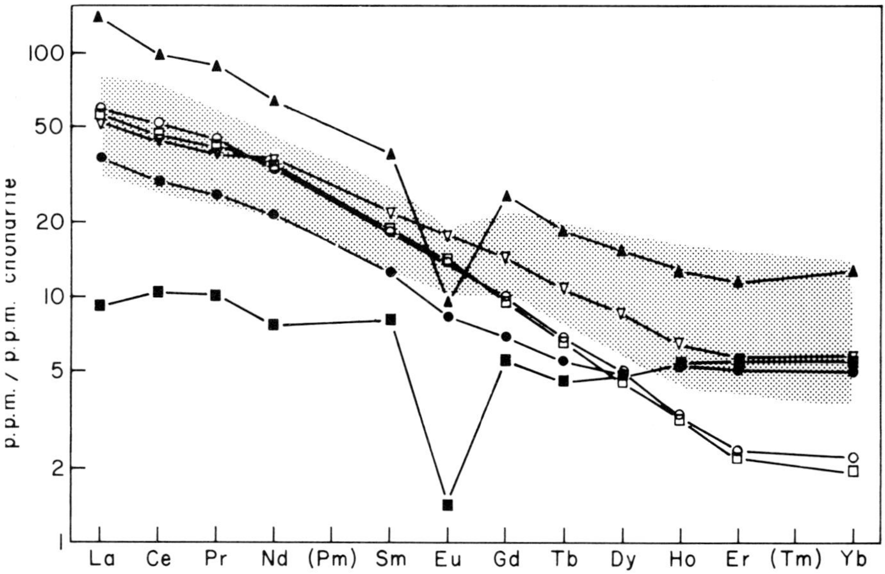

Fig. 8. Rare earth elements in El Salvador intrusive rocks. Data from Table 2, compared with the range of south-central Chilean andesites (Lopez-Escobar *et al.*, 1976, 1977); the more fractionated andesites from Tupungato and Marmolejo comprise the high light REE and low heavy REE limits of the andesite range, and are most similar to El Salvador porphyries. ▲——▲, Indio Muerto rhyolite; ▽——▽, "X" porphyry; ●——●, quartz porphyry; ■——■, quartz rhyolite; □——□, latite; ○——○, average "L" porphyry; stippled area indicates Chilean andesites.

La/Yb = 11–47). There are no europium anomalies. The "L" porphyry and the latite particularly show heavy REE depletion to 2–3 times chondritic values. The quartz porphyry appears to be the least fractionated of the porphyry suite. The Indio Muerto rhyolite is even more enriched than the porphyries (ΣREE = 165 p.p.m.), is strongly fractionated (La/Yb = 16.8), and has a strong negative europium anomaly. In contrast, the single quartz rhyolite sample has a distinctive and unusual pattern. It shows less REE enrichment than even mafic andesites and high-aluminium basalts from southern Chile, is very weakly fractionated (La/Yb = 2.6) and has a strong negative europium anomaly. The REE patterns of El Salvador porphyries are generally similar to those for south-central Chilean andesites, particularly Marmolejo and Tupungato, which represent the strongest light REE enrichment and heavy REE depletion, respectively, in the andesite range shown in Fig. 8.

Initial strontium isotope ratios are reported by Gustafson and Hunt (1975), and are tabulated in Table 3. The ratios are derived from Rb/Sr isochron determinations for the Indio Muerto rhyolite, ignimbrites apparently cogenetic with the Indio Muerto rhyolite, the quartz rhyolite and a group of mineral separates from mineralized as well as unmineralized porphyries. Initial strontium ratios for all rocks analysed are a remarkably uniform 0.7040–0.7042, though the Rb/Sr of individual samples analysed range from 0.14 to 4.1. There is no correlation of initial $^{87}Sr/^{86}Sr$ vs. Rb/Sr or SiO_2, unlike patterns seen in northern Chile and southern Peru, (James *et al.*, 1976; Klerkx *et al.*, 1977).

Table 3. Initial $^{87}Sr/^{86}Sr$ ratios of intrusive and volcanic suites from El Salvador, from isochron plots. From Gustafson and Hunt (1975); analyses by C. Brooks.

Rock suite	Rb	Sr	Rb/Sr	Initial $^{87}Sr/^{86}Sr$	Age (million years)
Indio Muerto rhyolite (6)	139	116	1.2	0.7042 ± 0.0002	50.4 ± 2.8
Indio Muerto volcanics (4)	138	118	1.2	0.7041 ± 0.0003	50.3 ± 3.2
Quartz rhyolite (6)	98	39	2.5	0.7040 ± 0.0003	45.4 ± 1.14
Porphyry suite (7 mineral separates)				0.7042 ± 0.0003	41.3 ± 1.1

2.4. Petrologic Interpretations

The major element character of porphyries and rhyolites at El Salvador is typical of calc–alkaline volcanics. The TiO_2 contents and MgO/FeO ratios are characteristically low. Compatible trace elements are strongly depleted, large ion lithophile (LIL) elements are enriched and there is strong REE enrichment and fractionation relative to most basaltic rocks. Relative to upper crustal materials, rubidium contents are low and initial $^{87}Sr/^{86}Sr$ ratios are very low.

No detailed petrologic modelling of El Salvador data has been attempted, but the arguments of Lopez-Escobar *et al.* (1976, 1977) regarding the origin of south-central Chilean andesites, especially Tupungato and Marmolejo, hold equally well. These authors have demonstrated that the combination of major and minor element evidence precludes interpreting these Chilean andesites as unmodified products of high-pressure partial melting of ocean floor basalt (eclogite). This conclusion holds by analogy for the El Salvador intrusives. Heavy REE depletion in both rock suites suggests partitioning between melt and garnet. Eclogite or garnet peridotite in or above the subduction zone are possible parents, though very small degrees of partial melting followed by subsequent fractional crystallization are apparently required. Significant metasomatic enrichment in incompatible elements of either primitive source material could allow larger degrees of subsequent partial melting. An eclogite source could have been either in the downgoing oceanic plate or in the lower and possibly recently-formed crust. The fact that amphibole can apparently produce similar REE fractionation either in a lower crustal source region or during subsequent fractional crystallization (Frey *et al.*, 1978), introduces considerable ambiguity into this argument, however.

The 9 million year history of magmatism at El Salvador is too great a time-span for the magma series to be the result of a single melting event. The constant and low $^{87}Sr/^{86}Sr$ ratio over this interval indicates either the same source (if a restricted volume) or a very homogenous source (if a large volume) for all intrusives in the district. This source region cannot have had a long history of radiogenic strontium addition, such as the older lower crust or ancient subcontinental mantle lithosphere proposed by Brooks *et al.* (1976) and Klerkx *et al.* (1977) as a source for volcanics in southern Peru and northern Chile. Also, El Salvador porphyries can have been affected by, at most, very minor contamination by old crustal material having low strontium or high cobalt, nickel or chromium contents. However, a significant amount of contamination with younger Andean volcanic material (at shallow levels of the crust) having low initial strontium ratios would be harder to distinguish. Quantitative limits on the amount of contamination permitted are indeterminate because of the unknown composition of the materials being mixed. Following the reasoning of Leeman (1976), the low Cr/Ni ratios of El Salvador rocks suggest dominance of clinopyroxene and/or amphibole or garnet rather than olivine in partial melting and subsequent fractional crystallization. Again an eclogite source is suggested, but many uncertainties preclude definitive modelling of the probably complex petrologic history of these rocks.

3. Panguna, Bougainville

The largest producing porphyry copper deposit in a western Pacific island arc setting is Panguna, on the island of Bougainville. It is at the western end of the Solomon Islands, a complex and presently active island arc (Blake and Miezitis, 1967; Coleman, 1966). A recent description of the deposit has been published by Baumer and Fraser (1975). The oldest exposed rocks are andesitic and basaltic lavas and pyroclastics of Lower Oligocene or Upper Miocene age. The copper deposit was formed 5–3.4 million years ago, during and following emplacement of a quartz diorite to granodiorite intrusive complex. The rock textures and the sequence and patterns of veining and alteration-mineralization assemblages show strong similarities with El Salvador and mark Panguna as also being a typical porphyry copper deposit. Ore reserves announced by Bougainville Copper Limited are 900 million tons of 0.48 per cent Cu. Panguna lacks the supergene enrichment largely responsible for the higher grade at El Salvador. It is characteristic of other western Pacific island arc deposits, having relatively low molybdenum (average 33 p.p.m.) and high gold (average 0.55 $g\,t^{-1}$) compared with the relatively high molybdenum and low gold contents of El Salvador, which is more typical of Cordilleran or continental margin porphyry copper types (Kesler, 1973).

The main mineralized centre occurs at the southern margin of the roughly 3.5 km diameter stock of Kaverong quartz diorite, and is intimately associated with what appear to be marginal intrusive phases of the main stock. For this study, trace element data (Table 5) were obtained from three samples, one each from fresh unmineralized Kaverong quartz diorite north of the pit, mineralized biotite granodiorite, and fresh (largely post-mineralization) Biuro granodiorite within the pit. Only the biotite granodiorite has undergone mineralization and alteration to a biotite–K-feldspar assemblage associated with chalcopyrite–bornite (similar to, but less intense than, alteration-mineralization of the "X" porphyry sample from El Salvador). As at El Salvador, comparison of analyses suggests that most compositional effects of the weak alteration-mineralization considered here are less than original magmatic variations within the intrusive complex (except for copper and sulphur), though the sparseness of the sampling precludes detailed analysis. The close similarity of the trace element character of the three intrusives indicates that this sampling can be taken as indicative of the general petrologic character of Panguna porphyries. The major element

Table 4. Major element content of Panguna intrusives and Bougainville volcanics.

	Kaverong quartz diorite	Biotite granodiorite	Biuro granodiorite	Bougainville andesites: average of ten
SiO_2	59.93	67.73	67.19	58.65
Al_2O_3	17.07	16.12	16.58	16.79
Fe_2O_3	3.35	1.09	1.78	4.09
FeO	2.76	1.48	1.37	2.35
MgO	2.41	1.19	1.11	2.98
CaO	6.12	2.95	4.27	6.94
Na_2O	4.09	4.97	4.70	4.21
K_2O	2.13	2.01	1.50	2.00
$H_2O(+)$	2.32	0.82	0.52	0.28
P_2O_5	0.29	0.18	0.19	—
TiO_2	0.63	0.33	0.33	0.65
MnO	0.04	0.00	0.02	0.16
S	0.01	0.17	0.00	—
CO_2	0.11	0.85	0.15	—
Cu	0.02	0.23	0.00	0.01
Total	101.29	100.13	99.72	99.09
$MgO/(MgO+FeO_T)$	0.29	0.33	0.27	0.33
K_2O/Na_2O	0.52	0.40	0.32	0.48

Panguna analyses from Mason (1975); Bougainville volcanics analyses from Taylor *et al.*, (1969*a*), average of nine andesites and one dacite.

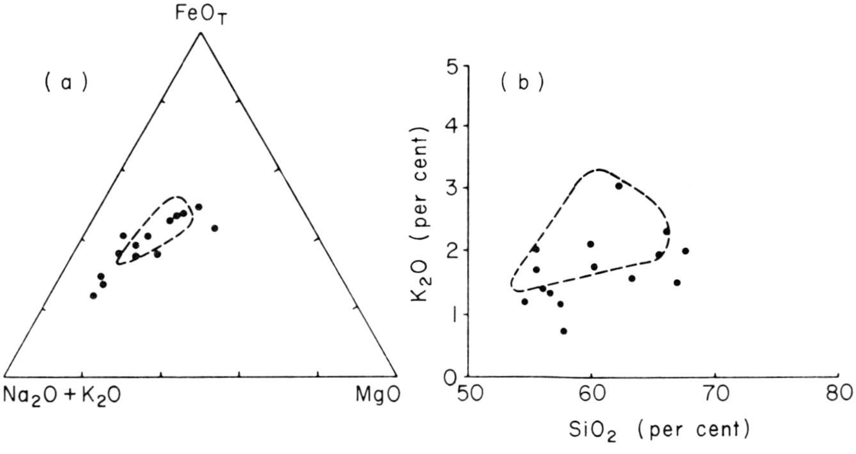

Fig. 9. Major element composition of Panguna intrusives compared with andesitic volcanics (including one dacite) from Bougainville. Data from Table 4 with additional analyses from Mason (1975); data on volcanics from Taylor *et al.* (1969*a*).

analyses (Table 4) and supplementary data on additional intrusives at Panguna presented in Figs 6, 7 and 9 are from Mason (1975). Comparative data for a suite of recent volcanics from Bougainville are from Taylor *et al.* (1969*a*).

3.1. Major Elements

Modern Bougainville "andesites" include low-silica and high-K varieties as well as normal andesites and dacites (Taylor, 1969). In a plot of K_2O vs. SiO_2 (Fig. 9(b), Panguna intrusive rocks spread across the range of composition of recent volcanics but tend to be lower in K_2O. On an FMA diagram (Fig. 9(a)) the intrusives coincide with the volcanic field but extend to lower FeO + MgO along the same Mg/Fe trend. The fact that total $Na_2O + K_2O$ contents of both intrusives and volcanics cover the same range of values (not plotted here) shows that this spread is not due to excessive total alkalis but to low FeO + MgO and low K_2O/Na_2O in the intrusive suite (Table 4). Whether or not this bias would disappear if a larger sampling was available is not known, though a similar but weaker bias is also suggested by the El Salvador data. The volcanics used for comparison on Bougainville are of much closer age and geographic distribution than those used for comparison with El Salvador. In both cases, however, the intrusive rocks are in most respects typical of volcanics in the same tectonic setting. Quartz dioritic rocks are more abundant at Panguna and on Bougainville than at El Salvador, and rhyolitic rocks are more abundant at El Salvador and in northern Chile than at Panguna (and also south-central Chile).

3.2. Minor Elements

In common with the El Salvador rocks, Panguna intrusives have similar to slightly depleted cobalt, nickel and vanadium contents when compared to their local volcanic suites. The chrome contents, however, are markedly higher than in local andesites, in contrast to El Salvador intrusives which show chrome depletion, relative to south-central Chilean andesites. The Ni/Co ratios are variable, and V/Ni ratios are generally high in both intrusives and volcanics on Bougainville relative to the Chilean rocks. LIL elements show essentially the same distribution in both intrusives and volcanics. The values are also very similar to those at El Salvador and in the Chilean volcanic suites. REE patterns for the three Panguna samples are remarkably similar to each other and to the volcanic suite (Taylor *et al.*, 1969*a*), though the intrusives tend to have slightly lower total REE contents than most volcanics. The patterns are typical of most circum-Pacific andesites. Compared to the El Salvador suite, Panguna intrusives have

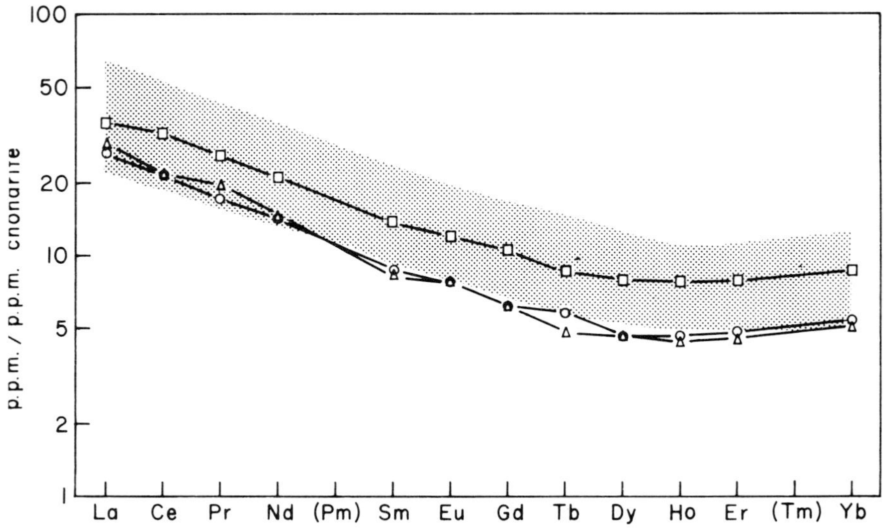

Fig. 10. Rare earth elements in Panguna intrusive rocks compared with the field of andesitic volcanics from Bougainville. Data from Table 4; volcanic data from Taylor *et al.* (1969*a*). □——□, Kaverong quarts diorite; △——△, biotite granodiorite; ○——○, Biuro granodiorite; stippled area shown Bougainville volcanics.

slightly lower total REE contents (45–67 p.p.m. vs. 61–98 p.p.m.) and are less fractionated (La/Yb = 6.2–7.4 vs. 11–47).

The $^{86}Sr/^{86}Sr$ ratios of the three Panguna samples range from 0.7033 to 0.7041 (Page, 1977, personal communication), comparable to the 0.7039 average for recent volcanics on Bougainville (Page and Johnson, 1974).

3.3. Petrologic Interpretations

The very high Cr/Ni ratios of the intrusives, following the line of reasoning used previously (Leeman, 1976), strongly indicate dominant partitioning between the melt and olivine rather than clinopyroxene or garnet. The REE patterns also indicate that garnet was not involved, a conclusion opposite to that reached for the El Salvador intrusives. Panguna melts therefore appear to have formed in a source region with quite different mineralogy to the source region of the El Salvador melts, one with either different bulk composition or at different temperature and pressure. Definition of these source regions is not possible at present, partially due to uncertainty about effects of subsequent fractional crystallization. The strontium isotope ratios, like those at El Salvador and most island-arc andesites, are significantly higher than ratios in "primitive" oceanic tholeiites. However, as pointed out by Gill and Compston (1973), the isotope arguments are not sufficiently constrained to allow determination of the reasons for this discrepancy. The crustal thickness and the history of Bougainville earlier than the Eocene when the oldest exposed rocks were formed are unknown, but comparison with neighbouring New Britain (Finlayson and Cull, 1973) suggest that the island is underlain by a substantial crustal thickness and has had a complex history.

4. Other Porphyry Copper Deposits

Single samples of three additional porphyry copper deposits in still different tectonic settings were analysed for minor elements (Table 5, Figs 6, 7 and 12) as a preliminary reconnaissance to see whether any suggestions of unique petrologic character could be found. The deposits were Atlas and Sipalay, on the islands of Cebu and Negros in the Republic of the Philippines, and Yerington in Nevada, USA.

Both Philippine deposits are related to porphyry complexes of probable Lower Tertiary age, intruded into what was probably an island-arc setting (Gervasio, 1966). Though the thickness and composition of the Philippine crust is now known, pre-Jurassic crystalline basement is exposed and a complex Mesozoic and Cenozoic depositional and deformational history is evident. The geology of both districts, which contain several centres of mineralization, has been documented in only a preliminary way (Santos-Yñigo, 1956; Kinkel *et al.*, 1956; Fernandez, 1966; Bryner, 1968). The deposits, both of which have been major producers for several years, are of particular interest in that they have contrasting molybdenum and gold values. While Atlas is typical of island-arc deposits with important gold but low molybdenum, Sipalay is nearly unique in this environment in being more like a Cordilleran deposit with low gold and high molybdenum. Rock and alteration types in the two deposits are in accord with this minor metal character. More felsic porphyries and more abundant K-feldspar alteration are exposed at Sipalay than at Atlas, where quartz diorite and only local K-feldspar alteration are more typical of island-arc deposits. The sample from the Lutopan Mine at Atlas is of a rather fresh and very poorly mineralized quartz diorite dike. The sample from Sipalay is of a fresh post-mineralization quartz porphyry which cuts all earlier veins and breccia within a similar looking quartz porphyry exposed in the small peripheral Baclao pit.

Yerington is a porphyry copper of Jurassic age in western Nevada which, though as yet only briefly described in the literature (Wilson, 1963; Howard, 1976), has been extensively studied by geologists of the Anaconda Company (work in preparation). It was apparently formed in a continental margin environment, and is one of a cluster of ore deposits related to cupolas on a

Table 5. Minor element content of Panguna intrusives, Bougainville volcanics and single intrusives from Atlas and Sipalay, Philippines, and Yerington, Nevada.

	Panguna intrusives			Bougainville Andesites	Other porphyry copper intrusives		
	Kaverong	Biotite granodiorite	Biuro granodiorite	Average of ten	Atlas	Sipalay	Yerington
U	0.8	0.5	0.5	0.9	0.5	0.33	2.1
Th	1.2	1.2	1.2	2.1	2.7	0.35	7.6
Ba	345	310	365	310	190	515	1400
Rb	40	12.9	25	39	29	20	61
Sr	617	653	700	734	447	805	553
La	11.2	9.2	8.4	10.9	13.4	3.0	23.5
Ce	26.1	17.5	17.5	19	31.6	7.1	50.7
Pr	3.01	2.28	2.0	2.9	3.0	0.90	6.8
Nd	12.7	8.92	8.5	13	12.5	4.18	27.8
Sm	2.65	1.58	1.68	2.8	2.6	0.97	4.5
Eu	0.87	0.56	0.56	0.89	0.75	0.34	1.12
Gd	2.74	1.60	1.62	2.6	2.40	0.92	2.95
Tb	0.42	0.24	0.28	0.54	0.35	0.13	0.30
Dy	2.58	1.54	1.51	3.8 (2)	1.96	0.84	1.49
Ho	0.57	0.32	0.34	0.57	0.40	0.19	0.30
Er	1.68	0.99	1.02	1.7	1.17	0.61	0.72

Yb	1.81	1.06	1.14	1.6	1.16	0.67	0.81
REE†	66.87	46.10	44.89	60.72‡	71.65	19.96	121.2
Hf	2.81	2.2	2.4	2.4	1.87	1.9	4.01
V	155	67	58	163	121	78	30
Co	13	10	4	17	11	6	7
Cr	68	104	94	24	21	6	11
Ni	6	7	5	8	6	8	10
Pb	3.6	4.6	3.2	3.7	2.4	5.2	5.7
Zn	42	34	34	—	57	27	32
Cu	237	2290	35	54	1200	165	728
K/Rb	442	1283	498	425	—	—	—
Rb/Sr	0.06	0.02	0.04	0.05	0.06	0.02	0.11
Ba/Rb	8.6	24.0	14.4	8.0	6.6	25.8	22.9
Th/U	1.5	2.4	2.4	2.3	5.4	1.1	3.6
Ni/Co	0.46	0.7	1.2	0.47	0.58	1.4	1.4
Cr/Ni	11.3	14.9	18.8	3.0	3.3	0.75	1.12
V/Ni	25.8	9.6	11.7	20.4	20.2	9.8	3.1
La/Yb	6.2	8.7	7.4	6.8	11.6	4.5	29.0
Eu*/Eu	1.0	1.1	1.0	—	0.93	1.1	0.95

† Includes estimate of Lu and Tm (both are minor).

‡ Estimate of Lu not included.

The Atlas sample is of a quartz diorite dike, Lutopan Mine. The Sipalay sample is of a later quartz porphyry in the Baclao pit. The Yerington sample is of late quartz monzonite porphyry—2.5. Analytical methods as in Table 2.

granodioritic to quartz monzonitic batholith which has been tilted and eroded to display the complex porphyry to batholith relationships. Gold and molybdenum contents at Yerington are very low. The sample analysed is of a largely post-mineralization quartz monzonite porphyry dike (QMP 2.5) from within the pit.

4.6. Major Elements

Kesler *et al* (1975) reviewed data for intrusives related to porphyry copper in the northern Caribbean and British Columbia island-arc associations as compared with continental deposits in the south-western USA. Figure 11 is

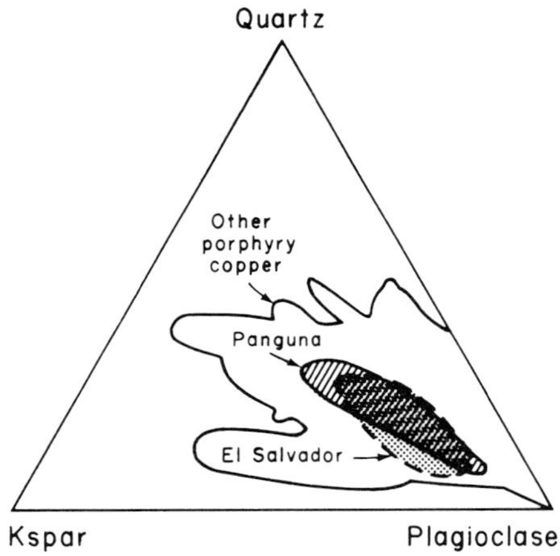

Fig. 11. Modal composition trends for porphyry intrusives from various porphyry copper deposits in North America, Northern Caribbean, Papua New Guinea and Solomon Islands compared with El Salvador and Panguna. Adapted from Fig. 4 of Kesler *et al.* (1975), with the addition of data from Mason (1975), and Gustafson and Hunt (1975).

taken from their compilation (Fig. 4) with the addition of many new analyses by Mason (1975) for intrusives associated with porphyry copper in island-arc and continental associations in Papua New Guinea and the Solomon Islands. Although Kesler *et al.* (1975) stress differences between the generally more K-rich granodioritic intrusives in continental environments as compared with less K-rich quartz diorites in island-arc settings, a generalization which conforms to widely recognised differences in igneous activity in these settings, the most important feature of this diagram is the continuity of copper porphyries over a wide composition range. Despite this wide range, the family of igneous textures of copper porphyries in all environments is remarkably similar, suggesting common histories of emplacement and crystallization. The fields of porphyry compositions from El Salvador and Panguna are largely coincident on this kind of a plot, even though the deposits formed in very different tectonic settings and with different minor element associations.

4.2. Minor Elements

Limited minor element data, except for metals, rubidium and strontium, are available for most porphyry copper deposits. The present sampling of the three deposits is much too small for interpretation other than to show a general similarity to El Salvador and Panguna within the broad range of values for both compatible and LIL elements (Figs. 6 and 7). The REE patterns (Fig. 12) show a similar wide range.

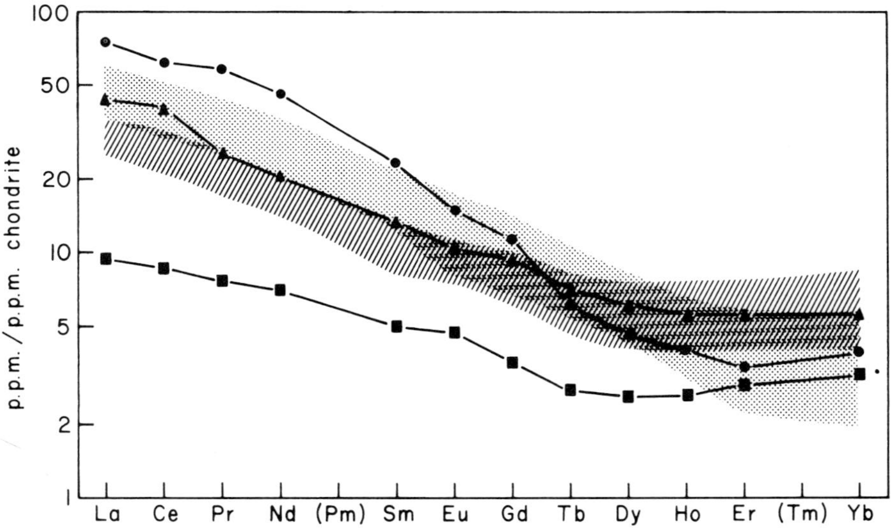

Fig. 12. Rare earth element data from porphyries at Atlas (▲——▲), Sipalay (■——■) and Yerington (●——●) compared with El Salvador and Panguna (hatched area). Data from Table 5.

Strontium isotope data are not available on these three deposits but are for many others. A recent review by Kesler *et al.* (1975) pointed out differences between low $^{87}Sr/^{86}Sr$ values (less than 0.705) for island-arc porphyries and higher values for deposits in continental settings. Figure 13 is a modification of their Fig. 6, showing new data for Panguna and El Salvador. Though Panguna fits this pattern, major discrepancies appear when Yandera and El Salvador ratios are considered. Yandera lies within

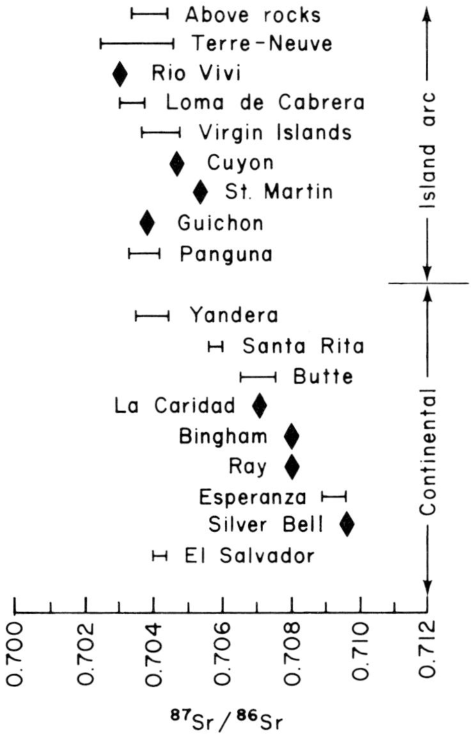

Fig. 13. Initial $^{87}Sr/^{86}Sr$ ratios for various porphyry copper intrusives. Modified from Fig. 6 of Kesler *et al.* (1975), with the addition of data on El Salvador (Gustafson and Hunt, 1975), Yandera (Page, 1977) and Panguna (R. W. Page, personal communication).

the New Guinea mobile belt to the south of the Ramu-Markham fault zone, probably over continental crust; El Salvador is certainly underlain by thick continental crust.

5. Petrologic Interpretations

The data presented here all confirm the long recognized fact that porphyry copper is associated with calc–alkaline volcanism. They fail to indicate any feature, other than the association with anomalous concentrations of sulphur and chalcophile metals, which distinguishes these intrusions from the broad range of intermediate calc–alkaline volcanics in a variety of tectonic environments. It is clear that different source regions (mineralogically and probably also in terms of composition, pressure and temperature) and fractional crystallization histories were involved in the formation of intrusive melts at Panguna and El Salvador. This is true even though detailed petrologic modelling could not provide unambiguous answers due to necessary assumptions as to the composition of possible source rocks, partitioning coefficients, and sequence of petrogenic events.

A broader sampling of deposits could only increase the diversity of petrologic character of this varied family of ore deposits. Porphyry copper magmas seem to have as diverse and complex origins as have "andesite" magmas in general, and in fact to be part of the overall "andesite problem". The problem of porphyry copper melts is probably a bit more restricted than that of andesites as a whole, because mineralized porphyries seem to be similar only to relatively mature andesites usually found in relatively late stages of volcanism. The deposits do not have affinities with the more mafic end of the andesitic suite, and are rarely associated with alkaline rocks. Porphyry copper deposits are not known in any of the more primitive arcs of the western Pacific (e.g. Tonga-Kermadek, New Hebrides or Mariannas arcs). The Chilean porphyry copper deposits are associated with late stages, rather than early stages, of the Andean orogeny, though deposits related to the culmination of the late Palaeozoic pre–Andean orogeny have recently been recognized in Argentina (Sillitoe, 1977).

As discussed in the introduction, the concept that all porphyry copper is related to active subduction is suspect, although a close relationship to active or past subduction is strongly indicated for most deposits. Andesites themselves are beginning to be more widely recognized as forming through processes other than active subduction and *via* pathways which are complex and diverse. Most recent models of andesite genesis fashioned under the twin constraints of major and minor element chemistry call on some sequential process of partial melting of deep-seated material which is itself neither simple primitive mantle or oceanic crust, but modified through incompatible element enrichment (e.g. Ringwood, 1974). This is followed in many cases by fractional crystallization at lower pressure. Some such complex process is envisioned as being necessary also for the formation of porphyry copper melts.

The actual source of sulphur and metals in porphyry copper deposits is even more elusive than the precise source of the melts. As was briefly considered in the introduction, geological evidence indicates that the bulk of sulphur and metals is carried by hydrothermal solutions derived from the same shallow magma chamber as the porphyritic intrusions. It is assumed that sulphur and metals were introduced into the shallow magma reservoir from the same deep source as the melt. Support for this assumption is derived from four different lines of reasoning.

1. Sulphur isotopes in most porphyry copper deposits and related hydrothermal veins show a narrow spread of $\delta(S^{34})$ values, close to the value

of meteoritic sulphur (Field, 1973). This so-called "magmatic" sulphur can be interpreted as having been derived from either an unfractionated or very well mixed source, probably in the upper mantle.

2. The tendency for minor molybdenum and gold contents of most porphyry copper deposits to reflect their continental or island arc tectonic setting also supports this contention, although we do not understand the causes of this effect or of the exceptions to the generalization, such as Sipalay.
3. Sulphur and chalcophile metals are probably more abundantly available in and above regions of subduction of oceanic crust than in any crustal source. The average content of oceanic crust is estimated at 800 g t^{-1} sulphur, 85 g t^{-1} copper and 1.5 g t^{-1} molybdenum (Lee and Yao, 1970). Given subduction of an oceanic crust 6 km thick at a rate of 3 cm y^{-1}, in just one million years there would be enough copper and molybdenum subducted along each 100 km section of oceanic trench to make roughly 1000 El Salvadors, or enough sulphur to make 100 El Salvadors if the concentration process were 100 per cent efficient.
4. As will be developed below, melts rising through the crust would be much more likely to lose sulphur and metals than to pick them up from their surroundings.

The major conclusion to be drawn from the preceding discussion is that porphyry copper melts which contain sulphur and metals can be derived from a variety of deep sources, probably in the upper mantle or lower crust, just as "andesitic" melts of different characters can be formed. No unique source region is required and, as indicated by the rare examples of deposits with most of the characteristics of porphyry copper in alkaline complexes (Barr *et al.*, 1976), even wide variations in magma source and evolution can be encompassed within the broad explanation of porphyry copper genesis. Porphyry coppers apparently share with calc–alkaline orogenic rock suites, broad source regions enriched in alkaline LIL and light REE, but some other factor must determine why some melts reach shallow reservoirs with enough sulphur and metals to form porphyry copper deposits while many others are barren. We are left with the conclusion that probably the only unique chemical characteristic of copper melts is the associated concentrations of sulphur and metals.

Sulphur is genetically, if not economically, a more critical factor than the metals. As Hunt (1977) has pointed out, porphyry copper deposits are really large sulphur anomalies with lower Cu/S ratios than most ordinary crustal rocks. For example, with up to 10^9 tons of sulphur and 10^7 tons of copper introduced into the El Salvador district, the ratio of Cu/S is approximately

1/100, and this ratio ranges in other deposits to about 1/10. The Cu/S ratio in average crustal rocks ranges from 1/5 to 1/20. The solubilities and mechanisms of solution of chalcophile elements are complex and not fully understood, especially at very high pressure. However, it is likely that magmatic processes which concentrate sulphur will also tend to concentrate copper, molybdenum and gold.

6. Proposed Model for Porphyry Copper Magma Genesis

The distribution of sulphur and chalcophile elements in igneous rocks of the upper crust (excluding surficial processes) is controlled by the evolution of aqueous fluids evolved from igneous melts. Understanding the history of the volatile components in these melts is a key to understanding the porphyry copper problem. Recent studies of volcanic gases and volatiles trapped in volcanic rocks (summarized and interpreted by Anderson, 1974, 1975; Wyllie, 1977) indicate that CO_2, SO_2 and chlorine as well as H_2O are major components of magmas derived from deep sources. At present, experimental information on the system andesite–H_2O–CO_2–S–Cl is too fragmentary, especially with regard to sulphur and chlorine, to be able to predict volatile content and behaviour over the large pressure–temperature–composition range of interest. However, data on andesite–H_2O–CO_2 and albite analogue systems provide some insights (Holloway, 1976; Mysen, 1976; Mysen *et al.*, 1975; Kadik and Eggler, 1975). Figures 14 and 15 summarize some of the pertinent experimental relationships. Figure 14 shows that the solubilities of H_2O and CO_2 in andesite increase with increasing pressure, and that there is a strong similarity of behaviour in albite and andesite systems. The solubility of CO_2 is about an order of magnitude lower than that of H_2O in both melts. Figure 15 shows the composition of a H_2O–CO_2 vapour phase in equilibrium with an albite melt, as a function of temperature and pressure. At high pressure, vapour in equilibrium with the melt is rich in CO_2, and with decreasing pressure there is a further gradual decrease in H_2O until a minimum is reached and then there is a rapid increase in the ratio $H_2O/(H_2O+CO_2)$ as pressure decreases beyond around 5 kbar.

The effect, as pointed out by Kadik and Eggler (1975) and Holloway (1976), is that felsic melts rising through the crust from high-pressure source regions, with not necessarily very high initial CO_2 contents, will generally become saturated with and evolve a CO_2-rich vapour phase throughout that rise. The initial vapour phase will have a low $H_2O/(H_2O+CO_2)$ ratio which will decrease until the 5 kbar minimum is passed and vapour then becomes rapidly more H_2O-rich. Thus, H_2O, which would otherwise be undersaturated and held in the melt, is lost with CO_2 over a large pressure (depth)

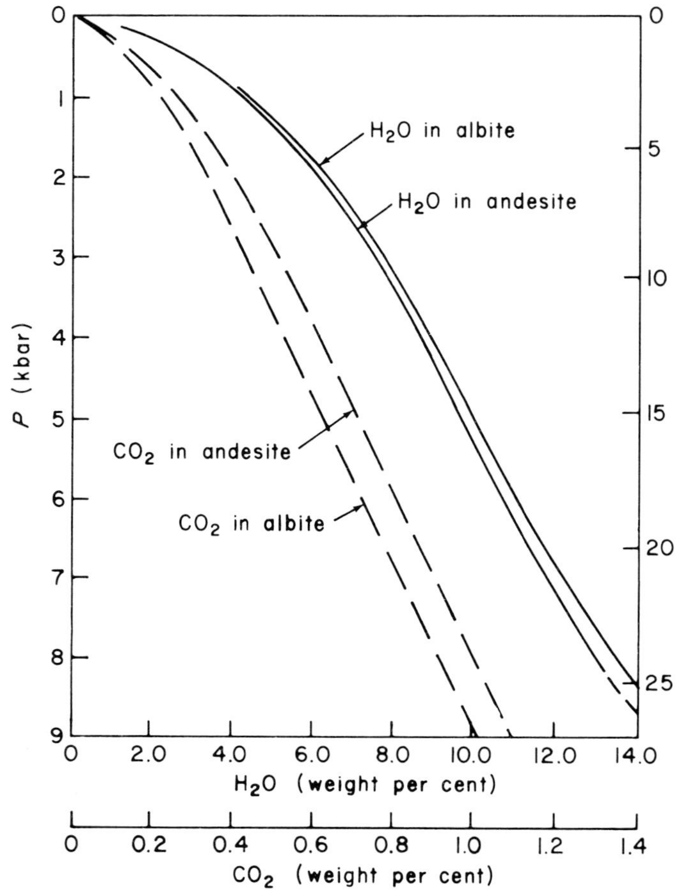

Fig. 14. Solubility of H_2O and CO_2 (weight per cent) in andesite and albite melts. Curves for H_2O in Mt Hood andesite at 1100 °C and for H_2O in albite at 40 °C above its liquidus (900–750 °C) are from Burnham (1967); curve for CO_2 in andesite at 1650 °C from Mysen *et al.* (1975); curve for CO_2 in albite at 1625 °C from Mysen (1976). Dashed lines are approximate extrapolations from higher pressure.

range. Burnham (1967) pointed out that because intermediate to felsic magmas reach shallow depths at temperatures below their dry melting temperature due to depression of the melting point by H_2O, saturation and loss of H_2O results in crystallization. Thus, he argued, porphyry copper magmas must have originated with relatively low H_2O contents to have avoided water saturation and freezing until shallow depths of emplacement were reached. Only in this case can volatile elements such as chlorine and sulphur, with associated chalcophile metals, be concentrated and evolved in

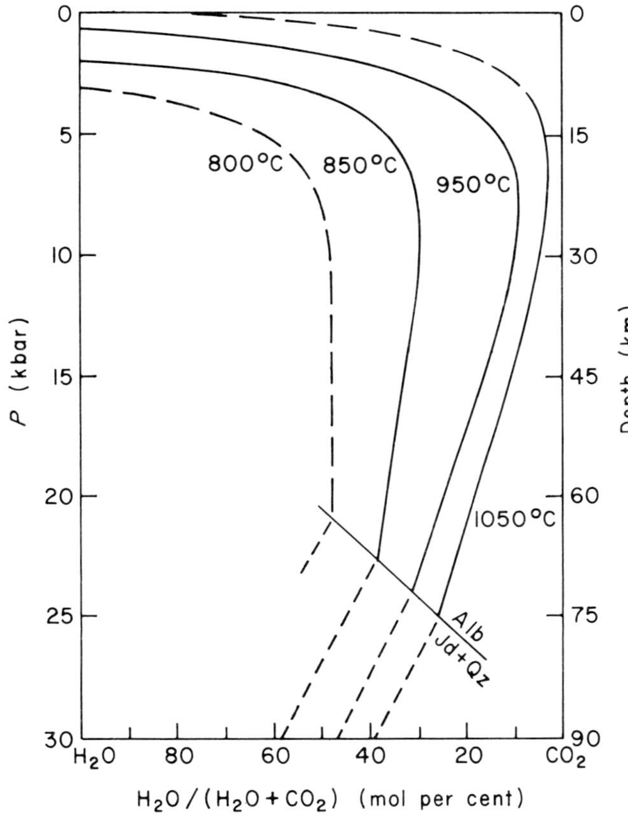

Fig. 15. Compositions (mol per cent) of H_2O–CO_2 vapour phase in equilibrium with an albite melt at 800, 850, 950 and 1050 °C; from Kadik and Eggler (1975).

shallow reservoirs. If, however, H_2O is partitioned out of a rising andesitic melt into a CO_2-rich phase at greater depth, it seems likely that chlorine and sulphur would also partition into this vapour phase and, in turn, carry their associated metals with them. Because alkalis and other major silicate components are much less soluble in CO_2 than in H_2O (Burnham, 1968) this vapour separation would probably not have a marked effect on the bulk composition of the melts which eventually reach the surface. Perhaps the tendency for porphyry copper intrusives to have a relatively low K/Na ratio is a consequence of CO_2 separation?

Figure 16 summarizes the proposed model. All melts generated in an orogenic belt which has had a long enough history of "conditioning" of the source area of calc–alkaline melts by volatile transport from an underlying subduction zone, are presumed to initially contain substantial amounts of

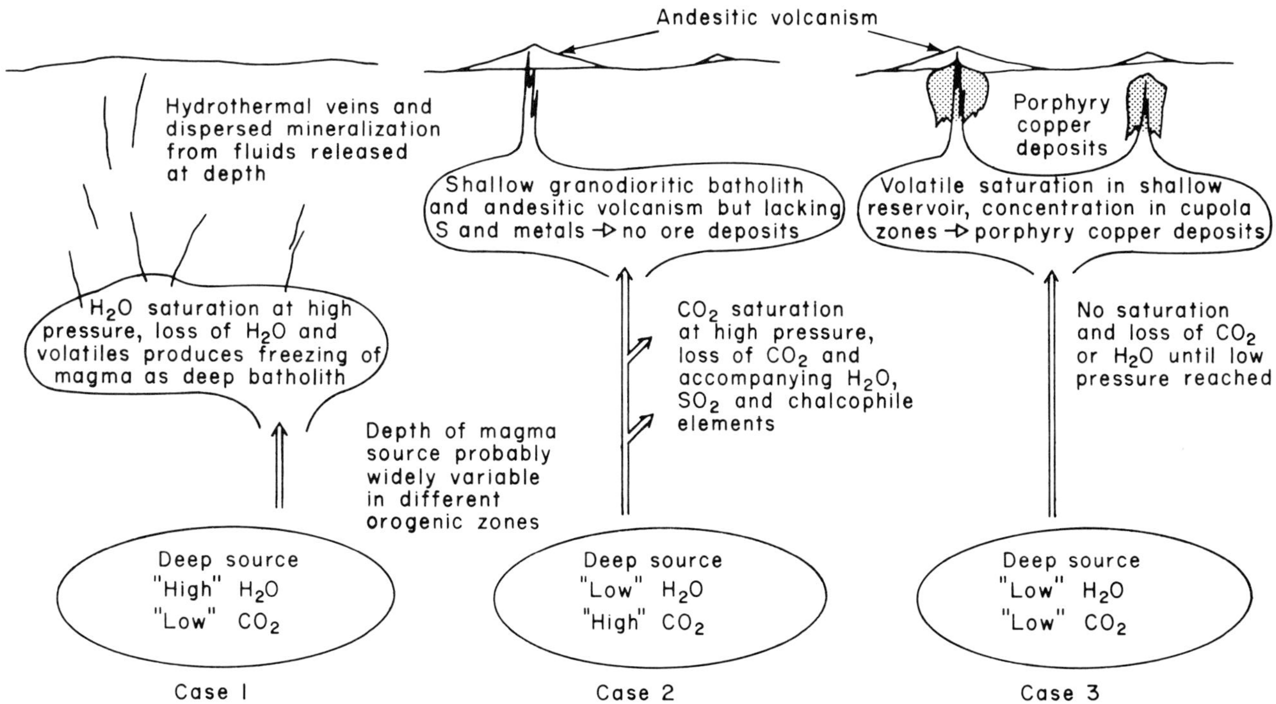

Fig. 16. Schematic model of the effect of initial H_2O and CO_2 contents of melts formed in a deep source on the volatile evolution as they rise through the crust, and on their ability to reach shallow reservoirs with sufficient metals and sulphur to form porphyry copper deposits.

sulphur, metals and associated volatiles. Variable H_2O and CO_2 contents are envisaged to result from inhomogeneities in the source region and variations in the degree of partial melting. Melts with too high an H_2O content saturate, freeze and lose all volatiles at depth, perhaps producing structurally controlled hydrothermal deposits but not porphyry copper. Melts with H_2O contents low enough to avoid deep H_2O saturation, but with enough CO_2 to evolve a vapour phase at depth, disperse their sulphur and metals throughout the deep crust. Only those melts with contents of both H_2O and CO_2 low enough to avoid any vapour saturation until the shallow reservoir is reached are capable of there evolving sulphur and metals to produce porphyry copper deposits in shallow cupolas.

There are several assumptions in this model which, on testing, may prove erroneous and require modification of the model. Though metals are apparently complexed with chlorine, at least at low pressure (Helgeson, 1964), and chlorine partitions strongly into an aqueous phase relative to melt (Kilinc and Burnham, 1972), making it reasonable to assume that extraction of H_2O with CO_2 at high pressure would also extract chlorine and complexed metals, it is not clear that sulphur would partition in the same way. It is conceivable that abortive porphyry copper melts might carry enough sulphur to first saturate at depth with a SO_2 or other sulphur vapour phase.

Preliminary results on the solubility of SO_2 in albite melts by Mysen and Arculus (in preparation) indicate a strong pressure coefficient of SO_2 solubility, and Anderson (1975) concludes that ocean floor basalts contain magmatic gas which is 60 per cent SO_2. However, there are too many unknowns regarding relative solubilities of sulphur in mixed volatile systems, oxidation state effects, and histories of evolution of volatiles prior to eruption of magmas to be able to make predictions on the basis of these indications. It is unlikely that melts in orogenic belts have anything like as much sulphur relative to H_2O and CO_2 as is indicated by oceanic basalts. Nonetheless, the partially independent behaviour of sulphur, which is linked to iron and oxidation states (Haughton *et al.*, 1974), and metals which are more directly linked to chlorine and H_2O, is probably a consideration which may partly control the wide variation in Cu/S ratios in different porphyry copper deposits.

Fluorite occurs in most molybdenite deposits of different types, including some porphyry coppers. Fluorite characterizes most porphyry molybdenum deposits (Climax, Colorado, being the classic example; Wallace *et al.*, 1968), where it, rather than the anhydrite characteristic of most porphyry coppers, fixes calcium during wallrock alteration. Fluorine, therefore, would appear to be as critical for distribution of molybdenite as chlorine is for copper. Hence, the relative abundance of molybdenum in different porphyry copper deposits would appear to be related to different histories of melt relative to the partitioning of chlorine, which strongly partitions into an aqueous phase, and fluorine, which strongly partitions into the melt relative to an aqueous phase (Burnham, 1968).

No systematic mineral association correlating with the abundance of gold in porphyry systems is apparent. There is a tendency for albite-rich alkali feldspar to proxy for K-feldspar in the more mafic, and commonly gold-rich, deposits in island arcs (and Woodall, 1975, has pointed out that sodic intrusives are commonly associated with gold deposits in many parts of the world), but the correlation has not been documented and there are certainly many exceptions. Whether the differences in minor metal ratios of porphyry coppers depend on differences in metal ratios at the source of the porphyry melts, on differences in abundance of chlorine, fluorine, sodium and other possible metal complexing agents, or on the differential partitioning of these elements with their associate metals during evolution of a H_2O–CO_2–(SO_2?) volatile phase remains an intriguing problem.

Clearly, more work on the andesite–H_2O–CO_2–Cl system would eventually lead to a better understanding of the processes of mineralization of the crust (e.g. Kadik and Eggler, 1975). It is the understanding of relationships in this complex system which would possibly lead to the ability to trace events within the shallow reservoir during crystallization of productive batholiths. These are the events which make ore, and are of most interest to geologists engaged in exploration. Prospectors and explorationists have always used signs of sulphide mineralization as the most reliable indicator of potential exploration targets. It is ironic that such signs may turn out to be the only geochemical clue to a magma which has had the "right" origin and history to make an ore body. However, even though bulk compositions do not distinguish between productive and barren intrusives, other parameters may. Abundant anhydrite formed at an early stage in many porphyry copper deposits indicates a relatively high oxidation state of sulphur in the magmatic fluids (Gustafson and Hunt, 1975; Field and Gustafson, 1976). This and the production of acid alteration may well be coupled through reactions such as

$$4SO_2 + 4H_2O \rightarrow 3SO_4^{2-} + S^{2-} + 8H^+.$$

Mason and McDonald (1978) report growth zoning toward increasing Mg/Fe ratios in hornblende phenocrysts in productive intrusives which is absent in barren intrusives. They interpret this zoning as indicating increasing oxidation states in late stages of crystallization. These features deserve further examination.

7. Time Evolutionary Aspects

The time pattern of global porphyry copper development has been considered by Meyer (1972) and Hunt (1977). They have shown that the great bulk of known porphyry copper apparently was emplaced in the last 75 million years, and that apparently diminishing amounts were emplaced back through time (Fig. 17). Very little porphyry copper mineralization is known to be more than about 450 million years old. It is at least partly true that the preponderance of young deposits reflects the fact that porphyry coppers are emplaced during late stages of tectonic evolution, characterized by uplift and erosion, so that chances of preserving these deposits diminish rapidly with age and deeper erosion. However, many shallow late-tectonic volcanic environments are preserved in the older geological record, but a decreasing number of them contain commercial concentrations of copper. In Queensland, Australia, more than 40 porphyry copper prospects in Paleozoic and Mesozoic intrusive centres have been found (Horton, 1978) but none of them are, as yet, of commerical tonnage and grade. Several deposits with porphyry copper affinity are known in the Archean of Canada and Western Australia, but all of very low copper grade. No deposits known to be older than Devonian are being mined anywhere, unless some of the mines in China are really Proterozoic (Ikonnikov, 1975), and unless the Archean deposit in the McIntire Mine, Ontario, is considered under a very broad definition to be a porphyry copper deposit. These exceptions to the overall pattern reinforce Meyer's (1972) point that "Locally almost any petrologic process of rock genesis is capable of producing a geochemical anomaly of ore grade, over a very wide range of geologic time". Though the detailed pattern of Fig. 17 probably reflects merely the incomplete state of

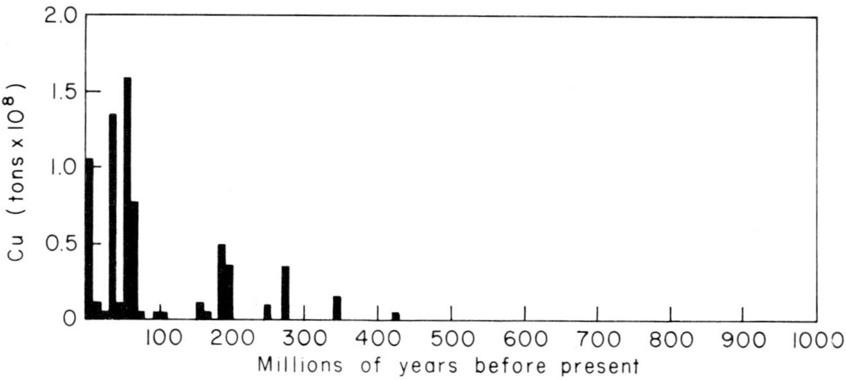

Fig. 17. Time pattern of the emplacement of copper in porphyry copper deposits of known age; from Hunt (1977).

dating known deposits and of discovery of unknown ones, the overall pattern is real.

If, as contended here, porphyry copper mineralization is simply a common manifestation of calc–alkaline volcanism, and as calc–alkaline volcanism is recognized to extend through the Proterozoic into the Archaen (Engel *et al.*, 1974), why are not porphyry copper deposits more abundant in the Precambrian? Is it due to a lack of subcrustal "conditioning" with alkalis, LIL elements, sulphur, etc., to provide adequately evolved source rocks in the absence of a subduction mechanism for such conditioning? At least one subeconomic deposit in south-west Africa, the Haib prospect, has clear porphyry copper characteristics, and occurs in a well documented calc–alkaline, orogenic volcanic-intrusive province 2000–1800 million years old (Reid, 1977). How unique is this province in the Proterozoic, and are there mechanisms other than subduction for producing thick differentiated calc–alkaline volcanic-intrusive sequences? Is the lack of porphyry copper in the Proterozoic due to perhaps a greater CO_2 content in subcrustal source regions, which led to bleeding off of the sulphur and metals before melts reached near-surface environments during this time? Much more work is required on the nature of Precambrian calc–alkaline volcanism before these questions can be answered, but the search promises to yield important insights into the evolution of the earth's crust and mineral wealth.

Acknowledgements

I am indebted to D. R. Sadler for production of most of the trace element data and for extensive assistance in the preparation of the manuscript; to E. Kiss, J. M. G. Shelley and P. H. Beasley for analytical assistance; to S. R. Taylor for the use of his analytical facility and for useful geochemical discussion; to R. J. Arculus for very helpful discussion and review of an early version of the manuscript; to F. A. Frey, R. J. Arculus and R. W. Johnson for access to unpublished manuscripts; to R. W. Page for unpublished strontium isotope data; to J. P. Hunt for permission to publish Fig. 17; and to K. E. Gibbs for editorial assistance.

References

Anderson, A. T. (1974). Chlorine, sulphur and water in magmas and oceans. *Bull. geol. Soc. Am.* **85**, 1485–1492.

Anderson, A. T. (1975). Some basaltic and andesitic gases. *Rev. Geophys. Space Phys.* **13**, 37–55.

Arculus, R. J. (1978). SO_2 in albite melts (in preparation).

Arculus, R. J. and Eggler, D. H. (1975). Solubility of carbon dioxide in melts of natural nephelinite, tholeiite and andesite composition. *Contr. Mineral. Petrol.* **53**, 227–239.

Barazangi, M. and Isacks, B. L. (1976). Spacial distribution of earthquakes and subduction of the Nazca plate beneath South America. *Geology* **4**, 686–692.

Barr, D. A., Fox, P. E., Northcote, K. E. and Preto, V. A. (1976). The alkaline suite porphyry deposits—a summary. In *Porphyry Deposits of the Canadian Cordillera* (A. Southerland-Brown, ed.), *Can. Inst. Min. Metall., spec. Vol.* **15**, 359–367.

Baumer, A. and Fraser, R. B. (1975). Panguna porphyry copper deposit, Bougainville. In *Economic Geology of Australia and Papua New Guinea*, Vol. 1. *Metals* (C. L. Knight, ed.), *Australasian Inst. Min. Metall. Monogr.* **5**, 855–866.

Blake, D. H., and Miezitis, Y. (1967). Geology of Bougainville and Buka Islands, New Guinea. *Bur. Miner. Resour. Australia Bull.* **93**.

Brooks, C., James, D. E. and Hart, S. R. (1976). Ancient lithosphere; its role in young continental volcanism. *Science, N.Y.* **193**, 1086–1094.

Bryner, L. (1968). Notes on the geology of porphyry copper deposits of the Philippines. *Mining Engng*, **19** (13), 12–23.

Burnham, C. W. (1968). Hydrothermal fluids at the magmatic stage. In *Geochemistry of Hydrothermal Ore Deposits* (H. L. Barnes, ed.), Holt, Rinehard and Winston, New York, pp. 34–76.

Coleman, P. J. (1966). The Solomon Islands as an island arc. *Nature, Lond.* **211**, 1249–1251.

Dewey, J. F. and Bird, J. M. (1970). Mountain belts and the new global tectonics. *J. geophys. Res.* **75**, 2625–2647.

Emmons, W. H. (1927). Relations of the disseminated copper ores in porphyry to igneous intrusives. *Am. Inst. Min. Metall. Engrs Trans.* **75**, 797–815.

Engel, A. E. J., Itson, S. P., Engel, C. G. and Stickney, D. M. (1974). Crustal evolution and global tectonics: a petrographic view. *Bull. geol. Surv. Am.* **85**, 843–858.

Fernandez, H. E. (1966). Origin of the mineralized breccia pipe of Baclao Mine, Sipalay, Negros Occidental. *Bull. Inst. Filip. Geol.* **1**, 47–54.

Field, C. W. (1973). Sulfur isotope abundances in hydrothermal sulfate-sulfide assemblages of the American Cordillera. *Geol. Soc. Am. Abstr., 1973 Annual Meeting*, p. 619 (abstract).

Field, C. W. and Gustafson, L. B. (1976). Sulfur isotopes in the porphyry copper deposit at El Salvador, Chile. *Econom. Geol.* **71**, 1533–1548.

Finlayson, D. M. and Cull, J. P. (1973). Time-term analysis of New Britain–New Ireland island arc structures. *Geophys. J. R. astronom. Soc.* **33**, 265–280.

Frey, F. A., Chappell, B. W. and Roy, S. D. (1978). Fractionation of rare-earth elements in the Tuolumne Intrusive series, Sierra Nevada batholith, California. *Geology* **6**, 239–242.

Gervasio, F. C. (1966). A study of the tectonics of the Philippine Archipelago. *Philippine Geologist* **20**(2), 51–75.

Gill, J. and Compston, W. (1973). Strontium isotopes in island arc volcanic rocks. In *The Western Pacific: Island Arcs, Marginal Seas, Geochemistry* (P. J. Coleman, ed.), University of Western Australia Press, Perth, pp. 483–496.

Guilbert, J. M. and Sumner, J. S. (1968). Distribution of porphyry copper deposits in the light of recent tectonic advances. In *Southern Arizona Guidebook* III, Arizona Geological Society, Tucson, pp. 97–112.

Guild, P. W. (1971). Metallogeny: a key to exploration. *Mining Engng* **23**, 69–72.
Gustafson, L. B. (1978). Some major factors of porphyry copper genesis. *Econom. Geol.* **73**, 600–607.
Gustafson, L. B. and Hunt, J. P. (1975). The porphyry copper deposit at El Salvador, Chile. *Econom. Geol.* **70**, 856–912.
Haughton, D. R., Roeder, P. L., and Skinner, B. J. (1974). Solubility of sulfur in mafic magmas. *Econom. Geol.* **69**, 451–467.
Helgeson, H. C. (1964). Complexing and Hydrothermal Ore Deposition, Pergamon, London.
Holland, H. D. (1972). Granites, solutions, and base metal deposits, *Econom. Geol.* **67**, 281–301.
Holloway, J. R. (1976). Fluids in the evolution of granitic magmas: consequences of finite CO_2 solubility. *Bull. geol. Soc. Am.* **87**, 1513–1518.
Horton, D. J. (1978). Porphyry-type copper–molybdenum mineralization belts in Eastern Queensland. *Econom. Geol.* **73**, 904–921.
Howard, K. L. Jr. (1976). Geology of the Yerington Mine, Lyon County, Nevada. *Econom. Geol.* **71**, 700 (abstract).
Hunt, J. P. (1977). Porphyry copper deposits. In *Volcanic Processes in Ore Genesis*, Special Publication No. 7, Geological Society, London, p. 98 (abstract).
Ikonnikov, A. B. (1975). Mineral resources of China. *Geol. Soc. Am. Microform Publ.* 2.
Isacks, B., Oliver, J. and Sykes, J. R. (1968). Seismology and the new blobal tectonics. *J. geophys. Res.* **73**, 5855–5900.
James, D. E. (1971). Place tectonic model for the evolution of the central Andes. *Bull. geol. Soc. Am.* **82**, 3325–3346.
James, D. E., Brookes, C. and Cuyubamba, A. (1976). Andean Cenozoic volcanism: magma genesis in the light of strontium isotopic composition and trace element geochemistry. *Bull. geol. Soc. Am.* **87**, 592–600.
Johnson, R. W., Mackenzie, D. E. and Smith, I. E. M. (1978). Delayed partial melting of subduction-modified mantle. *Tectonophysics* **46**, 197–216.
Kadik, A. A. and Eggler, D. H. (1975). Melt–vapour relations on the join $NaAlSi_3O_8$–H_2O–CO_2, *Carnegie Inst. Washington Yearbook* **74**, 479–484.
Kesler, S. E. (1973). Copper, molybdenum and gold abundances in porphyry copper deposits. *Econom. Geol.* **68**, 106–112.
Kesler, S. E., Jones, L. M. and Walker, R. L. (1975). Intrusive rocks associated with porphyry copper mineralization in island arc areas. *Econom. Geol.* **70**, 515–526.
Kilinc, I. A. and Burnham, C. W. (1972). Partitioning of chloride between a silicate melt and coexisting aqueous phase from 2 to 8 kilobars. *Econom. Geol.* **67**, 231–235.
Kinkel, A. R. Jr., Santos-Yñigo, L. and Samaniego, S. (1956). Geology of the copper deposits east of Sipalay, Negros Occidental. In *Copper Deposits of the Philippines*, Part I (A. R. Kinkel, *et al.*, eds), *Philip. Bur. Mines. spec. Projects Ser. Pub.* **16**, 160–169.
Klerkx, J., Deutsch, S., Pichler, H. and Zeil, W. (1977). Strontium isotopic composition and trace element data bearing on the origin of Cenozoic volcanic rocks of the central and southern Andes. *J. Volcanol. geotherm. Res.* **2**, 49–71.
Lee, T. and Yao, C. (1970). Abundance of chemical elements in the earth's crust and its major tectonic units. *Int. Geol. Rev.* **12**, 778–786.
Leeman, W. P. (1976). Petrogenesis of McKinney (Snake River) olivine, tholeiite in light of rare–earth element and Cr/Ni distributions, *Bull. geol. Soc. Am.* **87**, 1582–1586.

Lindgren, W., (1933). *Mineral Deposits*, McGraw-Hill, New York.

Livingston, D. E. (1973). A plate tectonic hypothesis for the genesis of porphyry copper deposits of the southern Basin and Range province. *Earth planet. Sci. Lett.* **20**, 171–179.

Lopez-Escobar, L., Frey, F. A. and Vergara, M. (1976). Andesites from central-south Chile: trace element abundances and petrogenesis. In *Proc. Symposium on Andean and Antarctic Volcanology Problems, IVACEI Special Series*, 725–761.

Lopez-Escobar, L., Frey, F. A. and Vergara, M. (1977). Andesites and high-alumina basalts from the central-south Chile high Andes: geochemical evidence bearing on their petrogenesis. *Contr. Mineral. Petrol.* **63**, 200–228.

Lowell, J. D. (1974). Regional characteristics of porphyry copper deposits of the Southwest. *Econom. Geol.* **69**, 601–617.

Lowell, J. D. and Guilbert, J. M. (1970). Lateral and vertical alteration-mineralization zoning in porphyry ore deposits. *Econom. Geol.* **65**, 378–408.

Mason, D. R. (1975). Geochemistry of intrusive rock suites and related porphyry copper mineralisation in the Papua New Guinea–Solomon Islands region. Ph.D. Thesis, Australian National University.

Mason, D. R. and McDonald, J. A. (1978). Intrusive rocks and porphyry copper occurrences of the Papua New Guinea–Solomon Islands region: a reconnaissance study. *Econom. Geol.* **73**, 857–877.

Meyer, C. (1972). Evolution of ore-forming processes with geologic time. Presidential address to the Society of Economic Geologists, San Francisco (unpublished).

Mitchell, A. H. G., and Garson, M. S. (1972). Relationships of porphyry copper and circum-Pacific tin deposits to palaeo–Benioff zones, *Inst. Min. Metall. Trans. B* **81**, 10–25.

Myson, B. O. (1976). The role of volatiles in silicate melts: solubility of carbon dioxide and water in feldspar, pyroxene and feldspathoid melts at 30 kb. and 1625°C. *Am. J. Sci.* **276**, 969–996.

Myson, B. O., Arculus, R. J. and Eggler, D. (1975). Solubility of carbon dioxide in melts of andesite, tholeite and olivine nephelinite composition to 30 kbar pressure. *Contr. Mineral. Petrol.* **53**, 227–239.

Nielsen, R. L. (1976). Recent developments in the study of porphyry copper geology: a review. In *Porphyry Deposits of the Canadian Cordillera* (A. Sutherland-Brown, ed.), Can. Inst. Min. Metall. *spec. Vol.* **15**, 487–500.

Page, R. W. (1976). Geochronology of igneous and metamorphic rocks in the New Guinea highlands. *Australian Bur. Miner. Resour. Bull.* 162.

Page, R. W. and Johnson, R. W. (1974). Strontium isotope ratios of Quaternary volcanic rocks from Papua New Guinea. *Lithos* **7**, 91–100.

Pichler, H., and Zeil, W. (1972). The Cenozoic rhyolite—andesite association of the Chilean Andes. *Bull. Volcanol.* **35**, 424–452.

Reid, D. L. (1977). Geochemistry of Precambrian igneous rocks in the Lower Orange River region. *Chamber of Mines, Precambrian Research Unit, Cape Town, Bulletin 22.*

Ringwood, A. E. (1974). The petrological evolution of island arc systems. *J. geol. Soc.* **130**, 183–204.

Sales, R. W. (1954). Genetic relations between granites, porphyrites, and associated copper deposits, *Mining Engng* **6**, 497–505.

Santos-Yñigo, L. (1956). Copper deposits of the Atlas Consolidated Mining and Development Corporation. In *Copper Deposits of the Philippines*, Part I (A. R. Kinkel *et al.*, eds). *Philip. Bur. Mines, spec. Projects Ser. Pub.* **16**, 142–159.

Sawkins, F. J. (1972). Sulphide ore deposits in relation to plate tectonics. *J. Geol.* **80**, 377–397.

Sheppard, S. M. F. and Gustafson, L. B. (1976). Oxygen and hydrogen isotopes in the porphyry copper deposit at El Salvador, Chile. Econom. Geol. **71**, 1549–1559.

Sillitoe, R. H. (1972). A plate tectonic model for the origin of porphyry copper deposits. *Econom. Geol.* **67**, 184–197.

Sillitoe, R. H. (1977). Permo-Carboniferous, upper Cretaceous and Miocene porphyry copper-type mineralization in the Argentinian Andes. *Econom. Geol.* **72**, 99–103.

Taylor, S. R. (1969). Trace element chemistry of andesites and associated calc–alkaline rocks. In *Proceedings of the Andesite Conference* (A. R. McBirney, ed.), *Oregon Dept Geol. Min. Res. Bull.* **65**, 43–64.

Taylor, S. R., Capp, A. C., Graham, A. L. and Blake, D. H. (1969*a*). Trace element abundances in andesites. II. Saipan, Bougainville and Fiji, *Contr. Mineral. Petrol.* **23**, 1–26.

Taylor, S. R., Kaye, M., White, A. J. R., Duncan, A. R. and Ewart, A. (1969*b*). Genetic significance of Co, Cr, Ni, Sc and V content of andesites, *Geochim. cosmochim. acta.* **33**, 275–286.

Taylor, S. R. and Gorton, M. J. (1977). Geochemical application of spark source mass spectrometry. III. Element sensitivity, precision and accuracy, *Geochim. cosmochim. acta* **33**, 275–286.

Wallace, S. R., Muncaster, N. K., Jonson, D. C., Mackenzie, W. B., Brookstrom, A. A. and Surface, V. E. (1968). Multiple intrusion and mineralization at Climax, Colorado. In *Ore Deposits in the United States 1933/1967* (J. D. Ridge, ed.), American Institute of Mining, Metallurgy and Petroleum Engineers, New York, pp. 605–640.

Wilson, J. R. (1963). Geology of the Yerington Mine. *Mining Congress J.*

Woodall, R. (1975). Gold in the Precambrian shield of Western Australia. In *Economic Geology of Australia and Papua New Guinea*, Vol. 1, *Metals* (C. L. Knight, ed.), *Australian Inst. Min. Metall. Monogr.* **5**, 175–184.

Wyllie, E. J. (1977). Mantle fluid compositions buffered by carbonates in peridotite–CO_2–H_2O. *J. Geol.* **85**, 187–207.

14. The Mechanical Behaviour of Rock under Crustal and Mantle Conditions

M. S. PATERSON

Abstract

A brief survey is given of the types of mechanical behaviour that might be expected in the crust and mantle in the light of recent experimental studies. Emphasis is placed on the variety of possible behaviour and the difficulties that arise out of this variety in making relevant extrapolations from the time scale of experiments to that of natural deformations. The particular case of olivine-rich rocks, of interest in mantle studies, is taken up in more detail, with a critical survey of available experimental data.

1. Introduction

The wide range in pressure and temperature as well as in mechanical and chemical environment in the crust and mantle raises the expectation that there will be a corresponding wide range of possible mechanical behaviour in the rocks in these regions. Experimental work has already revealed substantial diversity in the mechanical behaviour of rocks under some of these environmental conditions, such conditions as have already been attained in the laboratory, and further variety in behaviour may yet be revealed. However, it is not possible to explore all conditions in the laboratory and some extrapolation from the laboratory to the Earth is necessary, especially in the time scale and, to a lesser degree, in respect of pressure. Any extrapolation that is done must be guided by microstructural criteria, comparing laboratory observations with such evidence as remains microstructurally in the geologically deformed rocks of the relevant deformation processes. Therefore the elucidation of the deformation mechanisms is important as well as the study of the macroscopic mechanics of deformation. In this essay, an attempt is made, in the light of recent experimental studies, to survey the types of mechanical behaviour of rocks that may be involved in crust and mantle and to give some indication of the scope of our present knowledge on the rheology.

It is convenient first to consider separately the brittle and ductile fields of behaviour. However, these fields overlap considerably and some care is needed in distinguishing and defining what is brittle and what ductile. The realm of brittle behaviour is likely to be mainly the upper crust, where the pressure is at most a few hundred megapascals (a few kilobars) and the temperatures are low, but some brittleness may be important also under higher grade conditions if there is a pore fluid at a pressure not much below the lithostatic pressure. The occurrence of ductile flow at geologically realistic stress levels can be expected to depend mainly on temperature and to require, at least for silicate rocks, moderately high temperatures; however, it is difficult to be confident of predictions involving large extrapolations in strain rate or time and it is also currently thought that the presence of water may influence deformation rates in an important way. The brittle–ductile transition itself is a complex phenomenon, depending not only simply on pressure and temperature but also involving other environmental factors and boundary constraints, and its proper consideration brings in stability questions. We shall consider these topics in turn and then deal briefly with specific observations of likely relevance to flow in the mantle.

2. The Brittle Field

The concept of brittleness, that is, of the propensity to crack or to break into separate pieces, can be applied to rock on different scales. Primarily, the brittle field of rock behaviour is the field characterized by the occurrence of macroscopic brittle fracture in the absence of substantial amounts of preceding inelastic deformation. However, it is now clear that the macroscopic brittle fracture is not an isolated, singular event, but is the culmination of a complex development of effects at the microscopic scale which are themselves predominantly of a brittle nature. In a compressively loaded rock, these effects are detectable above about half the macroscopic fracture stress or peak load. In general they involve the formation, propagation or interaction of microcracks.

The microcrack is thus the elementary or fundamental entity involved in the brittle behaviour or rock. It is a crack of microscopic or submicroscopic dimensions that may be present initially in the rock or may be generated at points of local stress concentration during the loading. The detailed observation of microcracks is mainly of recent date. Optical microscopy gives a certain amount of information, especially about the larger microcracks and their distribution through the rock (see review by Simmons and Richter, 1976), but electron microscopy is beginning to reveal much more about the initial microcrack structure (Brace *et al.*, 1972; Sprunt and Brace, 1974; Hadley, 1976). From these studies it is now known that, at least in granitic rocks, the initial microcracks, or the elongate microcavities that act as nuclei for microcracking, are to a large extent of submicroscopic dimensions. They often follow grain boundaries and often have blunted ends which appear to represent the partial healing of former longer cracks; the ratio of length to width does not commonly exceed 100:1.

Of the pre-failure phenomena arising out of the activity of microcracks, dilatancy, acoustic emission and change in elastic wave velocity are those which have been studied in the most detail. Other notable effects include changes in permeability and in the electrical conductivity of water-saturated rock. A comprehensive review is given elsewhere (Paterson, 1978) and so only a few salient points will be touched on here. These are illustrated in Figs 1 and 2 in the case of Westerly granite in triaxial compression tests at room temperature.

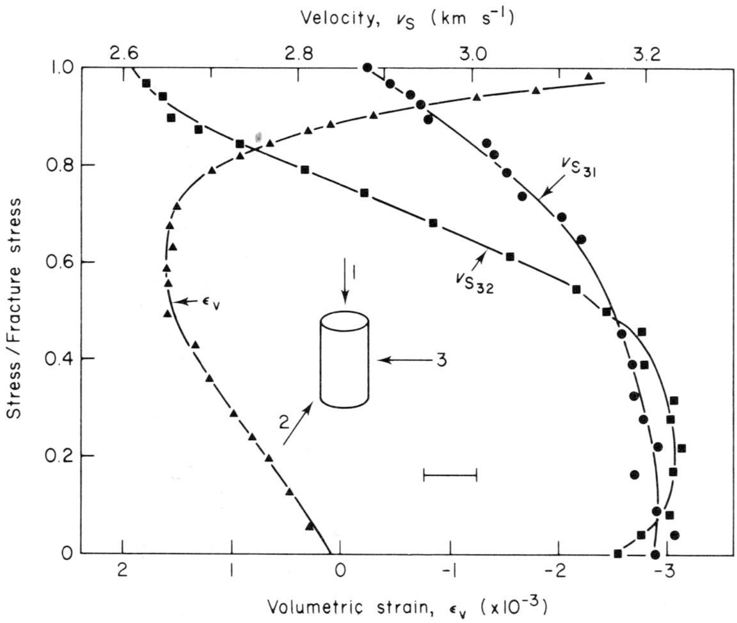

Fig. 1. Variation in volumetric strain $\varepsilon_v = -\Delta V/V$ and in shear wave velocities V_S in dry Westerly granite during loading up to fracture under confining pressure of 20 MPa. V_{S31} and V_{S32} are the velocities for propagation of shear waves normal to the loading axis, with vibration directions respectively parallel to and normal to this axis (after Bonner, 1974).

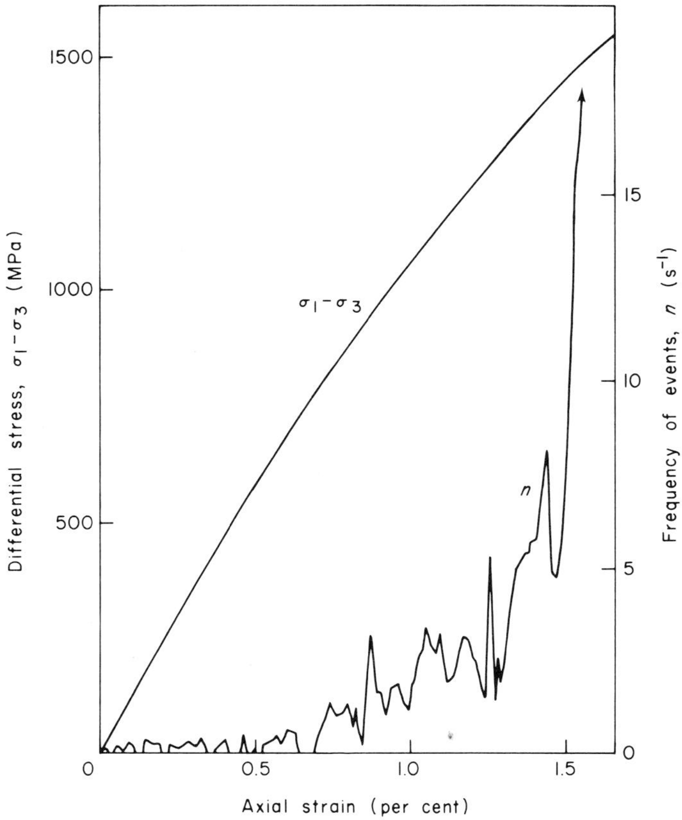

Fig. 2. Increase in the rate of acoustic emission in Westerly granite at 400 MPa confining pressure during loading along the stress–strain curve shown (after Scholz, 1968).

When the differential stress approaches about a half of the level needed to bring about macroscopic fracture in the granite (Fig. 1), the volumetric strain is seen to depart notably from linear elastic change, marking the onset of dilatancy, that is, of inelastic increase in volume corresponding to increase in void space within the rock. Correspondingly, the velocity of elastic waves also decreases markedly; shear wave velocities for propagation normal to the loading axis are shown in Fig. 1, but a similar effect is observed in the compressional wave velocity in the same direction. It is also observed that the rate of acoustic emission increases strikingly in the same stress range (Fig. 2). All three effects tend to become more marked with further increase in differential stress, the dilatancy and acoustic emission especially so just below the peak load.

These effects in the stressed granite are each consistent with a picture of proliferation and propagation of microcracks. The acoustic emission is probably a direct registration of microcracking events, either the initiation or the abrupt enlargement of microcracks, while the dilatancy represents volume increase due to the opening of the cracks. The changes in elastic wave velocities are consistent with theoretical predictions of the effects of introducing a substantial population of microcracks and they give information on the preferred orientation of the cracks; thus the birefringence, as seen in the difference in velocities for the two polarizations of shear waves shown in Fig. 1, is consistent with there being a predominance of cracks oriented more or less parallel to the direction of loading.

The microcracking development discussed so far, illustrated in granite but common to other rocks as well, is at first distributed fairly uniformly throughout the specimen. However, somewhere in the vicinity of the peak load, further development of cracking becomes concentrated in a localized zone, leading eventually to a macroscopic fracture appearing in this zone. The way in which the progression to macroscopic fracture occurs has not yet been fully worked out, but important new observations have been made recently, following the introduction of stiff and servo-controlled testing machines. (These machines permit the deformation to be arrested at any desired point in the unstable post-peak region of the load–deformation curve in order to observe the progress of fracture development.) Thus, in confined compression tests terminating in macroscopic shear fracture, it has been noted that near the peak load the microcrack population includes a high proportion of cracks that are inclined to the compression axis; however, the zonal concentration and linking-up of microcracks to form the macroscopic shear fracture is not normally evident until after the peak load is passed (Friedman, Perkins and Green, 1970; Wawersik and Brace, 1971).

The brittleness of rock thus has two aspects, different in scale and somewhat different in nature. The microcracking is one aspect, evident at the microscopic scale; this elementary process is essentially brittle in nature but is not necessarily or immediately expressed in macroscopic brittleness. The other aspect, the occurrence of a macroscopic brittle fracture, differs in being a more complex phenomenon; it involves a multiplicity of microcracks and its essential nature lies not so much in the individual crack as in the organization of the population of cracks and their interaction. Further consideration of these aspects of brittle behaviour will be taken up below in connection with the brittle–ductile transition.

3. The Ductile Field

The converse of brittleness is ductility or the capacity to flow; that is, the capacity to undergo permanent or inelastic change of shape without the intervention of macroscopic fracture. The inelastic deformation may occur by any of a considerable number of mechanisms or processes, depending on the nature of the material and on the physical environment. However, the ductile field of rock behaviour, like the brittle field, is defined in a macroscopic sense, independently of whatever the particular mechanism of flow may be.

The processes of deformation themselves can be viewed as involving, fundamentally, the rearrangement within the deforming body of some sort of elementary structural unit, the scale and nature of which may vary widely with different processes. For convenience we shall refer to these elementary units as the flow units. The concept of ductility is applicable at scales on which the material appears to be statistically homogeneous in structure, while the mechanism of deformation can only be discussed or studied at a finer scale at which the flow units are distinguishable. There are three general categories of flow units.

1. Clastic fragments or individual grains; for example, the fracture fragments that move relative to each other in cataclastic deformation, the sand grains or clay platelets that take part in soil flow, and, possibly, the crystal grains that move relative to each other in high-temperature "superplastic" creep of fine-grained polycrystalline materials.
2. Intracrystalline blocks, especially the glide packets that slide over each other in crystallographic slip through the movement of disolocations.
3. Individual atoms or molecules or the equivalent vacancies, the units in case of diffusional flow mechanisms.

The identification of the principal flow units and their pattern of relative movement, however, only gives the micro-geometry of the deformation. The dynamics of flow may be determined by additional factors, often by the slowest step in a sequence that includes auxiliary processes involved in accomodating the relative movement of the flow units, for example, the climb of dislocations in surmounting obstacles to their glide during deformation by slip. A full description of all vital aspects of a particular deformation mechanism may therefore be quite complex. Complexity is further compounded when the many different types of deformation mechanism in the ductile field are considered. Broadly, the mechanisms can be considered in three groups, corresponding in large part to the grouping of basic flow units given above, but rearranged slightly to reflect better the contrasts in temperature dependence of the flow stress. The three groups are:

1. *Cataclastic flow*. Involves distributed microfracturing and relative movement of the fragments and, because of the associated friction and dilatancy, tends to be highly sensitive to confining pressure but not to temperature.

2. *Crystal plasticity*. Slip processes within the grains, sometimes twinning. Comprises a wide variety of mechanisms, arising because of the many ways in which resistance to dislocation movement may arise, and covers a wide range in degrees of temperature dependence. The most widely studied experimentally.
3. *Diffusional and grain-sliding flow*. Change of shape by differential movement of the atoms, or by relative movement of the grains with accommodation by diffusion or minor crystal plasticity in order to retain coherence at grain boundaries. Tends to predominate at low stress levels and to have fairly high temperature dependence.

Although the contributions of the various mechanisms to the flow rate in a rock depend on many variables, under any given set of conditions one particular mechanism will tend to contribute substantially more than the others. Hence in a field of values of relevant physical variables, domains can be delimited within each of which a particular deformation mechanism is characteristic; such a field of domains is known as a "*deformation mechanism map*" (Ashby, 1972). Stress and temperature are the most commonly used variables for such maps. The boundaries of the domains can be calculated if laws relating strain rate to stress and temperature for deformation by each mechanism are assumed, as in examples for rocks given by Stocker and Ashby (1973), Rutter (1976) and White (1976). However, the construction of these maps has so far depended heavily on analogy with the known behaviour of metals and ceramics, and the experimental support from work on rocks is still rather meagre; a great deal of more detailed experimental work over a wider range of conditions is needed.

A complicating factor in characterizing the flow behaviour within a given mechanism field is that the resistance to flow is generally not constant but tends to increase or decrease with increasing strain. This work-hardening or work-softening reflects continued change in the structure of the material during the deformation. The change may be in the size of the flow units or in finer details of structure that determine the dynamics of the flow; for example, in cataclastic flow, there may be progressive breakdown into finer-scale elastic fragments and changes in their packing, or in crystal plasticity there may be changes in dislocation density, in the subgrain structure, or in the nature of the dislocation tangles that inhibit the motion of the mobile dislocations, or the grain boundaries themselves may migrate. However, there is a tendency to reach a stage eventually at which the rate of further change becomes negligible and the mechanical behaviour of the material can be said to have reached a *steady state*.

Although we have introduced it in structural terms, the steady state is primarily a macroscopic concept, here defined by the condition that the resistance to flow is independent of time, and hence of strain when the strain rate and other independent variables are fixed. Establishing steady-state conditions, using the strain independence as criterion, has been a common aim in experimental studies, although these have usually been restricted to simple coaxial deformations. Correspondingly, the extrapolation has only been attempted for steady-state flow, a restriction that is attractive on grounds of simplicity but which may also be a reasonable one in dealing with large-strain problems. However, the steady state, as defined in the above operational way, is a concept somewhat lacking in generality and care is needed in delimiting its applicability. Thus, behaviour that can be accepted as steady state to a sufficient approximation in a context of moderate strains, as in experiments, may not be acceptable as such over a very large range of strain. Further, the attainment of a steady state in a particular deformation path need not mean that the same state would eventually be reached along another deformation path under comparable conditions; that is, the existence of a steady state in particular circumstances does not necessarily mean that the behaviour under general strains can be characterized by an equation of state in which stress and strain rate are the only independent variables apart from environmental variables such as temperature, and in which there is the path independence implicit in any equation of state. This matter is one requiring detailed study before flow laws determined experimentally in simple deformations can be generalized and used with confidence in the solution of complex deformation problems. (For a broader and more sophisticated discussion of the phenomenology of flow laws, see Hart *et al.*, 1975; Hart, 1976; Poirier, 1976.)

Even in the experimental field itself there are some interesting and important questions concerning the establishment of flow laws applying to steady state conditions. Two such questions will now be discussed.

(1) How much strain is required for reaching the steady state? The answer is probably that the amount may be small or large depending on the nature of the deformation mechanism. One can distinguish two general ways in which a steady state may be reached. One way is through either the exhaustion of the possibility of further structural modification or the achievement of some sort of dynamic balance in which further structural modification is not needed for accommodating the relative movement of the flow units. (For example, with a sufficiently loose packing and the wearing off of sharp corners, there may be no need for further fragmentation in a cataclastic flow and a steady state could ensue.) In such a case, fairly large strains may be required to achieve a steady state. The other way in which a steady state may arise, although now only at relatively high temperatures, is through a balance between structural modification brought about by the strain itself and spontaneous restoration arising out of the effect of thermal fluctuations;

in this case, the steady state may be achieved after quite small strains, although not necessarily so. The distinction between the two ways of reaching a steady state thus lies mainly in the circumstance that the first involves factors that are all directly strain-dependent while the second involves a balance against a factor that depends directly only on time rather than on strain.

(2) What do we mean by a steady-state flow law and how do we determine it? A steady-state flow law is taken here to be a relationship between strain rate, stress and environmental parameters such as temperature and fugacity of mobile components, for which the strain rate at any given value of stress is the steady-state value that would eventually be attained in continuing deformation at this stress while the other independent variables are held constant. It should be noted that parameters characterizing the dynamic aspects of the structure do not appear as independent variables in a steady-state flow law. One can, of course, define other types of flow law, another conceptually important one being a constant-structure flow law, that is, a relationship involving not only strain rate, stress and the environmental variables but also involving a structure parameter as an independent parametric variable to be given specific values as desired; other quantities such as strain or similar aspects of specimen history now have to be manipulated so as to achieve the desired structure for each set of the other variables.

It follows that the ideal experimental procedure for determining the steady-state flow law is to carry out a set of constant strain rate or creep experiments, in each of which the specimen is deformed to a large enough strain to establish that the steady state has been reached (as indicated by attainment of constant stress or constant strain rate, respectively). One set of such experiments has to be carried out for each representative point in the field of values of environmental variables within which the flow law is to be established. Such a programme is, of course, a very onerous one if a large range of values of the independent variables is to be covered. The question then arises of whether one can use short-cut methods in which a multiplicity of stress, strain-rate points can be determined in the same experiment, for example, in a relaxation experiment. However, while such approaches can throw a good deal of light on the general character of the deformation properties, there cannot be a general expectation of success in obtaining thereby a steady-state flow law; success will depend on the rate of attainment of the steady state relative to the effective dwelling interval in strain or time at each stress, strain-rate point, and this will have to be judged or tested individually in each case. Thus, if a steady state is attainable with very small strains, or if the relevant structural factors characterizing the steady state are independent of the stress or strain rate at given values of the environmental variables, then a relaxation experiment can be used to derive a steady-state flow law (possibly the second situation occurs in Solenhofen limestone in the "superplastic" field; Schmid, Boland and Paterson, 1977). In the other extreme, where attainment of the steady state is very sluggish, the relaxation experiment may yield a constant-structure flow law. But in general the relaxation test may yield a flow law corresponding to some poorly-defined state between steady state and constant structure.

With appropriate experiments one can therefore establish empirically what steady-state flow laws are obeyed for particular rocks and demarcate the ranges of values of variables such as stress and temperature within which these laws hold; that is, establish "flow regime maps" for the rocks. Through accompanying microscopical and theoretical studies, one should eventually be able to convert these maps into the corresponding deformation mechanism maps. Such a programme represents one of the main thrusts of current work, with an emphasis on relatively high temperatures and low strain rates and on the achievement of steady-state conditions. This emphasis arises in part from the feeling that therein lies much potential geological interest and in part from this area being currently relatively tractable in terms of experiment and interpretation. However, it leaves aside such topics as transient behaviour, deformation under more complex stress states and the role of the deformation history, as well as the factors that govern the achievement of ductility in the first place. We now return to the latter question.

4. The Brittle–Ductile Transition

Brittleness or ductility is not an invariable property of a material but is dependent on the circumstances of test as well as on inherent properties of the material. In the case of rock, ductility may be induced by various means, the following being commonly effective, especially in combination:

1. increase in compressive mean stress, for example, through raising the confining pressure in a laboratory test (if pore fluid is present, its pressure must be allowed for, generally by simple subtraction, in specifying the "effective pressure");
2. increase in temperature;
3. decrease in strain rate;
4. introduction of a chemical weakening agent, notably water or hydroxyl ions.

It has been noted earlier that the macroscopic brittle fracture in a rock is the culmination of a complex development of microcracking; beyond a certain stage (in the vicinity of the peak load in a laboratory test), the microcracking activity becomes localized in a zone in which the macroscopic failure eventually occurs. However, the earlier part of the development, which took place in the steeply-rising part of the stress–strain curve, involves a fairly uniform distribution of microcracking activity. The central question in relation to the brittle–ductile transition is then: what determines whether localization of microcracking sets in leading to macroscopic failure? Or, conversely, how do the stabilizing factors mentioned above prevent this development in the ductile field? We first discuss the matter from the microscopic point of view and then the macroscopic.

When ductility is achieved solely by raising the pressure, one possibility is that the uniformly distributed proliferation of microcracking is promoted to the extent that large deformation by cataclastic flow becomes possible. This development would occur by virtue of a stabilizing influence of pressure on the propagation and interaction of microcracks that would prevent a localization of this activity, presumably through increase of energy absorption in friction, plastic work at crack tips and other restraints on crack mobility. However, this situation may rarely be realized in practice because the increase in pressure also markedly raises the threshold stress for initiating inelastic effects and so the stress may come to exceed the yield stress for the onset of crystal plasticity within the grains of the rock. At sufficiently high pressure, all cataclastic effects may be suppressed and the deformation occur wholly by crystal plasticity, but in many cases a component of cataclastic flow can still be expected (for further discussion and original references, see Paterson, 1978). The other factors listed above as promoting ductility do so mainly by lowering the flow stress for deformation by crystal plasticity or, under more extreme conditions, by possibly introducing diffusional or grain-sliding mechanisms. Where these factors predominate in determining the ductility, cataclastic effects will generally be negligible.

Macroscopically, particularly in the transitional and low-temperature ductile fields, the deformational stability of the material derives from a continuation of work hardening up to large strains; the work hardening ensures that localization of the deformation under uniform loading would lead to locally greater strength and hence to inhibition of further deformation there until all other regions had been similarly work hardened. However, when steady-state conditions are achieved, work hardening no longer occurs (by definition) and deformational stability of the material depends on the strain-rate sensitivity of its flow stress; localization of the deformation would now mean a locally higher strain rate, requiring a locally higher stress in violation of uniform loading. On the other hand, if significant intrinsic changes in the material accompanied the localization of deformation, for example, weakening due to syntectonic recrystallization or to near-adiabatic heating due to the straining itself, the deformational behaviour may again be unstable; in this way, local shear zones could develop even under uniform applied loading in the absence of brittle mechanisms.

5. Studies Relating to Flow in the Mantle

5.1. Experiments on Dunite

On the presumption that olivine is the most abundant mineral in the mantle, there have been several experimental studies of the deformation of olivine-rich rocks (usually dunite) carried out with a view to providing data for discussing flow in the mantle. The brittle–ductile transition of dunite has not been studied in detail as a function of pressure, temperature, strain rate and chemical evironment, but some salient observations have been made. Marginal ductility has been reported above about 500 MPa (5 kbar) confining pressure at room temperature (Byerlee, 1968), and substantial strains were achieved at 500 MPa pressure and strain rates of the order $10^{-4}\ s^{-1}$ by Turner *et al.* (1960) from 300 to 800 °C and by Raleigh (1968) at 800 and 1000 °C, in all cases with apparatus using a fluid medium and with nominally dry specimens. However, there was probably a component of cataclastic deformation in these cases, except perhaps at the highest temperatures (microcracking has been demonstrated by Boland and Hobbs (1973), in peridotite at temperatures up to about 950 °C at 300 MPa confining pressure); further, high-temperature steady-state flow was not achieved in these experiments. The only published measurements on approximately steady-state flow under a range of conditions are those of Carter and Ave'Lallemant (1970), Kirby and Raleigh (1973) and Post (1977), all made using solid-medium deformation apparatus and mainly made on Mt Burnet dunite. It is convenient to analyze these results in terms of an empirical flow law of the form

$$\dot{\varepsilon} = A\,\sigma^n e^{-Q/RT}, \tag{1}$$

which can be reasonably well fitted to the data, at least over limited ranges of stress and temperature. Here $\dot{\varepsilon}$ represents strain rate, σ differential stress, R the gas constant and T absolute temperature; A, n and Q are empirical constants.

First, we consider the experiments done under so-called "wet" conditions; that is, experiments in which the pressure medium was talc and the specimen was exposed to the water of dehydration of the talc in the vicinity of the specimen where the temperature is above the dehydration temperature (around 850 °C). The steady-state experiments of Carter and Ave'Lallemant were done at 1500 MPa confining pressure, mainly in the differential stress range 100–500 MPa. Those of Post were done at 500–1500 MPa confining pressure and mainly in the stress range 500–800 MPa, except for two experiments down to 250 MPa; Post also carried relaxation tests down to about 100 MPa stress. Both workers, in their analysis, included data from a number of runs in which stress-incrementing or temperature-incrementing was done after an initial steady-state run ("differential tests"). Probably as a result of the latter course and of working mainly in different stress regions, the conclusions drawn are somewhat different (but see Carter, 1976, for a revised analysis of the Carter and Ave'Lallemant data). However, at the point of best overlap of the studies, namely, at a strain rate of 10^{-5} s^{-1} and a temperature of 1100 °C, the agreement in stress level (250 MPa) is good and indicates that the difference in confining pressure is not having an important influence. Also, Post's data from constant strain-rate tests at 1100 °C give a value of n of between 4 and 5 which appears to agree with the n value for the corresponding data of Carter and Ave'Lallemant for the same temperature, as read from their original plot. A relaxation test of Post's at 1050 °C gives $n = 4.1$ for stresses down to about 250 MPa. On the other hand, Post's relaxation test gives $n = 3.4$ for stresses below 250 MPa and Carter and Ave'Lallemant deduce $n = 2.4$ when their lower stress and higher temperature data are included (or $n = 2.1$ according to Carter's revised analysis). There is therefore some evidence for a change in n from 4 or 5 to around 3 or less when the stress is reduced below 250–300 MPa, although clearly more data is needed before taking this conclusion seriously.

The temperature dependence of the flow of "dunite" under the "wet" conditions is given by the value of Q in equation (1). Q was found to be 392 ± 11 KJ mol^{-1} (94 kcal mol^{-1}) by Post and 334 ± 31 (revised to 226) by Carter and Ave'Lallemant. Since the span in $\ln \dot{\varepsilon}$ is two or three times the span in $\ln \sigma$, these figures should correspond more nearly to a determination of Q according to

$$Q = R\left(-\frac{\partial \ln \dot{\varepsilon}}{\partial(1/T)}\right)_{\sigma}$$

than according to

$$Q = nR\left(\frac{\partial \ln \sigma}{\partial(1/T)}\right)_{\dot{\varepsilon}}.$$

However, the lower Q values of Carter and Ave'Lallemant go with lower n values, so the differences in reported values of Q may be to some extent coupled with the differences in n through the second relationship. Again, it has to be borne in mind that much of the data comes from differential tests and may not accurately reflect steady-state conditions.

The experiments done under *"dry" conditions* do not present a very clear picture. In these experiments an anhydrous medium has been substituted for the talc that yields the water by dehydration in the "wet" experiments. Alsimag 222, a machinable porous ceramic consisting of forsterite and periclase (Post, 1977), has largely served as the anhydrous medium, although Post has inserted inner liners of glass or metal immediately adjacent to the specimen. The first striking aspect of the results is that the flow stresses reported for "dry" conditions are both much higher and more variable than those for "wet" conditions. Thus at 1100 °C, 10^{-5} s^{-1}, where a consistent flow stress of about 250 MPa is given for "wet" conditions, Carter and Ave'Lallemant report a flow stress for "dry" conditions of 650 MPa while Post's "dry" strengths are generally much higher; he has no experiment at 1100 °C, 10^{-5} s^{-1}, but reports the same flow stress of 650 MPa at temperatures of both 1100 °C and 1300 °C at a lower strain rate of about 10^{-6} s^{-1}. All the deduced values of n and Q are based on differential tests or relaxation tests, n being given as 4.8 (3.3 revised), 3.0 and 3, and Q as 500 (464), 420 and 527 kJ mol^{-1} by Carter and Ave'Lallemant, Kirby and Raleigh, and Post, respectively, these figures applying mainly in the stress range 100–500 MPa.

Aspects of the results for "dry" conditions that are problematical are as follows.

1. Because of the high strength of Alsimag and glass relative to talc in the cooler parts of the pressure vessel, the component of the "friction" correction in the force that relates to deformation of the pressure medium can be expected to be substantially higher under the "dry" conditions giving greater uncertainty in the corrected load. Moreover, the deformation of the medium may well not reach a steady state and so the normal procedure of allowing for "friction" during the "run-in" is likely to underestimate the correction, perhaps seriously, and so lead to an overestimate of the flow stress. The degree of "dryness" is said to be a matter of uncertainty. According to Post, steps taken to dry the pressure medium thoroughly and to dehydrate traces of serpentine in the specimens lead to higher flow stresses. Although Goetze (1978) has expressed doubts, Post's procedure of heating the dunite in a vacuum furnace at 950 °C in the presence of steel wool should in principle not affect the state of oxidation of the dunite; but confirmation of the absence of precipitated phases by electron microscopy has not been reported and is needed before accepting that the state of "dryness" is the only reason for Post's exceptionally high flow stresses.
2. In a large proportion of Post's "dry" tests, shear zones developed with various degrees of sharpness and he did not include results from these

tests in his analysis. Carter and Ave'Lallemant do not report the same phenomenon.

In view of the above problems, a conservative conclusion would be that the flow behaviour of dunite in the absence of water is not well known and there is really no uncontroversial evidence that dunite is markedly stronger in the absence of water than in its presence (the same comments as above apply to the work of Blacic, 1972, on single crystals of olivine). In fact the flow stress of "dry" Mt Burnet dunite measured in a creep experiment of Goetze and Brace (1972) in a gas apparatus was only slightly above 300 MPa at 1100 °C, $10^{-5}\ s^{-1}$, and preliminary experiments in the author's gas apparatus by Mr P. Chopra on oven-dried Anita Bay dunite give a slightly lower flow stress under the same conditions, both figures being similar to the flow stresses reported for Mt Burnet dunite under "wet" conditions in solid-medium apparatus.

Apart from the overall question of flow stress, two specific differences in behaviour between "wet" and "dry" conditions have been pointed to; namely, in the value of Q and in the recrystallization behaviour. The higher figure for "dry" conditions of Carter and Ave'Lallemant is based on differential tests between 1100 and 1350 °C and may be partly influenced by the fitting with a higher value of n, as discussed above, and partly by being in a somewhat higher temperature range than for the "wet" experiments. Post's figure for Q is frankly an "estimate", obtained by extrapolating from two lower figures obtained in temperature-raising experiments that were assumed to involve intermediate amounts of water. A higher value of Q for "dry" conditions can therefore at best be accepted only as a qualitative result until further studies are done.

A remarkable situation rises in recrystallization behaviour. Ave'Lallemant and Carter (1970) concluded that "the presence of externally released H_2O does not appear to affect the temperature of onset of recrystallization but does promote coarser grain size". Post reports that "recrystallization is observed in most of the deformed samples", but adds that while incipient recrystallization appears in "wet" experiments as low as 700 °C after several weeks, recrystallization is only observed at temperatures above 1200 °C in "dry" experiments. These and other experiments are consistent in pointing to a strong influence of water in increasing markedly the rate of recrystallization, probably even allowing an equilibrium grain size to be attained as one aspect of the steady state since Post notes an inverse correlation between recrystallized grainsize and stress (cf. also Mercier *et al.* 1977, and Ross *et al.* 1978). It would seem to be a reasonable speculation to attribute to the water a hydrolytic effect in aiding the breaking of Si—O bonds that must occur in completely reconstituting the crystal structure during recrystallization. However, such a hydrolytic role does not so obviously enter the deformation picture because the known slip systems, having Burgers vectors parallel to the crystallographic ones, can be conceived as operating by the motion of dislocations that need not cut any Si—O bonds, although such disruption may be needed during climb or in other accommodation mechanisms. Another indirect role of water may be through the crystallization if this affects the pertinent features of the steady-state structure and hence the level of flow stress.

The interpretations just suggested to account for some of the complexity in reported behaviour of olivine-rich rocks depart in several respects from what is probably the current consensus view, based on hydrolytic weakening of the grains, as represented by Carter (1976). Other alternative views are presented by Twiss (1976) and Goetze (1978). Some of these views may prove to be complementary, dealing with different regimes (as in the case to Twiss), but further experimental data for a wider range of conditions are needed for further progress.

5.2. Experiments on Olivine Single Crystals

Complementing the observations on the polycrystalline dunite, the flow behaviour of single crystals of Fo_{92} olivine has been studied in considerable detail by Phakey *et al.* (1972) and by Goetze and co-workers (Kohlstedt and Goetze, 1974; Kohlstedt *et al.* 1976; Durham and Goetze, 1977*a,b*; Durham *et al.* 1977; Goetze, 1978). The experiments of Phakey *et al.* were done in a talc pressure medium at 1000 MPa confining pressure, mostly at 10^{-5} s^{-1} strain rate and 800–1000 °C, and therefore at least partly under "wet" conditions. The experiments of Goetze and co-workers were done at atmospheric pressure in a H_2–CO_2 gas mixture with oxygen fugacity within the olivine stability field and at temperatures mostly in the range 1400–1600 °C. Extrapolating a little in temperature to give overlap at 1100 °C, 10^{-5} s^{-1} shows the two sets of measurements to be in agreement within experimental scatter, at a flow stress somewhat above 200 MPa for orientations favouring slip in the [100] direction and around 600 MPa for orientations favouring the [001] but not the [100] slip direction. Thus, the single crystal flow stresses for the softest orientations are not much below the flow stress of dunite measured under "wet" conditions, at 1100 °C, 10^{-5} s^{-1}, while the orientations unfavourable to slip in the [100] direction appear to be distinctly harder (orientations normal to the crystallographic axes being especially so).

There is therefore some basis, admittedly tenuous, for assuming that the flow stress of dunite at high temperatures will be similar in magnitude to that of olivine single crystals oriented favourable for [100] slip. There is still a problem of reconciling this proposition with the von Mises requirement of five independent slip systems, as has been discussed by Paterson (1974) and Durham and Goetze (1977*a*). Durham and Goetze suggest that any potential strain incompatibility between grains from limitation on the number of slip systems is accommodated by dislocation climb. However, Goetze (1978) has also argued that deformation of dunite at low stresses can be expected to proceed by a Coble creep mechanism rather than by crystal plasticity, so that the von Mises problem would not then arise.

A particularly interesting aspect of the work of Goetze and coworkers on the single crystals is that by using very high temperatures they have been able to study the flow behaviour in the lower stress range of 10–100 MPa, a range that may be of greater geophysical interest. In steady-state creep tests, they found for all orientations a flow law of the form (1) with $n = 3.6$ and $Q = 520$ kJ mol^{-1} (125 kcal mol^{-1}). These values of n and Q can probably be taken as setting upper limits for the corresponding parameters for polycrystalline olivine under these conditions (but see comment on Coble creep in previous paragraph).

5.3. Other Aspects

So far we have only discussed experiments on essentially single-phase material, carried out at pressures rather low relative to mantle pressures, and in which the deformation mechanisms have been predominantly intracrystalline slip. We now review very briefly what is known of behaviour outside these limits.

There are no direct experimental data on the influence of pressure on the high temperature flow behaviour of olivine-rich rocks. Presumably the influence is small compared with that of temperature since there are no obvious pressure effects in the experiments mentioned above, in which the confining pressure ranges from atmospheric to 1500 MPa. (A pressure effect corresponding to an activation volume of about 13 cm^3 mol^{-1} has recently been observed in experiments on dunite by Ross, Ave'Lallemant and Carter: N. L. Carter, private communication, 1978.) Stocker and Ashby have estimated an upper limit to the effect on the assumptions that it will appear as a modification of the value of Q through an effect of pressure in inhibiting diffusion and that the activation volume for diffusion will not be greater than 37 cm^3 mol^{-1}. However, Poirier and Vergobbi (1978) have made an alternative suggestion that the flow stress may be cross-slip controlled and that therefore the effect of increasing the pressure could even be to decrease the flow stress as has been observed in sodium chloride (Davis and Gordon, 1969*a*,*b*; Aladag *et al.*, 1970; Haasen *et al.*, 1970).

Departing from dunite itself, Carter and Ave'Lallemant (1970) have measured the flow stress in lherzolite under the same "wet" conditions as used for dunite and found it to be substantially weaker than dunite; the steady-state flow stress at 1100 °C, 10^{-5} s^{-1} strain rate for Snowy Mountains lherzolite appears to be roughly one half that for the Mt Burnet dunite. This observation would imply that the addition of pyroxenes to the olivine aggregate weakens it under these conditions. Further such studies are obviously desirable in view of their particular relevance to the mantle, as well as are studies on the role of partial melting for which there are so far no observations on ultrabasic rocks.

Although Post (1977) observed very little recrystallization in the absence of water, he did find that the "dry" dunite specimens had a strong tendency to form shear zones ("ductile faults") of varying degrees of sharpness and with varying rapidity, and that local recrystallization occurred in these zones, with an accompanying drop in strength (see also Blacic, 1972). Post attributes the weakening in part to enhanced grain-boundary sliding. Twiss (1976) has interpreted these observations in a more definite way as evidence of the onset of a superplastic mode of deformation involving the relative movement of the grains by grain-boundary sliding and he has adapted the model of Ashby and Verall (1973) to olivine, predicting a regime of superplastic flow with $n = 1$ at low stress levels, and deriving a value of Q of about 300 kJ mol^{-1} (71 kcal mol^{-1}) for this regime, using unpublished data of Post's. This topic is also one clearly calling for further experimental study since it has profound implications both for extrapolation of flow behaviour to the mantle and for explanation of the occurrence of mylonite zones (cf. Boullier and Gueguen, 1975; Gueguen and Boullier, 1976; and counter-argument by Mercier and Carter, 1975).

6. Conclusion

Experimental studies have greatly advanced our understanding of the fracture and flow properties of rock. In this paper we have re-examined some broader aspects of this understanding and discussed olivine-rich rocks in some detail. However, in applying the findings to the Earth, comparative microstructural observations are of paramount importance. They are essential both in discriminating between different mechanisms of deformation and in guiding the extrapolation of experimentally determined flow laws to natural conditions. Microstructural studies have already shown that the deformation processes being studied in the laboratory have much in common with the natural processes and that a basis exists for attempting this extrapolation, but further refinement in applications to the Earth can be expected as laboratory observations accumulate.

Acknowledgements

The author is indebted to his colleagues in the rock mechanics group at Australian National University for continually stimulating debate on the issues discussed in this paper. He is also indebted to Professor N. L. Carter for helpful comments.

References

Aladag, E., Davis, L. A. and Gordon, R. B. (1970). Cross slip and the plastic deformation of NaCl single and polycrystals at high pressure. *Phil. Mag.* **21**, 469–478.

Ashby, M. F. (1972). A first report on deformation-mechanism maps. *Acta metallurgica* **20**, 887–897.

Ashby, M. F. and Verrall, R. A. (1973). Diffusion-accommodated flow and superplasticity. *Acta metallurgica* **21**, 149–163.

Ave'Lallemant, H. G. and Carter, N. L. (1970). Syntectonic recrystallization of olivine and modes of flow in the upper mantle. *Bull. geol. Soc. Am.* **81**, 2203–2220.

Blacic, J. (1972). Effect of water on the experimental deformation of olivine. In *Flow and Fracture of Rocks* (H. C. Heard, I. Y. Borg, N. L. Carter and C. B. Raleigh, eds), Geophysics Monograph no. 16, American Geophysical Union, Washington, D.C., pp. 109–115.

Boland, J. N. and Hobbs, B. E. (1973). Microfracturing processes in experimentally deformed peridotite. *Int. J. Rock Mech. Min. Sci.* **10**, 623–626.

Bonner, B. P. (1974). Shear wave birefringence in dilating granite. *Geophys. Res. Lett.* **1**, 217–220.

Boullier, A. M. and Gueguen, Y. (1975). SP mylonites: origin of some mylonites by superplastic flow. *Contr. Mineral. Petrol.* **50**, 93–104.

Brace, W. F., Silver, E., Hadley, K. and Goetze, C. (1972). Cracks and pores—a closer look. *Science, N.Y.* **178**, 163–165.

Byerlee, J. D. (1968). Brittle–ductile transition in rocks. *J. geophys. Res.* **73**, 4741–4750.

Carter, N. L. (1976). Steady state flow of rocks. *Rev. Geophys. Space Phys.* **14**, 301–360.

Carter, N. L. and Ave'Lallemant, H. G. (1970). High temperature flow of dunite and peridotite. *Bull. geol. Soc. Am.* **81**, 2181–2202.

Davis, L. A. and Gordon, R. B. (1969*a*). Plastic deformation of alkali halide crystals at high pressure: work-hardening effects. *J. appl. Phys.* **40**, 4507–4513.

Davis, L. A. and Gordon, R. B. (1969*b*). On the deformation of alkali halide single crystals at high pressure. *Phys. Stat. Sol.* **36**, K133–K135.

Durham, W. B. and Goetze, C. (1977*a*). Plastic flow of oriented single crystals of olivine. 1. Mechanical data. *J. geophys. Res.* **82**, 5737–5753.

Durham, W. B. and Goetze, C. (1977*b*). A comparison of the creep properties of pure forsterite and iron-bearing olvine. *Tectonophysics* **40**, T15–T18.

Durham, W. B., Goetze, C. and Blake, B. (1977). Plastic flow of oriented single crystals of olivine. 2. Observations and interpretations of the dislocation structures. *J. geophys. Res.* **82**, 5755–5770.

Friedman, M., Perkins, R. D. and Green, S. J. (1970). Observation of brittle-deformation features at the maximum stress of Westerly granite and Solenhofen limestone. *Int. J. Rock Mech. Min. Sci.* **7**, 297–306.

Goetze, C. (1978). The mechanisms of creep of olivine. *Phil. Trans. R. Soc.* (in press).

Goetze, C. and Brace, W. F. (1972). Laboratory observations of high-temperature rheology of rocks. *Tectonophysics* **13**, 583–600.

Gueguen, Y. and Boullier, A. M. (1976). Evidence of superplasticity in mantle peridotites. In *The Physics and Chemistry of Minerals and Rocks*, NATO Advanced Study Institute, Newcastle upon Tyne, 1974 (R. G. J. Strens, ed), Wiley, London, pp. 19–33.

Haasen, P., Davis, L. A., Aladag, E. and Gordon, R. B. (1970). On the mechanism of stage III deformation in NaCl single crystals. *Scripta metallurgica* **4**, 55–56.

Hadley. K. (1976). Comparison of calculated and observed crack densities and seismic velocities in Westerly granite. *J. geophys. Res.* **81**, 3484–3494.

Hart, E. W. (1976). Constitutive relations for the nonelastic deformation of metals. *Trans Am. Soc. mech. Engrs, J. Engng Mater. Technol.* **98**, 193–202.

Hart, E. W., Li, C. Y., Yamada, H. and Wire, G. L. (1975). Phenomenological theory: a guide to constitutive relations and fundamental deformation properties. In *Constitutive Equations in Plasticity*, (A. S. Argon, ed.), MIT Press, Cambridge, Mass., pp. 149–197.

Kirby, S. H. and Raleigh, C. B. (1973). Mechanisms of high-temperature, solid-state flow in minerals and ceramics and their bearing on the creep behaviour of the mantle. *Tectonophysics* **19**, 165–194.

Kohlstedt, D. L. and Goetze, C. (1974). Low-stress high temperature creep in olivine single crystals. *J. geophys. Res.* **79**, 2045–2051.

Kohlstedt, D. L., Goetze, C. and Durham, W. B. (1976). Experimental deformation of single crystal olivine with application to flow in the mantle. In *The Physics and Chemistry of Minerals and Rocks*, NATO Advanced Study Institute, Newcastle upon Tyne, 1974 (R. G. J. Strens, ed.), Wiley, London, pp. 35–49.

Mercier, J.-C. C., Anderson, D. A. and Carter, N. L. (1977). Stress in the lithosphere: inferences from steady state flow of rocks. *Pure appl. Geophys.* **115**, 199–226.

Mercier, J.-C. and Carter, N. L. (1975). Pyroxene geotherms. *J. geophys. Res.* **80**, 3349–3362.

Paterson, M. S. (1974). Laboratory studies of flow in olivine-rich rocks. In *Advances in Rock Mechanics, Proceedings of the 3rd Congress of the International Society for Rock Mechanics, Denver 1974*, National Academy of Science, Washington, D.C., Vol. I, Part A, pp. 521–532.

Paterson, M. S. (1978). *Experimental Rock Deformation: The Brittle Field.* Springer-Verlag, Berlin, Heidelberg and New York (in press).

Phakey, P., Dollinger, G. and Christie, J. (1972). Transmission electron microscopy of experimentally deformed olivine crystals. In *Flow and Fracture of Rocks* (H. C. Heard, I. Y. Borg, N. L. Carter and C. B. Raleigh, eds), Geophysics Monograph no. 16, American Geophysical Union, Washington, D.C., pp. 117–138.

Poirier, J.-P. (1976). *Plasticité à Haute Température des Solides Cristallins*, Editions Eyrolles, Paris, Chapter 2.

Poirier, J.-P. and Vergobbi, B. (1978). Splitting of dislocations in olivine, cross-slip-controlled creep and mantle rheology. *Phys. Earth planet Interiors* **16**, 370–378.

Post, R. L. (1977). High-temperature creep of Mt. Burnet dunite. *Tectonophysics* **42**, 75–110.

Raleigh, C. B. (1968). Mechanisms of plastic deformation of olivine. *J. geophys. Res.* **73**, 5391–5406.

Ross, J. V., Ave'Lallemant, H. G. and Carter, N. L. (1978). Stress dependence of recrystallized grain and subgrain size in olivine. Personal communication, to be published.

Rutter, E. H. (1976). The kinetics of rock deformation by pressure solution. *Phil. Trans. R. Soc. A* **283**, 203–219.

Schmid, S. M., Boland, J. N. and Paterson, M. S. (1977). Superplastic flow in fine-grained limestone. *Tectonophysics* **43**, 257–291.

Scholz, C. H. (1968). Microfracturing and the inelastic deformation of rock in compression. *J. geophys. Res.* **73**, 1417–1432.

Simmons, G. and Richter, D. (1976). Microcracks in rocks. In *The Physics and Chemistry of Minerals and Rocks*, NATO Institute Newcastle-upon-Tyne, April 1974 (R. G. J. Strens, ed.), Wiley, London, pp. 105–137.

Sprunt, E. S. and Brace, W. F. (1974). Direct observation of microcavities in crystalline rocks. *Int. J. Rock Mech. Min. Sci.* **11**, 139–150.

Stocker, R. L. and Ashby, M. F. (1973). On the rheology of the upper mantle. *Rev. Geophys. Space Phys.* **11**, 391–426.

Turner, F. J., Griggs, D. T. and Heard, H. C. (1960). Deformation of rocks at 500° to 800 °C. In *Rock Deformation* (D. Griggs and J. Handin, eds.), *Geol. Soc. Am. Mem.* **79**, 39–104.

Twiss, R. J. (1976). Structural superplastic creep and linear viscosity in the earth's mantle. *Earth planet. Sci. Lett.* **33**, 86–100.

Wawersik, W. R. and Brace, W. F. (1971). Post-failure behaviour of a granite and a diabase. *Rock Mechanics* **3**, 61–85.

White, S. (1976). The effects of strain on the microstructures, fabrics, and deformation mechanisms in quartzites. *Phil. Trans. R. Soc. A* **283**, 69–86.

15. Earthquakes and Plate Tectonics

T. J. Fitch

Abstract

Seismicity on a global scale supports the plate tectonics hypothesis in that the higher levels of activity are concentrated at the plate margins where the higher slip rates are predicted; however, the connection between contemporary and geologic plate motion is difficult to quantify. Seismic slip rates have not been used to constrain geologically recent plate motions because, no matter how accurately determined, they typically differ by a factor of about two from geologically stable rates inferred from the magnetic anomalies in the sea-floor. These discrepancies, in the absence of other evidence, can be explained by uncertainty in the recurrence times for major earthquakes. However, a recurrence time implies at least quasi-periodic seismicity for some magnitude range that includes the major earthquakes. If such a simple representation is fit to major earthquake activity at the subduction zones marginal to the North Pacific times vary from decades to more than 1000 years. The corresponding convergence rates between the plates vary by less than a factor of two, which suggests that the recurrence times and the mechanisms of subduction generally depend as well on geologic factors not directly related to the plate velocities. Episodic activity in individual seismic zones and trends in the Middle East can be taken as typical of activity associated with plate convergence in continental regions. On a global scale, seismicity may also have an episodic character, in which case there might be corresponding variations in the plate motions.

There is a growing awareness that seismic slip is an inseparable part of plate boundary deformation which involves the frictional properties of the contact zone between the plates, the bulk and deformation properties of the lithosphere and underlying asthenosphere and the loading history. Where ground deformation has been monitored by geodetic techniques and/or by measured and accurately dated offsets, seismic slip is only a fraction, but usually the dominant fraction, of the accumulated slip between the plates. From this and other seismic evidence, stable sliding or aseismic fault slip accompanied by fore- and after-shocks is expected in the rupture zones of even moderate to small magnitude earthquakes. The start of the art with regard to plate margin deformation is such that the time dependence shown by even the more complete data sets can be satisfied by fundamentally different models, e.g. time-dependent fault slip vs. flow in the asthenosphere. Analyses of near-field seismic and geodetic data reveal large uncertainties in the temporal and spatial distributions of stress release. Stress drops of tens of bars to more than a kilobar have been reported for the same event. In spite of significant experimental errors, these seemingly inconsistent results suggest that large variations in tectonic stress exist within major fault zones. Notions such as this need to be tested with more complete data sets from broad-band near-field recordings and more extensive geodetic networks that are resurveyed every few years.

1. Introduction

The vast majority of earthquakes have shallow focal depths and are located near the slip zones between the twelve large plates shown in Fig. 1 and a

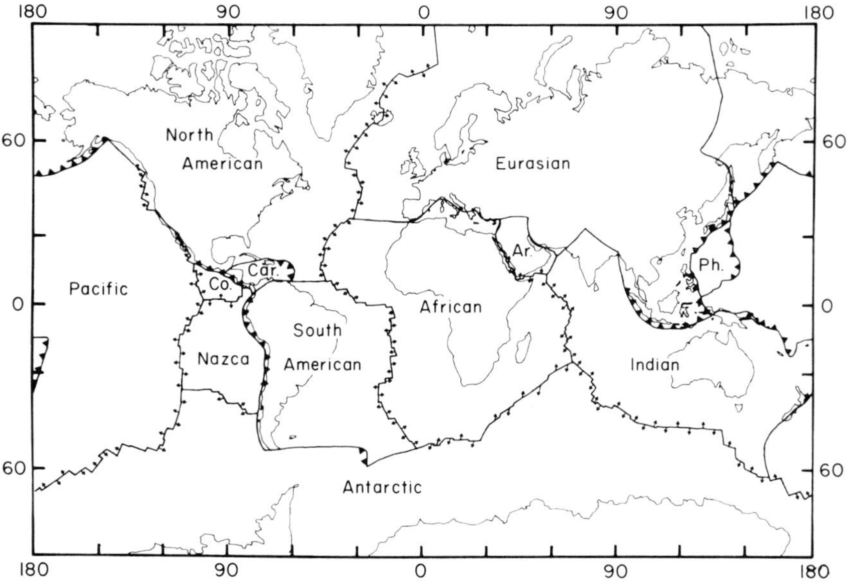

Fig. 1. The large plates. Convergent, transform and accretionary plate margins marked by saw-teeth, straight line segments and opposing arrows, respectively. (From Forsyth and Uyeda, 1975, with permission.)

number of smaller ones. In the following review of observations and hypotheses pertinent to earthquake occurrence, emphasis is placed on major earthquakes at non-accretionary plate margins because these events account for nearly all seismically related slip (Brune, 1968; Davies and Brune, 1971). Among the more topical problems given at least brief mention are possible relationships between global seismicity and plate dynamics, stress drops and apparent stress from earthquakes, tectonic stress near plate margins and within the plates, and mechanical properties of the slip zone between the plates. Solutions to some of these problems, or at least a more refined understanding of them, are expected in the near future as a result of national programs for earthquake prediction in the United States, Japan, the USSR and the Peoples' Republic of China.

Records of major earthquakes, $M_s > 7.5$, for the past 80 years (since the beginning of instrumental seismology) provide a data base for the analysis of global seismicity during a time interval that is sufficiently long to include recurrent events at some of the more active plate margins (Mogi, 1968*a*; Sykes, 1971). Historical earthquakes and prehistoric episodes of fault slip extend the seismic record in certain regions hundreds or even thousands of years. Ground movements that take place in time intervals greater than several hundred seconds occur aseismically and are inferred from repeated geodetic surveys as well as field observations. These data show that major earthquakes are but one stage in a deformation cycle that includes pre- and post-seismic ground movements as well as the coseismic movements (e.g. Fitch and Scholz, 1971).

Over geologic time intervals, millions of years, accumulated slip along the plate margins is well determined from the opening of the modern ocean basins. In much shorter time intervals, even shorter than the recurrence times for major earthquakes, which are 100's to 1000's years at the more active plate margins, activity appears to migrate over thousands of kilometers and to cross several plate margins. For example, Mogi (1968*a*) points out apparent migrations of large shallow earthquakes from Central America to Chile from 1940 to 1960 and from the Celebes (now Sulawesi) to Alaska from 1940 to 1964. This evidence suggests that at least major earthquakes along active plate margins are not independent events.

Interaction between static stress fields can account for some observed short-range effects such as slip on one section of the plate margin followed a short time later by slip on an adjacent section of the same margin (Mogi, 1969). The westward migration of shallow earthquakes along the Anatolian fault from 1939 to 1968 (Ambraseys, 1970) may be driven by such a mechanism. Interactions between remote sections of the same plate margin or other plate margins may result from the propagation of stress or plastic waves (Elsasser, 1967; Savage, 1971; Bott and Dean, 1973; Anderson, 1975). Such waves could impart a short-term episodic component of motion to the plates in addition to or in place of steady long-term motion driven by some form of thermal convection.

2. Plate Dynamics

A variety of seismic transmission effects support at least qualitatively a partition with depth of the Earth's near-surface material into cool elastic lithosphere and hot partially molten asthenosphere. The lithosphere is an efficient (high-Q) medium for the propagation of short period body waves p_n and s_n provided that the paths taken by these waves do not cross active tectonic regions (Molnar and Oliver, 1969). Similarly remnant lithosphere, slab-like bodies containing mantle seismic zones, is a high-Q medium for the propagation of short period P and S waves (Oliver and Isacks, 1967; Utsu, 1967).

The asthenosphere is usually taken to be a zone of anomalously low shear velocity (LVZ) in the depth range from about 50 to 200 km beneath ocean basins (e.g. Leeds *et al.*, 1974; Forsyth, 1975). Beneath shield regions the LVZ, if it exists, is thought to be less pronounced and the top of it displaced toward greater depth. For example, the Canadian shield model of Brune and Dorman (1963), which is a fit to surface wave dispersion, has a 4 per cent drop in the S wave velocity in the depth range 110–320 km. Beneath the west Australian shield surface, wave dispersion has not been able to resolve a LVZ (Goncz and Cleary, 1976).

The growth of the modern ocean basins implies several thousand kilometers of horizontal displacement of the major oceanic plates since the Cretaceous. Plate tectonics quantifies these displacements in terms of rigid body rotations relative to a given plate (McKenzie and Parker, 1967; Morgan, 1968; LePichon, 1968, 1970) or to the deep mantle (Morgan, 1973). Post-Miocene plate kinematics show that the plates composed almost entirely of oceanic lithosphere move faster than plates with a large fraction of continental lithosphere. The pronounced LVZ beneath the oceans may be conducive to high plate velocities.

The growth of the modern ocean basins as a dynamic process is poorly understood. Flow in the low velocity zone and in the deeper mantle is masked by the strength of the lithosphere (McKenzie, 1969; McKenzie and Weiss, 1975). One of the few places where thermal convection may be observable from short-wavelength observations is in the south-east Indian Ocean between Australia and Antarctica (Fig. 2). A 2000 km wide topographic low is centered on a section of the ridge called the Australian–Antarctic discordance by Weissel and Hayes (1974). They claim that asymmetric spreading occurred there between 10 and 4.7 million years ago with the north flank spreading about 1.0 cm y^{-1} faster than the south flank. Not only is the region depressed topographically when referred to depth–age relationships derived from thermal contraction models (Sclater and Francheteau, 1970), but the south and north flanks are topographically

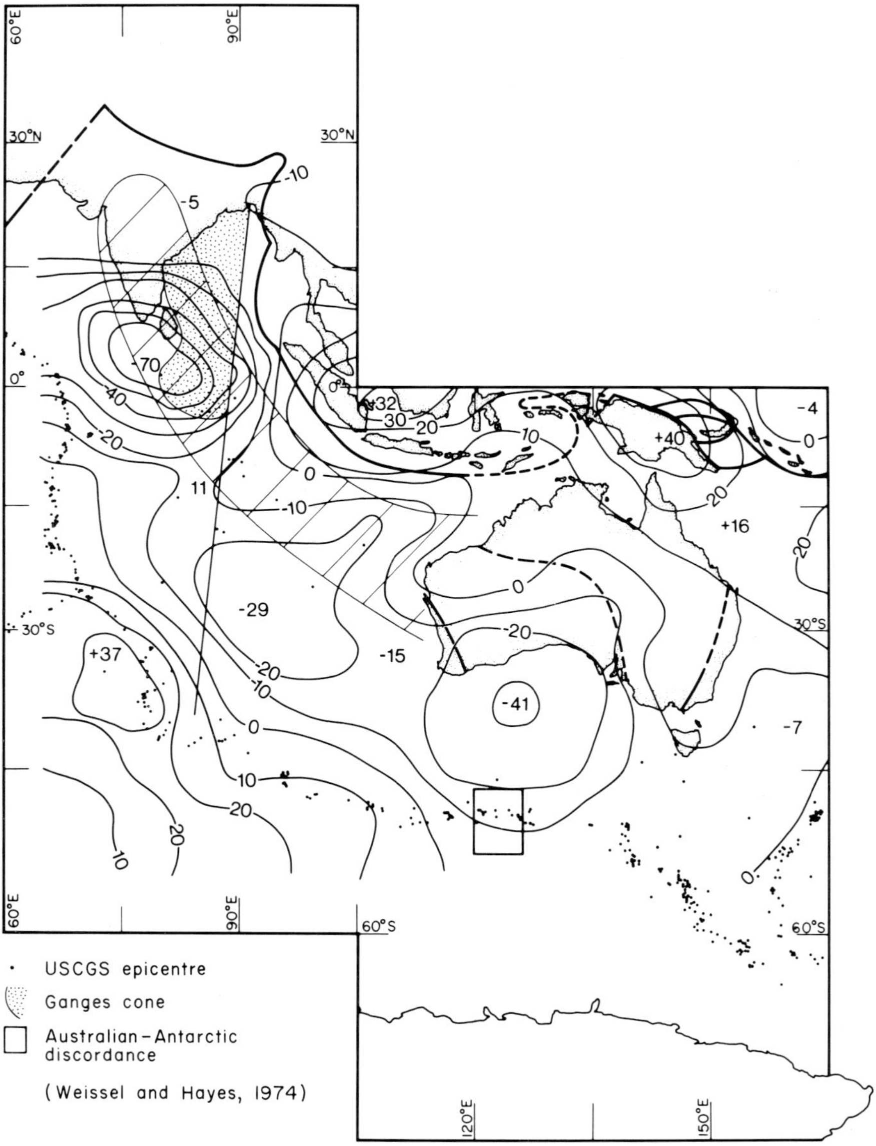

Fig. 2. Indian Ocean plate. Seismicity and global gravity anomalies. (From Fitch *et al.*, 1973, with modification.) Contours are free-air anomalies in milligals referred to an ellipsoid of flattening 1/298.25 (Kaula, 1972). Broad region of shallow earthquakes between southern India and western Australia is shown as a striped region. Seismic trends across Australia shown by solid and dashed curves. Northern plate margin shown by heavy solid and dashed curve.

asymmetrical as well with the south flank having the higher elevation. Weissel and Hayes explain these asymmetries by descending flow in mantle with hotter material flowing from the Antarctic side of the ridge.

Mass anomalies maintained by mantle flow are inferred from long-wavelength gravity anomalies derived from satellite observations (Gaposchkin and Lambeck, 1971; Kaula, 1972). Fig. 2 shows two of the stronger negative anomalies; one centered south-east of India and the other centered north of the previously mentioned Australian-Antarctic discordance. The analysis of the global gravity anomalies has not progressed significantly beyond the qualitative stage, whereby positive anomalies are correlated with active plate margins and negative ones with the interior of the plates generally, and in particular with the deep-sea basins (Kaula, 1972).

A simpler approach to the problem of plate dynamics is taken by Forsyth and Uyeda (1975). They inverted the plate velocities relative to the mantle for an *ad hoc* set of forces applied to the edges and bottoms of the plate. By setting the sum of torques on the plates to zero and solving for the relative strengths of the forces, they show that the dominant driving force for plate motion is a pull from the subducted parts of oceanic plates and that this driving force is balanced by resistance to subduction so that the downgoing slabs move with a terminal velocity.

The inertia of the plates can be ignored or, equivalently, the momentum of the plates is negligible. This can be demonstrated by the solution to a simple problem. Suppose the forces driving a plate of thickness H and density ρ were instantly turned off, how rapidly would the plate motion decay? If the driving forces were balanced solely by basal friction with an LVZ of thickness D and viscosity ν, then the frictional force is proportional to the plate velocity with the constant of proportionality ν/D per unit area of the plate. The solution to the problem is an exponentially decaying velocity:

$$V = V_0\,\mathrm{e}^{-(\nu/HD\rho)t}.$$

The time for the velocity to decay to e^{-1} of its initial value V_0 is $HD\rho/\nu$, which is about 3×10^{-7} s if H, D, ρ and ν are taken to be 100 km, 100 km, 3.4 g cm^{-3} and 10^{21} poise. Clearly the plates must be driven to maintain their motion regardless of order of magnitude uncertainty in the viscosity of the LVZ.

Forsyth and Uyeda emphasize the fact that the faster moving plates are being subducted from one edge. Plate rotation models show that oceanic plates, the Pacific, Philippine, Nazca and Cocos plates (Fig. 1) or ones with only a small fraction of continental material, the Indian Ocean plate, move at 6–9 cm y^{-1}, whereas the other plates move at 0–4 cm y^{-1}. McKenzie (1972) and Morgan (1973) point out that, if a pull on the downgoing slabs were a dominant driving force, then small plates attached to slabs would move faster than large plates because mantle drag would be proportional to the area of plate. This size–velocity correlation does not exist; the large Pacific plate is moving as fast as the smaller Cocos or Nazca plates.

To explain the lack of size–velocity correlation, Morgan and McKenzie prefer tight coupling between plates and mantle flow with plate edge effects having little influence on the motion. A more pronounced low-velocity zone for shear waves under oceans than under continents would then imply faster flow or less resistance to flow under oceans. The tight coupling arguments are consistent with the idea of scaled convection in the mantle, whereby the larger plates are identified with large-scale convection and their motions generate smaller-scale flow in the upper mantle (Richter and Parsons, 1975).

Forsyth and Uyeda offer an alternative explanation for the lack of a size–velocity correlation, whereby a pull on the downgoing slabs is balanced by resistance to subduction and coupling to flow in the mantle is loose. Their inversion scheme quantifies this argument but does not preclude Morgan and McKenzie's explanation because the tight coupling alternative is not represented in their force model. Furthermore, the major conclusions of Forsyth and Uyeda may be coincidental. Solomon *et al* (1977) have carried out a similar analysis of absolute plate velocities for the early Tertiary, 55 million years ago, and showed that at that time subducted oceanic plates did not move uniformly with respect to continental plates that were not being subducted. However, Jurdy (1977) showed that, if Antarctica is split into east and west parts, the early Tertiary plate motions are more compatible with recent plate motions in that the slower-moving plates are predominantly continental. An essential assumption in all these force models is that the forces on the plates can be inferred solely from the kinematics of the motion. This assumption may not be valid for the larger plates.

Episodic plate motion may be triggered by changes in the Earth's rotational velocity (variations in the length of the day) or by some other mechanism as yet undefined. A strong correlation between short-term variations in the length of day and an upsurge in global seismicity at about 1900 (Fig. 3) suggests a cause and effect relationship (Anderson, 1974). Lambeck and Cazenave (1973) have shown that changes in zonal wind patterns can account for the observed short-term changes in the length of the day. If the level of global seismicity is a measure of contemporary plate motion, then accelerated plate motion was triggered by changes in atmospheric circulation near the turn of the century.

This correlation and its possible implications are dependent on earthquake magnitude estimates from records made when seismology was in its

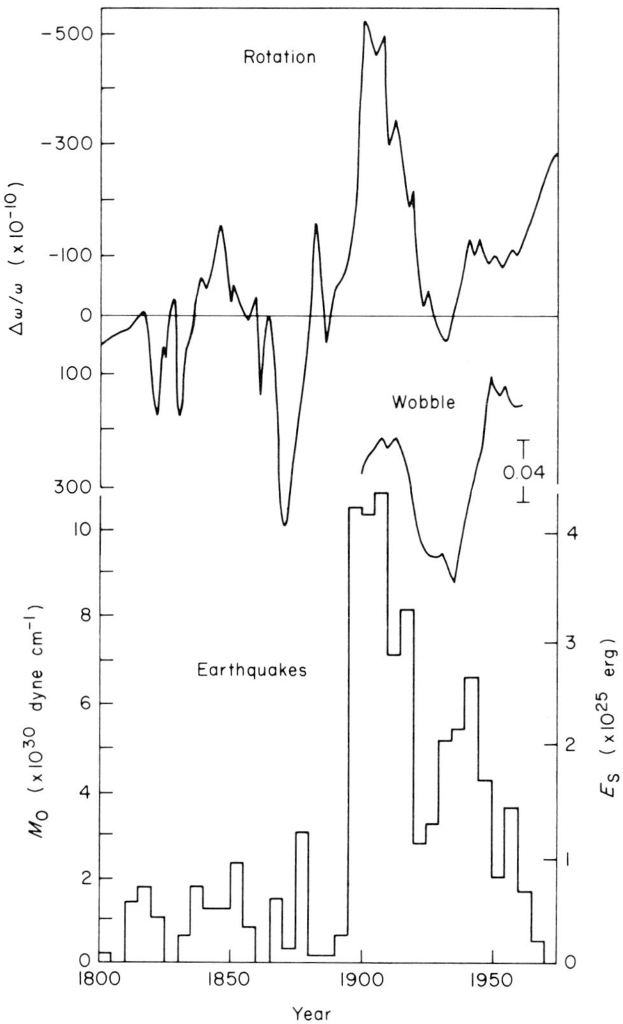

Fig. 3. Changes in rotation rate of the earth ($\Delta\omega/\omega$), Chandler wobble amplitudes, and 5-year means of earthquake energy (E_s) or moment (M_0). (From Anderson, 1974, with permission.)

infancy. Much of the early data were recorded on undamped Milne instruments. Kanamori and Abe (1978) have shown that Gutenberg's original magnitudes based on records from these instruments may be systematically overestimated by as much as 0.5 of a magnitude unit. Consequently, the turn of the century upsurge in seismicity may be an artefact of improper instrument calibration.

3. Plate Slip in Space and Time

Plate tectonics accounts for the present (Fig. 1) and the past (at least from the late Precambrian; McElhinny and McWilliams, 1977) distribution of continents and ocean basins by the accretion of oceanic lithosphere at active ridge crests and subduction of this lithosphere beneath island-arc and arc-like structures. The Post-Miocene history of the sea floor, inferred from magnetic anomaly patterns, is the basic standard by which to judge seismic slip rates and contemporary plate motion. The resolution of slip rates from magnetic anomalies is limited by the time interval between major reversals of the earth's dipole field. Since late Cretaceous anomaly #32, reversals in the field have occurred at irregular intervals ranging from a few hundred thousand to more than one million years, the last reversal occurred about 700 000 years ago (Table I in LePichon *et al.*, 1973).

Relative motion between the major plates since Late Pliocene anomaly #3 or Late Miocene anomaly #5 have been reported in a number of studies, each using slightly different data sets and assumptions. Table 1 summarizes

Table 1. Plate rotation and slip for the north Pacific region (Late Miocene to recent).

	Latitude °N	Longitude °E	Rate† (10^{-7} deg y^{-1})	Reference
Relative rotation				
Pacific–Eurasia	−67	138	−8.1	Le Pchon (1968, 1970)
	−67	114	−8.0	Morgan (1973)
	−63	90	−8.8	Chase (1972)
	−65.3±3.2	110.2±10.1	−9.1±0.9	Minster *et al.* (1974).
Pacific–North America	53	−47	6.0	Le Pichon (1968, 1970)
	54	−61	6.9	Morgan (1973)
	52	−73	7.9	Chase (1972)
	50.9±3.4	−66.3±7.3	7.5±0.8	Minister *et al.* (1974).
Absolute rotation (hot spot reference frame)				
Pacific	−67.3	120.6	−8.3	Minister *et al.* (1974)
Eurasia	38.1	−110.5	−1.2	Minister *et al.* (1974)
North America	−48.1	−82.1	−2.4	Minister *et al.* (1974)

	Slip rate (cm y^{-1})	Mode
Slip (with respect to Pacific plate)		
West North America	~5.5	Horizontal shear
East to west along Aleutian Trench	5.5 to 7.0	Convergence
Kamchatka to Kuril Trench	7.0 to 7.5	Convergence
Kuril to Japan Trench	7.5 to 9.0	Convergence

† Positive is clockwise rotation of first plate with respect to second or the mantle.

the results of these studies for the Pacific, North American and Eurasian plates. Anomaly #3 marks a change from predominantly normal polarity to a long period of reversed polarity that began 4.7 million years ago and lasted about 0.5 million years (Table I of LePichon *et al.*, 1973). Anomaly #5, which is 8.7 million years old, is recognized as a change from a period of reversed to one of normal polarity that lasted about 1.2 million years. Accumulated plate separation since anomaly #5 has been sufficiently small to allow the relative motions to be modeled as infinitesimal rotations (McKenzie and Parker, 1967; Morgan, 1968).

The accretion rate at most spreading centers lies in the range 0–10 cm y^{-1} (LePichon *et al.*, 1973). Variations as large as a factor of two or more in the accretion rate at a given spreading center occur in time intervals that are short by comparison with the lifetime of the center. For example, the opening of the North Atlantic basin slowed from 4.0–5.0 cm y^{-1} to about 2.0 cm y^{-1} about 53 million years ago (Pitman and Talwani, 1972). These variations reveal major changes in the forces controlling plate motion.

Equally dramatic changes have occurred in rates of subduction and horizontal shear; however, slip rates at non-spreading margins must be inferred from the opening of the ocean basins—directly if the non-spreading margin is joined to a spreading center or indirectly if it is not. For example, the San Andreas fault which separates the Pacific and North American plates in California is joined to active spreading centers in the Gulf of California where opening has taken place at a rate of 5.5 cm y^{-1} for the past 4.5 million years (Larson *et al.*, 1968). Slip rates for these plates prior to 4.5 million years ago must be inferred from the opening of the major ocean basins (Atwater and Molnar, 1973). Atwater and Molnar computed relative velocities of 4.0 cm y^{-1} between 29 and 21 million years ago, 1.3 cm y^{-1} between 21 and 10 million years ago and 4.0 cm y^{-1} from 10 to 4.5 million years ago.

Plate margin deformation is expected to be concentrated in an area about as wide as the plate thickness, which for old oceanic lithosphere is about 100 km. However, boundary conditions between the major plates may be conducive to partitioning of slip between a number of major fault zones. For example, convergence that is oblique to a subduction zone gives rise to a subparallel zone of horizontal shear on the continental side of the subduction zone (Fitch, 1972). The best example of this among active plate margins is in the Sumatran section of the western Sunda arc (Katili, 1971; Fitch, 1972).

Migration of a triple junction (McKenzie and Morgan, 1969), at least for a geologically brief period, may give rise to slip in a region hundreds of kilometers wide. For example, the Mid- to Late-Tertiary motion between the North America and Pacific plates can be represented by a north to south migration of a ridge, trench, transform junction along the western coast of North America (Atwater, 1970). The present rotation between these plates is consistent with simple shear along the San Andreas fault system. However, only a little over one half of the geologic slip rate of 5.5 cm y^{-1} (Larson *et al.*, 1968) can be accounted for by geodetic measurements of fault creep in central California (Savage and Burford, 1973) and only one (1906 San Francisco earthquake) of sixteen damaging earthquakes in California this century has occurred on the San Andreas fault (Richter, 1977). The missing slip may account for deformation in the coast ranges, beneath the continental shelf and possibly as far east as the Basin and Range Province.

The definitive evidence for the evolution of this plate margin comes from a one-sided pattern of magnetic anomalies in the Pacific basin adjacent to western North America (Atwater and Menard, 1970). Atwater (1970) argues that this pattern, shown in Fig. 4, requires a Mid-Tertiary subduction zone between North America and the now extinct Farallon plate (Fig. 5).

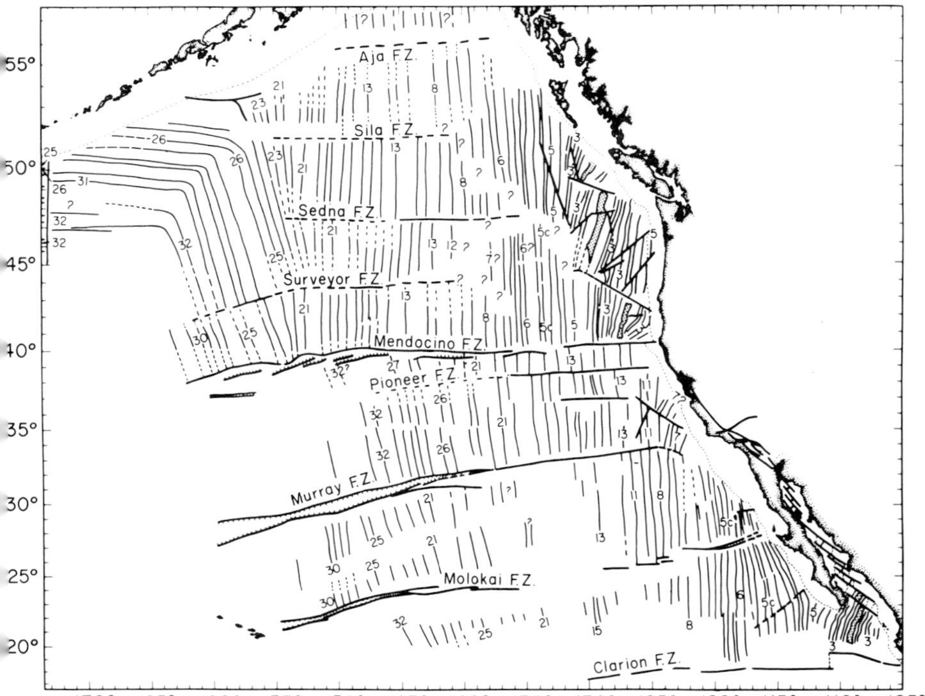

Fig. 4. Magnetic anomalies in the north-east Pacific. The ages of the numbered anomalies are given in Heirtzler *et al.* (1968); ages of the corresponding geologic epochs are given in Berggren (1969). (From Atwater and Menard, 1970, with permission.)

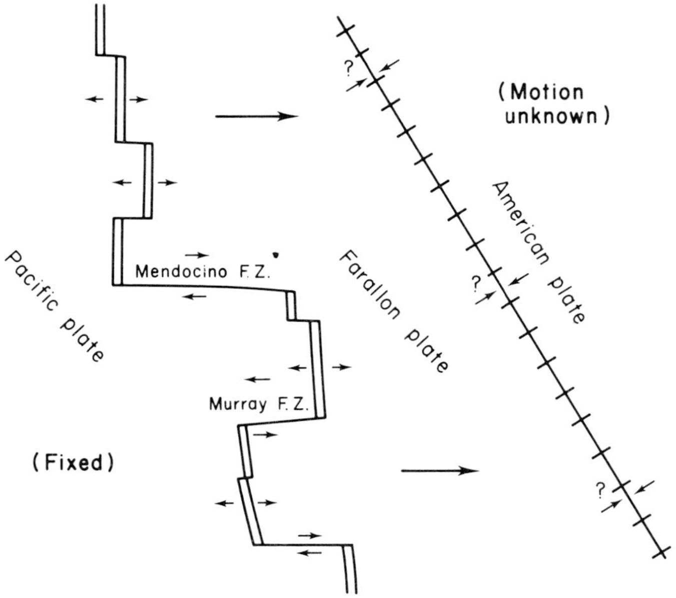

Fig. 5. Configuration of the American, Farallon and Pacific plate boundaries at the time of anomaly 21, about 53 million years ago. Single lines are transform faults, double lines are spreading centers, and hatched lines are zones of subduction (usually trenches). Large arrows show motions of plates with respect to the Pacific plate, which is arbitrarily held fixed. Small arrows show relative motions at points along plate boundaries. (From Atwater, 1970, with permission.)

The conversion of this subduction zone to a transform margin between the Pacific and North America plates gave rise to the San Andreas fault in California, the Queen Charlotte Islands fault off the coast of British Columbia and the Fairweather fault in south-eastern Alaska.

Perhaps the best preserved plate edge effects result from continental collision. About 40–50 million years ago collision between the Indian subcontinent and central Asia reduced the convergence rate between the Indian Ocean and Eurasia plates from 10 to 5 cm y^{-1} (Molnar and Tapponier, 1975). Since then the convergence rate between these plates has remained almost constant (Fig. 6) with the accumulated strain relieved by deformation north of the collision boundary in central Asia and China (Molnar and Tapponier, 1975). The apparent resistance to subduction comes presumably from the loss of negative buoyancy between the subducting plate and the asthenosphere; continental lithosphere being less dense than oceanic lithosphere.

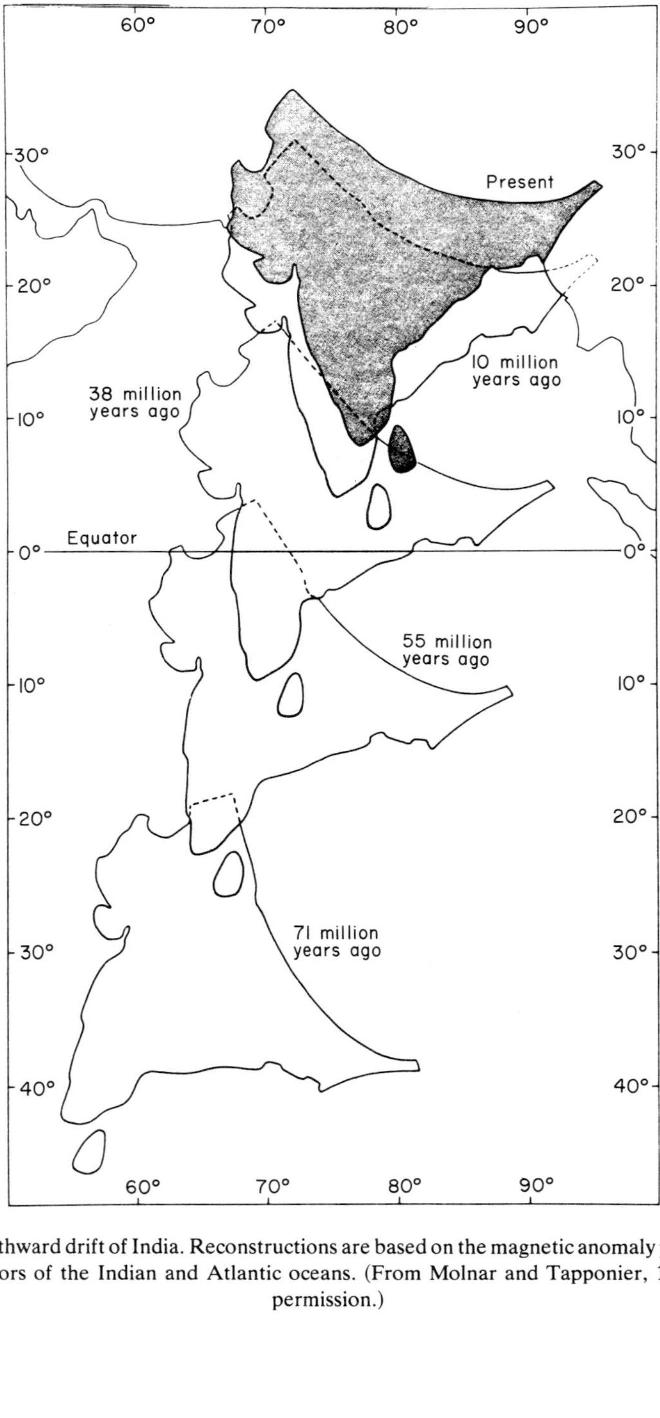

Fig. 6. Northward drift of India. Reconstructions are based on the magnetic anomaly patterns in the sea floors of the Indian and Atlantic oceans. (From Molnar and Tapponier, 1977, with permission.)

4. Earthquake Recurrence

Earthquakes at active plate margins are expected to show at least a vague periodicity if the relative motion between the plates is uniform for time intervals much longer than the time required to accumulate enough strain for a major earthquake. The history of major earthquakes along the convergent plate margins of the North Pacific limits recurrence times to at least a few decades and probably no longer than 1000 years.

Minimum recurrence times are inferred from locations of after-shocks of major earthquakes beneath the frontal arcs from the Aleutians to Japan (Fedotov, 1965; Mogi, 1968*a,b,c*; Sykes, 1971; Kelleher *et al.*, 1974). Few gaps and only minor overlap between after-shock zones are evident from maps representing the last 80 years of activity. This result is illustrated in Fig. 7—taken from Sykes (1971). Mogi (1968*b*) showed that a 30-year sample of this activity left large gaps between after-shock zones, which shows that recurrence times are generally greater than 30 years.

The maximum recurrence time is based on raised terraces in the focal region of the 1964 Great Alaskan earthquake that were dated by Plafker and Rubin (1967). Most of the region uplifted at the time of the earthquake had been subsiding slowly for a period of from 930 to 1360 years prior to the earthquake (Plafker and Rubin, 1967). The reversal in the sense of ground movement suggests that subsidence was caused by strain accumulation. During the subsidence interval there could not have been an earthquake of comparable size to the one in 1964.

The age spanned by the five older terraces on Middleton Island (one of which is shown in Fig. 8) suggests that long recurrence times are typical of the earthquake cycle in this region for the last 4400 years (Plafker and Rubin, 1967). However, the steps between the older terraces are from 6 to 9 m high, which is significantly higher than the 3.3 m terrace measured after the 1964 earthquake. If convergence between the plates at the eastern end of the Aleutian trench is uniform at a rate of 5.5 cm y^{-1} (Table 1), then 900 years of strain accumulation is equivalent to slip with a horizontal component of about 50 m. In Fig. 9(a) the maximum horizontal displacement inferred from triangulation surveys is about 20 m. This displacement yields a recurrence time for Great Alaskan earthquakes of about 400 years (Plafker, 1972). This time is probably a maximum because the field of horizontal displacements is referred to base stations that lie within the deformed region. Consequently, these displacements are to some extent underestimated. Some measure of the maximum horizontal slip is more appropriate for an estimate of recurrence time than a regional average value because average values are strongly influenced by the transition from the slipped to non-slipped part of the thrust zone, i.e. edge effects.

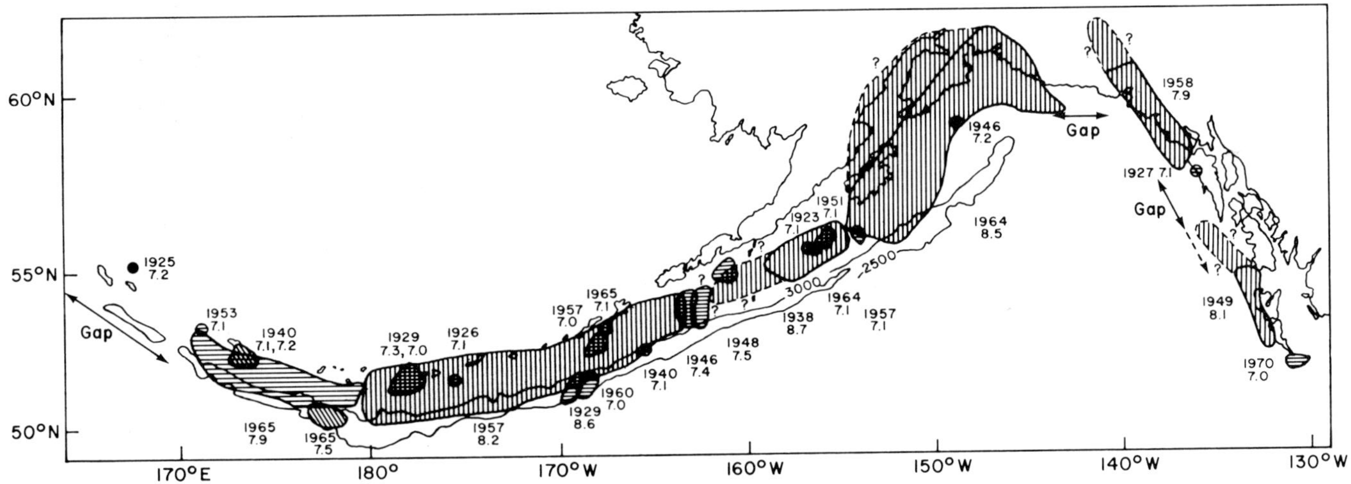

Fig. 7. After-shock areas of major and great earthquakes in the Aleutians and the north-east Pacific. Dates and magnitudes are indicated. Contours for 2500 and 3000 fathoms are shown in the Aleutian trench. Question marks denote uncertainties in the size of after-shock zones. (From Sykes, 1971, permission.)

Fig. 8. Muskeg-covered pre-earthquake marine terrace on Middleton Island (Fig. 9) at an altitude of 110–125 feet. It is one of five uplifted terraces on the island, and surf-cut rock platform exposed between the base of the sea cliff and the new high-tide level is a sixth terrace formed by uplift of about 11 feet in 1964. White specks are seagulls. Photograph taken near 7-foot tide stage, April 4, 1964. (Photo by George Plafker, U.S. Geological Survey; from Plafker, 1969, with permission.)

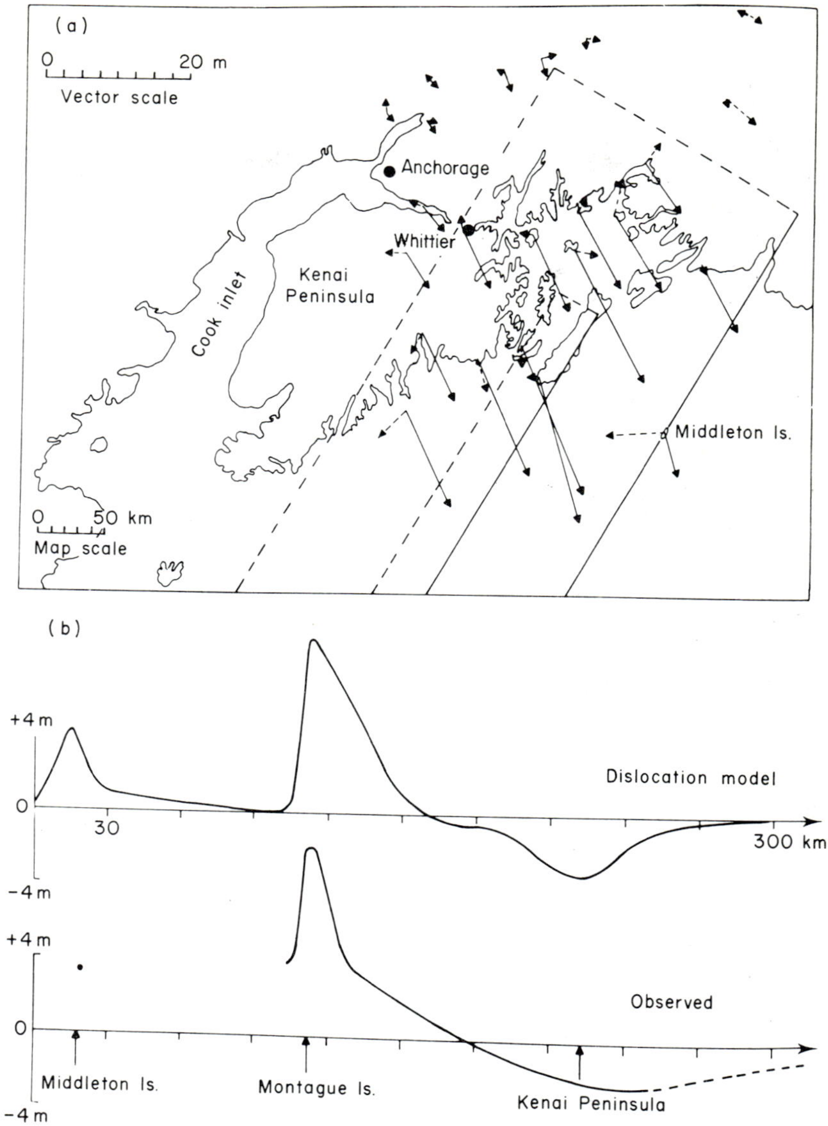

Fig. 9. (a) Horizontal displacement data for the "optimal" dislocation model. The large rectangle represents the projection of the main dislocation plane, while the smaller rectangle lying inside represents the projection of the subsidiary dislocation plane underlying Montague Island. The upper edges of the dislocation planes are shown by solid lines, the other edges by dashed lines. The dashed arrows (- - →) show the calculated horizontal displacement component (U_{th}); the solid arrows (→) show the discrepancy (U_{err}) between the observed and calculated displacement components. (b) Vertical displacements on a cross-section through Montague and Middleton Islands are shown for the calculated and the observed values. (From Hastie and Savage, 1970, with permission.)

More than half of the accumulated strain inferred from the terrace heights and the plate convergence rate is not represented by the coseismic movement in 1964. Further strain release is expected in a time interval that is short enough that the 1964 terraces, in effect, grow to heights comparable with the older terraces, or as Plafker (1972) points out the region could suffer one or more major earthquakes like the one in 1964. Post-seismic tilt of Middleton Island (Fig. 9(a)) suggests that at least local adjustments to the coseismic deformation have persisted for a period of ten years (Prescott and Libowski, 1977). This tilt is downward to the north-west, along an axis that nearly parallels the local trend of the Aleutian trench. Coseismic tilt from plate 1 of Plafker (1969) is in the opposite direction. The post-seismic measurements pertain to a time interval, 1974–1975, that is probably too short to average out experimental uncertainties effectively. Consequently, the inferred slip rate of about 1 m y^{-1} can not be taken as an accurate measure of post-seismic slip. No mention of the current height of the 1964 terrace can be found in the literature.

Historical records for the last 1200 years show that the part of south-west Japan that faces the Nankai trough (Fig. 10) has suffered a major off-shore earthquake every 100–200 years (Imamura, 1929, 1930; Matuzawa, 1964; Fitch and Scholz, 1971; Ando, 1975). The historic and recent events in this series have been correlated with uplifted marine terraces on peninsulas that jut into the Philippine Sea and a drowned coastal morphology away from the peninsulas (Watanabe, 1959; Yoshikawa, 1970). This deformation pattern is similar to that observed in the focal region of the 1964 Alaska earthquake. Repeated leveling and triangulation surveys carried out during the last 80 years by the Geographical Survey Institute of Japan reveal a sequence of ground movements the coseismic component of which resulted from under thrusting of the Philippine Sea plate beneath south-west Japan at the times of the 1944 Tonankai and the 1946 Nankaido earthquakes (Fitch and Scholz, 1971; Kanamori, 1972*a*).

Village records from medieval to recent times are sufficiently detailed with regard to uplift or subsidence of the coast and the areal extent of damage from the seismic sea waves (tsunamis) that Ando (1975) was able to estimate the lateral extent of each rupture in this series of underthrusts. In Fig. 11 (from Ando, 1975), this subduction zone is subdivided into four parts, with the more recent events assigned to one or more adjacent parts of the zone. In a similar vein, Mogi (1969) pointed out that after-shock zones of major earthquakes in this region and elsewhere often abut structural discontinuities lateral to the strike of the rupture.

Intraplate seismicity of south-western Japan is correlated with the megathrusts along the Nankai trough (Utsu, 1974; Shimazaki, 1976). Activity rises 50–30 years before a major off-shore earthquake and remains relatively high for about 10 years afterwards. In North-east Honshu, a similar correlation exists with major earthquakes associated with subduction along the Japanese trench (Fig. 10); however, unlike south-western Japan, the intraplate activity is concentrated in the 5 years before and 10 years after major off-shore earthquakes (Shimazaki, 1978). The recurrence rates for intraplate earthquakes in Japan (and probably elsewhere as well) are much greater than the corresponding rates for interplate earthquakes. Shimazaki (1976, 1978) argues that strain accumulation occurs at a much slower rate away from the plate margin. He proposes a model whereby intraplate strain accumulates slowly by some process associated with plate convergence,

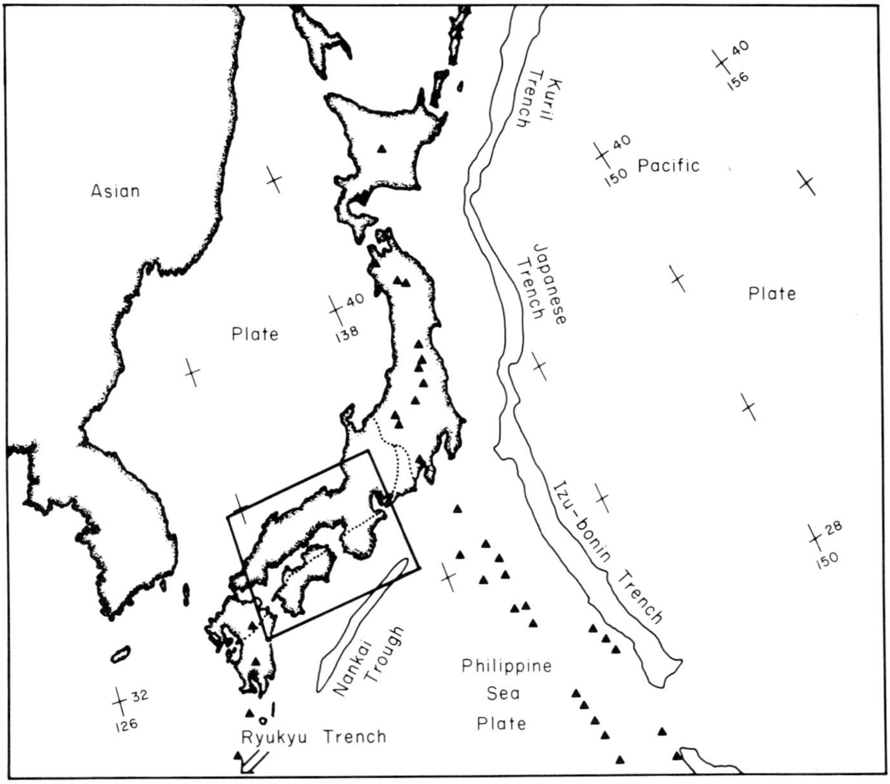

Fig. 10. Tectonic setting of Japan and adjacent areas. Vertical and horizontal movements at the time of the Nankaido earthquake of 1946 occurred within the square. The dotted curve maps the trace of the Fossa Magna (the north-striking line across central Honshu) and the Median Tectonic Line (Richter, 1958). Triangles represent terrestrial volcanoes that have been active historically and submarine volcanoes, of which some may not be presently active (Gutenberg and Richter, 1954). (From Fitch and Scholz, 1971).

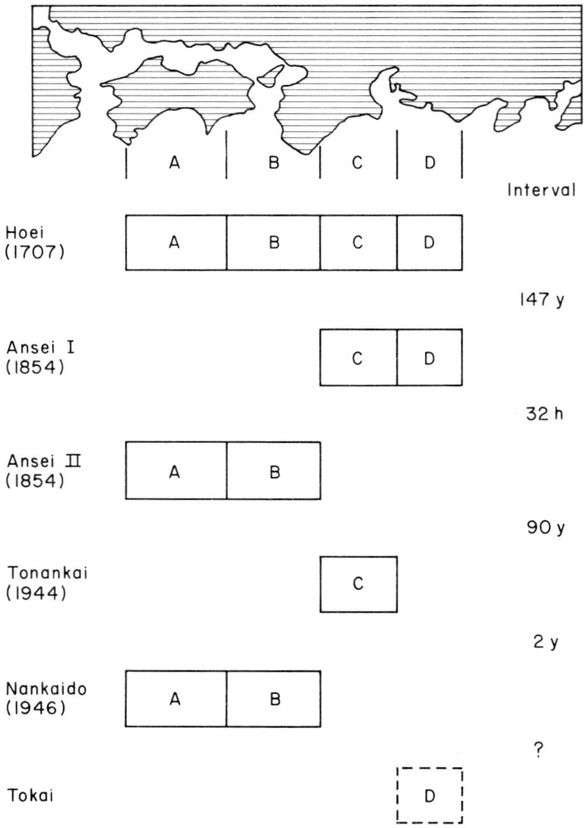

Fig. 11. Cyclic migration of Great Earthquake along the Nankai trough (Fig. 10). (From Ando, 1975, with permission.)

with short-term variations imposed by strain accumulation and release at the zone of underthrusting.

5. Earthquake Cycles

Figure 12 illustrates the classical elastic rebound theory of Reid (1910) for pure strike-slip and pure underthrust. This theory predicts a period of elastic strain accumulation followed by sudden and complete strain release at the time of an earthquake. This oversimplified view of the earthquake cycle can be modified to fit currently available geodetic and seismic data from active plate margins by allowing for time- and position-dependent friction between plates (Scholz *et al.*, 1969; Fitch and Scholz, 1971; Kanamori, 1973).

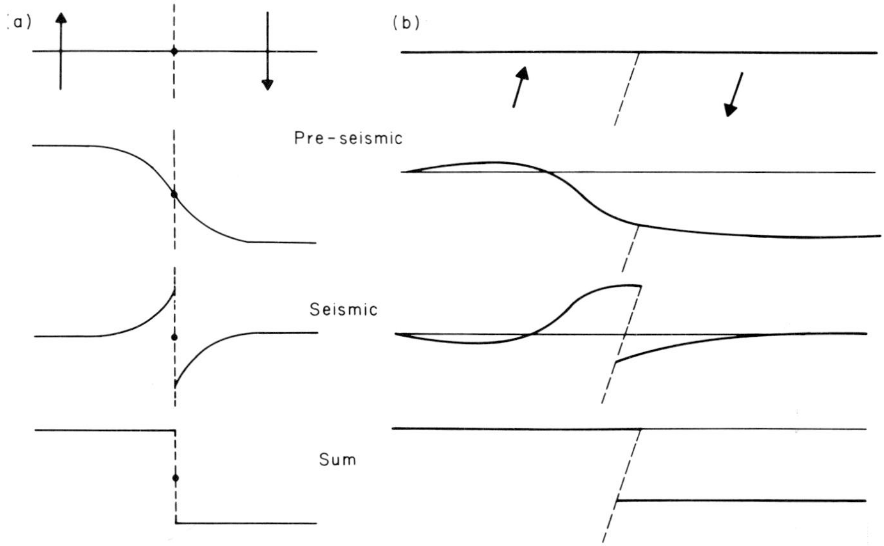

Fig. 12. Elastic rebound models of plate boundaries. (a) Transcurrent boundary. Horizontal deformation, on a greatly exaggerated scale, of a line initially normal to the fault trace (dashed line) is shown by solid curves. The preseismic curve gives deformation of such a line during a period of strain accumulation, the seismic shows deformation by an earthquake, and the sum is the total deformation for the entire earthquake sequence. The heavy arrows give the long-term motion of one side of the fault with respect to the other side. (b) Thrust boundary, vertical section. Heavy arrows indicate long-term motion on one side of the fault with respect to the other side. Deformation (vertical scale greatly exaggerated) of an initially undeformed free surface is shown by heavy solid curves. The fine lines in the three bottom drawings mark the position of the initially undeformed surface. (From Fitch and Scholz, 1971.)

Non-elastic effects such as stress relaxation in the underlying asthenosphere (Nur and Mavko, 1974) may be required to explain some of the longer-wavelength observations.

As models of fault slip become more complicated, more information is required of the data to distinguish between different models. This dilemma is apparent from contrasting interpretations of vertical movements of south-west Japan measured before and after the 1944 Tonankai and the 1946 Nankaido earthquakes, the largest of recent off-shore earthquakes (Fig. 13(a)). Observed and computed elevation changes in eastern Shikoku are compared in Fig. 14. Absolute ground movement between surveys is referred to a change in mean sea-level, where the leveling route passes a tide gauge station. The fault model represented by the solid circles is a planar dislocation buried in an elastic half space. Displacement fields generated by such a model are a linear function of slip on the dislocation surface. The appropriate equations are presented in Mansinha and Smylie (1971).

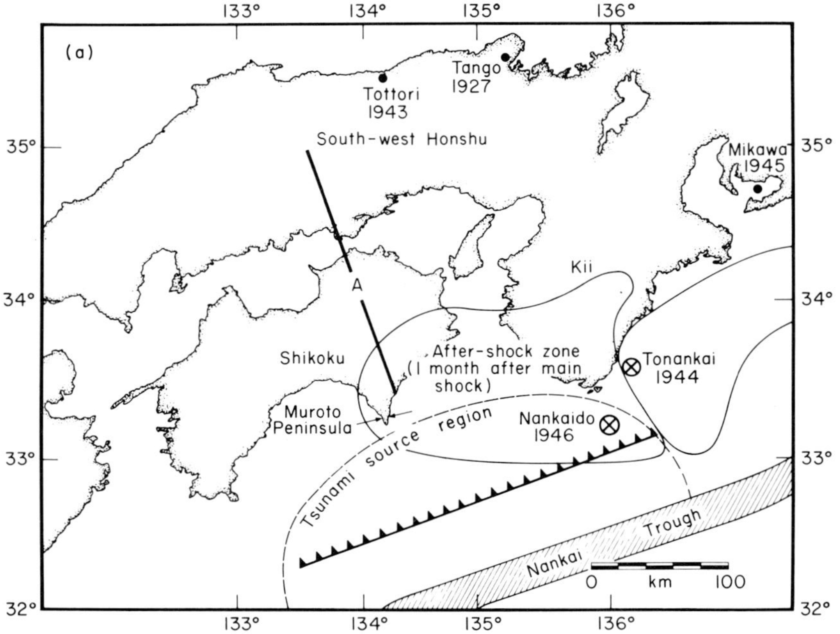

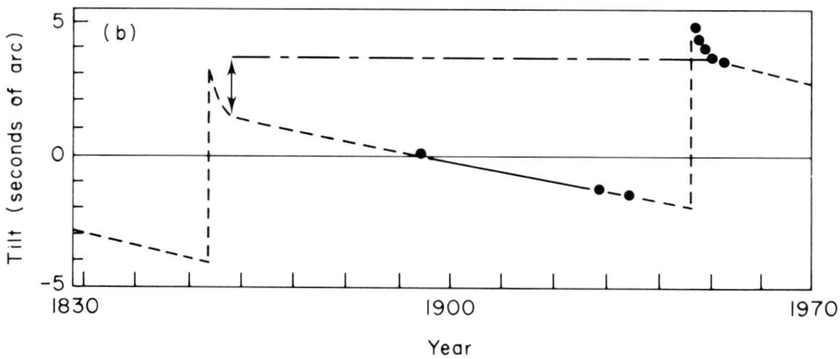

Fig. 13. (a) Recent major earthquakes in and near south-west Japan. After-shock zones are from Mogi (1968*b*). Inferred trace of Nankaido thrust is from Fitch and Scholz (1971). Tsunami source area pertaining to Nankaido rupture is from Hatori (1966). (b) Recent history of tilting at Muroto Point. Tilt of leveling lines projected on a vertical plane striking south-south-east near Muroto Point. North-north-west tilt is in the positive direction. A solid line or curve is used when the history of tilting is extrapolated between measured values. The length of the double-headed arrow is an estimate of the permanent tilt at the time of the Nankaido earthquake of 1946. Adapted from Okada and Nagata (1953). (From Fitch and Scholz, 1971.)

Unfortunately, there are an infinite number of these models that fit the elevation changes about as well as the one illustrated in Fig. 14. The principal cause of this uniqueness problem is the lack of observations on the continental slope where the megathrust either outcrops or reaches its shallowest depth. Slip of 5 m computed by Fitch and Scholz (1971) is nearly twice as much as Kanamorí (1972*a*) inferred from seismic evidence. Ando (1975) showed that horizontal deformation from repeated triangulation surveys in conjunction with the elevation changes are consistent with the seismic interpretation. Horizontal deformation computed in this way is referred to a base line that is assumed to be stationary between surveys. In Japan, intraplate deformation as well as deformation at the plate margins make the choice of a base line difficult at best. However, to prefer the Fitch and Scholz model to that of Ando's would imply a large and probably unacceptably large distortion of the base line.

The pre-seismic downwarp is nearly a mirror image of the coseismic uplift, aside from a scale factor that can be though of as a linear adjustment to the entire period of strain accumulation, 92 years. However, as Miyabe (1955) pointed out, and as can be seen in Fig. 14, significant deviations from a perfect mirror image relationship exist between the pre-seismic and coseismic movements. This relationship is improved by comparing the pre-seismic to the *sum* of coseismic and post-seismic movements. This improvement led Fitch and Scholz (1971) to the suggestion that the post-seismic uplift resulted from delayed slip on the surface that ruptured at the time of the earthquake and a possible extension of that surface to greater depth. To fit the observations with this model, backward slip is required on the shallow part of the slip surface and forward slip is required at depth.

Nur and Mavko (1974) claim that a reversal in sense between the coseismic and post-seismic movements can be explained by the relaxation of stress in the asthenosphere generated by sudden slip in an overlying elastic layer. Their computed post-seismic uplift is much broader than the observed uplift; however, this discrepancy may not invalidate the relaxation hypothesis because their two-dimensional model is overly simple. The model asthenosphere is assumed to have uniform viscosity, whereas the asthenosphere beneath south-western Japan is expected to be laterally heterogeneous with an inclined seismic zone extending to a depth of about 80 km (Kanamori, 1972*d*).

Repeated leveling surveys of Muroto peninsula (Fig. 13) after the Nankaido earthquake (Okada and Nagata, 1953) show that the rate of tilt diminished exponentially with a decay constant of about 5 years. The relaxation hypothesis of Nur and Mavko predicts a decay constant for post-seismic ground movement, τ, that is proportional to the ratio of the viscosity of the asthenosphere, ν, to twice the shear modulus, μ, of the

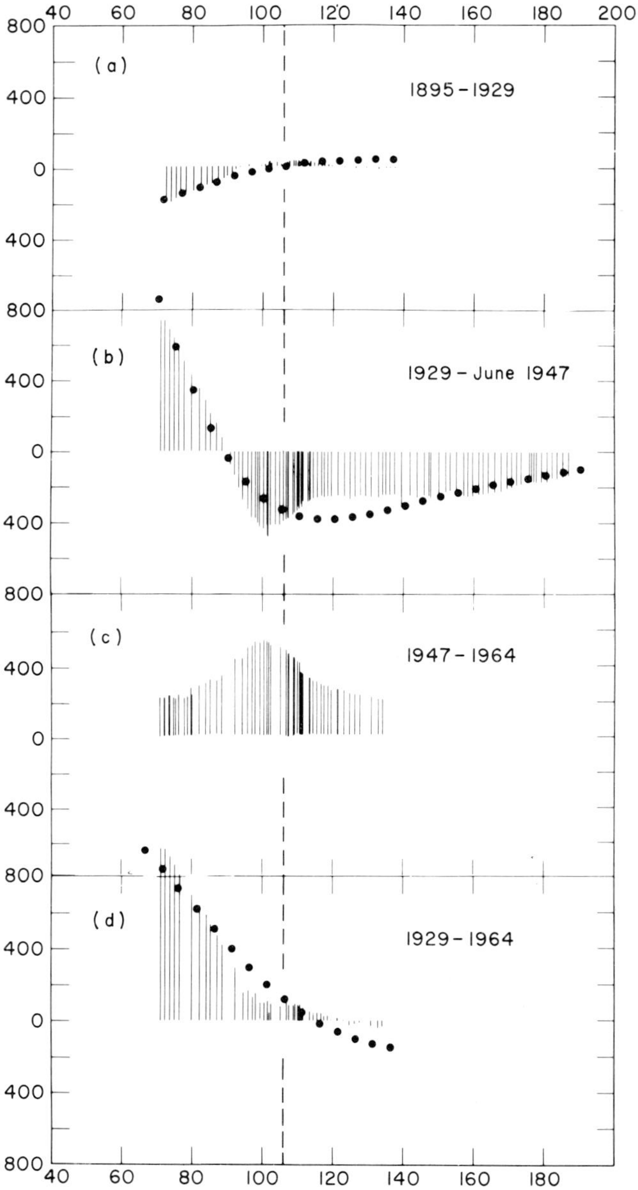

Fig. 14. Profile A. (a) Pre-seismic, (b) seismic, (c) post-seismic, and (d) the sum of seismic and post-seismic vertical displacements (in millimeters) on eastern Shikoku (Fig. 13(a)) are projected on to vertical planes perpendicular to the inferred trace of the megathrust. The abscissa gives the distance (in kilometers) from the trace of the megathrust. The dashed line represents the position of the hinge line for the marine terraces. The dotted curves are computed least-squates fits to the profiles. (From Fitch and Scholz, 1971.)

overlying elastic layer:

$$\tau = \alpha\nu/2\mu, \quad \text{where} \quad 1 < \alpha < 10.$$

If, following Nur and Mavko, α is taken to be 5 and μ is 8×10^{11} dyne cm^{-2}, then the viscosity of the asthenosphere is 5×10^{19} poise when the decay constant is 5 years. This viscosity is about an order of magnitude lower than that computed by Cathles (1975) from the rebound of eastern North America and Finnoscandia from glacial loading. Considering the contrast in tectonic environments and the simplicity of the models, order of magnitude agreement may be acceptably close.

Thatcher and Rundle's (1978) viscoelastic modeling of the Nankaido data is in greater harmony with the elastic models of Fitch and Scholz (1971) and Ando (1975) and, by inference, the viscoelastic solution computed by Nur and Mavko must be incorrect. Thatcher and Rundle argue that the co- and post-seismic fault movements initiate a viscous relaxation in the asthenosphere which pumps elastic strain energy into the lithosphere, thereby accounting for the pre-seismic downwarp. The post-seismic tilt in their model is caused by a downward extension of the coseismic rupture.

Evidence of delayed slip following the 1923 Great Kanto and the 1964 Great Alaska earthquakes is less ambiguous than in the case of the Nankaido earthquake. Scholz and Kato (1978) argue that reversed after-slip followed a combination of thrust and strike-slip movement at the time of the Great Kanto earthquake. The Kanto region includes Tokyo and is near the north end of the Nankai trough (Fig. 10), where the Philippine Sea, Eurasia and Pacific plates meet at a triple junction. Their evidence for reversed after-slip is elevation changes from repeated surveys across an inferred landward extension of the fault trace. The magnitude of the afterslip was 5–10 percent of that of the coseismic slip. A reversal in the direction of slip can be explained by a reversal in the stress field that drove the coseismic slip. This could happen if coseismic slip exceeded the forward slip required to exactly cancel the accumulated elastic strain. Slip in the sense of the pure underthrust should be more conducive to an overshoot than fault slip with a strike-slip component because to overshoot an underthrust does additional work against gravity whereas strike-slip requires additional work against the compliance of the confining medium. The gravitation force is less strong, consequently an underthrust favors an overshoot.

Repeated leveling surveys reveal post-seismic upheaval between Anchorage and Whittier (Fig. 9), a region of subsidence at the time of the 1964 Great Alaskan earthquake (Brown *et al.*, 1977). The apparent decay constant for this upheaval is about 5 years, just as in the case of post-seismic tilt of Muroto peninsula mentioned previously. Brown *et al.* account for this uplift by creep along a downdip extension of the slip surface that broke in 1964 and its apparent migration toward the Aleutian Trench.

Fault slip that occurs on a much shorter time scale, a few seconds to minutes, radiates the elastic body and surface waves recorded at seismological stations throughout the world. In such short time intervals, plate margin deformation is expected to differ from deformation that takes place in the years or even decades between successive geodetic surveys; however, systematic differences between geodetic and seismic models are difficult to resolve, at least from the megathrust models in Table 2. The most striking disparity between surveyed ground movements and movements estimated from radiated seismic energy pertains to the Nankaido region. The ground movement in western Shikoku (Fig. 13(a)) demonstrated beyond doubt that the rupture in 1946 extended more than 100 km south-west of the seismic focal region. Apparently this extension occurred without seismic expression.

A more complete model of a slip event includes not only the growth of the rupture but also the time dependence of slip at every position on the rupture surface. This dependence is usually taken to be independent of position. In particular, Kanamori (1972*b,c*) used the source time function, $G(t)$, where $G(t)=0$ for times less than the origin time and $G(t)=1-e^{-t/T}$ otherwise. The rise time of this function, T, governs the ratio of long-period to short-period energy radiated by the source. If T is more than a few minutes, most of the radiated energy will be outside the band pass of long-period seismographs and therefore seismically unobserved. Kanamori (1972*b,c*) estimates rise times from near field records and from the relative excitation of short- and long-period energy. He found that major off-shore earthquakes with rise times of about 100 s are correlated with the generation of large seismic sea waves (tsunamis). Most of the earthquakes, including the larger-magnitude events, had rise times of a few seconds to a few tens of seconds. This explains the lack of a correlation between earthquake magnitude, essentially a measure of 20 s energy, and the size of the tsunami.

The 1960 Great Chilean earthquake and its long period (300–600 s) precursor (Kanamori and Cipar, 1974) constitute the longest recorded episode of seismic fault slip. The precursor was recorded by a strain meter at the time of the P wave arrival from a major fore-shock 15 min before the main shock. The amplitudes of the long-period precursor were comparible with amplitudes from the main shock at similar periods. Consequently, the source, extended in time, is consistent with about twice the slip estimated for the main shock by itself. In Table 2 this is the earthquake for which geodetic and seismic slip have most nearly the same value. Apparently nearly the entire episode of fault slip was recorded seismically.

Table 2. Recent slip events at convergent plate margins: seismic vs. geodetic slip.

Event	Data	Area (km^2)	Dip (degrees)	Dip slip (m)	Strike slip (m)	M_0 (dyne cm^{-1})	Reference
1923 Great Kanto	P-wave first motions and long-period surface waves	130×70	34	2.1		7.6×10^{27}	Kanamori (1971*a*, 1973)
	Leveling, triangulation and mean sea levels	130×65	45	3	6		Ando (1971)
1946 Nankaido	P-wave first motions and long-period surface waves	120×80	10	3.1	—	1.5×10^{28}	Kanamori (1972*a*, 1973)
	Leveling, triangulation and mean sea levels	320×(50–30) 300×(120–70)	40–30 20–25	18–5 6–4	—		Fitch and Scholz (1971) Ando (1975)
1960 Chilean	Long-period surface waves from main shock and *precursor*	800×200	~10	30	—	6×10^{30}	Kanamori and Cipar (1974)
	Leveling and observed vertical deformation	100×(60–200)	~20	20–40	—		Plafker and Savage (1970)
1964 Alaskan	Long-period surface waves	600×~200	20	7	—	7.5×10^{29}	Kanamori (1970)
	Leveling and triangulation	600×204	3.7	12.2	9.7		Hastie and Savage (1970)
1973 Nemuro-Oki	P-wave first motions and long-period surface waves	100×60	27	1.6	—	6.7×10^{27}	Shimazaki (1974*b*)
	Leveling and mean sea levels	In agreement with seismic model but can only account for about one half of pre-seismic strain accumulation					Shimazaki (1974*a,b*)

6. Seismic Source

The seismic moment has replaced magnitude as the preferred measure of source strength because it is physically more meaningful and less vulnerable to errors resulting from the heterogeneity and anelasticity of the transmitting medium. Following Aki (1966), the moment for a shear dislocation or double couple can be written $M_0 = \mu \bar{u} S$, where μ, $\bar{u}$ and S are the shear modulus, mean slip and area of the slip surface, respectively. This definition of moment is independent of time and thus representative of the static source or the source at zero frequency. In practice, moments are inferred from the long-wavelength part of the recorded signal.

A point source can be expressed as a 3×3 tensor, $M_{0_{jk}}$, in which case the directional properties of the source are included as well as source strength. For the shear dislocation, Kostrov (1974) writes

$$M_{0_{jk}} = \mu \int_{\Sigma} (a_k(\mathbf{x}) n_j + a_j(\mathbf{x}) n_k) \, \mathrm{d}s,$$

where a_k and n_j are the direction cosines of slip (as a function of position on the slip surface Σ) and the normal to that surface, respectively. The trace of $M_{0_{jk}}$ is zero, which implies no expansion or contraction of the source region.

A volume change can radiate seismic energy directly or indirectly by the release of concomitant shear strain. Randall (1966, 1971) and McGarr (1976) show that shear stress drops, $\Delta\tau_{ij}$, distributed in a volume V, yield a moment,

$$M_{0_{jk}} = \int_{V} \Delta\tau_{jk} \, \mathrm{d}v.$$

The explosive or implosive component of a source is represented by a moment tensor with a non-zero trace, $M_{0_{jj}} \neq 0$.

The moment tensor can be generalized to include the effect of source time dependence. The time-dependent moment tensor was used by Gilbert (1970) to solve the problem of the excitation of free oscillations from a point source in a layered sphere. McCowan (1976) solved the analogous problem of surface wave excitation in a layered half space.

In contrast to the moment tensor, magnitude recognizes neither spatial nor temporal complexity of real or hypothetical sources. Body wave magnitude, m_b, is computed from logarithms of amplitudes of short-period P-waves weighted by the signal period, about 1 s (Richter, 1958). The surface wave magnitude, M_s, is computed from logarithms of amplitudes of 20 s Rayleigh waves (Richter, 1958). Magnitudes, either m_b or M_s contain large errors due to path effects that preferentially distort short- and medium-period waveforms by focusing, scattering, multipathing and amplitude diminution and dispersion from the Earth's anelasticity. Uncertainty in the better determinations of m_b is ± 0.3, which is equivalent to an amplitude uncertainty at the source of a factor of two (Basham, 1969; North, 1977*b*).

Magnitudes become less meaningful as the source strength increases. Earthquakes associated with major episodes of slip between plates, i.e. earthquakes with rupture zones hundreds of kilometers long and complex time functions, radiate short- and medium-period signals with large azimuthal and distance variations from source as well as transmission effects. The 1 and 20 s energy corresponds to a P-wave signal near the free surface with a wavelength of about 5 km and a Rayleigh wavelength of about 60 km, respectively. Radiation from a major earthquake in this period range, or equivalently in this range of wavelengths, is governed by the rise time and finite size of the source which will taper the high-frequency end of the amplitude spectrum (Fig. 15(a) and (b)).

The slope of the high-frequency end of the spectrum is governed by the discontinuities in the signal (Savage, 1972, from Bracewell, 1965). To illustrate this effect consider the Haskell (1964) source model (Fig. 15(a)). Slip begins at the origin and spreads instantaneously to a strip W wide and unilaterally propagates a distance L at a constant velocity v_r. The displacement time function is chosen to obey the following condition:

$$U = U_0 G(t - x/v_r),$$

where $G(t) = 0$ for $t < 0$ and $G(t) \to 1$ for large t. U_0 is the final slip. The previously mentioned time function with an exponential rise time satisfies these conditions. With this function, the slope of the far-field amplitude spectrum at high frequencies is ω^{-2} with the source length and finite rise time each providing a factor of ω^{-1}.

Higher-order discontinuities in the signal would impose even more rapid fall-off at high frequencies. For example, lateral propagation of the rupture to $\pm W/2$ in a finite time would change the asymptotic slope from ω^{-2} to ω^{-3} (Savage, 1972). Frasier and North (1978) show that short-period P-wave amplitudes from after-shocks of the 1965 Rat Island earthquake are more consistent with an ω^{-3} than an ω^{-2} slope. The amplitude spectra in Fig. 15(b) from Tucker and Brune (1977) are typical of spectra with a single corner frequency from after-shocks of the 1971 San Fernando, California earthquake. The high-frequency part of each spectrum above the noise level has a slope between ω^{-2} and ω^{-3}. Conservation of radiated seismic energy requires that the source amplitude spectrum have a slope of at least $\omega^{-1.5}$ at the high-frequency end.

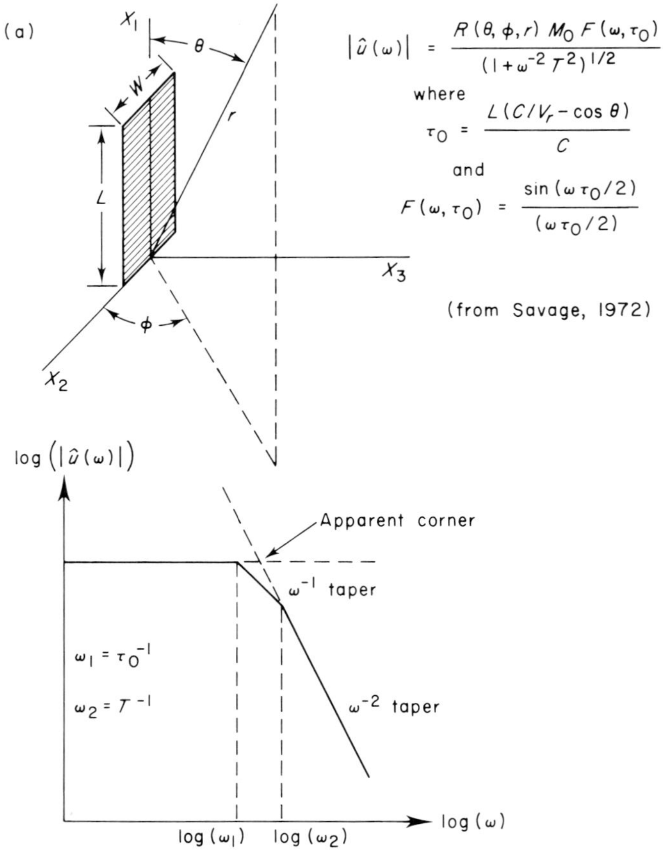

Fig. 15. (a) Haskell (1964) fault model and idealized displacement amplitude spectrum. T, C, V_r, $R(\theta, \phi, r)$ refers to the rise time for slip at the source, seismic phase velocity in the source region, rupture velocity, and radiation pattern, respectively. $\omega = 2\pi f$, where f is frequency in Hz.

Haskell's model, as well as other modern source models (e.g. Brune, 1970), is representative of the gross properties of the source. The physics of rupture propagation found in the fracture mechanics literature has not greatly influenced the seismic source models. In practice, source studies from teleseismic recordings, both long and short period, are limited by an imperfect separation of source and transmission effects, microseismic noise and the limited bandpass of the instruments. For example, if the hypothetical source represented in Fig. 15(a) was to be inferred

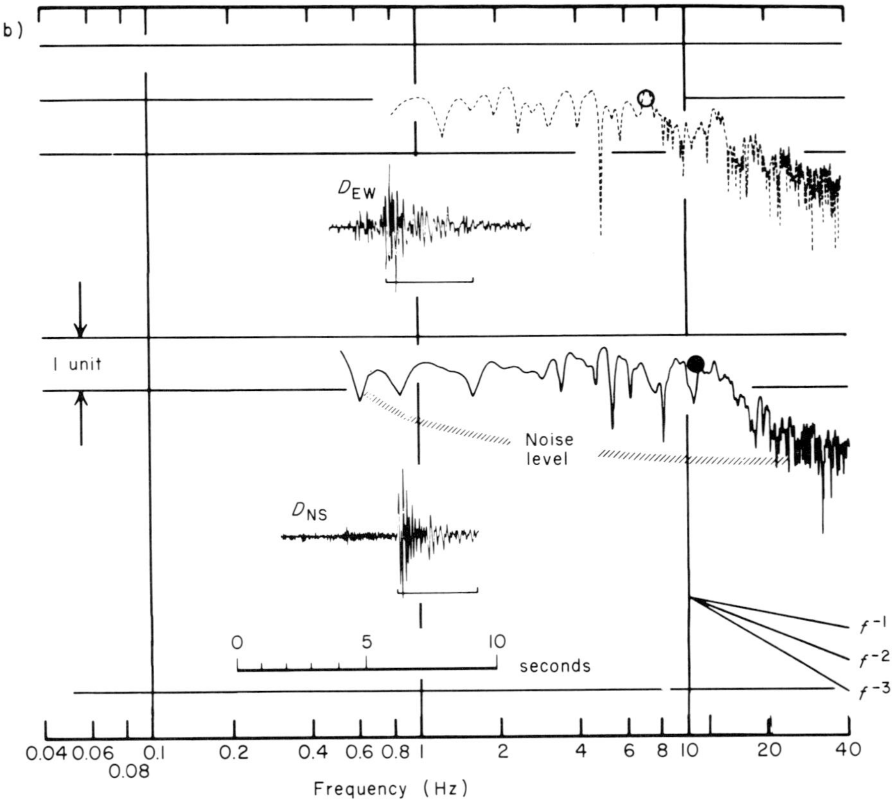

(b) S-wave amplitude spectra and seismograms from 1971 San Fernando earthquake. Data originally recorded digitally at 150 samples s^{-1} at a site within the source region east of the Pacoima Reservoir. Corners indicated by open and closed circles. Brackets beneath each seismogram identify the section of the record that was spectral analyzed. Spectra were corrected for instrument response and attenuation assuming a Q of 250 for S waves in the crust. (From Tucker and Brune, 1977, with permission.)

from real data, the spatial and temporal corners would likely appear as a single corner.

At long periods, the source amplitude spectrum from a shear dislocation is flat unless the source is underdamped (Haskell, 1964; Brune, 1970; Savage, 1972). The height of the flat part of the spectrum is proportional to the seismic moment with proportionality factors that are independent of source size and rise time (see, for example, Savage, 1972; Kostrov, 1974). Consequently, the moment or moment tensor can be determined uniquely from the long-period part of the seismogram. This is the case even with band-limited data, as long as spectral estimates are computed for periods longer than the period corresponding to the lowest-order discontinuity in the signal.

Moment estimates are usually reproducable within a factor of two with the better-determined moments uncertain by about ±30 percent (Kanamori and Anderson, 1975). Coseismic slip estimated at the area of the rupture can be inferred from either field observations or the size of the aftershock zone. Seismic slip rates comparable to geologic slip rates can be estimated by replacing moment by the sum of moments and rupture area by the area contact between plates and dividing the period of the observations, T:

$$\langle \dot{u} \rangle = \frac{\Sigma M_0}{Ts\mu}$$

(Brune, 1968). The sum of moments is estimated from an empirical moment magnitude relationship. Davies and Brune (1971) used $\log M_0 = M_s + 19.9$ for strike-slip earthquakes with $M_s > 7.5$. For dip-slip earthquakes, the moment from this equation was doubled to adjust for the relative excitation of surface waves from strike-slip and dip-slip sources. Because of the log relationship between moment and magnitude, nearly all seismic slip is accounted for by the major earthquakes. Uncertainty in the area of contact and in the effective shear modulus can be tolerated given errors of 100 percent or more in the accumulated moment.

Davies and Brune found that seismic slip and geologic slip taken from plate rotation models were in closer agreement than might have been expected considering the uncertainty in the accumulated moments. This could have been anticipated for the convergent margin of the Pacific basin, where after-shocks of major earthquakes from a record of seismicity for the last 80 years reveal few seismic gaps. Apparently slip took place nearly everywhere along these subduction zones at least once in this period, with the amount of slip comparable on the average with that predicted from plate kinematics. The largest discrepancy between seismic and geologic slip occurred in the New Guinea region, where the seismic rate of 2.3 cm y^{-1} is half an order of magnitude less than the geologic rate of 11.0 cm y^{-1}. Johnson and Molnar (1972) suggest that the Pacific and Indian Ocean plates in that region are decoupled from one another by a number of small intervening plates. Somehow the large convergence rate between the major plates is taken up in a buffer zone of small plates and is not fully represented by the recorded seismicity.

Seismic slip or regional seismic strain can account for less than one half of the geologic strain predicted for central Asia (Chen and Molnar, 1977). This conclusion is secure as long as the seismic strain is distributed throughout a crust of continental thickness, about 40 km, and the accumulated seismic moment is not drastically and systematically underestimated. The recurrence times for major earthquakes in central Asia are not well known; however, recorded seismicity in this region may be representative of the long-term average activity, even though this average might be a poor representation of activity in individual fault zones.

Farther west, in Iran, Turkey and the eastern Mediterranean region, accumulated seismic slip for the years 1963–1970 is more than an order of magnitude less than slip predicted from plate models (North, 1977*a*). Seismicity in individual fault zones of this region is characterized by periods of comparatively intense activity separated by periods of relative quiescence, each of which can last for hundreds of years (Ambraseys, 1971). These conclusions come from records of historic earthquakes that are thought to be nearly complete with regard to major shocks in the last 2000 years. Consequently, activity in any seven-year period such as the one pertaining to North's study has very little chance of being representative of long-term average activity represented in the plate models. Nevertheless a reasonable working hypothesis for continental plate boundaries is that crustal deformation generally, and in particular crustal shortening, is only partially representative of geologic strain between the plates. Here it is assumed that aseismic fault creep at crustal depths does not account for the missing geologic strain.

7. Stress Drop and Apparent Stress

Stress drop, the difference between the initial and final state of stress, is the difference in shear stress, $\Delta\sigma$, for the case of slip on a fault surface. The static stress drop is proportional to geologic strain, $\bar{u}/l$, where $\bar{u}$ and l are the average slip and the characteristic length, respectively. The coseismic or dynamic stress drop is proportional to the ratio of the seismic moment, M_0, to an effective volume, which is the product of area of slip, S, and the characteristic length, l. The proportionality factor, K, depends on the geometry of the slip surface and the sense of slip on that surface. For a circular fault, K is $7\pi/16$ (Eshelby, 1957; Keilis-Borok, 1959) and the characteristic length is the radius. K for long thin strike-slip and dip-slip faults is $2/\pi$ (Knopoff, 1958) and $8/3\pi$ (Starr, 1928), respectively, and l in both cases is the fault width. Here the K's have been simplified by setting Lamé's constant equal to the shear modulus.

Static stress drops from a number of strike-slip events in the western United States and Japan range from 20 to 100 bars (Chinnery, 1964). Kanamori and Anderson (1975) show that the average of coseismic stress drops from major shallow earthquakes is 60 bars. However, earthquakes at the plate margins have systematically lower stress drops than those located within the plates; average values for each group are 30 and 100 bars, respectively. Stress drops from smaller earthquakes show more scatter with values of less than 1 bar frequently reported (e.g. Wyss and Brune, 1968).

Coseismic stress drops are more vulnerable to errors in the effective volume than to errors in the seismic moment. Stress drop may be overestimated by more than 100 per cent if the rupture area, S, is taken from the size of the after-shock zone or from the extent of surface cracking and an assumed fault depth. Rupture area estimated in this way pertains to the area as $t \to \infty$, whereas the seismic moment is derived from band-limited information that pertains to no more than a few hundred seconds of fault slip. Microseismic noise will usually mask longer-period signals even if the recording system maintains sensitivity approaching the d.c. limit. Consequently, slow growth of the rupture during an after-shock series or during a period of stable sliding is not seismically recorded.

Underestimates of the rupture size and characteristic length are possible if these parameters are inferred from the junction of the long- and short-period parts of the amplitude spectrum, the corner frequency (Brune, 1970). For example, the hypothetical spectrum in Fig. 15(a) would yield an underestimate of the rupture length because the apparent corner frequency lies on the high-frequency side of ω_1, the corner frequency that is dependent on the fault length, L. In practice, amplitude spectra that have been azimuthally averaged to remove the source radiation pattern (Brune, 1970) reveal at most a single corner. The assumption that the apparent corner frequency is inversely proportional to the characteristic dimension of the fault yields an estimate of fault size that is uncertain by a factor of two or more (Hanks and Wyss, 1972). The corresponding estimate of stress drop is at best uncertain by a factor of about four (Hanks and Wyss, 1972).

The apparent stress, σ_a, is proportional to the radiated energy and is defined as the product of the average stress during rupture propagation, $(\sigma_0+\sigma_1)/2$ and the seismic efficiency, η. The tectonic stress is the stress at the onset of fault slip which will be greater than σ_0 if coseismic slip is preceded by a partial stress drop during an episode of stable sliding (Fig. 16). The seismic energy, $E_s = \sigma_a S\bar{u}$, and the moment–slip relationships can be combined to estimate the apparent stress $\sigma_a = E_s\mu/M_0$. For order of magnitude accuracy, E_s can be taken from an energy magnitude relationship of the form proposed by Richter (1958):

$$\log E_s = a + bM_s.$$

Kanamori and Anderson (1975) show that $\sigma_a \sim \Delta\sigma/2$ for major shallow earthquakes. By equating total energy to the sum of frictional and seismic

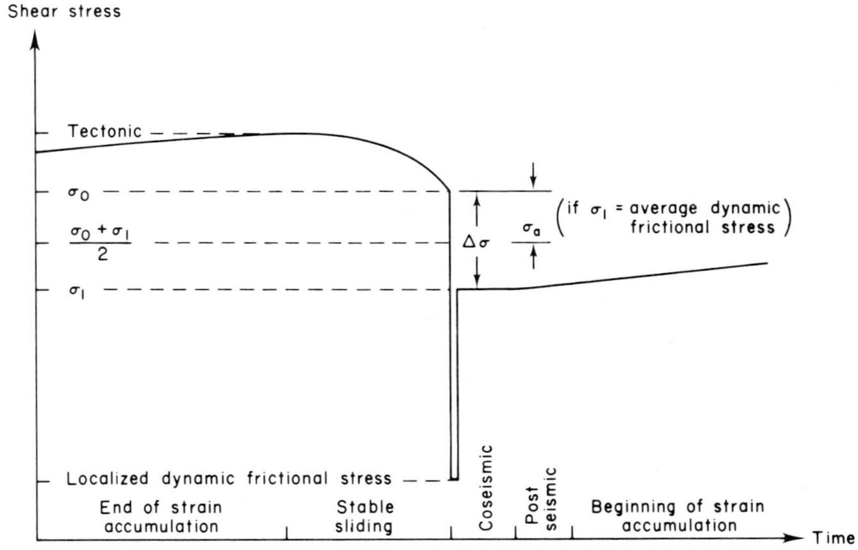

Fig. 16. Hypothetical stress history. The scales for both the stress and time axes are to be considered non-linear and shear stress is to refer to the mean stress over the rupture surface.

energy, it can be shown that the final stress, σ_1, and the frictional stress, σ_f, are equal when $\sigma_a = \Delta\sigma/2$. Savage and Wood (1971) show that $\sigma_a < \frac{1}{2}\Delta\sigma$ if, as Orowan (1960) proposed, frictional stress is greater than the final stress, σ_1, in order to bring fault slip to a halt.

The coseismic estimates of stress drop and apparent stress mentioned thus far come from far-field observations which average out accelerations and decelerations of the rupture that occur in time intervals that are short by comparison with periods representative of the long-period end of the recorded seismic spectrum. Near-field records are more sensitive to short-period variations in the rupture. For example, Hanks (1974) identified a large displacement pulse of short duration on a near-field accelerogram recording of the 1971 San Fernando earthquake. He argues that this pulse occurred at the time and place of rupture nucleation, in this case at a depth of about 12 km. The corresponding stress drop is in the range from 350 to 1400 bars. A schematic stress history for an event such as this is illustrated in Fig. 16.

Elastic models that fit the static vertical displacement from the San Fernando earthquake are in conflict with the dynamic model of Hanks in that maximum slip and stress drop are computed for that part of the rupture that passes through near surface sediments (Alwine and Jordan, 1973; Jungels and Frazier, 1973; McCowan *et al.*, 1977). These static models, unlike the simple models mentioned previously, have non-uniform slip on the fault surface. Conflicts also exist among the static models for this earthquake. The average stress drop along the slip surface of 290 bars, computed by McCowan *et al.* (1977), is an order of magnitude higher than the average value computed by Jungels and Frazier (1973). In both studies, two-dimensional finite element models were used to fit the observations. The complexity of these models precludes an easy assessment of the relative merits of each model.

The larger estimates of stress drop from the San Fernando earthquake are compatible with an empirical upper limit of a few hundred bars on teleseismic stress drops. Plate stresses from a variety of mechanisms, some of which are discussed in the next section, could be this large or larger. Near equality of plate stress and stress drop would imply either a vanishingly small sliding friction on the rupture surface or a non-equilibrium slip mechanism. Either possibility could arise from the loss of strength along the rupture front due to frictional heating. In an extreme case, frictional heating could melt part of the slip surface (McKenzie and Brune, 1972; Caldwell *et al.*, 1978), thereby promoting a total stress drop and, in an extreme case, a reversal in the stress field resulting from slip in excess of the equilibrium value. The best evidence for shear melting comes from glassy rocks found in a thrust zone in the Himalayas (Scott and Drever, 1954) and devitrification textures in some rocks found in ancient fault zones (e.g. Philpotts, 1964; Sibson, 1975). Sibson's field area in north-west Scotland is particularly significant in that it contains an exposure of the brittle/ductile transition.

A lower bound on tectonic stress in the San Fernando region can be taken as the upper bound on tectonic stress in the central California region of the San Andreas fault, where aseismic fault slip has been recorded since 1958 (Tocher, 1960; Nason, 1973). If the fault is stably sliding throughout its width, then the rate of heat release per unit length of fault is given by

$$E_{\mathrm{f}} = \sigma_0 \dot{u} W,$$

where σ_0 is the tectonic shear stress, $\dot{u}$ is the slip rate and W is the width of the slip surface, about 12–15 km if the width is equated to the maximum depth of earthquakes (Eaton *et al.*, 1970). The absence of a heat-flow anomaly limits the heat flow from frictional heating to 0.3 μcal cm^{-2} s^{-1}, which in turn limits the tectonic shear stress along this section of the fault (Brune *et al.*, 1969). This heat flow and 3.2 cm y^{-1} for the average slip rate (Savage and Burford, 1973) instead of the plate rotation rate of 5 cm y^{-1} (assumed by Brune *et al.*, 1969) yields an upper limit for tectonic stress of about 200 bars. If instead of stable sliding the motion occurs with stress drops that radiate seismic energy, then the sum of an upper limit for frictional stress and the stress drop yields an upper limit for tectonic stress. Stress drops of a few hundred bars would bring the upper limit of tectonic stress in central California to about 0.5 kbar.

Static and sliding friction measured in the laboratory, in principle, can be used to estimate lower bounds on tectonic stress before an earthquake. Friction experiments reveal a linear relationship between shear strength, τ, and the normal stress. σ_n, across the sliding surface called the Coulomb law of friction, $\tau = \tau_0 + n\sigma_n$, where τ_0 and n are the zero pressure strength and the coefficient of friction, respectively (Byerlee, 1967*a,b*; Brace, 1972). For common rock types, n ranges from 0.4 to 1.8 (Jaeger and Cook, 1969, pp. 57–61), with most experiments yielding about 0.6 independent of rock type, temperature and nature of the fault surface (Scholz, 1977). Consequently, kilobars of tectonic stress are required to overcome friction across fault surfaces that extend to depths greater than a few kilometers unless the coefficients of friction or the effective normal pressures are much lower than experiments would predict.

Scholz *et al.* (1969) argue that the coefficient of friction in the central section of the San Andreas fault near Parkfield is about 0.15, which implies a much smoother fault than those simulated in laboratory experiments. They estimated the roughness of the slip surface from a composite mechanism solution to after-shocks of the June 1966 Parkfield earthquake. The assumption was made, without justification, that signals from each after-shock pass through the same velocity structure.

The frictional strength of the fault could also be reduced by high pore pressure in the fault zone. Pore pressures have been incorporated into the broader context of a dilatancy theory of faulting (Nur, 1973 and Scholz *et al.*, 1973). The essential feature of this theory is that cracks open and possibly grow as tectonic stress builds toward a critical value. Laboratory experiments on crustal rocks show that a kilobar or more of deviatoric stress is required to observe dilatancy effects at pressure appropriate to the Earth's crust (Brace *et al.*, 1966). Kanamori and Anderson (1975) point out that if these laboratory results are applied directly to the Earth, then frictional stress in major fault zones is a large fraction of the tectonic stress because static and far-field stress drops have an upper limit of a few hundred bars. If one chooses 100 bars as the upper limit (as they did), then frictional stress is about 90 per cent of the tectonic stress.

8. Plate Stress

A variety of geological and geophysical observations reveal oriented tectonic stress fields near active plate margins. Bending moments give rise to graben tectonics on the seaward wall of deep sea trenches (Stauder, 1968), topographic rises seaward of deep trenches (Watts and Talwani, 1974) and horizontal compression near active subduction zones (Ichickawa, 1961, 1966; Stauder, 1973, 1975). Permanent flexures of the overriding plate are inferred from the coastal morphology of south-west Japan (Watanabe, 1959), south-central Alaska (Plafker, 1969), Chile (Plafker and Savage, 1970) and north-eastern New Guinea (Chappell, 1974). Bischke (1974) modeled a flexure hundreds of kilometers wide in the Shikoku region of south-western Japan (Fig. 9). He numerically deformed viscoelastic finite elements representative of the lithosphere and upper mantle to fit the observed vertical component of ground movement. Vertical loading of the downgoing plate suggestive of a positive density contrast between the underthrust oceanic plate and its surroundings was required to fit the observations. The lack of a mantle seismic zone beneath this part of Japan (Kanamori, 1972*d*) limits the source of the vertical loading to lithospheric depths, in this case about 80 km.

A topographic rise, typically a few hundred kilometers wide with an amplitude of a few hundred meters, is a continuous feature of the sea floor seaward of deep-sea trenches from the Aleutians to the Marianas (Watts and Talwani, 1974). In profile view, this rise has the shape of an overdamped sinusoid similar to that shown in Fig. 17. A thin elastic plate embedded in a fluid will take this shape if it is loaded from one end by a bending moment, M, that is either partly or solely the result of horizontal compression (Parsons and Molnar, 1976). If the applied moment is sufficiently small, the deformation can be described by vertical deflections, y, that obey the following equation

$$D\frac{\mathrm{d}^4 y}{\mathrm{d}x^4} - N\frac{\mathrm{d}^2 y}{\mathrm{d}x^2} + \rho g y = 0,$$

where ρ is the difference in density between the fluids above and below the plate and g is the acceleration of gravity (Parsons and Molnar, 1976), from LePichon *et al.*, 1973). In the equation, N is the horizontal force per unit length along the arc (positive for tension and negative for compression) and the flexural rigidity D is

$$EH(3/12)\,(1-\nu^2)\,,$$

where E, H and ν are the Young's modulus, thickness and Poisson's ratio, respectively, for the plate.

Hanks (1971) and Watts and Talwani (1974) showed that horizontal compressive stress of from 5 to 10 kbar would be sufficient to fit the observed shape of the rise with a deflection of about 3 km at the trench. Hanks (1977) argues that high compressive stress implies a basal shear stress of a few tens

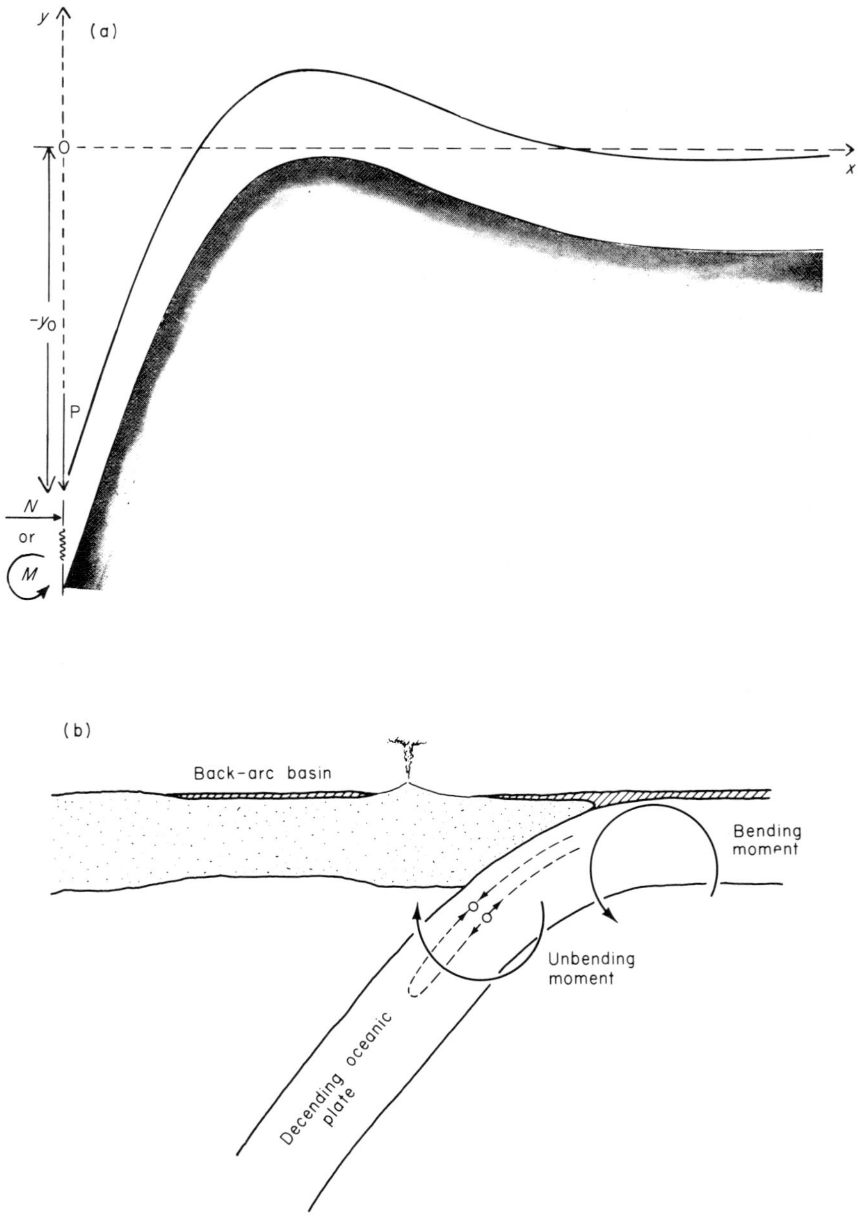

Fig. 17. (a) Simple geometry of the thin elastic plate model showing the boundary conditions applied at the origin, assumed to be the trench axis. The thickness of the plate is not to scale. (From Parsons and Molnar, 1976, with permission.) (b) Paired seismic zone and subduction induced bending and unbending of oceanic lithosphere.

of bars and that near an active plate margin it is this stress that is released at the times of major earthquakes. This model is consistent with the fault model of Savage and Burford (1973) in which the seismic stress drop is transmitted vertically from a near-surface brittle-elastic zone to an underlying plastic zone. The release of basal stress rather than stress on vertical planes accounts for teleseismic stress drops of a few tens of bars, which is in agreement with teleseismic observations.

Although a uniform basal shear stress for a large plate is an oversimplification, the magnitude of the proposed stress is reasonable in that the frictional heat generated is equivalent to no more than a few tenths of the mean heat flow of about 1.4 μcal cm^{-2} s^{-1} through the abyssal plains. This is the case for plate velocities as high as 10 cm y^{-1}. However, high horizontal compression can not be supported unequivocally and, in fact, is not required to support the rise (LePichon *et al.*, 1973; Caldwell *et al.*, 1976; Parsons and Molnar, 1976); all that is required is a moment with a magnitude of $Ny_0/2$ applied near the bottom of the trench.

The sinking force on the downgoing slab derived solely from a thermal contrast (McKenzie, 1969) may by itself provide the necessary moment. Parsons and Molnar (1976) estimate a moment of -2×10^{22} dyne cm^{-1} per unit length along the arc from the relationship $M = -\frac{1}{2}Bh \sin \theta$, where B is the sinking force per unit length and h and θ are the thickness of the elastic part of the plate, taken to be 30 km, and the dip of the plate, respectively. The elastic thickness is only a small fraction of the lithosphere thickness because strain in the descending plate near the trench exceeds the elastic limit for rocks, about 10^{-4}, by several orders of magnitude (LePichon *et al.*, 1973). Presumably the rock above the elastic part of the plate fails in tension (deviatoric) and that below fails in compression.

The lack of seismicity seaward of the deep-sea trenches generally, and in particular those in the north-west Pacific, can be explained in at least three ways. The near-surface part of the bent plate can behave elastically near the topographic rise. High compressive stress may keep this part of the plate from deforming by extensional failure (Hanks, 1971, 1977). Last, this part of the plate may be too broken up to transmit regional stress fields. In an extreme case, the latter alternative would have the horizontal and inclined parts of the plate mechanically decoupled. The elastic alternative seems the least plausible given the mechanical properties of the oceanic crust and the configurational strains. High compressive stress is not precluded by existing data.

Geophysical evidence appears to be more consistent with some form of decoupling. The near-surface structure from the bottom of the Japan trench to the seaward rise, inferred from seismic profiler records (Ludwig *et al*, 1966), shows an intensely faulted topography. Kanamori (1971*b*) argues that the 1933 Sanriku earthquake ruptured the entire Pacific plate adjacent to northern Honshu; however, the vertical extent of the rupture is not confirmed by accurate focal depth determinations. If this interpretation is correct and if similar events occur elsewhere sufficiently often, then the extreme version of the decoupling hypothesis would be favored. In this case, the rise would have to be supported by flow in the asthenosphere. Correlations between long-wavelength gravity anomalies and topography (Watts and Talwani, 1974; McKenzie, 1977) suggest that at least in part the rise is supported by an underlying flow.

Another Sanriku-type event, the Sumbawa earthquake of August 1977, occurred near the eastern end of the Java trench and was the largest seismic event since the 1964 Great Alaska earthquake (Stewart, 1978). Here again, the uncertainty in the depth of faulting has confounded the tectonic interpretation. This rupture either decoupled the horizontal and downgoing parts of the Indian Ocean plate or it was an indicator of high tectonic stress if the rupture was confined to a shallow depth within the lithosphere. Abe (1972*a*,*b*) argues that normal fault earthquakes adjacent to the Rat Island section of the Aleutian arc and the Peruvian section of the Andean subduction zone were decoupling events. His argument is supported by standard bulletin locations of after-shocks that generally yield unreliable depths for shallow earthquakes. To some extent this focal depth problem is unnecessary because depth sensitive phases such as pP and sP and the surface wave phase and amplitude spectra together can be used to constrain the depths of even shallow earthquakes (Chen and Forsyth, 1978). Locations with travel times calibrated for a particular region or locations relative to a master earthquake also retain depth sensitivity (Jackson and Fitch, 1978).

The elastic thickness of the Pacific plate subducted beneath northern Honshu can be equated to a separation of about 30–40 km between paired seismic zones in the depth range from 50 to 200 km (Hasegawa *et al.*, 1978; Uyeda, 1977). Paired seismic zones have also been resolved in this depth range beneath the Kuril (Veith, 1974) and Aleutian (Engdahl and Scholz, 1977) arcs. Plate stress consistent with focal mechanisms, schematically illustrated in Fig. 17(b), is the reverse of bending stress imposed by subduction, i.e. the upper part of the seismic zone is in down dip compression and the lower part is in down dip tension. This reversal suggests a recovery from subduction by unbending.

Shallow seismicity within the lithosphere containing the volcanic arcs is dominated by strike-slip and thrust activity. In Japan (Ichikawa, 1961, 1966) and Peru (Stauder, 1975), the stress axes from mechanism solutions to crustal earthquakes typically have nearly horizontal axes of maximum compressive stress oriented more or less normal to the direction of convergence. In south-west Japan, on the western side of the Kii peninsula, swarm activity correlates with the times of major off-shore earthquakes along the Nankai trough (Kanamori, 1972*d*). Similarly, dated ash deposits on Oshima volcano near the mouth of Sagami Bay (south of Tokyo) shows that major eruptions are correlated in time with the major earthquakes along this trough (Nakamura, 1964). Both the swarm and the volcanic activity fall off after a major earthquake and are renewed prior to the next one. Apparently the source regions for this activity concentrate tectonic stress. The effect is difficult to quantify because the results of whatever computations are undertaken depend critically on the shape of the source region, which is unknown (Kanamori, 1972*d*).

Horizontal compression in the vicinity of convergent plate margins may be only partly maintained by subduction. Hales (1969) has shown that gravitational sliding on a global scale can easily account for a few hundred bars of stress at the plate margins. He envisioned a plate of lithospheric thickness, about 100 km, sliding over a viscous low velocity zone (LVZ). The horizontal force per unit area required for steady-state laminar flow is $\nu V/D$, where ν and D are the viscosity of the LVZ (about 10^{21} poise) and its thickness (about 100 km), respectively. Equating this force to the down-dip component of force from the weight of the plate shows that an infinitesimal dip of the sliding surface is sufficient to move the plate at velocities comparable with the highest plate velocities (about 10 cm y^{-1}). Hales also points out that heat from the flow would tend to raise the temperature of LVZ, thereby encouraging more rapid plate motion by a reduction of the viscosity of the LVZ. Sykes and Sbar (1973) have suggested gravitational sliding as a possible driving mechanism for intraplate earthquakes. In eastern North America, axes of maximum compressive stress from shallow earthquakes are more or less horizontal and oriented normal to the crest of the North Atlantic ridge (Sykes and Sbar, 1973).

9. Summary

Plate motion during time intervals comparable with recurrence times for major earthquakes at active plate margins, a few decades to 1000 years, is at present unknown. Temporal variations in seismicity such as the build-up in activity near the turn of the century (Anderson, 1974) may be symptomatic of episodic plate motion, in which case the search for trigger mechanism should be encouraged. The spatial distribution of activity near active convergent and transform margins suggests that deformation is typically distributed over a zone a few hundred kilometers wide, and in the case of continental collision the zone may extend for more than 1000 km into what was the overriding plate prior to collision (Molnar and Tapponier, 1975).

Davies and Brune (1971) found that seismic slip at many convergent and transform margins was in close agreement with geologic slip, which is accurately known locally from field observations and globally from magnetic anomalies in the sea floor. This agreement may be fortuitous because accumulated seismic slip inferred from a moment–magnitude relationship applied to the 80-year sample of recorded seismicity is likely to be in error by a factor of two or more. However, this sample of activity at the more active plate margins may be representative of activity averaged over a much longer period.

Fore-shocks, stable sliding, premonitory slip, the non-zero rise time of seismic sources, after-shocks and post-seismic slip show that accumulated strain is released in varying amounts in periods ranging from a few seconds to a few months or even years. Post-seismic ground movements in the cases mentioned here and in Scholz (1972) diminish exponentially from the time of the main shocks. If these movements result solely from fault slip, a combination of forward and reverse slip is required, not necessarily on the same part of the slip surface that ruptured initially. Flow in an underlying asthenosphere may be required to account for some of the longer-wavelength observations.

Tectonic loading, whether regional, as in the case of subduction induced bending moments, or global, as in the case of gravitational sliding of an entire plate, probably does not exceed a few hundred bars to 1–2 kbar. Minimum values of tectonic stress inferred from geodetically and seismically determined stress drops lie in this range. However, different analyses of the geodetic and near-field seismic data from the San Fernando, California earthquake are inconsistent with regard to the distribution of stress drop along the slip surface (Alewine and Jordan, 1973; Jungels and Frazier, 1973; McCowan *et al.*, 1977; Hanks, 1974). These inconsistencies demonstrate that, even with the more complete data sets, source parameters for a variety of models fit the data nearly equally well. A severe uniqueness problem like this is to be expected because even the elastic layer parameters are not known accurately in most source regions and the correct parameterization of the fault zone is not known *a priori*. Hopefully, the national programs for earthquake prediction will result in models for fault slip that, at least in specific regions, diminish or remove these inconsistencies.

Acknowledgements

I am indebted to colleagues for preprints and comments with regard to this chapter and plate tectonic theory generally. In particular, Peter Molnar has pointed out simple yet compeling arguments such as the one showing that plate momentum is negligible.

This work was supported by the Defense Advanced Research Projects Agency.

References

Abe, K. (1972*a*). Lithospheric normal faulting beneath the Aleutian trench. *Phys. Earth planet. Interiors* **5**, 190–198.

Abe, K. (1972*b*). Mechanisms and tectonic implications of the 1966 and 1970 Peru earthquakes. *Phys. Earth planet. Interiors* **5**, 367–379.

Aki, K. (1966). Generation and propagation of G waves from the Niigata earthquake of June 16, 1964. Part 2. Estimation of earthquake moment, released energy and stress–strain drop from the G wave spectrum. *Bull. Earth Res. Inst.* **44**, 73–88.

Alewine, R. W. and Jordan, T. H. (1973). Generalized inversion of earthquake static displacement fields. *Geophys. J. R. astronom. Soc.* **35**, 357–380.

Ambraseys, N. N. (1970). Some characteristic features of the Anatolian fault zone. *Tectonophysics* **9**, 143–165.

Ambraseys, N. N. (1971). Value of historical records of earthquakes. *Nature, Lond.* **232**, 375–379.

Anderson, D. L. (1974). Earthquakes and the rotation of the Earth. *Science, N.Y.* **186**, 49.

Anderson, D. L. (1975). Accelerated plate tectonics. *Science, N.Y.* **187**, 1077–1079.

Ando, M. (1971). A fault-origin model of the Great Kanto Earthquake of 1923 as deduced from geodetic data. *Bull. Earth Res. Inst.* **49**, 19–32.

Ando, M. (1975). Source mechanisms and tectonic significance of historical earthquakes along the Naukai trough, Japan. *Tectonophysics* **27**, 119–140.

Atwater, T. (1970). Implications of plate tectonics for the Cenozoic tectonic evolution of western North America. *Bull. geol. Soc. Am.* **81**, 3513–3536.

Atwater, T. and Menard, H. W. (1970). Magnetic lineations in the northeast Pacific. *Earth planet. Sci. Lett.* **7**, 445–450.

Atwater, T. and Molnar, P. (1973). Relative motion of the Pacific and North American plates deduced from sea-floor spreading in the Atlantic, Indian and South Pacific Oceans. In *Proceedings of the Conference on Tectonic Problems of the San Andreas Fault System* (R. L. Kovach and A. Nur, eds), Stanford University Publications in Geological Science No. 8, Stanford University Press, Stanford, pp. 136–148.

Basham, P. W. (1969). Canadian magnitudes of earthquakes and nuclear explosions in southwestern North America, *Geophys. J. R. astronom. Soc.* **17**, 1–13.

Berggren, W. A. (1969). Cenozoic chronostratigraphy, planktonic foraminiferal zonation and the radiometric time scale. *Nature, Lond.* **224**, 1072–1075.

Bischke, R. E. (1974). A model of convergent plate margins based on the recent tectonics of Shikoka, Japan. *J. geophys. Res.* **79**, 4845–4858.

Bott, M. H. P. and Dean, D. S. (1973). Stress diffusion from plate boundaries. *Nature, Lond.* **243**, 339–341.

Brace, W. F. (1972). Laboratory studies of stick-slip and their application to earthquakes. *Tectonophysics* **14**, 189–200.

Brace, W. F., Paulding, B. W., Jr. and Scholz, C. (1966). Dilatancy in the fracture of crystalline rocks. *J. geophys. Res.* **71**, 3939–3954.
Bracewell, R. (1965). *The Fourier Transform and Its Applications*, McGraw-Hill, New York.
Brown, L. D., Reilinger, R. E., Holdahl, S. R. and Balazs, E. I. (1977). Post-seismic crustal uplift near Anchorage, Alaska. *J. geophys Res.* **82**, 3369–3378.
Brune, J. N. (1968). Seismic moment, seismicity and rate of slip along major fault zones. *J. geophys. Res.* **73**, 777–784 (1968).
Brune, J. N. (1970). Tectonic stress and spectra of seismic shear waves from earthquakes. *J. geophys. Res.* **75**, 4997–5009.
Brune, J. and Dorman, J. (1963). Seismic waves and earth structure in the Canadian shield. *Bull. seismol. Soc. Am.* **53**, 167–210.
Brune, J. N., Henyey, T. L. and Ray, R. F. (1969). Heat flow, stress and rate of slip along the San Andreas Fault, California. *J. geophys. Res.* **74**, 3821–3827.
Byerlee, J. D. (1967*a*). Frictional characteristics of granite under high confining pressure. *J. geophys. Res.* **72**, 3639–3648.
Byerlee, J. D. (1967*b*). Theory of friction based on brittle fracture. *J. appl. Phys.* **38**, 2928–2934.
Caldwell, R. K., Chinn, D. S., Moore, G. F. and Turcotte, D. L. (1978). Frictional heating on a fault zone with finite thickness. *Geophys J. R. astronom. Soc.* **52**, 525–530.
Caldwell, J. G., Haxby, W. F., Karig, D. E. and Turcotte, D. L. (1976). On the applicability of a universal elastic trench profile. *Earth planet. Sci. Lett.* **31**, 239–246.
Cathles, L. M., III (1975). *The Viscosity of the Earth's Mantle*, Princeton University Press, Princeton, N.J.
Chappell, J. (1974). Geology of coral terraces on Huon Peninsula, New Guinea: a study of Quaternary tectonic movements and sea level changes. *Bull. geol. Soc. Am.* **85**, 553–570.
Chase, C. G. (1972). The *N*-plate problem of plate tectonics. *Geophys. J. R. astronom. Soc.* **29**, 117–122.
Chen, T. and Forsyth, D. (1978). Structure of the lithosphere, a detailed study of two earthquakes seaward of the Tonga trench: implications for mechanical behavior of the oceanic lithosphere. *J. geophys. Res.* **83**, 4995–5004.
Chen, W.-P. and Molnar, P. (1977). Seismic moments of major earthquakes and the average rate of slip in Central Asia. *J. geophys Res.* **82**, 2945–2969.
Chinnery, M. A. (1964). The strength of the Earth's crust under horizontal shear stress. *J. geophys. Res.* **79**, 2085–2088.
Davies, G. F. and Brune, J. N. (1971). Regional and global fault slip rates from seismicity. *Nature phys. Sci.* **229**, 101–107.
Eaton, J. P., Lee, W. H. K. and Pakiser, L. C. (1970). Use of microearthquakes in the study of the mechanics of earthquake generation along the San Andreas fault in central California. *Tectonophysics* **9**, 259–282.
Elsasser, W. M. (1967). Convection and stress propagation in the upper mantle. In *The Application of Modern Physics to the Earth and Planetary Interiors* (S. K. Runcorn, ed.), Wiley-Interscience, New York, pp. 223–246.
Engdahl, E. R. and Scholz, C. H. (1977). A double Benioff zone beneath the central Aleutians: an unbending of the lithosphere. *Geophys. Res. Lett.* **4**, 473–476.
Eshelby, J. D. (1957). The determination of the elastic field of an ellipsoidal inclusion and related problems. *Proc. R. Soc. A* **241**, 376–396.

Fedotov, S. A. (1965). Regularities of the distribution of strong earthquakes in Kamchatka, the Kurile Islands and Northeastern Japan (in Russian). *Tr. Inst. Fiz. Zemli. Akad. Nauk SSSR* **36**, 66.

Fitch, T. J. (1972). Plate convergence, transcurrent faults, and internal deformation adjacent to southeast Asia and the Western Pacific. *J. geophys. Res.* **77**, 4432–4460.

Fitch, T. J. and Scholz, C. H. (1971). Mechanism of underthrusting in Southwest Japan: a model of convergent plate interactions. *J. geophys. Res.* **76**, 7260–7292.

Fitch, T. J., Worthington, M. H. and Everingham, I. B. (1973). Mechanics of Australian earthquakes and contemporary stress in the Indian Ocean Plate. *Earth planet. Sci. Lett.* **18**, 345–356.

Forsyth, D. W. (1975). The evolution of the upper mantle beneath mid-ocean ridges. In *Present State of Plate Tectonics* (J. Bonnin and R. S. Dietz, eds), *Tectonophysics* **38** (1–2), 89–118.

Forsyth, D. and Uyeda, S. (1975). On the relative importance of the driving forces of plate motion. *Geophys. J. R. astronom. Soc.* **43**, 163–200.

Frasier, C. W. and North R. G. (1978). Evidence for ω-cube scaling from amplitudes and periods of the Rat Island Sequence (1965). *Bull seismol. Soc. Am.* in press.

Gaposchkin, E. M. and Lambeck, K. (1971). Earth's gravity field to sixteenth degree and statin coordinates from satellite and terrestrial data. *J. geophys. Res.* **76**, 4855–4883.

Gilbert, F. (1970). Excitation of the normal modes of the Earth by earthquake sources. *Geophys. J. R. astronom. Soc.* **22**, 223–226.

Goncz, J. H. and Cleary, J. R. (1976). Variations in the structure the upper mantle beneath Australia. *Geophys. J. R. astronom. Soc.* **44**, 507–516.

Gutenberg, B. and Richter, C. F. (1954). *Seismicity of the Earth and Associated Phenomena*, Princeton University Press, Princeton, N.J.

Hales, A. L. (1969). Gravitational sliding and continental drift. *Earth planet. Sci. Lett.* **6**, 31–34.

Hanks, T. C. (1971). The Kuril Trench–Hokkaido Rise system: large shallow earthquakes and simple models of deformation. *Geophys. J. R. astronom. Soc.* **23**, 173–189.

Hanks, T. C. (1974). The faulting mechanism of the San Fernando earthquake. *J. geophys. Res.* **79**, 1215–1229.

Hanks, T. C. (1977). Earthquake stress drops, ambient tectonics stresses and stresses that drive plate motions. *Pure appl. Geophys.* **115**, 441–458.

Hanks, T. C. and Wyss, M. (1972). The use of body-wave spectra on the determination of seismic-source parameters. *Bull. seismol. Soc. Am.* **62**, 561–589.

Hasegawa, A., Umino, N. and Takagi, A. (1978). Double-planed structure of the deep seismic zone in the northeastern Japan arc. *Tectonophysics* **47**, 43–58.

Haskell, N. (1964). Total energy and energy spectral density of elastic wave radiation from propagating faults. *Bull. seismol. Soc. Am.* **54**, 1811–1842.

Hastie, L. M. and Savage, J. C. (1970). A dislocation model for the 1964 Alaska earthquake. *Bull. seismol. Soc. Am.* **60**, 1389–1392.

Hatori, T. (1966). Vertical displacement in a tsunami source area and the topography of the sea bottom. *Bull. Earth. Res. Inst.* **44**, 1449–1464.

Heirtzler, J. R., Dickson, G. O., Herron, E. M., Pitman, W. C., III and LePichon, X. (1968). Marine magnetic anomalies, geomagnetic field reversals and motions of the ocean floor and continents. *J. geophys. Res.* **73**, 2119–2136.

Ichikawa, M. (1961). Statistical investigation of earthquake mechanism in and near Japan during the period from 1950 to 1957. *Geophys. Mag.* **30**, 355–403.
Ichikawa, M. (1966). Mechanism of earthquakes in and near Japan 1950–1962, *Papers Meteor. Geophys., Meteor. Res. Inst., Japan. Meteor. Agency*, 201–229.
Imamura, A. (1929). On the chronic and acute earthtilting in the Kii peninsula. *Jap. J. Astron. Geophys.* **7**, 31–45.
Imamura, A. (1930). On the chronic and acute earthtilting in the southern part of Shikoku. *Jap. J. Astron. Geophys.* **8**, 29–37.
Jackson, J. and Fitch, T. J. (1978). Seismotectonic implications of relocated aftershock sequences in Iran and Turkey: an application of the master event technique, *Geophys. J. R. astronom. Soc.* in press.
Jaeger, J. C. and Cook, N. G. W. (1969). *Fundamentals of Rock Mechanics*, Methuen, London.
Johnson, T. and Molnar, P. (1972). Focal mechanisms and plate tectonics of the southwest Pacific. *J. geophys. Res.* **77**, 5000–5032.
Jungels, P. H. and Frazier, G. H. (1973). Finite element analysis of the residual displacements for an earthquake rupture: source parameters for the San Fernando earthquake. *J. geophys. Res.* **78**, 5062–5083.
Jurdy, D. M. (1977). An alternate model for absolute plate motions in the early Tertiary (abstract). *Trans. Am. Geophys. Union* **58**, 503.
Kanamori, H. (1970). The Alaska earthquake of 1964: radiation of long-period surface waves and source mechanism. *J. geophys. Res.* **75**, 5029–5039.
Kanamori, H. (1971*a*). Faulting of the Great Kanto earthquake of 1923 as revealed by seismological data. *Bull. Earth. Res. Inst.* **49**, 13–18.
Kanamori, H. (1971*b*). Seismological evidence for a lithospheric normal faulting—the Sanriku earthquake of 1933. *Phys. Earth planet. Interiors* **4**, 289–300.
Kanamori, H. (1972*a*). Tectonic implications of the 1944 Tonankai and the 1946 Nankaido earthquakes. *Phys. Earth planet. Interiors* **5**, 129–139.
Kanamori, H. (1972*b*). Determination of effective tectonic stress associated with earthquake faulting: the Tottori earthquake of 1943. *Phys. Earth planet. Interiors* **5**, 427–434.
Kanamori, H. (1972*c*). Mechanism of tsunami earthquakes. *Phys. Earth planet. Interiors* **6**, 346–359.
Kanamori, H. (1972*d*). Relation between tectonic stress, great earthquakes and earthquake swarms. *Tectonophysics* **14**, 1–12.
Kanamori, H. (1973). Mode of strain release associated with major earthquakes in Japan. *A. Rev. Earth planet. Sci.* **1**, 213–240 (1973).
Kanamori, H. and Abe, K. (1978). Re-evaluation of the turn-of-the-century seismicity peak using an undamped Milne seismograph (abstract). *Trans Am. Geophys. Union* **49**, 327.
Kanamori, H. and Anderson, D. L. (1975). Theoretical basis of some empirical relations in seismology. *Bull. seismol. Soc. Am.* **65**, 1073–1095.
Kanamori, H. and Cipar, J. J. (1974). Focal process of the Great Chilean earthquake May 22, 1960. *Phys. Earth planet. Interiors* **9**, 128–136.
Katili, J. A. (1971). A review of the geotectonic theories and tectonic maps of Indonesia. *Earth Sci. Rev.* **7**, 143–163.
Kaula, W. M. (1972). Global gravity and tectonics. In *The Nature of the Solid Earth* (E. C. Robertson, ed.), McGraw-Hill, New York.
Keilis-Borok, V. (1959). An estimation of the displacement in an earthquake source and of source dimensions. *Ann. Geofiz.* **12**, 205–214.

Kelleher, J., Savino, J., Rowlett, H. and McCann, W. (1974). Why and where great thrust earthquakes occur along island arcs. *J. geophys. Res.* **79**, 4889–4899.

Knopoff, L. (1958). Energy release in eartquakes. *Geophys. J. R. astronom. Soc.* **1**, 44–52.

Kostrov, V. V. (1974). Seismic moment and energy of earthquakes and seismic flow of rock. *Izv. Acad. Sci. USSR, Phys. Solid Earth* **1**, 23–44.

Lambeck, K. and Cazenave, A. (1973). The Earth's rotation and atmospheric circulation. I. Seasonal variations. *Geophys. J. R. astronom. Soc.* **32**, 79–94.

Larson, R. L., Menard, H. W. and Smith, S. M. (1968). Gulf of California: a result of sea-floor spreading and transform faulting. *Science, N.Y.* **161**, 781–784.

Leeds, A. R., Knopoff, L. and Kausel, E. G. (1974). Variations of upper mantle structure under the Pacific Ocean. *Science, N.Y.* **186**, 141–143.

LePichon, X. (1968). Sea-floor spreading and continental drift. *J. geophys. Res.* **73**, 3661–3697.

LePichon, X. (1970). Correction to paper by Xavier Le Pichon "Sea-floor spreading and continental drift". *J. geophys. Res.* **75**, 2793.

LePichon, X., Francheteau, J. and Bonnin, J. (1973). *Plate Tectonics*, Elsevier, Amsterdam.

Ludwig, W. J., Ewing, J. I., Ewing, M., Muraachi, S., Der, N., Asano, S., Hotta, H., Hayakawa, M., Asanuma, T., Ichikawa, K., and Noguchi, I. (1966). Sediments and structure of the Japan Trench. *J. geophys Res.* **71**, 2121–2137.

McCowan, D. W. (1976). Moment tensor representation of surface wave sources. *Geophys. J. R. astronom. Soc.* **44**, 595–599.

McCowan, D. W., Glover, P. and Alexander, S. S. (1977). A static and dynamic finite element analysis of the 1971 San Fernando, California earthquake. *Geophys. J. R. astronom. Soc.* **48**, 163–186.

McElhinny, M. W. and McWilliams, M. O. (1977). Precambrian geodynamics—a paleomagnetic view. *Tectonophysics* **40**, 137–160.

McGarr, A. (1976). Seismic moments and volume changes. *J. geophys. Res.* **81**, 1487–1494.

McKenzie, D. P. (1969). Speculations on the causes and consequence of plate motions. *Geophys. J. R. astronom. Soc.* **18**, 1–32.

McKenzie, D. P. (1972). Plate tectonics. In *The Nature of the Solid Earth* (E. C. Robertson, ed.), McGraw-Hill, New York, pp. 323–360.

McKenzie, D. P. (1977). Surface deformation, gravity anomalies and convection. *Geophys. J. R. astronom. Soc.* **48**, 211–238.

McKenzie, D. P. and Brune, J. N. (1972). Melting on fault planes during large earthquakes. *Geophys. J. R. astronom. Soc.* **29**, 65–78.

McKenzie, D. P. and Morgan, W. J. (1969). Evolution of triple junctions. *Nature, Lond.* **224**, 125–133.

McKenzie, D. P. and Parker, R. L. (1967). The North Pacific: an example of tectonics on a sphere. *Nature, Lond.* **216**, 1276–1280.

McKenzie, D. P. and Weiss, N. (1975). Speculations on the thermal and tectonic history of the Earth. *Geophys. J. R. astronom. Soc.* **42**, 131–174.

Mansinha, L. and Smylie, D. E. (1971). The displacement fields of inclined faults. *Bull. seismol. Soc. Am.* **61**, 1433–1440.

Matuzawa, T. (1964). *Study of Earthquakes*, Uno Shoten, Tokyo.

Minister, J. B., Jordan, T. H., Molnar, P. and Haines, E. (1974). Numerical modelling of instantaneous plate tectonics. *Geophys. J. R. astronom. Soc.* **36**, 541–576.

Miyabe, N. (1955). Vertical earth movements in Nankai district. *Bull. geogr. Surv. Inst.* **4**, 1–14.

Mogi, K. (1968*a*). Migration of seismic activity. *Bull. Earth Res. Inst.* **46**, 53–74.

Mogi, K. (1968*b*). Development of aftershock areas of great earthquakes. *Bull. Earth Res. Inst.* **46**, 175–203.

Mogi, K. (1968*c*). Some features of recent sesimic activity in and near Japan (1). *Bull. Earth Res. Inst.* **46**, 1225–1236.

Mogi, K. (1969). Relationship between the occurrence of great earthquakes and tectonic structures. *Bull. Earth Res. Inst.* **47**, 429–451.

Molnar, P. and Oliver, J. (1969). Lateral variations of attenuation in the upper mantle and discontinuities in the lithosphere. *J. geophys. Res.* **74**, 2648–2683.

Molnar P. and Tapponier, P. (1975). Cenozoic tectonics of Asia: effects of a continental collision. *Science, N.Y.* **189**, 419–426.

Molnar, P. and Tapponier, P. (1977). The collision between India and Eurasia. *Scient. Am.* **236**, 30–41.

Morgan, W. J. (1968). Rises, trenches, great faults and crustal blocks. *J. geophys. Res.* **73**, 1959–1982.

Morgan, W. J. (1973). Plate motions and deep mantle convection, In *Hess Volume* (R. Shagam, ed.), *Geol. Soc. Am. Mem.* **132**, 7–70.

Nakamura, K. (1964). Volcano-stratigraphic study of Oshima volcano, Izu. *Bull. Earth Res. Inst.* **42**, 649–728.

Nason, R. D. (1973). Fault creep and earthquakes on the San Andreas Fault. In *Proceedings of the Conference on Tectonic Problems of the San Andreas Fault System* (R. L. Kovach and A. Nur, eds), Standford University Publications in Geological Science No. 8, Stanford University Press, Stanford, pp. 275–285.

North, R. G. (1977*a*). Seismic moment, source dimensions and stress associated with earthquakes in the Mediterranean and Middle East. *Geophys. J. R. astronom. Soc.* **48**, 137–161.

North, R. G. (1977*b*). Station magnitude bias—its determination, causes and effects. Tech. Note 1977–24, Lincoln Laboratory, MIT, Lexington, Mass.

Nur, A. (1973). Role of pore fluids in faulting. *Phil. Trans. R. Soc. A* **274**, 297–304.

Nur, A. and Mavko, G. (1974). Postseismic viscoelastic rebound. *Science N.Y.* **183**, 204–206.

Okada, A. and Nagata, T. (1953). Land deformation of the neighborhood of Murato Point after the Naukaido great earthquake in 1946. *Bull. Earth Res. Inst.* **31**, 169–177.

Oliver, J. and Isacks, B. (1967). Deep earthquake zones, anomalous structures in the upper mantle and the lithosphere. *J. geophys. Res.* **72**, 4529–4275.

Orowan, E. (1960). Mechanism of seismic faulting. *Geol. Soc. Am. Mem.* **79**, 323–345.

Parsons, B. and Molnar, P. (1976). The origin of outer topographic rise associated with trenches. *Geophys. J. R. astronom. Soc.* **45**, 707–712.

Philpotts, A. R. (1964). Origin of pseudotachylites. *Am. J. Sci.* **263**, 1008–1035.

Pitman, W. C., III and Talwani, M. (1972). Sea-floor spreading in the North Atlantic. *Geol. Soc. Am. Bull.* **83**, 619–646.

Plafker, G. (1969). Tectonics of the March 1964 Alaska earthquake. *Geol. Survey prof. Paper 543-I.*

Plafker, G. (1972). Alaskan earthquake of 1964 and Chilean earthquake of 1960: Implications for arc tectonics. *J. geophys. Res.* **77**, 901–925.

Plafker, G. and Rubin, M. (1967). Vertical tectonic displacements in south-central Alaska during and prior to the great 1964 earthquake, *J. Geosci., Osaka City Univ.* **10**, 53–66.

Plafker, G. and Savage, J. C. (1970). Mechanism of the Chilean earthquakes of May 21 and 22, 1960. *Bull. geol. Soc. Am.* **81**, 1001–1030.

Prescott, W. H. and Lisowski, M. (1977). Deformation at Middleton Island, Alaska, during the decade after the Alaska earthquake of 1964. *Bull. seismol. Soc. Am.* **67**, 579–586.

Randall, M. J. (1966). Seismic radiation from a sudden phase transition. *J. geophys. Res.* **71**, 5297–5302.

Randall, M. J. (1971). Shear invariant and seismic moment for deep-focus earthquakes. *J. geophys. Res.* **76**, 4991–4992.

Reid, H. F. (1910). The mechanics of the earthquake. In *The California Earthquake of April 18, 1906*; *Report of the State Earthquake Investigation Committee*, Vol. 2, Carnegie Institute, Washington D.C.

Richter, C. F. (1958). *Elementary Seismology*, W. H. Freeman, San Francisco.

Richter, C. F. (1977). Letter of Acceptance for Medal of the Seismological Society of America 1976. *Bull seismol. Soc. Am.* **67**, 1244–1247.

Richter, F. M. and Parsons, B. (1975). On the interaction of two scales of convection in the mantle. *J. geophys. Res.* **80**, 2529–2541.

Savage, J. C. (1971). A theory of creep waves propagating along a transform fault. *J. geophys. Res.* **76**, 1954–1965.

Savage, J. C. (1972). Relation of corner frequency to fault dimensions. *J. geophys. Res.* **77**, 3788–3795.

Savage, J. C. and Burford, R. O. (1973). Geodetic determination of relative plate motion in central California. *J. geophys. Res.* **78**, 832–845.

Savage, J. C. and Wood, M. D. (1971). The relation between apparent stress and stress drop. *Bull. seismol. Soc. Am.* **61**, 1381–1388.

Scholz, C. H. (1972). Crustal movements in tectonic areas. In *Forerunners of Strong Earthquakes* (E. F. Savarensky and T. Rikitake, eds), *Tectonophysics* **14**, 201–217.

Scholz, C. H. (1977). A discussion of the state of shear stress on faults. In *Proceedings of the Second Conference on Experimental Studies of Rock Friction with Application to Earthquake Prediction*, U.S. Geological Survey, Menlo Park, Calif., pp. 289–289F.

Scholz, C. H. and Kato, T. (1978). The behavior of a convergent plate boundary: crustal deformation in the south Kanto district, Japan. *J. geophys. Res.* **83**, 783–797.

Scholz, C. H., Sykes, L. R. and Aggarwal, Y. P. (1973). Earthquake prediction: a physical basis. *Science, N.Y.* **181**, 803–810.

Scholz, C. H., Wyss, M., and Smith S. W. (1969). Seismic and aseismic slip on the San Andreas fault. *J. geophys. Res.* **74**, 2049–2069.

Sclater, J. G. and Francheteau, J. (1970). The implications of terrestrial heat flow observations on current tectonic and geochemical models of the crust and upper mantle of the earth. *Geophys. J. R. astronom. Soc.* **20**, 509–542.

Scott, J. S. and Drever, H. I. (1954). Frictional fusion along a Himalayan thrust. *Proc. R. Soc. Edinburgh* **65b**, 121–142.

Shimazaki, K. (1974*a*). Pre-seismic crustal deformation caused by an underthrusting oceanic plate, in Eastern Hakkaido Japan. *Phys. Earth planet. Interiors* **8**, 148–157.

Shimazaki, K. (1974*b*). Nemuro-Oki earthquake of June 17, 1973: a lithospheric rebound at the uppet half of the interface. *Phys. Earth planet. Interiors* **9**, 314–327.

Shimazaki, K. (1976). Intra-plate seismicity and inter-plate earthquakes: Historical activity in southwest Japan. *Tectonophysics* **33**, 33–42.

Shimazaki, K. (1978). Correlation between intraplate seismicity and interplate earthquakes in Tohoku, northeast Japan. *Bull. seismol. Soc. Am.* **68**, 181–192.

Sibson, R. H. (1975). Generation of pseudo tachylyte by ancient seismic faulting. *Geophys. Jl R. astonom. Soc.* **43**, 775–794.

Solomon, S. C., Sleep, N. H. and Jurdy, D. M. (1977). Mechanical models for absolute plate motions in the Early Tertiary. *J. geophys. Res.* **82**, 203–212.

Starr, A. T. (1928). Slip in a crystal and rupture in a solid due to shear. *Proc. Cambridge phil. Soc.* **24**, 489–500.

Stauder, W. S. J. (1968). Tensional character of earthquake foci beneath the Aleutian Trench with relation to sea-floor spreading. *J. geophys. Res.* **73**, 7693–7702.

Stauder, W. S. J. (1973). Mechanism and spatial distribution of Chilean earthquakes with relation to subduction of the Oceanic plate. *J. geophys. Res.* **78**, 5033–5061.

Stauder, W. S. J. (1975). Subduction of the Nazca plate under Peru as evidenced by focal mechanisms and by seismicity. *J. geophys. Res.* **80**, 1053–1064.

Stewart, G. S. (1978). Implications for plate tectonics of the Aug 19, 1977 Indonesian decoupling normal-fault earthquake (abstract). *Trans. Am. geophys. Union* **59**, 326.

Sykes, L. R. (1971). Aftershock zones of great earthquakes, seismicity gaps, and earthquake predictions for Alaska and the Aleutians. *J. geophys. Res.* **76**, 8021–8041.

Sykes, L. R. and Sbar, M. L. (1973). Intraplate earthquakes, Lithospheric stress and the driving mechanism of plate tectonics, *Nature, Lond.* **245**, 298–302.

Thatcher, W. and Rundle, J. B. (1978). A model for the earthquake cycle in underthrust zones. In *Seismic Gaps Symposium Proceedings*, U.S. Geological Survey, Menlo Park, Calif.

Tocher, D. (1960). Creep rate and related measurements of Vineyard, California. *Bull. seismol. Soc. Am.* **50**, 396–404.

Tucker, B. E. and Brune, J. N. (1977). Source mechanism and m_b–M_s analysis of aftershocks of the San Fernando earthquake. *Geophys. J. R. astronom. Soc.* **49**, 371–426.

Utsu, T. (1967). Anomalies in seismic wave velocity and attenuation with a deep earthquake zone 1. *J. Fac. Sci. Hokkaido Univ., Japan, Ser. 7* **3**, 1.

Utsu, T. (1974). Correlation between great earthquakes along the Nankai trough and destructive earthquakes in western Japan (in Japanese). *Rep. Coord. Comm. Earthquake Prediction, Geographical Surv. Inst.* **12**, 120–122.

Uyeda, S. (1977). Some basic problems in the trench-arc-back arc system. In *Island Arcs, Deep Sea Trenches and Back-Arc Basings* (M. Talwani and W. C. Pitman III, eds), Maurice Ewing Series No. 1, American Geophysical Union, Washington D.C., pp. 1–14.

Veith, K. F. (1974). The relationship of island arc seismicity to plate tectonics, Ph.D. theses, Southern Methodist University.

Watanabe, A. (1959). Geomorphology in relation to recent crustal movements in Japan. In *Proceedings of the I.G.U. Regional Conference in Japan, 1957*, pp. 215–221.

Watts, A. B. and Talwani, M. (1974). Gravity anomalies seaward of deep-sea trenches and their tectonic implications. *Geophys. J. R. astonom. Soc.* **36**, 57–90.

Weissel, J. K. and Hayes, D. E. (1974). The Australian–Antarctic discordance: new results and implications. *J. geophys. Res.* **79**, 2579–2587.

Wyss, M. and Brune, J. N. (1968). Seismic moment, stress, the source dimensions for earthquakes in the California–Nevada region. *J. geophys. Res.* **73**, 4681–4694.

Yoshikawa, T. (1970). On the relations between Quaternary tectonic movement and seismic crustal deformation in Japan. *Bull. Dept Geogr., Univ. Tokyo* **2**, 1–24.

16. The Present Status of the Geomagnetic Polarity Time Scale

I. McDOUGALL

Abstract

The geomagnetic polarity time scale for the last 4 million years was developed by combining palaeomagnetic polarity and K–Ar age data from subaerial volcanic rocks. Recently, changes have been recommended for the ^{40}K decay constants to more accurate and precise values, such that K–Ar ages on which the polarity time scale is based need to be increased by 2.67 per cent, compared with the previously reported ages. A revised time scale, using the newly recommended decay constants, and taking into account many new data, provides the following estimates for the age of epoch boundaries: Brunhes–Matuyama, 0.72 million years; Matuyama–Gauss, 2.47 million years; Gauss–Gilbert, 3.41 million years and Gilbert–epoch 5, 5.44 million years.

The polarity time scale for the last 3.3 million years was correlated with the pattern of linear marine magnetic anomalies, and by extrapolation yielded a predicted polarity time scale extending to about 80 million years ago. The changes in the ^{40}K decay constants require that the age of individual polarity inversions be increased; the increases range from 2.66 per cent at 5 million years to 2.52 per cent at 80 million years. Obviously, it is desirable to test the validity of the large extrapolations necessary to obtain this predicted polarity time scale, but lack of resolution in the K–Ar dating means that this is difficult to do for times earlier than 5 million years ago by direct measurement of individual volcanic rocks that are not in known stratigraphic relation to one another. However, detailed palaeomagnetic and K–Ar age measurements on thick sequences of basaltic lavas in Iceland demonstrate that direct testing of the extrapolations for the last 13 million years is possible. Results presently available show that at an age of 10 million years the predicted polarity time scale from the marine magnetic anomaly data is correct to within a few per cent.

1. Introduction

The geomagnetic polarity time scale has assumed considerable significance in many aspects of the earth sciences, ranging from its application to analyses of marine magnetic anomalies in terms of the seafloor spreading hypothesis, to providing a time scale for a wide variety of stratigraphic studies. The polarity time scale arose out of investigations designed to answer the question as to whether the occurrence of reversed remanent magnetization in rocks was because the geomagnetic field had opposite polarity to the present in earlier geological times, or because the rocks themselves had self-reversing properties such that they became magnetized antiparallel to the prevailing magnetic field.

The aim of the present contribution is to review the development of the polarity time scale, to assess the present state of knowledge as to the age of the boundaries between polarity intervals, to present a revised time scale, and to examine the extension of the polarity time scale to earlier geological periods.

2. History of Development of the Polarity Time Scale

Studies of magnetic directions recorded in volcanic rocks began in the nineteenth century, but the first published data showing that some lava flows and associated baked clays had directions of magnetization approximately antiparallel to the present magnetic field was given by Brunhes (1906), leading to speculation that the geomagnetic field had changed its polarity in the past. Mercanton (1926) recognized that this phenomenon should be registered in rocks world-wide, so that he obtained samples from Spitsbergen, Greenland, Iceland and Australia, and indeed found that some rocks had directions roughly opposite to the present field, whereas others had directions consistent with the present geomagnetic field. From stratigraphic studies in Japan and Korea, Matuyama (1929) demonstrated that early Quaternary lavas had reversed polarity whereas younger lavas had normal magnetic polarity. Similar observations were recorded by Roche (1951, 1956) from lavas in France, and by Opdyke and Runcorn (1956) on rocks from the United States, providing strong evidence for geomagnetic field reversal. Nagata *et al.* (1952), however, showed from laboratory studies that the Haruna dacite in Japan became magnetized in a direction opposite to the applied field on cooling through the Curie temperature, thus raising the question as to whether many of the rocks exhibiting reversed polarity also might have self-reversing properties.

If self-reversal was the correct explanation for reversed polarity observed in some rocks, then it would not be expected to be a time-dependent property, but rather would be related to some compositional or mineralogical factor. In contrast, as noted by Mercanton, if the reversed magnetic polarity was due to those rocks possessing it cooling in a reversed geomagnetic field, then the phenomenon should be found on a global scale, and all rocks of the same age should have the same polarity. It follows that rocks showing normal and reversed polarity should form a time sequence if the field reversal hypothesis is correct. Thus, to distinguish between the two hypotheses, it was necessary to determine the precise age of volcanic rocks whose magnetic polarities also had been measured. By the early 1960's, developments in the potassium–argon (K–Ar) isotopic dating method enabled quite young rocks to be dated with some precision, opening up the way for this problem to be tackled directly.

The first example of the combined use of magnetic polarity and K–Ar data was that of Rutten (1959), who concluded from studies on lavas from volcanoes near Rome that the present normal magnetic interval had extended from at least 0.47 million years ago, and that an earlier interval of normal polarity possibly existed about 2.4 million years ago. Subsequently, and especially from 1963 to 1966, a large amount of data on magnetic polarity and K–Ar ages of the same or closely related lavas was published, leading to the development of a polarity time scale for the last 4 million years and effectively proving that the field reversal hypothesis was correct.

In June 1963, Cox *et al.* (1963*a*) provided polarity and age data on nine samples, including three data points used by Rutten (1959), covering an age range from 0.47 to 3.2 million years ago. Cox *et al.* interpreted the results as indicating a relation between magnetic polarity and time, and suggested that a periodicity of magnetic polarity intervals of about 1 million years was discernible (Fig. 1). This was followed by virtually simultaneous publication in October, 1963 of papers by Cox *et al.* (1963*b*), who added ten new data points from lavas in the western United States, and McDougall and Tarling (1963), who provided more than 20 data points from polarity and K–Ar dating studies on lavas from Hawaii. The two time scales interpreted from these data (Fig. 1) were inconsistent with one another in one important respect: Cox *et al.* (1963*b*) suggested that the change from the penultimate normal to the last reversed interval occurred at about 1.85 million years ago, whereas McDougall and Tarling placed this boundary at 2.5 million years ago on the basis of reversed polarity for five lava flows with measured ages lying between 2.2 and 2.5 million years. Cox *et al.* (1964*a*) corrected two errors in their data published in October 1963, resulting in the two time scales coming into close agreement, as noted by McDougall and Tarling (1964), except for two data points. The results based upon some 40 age and polarity measurements covering the last 3.5 million years of time indicated normal polarity from the present to about 1 million years ago, a reversed interval from 1 to 2.5 million years ago, preceded by a normal polarity interval to at least $t = 3.2$ million years (Fig. 1). The two results that did not fit the simple pattern were a normal polarity lava dated at about 1.9 million

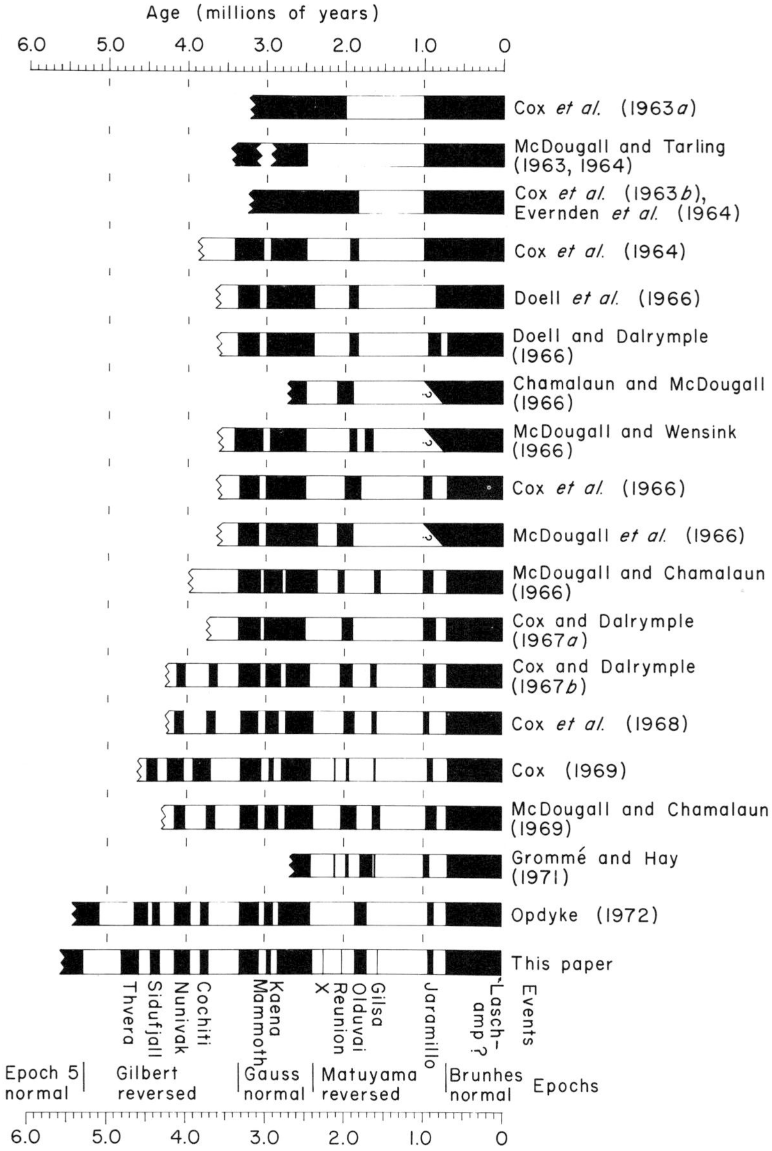

Fig. 1. Evolution of the polarity time scale from 1963 to the present. Black represents normal polarity, white represents reversed polarity.

years from Olduvai Gorge in East Africa (Grommé and Hay, 1963), and a reversely magnetized lava of age 2.95 million years from Hawaii (McDougall and Tarling, 1963), the latter authors suggesting from these data that the behaviour of the geomagnetic field may have been more complex than had been recognized up to this time.

Evernden *et al.* (1964) listed some of the available K–Ar and polarity data, and while no polarity log was published it is possible to deduce a time scale, which is virtually identical to that of Cox *et al.* (1963*b*).

Cox *et al.* (1964*b*) summarized all available results and included several unpublished results in their compilation. They emphasized that during the last 3.5 million years of geological time, changes in the polarity of the geomagnetic field had taken place at irregular intervals, and that within intervals of predominantly one polarity shorter intervals of opposite polarity, about 0.1 million years in duration, were recognizable (Fig. 1). The longer intervals were termed polarity epochs and the shorter intervals were called events. The epochs were named after pioneering workers in the field of geomagnetism (Brunhes, Matuyama and Gauss), and location names such as Olduvai and Mammoth were used to identify events.†

During 1966 a host of papers incorporating new data and providing compilations of previous results were published by Doell and Dalrymple (1966), Doell *et al.* (1966), Chamalaun and McDougall (1966), McDougall and Wensink (1966), Cox *et al.* (1966), McDougall *et al.* (1966) and McDougall and Chamalaun (1966). These results served to confirm the general outline of the polarity time scale, provided evidence for additional short events (Jaramillo, Gilsa and Kaena), and lead to refinement of the age of the boundaries between polarity intervals, as summarized in Fig. 1. By this time the evidence from the numerous magnetic polarity and K–Ar age measurements on lavas from diverse localities in both hemispheres was now so compelling in terms of its consistency that the field reversal hypothesis was regarded as proven. Likewise, it was evident that self-reversal was an unimportant factor in terrestrial rocks. Thus, in proving the field reversal hypothesis, a geomagnetic polarity time scale had been developed, and its potential applicability to a wide range of other geological studies was only just beginning to be recognized.

† The IUGS International Subcommission on Stratigraphic Classification recently prepared a series of draft proposals concerning magnetostratigraphic nomenclature (Hedberg, 1977). In these proposals it is recommended that the term polarity zone be used for magnetostratigraphic polarity units, the term chron be used for the geochronological equivalent, and the term chronozone be used for the chronostratigraphic equivalent. Under this scheme the terms chron and subchron are likely to replace the commonly used terms epoch and event for the divisions of the polarity time scale. As these changes have yet to be implemented the currently accepted terminology will be adhered to in the present paper.

In this respect two developments of major significance occurred during 1965 and 1966. First, Vine and Wilson (1965) attempted to correlate linear marine magnetic anomaly patterns with the polarity time scale on the basis of the hypothesis of Vine and Matthews (1963) and the seafloor spreading model of Hess (1962). While a good fit was not found at that time, refinements of the polarity time scale during 1966 enabled Vine (1966) to obtain an excellent correlation between the reversal history, as recorded in subaerial volcanic rocks, and the marine magnetic anomaly pattern observed across the Juan de Fuca Ridge, the Reykjanes Ridge and the East Pacific Rise. This heralded the remarkable developments of the following few years, leading to the concept of plate tectonics. Heirtzler *et al.* (1968) deduced a polarity time scale extending to about 79 million years ago by determining the spreading rate at the axial zone of the South Atlantic using the polarity time scale determined on land, and then extrapolating this rate linearly to estimate the age of the magnetic anomalies further removed from the axial zone.

The second important development at this time was the successful and accurate matching by Opdyke *et al.* (1966) and Ninkovich *et al.* (1966) of the reversal history recorded on land with magnetic polarity measurements on sediments in deep-sea sedimentary cores. Harrison and Funnell (1964) were the first to recognize that reversals were preserved in marine sediments, and Harrison (1966) made the first attempt to correlate these reversals with the polarity time scale.

The success of these correlations not only opened up vast new areas of investigation, but also served to convince those who may have still doubted that the field reversal hypothesis was correct.

3. Present Status of the Polarity Time Scale

Since 1966 there have been numerous studies published concerned with refinement and better definition of the polarity time scale for the last 4 million years of geological time. These studies have used the conventional approach of K–Ar dating and polarity measurement on volcanic rocks; in addition, data from deep-sea cores and from the interpretation of marine magnetic anomalies have been utilized. Cox and Dalrymple (1967*a*) were the first to employ statistical methods of analysis for estimating more objectively the age of polarity interval boundaries. Using this approach various revisions have been proposed by Cox (1969) and Dalrymple (1972), the latter providing a comprehensive listing of K–Ar and polarity results considered to be reliable. Watkins (1972) also summarized the history and development of the polarity time scale. In this section an appraisal of the present status of the polarity time scale will be given.

3.1. Epoch Boundaries

The definition as to which polarity change is regarded as marking an epoch boundary is to some extent arbitrary, but virtually all workers have utilized the boundaries suggested during the development of the polarity time scale. Beyond about 4 million years ago, polarity epochs have been defined by researchers dealing with the reversal chronology as recorded in deep-sea sediments.

The age of the Brunhes–Matuyama epoch boundary, the first authenticated field reversal in the geological past, appears to be quite well controlled at 0.69–0.70 million years (Cox and Dalrymple, 1967*a*; Cox, 1969; Dalrymple, 1972). This boundary has become a very important datum for the interpolation of ages in deep-sea sedimentary cores, particularly in relation to the detailed interpretation of palaeoclimates from oxygen isotope data (Shackleton and Opdyke, 1973). Thus it would be of considerable value to confirm the age of this boundary, and perhaps to determine it with greater precision. Such a study should preferably be done on a continuous sequence of fresh terrestrial lavas in which the polarity transition is recorded.

The lack of sufficient density of reliable polarity and K–Ar data points adjacent to the boundary between the Matuyama and Gauss epochs lead Dalrymple (1972, p. 117) to state that the age of this boundary "should be considered uncertain within the interval 2.2 to 2.5 million years". Cox (1969) previously had found that the uncertainty in the age of this boundary was so great that he resorted to deriving its estimated age of 2.43 million years by using marine magnetic anomaly data to interpolate from the relatively well-known ages of the Brunhes–Matuyama and the Gauss–Gilbert epoch boundaries. Resolution of this problem by direct K–Ar dating and polarity measurements on lavas was achieved by McDougall and Aziz-ur-Rahman (1972), who found a sequence of normal polarity lavas overlain by flows of reversed polarity on Norfolk Island in the southwest Pacific. Using the statistical approach of Cox and Dalrymple (1967*a*), an age of 2.41 ± 0.01 million years for the Matuyama–Gauss boundary was obtained, although an uncertainty in the order of 0.05 million years was considered to be more realistic in terms of the geology.

The age of the boundary between the Gauss and Gilbert epochs appears to be well controlled at 3.32 million years (Cox, 1969; Dalrymple, 1972). There is considerable variation in the estimates of the age of the boundary between the Gilbert epoch and epoch 5, which has been defined on the basis of polarity history observed in deep-sea cores (Foster and Opdyke, 1970). Opdyke (1972) suggested its age was about 5.1 million years from deep-sea core data, and Talwani *et al.* (1971) and Klitgord *et al.* (1975) gave ages of

5.18 and 5.25 million years for this boundary from analyses of marine magnetic anomaly data. Recently McDougall *et al.* (1976*a*, 1977) determined the age of the Gilbert-epoch 5 boundary at 5.34 million years, from direct K–Ar dating and polarity studies on two separate sequences of basalts in Iceland. It would appear that an age of 5.3 ± 0.1 million years is the best estimate for this boundary at the present time.

3.2. Age and Duration of Events

Bonhommet and Babkine (1967) reported essentially reversed directions of remanent magnetization on two lava flows at Laschamp and Olby in the Chaîne des Puys, France. Bonhommet (1970) and Bonhommet and Zahringer (1969) subsequently used these results to postulate the existence of a short interval of reversed polarity within the Brunhes epoch, called the Laschamp event, which was estimated to have an age within the range 8700–20 000 years. On the basis of new radiocarbon, thermoluminescence and K–Ar data on the lavas and associated materials given by Valladas *et al.* (1977) and Gillot and Cassignol (1977), Labeyrie and Gillot (1977) concluded that the age of the Laschamp event lies between 29 000 and 36 000 years. While it is still not clear whether the reversed polarity flows from the Chaîne des Puys reflect a global reversal of the geomagnetic field, it is possible that the Lake Mungo geomagnetic excursion, observed in aboriginal hearths at Lake Mungo, Australia (Barbetti and McElhinny, 1972, 1976), and dated at about 30 000 years, may be correlative with the Laschamp event.

In recent years a number of examples of intermediate to reversed directions of magnetization have been reported from sediments deposited during the Brunhes normal polarity epoch; see Verosub and Banerjee (1977) for a comprehensive summary. Many of these geomagnetic excursions appear to be of very short duration, and fully reversed directions are rather uncommon. In this context it is appropriate to mention the Blake event, dated at approximately 100 000 years, and discovered by Smith and Foster (1969) in deep-sea sediments from cores in the North Atlantic and Caribbean; additional information was given by Denham (1976) and Denham *et al.* (1977). These data indicate that at least locally the magnetic field appears to have been reversed. This event may also be recorded in sediments from a core in Lake Biwa, Japan (Yaskawa *et al.*, 1973; Kawaii *et al.*, 1972). As emphasized by Denham *et al.* (1977) and Verosub and Banerjee (1977) there is still a measure of doubt as to whether the Blake event and other geomagnetic excursions observed in sediments deposited during the younger part of the Brunhes epoch reflect behaviour of the geomagnetic field on a global scale. Many authors have pointed out the considerable difficulties associated with the interpretation of palaeomagnetic results from sediments because of problems caused by bioturbation, deformation and disturbance and secondary magnetization. It is becoming increasingly recognized that before a geomagnetic excursion recorded in sediments can be regarded as a global occurrence, consistent palaeomagnetic results are required from multiple sections of different lithology and sedimentation rates together with good age control. In summary, it would appear that none of the geomagnetic excursions or reversals postulated from the sedimentary record during Brunhes time are sufficiently well-documented to be included within the polarity time scale at the present time.

The recognition of the Jaramillo normal polarity event by Doell and Dalrymple (1966) resulted in the redefinition of the Brunhes–Matuyama boundary. Cox and Dalrymple (1967*a*) suggested that this event extended from 0.85 to 0.97 million years ago, and Opdyke (1972) estimated that it occurred during the period 0.87–0.92 million years ago, from the deep-sea core results.

Considerable uncertainty still exists as to the number, age and duration of normal polarity intervals (events) in the early part of the Matuyama reversed polarity epoch; the problem was reviewed by Watkins (1972), Dalrymple (1972), Opdyke (1972) and again briefly by Watkins *et al.* (1975). In summary it is clear from detailed K–Ar and polarity measurements on a sequence of lavas in Réunion that the Réunion normal polarity event has a mean age of 2.02 ± 0.02 million years with a duration of less than 50 000 years (McDougall and Watkins, 1973). Controversy still exists as to whether the Olduvai and Gilsá normal events are separate entities or one and the same event in the age range 1.5–1.9 million years ago. On the basis of new data, Watkins *et al.* (1975) suggested that the best estimate for the mean minimum age of the Gilsá event in its type locality in Iceland (McDougall and Wensink, 1966) is 1.58 ± 0.08 million years, but they were unable to provide an estimate of its duration. Grommé and Hay (1971) examined the age and polarity data from the type locality of the Olduvai event in Tanzania and were only able to state that "the span of time represented by these rocks might be as great as 1.6–1.9 m.y. or as little as 1.70–1.75 m.y.".

Polarity data from deep-sea sediment cores that penetrate through the early Matuyama are somewhat equivocal (Watkins, 1972; Opdyke, 1972). In some cores two normal events are identified; the younger normal polarity interval seems to be of shorter duration than the older, and may represent the Gilsá event. In the majority of cores, however, only a single normal polarity interval is identified, and there has been much discussion as to whether this represents the Olduvai or Gilsá events (see Watkins, 1972). Opdyke (1972) correlates it with the Olduvai event and shows that there is much variation in its estimated age derived from linear extrapolation of sedimentation rates between the Brunhes–Matuyama and Matuyama–Gauss boundaries in the cores. This is illustrated by the large uncertainties attached to his estimates of 1.71 ± 0.26 and 1.86 ± 0.19 million years for the ages of the younger and older boundaries, respectively, of this normal event, also yielding a mean duration of 0.15 ± 0.05 million years. A reassessment of all available K–Ar data from the type sequence at Olduvai Gorge was given by Curtis and Hay (1972). There is considerable scatter in the results, but Brock and Hay (1976) showed that nine samples from Tuff 1B, which is stratigraphically at about the middle of the normal polarity interval at Olduvai Gorge, have concordant K–Ar ages yielding a mean of 1.79 ± 0.03 million years. Thus the latest information from Olduvai itself is in good agreement with the deductions made by Opdyke (1972) from deep-sea cores.

As emphasized by Watkins *et al.* (1975), resolution of these difficulties concerning the Gilsá and Olduvai events is only likely to be possible by detailed polarity and dating studies on thick sequences of fresh subaerial lavas which were erupted during the early part of the Matuyama epoch. Nevertheless, it is suggested that the Gilsá event may be a separate normal polarity interval of short duration with a mean age of 1.58 ± 0.08 million years; and that the Olduvai event has duration and age limits as estimated by Opdyke (1972). The Réunion event seems well established at an age of 2.02 ± 0.02 million years. There is some evidence from marine magnetic anomaly data of another short normal polarity event (X), recognized by Heirtzler *et al.* (1968), and dated indirectly at about 2.17 ± 0.1 million years by Emilia and Heinrichs (1972). This normal polarity interval has yet to be positively identified in subaerial volcanic sequences.

In the Gauss normal polarity epoch two reversed polarity intervals (events) are recognized: the Mammoth event, defined and dated at 3.06 ± 0.10 million years by Cox *et al.* (1964*b*), but probably first recognized by McDougall and Tarling (1963); and the Kaena event, defined and dated at 2.80 ± 0.06 million years by McDougall and Chamalaun (1966). McDougall and Aziz-ur-Rahman (1972) summarized all available data, including results from deep-sea cores, and concluded that the Kaena event extended from 2.92 to 2.84 million years ago, and reported a more precise age of 2.87 ± 0.03 million years for the lava from the type locality in Oahu, Hawaii. They estimated that the Mammoth reversed interval occurred between 2.99 and 3.09 million years ago. Opdyke (1972) deduced very similar ages for these intervals.

Two normal polarity intervals were identified by Cox and Dalrymple (1967*b*) within the Gilbert reversed epoch; these were named the Cochiti event, dated at 3.7 million years, and the Nunivak event, dated at 4.1 million years. These events were regarded as correlatives of events "a" and "b" recognized in deep-sea cores by Hays and Opdyke (1967). Two older events also were found in marine sediments and called C_1 and C_2 by Foster and Opdyke (1970) with mean ages of about 4.37 and 4.56 million years, respectively (Opdyke, 1972). Both of these older intervals of normal polarity in the Gilbert epoch have subsequently been identified in a continuous sequence of lava flows in western Iceland by McDougall *et al.* (1977), and named the Sidufjall and Thvera events. The ages of the boundaries of the Sidufjall event are estimated to be at 4.33 and 4.45 million years, and the Thvera normal interval is estimated to extend from 4.60 to 4.80 million years ago. The uncertainty in the ages of these event boundaries may be as large as 0.1 to 0.15 million years.

3.3. A Revised Geomagnetic Polarity Time Scale

The preceding discussion permits the construction of a revised polarity time scale for the last 5.3 million years, shown in Fig. 2. The time scale is based almost entirely on primary K–Ar and magnetic polarity data from subaerial lava flows, but estimates of duration and age of the polarity events, and in particular the Olduvai event, are taken from studies of polarity in deep-sea cores as explained previously. Except for identification of the X normal polarity event in the early Matuyama, this time scale is quite independent of marine magnetic anomaly data.

The time scale is based upon K–Ar data calculated using the decay constants for ^{40}K proposed by Aldrich and Wetherill (1958). Beckinsale and Gale (1969) examined all available data from physical counting experiments on the decay of ^{40}K, and reported a revised set of decay constants which are of much higher precision and accuracy than those previously in use. Most workers have preferred to continue to use the decay constants suggested by Aldrich and Wetherill (1958) in order to prevent possible confusion in the literature, and because the revised decay constants produce a change of only about 2.6 per cent in the ages, which is within the precision quoted for many of the K–Ar ages. Recently the Subcommission on Geochronology of the International Commission on Stratigraphy (IUGS) proposed that in future all ages should be reported using a recommended set of decay constants, not only for ^{40}K but for other decay schemes as well (Steiger and Jäger, 1977). The recommended ^{40}K decay constants are based upon the summary of Beckinsale and Gale (1969) adjusted for the recently determined abundance of ^{40}K reported by Garner *et al.* (1975). The relevant values are: $\lambda_e + \lambda'_e = 0.581\ (\pm 0.004) \times 10^{-10}\ y^{-1}$; $\lambda_\beta = 4.962\ (\pm 0.009) \times 10^{-10}\ y^{-1}$; $^{40}K/K = 1.167\ (\pm 0.004) \times 10^{-4}$. Use of these values causes an increase in the age of all the boundaries for the last 5 million years of 2.67 per cent, and a revised time scale incorporating these changes is given on the right-hand side of Fig. 2. While the uncertainty associated with the age of many of the

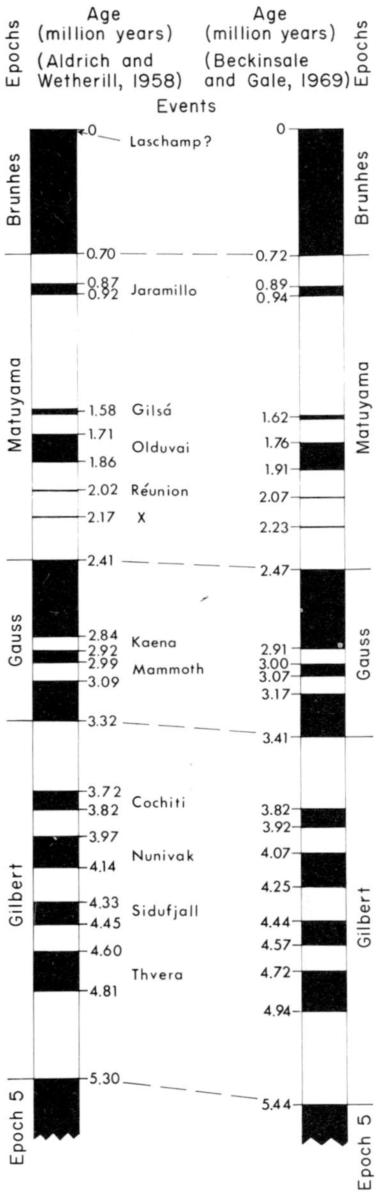

Fig. 2. Revised geomagnetic polarity time scales based mainly upon K–Ar and palaeomagnetic data on igneous rocks (see text). Scale on left-hand side derived using K–Ar ages calculated using ^{40}K decay constants proposed by Aldrich and Wetherill (1958), whereas time scale on right-hand side is adjusted to the recently recommended decay constants for ^{40}K (Steiger and Jäger, 1977) which are those suggested by Beckinsale and Gale (1969) modified to account for the revised abundance of ^{40}K. Black represents normal polarity, white represents reversed polarity.

boundaries between polarity intervals is larger than the change in age brought about by using the recommended decay constants for ^{40}K, it is obviously desirable to affect the change to minimize systematic errors. The effects, of course, are far reaching, in that, if the recommendations are followed, rates of seafloor spreading will have to be adjusted, and the polarity time scale derived by extrapolation from the known scale using marine magnetic anomaly data will likewise have to be revised. Because two partial decay constants are relevant to the calculation of K–Ar ages, the effect is not simply a matter of applying a fixed percentage correction. For example, an age of 80.0 million years using the earlier decay constants will be increased to 82.0 million years, i.e. by 2.52 per cent, when recalculated. Thus, while the changes are not great, they are nevertheless significant. The magnetic polarity time scale for the Late Cretaceous and Cainozoic recently proposed by La Brecque *et al.* (1977) can be recalculated readily in terms of the revised decay constants. The correction to be applied varies linearly, to a close approximation, from 2.66 per cent at 5.0 million years to 2.48 per cent at 100 million years.

4. Extension of the Polarity Time Scale

Using the then known polarity time scale for the past 3.5 million years, Heirtzler *et al.* (1968) produced a predicted time scale extending to about 79 million years ago, based upon analyses of linear marine magnetic anomalies. Such a time scale involves an extrapolation in the order of 20 times the length of the baseline, and, as emphasized by Heirtzler *et al.* (1968), the possibility of progressively larger systematic error with increasing age is very real. Thus it is of considerable importance to attempt to test the validity of this extrapolated time scale. To this end some studies have been made using palaeomagnetic and micropalaeontological data from deep-sea cores (Opdyke *et al.*, 1974; Theyer and Hammond, 1974), but there are many difficulties associated with this type of approach, including the problem of obtaining precise ages for the polarity intervals.

Likewise, the prospect of extending the polarity time scale to times earlier than 5 million years ago by K–Ar dating and polarity measurements on stratigraphically unrelated samples is very poor indeed. This is because of lack of resolution in the K–Ar dating, a point emphasized by Cox and Dalrymple (1967*a*). The analytical uncertainty associated with a K–Ar age generally is a percentage of the age. For example, taking a typical value of ±2 per cent for the precision of a K–Ar age, at 5 million years this is equivalent to ±0.1 million years, which is well within the range of durations of many of the shorter polarity intervals. It is clear, therefore, that attempts to construct a polarity time scale beyond an age of about 5 million years using the same methods as previously are likely to fail. This was shown to be the case by Dalrymple *et al.* (1967); they presented data on 39 samples with measured ages lying between 3.5 and 16 million years, but they were unable to distinguish any epochs or events, although three changes in polarity were identified. Dalrymple (1972, p. 117) regarded the prospects of extending the polarity time scale beyond 4.5 million years ago by these methods as "so dismal that any effort in this direction is probably futile".

The only approach that holds any prospect of success is to study thick sequences of volcanics, the detailed stratigraphy of which, together with magnetic polarity measurements, may provide the additional control required in the K–Ar dating. One of the few places on Earth, and perhaps the only place, where studies of this kind might yield useful results is Iceland. Here there has been virtually continuous eruption of basaltic lavas from about 13 million years ago to the present (Moorbath *et al.*, 1968), as a consequence of the growth of Iceland by crustal spreading astride the Mid-Atlantic Ridge (Bodvarsson and Walker, 1964). Recent cooperative studies with N. D. Watkins, L. Kristjansson and K. Saemundsson on some thick lava sequences in Iceland in fact have met with considerable success (McDougall *et al.*, 1976*a*,*b*, 1977), and two examples will be discussed briefly to illustrate this approach.

In the first example, a 3400 m thick succession of basaltic lavas in the Borgarfjördur region in western Iceland was studied following detailed stratigraphic mapping by Johannesson (1972) and Saemundsson and Noll (1974). The lava sequence was sampled systematically for magnetic polarity and K–Ar dating measurements and the results presented by McDougall *et al.* (1977). Figure 3 summarizes the results of the magnetic polarity and K–Ar dating, and it will be seen that the sequence was erupted from about 2 to 7 million years ago, during which time a considerable number of polarity changes are recorded. The succession extends from the early Matuyama through the Gauss, Gilbert epochs, epochs 5 and 6 and into epoch 7. The K–Ar age results are plotted against the stratigraphic height above the base of the sequence in Fig. 4, from which the general consistency of the ages can be verified, with only a few exceptions. Icelandic lavas are rather difficult to date by the K–Ar method because of their low potassium content, and particularly because many of the lavas show evidence of hydrothermal alteration brought about by burial metamorphism. No doubt some scatter in the K–Ar results is caused by these factors; this scatter effectively precludes the use of individual ages for precise control as to the age of the boundaries between polarity intervals. However, a remarkable and unexpected feature of these results is that the data conform to a linear regression model. A best fit straight line through the results, excluding the aberrant datum NP209, permits the derivation of the ages for the polarity interval boundaries as shown on the polarity log. The match of this polarity time scale, derived directly from K–Ar dating and polarity measurements, with those determined from extrapolations using magnetic anomaly data is excellent (Fig. 5), particularly the most recent versions of Talwani *et al.* (1971) and Klitgord *et al.* (1975). It should be noted that the ages given in Fig. 4 are based upon the Aldrich and Wetherill (1958) decay constants for ^{40}K, and need to be increased by an average of 2.66 per cent to conform with the recently recommended decay constants.

The second example relates to part of a 9 km thick lava sequence, originally mapped by Walker (1959), in eastern Iceland. Palaeomagnetic

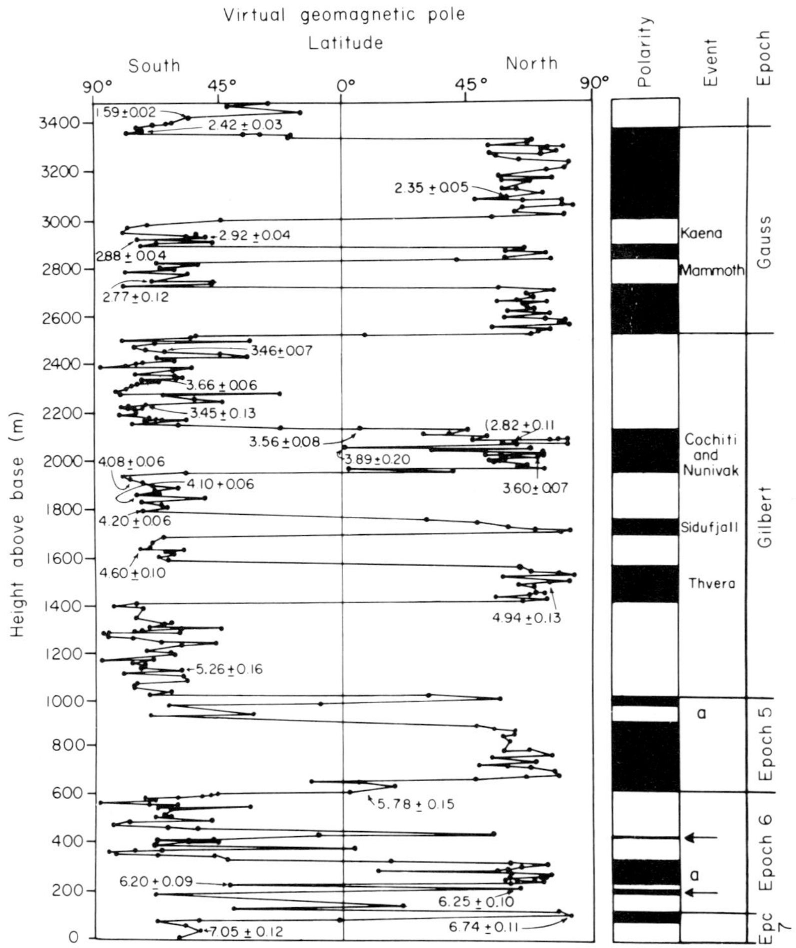

Fig. 3. Palaeomagnetic and K–Ar age data (in millions of years) from the lava sequence in Borgarfjördur region, western Iceland. Plot of latitude of virtual geomagnetic pole for each lava sampled against stratigraphic height above base of sequence. Interpretation of data in terms of polarity given on right, where black indicates normal polarity, white indicates reversed polarity. (After McDougall *et al.*, 1977.)

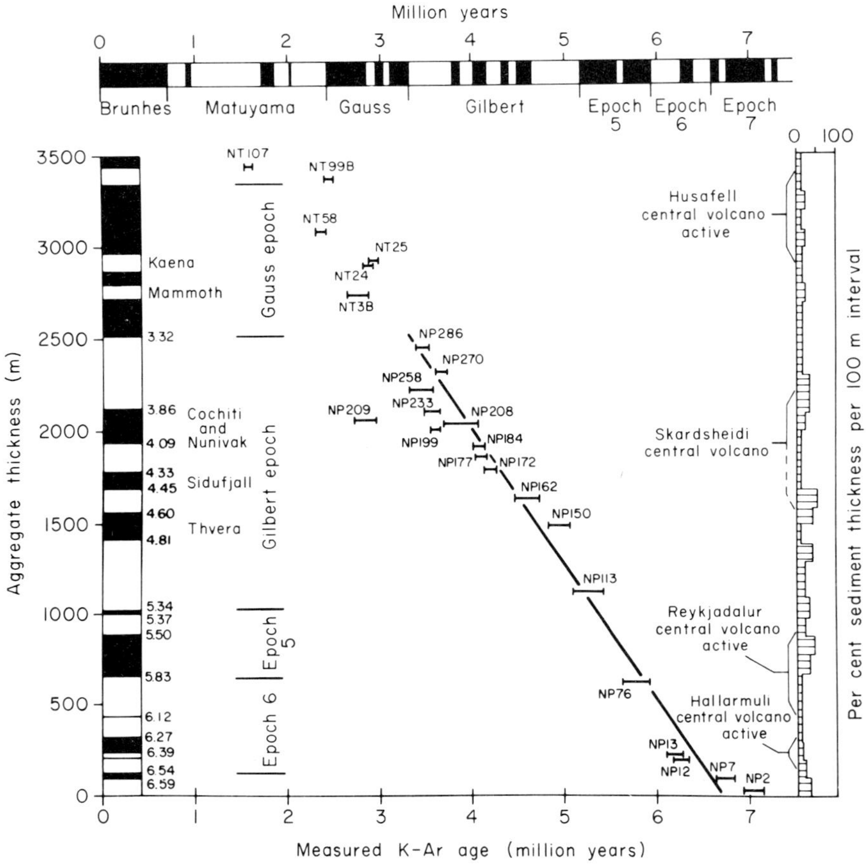

Fig. 4. The K–Ar ages from lavas of the Borgarfjördur area, western Iceland, are plotted against stratigraphic height above the base of the sequence. Polarity log (see Fig. 3) reproduced on left. Precision limits for K–Ar ages indicated by bars. Least squares fit regression line through data (excluding NP209) with Gilbert–Gauss epoch boundary taken as 3.32 million years. Ages on polarity log derived from regression. Polarity time scale after Talwani *et al.* (1971) given at top of diagram. See McDougall *et al.* (1977), from which diagram taken, for full details.

data were obtained on the whole of this thick succession under the direction of R. L. Wilson of the University of Liverpool, and published in polarity log form by Dagley *et al.* (1967). Recently full details of the stratigraphy and palaeomagnetism were given by Watkins and Walker (1977). Attempts were made by Dagley *et al.* (1967) to measure K–Ar ages on some of these lavas but were unsuccessful. Several of the sections comprising part of the overall sequence were re-collected for K–Ar dating, and the results were presented by McDougall *et al.* (1976*b*). The main aim of this latter study was to attempt to provide age control for a thick sequence of normal polarity

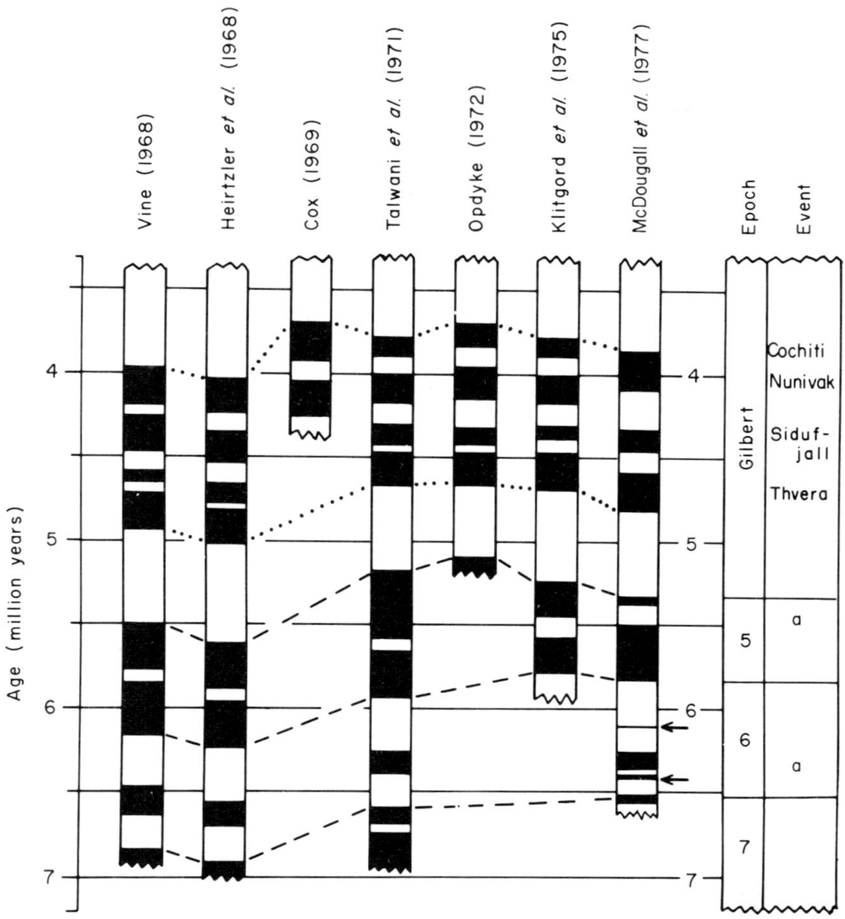

Fig. 5. Comparison of polarity time scale derived from study of Borgarfjördur sequence (polarity log at right) with scales proposed by previous workers. Black represnts normal polarity, white represents reversed polarity. (After McDougall *et al.*, 1977.)

lavas, which was thought to represent the on-land equivalent of marine magnetic anomaly 5 of Heirtzler *et al.* (1968). The results, summarized in Fig. 6, again show a linear correlation of measured K–Ar age with stratigraphic height. From the regression, estimates of the age of the polarity interval boundaries can be derived and are listed against the polarity log for the section on the left-hand side of Fig. 6. At the top of the diagram, two versions of the polarity time scale as deduced from the marine magnetic anomaly pattern are given. It will be seen that there is a good match between the observed polarity in the lava succession in eastern Iceland and the

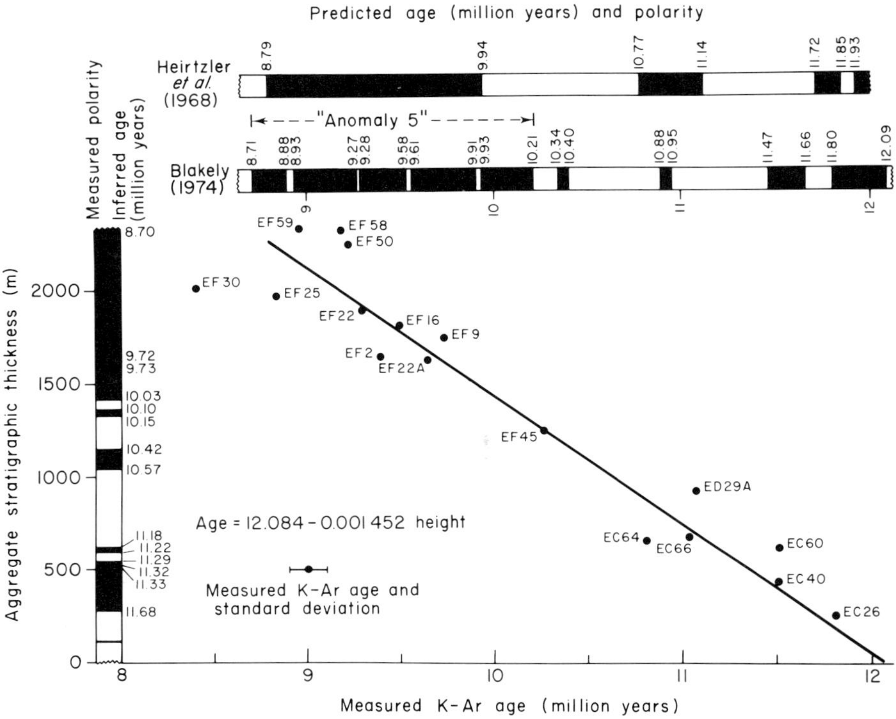

Fig. 6. Plot of K–Ar age against height above local base of lava sequence in Neskaupstadur region of eastern Iceland. Regression line through data points is indicated and ages of boundaries between polarity intervals given on polarity log for the sequence at left are derived from the regression. Polarity logs based upon analysis of marine magnetic anomalies by Heirtzler *et al.* (1968) and Blakely (1974) are given at the top of the diagram. See McDougall *et al.* (1976*b*), from which diagram taken, for full details.

marine magnetic profile data, especially that of Blakely (1974), such that there can be little doubt that we are dealing with the same part of the polarity time scale. Similarly there is good correlation of the ages as determined directly on land and from the extrapolations using the marine magnetic anomaly data. These results demonstrate that, at an age of about 10 million years, the extrapolated polarity time scale from the marine magnetic anomaly data is correct to within a few per cent. This result serves to emphasize that the assumptions underlying the seafloor spreading model are correct to a high order, giving greater confidence to the extrapolations of the polarity time scale to earlier times using marine magnetic anomalies.

These two examples illustrate that it is indeed possible in favourable situations to extend the polarity time scale directly by using K–Ar dating and magnetic polarity measurements on thick subaerial lava sequences. In addition, this approach can yield finer detail of the history of field reversal than by other methods—in the examples quoted, several additional short magnetic polarity intervals (events) have been recognized. The work in Iceland is as yet incomplete; measurements are being made on lava sequences covering much of the time interval from 3 to 13 million years ago, so that it is premature to attempt a complete synthesis of the data at present. Unless similar thick sequences of lavas of the right age can be found elsewhere on Earth, it seems somewhat unlikely that further extension of the polarity time scale by direct K–Ar dating and polarity measurement will be practicable.

Acknowledgements

It is most appropriate that this contribution appears in a volume dedicated to Professor J. C. Jaeger and Professor A. L. Hales because both have provided much encouragement and material assistance in studies the author has undertaken in relation to delineation of the polarity time scale. Professor Jaeger was instrumental in the author becoming involved in K–Ar dating studies in 1960, and gave much support and inspiration in the following years, especially in regard to elucidating the polarity time scale. The wise guidance provided by both Professor Jaeger and Professor Hales over the years is much appreciated; they have had a profound and lasting influence on the author.

In a review paper of this kind it is appropriate to acknowledge the important part played by colleagues—notably D. H. Tarling, F. H. Chamalaun, Aziz-ur-Rahman and N. D. Watkins. The more recent work in Iceland resulted from close collaboration between all those involved, especially N. D. Watkins, L. Kristjansson, K. Saemundsson, H. Johannesson and G. P. L. Walker. Financial support for the Iceland project was provided by the U.S. National Science Foundation (grant GA37178), the Icelandic authorities and the Australian National University.

References

Aldrich, L. T. and Wetherill, G. W. (1958). Geochronology by radioactive decay. *A. Rev. nuclear Sci.* **8**, 257–298.

Barbetti, M. and McElhinny, M. W. (1972). Evidence of a geomagnetic excursion 30 000 yr. B.P. *Nature, Lond.* **239**, 327–330.

Barbetti, M. and McElhinny, M. W. (1976). The Lake Mungo geomagnetic excursion. *Phil. Trans. R. Soc. A* **281**, 515–542.

Beckinsale, R. D. and Gale, N. H. (1969). A reappraisal of the decay constants and branching ratio of ^{40}K. *Earth planet. Sci. Lett.* **6**, 289–294.

Blakely, R. J. (1974). Geomagnetic reversals and crustal spreading rates during the Miocene. *J. geophys. Res.* **79**, 2979–2985.
Bodvarsson, G. and Walker, G. P. L. (1964). Crustal drift in Iceland. *Geophys. J. R. astronom. Soc.* **8**, 285–300.
Bonhommet, N. (1970). Discovery of a new event in the Brunhes period at Laschamp (France). In *Palaeogeophysics* (S. K. Runcorn, ed.), Academic Press, London, pp. 159–163.
Bonhommet, N. and Babkine, J. (1967). Sur la présence d'aimantations inversees dans la Chaine des Puys. *C.r. hebd Séanc. Acad. Sci., Paris B* **264**, 92–94.
Bonhommet, N. and Zahringer, J. (1969). Palaeomagnetism and potassium argon age determinations of the Laschamp geomagnetic polarity event. *Earth planet. Sci. Lett.* **6**, 43–46.
Brock, A. and Hay, R. L. (1976). The Olduvai event at Olduvai Gorge. *Earth planet. Sci. Lett.* **29**, 126–130.
Brunhes, B. (1906). Recherches sur la direction d'aimantation des roches volcaniques. *J. Phys.* **5**, 705–724.
Chamalaun, F. H. and McDougall, I. (1966). Dating geomagnetic polarity epochs in Réunion. *Nature, Lond.* **210**, 1212–1214.
Cox, A. (1969). Geomagnetic reversals. *Science, N.Y.* **163**, 237–245.
Cox, A. and Dalrymple, G. B. (1967*a*). Statistical analysis of geomagnetic reversal data and the precision of potassium–argon dating. *J. geophys. Res.* **72**, 2603–2614.
Cox, A. and Dalrymple, G. B. (1967*b*). Geomagnetic polarity epochs—Nunivak Island, Alaska. *Earth planet. Sci. Lett.* **3**, 173–177.
Cox, A., Doell, R. R. and Dalrymple, G. B. (1963*a*). Geomagnetic polarity epochs and Pleistocene geochronometry. *Nature, Lond.* **198**, 1049–1051.
Cox, A., Doell, R. R. and Dalrymple, G. B. (1963*b*). Geomagnetic polarity epochs: Sierra Nevada II. *Science, N.Y.* **142**, 382–305.
Cox, A., Doell, R. R. and Dalrymple, G. B. (1964*a*). Geomagnetic polarity epochs. *Science, N.Y.* **143**, 351–352.
Cox, A., Doell, R. R. and Dalrymple, G. B. (1964*b*). Reversals of the earth's magnetic field. *Science, N.Y.* **144**, 1537–1543.
Cox, A., Hopkins, D. M. and Dalrymple, G. B. (1966). Geomagnetic polarity epochs: Pribiloff Islands, Alaska. *Bull. geol. Soc. Am.* **77**, 883–910.
Cox, A., Doell, R. R. and Dalrymple, G. B. (1968). Radiometric time-scale for geomagnetic reversals. *Qt. J. geol. Soc.* **124**, 53–66.
Curtis, G. H. and Hay, R. L. (1972). Further geologic studies and K–Ar dating of Olduvai Gorge and Ngorongoro Crater. In *Calibration of Hominoid Evolution* (W. W. Bishop and J. A. Miller, eds), Scottish Academic Press, Edinburgh, pp. 289–302.
Dagley, P., Wilson, R. L., Ade-Hall, J. M., Walker, G. P. L., Haggerty, S. E., Sigurgeirsson, T., Watkins, N. D., Smith, P. J., Edwards, J. and Grasty, R. L. (1967). Geomagnetic polarity zones in Iceland. *Nature, Lond.* **216**, 25–29.
Dalrymple, G. B. (1972). Potassium-argon dating of geomagnetic reversals and North American glaciations. In *Calibration of Hominoid Evolution* (W. W. Bishop and J. A. Miller, eds), Scottish Academic Press, Edinburgh, pp. 107–134.
Dalrymple, G. B., Cox, A., Doell, R. R. and Grommé, C. S. (1967). Pliocene geomagnetic polarity epochs. *Earth planet. Sci. Lett.* **2**, 163–173.

Denham, C. R. (1976). Blake polarity episode in two cores from the Greater Antilles Outer Ridge. *Earth planet. Sci. Lett.* **29**, 422–434.

Denham, C. R., Anderson, R. F. and Bacon, M. P. (1977). Palaeomagnetism and radiochemical age estimates for late Brunhes polarity episodes. *Earth planet. Sci. Lett.* **35**, 384–397.

Doell, R. R. and Dalrymple, G. B. (1966). Geomagnetic polarity epochs: A new polarity event and the age of the Brunhes–Matuyama boundary. *Science, N.Y.* **152**, 1060–1061.

Doell, R. R., Dalrymple, G. B. and Cox, A. (1966). Geomagnetic polarity epochs: Sierra Nevada data, 3. *J. geophys. Res.* **71**, 531–541.

Emilia, D. A. and Heinrichs, D. F. (1972). Palaeomagnetic events in the Brunhes and Matuyama epochs identified from magnetic profiles reduced to the pole. *Marine Geophys. Res.* **1**, 436–444.

Evernden, J. F., Savage, D. E., Curtis, G. H. and James, G. T. (1964). Potassium–argon dates and the Cenozoic mammalian chronology of North America. *Am. J. Sci.* **262**, 145–198.

Foster, J. H. and Opdyke, N. D. (1970). Upper Miocene to Recent magnetic stratigraphy in deep-sea sediments. *J. geophys. Res.* **75**, 4465–4473.

Garner, E. L., Murphy, T. J., Gramlich, H. W., Paulsen, P. J. and Barnes, I. L. (1975). Absolute isotopic abundance ratios and the atomic weight of a reference sample of potassium. *J. Res. U.S. natn. Bur. Stand. A* **79**, 713–725.

Gillot, P. Y. and Cassignol, C. (1977). K/Ar dating of Laschamp magnetic event (abstract). Fifth European Colloquium of Geochronology, Cosmochronology and Isotope Geology, Pisa, September 1977.

Grommé, C. S. and Hay, R. L. (1963). Magnetizations of basalt of Bed 1, Olduvai Gorge, Tanganyika. *Nature, Lond.* **200**, 560–561.

Grommé, C. S. and Hay, R. L. (1971). Geomagnetic polarity epochs: age and duration of the Olduvai normal event. *Earth planet. Sci. Lett.* **10**, 179–185.

Harrison, C. G. A. (1966). The paleomagnetism of deep sea sediments. *J. geophys. Res.* **71**, 3033–4043.

Harrison, C. G. A. and Funnell, B. M. (1964). Relationship of palaeomagnetic reversals and micropalaeontology in two late Cenozoic cores from the Pacific Ocean. *Nature, Lond.* **204**, 566.

Hays, J. D. and Opdyke, N. D. (1967). Antarctic radiolaria, magnetic reversals, and climatic change. *Science, N.Y.* **158**, 1001–1010.

Hedberg, H. D. (1977). Preliminary report on magnetostratigraphic classification. Circular 55, International Subcommission on Stratigraphic Classification.

Heirtzler, J. R., Dickson, G. O., Herron, E. M., Pitman, W. C., III, and Le Pichon, X. (1968). Marine magnetic anomalies, geomagnetic field reversals, and motions of the ocean floor and continents. *J. geophys. Res.* **73**, 2119–2136.

Hess, H. H. (1962). History of ocean basins. In *Petrologic Studies: A Volume in Honor of A. F. Buddington* (A. E. Engel, H. L. James and B. F. Leonard, eds), Geological Society of America, Washington D.C., pp. 590–620.

Johannesson, H. (1972). The Tertiary lava pile between Nordurardal and Hvitarsida, western Iceland (in Icelandic). B.S. thesis, University of Iceland, Reykjavik, pp. 1–57.

Kawaii, N., Yaskawa, K., Nakajima, T., Torii, M. and Horie, S. (1972). Oscillating geomagnetic field with a recurring reversal discovered from Lake Biwa. *Proc. Jap. Acad.* **48**, 186–190.

Klitgord, K. D., Huestis, S. P., Mudie, J. D. and Parker, R. L. (1975). An analysis of near-bottom magnetic anomalies: sea-floor spreading and the magnetized layer. *Geophys. J. R. astronom. Soc.* **43**, 387–424.

Labeyrie, J. and Gillot, P. Y. (1977). Age of Laschamp event reevaluated. (Abstract). Fifth European Colloquium of Geochronology, Cosmochronology and Isotope Geology, Pisa, September 1977.

La Brecque, J. L., Kent, D. V. and Cande, S. C. (1977). Revised magnetic polarity time scale for Late Cretaceous and Cenozoic time. *Geology* **5**, 330–335.

McDougall, I. and Aziz-ur-Rahman (1972). Age of the Gauss-Matuyama boundary and the Kaena and Mammoth events. *Earth planet, Sci. Lett.* **14**, 367–380.

McDougall, I. and Chamalaun, F. H. (1966). Geomagnetic polarity scale of time. *Nature, Lond.* **212**, 1415–1418.

McDougall, I. and Chamalaun, F. H. (1969). Isotopic dating and geomagnetic polarity studies on volcanic rocks from Mauritius, Indian Ocean. *Bull. geol. Soc. Am.* **80**, 1419–1442.

McDougall, I. and Tarling, D. H. (1963). Dating of polarity zones in the Hawaiian Islands. *Nature, Lond.* **200**, 54–56.

McDougall, I. and Tarling, D. H. (1964). Dating geomagnetic polarity zones. *Nature, Lond.* **202**, 171–172.

McDougall, I. and Watkins, N. D. (1973). Age and duration of the Réunion geomagnetic polarity event. *Earth planet. Sci. Lett.* **19**, 443–452.

McDougall, I. and Wensink, H. (1966). Palaeomagnetism and geochronology of the Pliocene–Pleistocene lavas in Iceland. *Earth planet. Sci. Lett.* **1**, 232–236.

McDougall, I., Allsopp, H. and Chamalaun, F. H. (1966). Isotopic dating of the Newer Volcanics of Victoria, Australia, and geomagnetic polarity epochs. *J. geophys. Res.* **71**, 6107–6118.

McDougall, I., Watkins, N. D. and Kristjansson, L. (1976*a*). Geochronology and paleomagnetism of a Miocene–Pliocene lava sequence at Bessastadaá, eastern Iceland. *Am. J. Sci.* **276**, 1078–1095.

McDougall, I., Watkins, N. D., Walker, G. P. L. and Kristjansson, L. (1976*b*). Potassium–argon and paleomagnetic analysis of Icelandic lava flows: limits on the age of anomaly 5. *J. geophys. Res.* **81**, 1505–1512.

McDougall, I., Saemundsson, K., Johannesson, H., Watkins, N. D. and Kristjansson, L. (1977). Extension of the geomagnetic polarity time scale to 6.5 m.y.: K–Ar dating, geological and paleomagnetic study of a 3,500-m lava succession in western Iceland. *Bull. geol. Soc. Am.* **88**, 1–15.

Matuyama, M. (1929). On the direction of magnetization of basalt in Japan, Tyôsen, and Manchuria. *Japan Acad. Proc.* **5**, 203–205.

Mercanton, P. L. (1926). Inversion de l'inclinaison magnétique terrestre aux âges géologiques. *Terrestr. Magnet. atmos. Elect.* **31**, 187–190.

Moorbath, S., Sigurdsson, H. and Goodwin, R. (1968). K–Ar ages of the oldest exposed rocks in Iceland. *Earth planet. Sci. Lett.* **4**, 197–205.

Nagata, T., Uyeda, S. and Akimoto, S. (1952). Self-reversal of thermoremanent magnetism of igneous rocks. *J. Geomagnet. Geoelect.* **4**, 22–38.

Ninkovich, D., Opdyke, N. D., Heezen, B. C. and Foster, J. H. (1966). Paleomagnetic stratigraphy, rates of deposition and tephrachronology in North Pacific deep-sea sediments. *Earth planet. Sci. Lett.* **1**, 476–492.

Opdyke, N. D. (1972). Paleomagnetism of deep-sea cores. *Rev. Geophys. Space Phys.* **10**, 213–249.

Opdyke, N. D. and Runcorn, S. K. (1956). New evidence for reversal of the geomagnetic field near the Pliocene–Pleistocene boundary. *Science, N.Y.* **123**, 1126–1127.

Opdyke, N. D., Glass, B., Hays, J. D. and Foster, J. H. (1966). Paleomagnetic study of Antarctic deep-sea cores. *Science, N.Y.* **154**, 349–357.

Opdyke, N. D., Burckle, L. H. and Todd, A. (1974). The extension of the magnetic time scale in sediments of the central Pacific Ocean. *Earth planet. Sci. Lett.* **22**, 300–306.

Roche, A. (1951). Sur les inversions de l'aimantation rémanente des roches volcaniques dans les monts d'Auvergne. *C.r. hebd. Séanc. Acad. Sci., Paris* **233**, 1132–1134.

Roche, A. (1956). Sur la date de la dernière inversion du champ magnétique terrestre. *C.r. hebd. Séanc. Acad. Sci., Paris* **243**, 812–814.

Rutten, M. G. (1959). Paleomagnetic reconnaissance of mid-Italian volcanoes. *Geol. Mijnbouw* **21**, 373–374.

Saemundsson, K. and Noll, H. (1974). K/Ar ages of rocks from Húsafell, western Iceland, and the development of the Húsafell central volcano. *Jökull* **24**, 40–59.

Shackleton, N. J. and Opdyke, N. D. (1973). Oxygen isotope and palaeomagnetic stratigraphy of equatorial Pacific core V28-238: oxygen isotope temperatures and ice volumes on a 10^5 and 10^6 year scale. *Quaternary Res.* **3**, 39–55.

Smith, J. D. and Foster, J. H. (1969). Geomagnetic reversal in Brunhes normal polarity epoch. *Science, N.Y.* **163**, 565–567.

Steiger, R. H. and Jager, E. (1977). Subcommission on geochronology: Convention on the use of decay constants in geo- and cosmochronology. *Earth planet. Sci. Lett.* **36**, 359–362.

Talwani, M., Windisch, C. C. and Langseth, M. G., Jr. (1971). Reykjanes ridge crest: A detailed geophysical study. *J. geophys. Res.* **76**, 473–517.

Theyer, F. and Hammond, S. R. (1974). Palaeomagnetic polarity sequence and radiolarian zones, Brunhes to polarity epoch 20. *Earth planet. Sci. Lett.* **22**, 307–319.

Valladas, G., Gillot, P. Y., Poupeau, G. and Reyss, J. L. (1977). Thermoluminescence dating of recent volcanic rocks—Laschamp magnetic event dating (Abstract). Fifth European Colloquium of Geochronology, Cosmochronology and Isotope Geology, Pisa, Sept. 1977.

Verosub, K. L. and Banerjee, S. K. (1977). Geomagnetic excursions and their paleomagnetic record. *Rev. Geophys. Space Phys.* **15**, 145–155.

Vine, F. J. (1966). Spreading of the ocean floor: new evidence. *Science, N.Y.* **154**, 1405–1415.

Vine, F. J. (1968). Magnetic anomalies associated with mid-ocean ridges. In *The History of the Earth's Crust* (R. A. Phinney, ed.), Princeton University Press, Princeton, N.J., pp. 73–89.

Vine, F. J. and Matthews, D. H. (1963). Magnetic anomalies over ocean ridges. *Nature, Lond.* **199**, 947–949.

Vine, F. J. and Wilson, J. T. (1965). Magnetic anomalies over a young oceanic ridge off Vancouver Island. *Science, N.Y.* **150**, 485–489.

Walker, G. P. L. (1959). Geology of the Reydarfjördur area, eastern Iceland. *Qt. J. geol. Soc.* **114**, 367–393.

Watkins, N. D. (1972). Review of the development of the geomagnetic polarity time scale and discussion of prospects for its finer definition. *Bull. geol. Soc. Am.* **83**, 551–574.

Watkins, N. D. and Walker, G. P. L. (1977). Magnetostratigraphy of eastern Iceland. *Am. J. Sci.* **277**, 513–584.

Watkins, N. D., Kristjansson, L. and McDougall, I. (1975). A detailed paleomagnetic survey of the type location for the Gilsá geomagnetic polarity event. *Earth planet. Sci. Lett.* **27**, 436–444.

Yaskawa, K., Nakajima, T., Kawaii, N., Torii, M., Natsuhara, N. and Horie, S. (1973). Paleomagnetism of a core from Lake Biwa (1). *J. Geomagnet. Geoelect.* **25**, 447–474.

17. Pole positions and continental drift since the Devonian†

E. IRVING

Abstract

This chapter presents a series of 30 cartoons that describe, in a general and tentative fashion, the drift of the continents during the last 350 million years relative to each other and to the poles.

1. Historical

During the past century various schemes of continental drift have been proposed involving large-scale relative displacements of the continents. The essence of all such ideas was that the Earth once possessed either one primitive continent called *Pangaea*, or two, called *Laurasia* and *Gondwana*, and at some comparatively late time in geological history these primitive continents broke up and the fragments drifted into their present positions. An excellent account of the early evolution of ideas has been given by Hume (1948).

Alfred Wegener (see reference list) was the true prophet of continental drift. Although others (Snider, 1858; Fisher, 1882; Pickering, 1907; Taylor, 1910; Baker, 1912, 1932) had previously suggested it, it was he who first stated the concept boldly and at length, and who devoted a substantial part of his life to its development. Wegener was the first to attempt to integrate into the concept, much of what was then known about the Earth, and, with his father-in-law Wladimir Koppen, to draw a series of maps showing the development of the Earth's main surface features and the movements of the

† Contribution of Earth Physics Branch no. 717, Energy Mines and Resource, Ottawa, Canada.

geographical pole during the later Phanerozoic. Their reconstruction of the continents, with north-west Africa situated opposite the south-east coast of North America, and with an ocean called Tethys between Africa and Europe, is familiar to all. This reconstruction is referred to here as Pangaea A because it is the most recent assembly of the continental crust into a single supercontinent.

Wegener's other important contribution was to suggest a mechanism that appealed to processes operating today. Earlier advocates of continental drift had cited catastrophic causes, usually either the capture of the Moon, or its loss from the Pacific. Wegener thought that the Eötvös force moved the continents towards the equator, and that tidal drag impelled them westward. These forces are now considered inadequate, but Wegener should not be criticized for not providing a satisfactory dynamic model, because there is, even now, no generally acceptable mechanism for continental drift (see Chapters 8 and 15 of this book).

Wegener's ideas gained little support. Critics, unable or unwilling to deal with his work as a whole, proceeded to attempt to discredit it piecemeal, and, in the eyes of most scientists of the day, they succeeded, so much so that up until the mid-1960's continental drift remained outside the mainstream of research and teaching. The information necessary to convince the earth science community of the correctness of his concept was not forthcoming in his lifetime. That we now have this information in abundance is due, in no small measure, to John Jaeger and Anton Hales, who, along with a handful of other directors of major research groups, encouraged in the mid-1950's the palaeomagnetic studies that were to provide the first decisive physical measurements of relative continental displacements, namely, the discovery of the very large and indisputable differences between the Late Palaeozoic and Mesozoic palaeomagnetic poles obtained from the northern and southern continents.

By the mid-1950's polar differences of about 25° had already been discovered between Europe and North America, indicating that these continents were near together in the Late Paleozoic (Runcorn, 1956; Irving, 1956), but the differences were based on early observations with a limited sampling and experimental basis; magnetic (Creer 1958; As and Zidjerveld, 1958) and thermal (Wilson, 1961; Irving *et al.*, 1961) cleaning, which removed the unwanted secondary magnetizations in rocks, had not yet been developed. Although the result of Runcorn and Irving has since been amply confirmed, not many people at the time were convinced, partly because of the preliminary nature of the results, and partly because of the dominance of fixist ideas, which prevented people from considering the data seriously. However, in the late 1950's and the early 1960's more decisive results were produced in Australia at Jaeger's department in the Australian National

University, and in southern Africa. The work in southern Africa was first begun at the Bernard Price Institute, where Hales was director, and where Gough, an associate of Hales, also worked, and it was continued at the University College of Rhodesia and Nyasaland where Gough and his colleagues produced results of very great importance. These studies lead to the discovery of systematic polar differences of up to 90° among palaeomagnetic poles obtained from Australia and southern Africa and those from the northern hemisphere continents. Preliminary accounts of these large differences appeared in the late 1950's (Graham and Hales, 1957; Irving, 1957), but the main impact was felt when results were obtained from systematic studies of Mesozoic rocks, which could be dated either stratigraphically or by the newly-developed radiometric methods, and which had been subjected to magnetic and thermal cleaning. These polar differences between the northern and southern continents were significant beyond any possible technical or statistical error, and they demonstrated that relative continental motions of many thousands of kilometres had occurred during or since the Mesozoic (Gough *et al.* 1964; Irving *et al.*, 1963). The results were broadly consistent with Wegener's concept of the break-up of Pangaea into fragments that dispersed in the Mesozoic and Cenozoic. However, results from Upper Palaeozoic rocks yielded a distribution of palaeomagnetic poles that could not be reconciled with Wegener's simple concept, and this meant that there could have been an episode of continental drift comparable in magnitude to that envisaged by Wegener, but preceding it by about 150 million years (Irving and Green, 1958). Continental drift therefore, appeared to be a general feature of the Earth's behaviour, and not simply a process confined to the past 200 million years. These first quantitative measurements of "Wegenerian" continental drift, and of a pre-Wegenerian predominantly Carboniferous drift event, owe much to the foresight and leadership of Jaeger and Hales to whom this book is dedicated.

Because of their fragmentary nature, the results obtained in the late 1950's and early 1960's, although they indicated that the continents had been drifting relative to one another by thousands of kilometres at rates of the order of a centimetre per year since the Carboniferous, they did not, at the time, tell us much about the details of continental motions. During the last decade, results have been accumulating at an increasing tempo, and a reasonably coherent account can now be given. There are, however, many uncertain features, in particular, the width and nature of the Pre-Hercynian Ocean, little is known quantitatively about the motions of the tectonic blocks of eastern Asia and the Arctic region, and the description of events in the later Palaeozoic and Early Triassic is rudimentary; therefore the views expressed here on these questions are likely to be ephemeral.

2. Apparent Polar Wandering (APW)

A path of apparent polar wandering (Creer *et al.*, 1957) defines the variations in latitude and azimuth but not the longitude of a continent. The paths for North America, northern Eurasia and Gondwana are given in Figs 1, 2 and 3. The paths generally have a rather complex form, with bends (or hairpins) between which the polar trajectories (or tracks) are approximately straight. The tracks, in general, correspond to the motions between continents that occur at a time when the pattern of continental drift is approximately constant, and the hairpins mark the times of change in the pattern of continental motions. The North American path has hairpins at −200 million years (Triassic–Jurassic boundary) and −100 million years (mid-Cretaceous). The polar track between −200 and −300 million years is the palaeomagnetic signature in North American rocks of the opening of the middle Atlantic between Africa and North America during the Jurassic and Cretaceous, and the polar track between −60 million years and the present day is the signature of the opening of the northernmost Atlantic between Greenland and Norway during the Cenozoic. The rapid polar movement across Gondwana in the Early Carboniferous is, apparently, the palaeomagnetic signature of the closure of the "Pre-Hercynian" Ocean between Gondwana and Laurasia. The Gondwana path has a hairpin at −250 million years (end of the Palaeozoic) and one at −200 million years, and the track between them could correspond to a clockwise rotation of about 35° of Gondwana relative to Laurasia during the Triassic. The −200 million year hairpin in the Gondwana path marks the commencement of the opening of the northern Atlantic between Gondwana and Laurasia. These interpretations of the APW paths cannot, of course, be made by considering paths individually—their meaning only becomes clear when the motions of the continents are considered globally.

3. Procedures

When preparing the cartoons, pride-of-place has been given to the palaeomagnetic evidence. The latitudes and azimuths of the continents have been fixed by placing the palaeomagnetic poles (within their 95 per cent error circles) at the poles of the projection using Geuer's terrascope (Geuer, 1973). For times since −200 million years, relative longitudes have been taken from plate tectonic reconstructions using mainly the evidence from submarine magnetic anomalies (Francheteau, 1972; Phillips and Forsyth, 1972; Pitman and Talwani, 1972; McKenzie and Sclater, 1971). Maps for times prior to −200 million years have been made by working backwards, interval by interval, from the Early Jurassic, maintaining consistency with

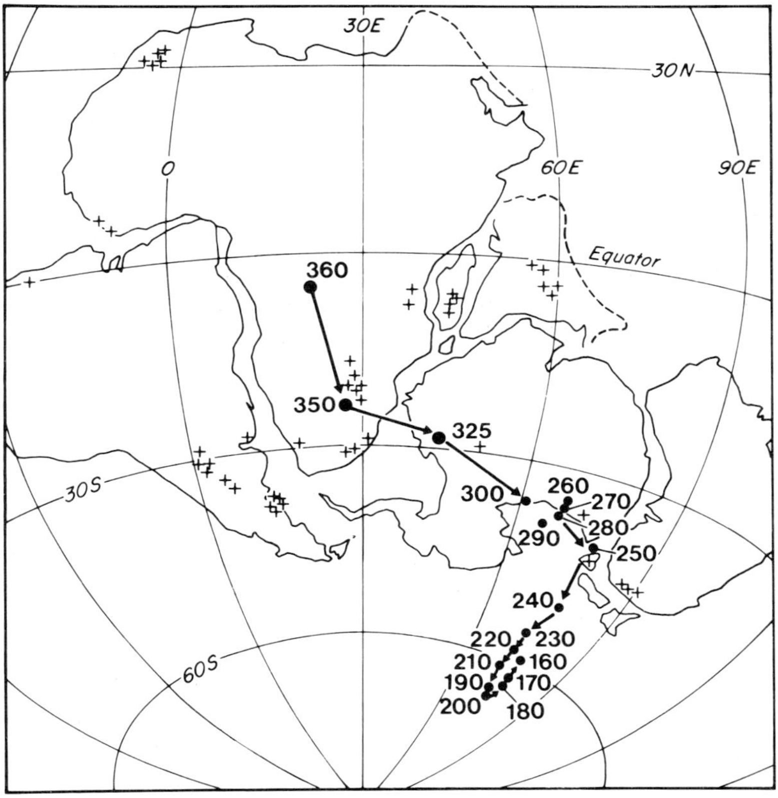

Fig. 1. APW relative to Gondwana for the Late Palaeozoic and Early Mesozoic. The poles from −300 to −160 million years have been obtained by statistical analysis. The calculations are made assuming that the geomagnetic field averaged to an axial geocentric dipole (AGD), and the integrity of the major continental blocks as described in the text. The palaeomagnetic poles from each continental block are first referred to a scale of millions of years, and overlapping averages at intervals of 10 million years calculated (Irving, 1977). The older poles are grouped into an interval of 40 million years duration and averaged to obtain the oldest point. The limits of the interval is then moved forward by 10° and a second average taken giving the next point and so on. Only every fourth value is independent. The mean 95 per cent errors of the poles are about 7°. The poles for the interval −360 to −325 million years are approximate poles obtained by visual inspection of sparse data from Australia and South America; the interpretation given here follows that given by Wensink (1975), Creer (1970), Hailwood (1974), McElhinny and Embleton (1974), and Morel and Irving (1978) and is speculative. Gondwana began to split in the Jurassic and a younger average path cannot be obtained. Results from the Speckled Sandstone of India are omitted. The values for Gondwana have been rotated relative to Africa using the reconstruction of Smith and Hallam (1970). Sampling localities are marked by crosses.

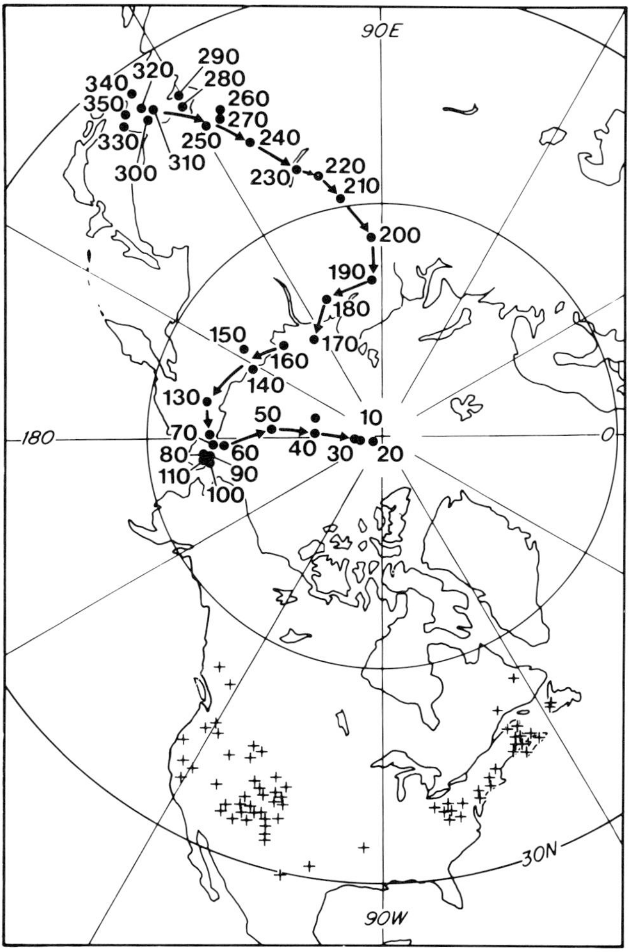

Fig. 2. Apparent polar wander (APW) relative to Laurentia since the Devonian. See caption to Fig. 1.

the palaeomagnetically determined latitudes and azimuths, but restricting relative motions to a minimum and disallowing major overlap between continents. Two assumptions are crucial to this procedure. The first is the assumption of minimum relative motion. This concept derives from the fact that the dynamics of modern plate motions are such as to minimise motions of the continental lithosphere, so that the velocities of plates bearing large continents are several times less than those of plates composed mostly of oceanic lithosphere (Minister *et al.* 1974). The second assumption is the

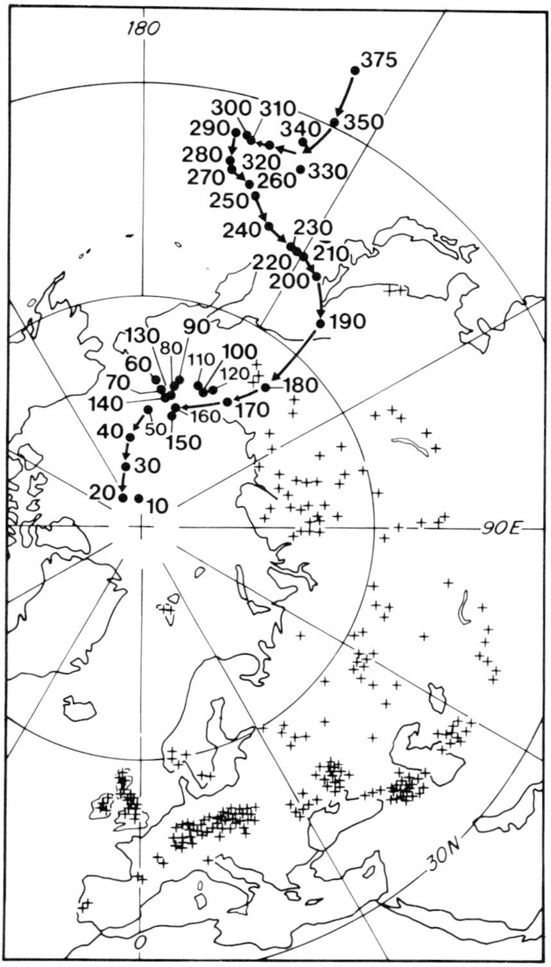

Fig. 3. APW relative to northern Eurasia since the Devonian. See caption to Fig. 1.

integrity of the three major continental blocks of Eurasia, Laurentia and Gondwana.

When preparing paleocontinental maps, the configuration of the ancient continental blocks must be estimated. We assume that the Siberian and Baltic Shields have had their present relative positions since the Devonian (see later), that the landmass of Laurentia has remained integral except for the western cordillera since the Devonian, and that the configuration for Gondwana given by du Toit (1937) and Smith and Hallam (1970) is essentially correct. The boundaries and names of the mini-blocks given by

Ziegler *et al.* (1977) have generally been followed. The manner in which Asia was assembled is one of the major outstanding problems (see, for example, Argand, 1922; Burrett, 1974) and probably will remain so until the present very inadequate palaeomagnetic coverage is greatly improved. The scheme for assembling Asia that is adopted here is very tentative indeed, and is as follows: Siberia and Baltica came together at the close of the Early Palaeozoic, Kazakhstania was added in the Carboniferous, North China in the Permian and Triassic, South China and the Malayan block in the middle of the Mesozoic, Kolyma in the Early Cretaceous, and, finally, India was added in the Eocene.

The reconstructions made in this way differ from those made by most other workers (e.g. Phillips and Forsyth, 1972; Smith and Briden, 1977), who have made their maps using evidence mainly from the oceans, such as the fit of the edges of continental shelves, transform faults, and marine magnetic anomalies. These authors then rotate the palaeomagnetic poles from all the land masses so that they conform with the plate-tectonic reconstruction and then average them to give a mean pole, which is then used to draw the global grid of latitude and (relative) longitude. By their procedure difference among palaeomagnetic poles from different continents for a given plate-tectonically derived reconstruction are assumed to be insignificant. In the procedure used here, the average poles for each of the continental block are calculated, and intercontinental polar differences are directly used to determine relative continental motions. The relative motions of continents implied by these intercontinent differences are not small—a thousand kilometres or more—and could prove to be critical for regional geological interpretations.

In order to focus attention on this important methodological problem and its consequences, two points of difference between the maps given below and those commonly to be found in the literature are noted. The maps produced by the methods of "classical" plate tectonics show the Arctic as a static region throughout the Mesozoic, the bordering continents of Laurentia and Eurasia being fixed relative to each other. But geologically the Mesozoic Arctic is far from static; there are fold belts, rifts, subsiding basins, and thick igneous sequences that in other contexts would be interpreted as evidence of plate motions. Consequently, in the cartoons given here, the Arctic is depicted as a mobile region during much of the Mesozoic. Second, in most contemporary maps, Wegner's Pangaea is regarded as integral from the Carboniferous to the Early Jurassic. On the contrary, I will argue that the continental blocks comprising Pangaea might have undergone spatial rearrangement during this interval. The arguments are based on the palaeomagnetic evidence, and the evolutionary scheme proposed has yet to be evaluated geologically.

4. The Duration of Pangaea A

It is now generally agreed that, at the beginning of the Jurassic, most of the continental crust was grouped into a single supercontinent, similar in all essentials to Wegener's Pangaea. But how long prior to the Jurassic did Pangaea A exist? An answer to this question can be given by comparing the APW paths with Pangaea A. If Pangaea A existed prior to the Jurassic, then the APW paths when plotted relative to this configuration of the continents should coincide. The paths will coincide only if the continental blocks have been correctly positioned within Pangaea.

In Fig. 4 the first attempt (Jaeger and Irving, 1957) to make this comparison is shown. At that time results were available from Australia, North America and Europe only. The poles for North America and Europe agreed reasonably well, but a group of Upper Palaeozoic poles from

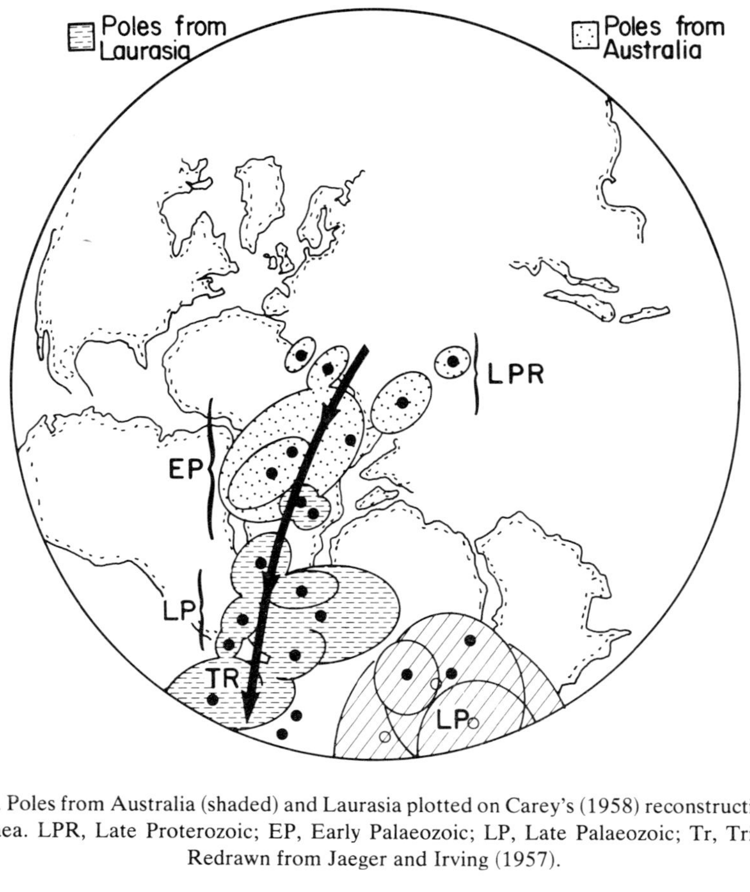

Fig. 4. Poles from Australia (shaded) and Laurasia plotted on Carey's (1958) reconstruction of Pangaea. LPR, Late Proterozoic; EP, Early Palaeozoic; LP, Late Palaeozoic; Tr, Triassic. Redrawn from Jaeger and Irving (1957).

Australia were discordant. Jaeger and Irving argued that if Australia were moved clockwise around the southern end of Pangaea toward South America the discordent poles would be brought on to the main curve; that is, Australia was incorrectly placed in Pangaea. This argument may not be so far fetched as it might seem because there is still much uncertainty about the placing of Australia and Antarctica within Gondwana.

In Fig. 5 the APW paths for North America and northern Eurasia are shown after closure of the north Atlantic using the fit of Bullard *et al.* (1965). The poles agree, within their errors, in the Late Carboniferous and Permian (−325 to −250 million years), and again in the Jurassic (−175 to −150 million years), and again in the Jurassic (− 175 to −150 million years). The

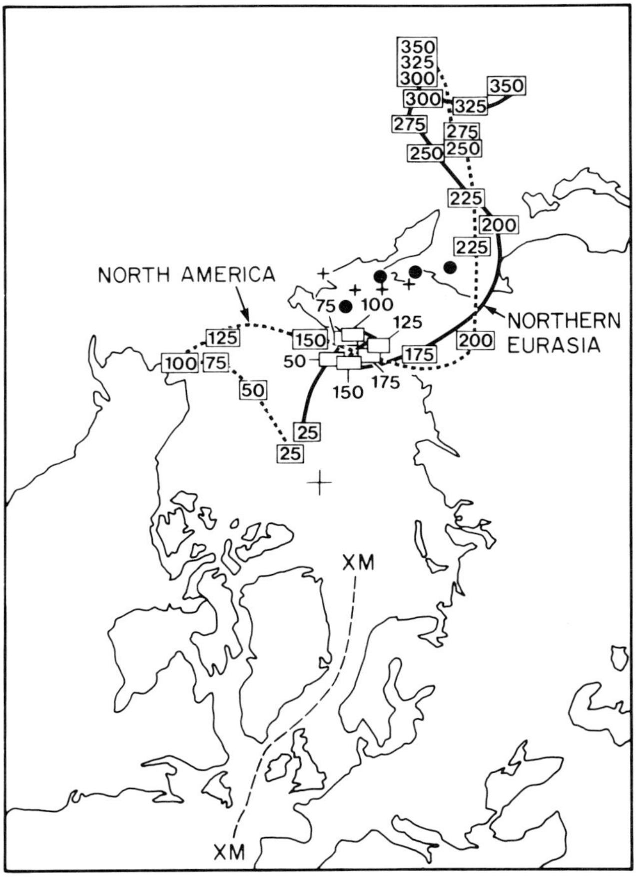

Fig. 5. APW paths for Eurasia and Laurentia compared after closure of the North Atlantic according to Bullard *et al* (1965).

poles at −125 million years are divergent, indicating that the northernmost Atlantic had commenced to open by the Early Cretaceous. The poles also differ in the Triassic, so that at −225 million years the paths begin to diverge, and at −200 million years there is a discrepancy of about 15°; the paths have a similar position but their timing is different. Evidently Europe and North America moved relative to one another during the Triassic (Roy, 1972).

The two APW paths of Fig. 5 are not, of course, grossly different from one another, and the mean path for the later Palaeozoic and earlier Mesozoic for Laurasia is given in Fig. 6 and compared with the path from Gondwana (Schmidt 1976). The paths are referred to Pangaea A. They differ by up to 20° between the Carboniferous and Triassic, but agree in the interval Late Triassic to Middle Jurassic. The comparisons of Figs 4, 5 and 6 suggest that

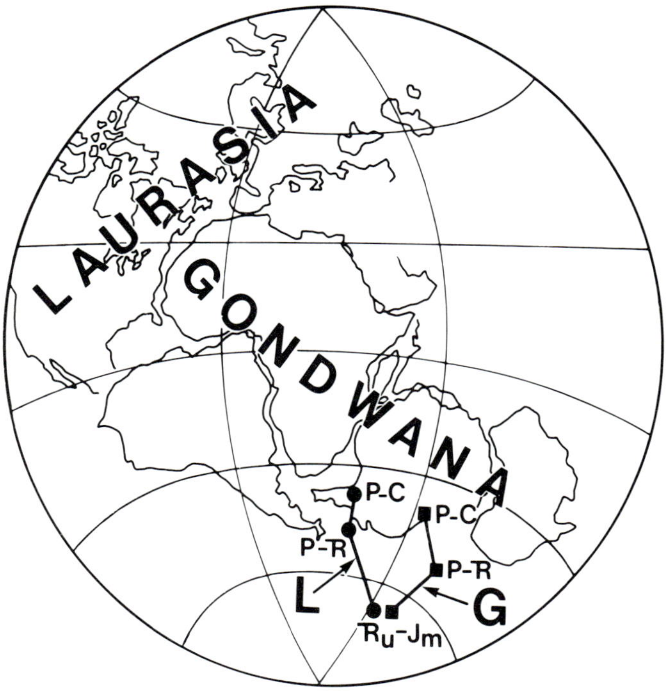

Fig. 6. Mean Late Palaeozoic and Early Mesozoic poles for Gonwana and Laurasia plotted relative to Pangaea A (Schmidt, 1976). The reconstruction of the continents around the Atlantic is that of Bullard *et al.* (1965), and around the Indian Ocean that of Smith and Hallam (1970). The latitude and longitude frame is the present geographical grid to which Africa is fixed. The results have been averaged in the following age groups: P–C (Permo-Carboniferous); P-Tr (Permo-Triassic), and Tru-Jm (Late Triassic to Middle Jurassic). Redrawn with permission of the author and the Australian National University.

Pangaea A was only a fleeting phenomenon, lasting no more than a few tens of millions of years during the latest Triassic and early Jurassic. If the poles for the Carboniferous, Permian and Triassic are unified and the relative longitudes of the various continental elements of Pangaea A retained, then North America and northern Gondwana would overlap by 15 or 20°, which is obviously absurd. This result, although not previously so well documented, has been known for many years, and several workers (Nairn, 1964; de Boer, 1965; Van Hilten, 1964; Irving, 1967) have tried to explain it by involving a large post-Triassic shear along Tethys—the "Tethys twist" of Van Hilten. De Boer (1965) and Van Hilten (1964) argued that the Tethys twist accounted for the Alpine deformation, but their ideas are no longer acceptable because the Early Jurassic palaeomagnetic results are now known to agree with Pangaea A (Figs 5 and 6); the motions needed to account for the discrepancy must have occurred before the Alpine Orogeny and before the creation of the western Tethys in the Late Triassic. McElhinny (1973, p. 263) may have been nearer the mark when he suggested that the evidence for the motions needed to explain the discrepancy must be sought in the Hercynian fold belts of Europe and north-west Africa.

5. Late Devonian Continents

The suggested reassembly for the Late Devonian (−360 million years) is shown in Figs 7 and 8. In the north there is a supercontinent, Laurasia, consisting of the Laurentian, Baltic-Russian and Siberian cratons and their

Figs 7 to 36. Continental drift since the Devonian. The latitudes of the major continents have been fixed palaeomagnetically using the poles of Figs 1, 2 and 3. This rule is relaxed for Figs 19–22 because it is not yet possible to provide satisfactory maps for the A–B transformation; the necessary details are unlikely to be revealed until the data became available to allow the use of a shorter averaging window, perhaps 20 million years. The longitude grid is arbitrary and may not correspond in pairs of maps of the same age. The latitudes may not agree exactly and this reflects the statistical errors (Irving, 1977). Not all maps are equally reliable and, on a relative reliability scale (A is good, B is moderate and C is fair), Figs 23–36 could be rated in category A, Figs 11–22 in category B and Figs 7–10 in category C. The symbols are as follows: A, B and C signify possible positions for continental fragments between Laurasia and Gondwana (see text); AA, Arctic Basin; AB, Arctic–Alaska block; BR, Broken Ridge; CP, Chukotsk Peninsula; EB, Eurasian Basin; EF, Ellsworth fold belt; FR, Franklinian fold belt; IR, Iran block; IT, Italian promontory; KB, Kolyma block; KG, Kerguelen-Gaussberg Ridge; KZ, Kazakhtania; LS, Labrador Sea; MB, Malayan block; NR, boreal rift in the Permian; 9R, 90° Ridge; S, Samfrau superfold belt; SC, South China block; SN, Chinese block; SP, preferred position of Spain and perhaps elements of central France and also central Europe; SR, Seychelles Ridge; SV, Sierra de la Ventana; TB, Tarim block; TG, Tasman foldbelt; TS, Tasman Sea; UR, Ural fold belt; XM, hypothetical zone of shear between Eurasia and Laurentia.

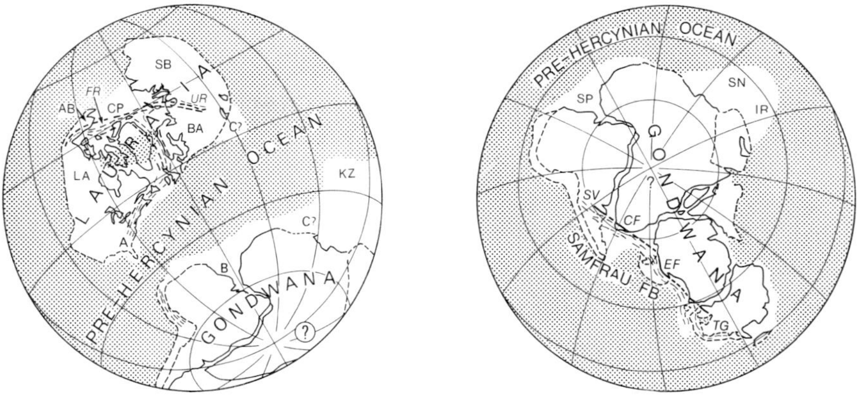

Figs 7, 8. 360 million years ago; Late Devonian.

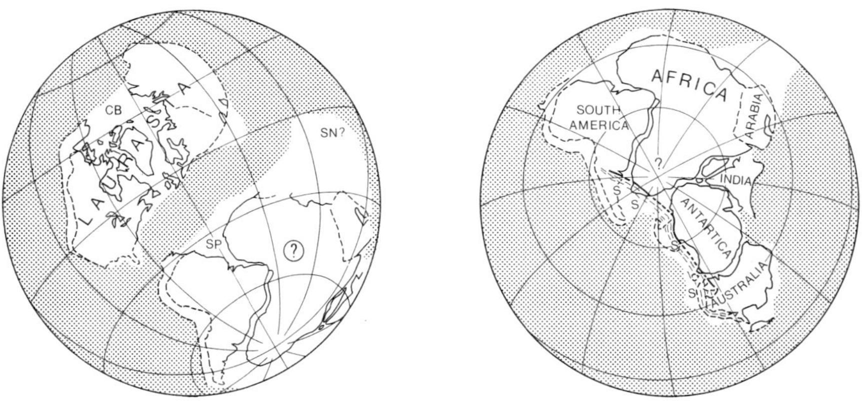

Figs 9, 10. 340 million years ago; Early Carboniferous.

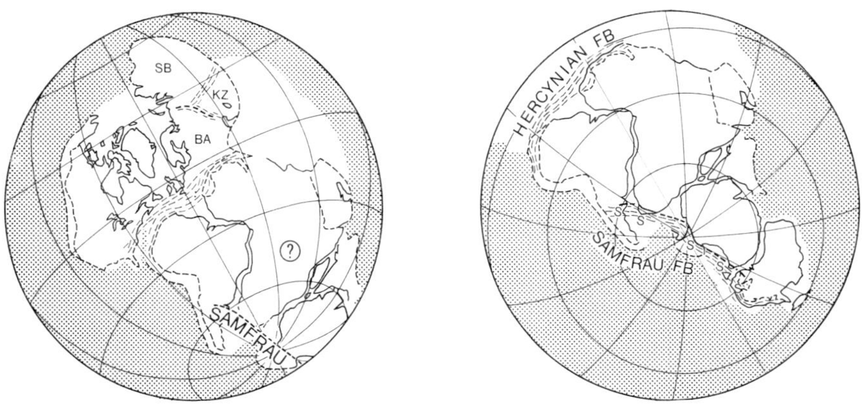

Figs 11, 12. 325 million years ago; Middle Carboniferous.

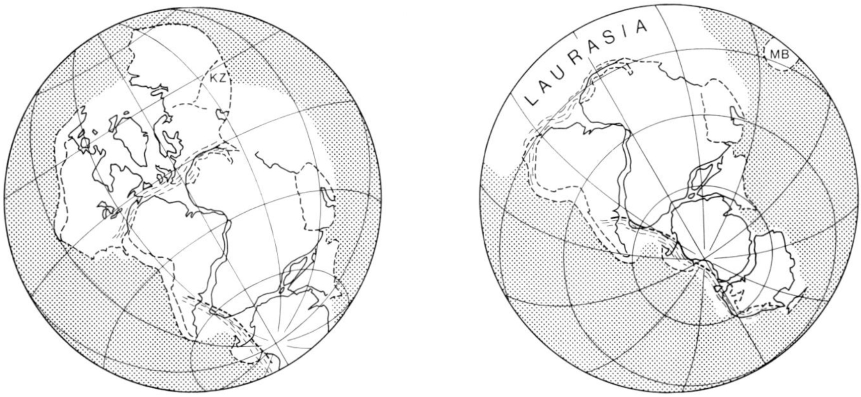

Figs 13, 14. 300 million years ago; Late Carboniferous.

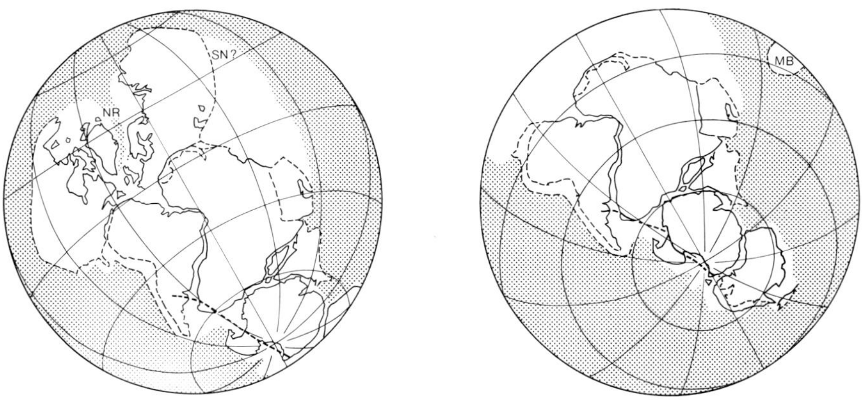

Figs 15, 16. 275 million years ago; Permian–Carboniferous boundary.

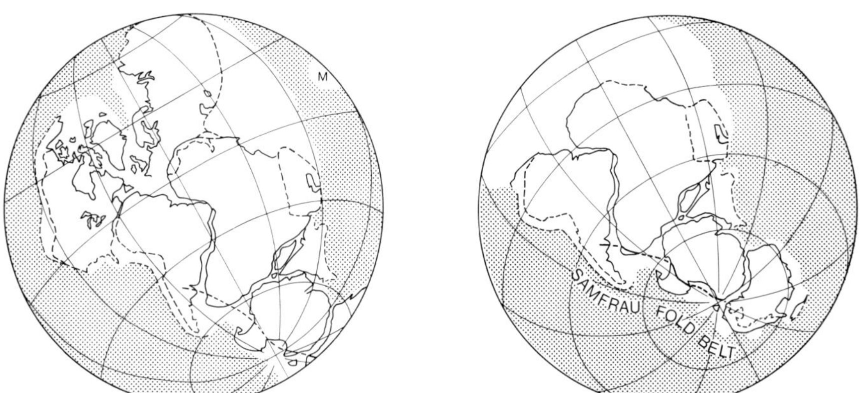

Figs 17, 18. 250 million years ago; Early Permian.

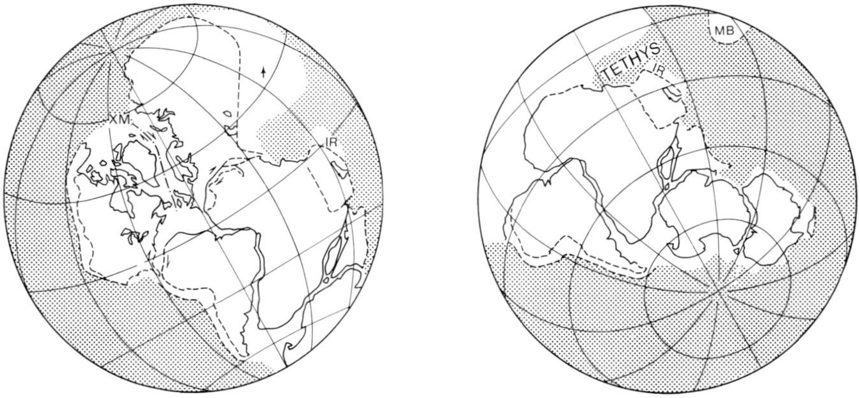

Figs 19, 20. 225 million years ago; Mesozoic-Palaeozoic boundary.

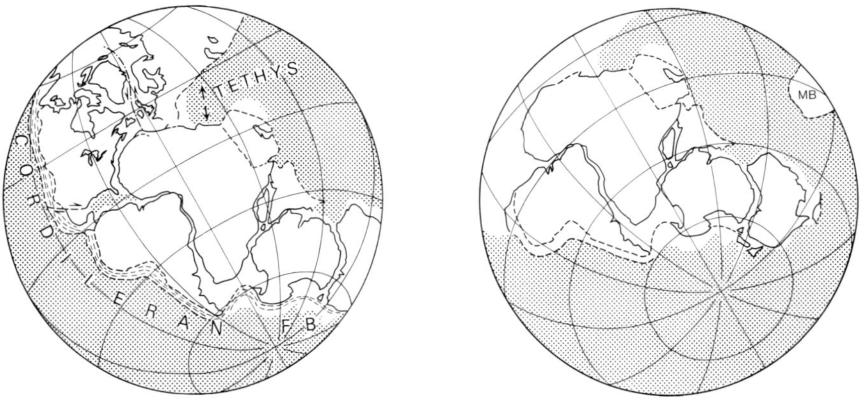

Figs. 21, 22. 200 million years ago; Late Triassic.

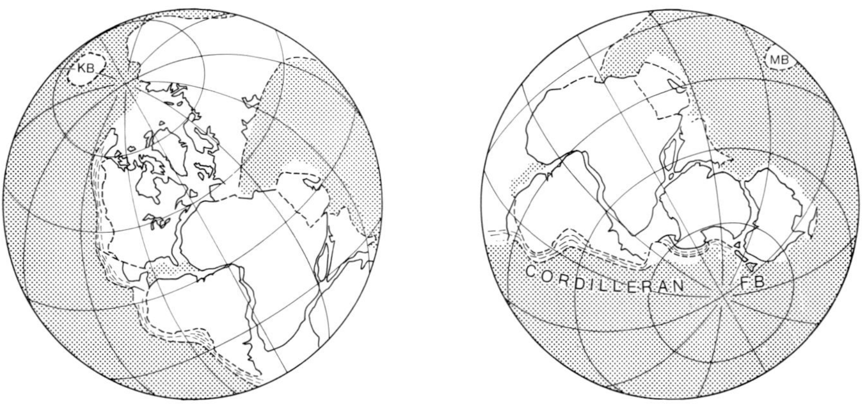

Figs 23, 24. 175 million years ago; Early Jurassic.

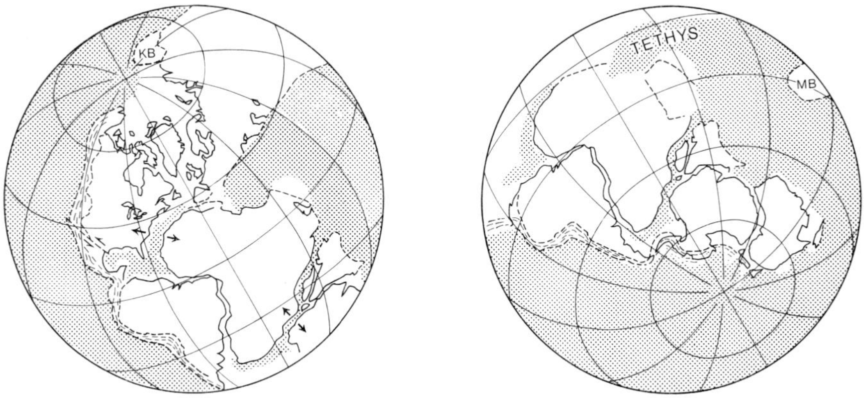

Figs 25, 26. 150 million years ago; Late Jurassic.

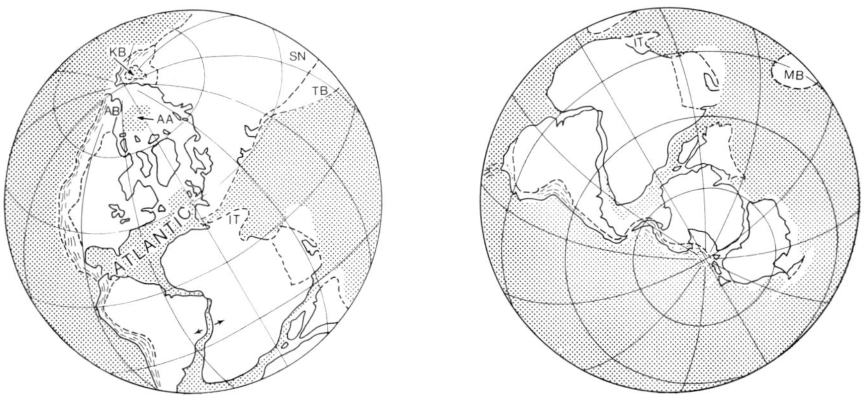

Figs 27, 28. 125 million years ago; Early Cretaceous.

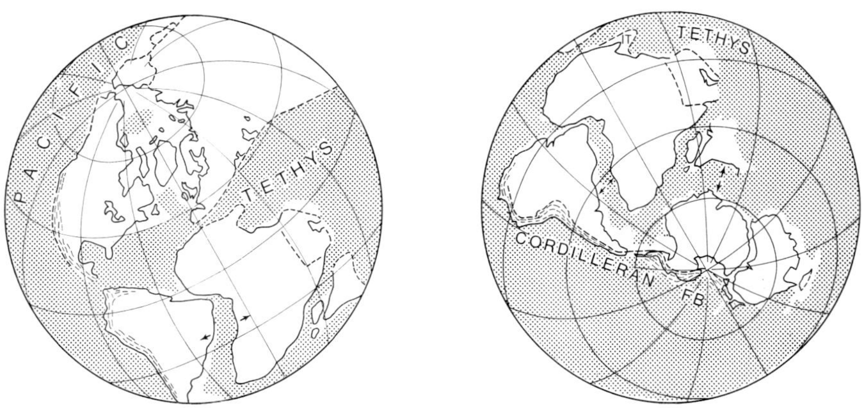

Figs 29, 30. 100 million years ago; Middle Cretaceous.

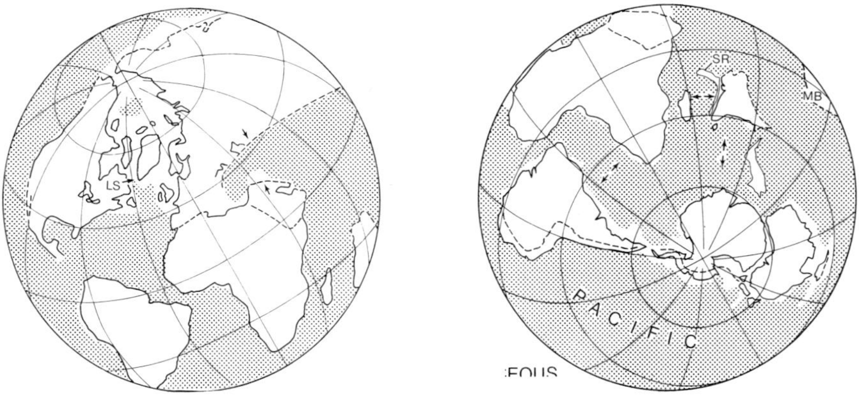

Figs 31, 32. 75 million years ago; Late Cretaceous.

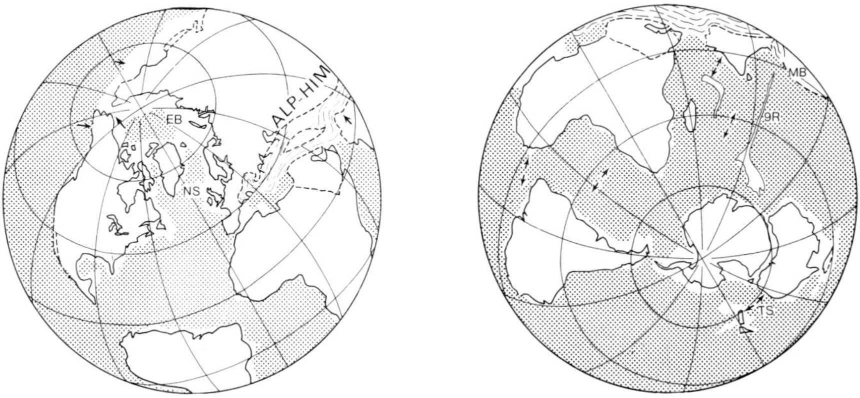

Figs 33, 34. 50 million years ago; Eocene.

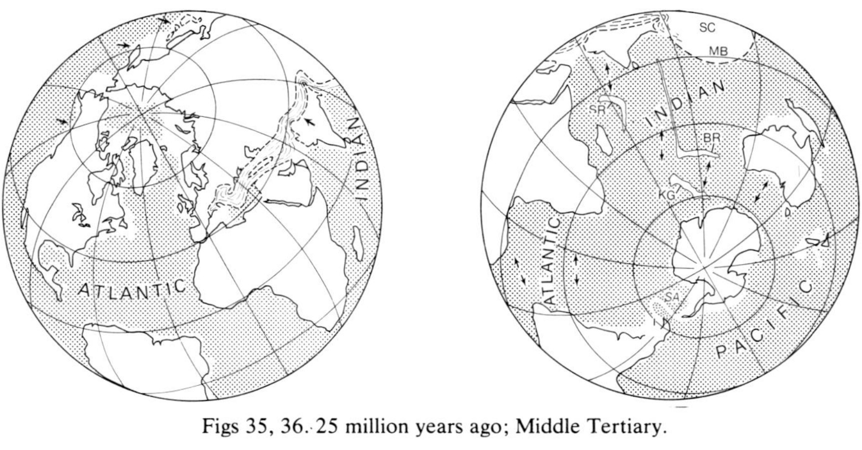

Figs 35, 36. 25 million years ago; Middle Tertiary.

bordering Early Palaeozoic foldbelts. Laurasia is in equatorial to mid-northerly latitudes. Svalbard adjoins Ellesmere Land and Greenland, and northern Alaska is rotated anticlockwise with respect to North America (Carey, 1958). Recently, palaeomagnetic results from Carboniferous rocks of the Brooks Range in northern Alaska have apparently demonstrated a 70° anticlockwise rotation of the northern half of Alaska (Newman *et al.*, 1977) Once this rotation is admitted one must ask how much, if any, of north-eastern Siberia has moved with northern Alaska? It is possible that the Chukotsk region, with or without the Kolyma block, may have been continuous with northern Alaska (Churkin, 1973). Alternatively, Kolyma may have been welded on to Asia separately (McElhinny, 1972). In this account, northern Alaska and Chukotsk (referred to as the Alaska–Arctic block AB by Newman *et al.* 1977) are assumed to have moved together and Kolyma to have been welded to Asia separately. Readers are referred to Sweeney *et al.* (1978) for further details.

The Ural mountains (UR) and their possible continuation westward into Novaya Zemblya, Svarlbard, north Greenland and the Franklinian (FR) fold belt of northern Laurasia, are of central importance in the evolution of Laurasia. This zone of deformation has apparently been formed by the marginal interaction of the four blocks (Laurentia, Baltica, Siberia and Kazakhstania) and consequently has a long and complex history governed by the form and timing of their relative motions. The fold belts west of, and including, the northern Urals could have been a product of the interaction of Laurentia, Siberia and Baltica, whereas the southern Urals have apparently been formed by the interaction of Kazakhstania and Baltica. The palaeomagnetic poles from the Devonian of Siberia and Baltica are in reasonable agreement, and therefore it is assumed here that there was no more than a narrow trough between them, so that the Franklinian and north Uralian troughs were either closed or narrow. Because of the longitude indeterminacy, this is not a rigorous solution, but only the simplest one arising out of the constraint of minimum motion. On the other hand the palaeomagnetic evidence indicates that Kazakstania did not form part of Laurasia in the Devonian, so the south Uralian trough apparently had not yet closed.

To the south of Laurasia there may have been an ocean about 2000–3000 km wide, beyond which there is the supercontinent of Gondwana (Du Toit, 1937), situated in middle to high southerly latitudes. The question marks in Figs 7 and 8 draw attention to the large uncertainties in the latitude determinations for Gondwana for Devonian and Early Carboniferous times (Fig. 1). The width of this pre-Hercynian ocean is therefore speculative. More palaeomagnetic data from the Devonian of Gondwana are needed before this question can be satisfactorily discussed.

In Gondwana the Samfrau superfold belt of Du Toit (1937) begins to develop, presumably as an ocean–continent interaction (Fig. 8). It extends for 6000 km from the Tasman foldbelt (TG) in Australia, through the Ellsworth (EF) and Cape (CF) fold belts, and Sierra de la Ventana (SV) of Buenos Aires to the central Andes. The position of the extreme south of South America is problematical; it has a Precambrian basement, and may have been a separate continental fragment welded on to Gondwana during the later Palaeozoic, so that the South American, and possibly also the southern African parts of Samfrau, may be a consequence of a continent–continent collision.

On either side of the Pre-Hercynian Ocean, in what is to become the zone of interaction between Laurasia and Gondwana during the mid-Carboniferous, the positions of many crustal elements are uncertain. For example, Spain and central Europe may have been situated in position B or C (McKerrow and Ziegler, 1972) of Fig. 7, and Florida and the Piedmont could have occupied positions A or B. The data from Spain in indecisive. The preferred position for Spain (SP) is between Africa and South America and elements of the Massif Central of France and central Europe may also have been near here on the south side of the Pre-Hercynian ocean. There are no relevant palaeomagnetic data from Florida and the Piedmont. Nor is very much known from palaeomagnetism about the positions of the Iran, Malayan and Chinese blocks and southern Indo-China. The Devonian and Early Carboniferous results from Kazakhstania do not agree with those from Siberia, and a more southerly situation is indicated in the Devonian. In order to provide continuity with subsequent maps, the North China block (SN) and Kazakhstania (KZ) have been arbitrarily attached to north-east Gondwana (Fig. 7).

6. Hercynian Orogeny and Pangaea B

At the Devonian–Carboniferous boundary, Laurasia and Gondwana are converging and they meet in the mid-Carboniferous (Figs 9 and 11). The relative longitudes for the middle and late Carboniferous were chosen to minimize subsequent relative motion, so that Africa is opposite Europe, and South America is opposite North America (Figs 11 and 13). This configuration is called Pangaea B. In principle, there are no constraints on longitude, so that Gondwana could have been further east, or it could have been situated west of North America (Van Hilten, 1964; Irving, 1967). These optional configurations are considered less likely because the subsequent motions needed to achieve the known Jurassic configuration would have been greater—a factor of two greater is the latter instance. In Gondwana the Samfrau superfold belt continues to develop.

By the end of the Carboniferous the fit between Laurasia and Gondwana tightens, but then apparently remains little changed until the end of the Permian; Pangaea B seems to have been a stable configuration lasting about 60 million years in the Late Carboniferous and Permian (Figs 13–18). Wrench-faults post-dating the main tectonism are characteristic of the Hercynian superfoldbelt (Arthaud and Matte, 1977) and indicate that some dextral shear between Gondwana and Laurasia that is undetectable palaeomagnetically, may have commenced by the end of the Carboniferous. Pangaea B provides for the possibility of continental collision during the Hercynian orogeny in central Europe, but does not provide for the continental collision that might have been responsible for the deformation of the Ouachita and Marathon uplift. Samfrau continues active. The Late Palaeozoic poles from Kazakhstania and Eurasia are now in agreement, indicating that the two are welded together (Fig. 13); it appears that Kazakhstania is thrust between Baltica and Siberia in much the same fashion as India will later be thrust between the Iran and South China blocks in the Tertiary. The Malayan block (MB, Fig. 14) is in equatorial latitudes and apparently separate from Gondwana (McElhinny *et al.*, 1974). A notable feature of Pangaea B is the presence of a narrow Permian rift in the north (NR); the palaeomagnetic evidence indicates that at the end of the Carboniferous, Laurentia and Eurasia move apart a little, forming the boreal rift (NR) (Figs 15 and 17).

The south pole migrated from central Africa in the Early Carboniferous, reaching eastern Antarctica in the Late Carboniferous (compare Figs 8 and 14). This motion was originally invoked to explain the migration of glacial and other facies, that becomes progressively younger from west to east in the Carboniferous (Du Toit, 1937; King 1961), and the palaeomagnetic evidence provides confirmation.

Pangaea B has been reconstructed on the premise of an axial geocentric dipole field, and the integrity of the major continental blocks as described above. Should these assumptions be incorrect then the concept of Pangaea B would be wrong. The first record of such a reconstruction for the Permian, with Africa adjoining Europe and Tethys closed, is in Irving (1964, Figs 11, 18), and Vine (1973) has also discussed it favourably. Recently, Westphal (1977) has made a spherical harmonic analysis of the Permian field separately for Gondwana and Laurasia, and found that for each there is, in addition to the dominant axial dipole, an important equatorial dipole component, analogous to the equatorial dipole of the geomagnetic field today. He then calculated the relative positions of Laurasia and Gondwana by first fitting the axial dipoles and then rotating Gondwana eastward relative to Laurasia to bring their equatorial dipoles together. The relative rotation required is 35°. Westphal's Permian reconstruction is less "tight" than the one given here, but the difference is not significant.

7. Pangaea B to A

At the Permian–Triassic boundary the palaeomagnetically determined latitudes now allows Gondwana to rotate anticlockwise, and the transformation from Pangaea B to A begins (Fig. 19). This is made possible by a northward movement of Laurentia along a zone (XM). The motion closes the boreal rift (MR, Fig. 15) between Laurentia and Eurasia. Major activity in the Samfrau ceases. By the Late Triassic the rotation of Gondwana relative to Laurasia stops, and the motions along XM reverses (Fig. 21). Laurentia is now locked to Gondwana, and both rotate anticlockwise relative to Eurasia, opening the Tethys and forming Pangaea A. Mexico is assumed to move north-west to Laurentia (Van der Voo *et al.*, 1976). During the transformation from Pangaea B to A it is as if Gondwana forces Laurentia northwards out of its path, and then, carrying Laurentia along with it, opens Tethys. The total relative motion along the site of the Hercynian superfold belt is about 3500 km.

The transformation is completed by the Late Triassic, but its time of commencement is poorly determined. Some shear may have begun in the Early Permian, and the time required may have been as long as 75 million years (average relative velocity 5 cm y^{-1}). The northward movement of Laurentia relative to Eurasia is about 1000 km, although the errors are such that the movement may be as much as 2000 km (Roy, 1972). Perhaps this motion was responsible for the opening of the Gulf of Mexico. Among the consequences of motion along XM might be the development of grabens, which, in the North Sea, contain oil, the formation of the Early Triassic Siberian Traps, the early opening of Rockall Trough, the dextral rotation of Novaya Zemblya, and the opening of the Kara Sea. During the Late Triassic, the Cordilleran superfold belt apparently developed as a continent–ocean interaction, and extends over 10 000 km from New Zealand to southern Alaska. The B to A transformation is marked by the movement of the pole away from Australia, which synchronizes with the disappearance from there of the last Gondwana glacials (Figs 18, 20 and 22).

Very fragmentary palaeomagnetic evidence indicates that the North Chinese block was in low latitudes during the Permian, and it could have formed a promontory attached to north-east Gondwana, but by the Triassic its latitude had apparently increased. It is therefore tentatively suggested that the North Chinese block undergoes sinistral shear relative to Eurasia at much the same time as Gondwana rotates dextrally. In this way the Chinese block is transferred during the Palaeozoic from Gondwana to Eurasia, and the eastern Tethys is formed at the end of the Palaeozoic.

8. Wegenerian Drift

By the beginning of the Jurassic the continents are assembled together as Wegener's Pangaea (Pangaea A, Figs 23 and 24), but it was short-lived, the world-wide occurrence of tholeiitic basalts and dolerites foreshadowing its early demise and the formation of the modern oceans. Observations, mainly from magnetic anomalies, now allow estimates of relative longitude to be made. By the Late Jurassic the North Atlantic opens perceptibly, and Mexico achieves its present position. A rift appears between eastern and western Gondwana (Figs 25 and 26). It is assumed that the South China block moves northwards and is attached to the Eurasian land mass. Malaya is still in equatorial latitudes, and could be welded on to Eurasia at this time.

In the Early Cretacous (Fig. 27) the pole positions are such that northern Laurentia and Eurasia move closer together as the northern Atlantic opens and the Alaska–Arctic block (AB) rotates anticlockwise opening the Amerasian Basin (AA). The Italian–Adriatic peninsula (IT) formed a recognizable promontary of Africa (Argand, 1922; Channel and Horath, 1976). During the Early Cretacous, Spain also begins to rotate anticlockwise, as Africa moves eastward relative to Europe (Williams, 1975). Riftin is initiated in the South Atlantic (Fig. 28). In the Middle Cretaceous, Tethys is open to the Atlantic (Fig. 29). During the Late Cretaceous the Labrador Sea (LS) opens, as Greenland rotates away from North America. Africa continues to move eastward relative to Europe, Spain completes its anticlockwise rotation, and Tethys closes to the west (Fig. 31). Africa with its Arabian and Italian promontories collides with Eurasia (Argand, 1922; Tapponier, 1977). The Pacific moves northward, presumably passing beneath the northern part of the cordilleran superfold belt (Francheteau, 1970; Harrison *et al.*, 1975). The South Atlantic has become a wide ocean (Fig. 32). There is apparently continuous land from New Guinea to South America, although the connection between South America and Antarctica may have been slender. Africa and India recede further away from Antarctica. Madagascar reaches its present position relative to Africa, India and the Seychelles Ridge (SR) separate from Madagascar, and India commences its rapid motion northward.

In the early Tertiary important changes begin (Figs 33 and 34). As the North Atlantic commences to open, north-eastern Eurasia moves away from northern North America, and the Eurasian (EB) Basin opens, completing the Arctic Ocean. This ocean is formed in two stages: in the early Cretaceous the Amerasian Basin forms as the Alaska–Arctic block migrates southward over the north Pacific subduction zone, and in the Early Tertiary the Eurasia Basin begins to open completing the Atlantic rift system. Svalbard, which is now part of the Eurasian plate, moves to the south-east. India leave the Seychelles Ridge (SR) and collides with Eurasia. The Carlsberg Ridge forms to the west, and the 90° Ridge (9R) to the east. Tethys closes to the west also, and the Alpine–Himalayan superfold belt forms. There was still more or less continuous continental crust connecting South America and New Guinea, but the Scotia Arc (SA) begins to form, initiating the separation between South America and Antarctica. New Zealand rifts from eastern Australia and the Tasman Sea (TS) is created.

By the middle of the Tertiary the main elements of modern physical geography emerge (Figs 35 and 36). Baffin Bay and the Labrador Sea cease to open. The Atlantic rift system, which now extends from the Eurasian Basin to the Antarctic, is fully developed. The Alpine–Himalayan superfold belt is established in its present form. In the south, Australia breaks away from Antarctica, and the Broken (BR) and Kerguelen-Gaussberg (KG) ridges rift apart. The Scotia arc is now fully developed, and South America and the Antarctic Peninsular sweep westward. The Red Sea opens and the East African Rift System forms.

9. Inner and Outer Tectonics

The tectonics of the outer skin of the Earth (outer tectonics) and of the lower mantle (inner tectonics) may be coupled by through mantle convection and the latter may control the behaviour of the core, and therefore the frequency spectrum of reversals of the geomagnetic field. There have been numerous discussions of the possible interactive behaviour of lithosphere, mantle and core (Hide, 1967; Cox, 1975; Irving and Pullaiah, 1976).

One of the features of the cartoons just described is that Pangaea B could have been the longest-lived continental configuration during the later Phanerozoic. It corresponds in time to the Late Palaeozoic quiet interval (the Kiaman of Irving and Parry, 1963) which is characterized by very few changes in polarity of the geomagnetic field and by reversed polarity. The Kiaman lasted for about 60 million years, from the Westphalian to the end of the Palaeozoic (Irving and Pullaiah, 1976). The coexistence of a long-lived supercontinent with a geomagnetically "stable" core probably has global significance. The extended existence of a Pangaea B could mean that there was little or no mantle convection for the last 60 million years of the Palaeozoic. This reasoning is more difficult to sustain for the shorter Mesozoic quiet intervals, it is possible that their shorter duration reflected changes that were less profound, and therefore less readily observable than for the Kiaman. The best established Mesozoic quiet interval is that in the

Cretaceous (−180 to −80 million years). It corresponds in time to the maximum extent of Laurasia and the closure of the western Tethys which brings Laurasia and Gondwana together, so that except for South America and eastern Gondwana the continents temporarily formed one land mass. There may also be a quiet interval in the Jurassic (−164 to −145 million years) which could correspond in time to Wegener's Pangaea (Pangaea A). Thus, the Cretaceous and Jurassic quiet intervals could also have coexisted with temporary assemblages of most of the continental crust. On the other hand, the very high frequency of reversals in the last 40 million years correlates with the maximum known dispersal of continents each with high average rates of motion.

Acknowledgements

I should like to acknowledge my deep indebtedness to John Jaeger in whose department I was a staff member for almost 10 years. He helped me greatly in my work, and it was under his general guidance, and with his unfailing support, that the Canberra palaeomagnetic laboratory, begun in late 1954, was made operational. I should also like to record my thanks to Anton Hales, who arrived in Canberra after I had left, but with whom, from time to time, I have had the privilege of many useful discussions, the most memorable of which was at the UGGI meeting in Rome 1954 when Keith Runcorn read our paper describing the first APW path, (Creer *et al.*, 1954), and when all of us, including both Jaeger and Hales, realized that we had within our grasp the means of measuring the drift of continents. I am also in debt to many colleagues at Montpellier for discussions that lead up to this paper, and to Kenneth Whitham of the Earth Physics Branch who first encouraged me to embark on a generalized review of this sort. Francette Brun, Ruth Decosse, Susanne Cumyn and Juan Geuer assisted in the preparation of the cartoons, and Richard Coles, Phillip Schmidt and Jack Sweeney read the manuscript and made helpful comments.

References

Argand, E. (1922). La tectonique de l'Asie. *C.r. Congr. geol. int.* **1**, 171–372.

Arthaud, F. and Matte, P. (1977). *Bull. geol. Soc. Am.* (in press).

As, J. A. and Zijdeveld, J. D. A. (1958). Magnetic cleaning of rocks in palaeo magnetic research. *Geophys. J. R. astronom. Soc.* **1**, 308–319.

Baker, H. B. (1912). The origin of continental forms. *Ann. Rep. Michigan Acad. Sci., Ann Arbor*, 116–141.

Baker, H. B. (1932). *The Atlantic Rift and its Meaning*, Detroit.

Bullard, E. C., Everett, J. E. and Smith, A. G. (1965). The fit of the continents around the Atlantic. *Phil. Trans. R. Soc. A* **258**, 41–51.

Burrett, C. F. (1974). Plate tectonics and the fusion of Asia. *Earth planet. Sci. Lett.* **21**, 181–189.
Carey, S. W. (1958). *A Tectonic Approach to Continental Drift.* University of Tasmania, Hobart, pp. 177–335.
Channel, J. E. T. and Horvath, F. (1976). African–Adriatic promontory as a paleogeographical premise to the Alpine orogeny and crustal movement in the Carpatho–Balkan region. *Tectonophysics* **35**, 71–101.
Churkin, M. (1973). Western boundary of the North American continental plate in Asia. *Bull. geol. Soc. Am.* **83**, 1027–1036.
Cox, A. (1975). The frequency of reversals and symmetry of the non-dipole field. *Rev. Geophys. Space Phys.* **13**, 35–51.
Creer, K. M. (1958). Preliminary palaeomagnetic measurements from South America. *Ann. Geophys.* **14**, 373–390.
Creer, K. M. (1970). A review of palaeomagnetism. *Earth Sci. Rev.* **6**, 369–466.
Creer, K. M. (1973). In *Applications of Continental Drift to the Earth Sciences* (D. H. Tarling and S. K. Runcorn, eds), Academic Press, London and New York, pp. 47–76.
Creer, K. M., Irving, E. and Runcorn, S. K. (1954). The directions of the geomagnetic field in Great Britain. *J. Geomag. Geoelec.* **6**, 163–168.
Creer, K. M., Irving, E. and Runcorn, S. K. (1957). Geophysical interpretation of palaeomagnetic directions from Great Britain. *Phil. Trans. Soc. A* **250**, 144–156.
De Boer, J. (1965). Paleomagnetic indications of megatectonic movements in the Tethys. *J. geophys. Res.* **70**, 931–944.
Du Toit, A. L. (1937). *Our Wandering Continents*, Oliver and Boyd, Edinburgh.
Fisher, O. (1882). On the physical cause of the ocean basins. *Nature, Lond.* **25**, 243–244.
Francheteau, J. (1972). Paleomagnetism and plate tectonics. Ph.D thesis, University of California, Scripps Institute of Oceanography ref. S10, 70–30.
Graham, K. T. W. and Hales, A. L. (1957). Palaeomagnetic measurements on Karroo Dolerites. *Adv. Phys. Phil. Mag.* **6**, 149–161.
Geuer, J. W. (1973). The terrascope. *Can. J. Earth Sci.* **10**, 1164–1169.
Gough, D. I., Opdyke, N. D. and McElhinny, M. W. (1964). The significance of paleomagnetic results from Africa. *J. geophys. Res.* **69**, 2509–2519.
Hailwood, E. A. (1974). Palaeomagnetism of the Missi norite (Morocco) and the Paleozoic reconstruction of Gondwanaland. *Earth planet. Sci. Lett* **23**, 376–386.
Hamilton, W. (1970). The Uralides and the motion of the Russian and Siberian platforms. *Bull. geol. Soc. Am.* **81**, 2553–2576.
Harrison, C. G. A., Jarrard, R. D., Vacquier, V. and Larson, R. C. (1975). Palaeomagnetism of Cretaceous Pacific seamounts. *Geophys. J. R. astronom. Soc.* **42**, 859–882.
Hide, R. (1967). Motions of the earth's core and mantle and variations of the geomagnetic field. *Science, N.Y.* **177**, 55–56.
Hume, W. F. (1948). *Terrestrial Theories.* Govt. Press, Cairo.
Irving, E. (1956). Palaeomagnetic and palaeclimatic aspects of polar wandering. *Geofis. pura appl.* **33**, 23–41.
Irving, E. (1957). Rock magnetism: a new approach to some palaeogeographic problems. *Phil. Mag. Suppl. Adv. Phys.* **6**, 194–218.
Irving, E. (1964). *Paleomagnetism and its Application to Geological and Geophysical Problems.* Wiley, New York and Chichester.

Irving, E. (1967). Palaeomagnetic evidence for shear along Tethys. *Systematics Assoc.*, No. 7, 59–76.
Irving, E. (1977). Drift of the major continental blocks since the Devonian. *Nature, Lond.* **270**, 304–309.
Irving, E. and Green R. (1958). Polar wandering relative to Australia. *Geophys. J. R. astronom. Soc.* **1**, 64–72.
Irving, E. and Parry, L. G. (1963). The magnetism of some Permian rocks from New South Wales. *Geophys. J. R. astronom. Soc.* **7**, 395–411.
Irving, E. and Pullaiah, G. (1976). Reversals of the geomagnetic field. *Earth Sci. Rev.* **12**, 35–64.
Irving, E., Robertson, W. A., Stott, P. M. (1963). The significance of the paleomagnetism from some Mesozoic rocks of eastern Australia. *J. geophys. Res.* **68**, 2313–2317.
Irving, E., Robertson, W. A., Stott, P. M., Tarling, D. H. and Ward, M. A. (1961). Treatment of partially stable sedimentary rocks showing planar distribution of directions of magnetization. *J. geophys. Res.* **66**, 1927–1933.
Jaeger, J. C. and Irving, E. (1957). Palaeomagnetism and the reconstruction of Gondwanaland. *C.r. 3rd Congr. Pacific Indian Ocean Sci. A, Tananarive*, 233–242.
King, L. C. (1961). Palaeoclimatology of Gondwanaland during the Palaeozoic and Mesozoic eras. In *Descriptive Palaeoclimatology* (A. E. M. Nairn, ed.), Interscience, New York, pp. 307–339.
Le Pichon, X., Sibuet, J. C. and Francheteau, J. (1977). The fit of the continents around the North Atlantic. *Tectonophysics* **38**, 169–209.
McElhinny, M. W. (1972). Palaemagnetic results from Eurasia. Implications of Continental Drift to the Earth Sciences (D. H. Tarling and S. K. Runcorn, eds), Academic Press, New York, pp. 77–86.
McElhinny, M. W. (1973). *Palaeomagnetism and Plate Tectonics.* Cambridge University Press, Cambridge.
McElhinny, M. W. and Embleton, B. J. J. (1974). Australian palaeomagnetism and the Phanerozoic plate tectonics of eastern Gondwana. *Tectonophysics* **22**, 1–29.
McElhinny, M. W., Haile, N. S., and Crawford, A. R. (1974). Palaeomagnetic evidence shows Malay Peninsula was not part of Gondwanaland. *Nature, Lond.* **252**, 641–645.
McKenzie, D. P. and Sclater, J. G. (1971). The evolution of the Indian Ocean since the Late Cretaceous. *Geophys. J. R. astronom. Soc.* **24**, 437–528.
McKerrow, W. S. and Ziegler, A. M. (1972). Palaeozoic Oceans, *Nature, Lond.* **240**, 92–94.
Minister, J. B., Jordan, T. M., Molnar, P. and Haines, E. (1974). Numerical modelling of instantaneous plate tectonics. *Geophys. J. R. astronom. Soc.* **36**, 541–576.
Morel, P. and Irving, E. (1978). Tentative paleocontinental maps for the Early Phanerozoic and Proterozoic. *J. Geol.* **86**, 535–561.
Nairn, A. E. M. (1964). Palaeomagnetic measurements on Karroo and post-Karroo rocks. *Overseas Geol. Min. Res. H.M. Stationery Office* **9**, 302–320.
Newman, G. W., Mull, C. G. and Watkins, N. D. (1977). Indications of Alaskan plate tectonics. In *Alaska Geological Society Symposium, April 1977*, pp. 16–19.
Phillips, J. D. and Forsyth, D. (1972). Plate tectonics paleomagnetism and the opening of the Atlantic. *Bull. geol. Soc. Am.* **83**, 1579–1600.
Pickering, W. H. (1907). The place of origin of the moon. *J. Geol.* **15**, 23–28.

Pitman, W. C. and Talwani, M. (1972). Sea-floor spreading in the North Atlantic. *Bull. geol. Soç. Am.* **83**, 619–646.
Roy, J. L. (1972). A pattern of rupture of the eastern North American-western Europe paleoblock. *Earth planet. Sci. Lett.* **14**, 103–114.
Runcorn (1956). Paleomagnetic comparisons between North America and Europe. *Proc. Geol. Assoc. Can.* **8**, 103–114.
Schmidt, P. W. (1976). The Late Palaeozoic and Mesozoic palaeomagnetism of Australia. Ph.D. thesis, Australian National University.
Smith, A. G. and Briden, J. C. (1977). *Palaeocontinental Maps for the Mesozoic and Cenozoic*, Cambridge University Press, Cambridge.
Smith, A. G., Briden, J. C. and Drewry, G. E. (1973). Phanerozoic world maps. In *Organisms and Continents through Time*, Palaeontological Association Special Paper No. 12, pp. 1–42.
Smith, A. G. and Hallam, A. (1970). The fit of the southern continents. *Nature, Lond.* **225**, 139–144.
Snider, A. (1858). *La Craption et ses Mystères Devoiles*, Paris.
Sweeney, J., Irving, E. and Geuer, J. (1978). Evolution of the Arctic Basin. *Pub. Earth Phys. Branch, Ottawa* **45**, no. 4, 91–100.
Tapponnier, P. (1977). Evolution tectonique du système alpin en métitfannée: poinconnement et écrasement rigide-plastique (in press).
Taylor, F. B. (1910). Bearing of the Tertiary Mountain belts on the origin of the earth's plan. *Bull geol. Soc. Am.* **21**, 179–226.
Van der Voo, R. and French (1974). Apparent polar wandering for the Atlantic bordering continents: Late Carboniferous to Eocene. *Earth Sci. Rev.* **10**, 99–119.
Van der Voo, S. R., Mauk, F. J. and French, R. B. (1976). *Geology* **2**, 177–178.
Van Hilten, D. (1964). Evaluation of some geotectonic hypotheses by paleomagnetism. *Tectonophysics* **1**, 3–71.
Vine, F. J. (1973). Organic diversity, palaeomagnetism, and Permian palaeogeography. In *Organisms and Continents through Time*, Palaeontological Association Special Paper No. 12, pp. 61–77.
Wegener, A. (1912). Die enstehung der kontinente. *Geol. Rundschau* **3**, 276–292.
Wegener, A. *Die Ensteting der Kontinente und Ozeane*, F. Vieweg, Germany, editions of 1915, 1920, 1922 (3rd ed. English tranl. by J. G. Skerl, Methuen, London, 1924), 1924 (4th ed. English trans. by J. Biram, Dover Publications, New York, 1966), 1936 (revised by K. Wegener).
Wensink, M. (1975). The structural history of the India-Pakistan subcontinent during the Phanerozoic. In *Progress in Geodynamics*, Royal Netherlands Academy of Arts and Science, The Hague, pp. 190–208.
Westphal, M. (1977). Configuration of the geomagnetic field and reconstruction of Pangaea in the Permian period. *Nature, Lond.* **267**, 136–137.
Williams, C. A. (1975). Sea-floor spreading in the Bay of Biscay and its relationship to the North Atlantic. *Earth planet. Sci. Lett.* **24**, 440–456.
Wilson, R. L. (1961). Thermal demagnetization of natural remanent magnetic moments of rocks. *Geophys. J. astronom. Soc.* **5**, 45–69.
Ziegler, A. M., Hansen, K. S., Johnson, M. E., Kelly, M. A. Scotese, C. R. and Van der Voo, R. (1977). Silurian continental distributions, paleogeography, climatology and biogeography. *Tectonophysics* **40**, 13–51.

Index